Advances in

GEOPHYSICS

VOLUME 48

Advances in
GEOPHYSICS

VOLUME 48

Advances in Wave Propagation in
Heterogeneous Earth

Series Editor
RENATA DMOWSKA
Division of Engineering and Applied Sciences
Harvard University
Cambridge, Massachusetts, USA

Guest Editors

RU-SHAN WU
Department of Earth Sciences
University of California
Santa Cruz, USA

VALERIE MAUPIN
Department of Geosciences
University of Oslo
Oslo, Norway

Amsterdam • Boston • Heidelberg • London • New York • Oxford
Paris • San Diego • San Francisco • Singapore • Sydney • Tokyo

Academic Press is an imprint of Elsevier
84 Theobald's Road, London WC1X 8RR, UK
Radarweg 29, PO Box 211, 1000 AE Amsterdam, The Netherlands
30 Corporate Drive, Suite 400, Burlington, MA 01803, USA
525 B Street, Suite 1900, San Diego, CA 92101-4495, USA

First edition 2007

Copyright © 2007 Elsevier Inc. All rights reserved

No part of this publication may be reproduced, stored in a retrieval system or transmitted in any form or by any means electronic, mechanical, photocopying, recording or otherwise without the prior written permission of the publisher

Permissions may be sought directly from Elsevier's Science & Technology Rights Department in Oxford, UK: phone (+44) (0) 1865 843830; fax (+44) (0) 1865 853333; Email: permissions@elsevier.com. Alternatively you can submit your request online by visiting the Elsevier web site at http://elsevier.com/locate/permissions, and selecting: Obtaining permission to use Elsevier material

Notice
No responsibility is assumed by the publisher for any injury and/or damage to persons or property as a matter of products liability, negligence or otherwise, or from any use or operation of any methods, products, instructions or ideas contained in the material herein. Because of rapid advances in the medical sciences, in particular, independent verification of diagnoses and drug dosages should be made

ISBN-13: 978-0-12-018850-5
ISBN-10: 0-12-018850-3

ISSN: 0065-2687

For information on all Academic Press publications
visit our website at books.elsevier.com

Printed and bound in USA

07 08 09 10 11   10 9 8 7 6 5 4 3 2 1

Working together to grow
libraries in developing countries

www.elsevier.com | www.bookaid.org | www.sabre.org

ELSEVIER   BOOK AID International   Sabre Foundation

PREFACE

Significant progress in our understanding of the Earth's structure and functioning is dependent on new and original observations. However, these observations cannot be interpreted in a quantitative way without tools to model them, and developing adequate modelling methods is also a prerequisite for progress. Seismological raw data in the 21st century are mostly three-component broadband recordings, and require advanced numerical tools to be modelled, especially if lateral variations in the model are accounted for in addition to the radial stratification of the Earth. Considerable progress has been made concerning modelling of elastic waves in laterally heterogeneous structures in recent decades, taking advantage of the development of computer power. The number of articles related to new developments of diverse methods is enormous and it can be very difficult for newcomers to get an overview of the different methods available, and to be able to find which method is most appropriate for his or her applications.

This volume aims at giving introductions to and basic reviews of the modelling methods for elastic waves in laterally heterogeneous structures that are most commonly used in contemporary seismology, or may have great potential for the future. We hope that it will serve as a bridge between method developers and potential users. We assume that the readers are familiar with the basics of theoretical seismology, presented in classical textbooks like *"Quantitative seismology"* by Aki and Richards (1980), and *"Seismic wave propagation in stratified media"* by Kennett (1983). This book will present methods that go beyond the classical tools. Previous books in the same spirit as this one have been very useful in spreading new methods and computational tools to a wider audience in seismology. The different chapters in Volumes 11 and 12 of *"Computational Physics"* (Bolt, 1972a, 1972b), in *"Seismological Algorithms: Computational Methods and Computer Programs"* (Doornbos, 1988), and in *"Seismic modelling of Earth structure"* (Boschi *et al.*, 1996), are references to which many seismologists turn back regularly. Except for ray theory, which has long been the dominant method to synthesize wavefields in three-dimensional structures, the previous monographs focused mainly on methodologies for calculating wavefields in one-dimensional structures. Modern computer power now makes the calculation of synthetic seismograms in three-dimensional structures accessible to a large community, and the Wave Propagation Commission of the IASPEI (International Association of Seismology and Physics of the Earth's Interior) thought it was timely to summarize the progress made in the field in a new monograph. This is the origin of this volume and many of the chapters are developed based on the progress reports organized by the Subcommission on Heterogeneity and Scattering of the Wave Propagation Commission. In organizing this book, we took into consideration also the progress of seismic modelling methods in exploration seismology and hope the book will be useful for all the global, earthquake, and exploration seismologists.

The book starts with new developments in the two most classical methods in seismology: ray theory on the one hand, a high-frequency asymptotic theory which has traditionally been used to model body waves, and mode methods on the other hand, which provides excellent modelling tools for low-frequency surface waves. Building upon their long histories, these two methods are under constant development, leading to wider domains of application and greater precision. Their respective domains of application tend to overlap, modes being used to calculate long-period body waves, and ray-theory being used to trace modes in laterally heterogeneous structures. S-waves in anisotropic media is a challenge to ray theory which is analyzed in detail in the chapter on ray theory. Mode coupling method is a powerful tool in modelling surface waves in laterally heterogeneous media and is discussed in detail in Chapter 2.

The two following chapters describe boundary methods, where the structure is separated into different units by boundaries, usually the interfaces of layers. Inside units, an analytical solution of the wave equation is known, and continuity conditions along the complex boundaries connect the solutions in different units. Used extensively in modelling the amplification in basin structures, the boundary integral and boundary element methods are reviewed in Chapter 3. In Chapter 4 the wave coupling by complex boundaries is posed as a global generalized reflection/transmission matrix method and the method is used to give a detailed demonstration of the formation of Love wave modes in laterally varying layered structures.

Wave scattering is a classical method to analyze the influence of 3-D volume heterogeneities and lateral variation of layered structures on the wavefield. The De Wolf approximation (multiple forward scattering–single backscattering approximation) and its implementation by the generalized screen-methods are ways to use scattering theory for a much wider range of models than single-scattering methods can handle. The computational cost is much lower than for the full-wave method, such as the finite difference or finite element method. The method is an extension of, but superior to, the classical Born and Rytov approximations. The fundamentals of the method and the different approximations are described in Chapter 5, with emphasis on dominantly vertical propagation. In Chapter 6, the same kind of approach is used for regional wave propagation, with dominant horizontal propagation. In the latter case, the treatment of free-surface, including irregular topography, becomes important.

Chapters 7 to 9 describe purely numerical methods. These methods have of course a high computational cost but can be used in a large variety of situations. The spectral element method described in Chapter 7 is a relatively new method in seismology, with few limitations except its computational cost. It is related to finite-element methods, which are not included in this book since they have not been found very well adapted to seismology and are expected to be replaced by spectral elements in the future. Finite-differences methods on the other hand have been used extensively in seismology, and the recent developments in elastodynamics are reviewed in Chapter 8. Chapter 9 is the only chapter in this book describing a method which is not based on continuum mechanics, but on a completely different discrete approach originating in fluid mechanics. Waves are simulated by quasi-particles carrying pressure on a discrete lattice. This approach is shown here to give very good results in the acoustic case, and may become an interesting new method when extended to elastodynamics.

The last chapter focuses on synthesis of envelopes of seismic traces rather than on the details of individual waveforms to analyze the statistical characteristics of the heterogeneities of the structures traversed by the waves. The validity of the approach is tested by comparison with sets of seismograms generated with finite-differences.

As opposed to *"Seismological Algorithms: Computational Methods and Computer Programs"* (Doornbos, 1988), this book is not directly associated with software. This is not because we underestimate the importance of publishing software. Rather than providing a technical guide on how to use software, we wish to focus on the fundamental understanding of the different methods. Websites, with their possibilities for easily updating the programs and their descriptions, are better suited nowadays to publishing software. The authors of Chapters 1 and 7 maintain websites with software related to their chapters. For several of the methods described in this book, programs are available at different websites, in particular the website of the SPICE European training network in computational seismology (http://www.spice-rtn.org). We encourage readers to consult this site to upload software and to view training material in computational seismology developed in the framework of this network.

We would like to thank the following scientists for their reviews of different chapters: Li-Yun Fu, Robert J. Geller, Matthew M. Haney, Robert W. Hobbs, Lianjie Huang, Einar Iversen, Michael Korn, Jean-Jacques Lévêque, Anatoli Levshin, Bertrand Maillot, Peter Mora, Chuck C. Mosher, Robert I. Odom, Luis Rivera, Arthur Rodgers, Francisco J. Sanchez-Sesma, Bernhard Schuberth, Geza Seriani, Hiroshi Takenaka and Jean Virieux.

We would also like to thank the IASPEI, in particular the ex-president, Brian Kennett, the ex-chairman of Wave Propagation Commission, Colin Thomson and the current president, Peter Suhadolc, for their support during this project.

This book also owes a lot to Renata Dmowska, the editor of the series *"Advances in Geophysics"*. We thank her for her tenacity in reminding us of deadlines and the help in the editing work, and hope she will forgive us for how much we put her patience to the test.

Valérie Maupin and Ru-Shan Wu
Oslo and Santa Cruz, January 9, 2006

REFERENCES

Aki, K., Richards, P.G. (1980). Quantitative Seismology. Freeman, New York.
Boschi, E., Ekström, G., Morelli, A. (Eds.) (1996). *Seismic modelling of Earth structure*. Editrice Compositori, Bologna.
Bolt, B.A. (Ed.) (1972a). *Seismology: Surface Waves and Free Oscillations. Methods in Computational Physics*, vol. 11. Academic Press, New York.
Bolt, B.A. (Ed.) (1972b). *Seismology: Body Waves and Sources. Methods in Computational Physics*, vol. 12. Academic Press, New York.
Doornbos, D.J. (Ed.) (1988). *Seismological Algorithms: Computational Methods and Computer Programs*. Academic Press, New York.
Kennett, B.L.N. (1983). Seismic Wave Propagation in Stratified Media. Cambridge Univ. Press, Cambridge.

CONTENTS

PREFACE v
CONTENTS ix
CONTRIBUTORS xix

Chapter 1

Seismic Ray Method: Recent Developments

VLASTISLAV ČERVENÝ, LUDĚK KLIMEŠ AND IVAN PŠENČÍK

1. Introduction 1
2. Seismic Ray Theories for Isotropic and Anisotropic Media 4
   2.1. Seismic Rays and Travel Times. Initial-Value Ray Tracing . . . 8
   2.2. Ray Histories, Two-Point Ray Tracing, Wavefront Tracing . . . 13
   2.3. Dynamic Ray Tracing. Paraxial Ray Methods 29
   2.4. Ray-Theory Amplitude 39
   2.5. Effects of Interfaces 42
   2.6. Ray-Theory Green Function 45
   2.7. Ray-Theory Seismograms 49
3. Ray-Theory Perturbations 51
   3.1. Perturbation Parameters 52
   3.2. Perturbation of Travel Time 53
   3.3. Optimizing Model Updates During Linearized Inversion of Travel Times 55
4. Coupling Ray Theory for S Waves 56
   4.1. Coupling Equation for S Waves 57
   4.2. Coupling-Ray-Theory S-Wave Propagator Matrix 58
   4.3. Quasi-Isotropic Approximations of the Coupling Ray Theory . . 59
   4.4. Numerical Examples 62
5. Summation of Gaussian Beams and Packets 68
   5.1. Gaussian Beams 68
   5.2. Gaussian Packets 74
   5.3. Optimization of the Shape of Gaussian Beams or Packets . . . 81
   5.4. Asymptotic Summation of Gaussian Beams and Packets 82
   5.5. Decomposition of a General Wavefield into Gaussian Packets or Beams 94
   5.6. Sensitivity of Waves to Heterogeneities 95
   5.7. Migrations 96
6. Ray Chaos, Lyapunov Exponents, Models Suitable for Ray Tracing . 97

	6.1. Lyapunov Exponents and Rotation Numbers	97
	6.2. Models Suitable for Ray Tracing	106
7.	Other Topics Related to the Ray Method	110
	7.1. Higher-Order Ray Approximations	110
	7.2. Direct Computation of First-Arrival Travel Times	112
	7.3. Ray Method with Complex Eikonal	114
	7.4. Hybrid Methods	116
	7.5. Several Other Extensions of the Ray Method	118
	Acknowledgements	118
	References	118

Chapter 2

Introduction to Mode Coupling Methods for Surface Waves

Valérie Maupin

1.	Introduction	127
2.	Modes and Mode Coupling	128
3.	Local Modes	131
	3.1. A Simple Example of Local Mode Coupling: Welded Spaces	131
	3.2. Local Mode Coupling: The General Case	136
	3.3. Surface Wave Ray Tracing	139
4.	Reference Modes	143
	4.1. Two-Dimensional Method of Kennett	144
	4.2. Multiple-Scattering	146
	4.3. Single-Scattering Approximation	148
5.	Conclusion and Perspectives	152
	References	153

Chapter 3

Boundary Integral Equations and Boundary Elements Methods in Elastodynamics

Michel Bouchon and Francisco J. Sánchez-Sesma

1.	Introduction	157
2.	The Direct BIE	159
3.	The Indirect BIE	161
4.	The Calculation of the Green's functions	166

	4.1. Approximate Representations	166
	4.2. 2-D Green's Functions	167
	4.3. 2.5-D Green's Functions	168
	4.4. 3-D Green's Functions	170
	4.5. Non-Homogeneous Media Green's Functions	170
5.	Discretization and Inversion	173
6.	Time Domain Implementation	174
7.	Other Boundary Methods	176
8.	Hybrid Methods	179
9.	Domains of Application	179
10.	Concluding Remarks	183
	Acknowledgements	183
	References	183

Chapter 4

Generation and Propagation of Seismic SH Waves in Multi-Layered Media with Irregular Interfaces

XIAO-FEI CHEN

1. Introduction	191
2. Seismic SH Wave Generation and Propagation in Multi-Layered Media with Irregular Interfaces	192
2.1. Integral Equation and its Discrete Wave Number Representation	193
2.2. Simultaneous Matrix Equations	196
2.3. Global Generalized Reflection/Transmission Matrices	201
2.4. Solution Synthesis	205
2.5. Numerical Validation by the Analytical Solution	210
2.6. Comparisons with Other Existing Methods	215
2.7. More Numerical Examples	222
3. Love Wave in Multi-Layered Media with Irregular Interfaces	233
3.1. Modal Solution of Love Wave in an Arbitrarily Irregular Multi-Layered Medium	234
3.2. Excitation Formula of Love Wave in Irregular Multi-Layered Media	241
3.3. Comparisons with Classic Theory of Love Wave in Laterally Homogeneous Multi-Layered Media	246
4. Concluding Remarks	250
Acknowledgements	251
Appendix A	252
Appendix B	255

Appendix C 257
Appendix D 259
References 261

Chapter 5

One-Way and One-Return Approximations (de Wolf Approximation) for Fast Elastic Wave Modeling in Complex Media

RU-SHAN WU, XIAO-BI XIE AND XIAN-YUN WU

1. Introduction 266
2. Born, Rytov, De Wolf Approximations and Multiple Scattering Series   267
   2.1. Born Approximation and Rytov Approximation: Their Strong and Weak Points 268
   2.2. De Wolf Approximation 272
   2.3. The De Wolf Series (DWS) of Multiple Scattering 275
3. A Dual-Domain Thin-Slab Formulation for One-Return (MFSB) Synthetics 277
   3.1. The Case of Scalar Media 278
   3.2. The Case of Acoustic Media 280
   3.3. The Case of Elastic Media 282
   3.4. Implementation Procedure of the One-Return Simulation 286
4. Fast Algorithm of the Elastic Thin-Slab Propagator and Some Practical Issues 287
   4.1. Fast Implementation in Dual Domains 287
   4.2. Incorporation of Boundary Transmission/Reflection into the One-Return Method 290
   4.3. Treatment of Anelasticity: The Q-Factor 290
   4.4. Numerical Tests for the Elastic Thin-Slab Method 291
5. The Screen Approximation 297
   5.1. Screen Propagators for Acoustic Media 297
   5.2. Screen Approximation for Elastic Media: Elastic complex Screen Method 303
   5.3. Complex Screen Method with First-Order Corrections 304
6. Reflected Wave Field Modeling Using Thin-Slab Method 306
   6.1. Reflection Coefficients of Sedimentary Interfaces 307
   6.2. Reflections from a Dipping Sandstone Reservoir 309
   6.3. Reservoir Reflections with Scattering and Attenuation from a Heterogeneous Overburden 311
   6.4. Thin Layer AVO Response 313
   6.5. AVO Response in Laterally Varying Media 316

7. Conclusions	318
Acknowledgements	318
References	319

Chapter 6

Simulation of High-Frequency Wave Propagation in Complex Crustal Waveguides Using Generalized Screen Propagators

RU-SHAN WU, XIAN-YUN WU AND XIAO-BI XIE

1. Introduction	324
2. A Brief Description of the Generalized Screen Propagator for Guided Waves	328
3. SH Wave Case	331
3.1. Half-Space Screen Propagator	331
3.2. Treatment of the Moho Discontinuity	334
3.3. Numerical Verifications and Simulation Examples	335
3.4. Application to Energy Partition and Attenuation in Crustal Waveguide with Random Heterogeneities	338
3.5. SH-Waves in Crustal Waveguides with Irregular Surface Topography	343
4. P-SV Case	350
5. Conclusion	356
Acknowledgements	358
References	358

Chapter 7

Spectral-Element Analysis in Seismology

EMMANUEL CHALJUB, DIMITRI KOMATITSCH, JEAN-PIERRE VILOTTE, YANN CAPDEVILLE, BERNARD VALETTE AND GAETANO FESTA

1. Introduction	365
2. The Wave Equation in Regional and Global Seismology	367
2.1. General Framework	368
2.2. Reference Configuration	368
2.3. The Gravito-Elastodynamic Equations	369
2.4. Potential Formulation in Fluid Regions	371
2.5. Initial and Boundary Conditions	374

    2.6. Coupling Boundaries 375
    2.7. Weak Formulation 377
3. Spectral-Element Approximation 379
    3.1. Spatial Discretization 379
4. Time Discretization 389
    4.1. Newmark Time Stepping Method 389
    4.2. Rotation 390
    4.3. Self-Gravitation 391
    4.4. Attenuation 393
    4.5. Implementation of a Perfectly Matched Layer 393
5. Implementation 394
    5.1. Choice of Parameters for Accurate Calculations 394
    5.2. Parallel Implementation 395
6. Applications 396
    6.1. 3-D Ground Motion in Alpine Valleys 396
    6.2. 3-D Spectral-Element Reference Solution 402
    6.3. 3-D Modeling in the D'' region 403
7. Conclusions and Perspectives 406
    Acknowledgements 407
    Appendix A. Modified Hooke's Law 407
        A.1. Intrinsic Attenuation 407
        A.2. Fictitious Attenuation in Absorbing Layers 409
    Appendix B. DtN Operators 410
        B.1. Dynamic Coupling 410
        B.2. Static Coupling 412
        B.3. Ocean Load Approximation 412
    References 413

Chapter 8

The Finite-Difference Time-Domain Method for Modeling of Seismic Wave Propagation

PETER MOCZO, JOHAN O.A. ROBERTSSON AND LEO EISNER

1. Introduction 421
    1.1. The Grid 423
    1.2. The FD Approximations to Derivatives 424
    1.3. Basic Properties of the FD Equations and their Solution 426
    1.4. Explicit and Implicit FD Schemes 427
    1.5. Homogeneous and Heterogeneous FD Schemes 427
    1.6. The Equation of Motion, Hooke's Law, and FD Schemes . . . 428

1.7. Algorithms for Enhancing Computational Performance	432
2. Optimally Accurate FD Operators	433
2.1. General Criterion for Optimally Accurate FD Operators	434
2.2. Optimally Accurate FD Operators for 1-D problem in a Homogeneous Medium	437
3. Complex Viscoelastic Media with Material Discontinuities	439
3.1. A Material Discontinuity in the Elastic Medium	439
3.2. Incorporation of the Realistic Attenuation	448
4. Anisotropic Media	464
5. Free Surface	468
5.1. Traction-Free Boundary Condition	468
5.2. Planar Free Surface	469
5.3. Free-Surface Topography	473
6. Wavefield Excitation	481
6.1. Direct Modeling of the Point Sources	481
6.2. Introducing the Source Wavefield Along a Boundary Internal to the Grid	483
6.3. Dynamic Modeling of Earthquake Rupture	486
6.4. Non-Reflecting Boundaries	495
7. Memory Optimization and Parallelization	501
7.1. Memory Optimization	501
7.2. Parallelization	506
Acknowledgements	507
References	508

Chapter 9

A Lattice Boltzmann Approach to Acoustic-Wave Propagation

Lianjie Huang

1. Introduction	517
2. The History of the Use of Discretization	520
3. Lattice Gas Approach	522
3.1. Particle Movement	522
3.2. Particle Collision Rules	522
3.3. Microdynamical Equation	524
3.4. Conservation Relations	525
4. Lattice Boltzmann Approaches	525
4.1. I: A Fermi–Dirac-Type Approach	525
4.2. II: A Bose-Like Approach	528

5. A Lattice Boltzmann Method for Modeling Acoustic Waves: Phononic Lattice Solid by Interpolation 528
   5.1. Transportation Step 529
   5.2. Transmission and Reflection Steps 529
   5.3. Collision Step 531
   5.4. The First-Order Chapman–Enskog Expansion of the Quasi-Particle Number Density N_α 533
   5.5. The Acoustic-Wave Equation 534
6. Absorbing Boundary Condition 536
   6.1. Zero-Valued Reflection Coefficients at Absorbing Boundaries . . 536
   6.2. Viscous Absorbing Boundary Layers 537
   6.3. Incidence Angle Dependence of the Absorbing Boundary Condition 539
7. Free-Surface Reflections 540
   7.1. Numerical Simulations of Free-Surface Reflections 544
   7.2. Numerical Simulations of Reflections From Free-Surface Topography 545
8. Acoustic Waves in Strongly Heterogeneous Media 546
   8.1. Wave Propagation in a Finely Layered Medium 548
   8.2. Wave Propagation in a Medium with Aligned Heterogeneities . . 549
   8.3. Wave Propagation in a Medium with a Circular Empty Bubble . 550
   8.4. Wave Propagation in a Medium with Random Empty Bubbles . 553
9. Conclusions 554
   Acknowledgements 556
   References 556

Chapter 10

Synthesis of Seismogram Envelopes in Heterogeneous Media

HARUO SATO AND MICHAEL C. FEHLER

1. Introduction 562
2. Scalar Wave Equation in 2-D Random Media 567
3. Finite Difference Simulation in Gaussian-Type Random Media 568
   3.1. Realization of Random Media 568
   3.2. Finite Difference Simulation 569
   3.3. FD Envelopes as Reference Envelopes 571
4. Markov Approximation for Gaussian-Type Random Media 571
   4.1. Coherent Wave Field 572
   4.2. Two-Frequency Mutual Coherence Function 574
   4.3. Quasi-Monochromatic Waves 574

4.4. Analytic Solution for the Case of Gaussian ACF 576
4.5. Markov Envelopes in Gaussian-Type Random Media 578
5. Markov Approximation for von Kármán-Type Random Media 578
   5.1. FD Envelopes as Reference Envelopes 579
   5.2. Markov Envelopes 580
   5.3. Comparison of Markov Envelopes with Reference Envelopes . . 582
6. Radiative Transfer Theory for von Kármán-Type Random Media of $\kappa = 0.1$ 582
   6.1. Momentum Transfer Scattering Coefficient 583
   6.2. Radiative Transfer Integral Equation for Isotropic Scattering Process 584
7. Hybrid Envelope Synthesis for von Kármán-Type Random Media Having $\kappa = 0.1$ 588
8. Overview of Envelope Studies 589
   8.1. Envelope Broadening Studies 589
   8.2. Diffusion Studies 590
   8.3. Three-Component Envelope Studies 590
   8.4. Envelopes of Teleseismic Seismograms 591
9. Conclusion and Discussion 591
   Acknowledgements 593
   References 593

SUBJECT INDEX 597

CONTRIBUTORS

Numbers in parentheses indicate the pages on which the authors' contributions begin

BOUCHON, MICHEL (157) Université Joseph Fourier, IRIGM, BP 53, F-38041 Grenoble, France

CAPDEVILLE, YANN (365) Institut de Physique du Globe, 4 place Jussieu, 75252 Paris Cedex 05, France

CHALJUB, EMMANUEL (365) Laboratoire de Géophysique Interne et de Tectonophysique, BP 53, 38041 Grenoble Cedex 9, France

CHEN, XIAO-FEI (191) Laboratory of Computational Geodynamics, Department of Geophysics, School of Earth and Space Sciences, Peking University, Beijing 100871, China

ČERVENÝ, VLASTISLAV (1) Department of Geophysics, Faculty of Mathematics and Physics, Charles University, Ke Karlovu 3, 121 16 Praha 2, Czech Republic

EISNER, LEO (421) Schlumberger Cambridge Research Ltd., Cambridge, United Kingdom

FEHLER, MICHAEL C. (561) Los Alamos National Laboratory, Los Alamos, New Mexico, USA

FESTA, GAETANO (365) Institut de Physique du Globe, 4 place Jussieu, 75252 Paris Cedex 05, France

HUANG, LIANJIE (517) Geophysics Group, Mail Stop D443, Los Alamos National Laboratory, Los Alamos, NM 87545, USA

KLIMEŠ, LUDĚK (1) Department of Geophysics, Faculty of Mathematics and Physics, Charles University, Ke Karlovu 3, 121 16 Praha 2, Czech Republic

KOMATITSCH, DIMITRI (365) Seismological Laboratory, California Institute of Technology, 1200 East California Boulevard, Pasadena, California 91125, USA and Now at: Laboratoire de Modélisation et d'Imagerie en Géosciences, CNRS UMR 5212 and Magique 3D INRIA Futurs, Université de Pau et des Pays de l'Adour, BP 1155, 64013 Pau Cedex, France

MAUPIN, VALÉRIE (127) Department of Geosciences, University of Oslo, PO Box 1047 Blindern, 0316 Oslo, Norway

MOCZO, PETER (421) Faculty of Mathematics, Physics and Informatics, Comenius University, Bratislava, Slovak Republic

PŠENČÍK, IVAN (1) Geophysical Institute, Academy of Sciences of Czech Republic, Boční II 1401, 141 31 Praha 4, Czech Republic

ROBERTSSON, JOHAN O.A. (421) Western Geco Oslo Technology Center, Schlumberger House, Asker, Norway

SATO, HARUO (561) Tohoku University, Sendai, Japan

SÁNCHEZ-SESMA, FRANCISCO J. (157) Institute of Geophysics, Mexico National Autonomous University (UNAM), Mexico

VALETTE, BERNARD (365) Laboratoire de Géophysique Interne et de Tectonophysique, IRD, Université de Savoie, 73376 Le Bourget-du-Lac Cedex, France

VILOTTE, JEAN-PIERRE (365) Institut de Physique du Globe, 4 place Jussieu, 75252 Paris Cedex 05, France

WU, RU-SHAN (265, 323) Modeling and Imaging Laboratory, Institute of Geophysics and Planetary Physics, University of California, Santa Cruz, California, USA

WU, XIAN-YUN (265, 323) Modeling and Imaging Laboratory, Institute of Geophysics and Planetary Physics, University of California, Santa Cruz, California, USA

XIE, XIAO-BI (265, 323) Modeling and Imaging Laboratory, Institute of Geophysics and Planetary Physics, University of California, Santa Cruz, California, USA

SEISMIC RAY METHOD: RECENT DEVELOPMENTS

VLASTISLAV ČERVENÝ[1], LUDĚK KLIMEŠ[1] AND IVAN PŠENČÍK[2]

[1]*Department of Geophysics, Faculty of Mathematics and Physics, Charles University, Ke Karlovu 3, 121 16 Praha 2, Czech Republic*
[2]*Geophysical Institute, Academy of Sciences of Czech Republic, Boční II 1401, 141 31 Praha 4, Czech Republic*

ABSTRACT

The seismic ray method has found broad applications in the numerical calculation of seismic wavefields in complex 3-D, isotropic and anisotropic, laterally varying layered structures and in the solution of forward and inverse problems of seismology and seismic exploration for oil. This chapter outlines the basic features of the seismic ray method, and reviews its possibilities and recent extensions. Considerable attention is devoted to ray tracing and dynamic ray tracing of S waves in heterogeneous anisotropic media, to the coupling ray theory for S waves in such media, to the summation of Gaussian beams and packets, and to the selection of models suitable for ray tracing.

1. INTRODUCTION

The seismic ray method is based on the approximate high-frequency solution of the elastodynamic equation. It leads to the decomposition of the wavefield into independent contributions called elementary waves, which propagate along rays. These elementary waves may represent various seismic body waves propagating in heterogeneous, isotropic or anisotropic, layered and block structures, such as direct, reflected, converted or multiply reflected/transmitted waves. The great advantage of the ray method is that the individual elementary waves can be treated separately.

In the ray-theory approximation considered in this chapter, elementary waves are specified by their travel times $\tau(x_i)$ and complex-valued vectorial amplitudes $\mathbf{U}(x_i)$, as functions of Cartesian coordinates x_i.

Travel time $\tau(x_i)$ satisfies a nonlinear partial differential equation of the first order, called the eikonal equation. The explicit form of the eikonal equation is obtained from the elastodynamic equation. The eikonal equation is solved for $\tau(x_i)$, usually by the method of characteristics. The characteristics of the eikonal equation define the rays. Mathematically, the characteristics of the eikonal equation are represented by a system of nonlinear ordinary differential equations of the first order, called the ray tracing system, see Section 2.1. Conventional numerical methods can be used to solve this system with the required accuracy, so that the computation of rays (as an initial-value problem) is straightforward. In addition to the ray trajectory, we also obtain the travel times $\tau(x_i)$ and the Cartesian components $p_i = \partial \tau / \partial x_i$ of slowness vector **p** at each point of the ray. The procedure of solving the ray tracing system is called ray tracing, and the relevant travel times are called ray-theory travel times. At interfaces, the ray tracing system must be supplemented by Snell's law.

In applications, the whole system of rays is usually needed, not just one single ray. Mostly, a two-parametric system of rays, corresponding to a system of wavefronts of a selected elementary wave, is considered. The ray-parameter domain can be decomposed into regions of equal ray histories, see Section 2.2. These regions can then be sampled, e.g., by triangulation. The introduction of ray histories and the division of the ray-parameter domain into regions of equal ray histories enables a highly accurate and reliable 3-D two-point ray tracing algorithm and very successful wavefront tracing methods to be designed. The concept of ray histories is of principal importance in the asymptotic summation of Gaussian beams or Gaussian packets, and has facilitated the introduction of calculating travel times and other quantities on dense rectangular grids of points by interpolation within ray cells, which is of crucial importance in prestack seismic migrations and non-linear determination of seismic hypocentres.

In Section 2.3, dynamic ray tracing and paraxial ray methods are discussed. Dynamic ray tracing is a very powerful procedure, which is based on the solution of a system of linear ordinary differential equations of the first order along the ray. As the system is linear, its solution is simple. Mostly, the system is expressed either in global Cartesian coordinates, see Section 2.3.1, or in ray-centred coordinates, see Section 2.3.2. Using dynamic ray tracing, it is also possible to determine the DRT (dynamic ray tracing) propagator matrix along the ray, see Section 2.3.3. The DRT propagator matrix can be used to calculate geometrical spreading along the ray, needed in the computation of amplitudes. The DRT propagator matrix can also be used to determine approximately ray-theory travel times and other relevant quantities not only along a ray, but also in its "quadratic" vicinity. We then speak of the paraxial approximation of the ray method (shortly paraxial ray method), paraxial rays, paraxial travel times, etc., see Sections 2.3.4–2.3.6.

Sections 2.4–2.6 are devoted to the computation of the complex-valued vectorial amplitude $\mathbf{U}(x_i)$. The scalar complex-valued amplitude $A(x_i)$ satisfies the transport equation. The transport equation can be simply solved along rays. The transport equation then reduces to an ordinary differential equation of the first order, which can be solved analytically in terms of geometrical spreading. Geometrical spreading is calculated along the ray using dynamic ray tracing. To obtain the final expressions for vectorial complex amplitudes, or for the Green function, we must also determine the amplitude contributions at the initial point of the ray (for example, the radiation pattern of the source), at the end point of the ray (for example, the conversion coefficients at the receiver), and at the points of reflection/transmission (R/T) coefficients.

Once travel time $\tau(x_i)$ and the vectorial complex-valued amplitude $\mathbf{U}(x_i)$ at a given spatial point x_i are known, the particle ground motion for an arbitrary high-frequency signal can be easily computed using the Fourier transform, see Section 2.7. Finally, considering selected elementary waves arriving at receiver points distributed along a profile, we obtain the whole ray-theory seismogram section.

Simple procedures of solving approximately both forward and inverse kinematic problems of heterogeneous, isotropic or anisotropic media are based on the perturbation theory. See Section 3 for a description of the basic principles of the perturbation theory for travel times.

For finite frequencies, the ray method is only approximate. Its accuracy, however, is sufficient to solve many important 3-D wave propagation problems, which can hardly be treated by any other means. Various generalizations and extensions of the ray method, which increase its accuracy and can be applied in situations where the standard zero-order approximation of the ray method is not applicable, have been proposed. Many of them use ray fields as a framework. Several generalizations are described or reviewed to a greater or smaller extent in Sections 4–5 and 7. Section 4 is devoted to the coupling ray theory for S waves. This theory plays a fundamental role in the propagation of S waves in weakly anisotropic media, and in anisotropic media close to S-wave singularities. A detailed treatment of Gaussian beams and Gaussian packets can be found in Section 5. The summation of Gaussian beams and Gaussian packets overcomes the difficulties of the ray theory with caustics.

The ray method can be applied to isotropic, anisotropic and weakly anisotropic 3-D structures, with 3-D variations of elastic moduli and density, with curved interfaces, for arbitrary source-receiver configurations, and for very general types of seismic body waves. It is, however, limited to smooth models (smooth distribution of elastic moduli and density inside layers and blocks, smooth interfaces). For more details on the construction and smoothing of 3-D models for the application of ray methods, see Section 6.

Unless otherwise stated, we consider only perfectly elastic media. Moreover, we do not consider inhomogeneous waves, with complex-valued travel times and complex-valued rays. The waves in viscoelastic models are still the subject of research.

The seismic ray method has been described in great detail in several textbooks and papers. See, for example, Babich and Buldyrev (1972), Červený et al. (1977), Hanyga et al. (1984), Bleistein (1984), Virieux (1996), Dahlen and Tromp (1998), Bleistein et al. (2001), Červený (2001), Chapman (2002, 2004). Therefore, the well-known properties of the ray method are presented here without detailed derivation, just for the sake of completeness. This concerns mostly Section 2, for which the book by Červený (2001) can be considered a general reference. More space is, however, devoted to several recently intensively studied, important concepts such as the coupling ray theory, seismic wave modelling with the use of Gaussian beams or ray chaos, see Sections 4–6.

The chapter does not contain a systematic and complete bibliography. This would exceed considerably the admissible extent of the chapter. Exceptions are Sections 4–6, which contain more extensive lists of recent references. More references on the ray theory itself can be found in the books and papers mentioned in the above paragraph.

The ray method has also been used in other branches of physics, particularly in the electromagnetic theory. Although we consider the elastic waves described by the *elastodynamic equation*, most results are applicable to any non-dissipative hyperbolic second-order partial differential equations (wave equations), including the system of Maxwell equations. Most wave propagation problems discussed here can thus be applied with minor modifications in the electromagnetic theory. The basic principles of the summation of Gaussian beams and packets are independent of a particular selection of the wave equation. References relating to the electromagnetic ray theory, however, are given here only exceptionally. For general references, see the books by Babich and Buldyrev (1972), Felsen and Markuvitz (1973), Kravtsov and Orlov (1980).

To express the equations in this chapter in a concise form, we mostly use the component notation for vectors and matrices, and exceptionally the matrix notation. In the component notation, the upper-case indices $(A, B, C, \ldots)$ take the values of 1 and 2, the lower-case indices $(a, b, c, \ldots)$ the values 1, 2 and 3. In Section 3, the Greek lower-case subscripts $(\alpha, \beta, \gamma, \ldots)$ index the perturbation parameters, and the number of their values is given by the number of perturbation parameters. In Section 5, the Greek lower-case subscripts $(\alpha, \beta, \gamma, \ldots)$ index the space–time coordinates, and take the values 1, 2, 3 and 4. The Einstein summation convention is used throughout the chapter.

There are several ray tracing software packages around the world, available as commercial products as well as free software. In this chapter, we briefly mention only the software packages used to calculate the numerical examples presented hereinafter. These software packages are available from the WWW pages "http://sw3d.mff.cuni.cz" of the consortium project "Seismic Waves in Complex 3-D Structures" (SW3D), where more information on the seismic ray method and additional references may also be found.

2. Seismic Ray Theories for Isotropic and Anisotropic Media

Various approaches have been proposed to derive basic equations of the ray method. In the classical ray method, the displacement vector $\mathbf{u}(x_i, t)$ of an arbitrary time-harmonic elementary wave propagating in a heterogeneous isotropic elastic medium is represented by an *asymptotic ray series* in inverse powers of frequency $\omega > 0$:

$$\mathbf{u}(x_i, t) = \exp\bigl[-i\omega(t - \tau)\bigr]$$
$$\times \bigl[\mathbf{U}^{(0)} + \mathbf{U}^{(1)}/(-i\omega) + \mathbf{U}^{(2)}/(-i\omega)^2 + \cdots\bigr]. \tag{1}$$

Here $\tau = \tau(x_i)$ is the real-valued travel time, $\mathbf{U}^{(n)} = \mathbf{U}^{(n)}(x_i)$, $n = 0, 1, 2, \ldots$ are complex-valued vectorial amplitude coefficients of the ray series, t is time and x_i are Cartesian coordinates. Surfaces $\tau(x_i) = $ const. are called wavefronts. Quantities τ and $\mathbf{U}^{(n)}$ are assumed to be frequency independent. Alternatively, the ray series can be expressed for high-frequency transient waves or for waves discontinuous along the wavefronts. The ray series method for isotropic media was proposed by Babich (1956) and independently by Karal and Keller (1959). For anisotropic media, it was generalized in the pioneering work of Babich (1961).

Inserting (1) into the elastodynamic equation yields a basic recurrence system of equations of the ray method, which can be used to determine successively explicit equations for τ and $\mathbf{U}^{(0)}$, $\mathbf{U}^{(1)}$, ... Formally, the derivation of these equations is simple, see Červený (2001, Section 5.7). The successive computation of higher-order amplitude coefficients $\mathbf{U}^{(n)}$, $n = 1, 2, \ldots$, is, however, complicated as it requires the knowledge of higher-order spatial derivatives of functions describing the model, up to the $(2n + 2)$nd order. The higher n, the higher-order spatial derivatives are needed. Consequently, the computation of higher-order amplitude coefficients of the ray series in realistic structures is very unstable since it is sensitive to the fine details of the model. These are the main reasons why only the *zero-order approximation* of the ray method

$$\mathbf{u}(x_i, t) = \mathbf{U}(x_i) \exp\bigl\{-i\omega\bigl[t - \tau(x_i)\bigr]\bigr\} \tag{2}$$

has been used in most seismological applications. Fortunately, the zero-order approximation (2) is often sufficient. In (2), we have dropped the superscript (0) of $\mathbf{U}(x_i)$, and have called $\mathbf{U}(x_i)$ the *complex-valued vectorial amplitude*. It is again assumed that $\tau(x_i)$ and $\mathbf{U}(x_i)$ in (2) are frequency independent.

In this section, we use the zero-order approximation (2) of the ray method systematically, and explain how travel time $\tau(x_i)$ and vectorial amplitude $\mathbf{U}(x_i)$ can be determined from the elastodynamic equation.

For heterogeneous anisotropic perfectly elastic media, the source-free elastodynamic equation reads

$$(c_{ijkl} u_{k,l})_{,j} = \varrho \ddot{u}_i, \tag{3}$$

where $c_{ijkl}(x_n)$ are the real-valued elastic moduli, and $\varrho(x_n)$ is the density. In this chapter, we consider only stationary non-dissipative elastic media, in which c_{ijkl} and ϱ are time-independent.

Inserting (2) into (3) yields
$$-N_i(\mathbf{U}) + i\omega^{-1} M_i(\mathbf{U}) + \omega^{-2} L_i(\mathbf{U}) = 0, \tag{4}$$
where
$$N_i(\mathbf{U}) = c_{ijkl} p_j p_l U_k - \varrho U_i,$$
$$M_i(\mathbf{U}) = c_{ijkl} p_j U_{k,l} + (c_{ijkl} p_l U_k)_{,j},$$
$$L_i(\mathbf{U}) = (c_{ijkl} U_{k,l})_{,j}. \tag{5}$$

Here p_i are Cartesian components of slowness vector $\mathbf{p}$, given by the relation
$$p_i = \frac{\partial \tau}{\partial x_i} = \frac{n_i}{\mathcal{C}}. \tag{6}$$

n_i denote the Cartesian components of unit vector $\mathbf{n}$, perpendicular to the wavefront, and $\mathcal{C} = \mathcal{C}(x_i, n_j)$ is the phase velocity. Obviously, the phase velocity depends both on position and direction of propagation $\mathbf{n}$.

Equations (3)–(5) remain valid for heterogeneous isotropic media; we only take into account that $c_{ijkl} = \lambda \delta_{ij}\delta_{kl} + \mu(\delta_{ik}\delta_{jl} + \delta_{il}\delta_{jk})$, where λ and μ are Lamé's elastic moduli.

As travel time $\tau(x_i)$ and vectorial amplitude $\mathbf{U}(x_i)$ are assumed to be frequency-independent, N_i, M_i and L_i are also frequency-independent. Since Eq. (4) should be satisfied for any frequency ω, N_i, M_i and L_i should equal zero:
$$a_{ijkl} p_j p_l U_k - U_i = 0, \tag{7}$$
$$a_{ijkl} p_j U_{k,l} + \varrho^{-1}(\varrho a_{ijkl} p_l U_k)_{,j} = 0, \tag{8}$$
$$(\varrho a_{ijkl} U_{k,l})_{,j} = 0, \tag{9}$$
where
$$a_{ijkl} = \frac{c_{ijkl}}{\varrho} \tag{10}$$
are density-normalized elastic moduli. The first two equations, (7) and (8), are sufficient to find explicit equations for $\tau(x_i)$ and $\mathbf{U}(x_i)$. Their solutions coincide with the zero-order term of the ray series. Note that (8) is satisfied only partially, and (9) is not satisfied at all.

Equation (7) can be expressed in the form
$$(\Gamma_{ik} - \delta_{ik})U_k = 0, \quad i = 1, 2, 3, \tag{11}$$
where
$$\Gamma_{ik}(x_m, p_n) = a_{ijkl} p_j p_l. \tag{12}$$

In the seismic ray method, the 3×3 matrix $\Gamma_{ik}(x_m, p_n)$ is usually called the *Christoffel matrix*, although it does not strictly correspond to the common definition of the Christoffel matrix, in which the slowness vector $\mathbf{p}$ in (12) is replaced by a real-valued unit vector **n**, oriented in the direction of **p**. In this chapter, arguments p_i of the Christoffel matrix always represent the components of the slowness vector. Note that $\Gamma(x_m, p_n)$ is a symmetric positive definite matrix in non-dissipative elastic media, and its elements are homogeneous functions of the second degree in p_i.

The Christoffel matrix (12) has three eigenvalues $G_m(x_k, p_n)$ and three relevant eigenvectors $\mathbf{g}^{(m)}(x_k, p_n)$, $m = 1, 2, 3$. They correspond to the three elementary waves propagating in heterogeneous anisotropic media, namely S1, S2 and P. Since matrix Γ is symmetric and positive definite, all the three eigenvalues G_1, G_2, G_3 are real-valued and positive. Moreover, they are homogeneous functions of the second degree in p_i. The eigenvalues of the Christoffel matrix generally differ, and can be calculated by solving the appropriate algebraic equation of the third degree. If two of them coincide, we speak of a degenerate case. Eigenvectors $\mathbf{g}^{(m)}$ are considered to be normalized (unit) vectors. For **n** given, $\mathbf{g}^{(m)}$ represent three mutually perpendicular unit vectors. Note that eigenvalue G_m and eigenvector $\mathbf{g}^{(m)}$ satisfy the relation $G_m = \Gamma_{ik} g_i^{(m)} g_k^{(m)}$ (no summation over m).

Let us consider the mth elementary wave. It follows from (11) that eigenvalue G_m should satisfy the equation

$$G_m(x_i, p_j) = 1, \tag{13}$$

and that vectorial amplitude **U** is expressed in terms of the unit real-valued eigenvector $\mathbf{g}^{(m)}$ of the Christoffel matrix (12) as follows:

$$\mathbf{U} = A\mathbf{g}^{(m)}. \tag{14}$$

Here $A = A(x_i)$ is a complex-valued frequency-independent *scalar amplitude*. Equation (13) is a nonlinear partial differential equation of the first order for travel time $\tau(x_i)$, called the *eikonal equation* for heterogeneous anisotropic media.

As G_m is a homogeneous function of the second degree in p_i, and $p_i = n_i/\mathcal{C}$, we obtain $G_m(x_i, p_j) = \mathcal{C}^{-2} G_m(x_i, n_j)$. Using (13) we then obtain

$$\mathcal{C}(x_i, n_j) = [G_m(x_i, n_j)]^{1/2}. \tag{15}$$

Here $\mathcal{C}(x_i, n_j)$ is the phase velocity of the mth elementary wave, at position **x** and direction **n**. Note that $G_m(x_i, n_j)$ in (15) actually represents an eigenvalue of matrix $a_{ijkl} n_j n_l$.

Let us now discuss Eq. (8). Multiplying Eq. (8) by $g_i^{(m)}$, we obtain the partial differential equation of the first order for the complex-valued scalar amplitude $A(x_i)$:

$$2\mathcal{U} \cdot \nabla(\sqrt{\varrho} A) + (\sqrt{\varrho} A) \nabla \cdot \mathcal{U} = 0, \tag{16}$$

where

$$\mathcal{U}_i = a_{ijkl} p_l g_k^{(m)} g_j^{(m)} \tag{17}$$

(no summation over m). Vector $\mathcal{U}$, with Cartesian components $\mathcal{U}_i$ given by (17), is the ray-velocity vector corresponding to eikonal equation (13). Its orientation is generally different from **p** and $\mathbf{g}^{(m)}$. Equation (16) represents one possible form of the *transport equation*.

The three derived equations, namely eikonal equation (13), Eq. (14) for the vectorial amplitude **U** and the transport equation (16) play a basic role in the seismic ray method, and will be discussed in detail in the following sections.

In heterogeneous isotropic media, the eigenvalues of the Christoffel matrix (12) can be expressed analytically:

$$G_1(x_i, p_j) = G_2(x_i, p_j) = \beta^2(x_i) p_n p_n,$$
$$G_3(x_i, p_j) = \alpha^2(x_i) p_n p_n. \tag{18}$$

Here G_1 and G_2 correspond to S waves, with velocity $\beta = \sqrt{\mu/\varrho}$, and G_3 to P waves, with velocity $\alpha = \sqrt{(\lambda + 2\mu)/\varrho}$. The two S-wave eigenvalues coincide, $G_1 = G_2$, so that the case of S waves in heterogeneous isotropic media is degenerate. The phase velocities of P and S waves do not depend on the direction of propagation, $\mathcal{C}(x_m) = V(x_m)$, where $V = \alpha$ for P waves, and $V = \beta$ for S waves. Slowness vector **p** and ray-velocity vector $\mathcal{U}$ are parallel, with $\mathcal{U} = V^2 \mathbf{p}$. The eigenvector $\mathbf{g}^{(3)}$ of the P wave equals **n**. The eigenvectors $\mathbf{g}^{(1)}$ and $\mathbf{g}^{(2)}$ of S waves are mutually perpendicular and perpendicular to **p**. As $\mathbf{g}^{(1)}$ and $\mathbf{g}^{(2)}$ correspond to the degenerate case, they are not specified uniquely. The condition how $\mathbf{g}^{(I)}$ vary along the ray should be supplemented. Using Eq. (8), we thus arrive at the "decoupling" condition,

$$p_i g_{j,i}^{(I)} = a p_j, \tag{19}$$

where a is a scalar real-valued quantity that should be chosen to guarantee that $\mathbf{g}^{(I)}$ are unit vectors.

2.1. Seismic Rays and Travel Times. Initial-Value Ray Tracing

The eikonal equation is a nonlinear partial differential equation of the first order for travel time $\tau(x_i)$ and can be derived from the elastodynamic equation. It can be expressed in Hamiltonian form

$$\mathcal{H}(x_i, p_j) = 0, \tag{20}$$

where $\mathcal{H}$ is the Hamiltonian or Hamilton function, x_i are Cartesian coordinates and $p_i = \partial \tau / \partial x_i$ are Cartesian components of slowness vector **p**. We consider Hamiltonians which are homogeneous functions of the second degree in p_i (with a possible additional constant). For example, for elementary waves propagating in a heterogeneous isotropic medium, we use the Hamiltonian

$$\mathcal{H}(x_i, p_j) = \frac{1}{2}[V^2(x_i) p_k p_k - 1], \tag{21}$$

following from (13) and (18). Here $V(x_i)$ is the spatially variable velocity of P or S waves. In heterogeneous anisotropic media, we use the Hamiltonian given by the relation

$$\mathcal{H}(x_i, p_j) = \frac{1}{2}[G_m(x_i, p_j) - 1], \qquad (22)$$

following from (13). Here $G_m(x_i, p_j)$ is a selected eigenvalue of the 3×3 Christoffel matrix (12). For S waves in heterogeneous anisotropic media, we also use the averaged Hamiltonian (Bakker, 2002)

$$\mathcal{H}(x_i, p_j) = \frac{1}{4}[G_1(x_i, p_j) + G_2(x_i, p_j) - 2] \qquad (23)$$

of both S waves, which enables the *common S-wave rays* to be traced. These rays are used in the coupling ray theory, see Section 4.

If eigenvalue G_m coincides with one of the two remaining eigenvalues, eigenvector $\mathbf{g}^{(m)}$ cannot be uniquely determined. We then speak of the degenerate case and degenerate Christoffel matrix. The two elementary waves are coupled, and the polarization of the resulting wave is generally nonlinear, see Section 2.4.3.

In realistic cases, the P-wave eigenvalue G_3 is well separated from the S-wave eigenvalues G_1 and G_2. This means that the degenerate case does not exist for P waves. Consequently, the algorithms proposed for P waves in anisotropic media can be used quite universally also in isotropic and weakly anisotropic media.

For S waves, the situation is considerably more complex. There are two different degenerate cases for S waves: (a) Eigenvalues G_1 and G_2 coincide *locally* along certain lines or at certain points in anisotropic media. We then speak of S-wave singularities. (b) Eigenvalues G_1 and G_2 coincide *globally* in isotropic media, see (18). In both these degenerate cases, the S1 and S2 waves are coupled, locally or globally, and propagate as a single wave. The algorithms for S waves propagating in heterogeneous anisotropic media cannot be used for S waves propagating in heterogeneous isotropic media. Consequently, there are two main ray theories for S waves: the first for the computation of S waves in heterogeneous anisotropic media, based on Eq. (22), and called the anisotropic ray theory for S waves, and the second for the computation of S waves in heterogeneous isotropic media, based on Eq. (21), and called the isotropic ray theory for S waves. None of these ray theories is valid for S waves in heterogeneous weakly anisotropic media, and a different approach, based on the coupling ray theory for S waves, must be used. For the coupling ray theory for S waves, see Section 4.

The eikonal equation can be solved by the method of characteristics, which represent rays. The parameter, specifying the points along the characteristics, depends on the selected Hamiltonian. For Hamiltonians which are homogeneous functions in p_i (with a possible additional constant), the parameter has the physical meaning of travel time τ. Consequently, we consider consistently with

Eqs. (21), (22) and (23) travel time τ to be the parameter along the ray. The characteristic of the eikonal equation is then described by six ordinary differential equations in x_i and p_i,

$$\frac{dx_i}{d\tau} = \frac{\partial \mathcal{H}}{\partial p_i}, \qquad \frac{dp_i}{d\tau} = -\frac{\partial \mathcal{H}}{\partial x_i}. \tag{24}$$

Thus, travel time τ is a parameter along the characteristic curve, and need not be computed additionally by quadratures. Equations (24) also introduce two vectors which play an important role in the seismic ray method: ray-velocity vector

$$\mathcal{U}(\tau) = \frac{d\mathbf{x}}{d\tau}, \tag{25}$$

and vector

$$\eta(\tau) = \frac{d\mathbf{p}}{d\tau}. \tag{26}$$

The determination of the derivatives of the Hamiltonian in ray tracing system (24) may require the calculation of the eigenvectors of the Christoffel matrix. Whereas there are no problems with Hamiltonians (21) or (23), and with Hamiltonian (22) for a P wave, the determination of the phase-space derivatives of Hamiltonian (22) for an S wave may be difficult.

The system of six ordinary differential equations (24) represents the ray tracing system (also sometimes called the kinematic ray tracing system, to distinguish it from the dynamic ray tracing system). If the initial conditions $x_i = x_{i0}$ and $p_i = p_{i0}$ are specified for $\tau = \tau_0$ in such a way that $\mathcal{H}(x_{i0}, p_{j0}) = 0$, see (20), ray tracing system (24) can be solved by conventional numerical methods (predictor-corrector, for example). We then speak of initial-value ray tracing and of initial-value rays. Relation (20) must be satisfied along the whole ray and can be used to check and control the accuracy of ray tracing.

For heterogeneous isotropic media, the ray tracing system reads, see (21) and (24),

$$\frac{dx_i}{d\tau} = V^2 p_i, \qquad \frac{dp_i}{d\tau} = -\frac{1}{2} p_k p_k \frac{\partial V^2}{\partial x_i}. \tag{27}$$

As $p_k p_k = 1/V^2$ along the ray, see (21), the r.h.s. of the second equation of (27) can also be replaced by $-\partial \ln V / \partial x_i$. Isotropic ray tracing system (27) can be used quite safely for both P and S waves, even if two rays with different initial conditions intersect (multipathing) or are tangent (caustics) at some points of the rays.

Ray tracing system (27) for heterogeneous isotropic media has been used to compute rays in many computer packages. Let us mention here, for example, the 2-D ray tracing package SEIS, see Červený and Pšenčík (1984), and the 3-D ray tracing package CRT (complete ray tracing), see Červený et al. (1988).

For heterogeneous anisotropic media, the ray tracing system can be expressed as follows:

$$\frac{dx_i}{d\tau} = a_{ijkl} p_l g_j^{(m)} g_k^{(m)}, \qquad \frac{dp_i}{d\tau} = -\frac{1}{2}\frac{\partial a_{jkln}}{\partial x_i} p_k p_n g_j^{(m)} g_l^{(m)}. \qquad (28)$$

Here $g_i^{(m)}$ is the eigenvector of the Christoffel matrix corresponding to the considered wave. No summation over (m) is applied. Ray tracing system (28) has found useful applications in heterogeneous anisotropic media. Anisotropic ray tracing system (28) is used, for example, in the computer package CRT (Červený et al., 1988).

Ray tracing system (28) can also be expressed in an alternative form. We define subdeterminants

$$D_{ij} = \frac{1}{2}\varepsilon_{ikl}\varepsilon_{jrs}(\Gamma_{kr} - \delta_{kr})(\Gamma_{ls} - \delta_{ls}) \qquad (29)$$

of the Christoffel matrix. Here ε_{ijk} is the Levi-Civita symbol ($\varepsilon_{123} = \varepsilon_{312} = \varepsilon_{231} = 1$, $\varepsilon_{321} = \varepsilon_{213} = \varepsilon_{132} = -1$, $\varepsilon_{ijk} = 0$ otherwise). It is not difficult to prove that, for a selected elementary wave for which $G_m = 1$ and $D_{ss} \neq 0$, the factor D_{ij}/D_{ss} can be expressed in terms of the eigenvector of the Christoffel matrix corresponding to the considered wave as follows:

$$\frac{D_{ij}}{D_{ss}} = g_i^{(m)} g_j^{(m)}. \qquad (30)$$

No summation over (m) is applied. Using this relation, the anisotropic ray tracing system (28) yields, for D_{ss} sufficiently different from zero, ray tracing system

$$\frac{dx_i}{d\tau} = a_{ijkl} p_l \frac{D_{jk}}{D_{ss}}, \qquad \frac{dp_i}{d\tau} = -\frac{1}{2}\frac{\partial a_{jkln}}{\partial x_i} p_k p_n \frac{D_{jl}}{D_{ss}}. \qquad (31)$$

The mode of the wave we wish to compute is specified by the initial conditions x_{i0}, p_{j0}, which must satisfy Eq. (20) at the initial point, corresponding to the relevant wave. Anisotropic ray tracing system (31) is used, for example, in the computer package ANRAY (Gajewski and Pšenčík, 1987, 1990).

Anisotropic ray tracing system (28) or (31) can be used quite universally for P waves, even for P waves in heterogeneous isotropic and weakly anisotropic media. For S1 and S2 waves, however, the situation is more complicated. The anisotropic ray tracing systems do not work universally for these waves, but may fail in degenerate cases.

Let us first consider an elementary wave, corresponding to eigenvalue G_m of the Christoffel matrix (12), well separated from the other two eigenvalues. In this *non-degenerate case*, we may identify eigenvalue G_m by its unit value, or we may identify the corresponding eigenvector $\mathbf{g}^{(m)}$ by its continuity along the ray. Eigenvector $\mathbf{g}^{(m)}$ can then be inserted into ray tracing equations (28). In this case, ray tracing equations (28) may be replaced by ray tracing equations (31).

In the *vicinity of S-wave singularities*, eigenvalue G_m of the Christoffel matrix (12) cannot be safely identified by its unit value. We must thus identify eigenvector $\mathbf{g}^{(M)}$ by means of its continuity along the ray (Vavryčuk, 2001, Section 5.1.1), if we are performing ray tracing in an anisotropic medium. In this case, we should use Eq. (28) for ray tracing and avoid Eq. (31).

If the segment of the ray in the vicinity of the S-wave singularity is sufficiently short, we may simply compare the eigenvectors in front of and beyond the vicinity of the S-wave singularity (Vavryčuk, 2001, Section 5.1.2; Bulant and Klimeš, 2002, Eqs. (18) and (37)), and possibly interpolate the eigenvectors.

If the segment of the ray in the vicinity of the S-wave singularity (singular region) is too long, we must continue vectors $\mathbf{g}^{(M)}$ along the ray so that they do not rotate about vector $\mathbf{g}^{(3)}$,

$$\frac{dg_i^{(1)}}{d\tau} g_i^{(2)} = 0. \tag{32}$$

This condition follows from Eq. (8), exactly for S waves in isotropic media, see (19), and approximately for S waves in very weakly anisotropic media. For the continuation, we may use, e.g., equation

$$\frac{dg_i^{(M)}}{d\tau} = \frac{g_i^{(3)}}{G_M - G_3} g_j^{(3)} \left(\frac{\partial \Gamma_{jk}}{\partial x_l} \mathcal{U}_l + \frac{\partial \Gamma_{jk}}{\partial p_l} \eta_l \right) g_k^{(M)}, \tag{33}$$

where no summation over $M = 1, 2$ is applied. Equation (33) follows from condition (32) and from the derivative of equation $\Gamma_{ik} g_k^{(M)} = G_M g_i^{(M)}$ along the ray. We then compare the eigenvectors determined in front of the singular region and continued through the singular region with the eigenvectors determined beyond the singular region. If the difference between the eigenvectors is not negligible, we should terminate tracing the ray.

Solving the ray tracing system (24) for given initial conditions x_{i0} and p_{i0}, we obtain not only the ray trajectory (specified by $x_i(\tau)$), but also slowness vector $\mathbf{p}(\tau)$ (perpendicular to the wavefront) at any point of the ray, and the relevant phase velocity $\mathcal{C} = (p_i p_i)^{-1/2}$ (i.e., the velocity of propagation of the wavefront in the direction of its normal). We also obtain two other important vectors $\mathcal{U}$ and η at any point of the ray, see (25) and (26). Physically, $\mathcal{U}$ is the ray-velocity vector, tangent to the ray, whose magnitude is equal to velocity $\mathcal{U}$ of propagation of the wavefront along the ray. This vector has the direction of the time-averaged energy flux. In the seismological literature, $\mathcal{U}$ is also often called the group velocity vector. In isotropic media, $\mathcal{U}$ and $\mathbf{p}$ are parallel, but not in anisotropic media. In other words, the time-averaged energy flux is not perpendicular to the wavefronts in anisotropic media. Both $\mathcal{U}$ and $\mathbf{p}$ play an important role in the seismic ray method, particularly in dynamic ray tracing, and satisfy the well-known relation $\mathcal{U} \cdot \mathbf{p} = 1$.

Similarly as isotropic ray tracing system (27), anisotropic ray tracing systems (28) and (31) can be used safely outside S-wave singularities for any wave mode if the rays with different initial values intersect (multipathing) or pass through caustic points.

In media composed of layers or blocks, ray tracing must be supplemented by the prescription specifying which of the generated waves should be considered after reflection/transmission. We refer to this prescription as the *code of the elementary wave*. Every elementary wave is thus specified by its code. This distinguishes it from other elementary waves.

At boundaries of layers or blocks, formulae for the transformation of quantities, calculated along the ray, during reflection/transmission must be applied, see Section 2.5.

As mentioned above, the ray method requires sufficiently smooth models. Once heterogeneities in the model exceed a certain limit, the rays may have a chaotic character, particularly for large travel times (long rays). For more details on chaotic rays, see Section 6.

2.2. Ray Histories, Two-Point Ray Tracing, Wavefront Tracing

In applications, the whole system of rays is usually needed, not just one single ray. Mostly, a two-parametric system of rays, corresponding to a system of wavefronts of a selected elementary wave (orthonomic system of rays), is considered. We also call this system the ray field. We denote the two parameters γ_1 and γ_2, and refer to them as the *ray parameters*. For example, for an elementary wave generated by a point source, the two ray parameters may be chosen as the take-off angles of initial slowness vectors at the source. Similarly, for a wave generated at an initial surface, the ray parameters can be chosen as curvilinear coordinates of initial points of rays along the initial surface. The properties of ray fields, expressed in terms of ray parameters, play a very important role in the seismic ray method and its applications. The two ray parameters γ_1, γ_2 may be supplemented by a third parameter $\gamma_3 = \tau$, which specifies uniquely the position of any point on the ray. We then speak of *ray coordinates* $\gamma_1, \gamma_2, \gamma_3$. See Červený (2001, Section 3.10).

The introduction of ray histories (Bulant, 1996; Vinje *et al.*, 1996a, p. 824) and the division of the ray-parameter domain into regions of equal ray histories brought about a considerable simplification and generalization of the 2-D two-point ray tracing algorithm originally proposed by Červený and Pšenčík (1984) and enabled a highly accurate and reliable 3-D two-point ray tracing algorithm (Bulant, 1996) to be designed. The concept of ray histories has facilitated the introduction of calculating travel times and other quantities on dense rectangular grids of points by interpolation within ray cells (Vinje *et al.*, 1993a; Vinje *et al.*, 1996a, Section 2.4; Bulant and Klimeš, 1999), which is of crucial importance in prestack seismic migrations and nonlinear determination of seismic hypocentres. The relevant algorithms are simultaneously used in very successful wavefront tracing methods, and allow diffracted edge waves to be calculated. Note that during recent years, ray histories have also become very important in 3-D computer graphics and animation, where, however, the rays are straight, which considerably simplifies the determination of possible boundaries between ray histories compared to seismology.

The definition of ray histories allows the ray-parameter domain to be decomposed into regions of equal ray histories. These regions can then be sampled, e.g., by triangulation. This procedure is, in general, called here *controlled initial-value ray tracing*. Controlled initial-value ray tracing has many applications. It is the basis of boundary-value (two-point) ray tracing, it is applied in wavefront tracing, it serves as the preprocessor for interpolation within ray cells, and it is of principal importance in the asymptotic summation of Gaussian beams or Gaussian packets.

2.2.1. Model and Ray Histories

The travel time of an elementary wave is multivalued with respect to spatial coordinates. It thus cannot globally be defined as the function of spatial coordinates x_i. It has to be parametrized by ray coordinates γ_a, composed of two ray parameters γ_1, γ_2 and of parameter γ_3 along the rays. The spatial distribution of travel time can be described by the projection of ray coordinates onto spatial coordinates and travel time,

$$\gamma_a \to x_i(\gamma_a), \tau(\gamma_a). \tag{34}$$

We may then say that travel time τ is a "continuous function" if both coordinates x_i and travel time τ are continuous functions of the ray coordinates, and that the nth derivative of travel time τ is a "continuous function" if both the nth derivatives of coordinates x_i and the nth derivative of travel time τ with respect to ray coordinates are continuous functions of the ray coordinates. We may define the continuity of other quantities calculated along the rays analogously.

If the nth derivatives of the travel time are to be continuous, the nth derivatives of the density-normalized elastic moduli (i.e. of the propagation velocity in isotropic media) must also be continuous. If the amplitudes are to be continuous, the second derivatives of the density-normalized elastic moduli must also be continuous.

For the calculation of the nth derivatives of travel time, a *smooth model* (smooth velocity model, smooth macromodel) thus means the model with the continuous nth derivatives of the density-normalized elastic moduli. The minimum requirement on a smooth model for the zero-order ray theory are the continuous second derivatives of the density-normalized elastic moduli.

The density-normalized elastic moduli may be discontinuous only along "*interfaces*". An interface of the $(n + 1)$th order means that the nth derivatives of the density-normalized elastic moduli are discontinuous. Interfaces thus divide the model into *blocks* (e.g., a layer or a salt dome). The variation of the density-normalized elastic moduli within each block should be smooth in the same sense as in the smooth model (Gjøystdal *et al.*, 1985; Červený *et al.*, 1988). A model composed of these blocks, which are separated by interfaces, may be called the *block model* (block velocity model, block macromodel).

The interfaces may be specified in many ways. Here we assume that they are composed of one or several smooth surfaces or of parts thereof. The surfaces forming the interfaces may be defined implicitly, as the zero isosurfaces of given functions. If we wish the nth derivatives of travel time to be continuous within the block beyond the interface, we require the $(n - m)$th derivatives of the function describing the interface of the $(m + 1)$th order to be continuous.

In models with interfaces, not all rays pass through an equal sequence of blocks and interfaces. The continuity or smoothness of the travel time or of other quantities calculated along the rays is then violated. We thus have to introduce, for each elementary wave (specified by the code of the elementary wave), the *ray histories*. Rays of the same ray history pass through an equal sequence of blocks and interfaces. Each sequence of blocks and interfaces encountered during ray tracing thus defines the corresponding ray history. The ray histories may be indexed by integers, and may thus be understood to be the integer-valued functions of ray coordinates (similarly as, e.g., the KMAH index defined in Section 2.6.3). This integer-valued function of ray coordinates is called the *history function*.

If we perform ray tracing from a given source to a given reference surface (e.g., two-point ray tracing), we are usually interested in the complete ray history from the source to the reference surface. In this case, the ray history depends on two ray parameters γ_A. If the spatial distribution of the results of ray tracing is required (e.g., interpolation within ray cells, or wavefront tracing), the ray history depends on three ray coordinates γ_a, and is determined along the ray gradually, starting from the source. The history function dependent on three ray coordinates γ_a is discontinuous for ray parameters, whose images $x_i(\gamma_a)$ are the points of intersection of rays with interfaces.

Travel time is a smooth function for a single ray history. At the boundary between ray histories, travel time or its derivatives may be discontinuous. Analogously for the spatial coordinates $x_i(\gamma_a)$ of rays and for many other quantities calculated along the rays. Geometrical spreading may even approach infinity at the boundary between ray histories. Whereas travel time is continuous at the caustics, some other quantities are not (e.g., the KMAH index, amplitude, second derivatives of travel time). If we are interested in the quantities discontinuous at the caustics, we may optionally wish to include also the KMAH index into the definition of ray histories. In this case, the ray history including the KMAH index is determined both by the ray history without the KMAH index and by the KMAH index. The KMAH index is defined in Section 2.6.3, and dynamic ray tracing (Section 2.3) has to be applied to determine it. The history function thus may or may not include the KMAH index.

Since the continuity and smoothness is violated at the boundaries between ray histories, neither interpolation, nor paraxial ray approximation, nor perturbation expansion can be applied across these boundaries. That is why the boundaries between ray histories should be carefully determined, and why methods such as interpolation, paraxial ray approximation or perturbation expansion can be applied just within a single ray history. The determination of the boundaries between ray histories is thus the basic and most involved step in applying such methods. Interpolation, paraxial ray approximation or perturbation expansion on their own are usually just relatively easy and routine subsequent steps of the application.

2.2.2. Controlled Initial-Value Ray Tracing

Rays taking off from the source are parametrized by ray parameters γ_A. The given 2-D set of all ray parameters (γ_1, γ_2) is the *ray-parameter domain*. To simplify the description of the ray-parameter domain, it is often assumed that the ray-parameter domain is the image of the rectangular *normalized ray-parameter domain* in the given mapping $\gamma_A = \gamma_A(\eta_B)$, which projects *normalized ray parameters* η_B onto ray parameters γ_A. The normalized ray-parameter domain is usually selected as a unit square. For this reason, no description of axes is used in the figures of the normalized ray-parameter domain.

The history function is an integer-valued function of two ray parameters, and describes the complete ray history from the source to the reference surface, or to the endpoint of the ray if no reference surface is specified.

Controlled initial-value ray tracing consists in dividing the ray-parameter domain into regions of the same value of the history function, and in sampling these regions. We briefly describe here the algorithm of controlled initial-value ray tracing based on the triangulation of the ray-parameter domain of a given elementary wave according to Bulant (1996, 1999). The algorithm is designed for general 3-D models composed of heterogeneous blocks separated by curved interfaces. The algorithm is independent of the type and shape of the source (point source, curved linear source, curved surface source) and of its parametrization.

Initial-value ray tracing assigns the integer value of the history function to every ray. Initial-value ray tracing also assigns unique coordinates ξ_K of incidence to the rays incident at the reference surface. These rays are referred to as *successful rays*.

The ray-parameter domain should be divided into regions of the same value of the history function. If defining the ray-parameter domain as the image of the normalized ray-parameter domain, this division is performed in the normalized ray-parameter domain.

The division of the ray-parameter domain into regions of the same value of the history function may be carried out by triangulating these regions. We refer to the narrow belts between the individual triangulated regions as *demarcation belts*. The demarcation belts between the different ray histories correspond to the numerical uncertainty of the boundaries between the ray histories. The demarcation belts must thus be kept reasonably narrow.

SEISMIC RAY METHOD: RECENT DEVELOPMENTS

The ray-parameter domain is divided into regions of equal value of the history function by means of triangulating the ray-parameter domain. The triangulation is performed iteratively.

First, the ray-parameter domain is covered by *basic triangles*. The size of the basic triangles is measured with respect to the *ray-parameter metric tensor*. The ray-parameter metric tensor may specify the metric in the ray-parameter domain (e.g., the metric on the unit sphere for a point source). This ray-parameter metric tensor must be defined for all ray parameters, including rays, which do not leave the source. This ray-parameter metric tensor should also be reasonably smooth over all ray histories. The basic triangles are roughly equilateral with respect to the ray-parameter metric.

A *homogeneous triangle* is a triangle whose apexes are formed by rays with equal value of the history function. All "inhomogeneous" triangles have to be divided into homogeneous triangles and into narrow demarcation belts along the boundaries between different ray histories.

In the first step of the division, all *boundary rays* have to be found on the sides of the inhomogeneous triangle. The boundary rays are pairs of rays with very slightly different ray parameters, but with different ray histories. Their maximum distance in the ray-parameter domain is given a priori. The pairs of boundary rays serve to separate the regions of equal value of the history function. The boundary rays are sought using the method of halving intervals, which requires the auxiliary rays, which are not further used in the triangulation, to be traced, see Fig. 1. During this step, narrow strips of new ray histories can often be found along the

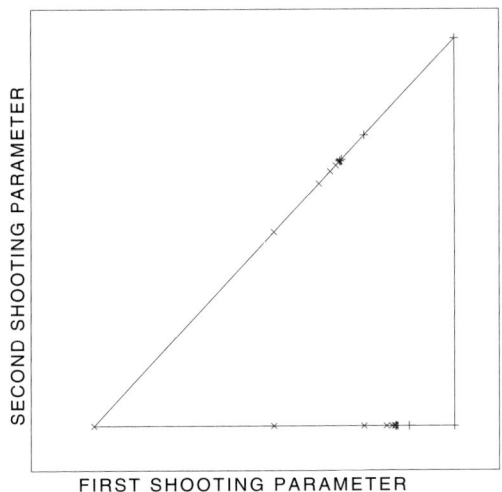

FIG. 1. Searching for boundary rays on the sides of the inhomogeneous triangle by halving intervals. Rays are symbol-coded according to their history. The three vertices of the triangle are the basic rays. Note that the distance between the two rays in each pair of boundary rays is very small.

sides of the basic triangles. In this way, several pairs of boundary rays may be found along one side of the triangle.

In the second step, the boundary between one ray history and the other ray histories is traced, see Fig. 2. For the description of tracing the boundary, refer to Bulant (1996, Section 3.3). In this way, we obtain *homogeneous polygons*, which should then be divided into homogeneous triangles.

In the third step, the homogeneous polygons should be divided into homogeneous triangles. The division of homogeneous polygons into triangles, not too different from equilateral, requires tracing several new rays, situated both on the long sides of the homogeneous polygons and inside the homogeneous polygons, see Fig. 3.

If a ray of a different ray history is identified in a homogeneous triangle, the ray together with the vertices of the triangle is used to create three new inhomogeneous triangles. The algorithm then reverts to the division of these three triangles into homogeneous triangles.

If tracing anisotropic-ray-theory S-wave rays, it is necessary to check also the continuity of the eigenvectors of the Christoffel matrix with respect to ray parameters, both at the initial points of rays and at the endpoints of rays. The discontinuities of the eigenvectors should be demarcated analogously to the boundaries between ray histories.

We would appreciate if the ray-parameter metric tensor could also take into account geometrical spreading or another quantification of the complexity of the system of rays. Unfortunately, this might create considerable discontinuities of the

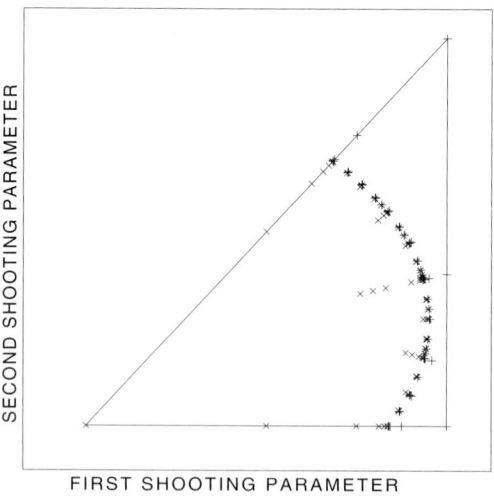

FIG. 2. The boundary between the two ray histories is demarcated by pairs of boundary rays. All the boundary and basic rays of the same ray history create a homogeneous polygon. Here are two homogeneous polygons, which will be divided into homogeneous triangles in the next step.

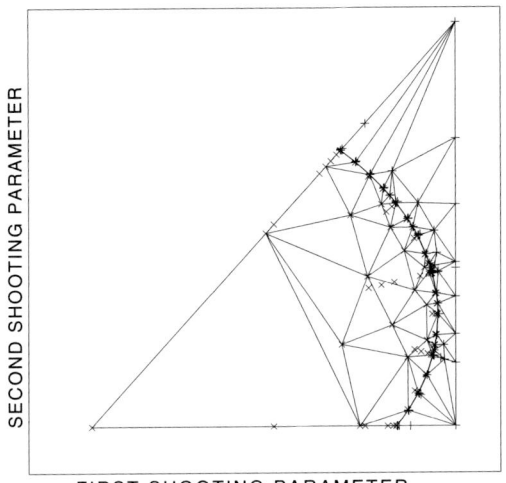

FIG. 3. Both homogeneous polygons are covered by homogeneous triangles. Each homogeneous triangle is formed by three rays of equal ray history. Several new rays are traced in order to create homogeneous triangles "not too different" from equilateral.

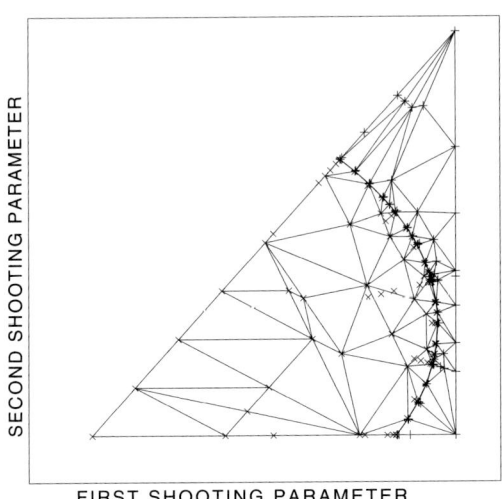

FIG. 4. Again the same basic triangle as in Fig. 3. Now all the homogeneous triangles corresponding rays, which are too large on the reference surface, are divided into smaller ones. This means that the secondary ray-parameter metric tensor was defined so that it measures the distances between the points of intersection of rays with the reference surface.

ray-parameter metric tensor at the boundaries between ray histories. That is why we also define the *secondary ray-parameter metric tensor*, which is smooth only within the individual regions of equal value of the history function. The secondary ray-parameter metric tensor is used to measure the size of homogeneous triangles. If a homogeneous triangle is large with respect to the secondary ray-parameter metric tensor, it is divided into smaller triangles, see Fig. 4.

2.2.3. Two-Point Ray Tracing

The algorithm for determining all two-point rays of a given elementary wave by means of the shooting method, proposed by Bulant (1996), is based on the controlled initial-value ray tracing. The algorithm is designed for general 3-D models composed of heterogeneous blocks separated by curved interfaces. The algorithm is independent of the type and shape of the source (point source, curved linear source, curved surface source) and of its parametrization.

Solving the *two-point ray tracing* problem means finding all rays of a given elementary wave, which take off from the source and pass through the receiver. Note that we have used the term *two-point ray* in place of the term *boundary-value ray*, which would be more appropriate because the *source* is general and is not limited to a point. The *receiver* is a given point, situated at the *reference surface*. This receiver need not exactly coincide with the actual receiver, it may represent the projection of the actual receiver onto the reference surface. The actual receivers should be situated in the vicinity of the reference surface. The receivers at the reference surface are parametrized by two reference coordinates ξ_K along the reference surface. The reference coordinates may be general functions of model coordinates x_i, $\xi_K = \xi_K(x_i)$.

The history function for two-point ray tracing is an integer-valued function of two ray parameters, and describes the complete ray history from the source to the reference surface. The history function for two-point ray tracing should include the KMAH index in order to catch all two-point rays on a triplication of travel time. The demarcation belts between different ray histories must be kept reasonably narrow, because all two-point rays are sought within the individual triangulated regions only, and the two-point rays situated inside the demarcation belts are not found.

Once the ray-parameter domain is decomposed, with the given accuracy, into regions of the same value of the history function, the two-point rays may be found iteratively within the individual regions corresponding to successful rays.

In two-point ray tracing, the secondary ray-parameter metric tensor may take into account, e.g., geometrical spreading at the reference surface.

For the coverage of the normalized ray-parameter domain by rays of various ray histories, traced in the model depicted in Fig. 5, refer to Fig. 6. The arrangement of the individual homogeneous subdomains is quite complicated. Some regions of equal ray histories are disconnected, because the rays reflect from the

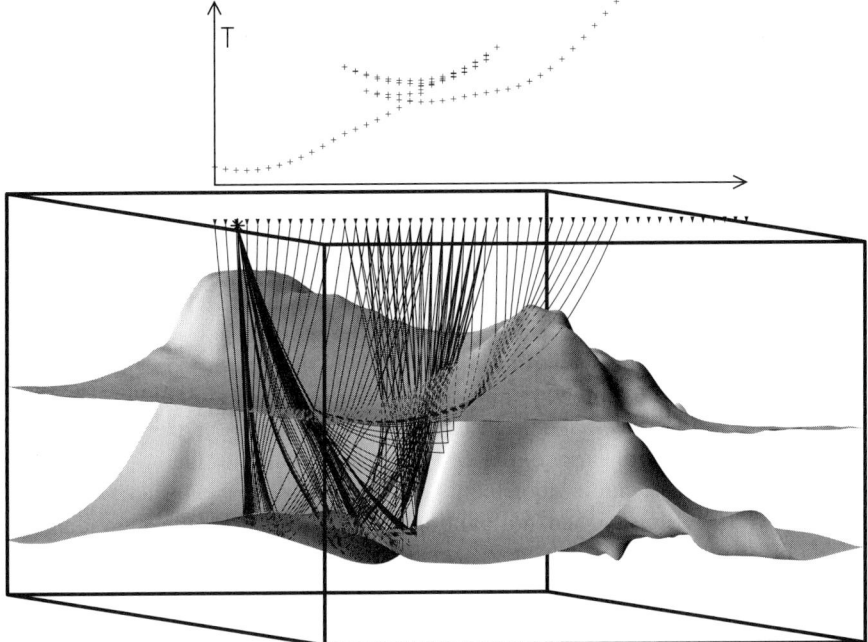

FIG. 5. The three-layer model used for two-point ray tracing (Bulant, 1999), with two-point rays reflected from the lower interface. The asterisk represents the point source, the small triangles are the receivers. The time curve above the model represents the corresponding two-point travel times. Only each fourth receiver and each fourth two-point ray indicated in Fig. 6 has been plotted.

disconnected parts of the lower interface. The small separation of neighbouring two-point rays indicates large geometrical spreading.

The two-point rays of the wave diffracted at the edge of a lenticular inclusion, traced using the described algorithm, are displayed in Fig. 7.

2.2.4. Other Applications of Controlled Initial-Value Ray Tracing

In addition to two-point ray tracing, controlled initial-value ray tracing may also be applied to generate discrete systems of rays, e.g., for the interpolation within ray cells, for the asymptotic summation of Gaussian beams, or for the asymptotic summation of Gaussian packets.

Controlled initial-value ray tracing for the asymptotic summation of Gaussian beams differs from two-point ray tracing by no receivers specified along the reference surface.

In controlled initial-value ray tracing for the asymptotic summation of Gaussian packets or for the interpolation within ray cells, no reference surface is usually

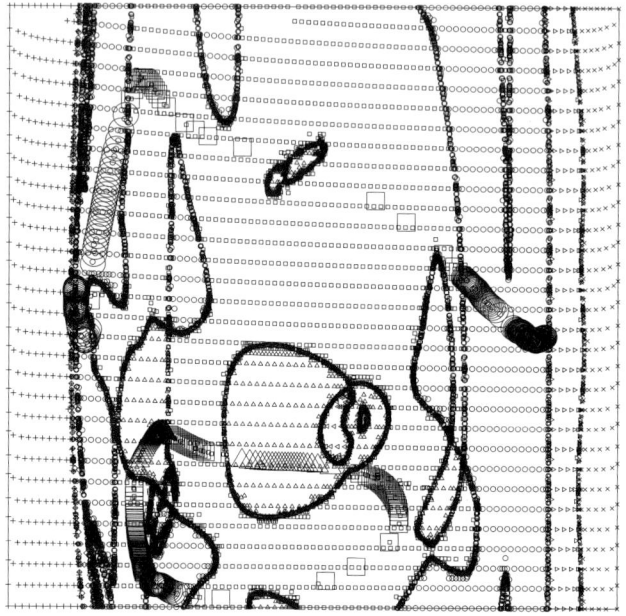

FIG. 6. The coverage of the normalized ray-parameter domain by rays shot during the computation of the two-point rays in the model shown in Fig. 5. The rays are symbol-coded according to the ray history. Two-point rays are plotted as larger symbols. The basic triangles are roughly equilateral with respect to the ray-parameter metric rather than with respect to the Euclidean metric on this sheet of paper.

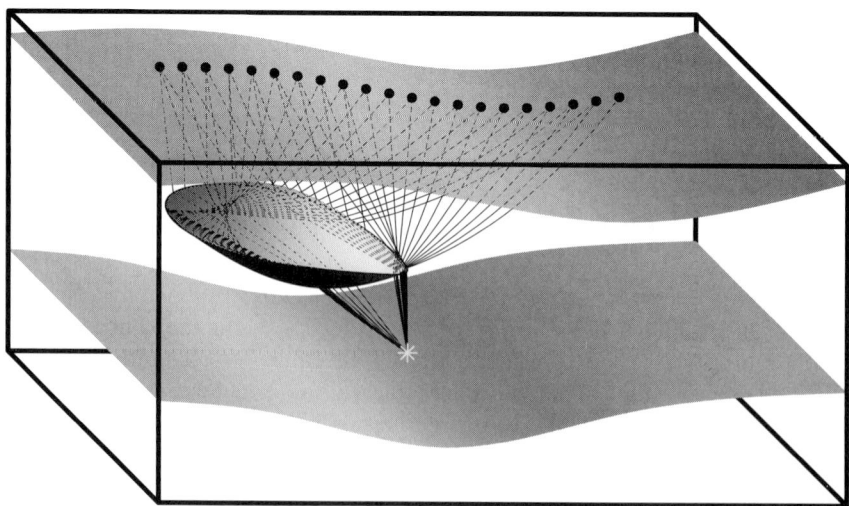

FIG. 7. Two-point rays diffracted at the edge of the lenticular inclusion.

specified. If no reference surface is specified, the history function describes the complete ray history from the source to the endpoint of the ray.

The history function for controlled initial-value ray tracing usually does not include the KMAH index, because the asymptotic summation of Gaussian beams, the asymptotic summation of Gaussian packets, and usually also the interpolation within ray cells do not require sorting rays according to the KMAH index.

In controlled initial-value ray tracing for the interpolation within ray cells, the secondary ray-parameter metric tensor may take into account geometrical spreading along the whole ray.

In controlled initial-value ray tracing for the asymptotic summation of Gaussian beams, the secondary ray-parameter metric tensor should control the discretization error, see Eq. (171). The secondary ray-parameter metric tensor then corresponds to the shape of Gaussian beams optimized for the propagation from the source to the reference surface.

In controlled initial-value ray tracing for the asymptotic summation of Gaussian packets, the secondary ray-parameter metric tensor should also be used to control the discretization error. In this case, the secondary ray-parameter metric tensor may take into account metric tensors (171) corresponding to the shapes of Gaussian packets optimized for the propagation from the source to various points along the ray.

2.2.5. Wavefront Tracing

In *wavefront tracing* (Vinje *et al.*, 1993a; Vinje *et al.*, 1993b, 1996a, 1996b; Lambaré *et al.*, 1996; Gjøystdal *et al.*, 2002), a system of wavefronts with a given step in travel time is considered for each elementary wave. The rays are then traced from the source to the first wavefront, from the first wavefront to the second wavefront, and so on. The ray histories for tracing a ray to individual wavefronts then differ. The ray history for tracing to a wavefront corresponds to the propagation from the source to this wavefront.

The triangulation of the ray-parameter domain for tracing to the first wavefront may be carried out analogously as in controlled initial-value ray tracing. When tracing from the first wavefront to the second wavefront, the homogeneous triangles may be divided. There are two different reasons for this division:

(a) Rays of an equal ray history on the first wavefront are of different ray histories on the second wavefront. A homogeneous triangle then becomes inhomogeneous, and has to be divided.
(b) The evolution of the secondary ray-parameter metric tensor makes the homogeneous triangle too large on the second wavefront. The homogeneous triangle then has to be divided into smaller homogeneous triangles (Vinje, 1997).

When tracing from the first wavefront to the second wavefront, new rays required for the division of the triangles on the second wavefront are not shot from the source. They are shot from the first wavefront instead. The initial values for the new rays are determined by the interpolation within homogeneous triangles corresponding to the first wavefront. The interpolation of the initial coordinates should be performed along the spatial boundary of the ray cell rather than along the wavefront, if the interpolation within ray cells is to be applied, see Section 2.2.6. Analogously for tracing to subsequent wavefronts.

Wavefront tracing is especially suitable for interpolating within ray cells. In this application, it has a great advantage over controlled initial-value ray tracing. In controlled initial-value ray tracing, the ray history corresponds to the whole ray. The narrow demarcation belts between different ray histories then create demarcation volumes which always extend from the source, even if the ray histories begin to differ later on. These demarcation volumes then form gaps inside the volume corresponding to equal ray history. This problem does not occur in wavefront tracing.

Another, but less important advantage of wavefront tracing is the smaller total length of traced rays than in controlled initial-value ray tracing. In particular, the auxiliary rays required during the determination of boundaries between ray histories are traced only between two neighbouring wavefronts. The difference in the total length of useful rays (vertices of homogeneous triangles) is less pronounced, and is partly reduced by more expensive numerical integration. Because of short ray elements between wavefronts, wavefront tracing requires the application of Runge–Kutta methods, which are roughly twice more expensive than predictor–corrector methods recommended for two-point ray tracing and controlled initial-value ray tracing (Červený et al., 1988, Section 5.8). Let us emphasize that the most important factor, influencing the efficiency of both controlled initial-value ray tracing and wavefront tracing by an order of magnitude, is the efficiency of determining the boundaries between ray histories. The second important factor is a good selection of the secondary ray-parameter metric tensor. The total length of traced rays is not too important.

Wavefront tracing is thus an excellent method for calculating travel times and other quantities. The reliability and efficiency of wavefront tracing strongly depends, especially in block models with interfaces, on detailed numerical algorithms for updating the triangulation of the ray-parameter domain in proceeding from one wavefront to the next.

2.2.6. Interpolation Within Ray Cells

Each homogeneous triangle in the ray-parameter domain, created during controlled initial-value ray tracing, generates a *ray tube* limited by the corresponding three rays. Each traced ray is recorded as a set of points of two types. First, there are the points on the ray stored with a given travel-time step; we refer to them as *time points*. Second, there are the points of interaction of the ray with interfaces and other surfaces (reflection, transmission or termination); we refer to them as *interaction points*. A ray tube is thus represented as a set of the above points on the three rays, which form the ray tube, see Fig. 8.

For interpolation, the ray tube is decomposed into *ray cells*. *Regular ray cells* are defined by six points on the rays. Both the bottom and the top of the regular ray cell are formed by triangles defined by three points on the rays. These are usually the time points of the same travel-time level. *Degenerate ray cells* are formed by five or four points. They occur mainly at a point source, where the bottom of a ray cell is formed by a single point, in front of and beyond interfaces, where one or two points of the top of a ray cell coincide with the corresponding point(s) of the bottom, and in front of the end surface where the ray tube terminates.

We start the decomposition at the source, taking the first points on the rays as the bottom of the first ray cell. The bottom is formed by a single point for point sources, by a line segment for straight-line sources, or by a triangle for curved-line sources and surface sources. We then proceed along the rays to the second points on the rays, taking the second points simultaneously as the top of the first ray cell and as the bottom of the second ray cell (if all the second points are time points). We continue this procedure along the rays and create regular ray cells, until we reach the first interaction point on anyone of the rays. Here we suspend proceeding along this ray, and we proceed only along the remaining two rays (or one ray), until we reach the interaction points on all the three rays. We may thus create one or more degenerate ray cells, until the three interaction points form the

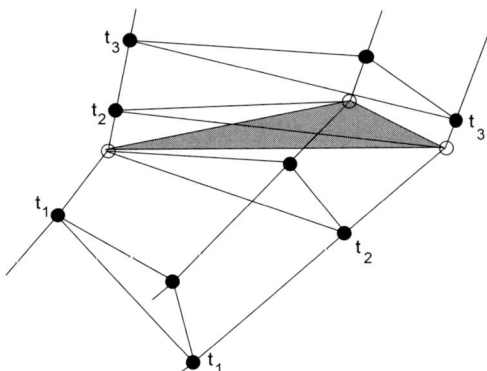

FIG. 8. Decomposition of a ray tube into ray cells. Three rays, forming a ray tube, cross an interface. Bullets represent the points at three travel-time levels t_1, t_2 and t_3 ("time points"). Circles are the points of intersection of the rays with the interface ("interaction points"). Starting from the bottom of the figure, we can see one regular ray cell formed by six points, then two degenerate cells formed by five and four points, and again a regular ray cell. The bottom of the first ray cell and the top of the fourth cell are formed by triangles which approximate wavefronts at travel-time levels t_1 and t_3. The shaded triangle, which is the top of the second cell and the bottom of the third cell, approximates the interface.

top of the last created ray cell. If they lie at the end surface, the decomposition of the ray tube into ray cells is completed. If they are located at an interface, we continue the process of creating ray cells beyond this interface. We may again have to suspend proceeding along some ray(s) and we may create some degenerate ray cell(s), until we get time points of equal travel time at the top of the recently created ray cell. We can then continue with creating regular ray cells with equal travel time at their bottoms and tops. See Fig. 8.

Let us note that ray cells in wavefront tracing are analogous to ray cells created during controlled initial-value ray tracing. The only difference is that the ray tube can be split into several ray tubes in wavefront tracing. This splitting should always be realized along the planar top triangle of the last cell before splitting in order to assure perfect contact of the ray cells.

In decomposing ray tubes into ray cells, we must ensure, that the sides of all the ray cells are equal for neighbouring ray tubes. Otherwise, some gridpoints of a receiver grid might be considered to be located within none, or within both of the neighbouring ray cells.

By determining a ray cell we understand the algorithm for the decision, whether an arbitrary point lies within the ray cell, or not. Assume a ray cell defined by six points on three rays: points $\mathbf{B}_i$ at the bottom, and points $\mathbf{C}_i$ at the top of the ray cell, see Fig. 9. We introduce a local coordinate system parametrized by parameters w_1, w_2, w_B. The coordinates of arbitrary point $\mathbf{X}$ are then given as

$$\mathbf{X} = \mathbf{A}_3 + w_1(\mathbf{A}_1 - \mathbf{A}_3) + w_2(\mathbf{A}_2 - \mathbf{A}_3), \tag{35}$$

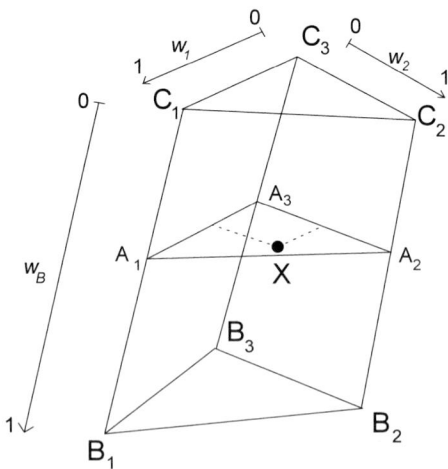

FIG. 9. A ray cell is defined by six points $\mathbf{B}_1, \mathbf{B}_2, \mathbf{B}_3, \mathbf{C}_1, \mathbf{C}_2, \mathbf{C}_3$, situated on three rays. Point $\mathbf{X}$ is examined, whether it is located within the ray cell. Its position may be expressed using the local coordinate system parametrized by parameters w_1, w_2, w_B. If all of the quantities w_1, w_2, w_B, $1 - w_1 - w_2$ are in interval $\langle 0, 1 \rangle$, point $\mathbf{X}$ lies within the ray cell.

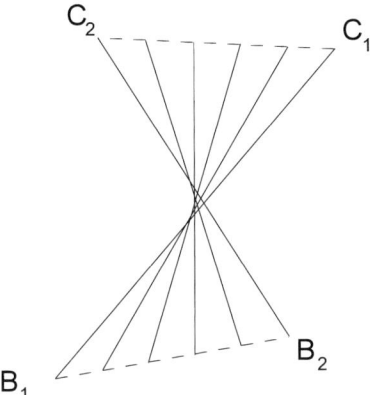

FIG. 10. A 2-D ray cell touches a caustic. The lines approximate rays, the dashed lines represent the bottom and the top of the cell. Each point of a small area in the middle of the cell lies on two rays. For such points we get two (two or three in 3-D) different sets of local coordinates, and we then obtain two (two or three in 3-D) different values of interpolated quantities.

where

$$\mathbf{A}_i = \mathbf{C}_i + w_B(\mathbf{B}_i - \mathbf{C}_i), \quad i = 1, 2, 3. \tag{36}$$

Parameters w_1, w_2 are the barycentric coordinates in triangle $\mathbf{A}_1\mathbf{A}_2\mathbf{A}_3$. A degenerate ray cell is treated as a special case of a regular ray cell with coinciding vertices. Equation (35) is the parametric description of planar triangle $\mathbf{A}_1\mathbf{A}_2\mathbf{A}_3$. Equations (36) are parametric descriptions of straight-line segments $\mathbf{B}_1\mathbf{C}_1$, $\mathbf{B}_2\mathbf{C}_2$ and $\mathbf{B}_3\mathbf{C}_3$, approximating the curved ray segments. Equations (35) and (36) may also be expressed as

$$\mathbf{X} = w_B \sum_{i=1}^{3} w_i \mathbf{B}_i + w_C \sum_{i=1}^{3} w_i \mathbf{C}_i \tag{37}$$

with $w_C = 1 - w_B$, $w_3 = 1 - w_1 - w_2$.

Point $\mathbf{X}$ lies within the ray cell if all values w_1, w_2, w_3, w_B are within interval $\langle 0, 1 \rangle$. Thus, for each receiver, which might be located within the selected ray cell, we compute its local coordinates w_1, w_2, w_B and quantities w_3, w_C. If all of them display values from interval $\langle 0, 1 \rangle$, point $\mathbf{X}$ is within the ray cell, and the values of all required quantities will be determined by interpolation within the ray cell.

Local coordinate w_B is the root of the cubic equation (Bulant and Klimeš, 1999, Eq. (6)). For each root w_B from interval $\langle 0, 1 \rangle$, local coordinates w_1 and w_2 are the unique solutions of the linear equation (Bulant and Klimeš, 1999, Eq. (17)). If the ray cell touches a caustic, two or three (in 3-D) different sets of values w_1, w_2, w_3, w_B may be found within interval $\langle 0, 1 \rangle$ for the same point $\mathbf{X}$, see Fig. 10. Different coordinates w_1, w_2, w_B of the same point $\mathbf{X}$ correspond to different

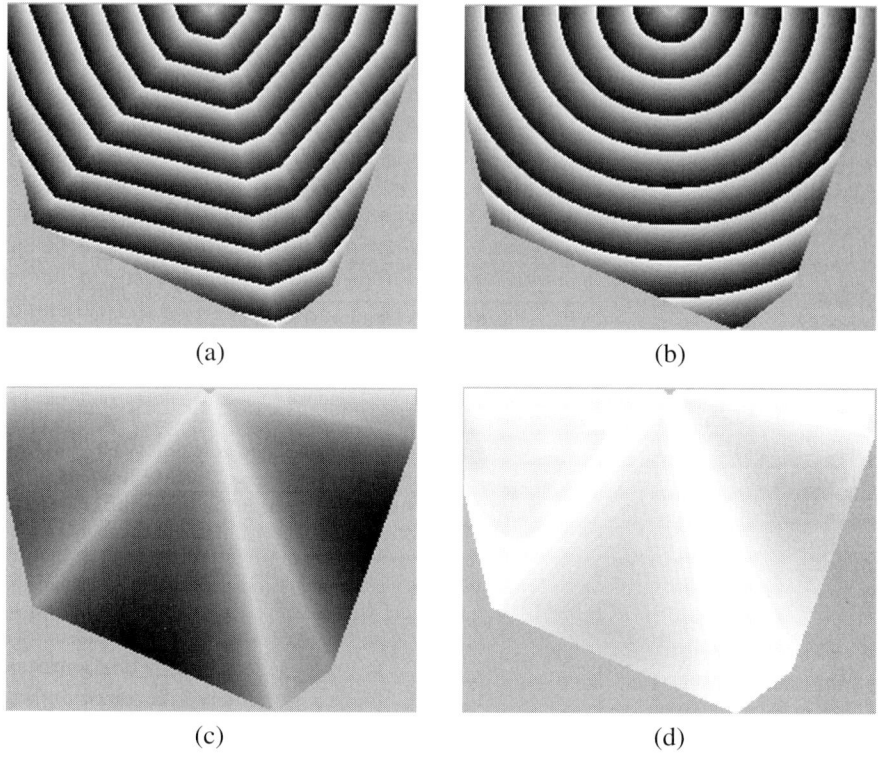

FIG. 11. (a) Travel times in a homogeneous model interpolated bilinearly within wide 3-D ray tubes. One grey-scale cycle from white to dark grey corresponds to 0.3 s. The solid light-grey region is not covered by ray cells. (b) Travel times in a homogeneous model calculated using the bicubic interpolation within the same wide 3-D ray tubes as in (a). The maximum travel time is about 2.7 s. The solid light-grey region is not covered by ray cells. (c) Differences of the bilinearly interpolated travel times from the exact solution. The maximum difference is about 0.308 s. The grey scale is the same as in (a). The solid light-grey region is not covered by ray cells. (d) Differences of the bicubically interpolated travel times from the exact solution. The maximum difference is about 0.017 s. The grey scale is the same as in (a). The solid light-grey region is not covered by ray cells.

ray parameters γ_1, γ_2, and the interpolation should be performed for each set of coordinates w_1, w_2, w_B.

In each ray cell, travel time and other quantities are interpolated as functions of local coordinates w_1, w_2, w_B. Since both the travel time and its gradient are known at all vertices of a ray cell, the travel time can be approximated by bicubic interpolation (Bulant and Klimeš, 1999, Eqs. (25)–(27) and (35)), see Fig. 11b. The application of the second derivatives of travel time during interpolation should be avoided because they may be infinite. Other quantities (e.g., amplitudes) may be approximated by bilinear interpolation (Bulant and Klimeš,

1999, Eqs. (18) and (19)). Refer to Figs. 11a–d for the comparison between bicubic and bilinear interpolation of travel time.

The bicubic interpolation fits the values and the first spatial partial derivatives at the six vertices of the ray cell. The functional values at each cell face depend on the functional values and derivatives at the vertices of the face only. The interpolated values are thus continuous across the cell faces. The gradient of the interpolated function is continuous at the triangular cell faces and along the cell edges, but may be discontinuous at the tetragonal cell faces separating ray tubes. The bicubic interpolation fits linear and quadratic functions exactly.

The described algorithm incorporates both the decision, whether a receiver lies in the ray cell, and the interpolation to the receiver. It is applicable to all ray cells formed by six, five or four points. All the quantities computed along rays may be interpolated, using the values of the quantities only at the vertices of the corresponding ray cell.

2.3. Dynamic Ray Tracing. Paraxial Ray Methods

Ray tracing can be used to compute the ray as a spatial trajectory, and to determine travel time τ, slowness vector **p**, ray-velocity vector $\mathcal{U}$ and vector η at any point of the ray. These quantities, however, are not known in the vicinity of the traced ray. To find them in the vicinity of the traced ray, it would be necessary to compute new rays. In addition, geometrical spreading and ray amplitudes cannot be computed by single ray tracing, without considering vicinity rays.

These disadvantages can be reduced by an additional procedure, called *dynamic ray tracing* (DRT), or also "paraxial ray tracing". DRT consists in solving a system of linear ordinary differential equations of the first order along a selected central ray (DRT system). It yields the derivatives of coordinates of points forming the ray and of the corresponding slowness vectors with respect to, e.g., ray parameters. DRT was originally used to determine geometrical spreading, needed to calculate the amplitudes along the ray. Recently, however, the number of applications of DRT has increased so much that DRT should be a part of any ray tracing computer package. It can be used to determine the DRT propagator matrix, which can be applied to compute all kinematic quantities not only along a single ray, but also in some vicinity of the ray. We call this vicinity the paraxial vicinity of the ray, and speak of paraxial rays, paraxial travel times, paraxial slowness vectors, etc. The DRT propagator matrix can also be used to solve various paraxial boundary-value ray tracing problems of a general four-parametric system of rays.

The DRT system can be expressed in many forms and in various coordinate systems. The most common coordinate systems, in which the DRT system is expressed, are general Cartesian coordinate systems and ray-centred coordinate systems connected with the central ray. The DRT in Cartesian coordinates is described in Section 2.3.1, and in ray-centred coordinates in Section 2.3.2. In Section 2.3.3, the computation of DRT propagator matrices is discussed, and in

Sections 2.3.4 and 2.3.5, the application of DRT propagator matrices in the computation of higher-order partial derivatives of the travel time field is described. Finally, Section 2.3.6 contains a list of some other applications of DRT propagator matrices.

2.3.1. Dynamic Ray Tracing in Cartesian Coordinates

In global Cartesian coordinates, DRT is designed to compute six quantities Q_n and P_n ($n = 1, 2, 3$) along the ray,

$$Q_n = \frac{\partial x_n}{\partial \gamma}, \qquad P_n = \frac{\partial p_n}{\partial \gamma}, \tag{38}$$

where γ is a chosen ray coordinate (e.g., initial take-off angle) or some initial parameter of the ray (e.g., x_{i0}, p_{j0}), or some other parameter affecting the ray. The DRT system for Q_n and P_n is then simply obtained by taking the partial derivatives of (24) with respect to γ:

$$\frac{dQ_n}{d\tau} = A_{nm}Q_m + B_{nm}P_m,$$
$$\frac{dP_n}{d\tau} = -C_{nm}Q_m - D_{nm}P_m. \tag{39}$$

Here

$$A_{nm} = \frac{\partial^2 \mathcal{H}}{\partial p_n \partial x_m}, \qquad B_{nm} = \frac{\partial^2 \mathcal{H}}{\partial p_n \partial p_m},$$
$$C_{nm} = \frac{\partial^2 \mathcal{H}}{\partial x_n \partial x_m}, \qquad D_{nm} = \frac{\partial^2 \mathcal{H}}{\partial x_n \partial p_m} \tag{40}$$

and $\mathcal{H}$ denotes the Hamiltonian, which is a homogeneous function of the second degree in p_i (with a possible additional constant). The variable τ along the ray again represents the travel time.

Thus, DRT system (39) consists of six linear ordinary differential equations. At the initial point of the ray, the initial values of Q_n and P_n, corresponding to the eikonal equation (20), should be chosen so that they satisfy the constraint relation:

$$\mathcal{U}_i P_i - \eta_i Q_i = 0 \tag{41}$$

resulting from the differentiation of (20) with respect to γ. Constraint relation (41) then remains valid along the whole ray.

Two solutions of the DRT system (39) are known analytically. The first of them is the so-called ray-tangent solution,

$$Q_i = \mathcal{U}_i, \qquad P_i = \eta_i, \tag{42}$$

and the second the non-eikonal solution

$$Q_i = \tau \mathcal{U}_i, \qquad P_i = p_i + \tau \eta_i. \tag{43}$$

The ray-tangent solution (42) satisfies the constraint relation (41), but the non-eikonal solution does not. Consequently, solution (43) does not correspond to eikonal equation (20). It may, however, be useful in the construction of the DRT propagator matrix, see Section 2.3.3.

DRT system (39) can be used both for heterogeneous isotropic and anisotropic media, only Hamiltonian $\mathcal{H}$ should be specified correspondingly. For *heterogeneous isotropic media*, the Hamiltonian is given by (21). DRT system (39) is then simple:

$$\frac{dQ_n}{d\tau} = (\partial V^2/\partial x_i) p_n Q_i + V^2 P_n,$$
$$\frac{dP_n}{d\tau} = -\frac{1}{2} V^{-2} (\partial^2 V^2/\partial x_i \partial x_n) Q_i - (\partial V^2/\partial x_n) p_i P_i. \tag{44}$$

For *heterogeneous anisotropic media*, the Hamiltonian is given by (22). The expressions for A_{nm}, B_{nm}, C_{nm} and D_{nm} are then more complicated than in isotropic media. For this reason, we do not present them here; they can be found in the ray theory literature. See, e.g., the detailed derivation in Červený (2001, Section 4.14.1). Gajewski and Pšenčík (1990) have used the relevant DRT system in computer package ANRAY.

Let us briefly discuss the physical meaning of the computed quantities Q_n and P_n, given by (38). Quantities Q_n and P_n show how point x_n and slowness vector p_n at the point on the ray change in the vicinity of the ray, when parameter γ is changed. Thus, using the results of DRT, we can approximately compute various ray-theory quantities (rays, travel times, slowness vectors, etc.) in the "paraxial" vicinity of the ray. We then speak of paraxial rays, paraxial travel times, paraxial slowness vectors, etc.

2.3.2. Dynamic Ray Tracing in Ray-Centred Coordinates

The DRT in ray-centred coordinates for heterogeneous isotropic media was introduced to seismology by Popov and Pšenčík (1978a, 1978b). In this section, we first introduce the ray-centred coordinate system in the form which can be used both for isotropic and anisotropic media. Thereafter, we shall discuss the relevant DRT systems.

Let us consider an arbitrarily selected ray Ω and call it the central ray. Along it, let us introduce the ray-centred coordinates q_1, q_2, q_3, in which central ray Ω is the q_3-coordinate axis. Coordinate q_3 may be an arbitrary monotonic variable along the ray. We introduce q_3 by the relation

$$q_3 = q_{30} + \int_{\tau_0}^{\tau} w(\tau) \, d\tau, \tag{45}$$

where the integration is taken along ray Ω. Consequently, q_3 equals τ for $w(\tau) = 1$, q_3 equals the arclength for $w(\tau) = \mathcal{U}(\tau)$, etc. Ray-centred coordinates q_1 and

q_2 are introduced as Cartesian coordinates in a plane tangent to the wavefront, intersecting central ray Ω at the point specified by q_3. At points along ray Ω, coordinates q_1 and q_2 are zero. The mutual relation between the global Cartesian coordinates x_i and the ray-centred coordinates q_k may be expressed as follows:

$$x_i(q_n) = x_i^\Omega(q_3) + H_{iN}(q_3) q_N. \tag{46}$$

Here $x_i = x_i^\Omega(q_3)$ represents the equation of the central ray (along which $q_1 = q_2 = 0$). The 3×3 transformation matrices from ray-centred to Cartesian coordinates and back are denoted by $H_{im} = \partial x_i / \partial q_m$ and $\bar{H}_{mi} = \partial q_m / \partial x_i$, and satisfy relations

$$H_{im} \bar{H}_{mj} = \delta_{ij}, \qquad \bar{H}_{mi} H_{in} = \delta_{mn}. \tag{47}$$

Elements H_{i3} and $\bar{H}_{3i}$ are given by simple explicit relations:

$$H_{i3} = w^{-1} \mathcal{U}_i, \qquad \bar{H}_{3i} = w p_i. \tag{48}$$

Note that H_{i1} and H_{i2} represent the ith Cartesian components of basis vectors $\mathbf{H}_1$ and $\mathbf{H}_2$, situated in the plane tangent to the wavefront at Ω.

In heterogeneous anisotropic media, the ray-centred coordinate system is non-orthogonal. Vectors H_{i1}, H_{i2} and H_{i3} represent contravariant basis vectors (with $H_{i3} = w^{-1} \mathcal{U}_i$ tangent to the ray, and H_{i1}, H_{i2} tangent to the wavefront), and vectors $\bar{H}_{1i}$, $\bar{H}_{2i}$ and $\bar{H}_{3i}$ covariant basis vectors (with $\bar{H}_{3i} = w p_i$, perpendicular to the wavefront, and $\bar{H}_{1i}$, $\bar{H}_{2i}$ perpendicular to the ray).

An efficient algorithm has been proposed to compute the contravariant basis vectors $\mathbf{H}_1$ and $\mathbf{H}_2$ along ray Ω, which is the same for isotropic and anisotropic media (Klimeš, 2006c, Section 5.4). They can be determined using a simple ordinary differential equation

$$\frac{d H_{IM}}{d\tau} = -\frac{(H_{lM} \cdot \eta_l) p_i}{p_k p_k}. \tag{49}$$

Vectors $\boldsymbol{\eta}$ and $\mathbf{p}$ are known from ray tracing, see (26) for $\boldsymbol{\eta}$. The ordinary differential equation (49) does not depend on w, and may be used for any monotonic coordinate q_3. At the initial point of ray Ω, we choose $\mathbf{H}_1$, $\mathbf{H}_2$ as mutually perpendicular unit vectors, perpendicular to slowness vector $\mathbf{p}$. Then $\mathbf{H}_1$ and $\mathbf{H}_2$, given by (49), have the same property along the whole ray. Actually, it is sufficient to calculate only one of the basis vectors $\mathbf{H}_1$, $\mathbf{H}_2$, say $\mathbf{H}_1$; the second may then be determined from the first and from the slowness vector. The determination of the remaining elements of the transformation matrices is straightforward,

$$\mathbf{H}_2 = \frac{\mathbf{p} \times \mathbf{H}_1}{|\mathbf{p} \times \mathbf{H}_1|}, \qquad \bar{\mathbf{H}}_1 = \frac{\mathbf{H}_2 \times \mathbf{H}_3}{\mathbf{H}_3 \cdot (\mathbf{H}_1 \times \mathbf{H}_2)},$$

$$\bar{\mathbf{H}}_2 = \frac{\mathbf{H}_3 \times \mathbf{H}_1}{\mathbf{H}_3 \cdot (\mathbf{H}_1 \times \mathbf{H}_2)}, \tag{50}$$

see (47) and (48).

In heterogeneous isotropic media, the contravariant and covariant basis vectors coincide, because we have chosen vectors H_{i1} and H_{i2} unit and mutually perpendicular.

The numerical solution of (49) is easy, fast and stable, and may be used both for heterogeneous isotropic and anisotropic media. Once the basis vectors of the ray-centred coordinate system are known along the ray, the Cartesian and ray-centred coordinates of any point in the vicinity of central ray Ω can be simply mutually recalculated.

There are many other possibilities of introducing the ray-centred basis vectors and relevant ray-centred coordinates than (49). For example, we could use an equation analogous to (49) directly for the covariant basis vectors $\bar{H}_{i1}$ and $\bar{H}_{i2}$, perpendicular to the ray. However, Eq. (49) with (50) probably yields the simplest algorithm to calculate the basis vectors along the ray. The advantages are that both $\mathbf{p}$ and $\eta = d\mathbf{p}/d\tau$ are known from ray tracing and that the same system (49) may be used both for heterogeneous anisotropic and isotropic media.

Similarly as in (38), we introduce six quantities $Q_n^{(q)}$ and $P_n^{(q)}$ ($n = 1, 2, 3$) along the ray,

$$Q_n^{(q)} = \frac{\partial q_n}{\partial \gamma}, \qquad P_n^{(q)} = \frac{\partial(\partial \tau/\partial q_n)}{\partial \gamma}, \tag{51}$$

where γ is a chosen ray parameter. Superscript (q) is used to remind the reader that ray-centred coordinates q_i are being used. A great advantage of the DRT system in ray-centred coordinates is that the system consisting of six equations for $Q_n^{(q)}$ and $P_n^{(q)}$ ($n = 1, 2, 3$) may be decomposed into two independent systems. The first is composed of four equations for $Q_N^{(q)}$ and $P_N^{(q)}$ ($N = 1, 2$), and the second of two equations for $Q_3^{(q)}$ and $P_3^{(q)}$. The second system can be solved analytically, and is of no interest here. Consequently, the DRT system in ray-centred coordinates is reduced to four equations for $Q_N^{(q)}$ and $P_N^{(q)}$:

$$\frac{dQ_N^{(q)}}{d\tau} = A_{NM}^{(q)} Q_M^{(q)} + B_{NM}^{(q)} P_M^{(q)},$$

$$\frac{dP_N^{(q)}}{d\tau} = -C_{NM}^{(q)} Q_M^{(q)} - D_{NM}^{(q)} P_M^{(q)}. \tag{52}$$

Here

$$A_{NM}^{(q)} = \bar{H}_{Ni} H_{jM} A_{ij} - d_{NM}, \qquad B_{NM}^{(q)} = \bar{H}_{Ni} \bar{H}_{Mj} B_{ij},$$
$$C_{NM}^{(q)} = H_{iN} H_{jM}(C_{ij} - \eta_i \eta_j), \qquad D_{NM}^{(q)} = H_{iN} \bar{H}_{Mj} D_{ij} - d_{MN}, \tag{53}$$

and

$$d_{NM} = \bar{H}_{Ni} \frac{dH_{iM}}{d\tau} = -(\bar{H}_{Ni} p_i)(H_{lM} \eta_l)/(p_k p_k). \tag{54}$$

The 3×3 matrices A_{ij}, B_{ij}, C_{ij}, D_{ij} are given by (40), η_i by (26). Basis vector H_{i1} is calculated along the ray using the ordinary differential equation (49), H_{i2} with $\mathbf{H}_2 = \mathcal{C}(\mathbf{p} \times \mathbf{H}_1)$, and basis vectors $\bar{H}_{Mi}$ are calculated from H_{iM} using (47). For a detailed derivation and discussion see Klimeš (1994, 2006b, 2006c).

Thus, the DRT system in ray-centred coordinates consists of four equations only, and is not more complicated than the DRT system in Cartesian coordinates. The only additional thing we must do is solve (49) for $\mathbf{H}_1$. Basis vectors $\mathbf{H}_1$, $\mathbf{H}_2$ are, however, useful even in other applications. DRT system (52) in ray-centred coordinates is used in 3-D computer package CRT (Červený et al., 1988; Klimeš, 2006b).

In *heterogeneous anisotropic media*, the ray-centred coordinate system is curvilinear and non-orthogonal, as the ray is not perpendicular to the wavefront.

In *heterogeneous isotropic media*, the dynamic ray tracing system in ray-centred coordinates is considerably simpler. and the ray-centred coordinate system is orthogonal. The DRT system in ray-centred coordinates reads,

$$\frac{dQ_N^{(q)}}{d\tau} = V^2 P_N^{(q)}, \qquad \frac{dP_N^{(q)}}{d\tau} = -V^{-1} V_{NK}^{(q)} Q_K^{(q)}. \tag{55}$$

Here

$$V_{NK}^{(q)} = \frac{\partial^2 V}{\partial q_N \partial q_K} = H_{iN} H_{jK} \frac{\partial^2 V}{\partial x_i \partial x_j}. \tag{56}$$

DRT system (55) in ray-centred coordinates is used in 2-D computer package SEIS, see Červený and Pšenčík (1984), and in 3-D computer package CRT, see Červený et al. (1988).

The DRT system (52) can be slightly modified if we use wavefront orthonormal basis vectors instead of the ray-centred coordinate system. See (Červený, 2001, Section 4.2.2).

2.3.3. DRT Propagator Matrix

The dynamic ray tracing system consists of linear ordinary differential equations of the first order. Consequently, it allows the fundamental matrix, consisting of linearly independent solutions of the DRT system, to be introduced. This can be done by specifying the fundamental matrix by the identity matrix at an arbitrarily selected point $\tau = \tau_0$ of the ray. We call this fundamental matrix the *DRT propagator matrix*, and denote it $\mathbf{\Pi}(\tau, \tau_0)$. Clearly, the DRT propagator matrix $\mathbf{\Pi}(\tau, \tau_0)$ depends on variable τ and on the initial time τ_0. Once a routine for the solution of the DRT system is available, the computation of the DRT propagator matrix is easy: we merely perform dynamic ray tracing along the relevant ray, starting from τ_0, with the initial conditions specified by the identity matrix.

As the DRT system in Cartesian coordinates consists of six linearly independent equations, the DRT propagator matrix $\mathbf{\Pi}(\tau, \tau_0)$ in Cartesian coordinates is

6×6. Analogously, the DRT propagator matrix $\Pi(\tau, \tau_0)$ in ray-centred coordinates is 4×4. The DRT propagator matrices have several interesting and important properties, valid both for 6×6 and 4×4 matrices; only the size of the matrices involved is different. For this reason, we will discuss only the properties of the 4×4 DRT propagator matrix $\Pi(\tau, \tau_0)$ in ray-centred coordinates; the properties of the 6×6 DRT propagator matrix in Cartesian coordinates are quite analogous.

(a) If the DRT propagator matrix $\Pi(\tau, \tau_0)$ is known, the solution of the DRT system can be easily found at point τ of the ray for arbitrary initial conditions at τ_0:

$$\begin{pmatrix} \mathbf{Q}^{(q)}(\tau) \\ \mathbf{P}^{(q)}(\tau) \end{pmatrix} = \Pi(\tau, \tau_0) \begin{pmatrix} \mathbf{Q}^{(q)}(\tau_0) \\ \mathbf{P}^{(q)}(\tau_0) \end{pmatrix}. \tag{57}$$

(b) The DRT propagator matrix $\Pi(\tau, \tau_0)$ is symplectic,

$$\Pi^T(\tau, \tau_0) \mathbf{J} \Pi(\tau, \tau_0) = \mathbf{J}, \quad \text{where } \mathbf{J} = \begin{pmatrix} \mathbf{0} & \mathbf{I} \\ -\mathbf{I} & \mathbf{0} \end{pmatrix}. \tag{58}$$

Here $\mathbf{I}$ and $\mathbf{0}$ are 2×2 identity and null matrices. It follows from (58) that $\det \Pi(\tau, \tau_0) = 1$, so that $\Pi(\tau, \tau_0)$ is regular along the whole ray.

(c) The DRT propagator matrix $\Pi(\tau, \tau_0)$ satisfies the chain rule,

$$\Pi(\tau, \tau_0) = \Pi(\tau, \tau_1) \Pi(\tau_1, \tau_0), \tag{59}$$

where τ_1 specifies an arbitrary point on the ray. Equation (59) also implies

$$\Pi^{-1}(\tau_1, \tau_0) = \Pi(\tau_0, \tau_1), \tag{60}$$

for points τ_0 and τ_1 arbitrarily situated on the ray.

(d) The DRT propagator matrix can also be defined for rays interacting with interfaces (reflections, transmissions). We merely introduce the *interface propagator matrix* $\Pi(\tau^r, \tau^{inc})$, where τ^{inc} corresponds to the point of incidence, and τ^r to the corresponding point of reflection/transmission (at the same interface), $\tau^r = \tau^{inc}$. For the point of incidence τ^{inc} situated between τ_0 and τ, we then modify (59) as follows:

$$\Pi(\tau, \tau_0) = \Pi(\tau, \tau^r) \Pi(\tau^r, \tau^{inc}) \Pi(\tau^{inc}, \tau_0). \tag{61}$$

Thus, it is only necessary to insert the interface propagator matrix $\Pi(\tau^r, \tau^{inc})$ into the proper place of the chain. For more details on 6×6 interface propagator matrices in Cartesian coordinates refer to Farra and Le Bégat (1995) and Moser (2004). Of course, (61) may be further chained.

In practical applications, the 4×4 DRT propagator matrix $\Pi(\tau, \tau_0)$ in ray-centred coordinates is more useful than the 6×6 propagator matrix $\Pi(\tau, \tau_0)$ in Cartesian coordinates. The reason is that the general system of rays in heterogeneous media (without structural interfaces) is four-parametric, and the 4×4 DRT propagator matrix strictly corresponds to this system of rays. Contrary to this, the 6×6 DRT propagator matrix in Cartesian coordinates includes two redundant solutions (ray-tangent and non-eikonal) which must be excluded if we wish to apply it to a four-parametric system of rays. Consequently, the solution of various boundary-value problems for a four-parametric system of rays is more straightforward if we use the 4×4 DRT propagator matrices.

Let us briefly comment on determining the relative geometrical spreading from the 4×4 DRT propagator matrix in ray-centred coordinates. It is common to express the 4×4 DRT propagator matrix $\mathbf{\Pi}(\tau, \tau_0)$ in ray-centred coordinates in the following form:

$$\mathbf{\Pi}(\tau, \tau_0) = \begin{pmatrix} \mathbf{Q}_1(\tau, \tau_0) & \mathbf{Q}_2(\tau, \tau_0) \\ \mathbf{P}_1(\tau, \tau_0) & \mathbf{P}_2(\tau, \tau_0) \end{pmatrix}. \tag{62}$$

The 2×2 submatrix $\mathbf{Q}_2(\tau, \tau_0)$ can be used to determine the relative geometrical spreading $\mathcal{L}(\tau, \tau_0)$ of the Green function,

$$\mathcal{L}(\tau, \tau_0) = \left|\det \mathbf{Q}_2(\tau, \tau_0)\right|^{1/2}. \tag{63}$$

It is not difficult to show that the relative geometrical spreading is reciprocal, $\mathcal{L}(\tau, \tau_0) = \mathcal{L}(\tau_0, \tau)$.

2.3.4. Second-Order Spatial Derivatives of Travel Time

The first-order partial travel-time derivatives with respect to Cartesian coordinates x_i (components of slowness vector $\mathbf{p}$) are known along the ray from ray tracing, see (24). To compute the second-order and higher-order derivatives, DRT is needed.

The second-order partial derivatives of the travel-time field are obtained directly as a by-product of DRT. If we denote by $\mathbf{M}$ the 2×2 matrix of the second derivatives of the travel time field with respect to ray-centred coordinates q_1, q_2, we obtain

$$\mathbf{M} = \mathbf{P}^{(q)}(\mathbf{Q}^{(q)})^{-1}. \tag{64}$$

The 2×2 matrices $\mathbf{Q}^{(q)}$ and $\mathbf{P}^{(q)}$ are defined by their components

$$Q^{(q)}_{NK} = \frac{\partial q_N}{\partial \gamma_K}, \quad P^{(q)}_{NK} = \frac{\partial(\partial \tau/\partial q_N)}{\partial \gamma_K}, \tag{65}$$

and can be calculated using DRT system (52). Analogously to (64), we can also introduce the 3×3 matrix $\mathbf{N} = \mathbf{PQ}^{-1}$ of the second derivatives of the travel-time field with respect to Cartesian coordinates x_i and the 3×3 matrices $\mathbf{P}$ and $\mathbf{Q}$ calculated by DRT system (39).

Matrix (64) satisfies the nonlinear matrix Riccati equation

$$\frac{d\mathbf{M}}{d\tau} = -\mathbf{MB}^{(q)}\mathbf{M} - \mathbf{MA}^{(q)} - \mathbf{D}^{(q)}\mathbf{M} - \mathbf{C}^{(q)} \tag{66}$$

following from DRT system (52). In isotropic media, this Riccati equation reads

$$\frac{d\mathbf{M}}{d\tau} = -V^2 \mathbf{M}\mathbf{M} - V^{-1}\mathbf{V}^{(q)}. \tag{67}$$

Here V is the velocity and $\mathbf{V}^{(q)}$ the 2×2 matrix (56). The disadvantage of the Riccati DRT system is that it is nonlinear, whereas the DRT system (52) is linear. The Riccati DRT system is, however, broadly used in the theory of Gaussian beams, see Section 5.1.

2.3.5. Third-Order and Higher-Order Spatial Derivatives of Travel Time

Once DRT has been performed, the third-order and higher-order partial derivatives of travel time can be computed by simple numerical quadratures along rays. The general form of the equations for the third-order and higher-order partial derivatives of travel time was suggested by Babich *et al.* (1985). The explicit equations for the third-order and higher-order partial derivatives of travel time in both isotropic and anisotropic smooth models without interfaces were derived by Klimeš (2002a). The equations are derived for a general Hamiltonian. The derivatives of travel time are expressed in the form of integrals along the ray, with the appropriate initial values. The integrands are composed of partial derivatives of travel time of orders lower than the one being computed, and of the phase-space partial derivatives of the Hamiltonian. The derivatives of the Hamiltonian contain partial derivatives of density-normalized elastic moduli up to the order corresponding to that of the calculated derivatives of travel time.

Examples of possible applications of the third-order and higher-order partial derivatives of travel time are:

(a) Estimation of the *accuracy of travel-time interpolation within ray cells*, and the control of the accuracy by optimizing the size of ray cells in wavefront tracing and in controlled initial-value ray tracing followed by travel-time interpolation.
(b) Estimation and control of the *accuracy of the paraxial ray approximation*.
(c) Estimation of the *accuracy of the paraxial approximation of Gaussian beams*.
(d) Estimation of the *accuracy of the paraxial expansions of Fresnel edge waves* approximating the diffractions.
(e) Application to the *local and uniform asymptotic approximations* in heterogeneous media. They may be useful in deriving general high-frequency asymptotic approximations of the Maslov and Gaussian-beam integrals with respect to the ray parameters, such as the Airy approximations (third-order Taylor expansion of travel time with respect to ray parameters) and the Pearcey approximations (fourth-order Taylor expansion of travel time with respect to ray parameters).

(f) Calculating or approximating the *higher-order terms in the ray series*.
(g) Determination of the *dependence of the smoothness of travel time on the smoothness of the model*. For example, continuous and finite nth velocity derivatives ($n \geq 2$) imply continuous and finite nth derivatives of travel time off caustics.
(h) The third- and higher-order spatial derivatives of travel time are required when calculating the *third- and higher-order derivatives of travel time with respect to perturbation parameters*.

Note that the expressions for the partial derivatives of the vectorial amplitude may also be derived (e.g., Klimeš, 2006a).

2.3.6. Applications of DRT Propagator Matrix. Paraxial Ray Methods

The DRT propagator matrix has found many applications in the ray method, in heterogeneous both isotropic and anisotropic layered media. A list of several of them follows:

1. Paraxial ray theory. Computation of paraxial rays, paraxial travel times, and paraxial slowness vectors in the vicinity of the ray (Červený, 2001, Section 4.6).
2. Computation of the curvature of the wavefront along the ray (see Section 2.3.4).
3. Computation of the ray Jacobian, geometrical spreading, relative geometrical spreading, and phase shift due to caustics (see Sections 2.4 and 2.6).
4. Computation of the third-order and higher-order spatial derivatives of travel time (see Section 2.3.5).
5. Computation of the second-order and higher-order perturbations of travel time (see Section 3.2).
6. Computation of the perturbations or spatial derivatives of amplitude (Klimeš, 2006a).
7. Computation of paraxial Fresnel volumes (Červený and Soares, 1992).
8. Investigation of chaotic rays and evaluation of Lyapunov coefficients (see Section 6).
9. Construction of surface-to-surface propagator matrices (Hubral *et al.*, 1992a, 1992b; Schleicher *et al.*, 1993; Sun, 2004).
10. Two-point ray tracing.
11. Various boundary-value problems for the four-parametric system of paraxial rays.
12. Investigation of reciprocity relations along a ray (see Sections 2.6.2 and 2.6.3).
13. Determination of geometrical spreading from travel time measurements (Červený, 2001).
14. Factorization of geometrical spreading. Fresnel-zone matrices.

15. Computation of Gaussian beams and Gaussian packets (see Section 5).
16. Maslov method (see Section 5.4.6).
17. Numerous applications in ray perturbation theory (Farra, 1999).
18. Isochrone perturbation (Iversen, 1996) and computation of the curvature of an isochrone (Iversen, 2001a, Eqs. (4), (9) and (10)). "Velocity rays" (Iversen, 2001c).
19. Point to curve ray tracing (Hanyga and Pajchel, 1995; Hanyga, 1996) and its applications, e.g., "isochrone rays" (Iversen, 2001b, Eqs. (17) and (18), 2004, Eqs. (37)–(40)).
20. Applications in migration methods (Tygel et al., 1994; Sun, 2004).
21. Kirchhoff–Helmholtz integrals (Tygel et al., 1994; Chapman, 2004).
22. Applications in various diffraction problems of elementary waves. Edge waves and tip waves (Klem-Musatov and Aizenberg, 1984; Klem-Musatov, 1994; Hanyga et al., 2001). For more details and many references see Červený (2001, Chapter 4).

2.4. Ray-Theory Amplitude

Once the ray tracing and dynamic ray tracing have been performed, and the eigenvector $\mathbf{g}^{(m)}(x_i)$ of the Christoffel matrix and geometrical spreading have been determined along the ray, it is not difficult to compute the vectorial complex-valued amplitude $\mathbf{U}(x_i)$ of an arbitrary elementary wave. It is, however, still necessary to take into account the conditions at the initial point of the ray, at the end point of the ray, and at points of contact of the ray with interfaces (or with the Earth's surface). These conditions are known theoretically for most situations of practical interest, and many of them are included in the available ray program packages.

For this reason, we shall be brief. (a) The initial point of the ray often represents a point source. The most important point sources are single-force and moment-tensor point sources, including the explosive point source. The source may be situated in a smooth medium, at the surface of the Earth, or at an interface. Ray-theory expressions are known even for a line source (arbitrarily curvilinear) and for a surface source (arbitrarily curved). (b) The most common conditions for the receiver point are expressed in terms of conversion coefficients. The conversion coefficients may correspond to a receiver situated at the Earth's surface or at an interface. (c) The algorithms for evaluating the (complex-valued) reflection/transmission coefficients at interfaces and at the Earth's surface are well known for both isotropic and anisotropic media. Displacement R/T coefficients, or normalized R/T coefficients (normalized with respect to the energy flux across the interface) are mostly used. A useful property of the normalized R/T coefficients is that they satisfy a reciprocity relation.

The vectorial complex-valued amplitude $\mathbf{U}(x_i)$ of an elementary wave propagating in a heterogeneous anisotropic medium is expressed in terms of the eigenvector $\mathbf{g}^{(m)}(x_i)$ of the Christoffel matrix and scalar amplitude $A(x_i)$, see (14).

In Section 2.4.1, we shall discuss the determination of scalar amplitude A, in Section 2.4.2 the behaviour of $\mathbf{U}$ in regular regions (non-degenerate case), and in Section 2.4.3 the behaviour of $\mathbf{U}$ in the vicinity of S-wave singularities and in isotropic media.

2.4.1. Determination of the Scalar Amplitude

In the zero-order approximation of the ray method, scalar amplitude A is a solution of transport equation (16), in which $\mathcal{U}$ given by (17) coincides with the ray-velocity vector (25) corresponding to Hamiltonian (22). For other Hamiltonians, e.g., for Hamiltonian (23), we may define reference scalar amplitude A as a solution of transport equation (16) with $\mathcal{U}$ given by (25).

The transport equation can be simply solved along the ray in terms of the ray Jacobian J. We define J by the following expression

$$J = \left(\frac{\partial \mathbf{x}}{\partial \gamma_1} \times \frac{\partial \mathbf{x}}{\partial \gamma_2} \right) \cdot \mathbf{t}, \tag{68}$$

where $\mathbf{t} = \mathcal{U}/\mathcal{U}$ is the unit vector tangent to the ray, and the derivatives are taken along the wavefront. Alternatively, we can define ray Jacobian J as the Jacobian $J^{(T)}$ of the transformation from ray coordinates $\gamma_1, \gamma_2, \gamma_3 = \tau$ to Cartesian coordinates x_1, x_2, x_3, normalized with respect to ray velocity $\mathcal{U}$. It is possible to prove that

$$\nabla \cdot \mathcal{U} = (J\mathcal{U})^{-1} \mathrm{d}(J\mathcal{U})/\mathrm{d}\tau, \tag{69}$$

where the derivative with respect to τ is taken along the ray. Transport equation (16), applied along the ray, then yields

$$\frac{\mathrm{d}(\sqrt{\varrho J \mathcal{U}} A)}{\mathrm{d}\tau} = 0. \tag{70}$$

This can be simply solved for A:

$$A(\tau) = \left[\frac{\varrho(\tau_0)\mathcal{U}(\tau_0)J(\tau_0)}{\varrho(\tau)\mathcal{U}(\tau)J(\tau)} \right]^{1/2} A(\tau_0). \tag{71}$$

Ray Jacobian $J(\tau)$ may be positive, negative or zero at different points of the ray. Points of the ray where $J(\tau) = 0$ are called caustic points. At these points, the ray method fails; it yields infinite amplitudes. Since the ray Jacobian may be negative, the square-root factor in (71) may introduce a phase shift. Consequently, it is useful to express Eq. (71) in the following form:

$$A(\tau) = \left[\frac{\varrho(\tau_0)\mathcal{U}(\tau_0)|J(\tau_0)|}{\varrho(\tau)\mathcal{U}(\tau)|J(\tau)|} \right]^{1/2} A(\tau_0) \exp[iT^C(\tau, \tau_0)]. \tag{72}$$

Here $T^C(\tau, \tau_0)$ is called the phase shift due to caustics. For more details, see Section 2.6.3. The non-negative function $\sqrt{|J(\tau)|}$ is often called the geometrical spreading. However, this terminology has not been firmly established.

Thus, we express scalar amplitude $A(\tau)$ in terms of $|J(\tau)|$ and $T^C(\tau, \tau_0)$, which can be both determined by dynamic ray tracing.

Often, some alternative quantities instead of the ray Jacobian are used, e.g., Jacobian $J^{(T)}$, or the scalar surface element $\Omega^{(T)}$, cut out of the wavefront by the ray tube and normalized with respect to $d\gamma_1 d\gamma_2$. The relation between J, $J^{(T)}$ and $\Omega^{(T)}$ is as follows:

$$J\mathcal{U} = J^{(T)} = \mathcal{C}\Omega^{(T)}. \tag{73}$$

Equations (71) and (72) for $A(\tau)$ can then be expressed in terms of $J^{(T)}$ or $\Omega^{(T)}$.

Equations (71) and (72) require a modification for a point or a line source at τ_0, for a ray crossing the structural interface, and for an initial and/or end point of the ray, situated at a structural interface or at the Earth's surface. For these situations, see Červený (2001).

2.4.2. Polarization in a Non-Degenerate Case

Let us consider an elementary wave, propagating in a heterogeneous isotropic or anisotropic medium, corresponding to eigenvalue G_m of the Christoffel matrix (12), well separated from the other two eigenvalues. *Polarization vector* **g** then coincides with the relevant eigenvector $\mathbf{g}^{(m)}$ of the Christoffel matrix, and the vectorial amplitude is given by (14).

At any point of the ray, at which the slowness vector is known and at which the Christoffel matrix is not degenerate, polarization vectors can be determined uniquely by conventional methods. The three polarization vectors $\mathbf{g}^{(1)}$, $\mathbf{g}^{(2)}$ and $\mathbf{g}^{(3)}$, corresponding to one wavefront normal, are mutually orthogonal.

Equation (14) for the amplitude vector **U** can be used only if the eigenvalue G_m, corresponding to the elementary wave under consideration, is sufficiently different from the others. This applies to P waves propagating in both isotropic and anisotropic media, and to non-degenerate S waves in anisotropic media. As we can easily see from (14), in this case the elementary wave under consideration is linearly polarized.

2.4.3. Polarization in the Vicinity of S-Wave Singularities or in Isotropic Media

For S waves propagating in isotropic media, in weakly anisotropic media, and for S waves propagating close to S-wave singularities in heterogeneous anisotropic media, the complex-valued amplitude vector **U** must be considered in the form

$$\mathbf{U} = B\mathbf{g}^{(1)} + C\mathbf{g}^{(2)}. \tag{74}$$

Here $\mathbf{g}^{(1)}$ and $\mathbf{g}^{(2)}$ are eigenvectors of the Christoffel matrix (12), corresponding to eigenvalues G_1 and G_2, and B and C are two complex-valued scalar quantities. As we can see from (74), the elementary wave under consideration is elliptically polarized if B and C are "out of phase". This applies globally to S waves in heterogeneous isotropic and weakly anisotropic media, and locally to S waves propagating in heterogeneous anisotropic media close to S-wave singularities. In heterogeneous weakly anisotropic media or in the vicinity of an S-wave singularity, B and C are frequency-dependent. They can be determined from two coupled transport equations. More details are given in Section 4.

For S waves propagating in heterogeneous isotropic media, the two transport equations for the complex-valued scalar amplitudes B and C in (74) are generally coupled. They can be decoupled only in special cases. Decoupling condition (19) for isotropic media can be expressed in the form of an ordinary differential equation of the first order, which must be satisfied along the ray:

$$\frac{d\mathbf{g}^{(M)}}{d\tau} = \frac{-(\mathbf{g}^{(M)} \cdot \boldsymbol{\eta})\mathbf{p}}{\mathbf{p} \cdot \mathbf{p}}. \tag{75}$$

This equation is a special case of Eq. (33). Here η is given by (26) and is known from ray tracing. In the terminology of Riemannian geometry, we say that vector $V\mathbf{g}^{(M)}$ (where velocity V equals α or β) is transported parallely along the ray. If $\mathbf{g}^{(1)}$ and $\mathbf{g}^{(2)}$ are chosen to be mutually perpendicular unit vectors perpendicular to $\mathbf{p}$ at the initial point of the ray, Eq. (75) guarantees that they have the same property along the whole ray. Consequently, $\mathbf{g}^{(1)}$ and $\mathbf{g}^{(2)}$ are always perpendicular to the ray and do not rotate around the ray as the wave progresses. As $\mathbf{g}^{(1)}$, $\mathbf{g}^{(2)}$ and $\mathbf{p}$ are always orthogonal, it is sufficient to compute just $\mathbf{g}^{(1)}$ (or $\mathbf{g}^{(2)}$) from (75), the second vector $\mathbf{g}^{(2)}$ (or $\mathbf{g}^{(1)}$) can be simply determined using the orthogonality condition.

At interfaces between isotropic media, the polarization vectors of reflected or transmitted S waves can be chosen as two arbitrary, mutually perpendicular unit vectors, perpendicular to the ray of a generated wave. The corresponding R/T coefficients, however, must be adjusted to this choice.

2.5. Effects of Interfaces

Reflection or transmission of a wave at an interface of the first order (across which parameters of the medium change abruptly) leads to abrupt changes of most quantities calculated along a ray. Only the ray path itself and travel time remain continuous. Other quantities like the direction of the ray, the direction of the corresponding slowness vector, values of all the quantities calculated in dynamic ray tracing and, consequently, geometrical spreading and the ray amplitude change discontinuously. Let us briefly describe the transformation of the individual quantities.

Due to the continuity of the travel times of incident and generated waves along the interface, specifically due to the continuity of their first-order derivatives, the slowness vectors of the incident and generated waves must have the same projection onto the plane tangent to the interface at the point of incidence. This equality of the tangential projections of the slowness vectors is the expression of Snell's law. It holds in isotropic as well as anisotropic media. The determination of the slowness vector of a generated wave thus reduces to the determination of the component of the slowness vector along the normal v_i to the interface. The slowness vector $p_i^{(m)}$ of a generated wave can thus be expressed as

$$p_i^{(m)} = [p_i - v_i(v_k p_k)] + \xi^{(m)} v_i. \tag{76}$$

Here p_i is the slowness vector of the incident wave, v_i is the unit normal to the interface and $\xi^{(m)}$ is the sought component of the slowness vector along normal v_i. The term in the square brackets is the tangential vectorial component of the slowness vector, which is the same for incident and generated waves. The slowness vector of any generated wave propagating in an anisotropic medium must satisfy the condition of solvability of the Christoffel equation

$$\det(a_{ijkl} p_j^{(m)} p_l^{(m)} - \delta_{ik}) = 0. \tag{77}$$

Here a_{ijkl} again denotes the tensor of density-normalized elastic moduli of the medium in which the generated wave propagates. By inserting (76) into (77) we arrive at the sixth-order polynomial equation for $\xi^{(m)}$. Since the coefficients of the polynomial equation are real-valued, its complex roots appear in complex conjugate pairs. Real roots correspond to subcritical incidence, complex roots indicate supercritical incidence of the incident wave. One equation (77), with parameters a_{ijkl} specifying the medium in which incident and reflected waves propagate, holds for the three reflected waves; another equation (77), with parameters a_{ijkl} specifying the medium in which transmitted waves propagate, holds for the three transmitted waves. Since each equation has six roots, only three of them have physical meaning and the remaining three must be eliminated. Of the real-valued roots only those are acceptable for which the corresponding ray-velocity vectors (parallel to the time-averaged energy flux vectors) point into the medium, in which the generated wave propagates. Of the complex conjugate roots only those are acceptable, which correspond to inhomogeneous waves (waves with complex-valued slowness vector, whose real and imaginary parts make a non-zero angle) decreasing exponentially with distance from the interface. The limiting case, in which real-valued roots become complex-valued, is the case of critical incidence, in which the ray-velocity vector is parallel to the interface. Note that in anisotropic media, the slowness vector of the wave, generated by a critically incident wave, is generally not parallel to the interface. Since the ray-velocity vector and the slowness vector have generally different directions in anisotropic media, it may happen that the slowness and ray-velocity vectors of a generated wave point to opposite sides of the interface.

If the generated wave propagates in an isotropic medium, condition (77) simplifies. The slowness vector must satisfy the corresponding eikonal equation, from which its component along the normal to the interface can be determined. In this case, two roots are obtained for a single wave and one of them must be eliminated. The conditions of elimination are the same as in the case of an anisotropic medium. Since in an isotropic medium the ray-velocity vector and the slowness vector are parallel, the ray-velocity vector can be substituted by the slowness vector in the elimination process.

To determine slowness vectors of generated waves in an anisotropic medium, a polynomial equation of the sixth order must be solved numerically. It yields the normal components of slowness vectors of all waves generated on the relevant side of the interface. In isotropic media, simple analytic formulae exist for the determination of the slowness vector of a given generated wave alone.

Equation (76) clearly indicates that the slowness vectors of generated waves are situated in the plane of incidence defined by the slowness vector of the incident wave and the normal to the interface. In anisotropic media, the ray-velocity vectors may point out of the plane of incidence. In isotropic media, the ray-velocity vectors of all generated waves are situated in the plane of incidence.

The transformation relations at an interface for the solutions of the dynamic ray tracing system also follow from the continuity of the travel times of generated and incident waves. In this case, second-order travel-time derivatives are used, which means that the transformation is affected by the first-order derivatives of the parameters of the medium and by the curvature of the interface at the point of incidence. Due to this dependence, the transformation relations for dynamic ray tracing must be applied even at interfaces of the second order, i.e. interfaces, at which gradients of elastic parameters change abruptly while the parameters themselves are continuous.

The ray amplitudes of generated waves at the point of reflection/transmission are obtained from the ray amplitude of the incident wave by multiplying it by the corresponding reflection or transmission coefficients. In the zero-order ray approximation, the coefficients correspond to the plane-wave reflection/transmission coefficients. Mostly displacement reflection/transmission coefficients are used. Normalized coefficients, which have the convenient property of reciprocity, see Section 2.6.2, can be used instead of them. The coefficients depend on the parameters of the medium, the density and the angle of incidence. In addition to the coefficients, which are applied at the points of reflection or transmission at interfaces, the conversion coefficients should be used at sources and receivers when these are situated at interfaces.

2.6. Ray-Theory Green Function

2.6.1. Elementary Ray-Theory Green Function

A very important role in various applications in seismology and seismic exploration for oil is played by the elastodynamic Green function. The elastodynamic Green function $G_{in}(R, t, S, t_0)$ represents the ith Cartesian component of the displacement vector at location R and time t due to a single-force point source situated at location S and oriented along the nth Cartesian axis, with the time dependence $\delta(t - t_0)$. Quite analogously, we can introduce the elementary ray-theory elastodynamic Green function. Two differences are that the elementary ray-theory elastodynamic Green function does not correspond to the complete wavefield, but to the wavefield of an arbitrarily selected elementary wave, and that it is not exact, but only represents its zero-order ray approximation. The ray method can be used to obtain a surprisingly simple expression of the elementary ray-theory elastodynamic Green function for any selected elementary wave (e.g., any multiply reflected/transmitted wave), propagating in an arbitrary 3-D heterogeneous anisotropic or isotropic layered structure along any ray connecting the point source S and the receiver at R. The Fourier transform of Green function $G_{in}(R, t, S, t_0)$ with respect to $t - t_0$ reads

$$G_{in}(R, S, \omega) = \frac{g_n^{(m)}(S) g_i^{(m)}(R) \exp[i T^G(R, S) + i\omega\tau(R, S)]}{4\pi [\varrho(S)\varrho(R)\mathcal{C}(S)\mathcal{C}(R)]^{1/2} \mathcal{L}(R, S)} \mathcal{R}^C \qquad (78)$$

(no summation over m). Here $g_i^{(m)}$ is the ith Cartesian component of the relevant unit eigenvector $\mathbf{g}^{(m)}$ of the Christoffel matrix, $\mathcal{C}$ is the phase velocity, and ϱ the density. Further, $\tau(R, S)$ is the travel time from S to R, $T^G(R, S)$ the complete phase shift due to caustics along the ray from S to R (see Section 2.6.3), $\mathcal{L}(R, S)$ the relative geometrical spreading (63) along the ray from S to R, and $\mathcal{R}^C$ the product of all normalized displacement R/T coefficients at all points of contact of the ray with interfaces or with the Earth's surface, between S and R.

The factor $[\mathcal{C}(S)\mathcal{C}(R)]^{1/2}$ in the denominator of (78) corresponds strictly to the definition of the relative geometrical spreading given by (63). If the definition of the relative geometrical spreading is changed, the factor may also change, for example, to $[\mathcal{U}(S)\mathcal{U}(R)]^{1/2}$. See the detailed discussion in Schleicher et al. (2001).

Equation (78) is valid for points S and R, situated in the smooth parts of the model. For points S and R situated on an interface and/or the Earth's structure, eigenvectors $g_n^{(m)}(S)$ and $g_i^{(m)}(R)$ must be replaced by more complicated expressions.

The elementary ray-theory elastodynamic Green function may be used in similar applications as the exact elastodynamic Green function. Usually it is sufficient to consider several elementary ray-theory Green functions, corresponding to the most energetic elementary waves propagating along different rays from point S to point R (direct P, S1 and S2 waves, various primary reflected waves, etc.), and to form the ray-theory Green function in this way.

2.6.2. Reciprocity of the Ray-Theory Green Function

It is possible to prove that the elementary ray-theory elastodynamic Green function is reciprocal in the following sense:

$$G_{in}(R, S, \omega) = G_{ni}(S, R, \omega). \tag{79}$$

Here $G_{ni}(S, R, \omega)$ corresponds to the backward propagation along the same ray path, from R to S. The reciprocity relation (79) is valid for any multiply reflected (possibly converted) elementary wave propagating in a 3-D anisotropic laterally varying layered structure with any number of reflection/transmission points. Similarly, reciprocity (79) remains valid for the wavefield, composed of any number of elementary waves propagating from S to R along different ray trajectories. Reciprocity relation (79) follows simply from (78), if we take into account that travel time $\tau(R, S)$, relative geometrical spreading $\mathcal{L}(R, S)$, the complete phase shift due to caustics $T^G(R, S)$ and the product of normalized R/T displacement coefficients $\mathcal{R}^C$ are reciprocal. The reciprocity of travel time $\tau(R, S)$ is obvious. The reciprocity of $\mathcal{L}$ follows from the symplecticity of the DRT propagator matrix, and the reciprocity of $\mathcal{R}^C$ was proved by Chapman (1994). It should be noted that the product of the conventional displacement R/T coefficients is not reciprocal. For the reciprocity of the phase shift due to caustics see Section 2.6.3. See also Kendall et al. (1992).

Note that reciprocity relation (79) is valid only for the single force point source. It is not valid for other types of point sources, e.g., for the explosive source (centre of dilatation).

2.6.3. Phase Shift of the Green Function Due to Caustics. KMAH Index

In heterogeneous anisotropic media, the complete phase shift due to caustics $T^G(R, S)$ is given by the relation,

$$T^G(R, S) = T^C(R, S) + \frac{1}{2}\pi\sigma_0(S) = -\frac{1}{2}\pi\big(k(R, S) - \sigma_0(S)\big). \tag{80}$$

Here $\sigma_0(S)$ corresponds to the phase shift of the point source, and $T^C(R, S)$ is the phase shift due to caustics outside the source, between S and R. Phase shift $\sigma_0(S)$ depends on the form of the ray-velocity surface at source S and on the direction of the ray at S. For rays corresponding to the forward branch of the ray-velocity surface, $\sigma_0 = 0$ (similarly as in isotropic media). For rays corresponding to the reverse branch of the ray-velocity surface, $\sigma_0 = 1$ or even $\sigma_0 = 2$ (Pšenčík and Teles, 1996). The quantity $k(R, S)$, called the KMAH index, is an integer, changing only when the ray under consideration passes through a caustic point.

In isotropic media, the slowness surface is always convex, so that $\sigma_0(S) = 0$. Moreover, the KMAH index $k(R, S)$ is always positive. The increment of the KMAH index equals unity at a line caustic, and two at a point caustic. The caustic points, and consequently $k(R, S)$, can be found as a by-product of DRT.

We can also introduce the KMAH index for anisotropic media $k^G(R, S)$ by the relation

$$k^G(R, S) = k(R, S) - \sigma_0(S). \tag{81}$$

In this section, we consider the KMAH index in the form of (81). It is possible to prove that the KMAH index (81) is reciprocal in the following sense

$$k^G(R, S) = k^{G-}(S, R). \tag{82}$$

Here $k^G(R, S)$ corresponds to the forward ray, and $k^{G-}(S, R)$ to the reverse ray. For more details refer to Klimeš (1997, 2006d), Chapman (2004, p. 176).

Unlike in isotropic media, where the increment of the KMAH index is always positive, the increment of the KMAH index of S waves in anisotropic media may be either positive, or negative, depending on the convexity or non-convexity of the slowness surface. Lewis (1965) derived a general phase-shift rule for a point caustic, expressed in terms of the signature of the matrix of the second derivatives of travel time. Garmany (2000) expressed a general phase-shift rule for a line (simple) caustic in terms of the second derivatives of the eigenvalue of the Christoffel matrix with respect to the slowness vector. Bakker (1998) derived equations for the phase shift corresponding to a general wavefield, due to both line (simple) and point caustics in 3-D anisotropic media. Bakker's rules are expressed in terms of the second derivatives of the eigenvalue of the Christoffel matrix with respect to the slowness vector.

In the following section, a simple example is given to illustrate simultaneously the positive and negative phase shifts in anisotropic media, and the reciprocity (82) of the phase shift of the Green function.

2.6.4. Example of Phase Shifts

Assume a 2-D homogeneous anisotropic medium with a triplication on the S-wave ray-velocity surface. Select a ray corresponding to the reverse branch of the ray-velocity surface. Assume that the homogeneous anisotropic medium is separated from the homogeneous isotropic medium by a planar interface perpendicular to the selected ray. For the sake of simplicity, we may assume that the (phase and ray) velocities in the direction of the selected ray are similar (or equal) in both media.

The slowness vectors of the paraxial rays, propagating in the anisotropic medium to one side of the selected ray, point to the other side of the selected ray, see Fig. 12. On entering the isotropic medium, the paraxial rays are refracted to the direction of the slowness vector and cross the selected ray at the caustic.

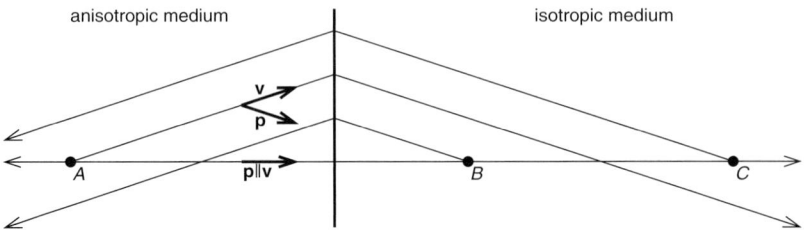

FIG. 12. The selected ray and paraxial rays from points A, B, C in the discussed example. In the anisotropic halfspace, paraxial ray-velocity vector $\mathbf{v}$ and slowness vector $\mathbf{p}$ are deflected in opposite directions.

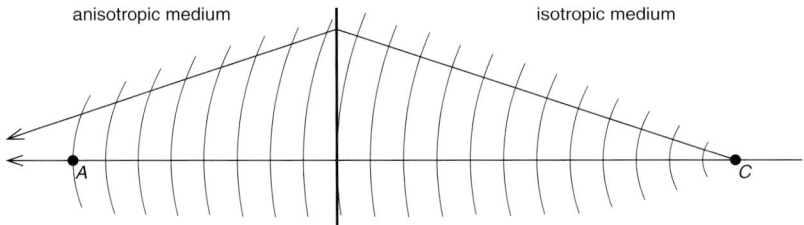

FIG. 13. $C \to A$: There is no caustic along the ray from point C to point A. As the KMAH index is defined zero at the point source in the isotropic medium, the resulting KMAH index at point A is 0.

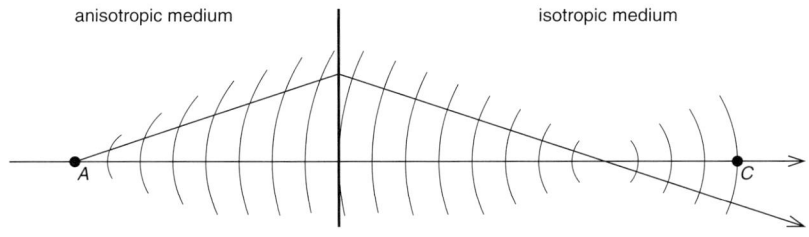

FIG. 14. $A \to C$: The initial KMAH index at point A on the reverse wavefront branch equals -1, see (81) with $\sigma_0 = 1$. There is a single caustic along the ray from point A to point C, located in the isotropic medium, causing the KMAH index to increase by $+1$. The resulting KMAH index at point C is then 0.

Consider three points on the selected ray: Point A in the anisotropic medium, point B situated in the isotropic medium between the interface and the caustic corresponding to the point source at A, and point C situated in the isotropic medium beyond the caustic corresponding to the point source at A. The paraxial rays from all 3 points are depicted in Fig. 12. The paraxial rays corresponding to the same paraxial slowness vector are parallel in each of the halfspaces.

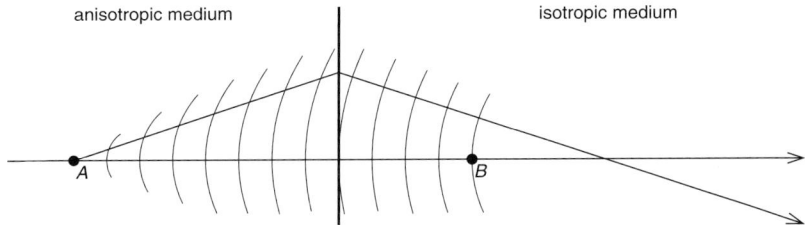

FIG. 15. $A \to B$: The initial KMAH index at point A is again -1, but there is no caustic along the ray from point A to point B. The resulting KMAH index at point B is now -1.

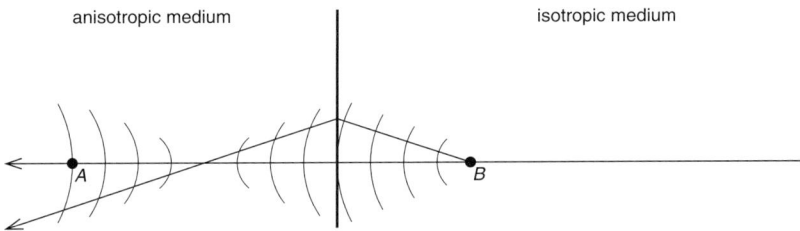

FIG. 16. $B \to A$: The initial KMAH index at point B is 0, as for $C \to A$. There is a single caustic along the ray from point B to point A, located in the anisotropic medium on the reverse branch of the ray-velocity surface, causing the KMAH index to decrease by 1. The resulting KMAH index at point A is then -1.

Accumulation of the phase shift along the four rays between points A, B and C is shown in Figs. 13–16.

2.7. Ray-Theory Seismograms

For transient elementary waves propagating in a heterogeneous, anisotropic or isotropic, perfectly elastic, layered structure, the elementary ray seismogram is given by the formula

$$u_i(x_n, t) = \text{Re}\{U_i(x_n) F[t - \tau(x_m)]\}. \tag{83}$$

Here $F(\zeta)$ is the analytical signal, given by relation

$$F(\zeta) = x(\zeta) + ig(\zeta). \tag{84}$$

The real-valued function $x(\zeta)$ may be arbitrary, and may represent, for example, the source-time function. For simplicity, we shall refer to it as the input signal. The real-valued function $g(\zeta)$ is the Hilbert transform of $x(\zeta)$. Once the input signal $x(\zeta)$ is known, its Hilbert transform and the analytical signal $F(\zeta)$ can be determined. Two simple examples of the analytical signals are as follows: (a) For

$x(\zeta) = \delta(\zeta)$, where $\delta(\zeta)$ is the Dirac delta function, $F(\zeta) = \delta(\zeta) - i/\pi\zeta$. (b) For $x(\zeta) = \cos(\zeta)$, the analytical signal is $F(\zeta) = \exp[-i\zeta]$.

We call $x(\zeta)$ the causal input signal if $x(\zeta) = 0$ for $\zeta < 0$. It is well known that the Hilbert transform $g(\zeta)$ of a causal input signal $x(\zeta)$ is not causal. Thus, the elementary ray-theory signal $u_i(x_n, t)$ *is in principle non-causal*, even when the input signal $x(\zeta)$ is causal.

The situation is simple if $U_i(x_n)$ is real-valued. Equation (83) then yields

$$u_i(x_n, t) = U_i(x_n) x(t - \tau(x_i)), \tag{85}$$

and the form of the elementary ray-theory signal $u_i(x_n, t)$ remains preserved along the whole ray. Moreover, $u_i(x_n, t)$ is causal along the whole ray if $x(\zeta)$ is causal.

The situation is quite different for complex-valued $U_i(x_n)$. Note that $U_i(x_n)$, real-valued at the initial point of the ray, may become complex-valued in other parts of the ray. This is mainly due to the caustic points along the ray, and due to the supercritical reflections at interfaces. Thus, without considering complex-valued $U_i(x_n)$, the ray method would be incomplete.

Let us now briefly discuss certain important properties of elementary ray-theory signals (83), valid both for $U_i(x_n)$ real-valued and complex-valued and for non-dissipative media.

1. Analytical signal $F(\zeta)$ is preserved along the whole ray.
2. The normalized envelope of the elementary ray-theory signal (83) is preserved along the whole ray. The normalization is taken with respect to the maximum value of the envelope. Thus, the normalized envelope is the same at the initial point of the ray, at interfaces, at caustic points, and at the receiver. It is always equal to the normalized envelope of the analytical signal.
3. Although the elementary ray-theory signal is not, in principle, causal, we may say that it is *effectively causal*, particularly for smooth, absolutely integrable input signals $x(\zeta)$. The explanation of effective causality is as follows: Assume that the ray-theory travel time corresponds to the maximum of the normalized envelope, and denote it at a given point of the ray by t_{max}. The effective first arrival travel time t_{min} may then be taken at a time, for which the value of the normalized envelope is negligibly small, for example 0.001. Property 2 then guarantees that $t_{max} - t_{min}$ is preserved along the whole ray, including the interfaces, caustic points, etc.

    It might seem useful to consider very narrow causal input signals $x(\zeta)$ in the ray method, for example signals close to $x(\zeta) = \delta(\zeta)$. For such signals, however, the envelope is broad (due to the broad Hilbert transform), and $t_{max} - t_{min}$ is large. Thus, the effect of non-causality is particularly strong in this case. A very suitable option is to use smooth, non-causal, high-frequency input signal $x(\zeta)$, for example the Gabor signal. Fortunately, the three-parametric Gabor signal can easily simulate most signals $x(\zeta)$ known from seismological and seismic exploration records. The effects of changing phase shifts of $U_i(x_n)$ on the ray-theory signals $u_i(x_n, t)$ are then expressed only in the changes of $u_i(x, t)$ under its fixed envelope.

3. Ray-Theory Perturbations

Perturbation methods play an important role in ray methods. They can be used for fast but approximate solution of forward problems in complicated models. Perturbation methods play an equal or even more important role in inverse problems.

There is a great variety of perturbation methods in use, most of them being based on the following assumptions. We assume that a model, in which we wish to study wave propagation, differs only little from another model called the *background* or *reference model*. The solution in the perturbed model can then be sought in the form of a power series in the deviations of the perturbed and reference models. If only the first term of the series is considered, which is often the case, we speak of the *first-order perturbation expansion*. The perturbed and reference models may both be isotropic or anisotropic, or the reference model may be isotropic and the perturbed model anisotropic. Various quantities may be sought in this way.

Most often the quantity sought is the travel time. The corresponding first-order formula for the isotropic reference and perturbed models has been known for a long time, see Aki and Richards (1980). For the anisotropic perturbed model, the formulae for the first-order travel-time perturbations can be found in Červený (1982), Červený and Jech (1982), Hanyga (1982), Jech and Pšenčík (1989), Chapman and Pratt (1992). In these approaches a reference ray is traced in the reference isotropic or anisotropic medium, and the first-order travel-time correction is sought by integration along the reference ray. Even higher-order travel-time perturbations can be calculated. Farra (1999) proposed a procedure for calculating the second-order travel-time perturbations. Her procedure is based on the knowledge of the first-order perturbations of a reference ray. Recently, developing ideas of Babich et al. (1985), Klimeš (2002a) proposed a procedure for calculating higher-order perturbations (or derivatives) of the travel time along a reference ray in the reference medium. Refer to Section 3.2 for more details.

The perturbation theory can be used not only to determine the travel time, but the whole wavefield. Farra and Madariaga (1987) proposed such an approach for an isotropic reference and perturbed medium. The approach consists in computing perturbed rays and related dynamic ray tracing by integrating along a reference ray. The approach was later extended also to anisotropic media, see, e.g., Farra (1989), Nowack and Pšenčík (1991). In all the above-mentioned approaches, it is necessary to trace rays in the reference medium, and then to calculate the perturbations of the sought quantities along these rays. Recently, Pšenčík and Farra (2005) and Farra (2005) proposed an alternative approach to calculating rays in heterogeneous weakly anisotropic media, in which the deviation of anisotropy from isotropy is considered to be small. In ray tracing system (24), they substituted the exact Hamiltonian by its first-order approximation with respect to the deviation of anisotropy from isotropy. In this way, their ray tracing yields "first-order rays" and corresponding first-order travel times without calculating the reference rays. If it becomes necessary to increase the accuracy of the calculated travel times, the second-order travel-time correction can be easily calculated by integrating along the first-order rays. In weakly anisotropic media, this ray tracing represents a natural generalization of ray tracing for isotropic media. Indeed, for vanishing anisotropy, the first-order ray tracing reduces to exact "isotropic" ray tracing. The first-order ray tracing equations are, as in isotropic media, different for P and S waves. In each case, the ray tracing equations depend on 15 of the 21 weak anisotropy parameters, because there are 6 parameters related only to P waves, 6 parameters related only to S waves, and 9 parameters affecting both P and S waves. Weak anisotropy parameters may be chosen in various ways. For example, Pšenčík and Gajewski (1998) and Farra and Pšenčík (2003) use a generalization of Thomsen's (1986) linearized parameters. During numerical tests using a configuration and models typical for seismic exploration, it was found that the relative errors of the travel times including the second-order correction were well under 0.05% for anisotropy of about 8%, and they did not exceed 0.3% for anisotropy of about 20%.

A possibility of how to calculate the whole wavefield using the perturbation theory is the *generalized Born scattering*, see Coates and Chapman (1991), Chapman (2004). This approach is based on the first-order Born approximation, in which the exact Green function in the reference medium is substituted by the ray-theory Green function. The errors caused by this substitution are introduced into the scattering integral so that scattering occurs not only from perturbations of the medium, but also from the errors.

In this chapter, we concentrate on the perturbations of travel time only. Similar formulae as for travel time may also be derived for other ray-theory quantities.

3.1. Perturbation Parameters

We assume that the Hamiltonian is a function of *phase-space coordinates*, composed of spatial coordinates x_i and slowness-vector components p_j, and of any number of *perturbation parameters* f_κ,

$$\mathcal{H} = \mathcal{H}(x_i, p_j, f_\kappa). \tag{86}$$

The travel time and other quantities calculated in the model using this Hamiltonian are then functions of *spatial coordinates* x_i and of perturbation parameters f_κ,

$$\tau = \tau(x_i, f_\kappa). \tag{87}$$

In this section, we index the perturbation parameters by lower-case Greek subscripts. Perturbation parameters are the parameters to be perturbed. They may have various meanings. For example, perturbation parameters f_κ may parametrize the model. Parameters f_κ may be the B-spline coefficients of the functions describing the model. However, the applicability of perturbation parameters is more general. For example, there are two S waves with the two respective Hamiltonians corresponding to the anisotropic ray theory in anisotropic media. Instead of selecting one of these two Hamiltonians, we may consider a one-parametric set of Hamiltonians, parametrized by a parameter f_μ, such that we obtain the Hamiltonian corresponding to one S wave for $f_\mu = -1$, the Hamiltonian corresponding to the other S wave for $f_\mu = +1$, and the averaged Hamiltonian of both S waves corresponding to the anisotropic common ray tracing (Bakker, 2002) for $f_\mu = 0$. Another example: instead of using Hamiltonian $\mathcal{H} = \mathcal{H}(x_i, p_j, f_\kappa)$ and perturbation expansion from $f_\kappa = f_\kappa^1$ to $f_\kappa = f_\kappa^2$, we may define a new one-parametric Hamiltonian $\tilde{\mathcal{H}} = \tilde{\mathcal{H}}(x_i, p_j, f)$,

$$\tilde{\mathcal{H}}(x_i, p_j, f) = \mathcal{H}(x_i, p_j, f_\kappa^1) + [\mathcal{H}(x_i, p_j, f_\kappa^2) - \mathcal{H}(x_i, p_j, f_\kappa^1)]f \quad (88)$$

and apply the perturbation expansion from $f = 0$ to $f = 1$.

We shall refer to the partial derivatives with respect to perturbation parameters f_κ as the *perturbation derivatives*.

We denote the partial derivatives with respect to spatial coordinates x_i by lower-case Roman subscripts following a comma, and the *perturbation derivatives by lower-case Greek subscripts* following a comma. In phase space, we denote the partial derivatives with respect to components p_j of the slowness vector by lower-case Roman superscripts following a comma. For instance,

$$\mathcal{H}_{,ij\ldots n\alpha\ldots\nu}^{,ab\ldots f} = \frac{\partial}{\partial f_\alpha} \cdots \frac{\partial}{\partial f_\nu} \frac{\partial}{\partial x_i} \frac{\partial}{\partial x_j} \cdots \frac{\partial}{\partial x_n} \frac{\partial}{\partial p_a} \frac{\partial}{\partial p_b} \cdots \frac{\partial}{\partial p_f} \mathcal{H} \quad (89)$$

denotes the perturbation derivatives of the partial *phase-space derivatives* of the Hamiltonian. The perturbation derivatives are taken at fixed phase-space coordinates x_i, p_j, and the partial phase-space derivatives are calculated at fixed perturbation parameters f_κ. Similarly,

$$\tau_{,ij\ldots n\alpha\ldots\nu} = \frac{\partial}{\partial f_\alpha} \cdots \frac{\partial}{\partial f_\nu} \frac{\partial}{\partial x_i} \frac{\partial}{\partial x_j} \cdots \frac{\partial}{\partial x_n} \tau \quad (90)$$

denotes the perturbation derivatives of the *spatial derivatives* of travel time τ. The perturbation derivatives are taken at fixed spatial coordinates x_i, and the partial spatial derivatives are calculated at fixed perturbation parameters f_κ. Analogously for other quantities, e.g., amplitude $A = A(x_i, f_\kappa)$.

3.2. Perturbation of Travel Time

The first-order perturbation derivatives of travel time are very important in (a) linearized inversion of travel times; (b) approximate solution of the complex-valued eikonal equation in the vicinity of real-valued rays; (c) common ray approximations of the coupling ray theory. In all these methods, the error of the first-order perturbation expansion should be estimated and controlled. The error of the first-order perturbation expansion can be approximated by the neglected second-order term. It is thus of principal importance in all the above applications to estimate the second-order perturbation derivatives of travel time. A brief note on the application of the second-order perturbation derivatives of travel time to the linearized inversion of travel times is given in Section 3.3. For estimating the errors of common ray approximations of the coupling ray theory using the second-order perturbation derivatives of travel time, refer to Klimeš and Bulant (2004, 2006).

The perturbation derivatives of the travel time of all orders can be calculated by simple numerical quadratures along the unperturbed ray. The perturbation derivatives of the spatial derivatives of travel time can also be calculated by analogous numerical quadratures along the unperturbed ray. As regards the general equations for the perturbation derivatives of all orders, refer to Klimeš (2002a). Here we only present the equations for the first-order and second-order perturbation derivatives of travel time, expressed in terms of a general Hamiltonian.

The individual perturbation derivatives of travel time depend on the form of the Hamiltonian. For different Hamiltonians, we obtain different perturbation expansions of travel time. Some of these perturbation expansions may be more accurate for particular applications than other perturbation expansions (Klimeš, 2002a).

3.2.1. First-Order Perturbation Derivatives of Travel Time

Equation

$$\tau_{,\alpha}(\gamma_3) = \tau_{,\alpha}(\gamma_3^0) + \int_{\gamma_3^0}^{\gamma_3} d\gamma_3 \, (-\mathcal{H}_{,\alpha}) \tag{91}$$

for the first-order perturbation derivatives of travel time is well-known (e.g., Červený, 2001, Eq. (3.9.6)). The integration variable γ_3 along the ray is determined by the form of the Hamiltonian. For the Hamiltonians considered in this chapter, the integration variable is the travel time, $\gamma_3 = \tau$.

3.2.2. First-Order Perturbation Derivatives of the Travel-Time Gradient

The first-order perturbation derivatives of the travel-time gradient may be calculated using relation

$$\tau_{,i\alpha} = T_{a\alpha} Q_{ai}^{-1}, \tag{92}$$

where Q_{ai}^{-1} are the elements of the matrix inverse to matrix $Q_{ia} = \partial x_i / \partial \gamma_a$ of geometrical spreading in Cartesian coordinates. The perturbation derivatives of the covariant spatial derivatives $T_{a\alpha}$ of travel time with respect to ray coordinates can be obtained by quadrature

$$T_{a\alpha}(\gamma_3) = T_{a\alpha}(\gamma_3^0) + \int_{\gamma_3^0}^{\gamma_3} d\gamma_3 \left(-\mathcal{H}_{,i\alpha} Q_{ia} - \mathcal{H}_{,\alpha}^i P_{ia}\right). \tag{93}$$

The matrix $P_{ia} = \partial p_i / \partial \gamma_a$ of the transformation from ray coordinates to the components of the slowness vector is calculated, together with the matrix Q_{ia} of geometrical spreading, by dynamic ray tracing.

3.2.3. Second-Order Perturbation Derivatives of Travel Time

The second-order perturbation derivatives of travel time can be calculated by quadrature

$$\tau_{,\alpha\beta}(\gamma_3) = \tau_{,\alpha\beta}(\gamma_3^0) + \int_{\gamma_3^0}^{\gamma_3} d\gamma_3 \, K_{\alpha\beta}, \tag{94}$$

where the integration kernel is given by

$$K_{\alpha\beta} = -\mathcal{H}_{,\alpha\beta} - T_{a\alpha} Q_{ar}^{-1} \mathcal{H}_{,\beta}^{,r} - T_{a\beta} Q_{ar}^{-1} \mathcal{H}_{,\alpha}^{,r}$$
$$- T_{a\alpha} Q_{ar}^{-1} T_{b\beta} Q_{bs}^{-1} \mathcal{H}^{,rs}. \tag{95}$$

Quadrature (94) is applicable only to ray segments with the regular matrix Q_{ia} of geometrical spreading. Along ray segments with singular matrix Q_{ia} but regular matrix P_{ia}, Eq. (94) may be replaced by equation

$$\tau_{,\alpha\beta}(\gamma_3) = \tau_{,\alpha\beta}(\gamma_3^0) + \left[T_{a\alpha} P_{ar}^{-1} T_{b\beta} Q_{br}^{-1}\right]_{\gamma_3^0}^{\gamma_3} + \int_{\gamma_3^0}^{\gamma_3} d\gamma_3 \, K_{\alpha\beta}^{\text{caust}}, \tag{96}$$

where the integration kernel is given by

$$K_{\alpha\beta}^{\text{caust}} = -\mathcal{H}_{,\alpha\beta} + T_{a\alpha} P_{ar}^{-1} \mathcal{H}_{,r\beta} + T_{a\beta} P_{ar}^{-1} \mathcal{H}_{,r\alpha}$$
$$- T_{a\alpha} P_{ar}^{-1} T_{b\beta} P_{bs}^{-1} \mathcal{H}_{,rs}. \tag{97}$$

3.3. Optimizing Model Updates During Linearized Inversion of Travel Times

The size of the individual iterations in the linearized inversion of travel times should be controlled by means of the second-order perturbation derivatives of travel time. This control can be achieved by including the square of the relevant Sobolev norm of the model update into the objective function for the linearized inversion of travel times. The Sobolev scalar product of two functions is a linear combination of the L_2 Lebesgue scalar products of the zero-order, first-order, second-order or higher-order derivatives of the functions, see, e.g., Tarantola (1987).

Since the second-order perturbation derivatives of travel time are proportional to the square of the first-order perturbation derivatives of the slowness or velocity gradient (Klimeš, 2002a), the error of a single iteration of the linearized inversion of travel times is roughly proportional to the square of the Sobolev norm composed of the first derivatives of the slowness or velocity update. We should thus include the square of the Sobolev norm composed of the first derivatives of the slowness or velocity update into the objective function (Klimeš, 2002c).

The weighting factor of this Sobolev norm in the objective function may iteratively be adjusted. For this purpose, we only need to calculate the homogeneous second-order perturbation derivative of travel time in the direction of the model update, and the mixed second-order perturbation derivative of travel time in the direction of the model update and in the direction of the derivative of the model update with respect to the weighting factor. The calculation of these second-order perturbation derivatives of travel time costs 4 numerical quadratures along each two-point ray. The results are then used to correct the estimate of the weighting factor and to calculate a new model update, new second-order perturbation derivatives of travel time, and a new weighting factor, until reaching the nearly-optimum size of the model update for the rapid convergence of the travel-time inversion.

4. Coupling Ray Theory for S Waves

As mentioned in Section 2, there are two different high-frequency asymptotic ray theories: the *isotropic ray theory* assuming equal velocities of both S waves and the *anisotropic ray theory* assuming both S waves strictly decoupled. In the isotropic ray theory, the S-wave polarization vectors do not rotate about the ray, whereas in the anisotropic ray theory they may rotate rapidly about the ray, see Fig. 17. Thomson *et al.* (1992) demonstrated analytically that the high-frequency asymptotic error of the anisotropic ray theory is inversely proportional to the square root or higher-order root of the frequency if a ray passes through the point of equal S-wave eigenvalues of the Christoffel matrix.

In weakly anisotropic models, at moderate frequencies, the S-wave polarization vector tends to remain unrotated about the ray, but is partly attracted by the rotation of the eigenvectors of the Christoffel matrix. The intensity of the attraction increases with frequency. This behaviour of the S-wave polarization vector is described by the *coupling ray theory* proposed by Coates and Chapman (1990) and (in the quasi-isotropic approximation) by Pšenčík (1998). The coupling ray theory is applicable at all degrees of anisotropy, from isotropic models to considerably anisotropic ones. The frequency-dependent coupling ray theory is the generalization of both the zero-order isotropic and anisotropic ray theories and provides continuous transition between them. The numerical algorithm for calculating the frequency-dependent complex-valued S-wave polarization vectors of the coupling ray theory has been designed by Bulant and Klimeš (2002).

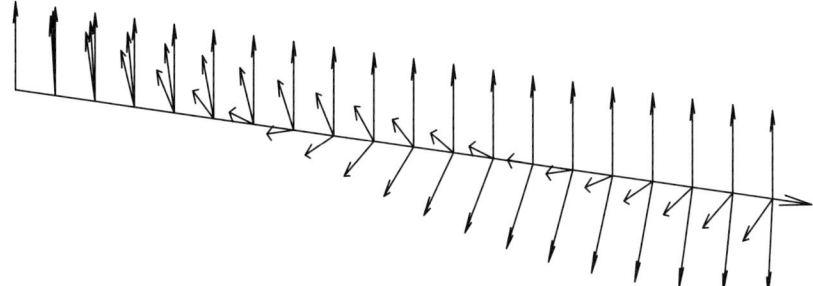

FIG. 17. S-wave polarization vector along a ray: Isotropic-ray-theory polarization vector points upwards. Anisotropic-ray-theory polarization vector points downwards on the right-hand side. Coupling-ray-theory polarization vector points between the isotropic-ray-theory and anisotropic-ray-theory polarization vectors.

The results of the isotropic ray theory, anisotropic ray theory and coupling ray theory have been compared with the exact solution in the "simplified twisted crystal" model designed by Vavryčuk (1999), analytically by Klimeš (2004a) and numerically by Bulant et al. (2004). The quasi-isotropic approximation of the coupling ray theory has been numerically compared with the isotropic ray theory, anisotropic ray theory and reflectivity method in a more realistic 1-D model by Pšenčík and Dellinger (2001).

4.1. Coupling Equation for S Waves

Assume a *reference ray* in phase space, parametrized by reference travel time τ, with reference slowness vectors $p_i(\tau)$ known at all its points $x_j(\tau)$. Using the reference slowness vector, we can calculate the reference Christoffel matrix

$$\Gamma_{jk}(\tau) = p_i(\tau) a_{ijkl}(x_m(\tau)) p_l(\tau) \tag{98}$$

and its eigenvectors $g_i^{(1)}(\tau), g_i^{(2)}(\tau), g_i^{(3)}(\tau)$ along the reference ray, see (12). Assume that eigenvectors $g_i^{(1)}(\tau)$ and $g_i^{(2)}(\tau)$ correspond to S waves and that they vary continuously along the reference ray. Continuity is not required in regions where the corresponding two eigenvalues are equal. Let us denote by $\tau_1(\tau)$ and $\tau_2(\tau)$ the travel times corresponding to eigenvectors $g_i^{(1)}(\tau)$ and $g_i^{(2)}(\tau)$, respectively. They may be approximated by quadratures along the unperturbed reference ray,

$$\frac{d\tau_1}{d\tau} = [\Gamma_{jk} g_j^{(1)} g_k^{(1)}]^{-\frac{1}{2}}, \qquad \frac{d\tau_2}{d\tau} = [\Gamma_{jk} g_j^{(2)} g_k^{(2)}]^{-\frac{1}{2}} \tag{99}$$

(Klimeš, 2002a, Eqs. (43) and (65)). The time-harmonic coupling-ray-theory solution u_i of the elastodynamic equation (3) may then be expressed, for S waves, as the linear combination

$$u_i = A \sum_{M=1}^{2} g_i^{(M)} a_M \exp[i\omega(\bar{\tau} - t)] \tag{100}$$

of the time-harmonic anisotropic-ray-theory solutions, where

$$\bar{\tau}(\tau) = \frac{1}{2}[\tau_1(\tau) + \tau_2(\tau)] \tag{101}$$

is the average travel time, and $A = A(\tau)$ is the complex-valued scalar amplitude in the high-frequency approximation (Červený, 1972), corresponding to the system of reference rays. The coupling-ray-theory (Coates and Chapman, 1990) equation for complex-valued amplitude factors $a_M = a_M(\tau)$ reads (Bulant and Klimeš, 2002, Eq. (9))

$$\frac{d}{d\tau}\begin{pmatrix} a_1 \\ a_2 \end{pmatrix} = \left[\begin{pmatrix} 0 & 1 \\ -1 & 0 \end{pmatrix}\frac{d\varphi}{d\tau} + \begin{pmatrix} i & 0 \\ 0 & -i \end{pmatrix}\frac{\omega}{2}\frac{d(\tau_1 - \tau_2)}{d\tau}\right]\begin{pmatrix} a_1 \\ a_2 \end{pmatrix}, \tag{102}$$

where

$$\frac{d\varphi}{d\tau} = \frac{dg_k^{(1)}}{d\tau} g_k^{(2)} = -g_k^{(1)} \frac{dg_k^{(2)}}{d\tau} \tag{103}$$

is the angular velocity of the eigenvector rotation.

4.2. Coupling-Ray-Theory S-Wave Propagator Matrix

Propagator matrix $\mathbf{\Pi}^g$ of Eq. (102), defined as

$$\Pi^g_{MN}(\tau, \tau_0) = \frac{\partial a_M(\tau)}{\partial a_N(\tau_0)}, \tag{104}$$

is a complex-valued 2×2 matrix satisfying equation (Bulant and Klimeš, 2002, Eq. (11))

$$\frac{d}{d\tau}\mathbf{\Pi}^g = \left[\begin{pmatrix} 0 & 1 \\ -1 & 0 \end{pmatrix}\frac{d\varphi}{d\tau} + \begin{pmatrix} i & 0 \\ 0 & -i \end{pmatrix}\frac{d\psi}{d\tau}\right]\mathbf{\Pi}^g, \tag{105}$$

directly following from Eq. (102). Here

$$\psi(\tau) = \frac{1}{2}\omega[\tau_1(\tau) - \tau_2(\tau)]. \tag{106}$$

Note that propagator matrix $\mathbf{\Pi}^g$ is symplectic and unitary.

It is difficult to integrate equation (105) by the Runge–Kutta or another numerical method that requires derivative $d\varphi/d\tau$ along the reference ray to be calculated, because this derivative is undefined in the singular regions, in which the two eigenvalues of Christoffel matrix (98) are equal. The method of calculating propagator matrix $\mathbf{\Pi}^g$, suitable for Eq. (105), has been proposed by Červený (2001) and Bulant and Klimeš (2002), with emphasis on numerical implementation. The method does not require the calculation of the angular velocity $d\varphi/d\tau$ of the rotation of the eigenvectors of the Christoffel matrix along the reference ray and does not require $d\varphi/d\tau$ to be smooth or finite along the reference ray.

The proposed method of solving Eq. (105) takes advantage of the chain rule. Since Π^g is a propagator matrix satisfying the chain rule, it may be numerically calculated as the product of propagator matrices Π^g corresponding to reasonably small segments of the reference ray (Červený, 2001). Frequency-dependent propagator matrices along the individual small ray segments may be approximated by the method of mean coefficients (Červený, 2001). The accuracy of the proposed algorithm of numerical integration of the coupling equation has been estimated by Bulant and Klimeš (2002). The estimate enables the integration step to be controlled, so that the relative error in the wavefield amplitudes due to the integration is kept below a given limit, which is of principal importance for numerical applications.

4.3. Quasi-Isotropic Approximations of the Coupling Ray Theory

The coupling ray theory by Coates and Chapman (1990) is applicable at all degrees of anisotropy, but it is often replaced by various *quasi-isotropic approximations*. There are many frequently used quasi-isotropic approximations of the coupling ray theory (Klimeš and Bulant, 2004), which impair the accuracy of the coupling ray theory both with increasing frequency and increasing degree of anisotropy. For example, the reference ray may be calculated in different ways (Bakker, 2002; Klimeš and Bulant, 2004, 2006), the Christoffel matrix may be approximated by its quasi-isotropic projections onto the plane perpendicular to the reference ray and onto the line tangent to the reference ray (Pšenčík, 1998), travel times corresponding to the anisotropic ray theory may be approximated in several ways, e.g., by the first-order quasi-isotropic perturbation with respect to the density-normalized elastic moduli (Pšenčík, 1998), etc. Most of these quasi-isotropic approximations can be avoided with minimum effort (Bulant and Klimeš, 2002, 2004), except for the *common ray approximation* for S waves.

The effects of the quasi-isotropic approximation of the Christoffel matrix, the quasi-isotropic projection of the Green function and the quasi-isotropic perturbation of travel times have been demonstrated by numerical comparison with the exact solution for the one-way plane-wave propagator matrix in a simple 1-D anisotropic "oblique twisted crystal" model by Bulant and Klimeš (2004). The effect of the quasi-isotropic projection of the Green function has been numerically demonstrated by Bulant and Klimeš (2002) on the example of seismograms calculated by the coupling ray theory in a more realistic 1-D anisotropic model. Klimeš and Bulant (2004) studied the effects of the anisotropic common ray approximation and the less accurate isotropic common ray approximation, and compared them with the effects of the quasi-isotropic approximation of the Christoffel matrix, of the quasi-isotropic projection of the Green function, and of the quasi-isotropic perturbation of travel times in three 1-D anisotropic models of differing degree of anisotropy.

When using any of these quasi-isotropic approximations of the coupling ray theory, the errors due to the applied quasi-isotropic approximation should be calculated and checked.

4.3.1. Selection of the Reference Ray

The isotropic ray theory is always the limiting case of the coupling ray theory for decreasing anisotropy at a fixed frequency. On the other hand, the high-frequency limit of the coupling ray theory at a fixed anisotropy depends on the choice of the reference ray, and even on the choice of the *system* of reference rays, because the amplitudes are determined by the paraxial reference rays.

From the point of view of the high-frequency asymptotic validity, the frequency-independent reference ray is best represented by the *anisotropic-ray-theory reference ray*, provided that we choose the initial condition for the polarization vector in the coupling equation given by the eigenvector of the Christoffel matrix corresponding to the reference ray. The anisotropic-ray-theory travel time corresponding to the selected polarization vector is then exact, and only the difference between the two anisotropic-ray-theory S-wave travel times is approximate. The coupling ray theory may then also be used at high frequencies because the approximate travel-time difference only influences the coupling due to low-frequency scattering. The coupling ray theory then correctly converges to the anisotropic ray theory for high frequencies. For other choices of reference rays, the high-frequency limit of the coupling ray theory at a fixed anisotropy is incorrect, although the differences may be small at the finite frequencies under consideration. Note that the anisotropic-ray-theory reference ray can be traced only if the eigenvectors of the Christoffel matrix vary continuously along the whole ray (Vavryčuk, 2001).

In the *common ray approximation*, only one reference ray is traced for both anisotropic-ray-theory S waves, and both S-wave anisotropic-ray-theory travel times are approximated by the perturbation expansion from the common reference ray. The common ray approximation thus considerably simplifies the coding of the coupling ray theory and numerical calculations, but may introduce errors in travel times due to the perturbation. These travel-time errors can deteriorate the coupling-ray-theory solution at high frequencies. The travel-time errors due to the common ray approximations can be calculated using the equations proposed by Klimeš and Bulant (2004, 2006).

In the *anisotropic common ray approximation*, the common reference ray is traced using the averaged Hamiltonian of both anisotropic-ray-theory S waves

(Bakker, 2002; Klimeš, 2006b). This is probably the best common ray approximation (Klimeš and Bulant, 2004, 2006).

In the less accurate *isotropic common ray approximation*, the reference ray is traced in the reference isotropic model. The reference isotropic model may be selected in different ways, yielding quasi-isotropic approximations of differing accuracies.

In the common ray approximations, the S-wave travel times are usually approximated by the first-order perturbation expansion from the common reference ray. The errors of S-wave travel times may then be approximated by second-order terms in the perturbation expansion. A method of estimating the errors due to the isotropic common ray approximation and the anisotropic common ray approximation has been proposed and numerically demonstrated by Klimeš and Bulant (2004, 2006) and Bulant and Klimeš (2006). The method is based on the equations for the second-order perturbations of travel time derived by Klimeš (2002a).

The accuracy of the anisotropic common ray approximation can be studied along isotropic common rays, without tracing the anisotropic common rays. If the error of the isotropic common ray approximation exceeds an acceptable limit, we can immediately decide whether the anisotropic common ray approximation (Bakker, 2002; Klimeš, 2006b) would be sufficiently accurate, or whether the anisotropic-ray-theory rays should be traced as reference rays for the coupling ray theory. The numerical results by Klimeš and Bulant (2004) and by Bulant and Klimeš (2006) demonstrate that the anisotropic common ray approximation by Bakker (2002) and Klimeš (2006b) is worth coding and applying.

4.3.2. Quasi-Isotropic Projection of the Green Function

The coupling-ray-theory solution (100) may be approximated by its projection

$$\tilde{u}_i = h_{iM} h_{mM} u_m \qquad (107)$$

onto the reference S-wave polarization plane, given by two orthonormal reference polarization vectors h_{k1}, h_{k2}. This approximation may simplify the modification of existing isotropic ray tracing codes for the coupling ray theory. The error of this approximation is obvious and simple to calculate.

4.3.3. Quasi-Isotropic Approximation of the Christoffel Matrix

The Christoffel matrix may be approximated by its projections onto the reference S-wave polarization plane and onto the reference P-wave polarization line. Denote the polarization vectors of the isotropic ray theory, or the reference polarization vectors in general, by h_{k1}, h_{k2} and h_{k3}. If the Christoffel matrix is approximated by its projections onto plane h_{j1}, h_{k2} and onto vector h_{l3}, namely,

$$\tilde{\Gamma}_{jk} = h_{jM}h_{mM}\Gamma_{mn}h_{nN}h_{kN} + h_{j3}h_{m3}\Gamma_{mn}h_{n3}h_{k3}$$
$$= \Gamma_{jk} - (h_{jM}h_{k3} + h_{j3}h_{kM})h_{mM}\Gamma_{mn}h_{n3}, \tag{108}$$

then eigenvectors $g_k^{(1)}$ and $g_k^{(2)}$ become located in plane h_{j1}, h_{k2} as in the zero-order quasi-isotropic approximation of Pšenčík (1998). This approximation includes the quasi-isotropic projection of the Green function.

The quasi-isotropic approximation of the Christoffel matrix usually generates much greater error than the other quasi-isotropic approximations. The alteration of the coupling-ray-theory seismograms due to the quasi-isotropic approximation of the Christoffel matrix has been demonstrated by Bulant and Klimeš (2002).

4.3.4. Quasi-Isotropic Perturbation of Travel Times

The anisotropic-ray-theory travel times used in the coupling ray theory should be calculated by the numerical quadrature of Eq. (99) along the reference ray. In the quasi-isotropic perturbation of travel times, the anisotropic-ray-theory travel times are calculated from the reference travel time by the first-order perturbation with respect to the density-normalized elastic moduli,

$$\frac{d\tau_1}{d\tau} \approx \left(\Gamma_{jk}^0 g_j^{(1)} g_k^{(1)}\right)^{-\frac{1}{2}} - \frac{1}{2}(\Gamma_{jk} - \Gamma_{jk}^0)g_j^{(1)}g_k^{(2)}\left(\Gamma_{jk}^0 g_j^{(1)} g_k^{(1)}\right)^{-\frac{3}{2}}. \tag{109}$$

If the isotropic common ray approximation and the quasi-isotropic approximation of the Christoffel matrix are also applied, then

$$\Gamma_{jk}^0 g_j^{(1)} g_k^{(1)} = 1, \tag{110}$$

and Eq. (109) becomes

$$\frac{d\tau_1}{d\tau} \approx \frac{3}{2} - \frac{1}{2}\Gamma_{jk}g_j^{(1)}g_k^{(1)}, \tag{111}$$

as in the quasi-isotropic approximation of Pšenčík (1998). Analogously, we obtain $d\tau_2/d\tau$. The quasi-isotropic perturbation of travel times leads to an erroneous time shift in coupling-ray-theory seismograms, but has a negligible impact on the amplitudes.

4.4. Numerical Examples

The vertically heterogeneous 1-D anisotropic model QI was used by Pšenčík and Dellinger (2001, model WA rotated by 45°) for comparison of the coupling-ray-theory synthetic seismograms with the reflectivity method. The density-normalized elastic moduli a_{ijkl} in both model QI and in the reference isotropic model QI0 are interpolated linearly with depth. For the discussion and description of this model refer to Pšenčík and Dellinger (2001).

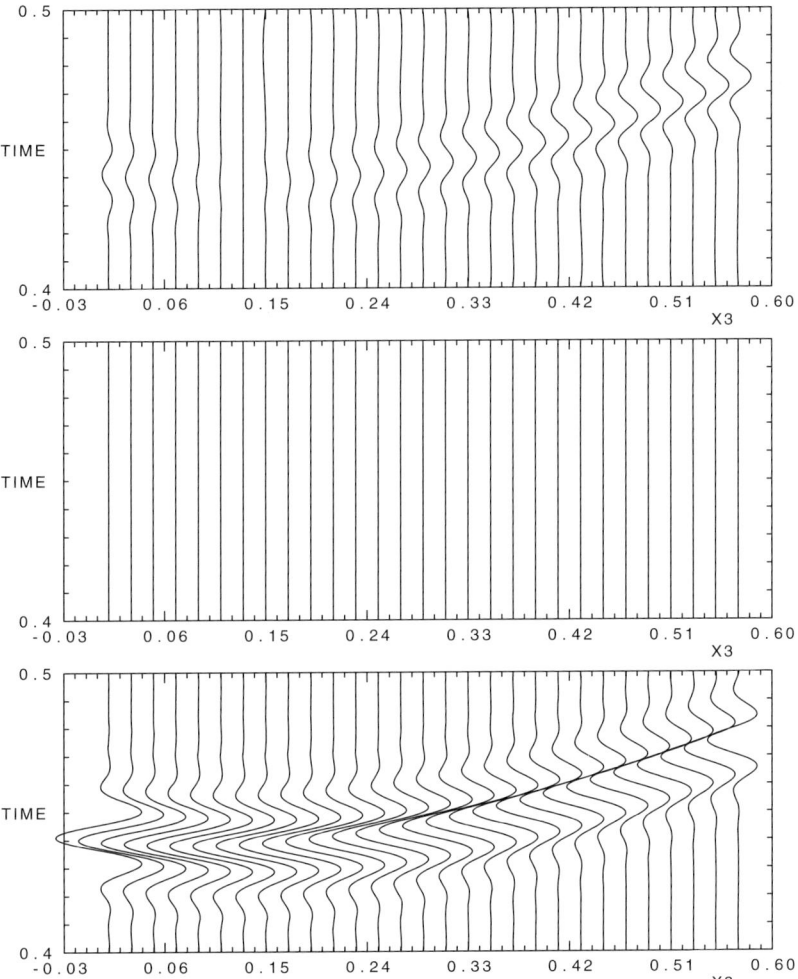

FIG. 18. Isotropic-ray-theory seismograms in model QIH (actually calculated in isotropic model QI0). From top to bottom: the first (radial) component, the second (transverse) component, the third (vertical) component.

The differences of the elastic moduli of model QIH from the elastic moduli of the reference isotropic model QI0 are exactly twice smaller than the differences of model QI. The differences of the elastic moduli of model QI2 from the elastic moduli of the reference isotropic model QI0 are exactly twice larger than the differences of model QI. The differences of the elastic moduli of model QI4 from the elastic moduli of the reference isotropic model QI0 are exactly 4 times larger

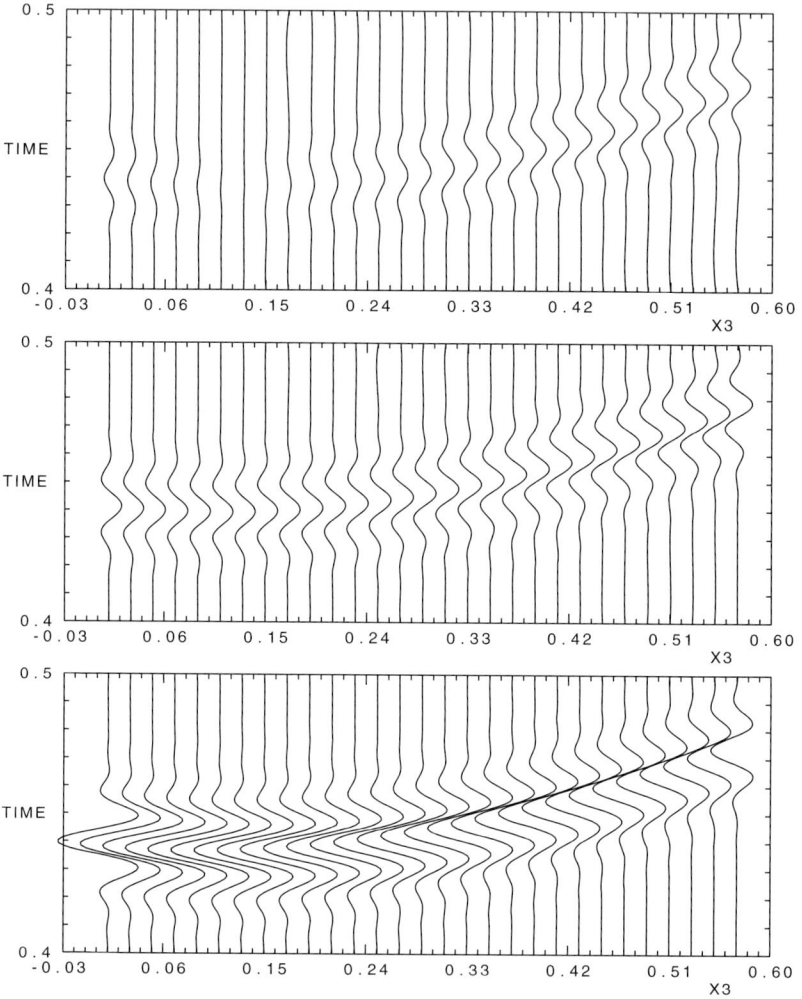

FIG. 19. Anisotropic-ray-theory seismograms in model QIH. From top to bottom: the first (radial) component, the second (transverse) component, the third (vertical) component.

than the differences of model QI. For a description of models QI0, QI, QI2 and QI4 refer to Klimeš and Bulant (2004) and to Bulant and Klimeš (2006).

The synthetic seismograms were calculated at 29 receivers located in a vertical well drilled 1 km from a vertical force. The source time function is the Gabor signal with reference frequency 50 Hz, band-pass filtered by a cosine filter.

Isotropic-ray-theory, coupling-ray-theory and anisotropic-ray-theory seismograms in model QIH are compared in Figs. 18–20. The most pronounced differences can be observed in polarization. They are clearly visible on the second

SEISMIC RAY METHOD: RECENT DEVELOPMENTS 65

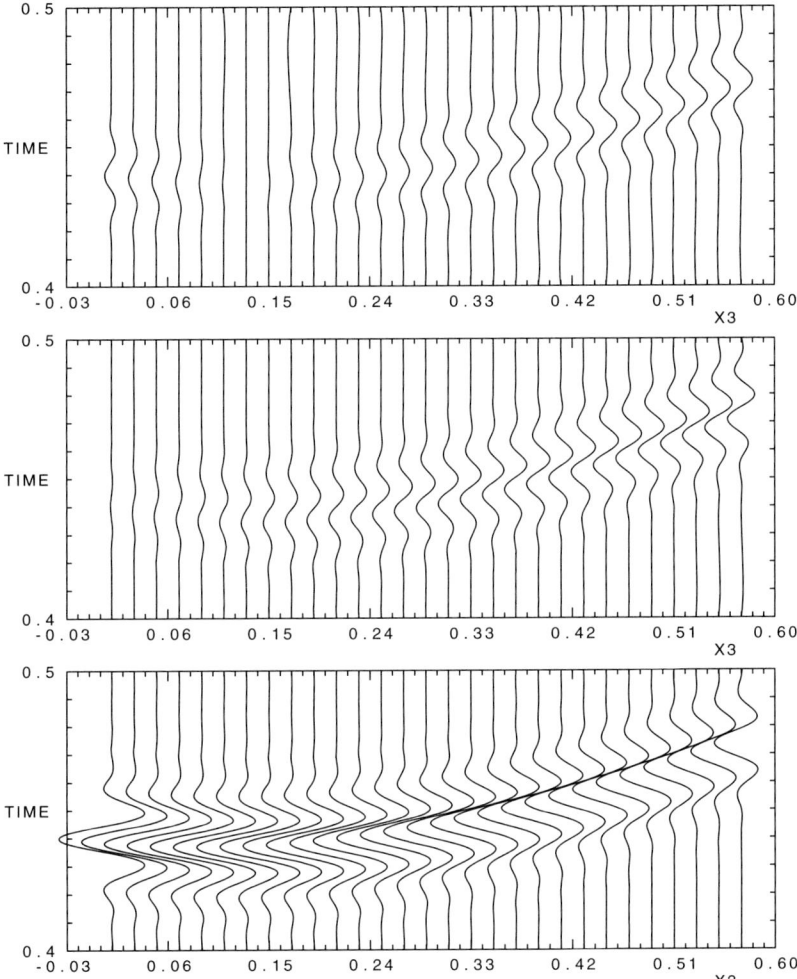

FIG. 20. Coupling-ray-theory seismograms in model QIH. From top to bottom: the first (radial) component, the second (transverse) component, the third (vertical) component. Compare the second (transverse) component with Figs. 18 and 19.

(transverse) component. A small time shift of the isotropic-ray-theory seismograms with respect to the anisotropic-ray-theory and coupling-ray-theory seismograms could be corrected by a better selection of the reference isotropic model.

The second (transverse) component of the coupling-ray-theory seismograms in all 4 anisotropic models QIH, QI, QI2 and QI4 are compared in Fig. 21. For weak anisotropy, the change of polarization with increasing anisotropy is indicated by a clear increment of the transverse amplitudes in the two upper models.

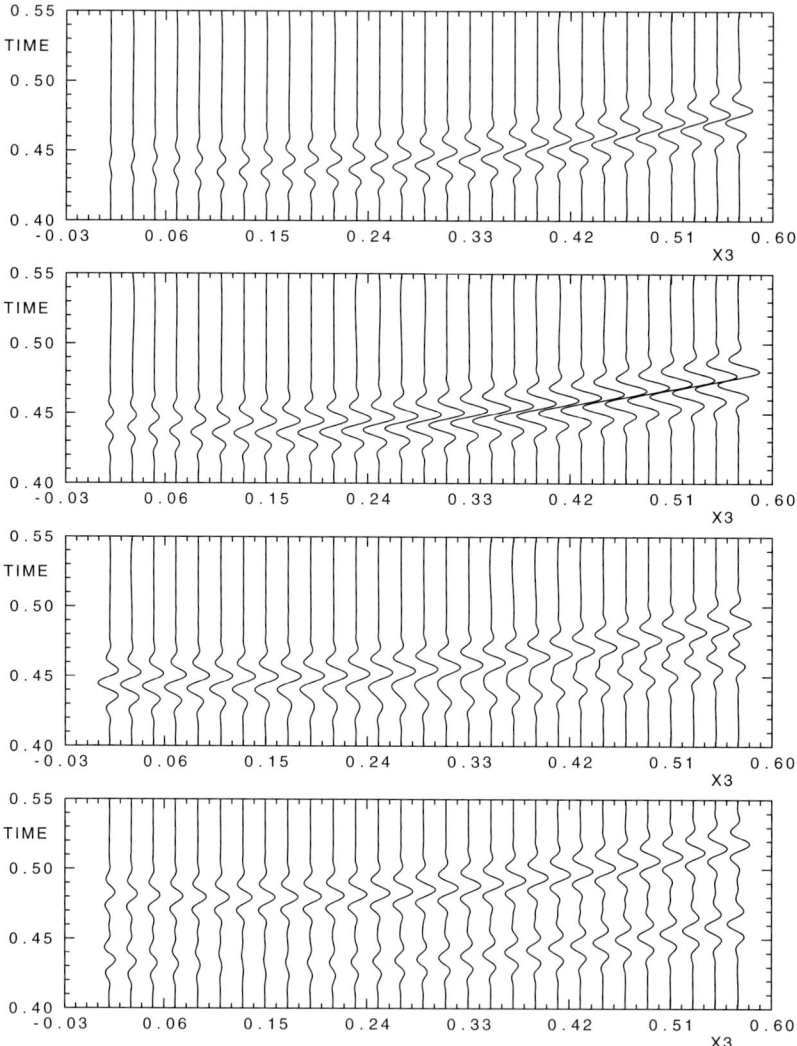

FIG. 21. Coupling-ray-theory seismograms in models QIH, QI, QI2 and QI4 (from top to bottom). Only the second (transverse) component is shown. This component vanishes in the reference isotropic model QI0 (Fig. 18). See the change of polarization indicated by clear increment of transverse amplitudes in the two upper models, and the clear development of S-wave splitting in the two bottom models.

The clear development of S-wave splitting, if anisotropy is increased further, can be observed in the two bottom models.

Figure 22 shows the comparison of the synthetic seismograms computed by the quasi-isotropic approximation of the coupling ray theory and by the Chebyshev

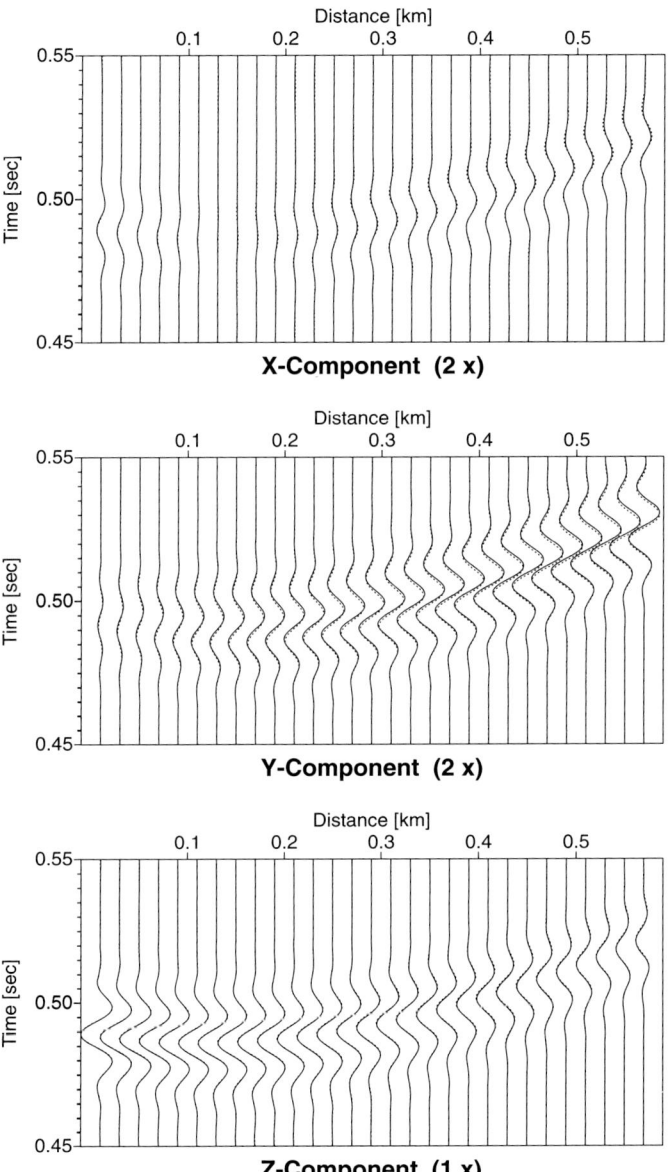

FIG. 22. Comparison of the seismograms computed using the quasi-isotropic approximation of the coupling ray theory (dotted line) with the Chebyshev spectral method (solid line) in model QI. From top to bottom: the first (radial) component (twice enlarged), the second (transverse) component (twice enlarged), the third (vertical) component. For better viewing, the first two components are enlarged twice with respect to the third. Slight discrepancies are mostly due to the quasi-isotropic approximation of the Christoffel matrix.

spectral method (Tessmer, 1995) in model QI. Slight discrepancies are mostly due to the quasi-isotropic approximation of the Christoffel matrix (Bulant and Klimeš, 2002; Klimeš and Bulant, 2004).

5. Summation of Gaussian Beams and Packets

Gaussian beams and packets may serve as building blocks of a wavefield. The summation of Gaussian beams (Červený et al., 1982; Popov, 1982; Červený, 1985; Nowack, 2003) and packets overcomes the problems of the standard ray theory with caustics. The summation of Gaussian beams and packets is considerably comprehensive and flexible. It may be formulated in many ways, including the Maslov method and its various generalizations as special cases. The form of the summation depends primarily on the specification of the wavefield, also comprising the system of Gaussian packets scattered from Gabor functions forming medium perturbations.

We shall now concentrate on the representation of non-directional wavefields without pronounced diffractions rather than on the decomposition of directional beams or of diffracted waves. We thus omit the higher-order Gaussian beams (Bessel–Gaussian beams, Hermite–Gaussian beams, Laguerre–Gaussian beams) and diffracted Gaussian beams.

A Gaussian beam or packet is expressed in terms of a complex-valued vectorial amplitude and complex-valued travel time (phase function, eikonal). In this Section 5, we prefer the term "phase function" specifically for Gaussian packets, otherwise we shall use the term "travel time". The travel time depends on the wave equation through the eikonal equation, whereas the amplitude depends on the wave equation through the transport equation.

In Section 5, we consider only isotropic media. Asymptotic summations of Gaussian beams or packets are performed in the frequency domain, because the frequency domain is the natural representation in time-independent media.

5.1. Gaussian Beams

A Gaussian beam is a high-frequency asymptotic time-harmonic solution of the elastodynamic equation, with an approximately Gaussian profile perpendicularly to the central ray. Refer to Fig. 24 for an illustration. A Gaussian beam represents a bundle of complex-valued rays concentrated in the vicinity of a real-valued central ray. A paraxial Gaussian beam is a Gaussian beam approximated by the paraxial ray approximation in the vicinity of the central ray, or in the vicinity of a selected reference point. The paraxial ray approximation consists in the second-order Taylor expansion of travel time at a point on the central ray and in a constant amplitude.

The evolution of the shape of a Gaussian beam along the central ray is determined by the respective ordinary differential equations. The travel time and its first derivatives along the real-valued central ray are determined, together with the central ray, by ray tracing and are real-valued. The complex-valued second derivatives of travel time, which control the shape of the Gaussian beam, are determined by dynamic ray tracing, see Section 2.3. Dynamic ray tracing is also required for determining the complex-valued amplitude of the Gaussian beam.

Infinitely broad paraxial Gaussian beams (with real-valued second derivatives of travel time) correspond to the paraxial approximation of time-harmonic ray-theory wavefields. If used in a superposition, they must differ from the paraxial approximation of the decomposed ray-theory wavefield. Ribbon Gaussian beams are paraxial Gaussian beams infinitely broad only in a single *singular direction* perpendicular to the central ray. If used in a superposition, they correspond to the paraxial approximation of the decomposed ray-theory wavefield in the singular direction. If the ribbon Gaussian beams used in a superposition are infinitely broad in all directions, they coincide with the paraxial approximation of the decomposed ray-theory wavefield only in the singular direction, but must differ from it in other directions perpendicular to the central ray. Note that infinitely broad Gaussian beams correspond to the standard Maslov method. However, these infinitely broad Gaussian beams are not recommended for the frequency-domain superposition integrals.

5.1.1. Gaussian Beam

A Gaussian beam is an approximate time-harmonic high-frequency asymptotic solution of the elastodynamic equation (or other hyperbolic wave equation), and may be expressed in the form of

$$\mathbf{u}^{GB}(x_i, t) = \mathbf{U} \exp[i\omega(\tau - t)], \tag{112}$$

where both the complex-valued vectorial amplitude $\mathbf{U} = \mathbf{U}(x_i)$ and the complex-valued travel time $\tau = \tau(x_i)$ are smooth functions of spatial coordinates x_i.

The complex-valued travel time satisfies the eikonal equation

$$V^2 \tau_{,i} \tau_{,i} = 1. \tag{113}$$

Introducing slowness vector

$$p_i = \tau_{,i} \tag{114}$$

and Hamiltonian (21), the eikonal equation may be solved by the method of characteristics. The corresponding Hamilton ray tracing equations for frequency-independent propagation velocity $V = V(x_k)$ are

$$\frac{dy_i}{d\gamma_3} = [V(y_k)]^2 p_i, \quad \frac{dp_i}{d\gamma_3} = -[V(y_n)]^{-1} V_{,i}(y_k), \tag{115}$$

see (27). Here we have denoted the coordinates by y_i instead of x_i in order to be able to distinguish the points of the ray from general coordinates x_i. We have also denoted the travel time along the ray γ_3 in order to distinguish it from the complex-valued travel time $\tau = \tau(x_i)$ in Eq. (112).

Since the Hamiltonian is real-valued along the real-valued rays, the imaginary part of the travel time is constant. We shall refer to the ray, along which the imaginary part of the travel time is minimum, as the central ray. We shall assume that the value of the imaginary part of the travel time along the central ray is zero,

$$\text{Im}\{\tau[y_i(\gamma_3)]\} = 0. \tag{116}$$

Since (116) corresponds to the minimum of the imaginary part of τ, inequality

$$\text{Im}[\tau(x_i)] \geq 0 \tag{117}$$

is the consequence of option (116). The Gaussian beam is thus concentrated close to its central ray.

5.1.2. Paraxial Gaussian Beam

We shall refer to the approximation of a Gaussian beam, obtained by substituting the exact travel time τ by its second-order Taylor expansion, as the paraxial approximation of the Gaussian beam or, briefly, as the *paraxial Gaussian beam*. We shall refer to Gaussian beam (112) with the exact travel time τ, satisfying eikonal equation (113) as the *strict Gaussian beam*.

The paraxial approximation of a strict Gaussian beam centred at point $y_i(\gamma_3)$, is

$$\mathbf{u}^{\text{PGB}}(x_i, t) = \mathbf{U} \exp\{i\omega[p_i^R r_i + \tfrac{1}{2} N_{ij} r_i r_j - t]\}, \tag{118}$$

where

$$r_i = x_i - y_i(\gamma_3) \tag{119}$$

is the difference of the spatial coordinates from reference point y_i. p_i^R is the ray-theory slowness vector at the reference point, and

$$N_{ij} = \tau_{,ij} \tag{120}$$

are the second derivatives of the travel time at the reference point with respect to Cartesian coordinates. Quantities $\mathbf{U}$, p_i and N_{ij} on the right-hand side of Eq. (118) are taken at the reference point $y_i(\gamma_3)$ of the Gaussian beam.

The complex-valued second partial derivatives N_{ij} of the travel time with respect to Cartesian coordinates x_i at reference point y_i determine the shape of the Gaussian beam. The imaginary part of the 3×3 matrix N_{kl} has rank 2 and should be chosen positive-semidefinite. The evolution of the complex-valued second derivatives N_{ij} of the travel time along the spatial central ray is determined by the quantities calculated by dynamic ray tracing. Dynamic ray tracing is also required for determining the complex-valued amplitude of the Gaussian beam.

The difference between the strict Gaussian beam and its paraxial approximation is of the asymptotic order $\omega^{-\frac{1}{2}}$. A single paraxial Gaussian beam is then a worse solution to the equations of motion than the strict Gaussian beam by the asymptotic order $\omega^{\frac{1}{2}}$. However, the differences of order $\omega^{-\frac{1}{2}}$ between the paraxial and strict Gaussian beams are odd functions of the coordinates and so vanish in the integration over the space in the asymptotic summation of Gaussian beams. The error of the 2-parametric (1-parametric in 2-D models) time-harmonic superposition of paraxial Gaussian beams is thus of the asymptotic order ω^{-1}, i.e. of the same order as the error of the superposition of the strict Gaussian beams.

As the travel time, the amplitude can also be replaced by its Taylor expansion. The approximation of the amplitude up to the nth order introduces an error of the asymptotic order $\omega^{-\frac{n+1}{2}}$. Since the error of the second-order expansion of the travel time is of order $\omega^{-\frac{1}{2}}$, the amplitude can be replaced by the constant in the paraxial approximation.

5.1.3. Equations for the Second Derivatives of Travel Time in Cartesian Coordinates

Differentiating eikonal equation (113) with respect to spatial coordinates and considering (120), we obtain relation

$$V^2 p_k N_{ki} = -V^{-1} V_{,i}. \tag{121}$$

Differentiating eikonal equation (113) twice with respect to spatial coordinates, and inserting it together with ray tracing equations (115) into equation $d\tau_{,jk}/d\gamma_3 = \tau_{,ijk} \, dy_k/d\gamma_3$, we arrive at the Riccati equation

$$\frac{dN_{jk}}{d\gamma_3} = -V^2 N_{ij} N_{ik} - V^{-1} V_{,jk} - V^{-2} V_{,j} V_{,k} - 2V V_{,j} N_{ik} p_i$$
$$- 2V V_{,k} N_{ij} p_i. \tag{122}$$

Inserting (121) into (122), we obtain the Riccati equation for the 3×3 matrix N_{ij},

$$\frac{dN_{jk}}{d\gamma_3} = -V^2 N_{ij} N_{ik} - V^{-1} V_{,jk} + 3V^{-2} V_{,j} V_{,k}. \tag{123}$$

Equation (123) describes the evolution of the shape of a Gaussian beam in general Cartesian coordinates. Note that Eq. (123) can also be transformed into the system of two linear ordinary differential equations for two 3×3 matrices.

5.1.4. Equations for the Second Derivatives of Travel Time in Ray-Centred Coordinates

The second-order covariant derivatives M_{mn} of the travel time in ray-centred coordinates are related to the partial derivatives N_{mn} in Cartesian coordinates through equation

$$N_{jk} = H_{jm} H_{kn} M_{mn}. \tag{124}$$

Riccati equation (123) for matrix N_{ij} is equivalent to the Riccati equation for 3×3 matrix M_{mn}. Here H_{im} is the ith component of mth basis vector of the ray-centred coordinate system, see Section 2.3.2.

The derivatives of the basis vectors of the ray-centred coordinate system along the ray satisfy relations, see (49),

$$\frac{dH_{jM}}{d\gamma_3} H_{jN} = 0, \qquad \frac{dH_{j3}}{d\gamma_3} H_{jN} = -V_N^{(q)},$$

$$\frac{dH_{jM}}{d\gamma_3} H_{j3} = V_M^{(q)}, \qquad \frac{dH_{j3}}{d\gamma_3} H_{j3} = 0, \tag{125}$$

where

$$V_m^{(q)} = V_{,i} H_{im} \tag{126}$$

is the velocity gradient in ray-centred coordinates.

Equation (121), transformed using (124) into ray-centred coordinates, reads

$$M_{m3} = M_{3M} = -V^{-2} V_m^{(q)}. \tag{127}$$

Equation (123), transformed using (124) into ray-centred coordinates, reads

$$\frac{dM_{mn}}{d\gamma_3} = -M_{jn} \frac{dH_{ij}}{d\gamma_3} H_{im} - M_{mk} \frac{dH_{ik}}{d\gamma_3} H_{in} - V^2 M_{km} M_{kn}$$
$$- V^{-1} V_{mn}^{(q)} + 3 V^{-2} V_m^{(q)} V_n^{(q)}, \tag{128}$$

where

$$V_{mn}^{(q)} = V_{,ij} H_{im} H_{jn} \tag{129}$$

are the covariant velocity derivatives in ray-centred coordinates. Inserting (125) for the derivatives of the basis vectors, Eq. (128) may be expanded into the equation for the 2×2 submatrix M_{MN} and equations for matrix elements M_{M3} and M_{33}. The Riccati equation for the 2×2 submatrix M_{MN}, with (127) inserted for M_{M3}, reads

$$\frac{dM_{MN}}{d\gamma_3} = -V^2 M_{MK} M_{KN} - V^{-1} V_{MN}^{(q)}. \tag{130}$$

The equations for matrix elements M_{M3} and M_{33} need not be used because their solution is given by Eq. (127).

The covariant velocity derivatives $V_i^{(q)}$ and $V_{ij}^{(q)}$ in ray-centred coordinates are defined by Eqs. (126) and (129).

5.1.5. Solving the Equation for a Gaussian Beam

To express equation (130) in matrix notation, we introduce matrices

$$\mathbf{M} = \begin{pmatrix} M_{11} & M_{12} \\ M_{21} & M_{22} \end{pmatrix}, \quad \mathbf{V}^{(q)} = \begin{pmatrix} V_{11}^{(q)} & V_{12}^{(q)} \\ V_{21}^{(q)} & V_{22}^{(q)} \end{pmatrix}. \tag{131}$$

Equation (130) in matrix notation has form (67), and its solution may be expressed as

$$\mathbf{M} = \mathbf{M}^{\mathrm{T}} = \mathbf{P}^{(q)}(\mathbf{Q}^{(q)})^{-1}, \tag{132}$$

see (64). Complex-valued matrices $\mathbf{Q}^{(q)}$ and $\mathbf{P}^{(q)}$ satisfy dynamic ray tracing equations (55). Introducing the DRT propagator matrix (62), matrices $\mathbf{Q}^{(q)}$ and $\mathbf{P}^{(q)}$ may be expressed in terms of their initial values $\mathbf{Q}^0$ and $\mathbf{P}^0$ as

$$\begin{aligned} \mathbf{Q}^{(q)} &= \mathbf{Q}_1 \mathbf{Q}^{(q)0} + \mathbf{Q}_2 \mathbf{P}^{(q)0}, \\ \mathbf{P}^{(q)} &= \mathbf{P}_1 \mathbf{Q}^{(q)0} + \mathbf{P}_2 \mathbf{P}^{(q)0}, \end{aligned} \tag{133}$$

with

$$\mathbf{P}^{(q)0} = \mathbf{M}^0 \mathbf{Q}^{(q)0}. \tag{134}$$

Matrix (132) may then be expressed as

$$\mathbf{M} = (\mathbf{P}_1 + \mathbf{P}_2 \mathbf{M}^0)(\mathbf{Q}_1 + \mathbf{Q}_2 \mathbf{M}^0)^{-1}. \tag{135}$$

5.1.6. Amplitude of a Gaussian Beam

The vectorial amplitude $\mathbf{U}$ of the Gaussian beam may be expressed in terms of the complex-valued scalar amplitude A and unit polarization vector $\mathbf{g}$,

$$\mathbf{U} = A\,\mathbf{g}. \tag{136}$$

The unit polarization vector $\mathbf{g}$ is identical to the analogous ray-theory polarization vector.

The complex-valued scalar amplitude evolves along the central ray according to equation

$$A = A^0 \sqrt{\frac{V^0 \varrho^0}{V \varrho \det(\mathbf{Q}_1 + \mathbf{Q}_2 \mathbf{M}^0)}}. \tag{137}$$

Here velocity V and density ϱ correspond to the elastodynamic equation, and should be replaced by the respective material parameters for other wave equations.

5.2. Gaussian Packets

Gaussian packets, also called (space–time) Gaussian beams (Ralston, 1983), quasiphotons (Babich and Ulin, 1981) or coherent states (Combescure *et al.*, 1999), are high-frequency asymptotic space–time solutions of the elastodynamic equation. A Gaussian packet has an approximately Gaussian profile in all spatial directions and in time. A Gaussian packet is concentrated to a real-valued space–time ray, as a Gaussian beam to a spatial ray. In a stationary medium, a Gaussian packet propagates along its real-valued spatial central ray. A paraxial Gaussian packet is a Gaussian packet approximated by the paraxial ray approximation in the vicinity of its central point. The time-dependent central point is the spatial position of the maximum of the envelope of the Gaussian packet. The central point of a Gaussian packet moves along the spatial central ray according to the ray tracing equations. The shape of a Gaussian packet is determined by the second derivatives of the phase function.

The equations for the evolution of a Gaussian packet were derived by Babich and Ulin (1981) and Ralston (1983). The phase function and its first derivatives along the real-valued central ray are determined by ray tracing and are real-valued. The evolution of the complex-valued second derivatives of the phase function, which control the shape of the Gaussian packet, along the central ray is determined by the quantities calculated by dynamic ray tracing with additional quadratures along the central ray. Dynamic ray tracing is also required for determining the complex-valued amplitude of the Gaussian packet.

The evolution equations are required to propagate Gaussian packets constituting a general time-domain wavefield (Žáček, 2006b), and to optimize the shape of Gaussian packets (Žáček, 2006a). The evolution equations for the shape of Gaussian packets are unnecessary for the asymptotic decomposition of the time-harmonic wavefield specified in terms of amplitude and phase (Klimeš, 1984b, 1989b), but may be applied to the optimization of the shape of Gaussian packets used in this asymptotic decomposition.

Infinitely long paraxial Gaussian packets correspond to the paraxial approximations of Gaussian beams. Infinitely broad paraxial Gaussian packets correspond to the paraxial approximations of space–time ray-theory wavefields.

5.2.1. Gaussian Packet and the Space–Time Eikonal Equation

A Gaussian packet is an approximate space–time high-frequency asymptotic solution of the corresponding hyperbolic wave equation, and may be expressed in the form of

$$\mathbf{u}^{GP}(x_\kappa) = \mathbf{U} \exp(i\omega\tau), \tag{138}$$

where both the complex-valued vectorial amplitude $\mathbf{U} = \mathbf{U}(x_\alpha)$ and the complex-valued phase function $\tau = \tau(x_\alpha)$ are smooth functions of the four space–time coordinates x_α. Hereinafter, the lower-case Greek subscripts $_{\alpha,\beta,\ldots}=_{1,2,3,4}$ will correspond to four space–time coordinates: spatial coordinates x_i with $_i=_{1,2,3}$, and time x_4.

The complex-valued phase function satisfies the space–time eikonal equation

$$V^2 \tau_{,i} \tau_{,i} - \tau_{,4} \tau_{,4} = 0. \tag{139}$$

Introducing space–time slowness vector

$$N_\alpha = \tau_{,\alpha} \tag{140}$$

and Hamiltonian $\mathcal{H} = \mathcal{H}(x_\alpha, N_\alpha) = V^2(x_k) N_i N_i - N_4 N_4$, the eikonal equation may be solved by the method of characteristics. The corresponding Hamilton space–time ray tracing equations for time-independent propagation velocity $V = V(x_k)$ are

$$\frac{dy_i}{d\gamma_3} = [V(y_k)]^2 N_i, \qquad \frac{dy_4}{d\gamma_3} = -N_4,$$

$$\frac{dN_i}{d\gamma_3} = -[V(y_n)]^{-1} V_{,i}(y_k), \qquad \frac{dN_4}{d\gamma_3} = 0. \tag{141}$$

Here again we have denoted the space–time coordinates by y_α instead of x_α in order to be able to distinguish the points of the space–time ray from general space–time coordinates x_α.

The imaginary part of phase function τ must have, at a fixed time $x_4 = y_4(\gamma_3)$, a strict minimum over space at some point $x_i = y_i(\gamma_3)$. The imaginary part of the phase function can be approximated in the vicinity of its minimum y_i by the quadratic form of spatial distances $[x_i - y_i(\gamma_3)]$. For sufficiently high frequencies, the amplitude profile of the Gaussian packet is thus *nearly* Gaussian in space and concentrated close to point $y_i(\gamma_3)$. We shall refer to point $y_i(\gamma_3)$ as the central point of the Gaussian packet.

Since the Hamiltonian is zero along the space–time rays, the complex-valued phase function is constant. In this way, the minimum of the imaginary part of τ, which yields the maximum of the high-frequency Gaussian packet, is identical with the space–time ray passing through point $y_i(\gamma_3^0)$ at time $y_4(\gamma_3^0)$ (Norris et al., 1987). We shall assume that the value of the phase function at this space–time ray is zero,

$$\tau[y_\alpha(\gamma_3)] = 0. \tag{142}$$

Since (142) corresponds to the minimum of the imaginary part of τ, inequality

$$\text{Im}[\tau(x_\alpha)] \geq 0 \tag{143}$$

is the consequence of option (142).

The Gaussian packet is thus concentrated close to the space–time ray as the Gaussian beam is concentrated close to the spatial ray, and is also called a (space–time) Gaussian beam (Ralston, 1983). Thus, in a time-independent perfectly elastic model, the Gaussian packet moves along the spatial ray with the ray velocity. The Gaussian packet remains approximately Gaussian in space at all times $x_4 = y_4(\gamma_3)$. At any time, the principal part of the energy of the Gaussian packet is effectively concentrated in some vicinity of the central point $y_i(\gamma_3)$ of the Gaussian packet. This vicinity is called the *effective region* of the Gaussian packet. At a specified time, the effective region of a Gaussian packet is limited in space and, for a sufficiently high frequency, it is concentrated close to its central point. In space–time, however, the effective region of a Gaussian packet is not limited.

We select

$$N_4 = -1. \tag{144}$$

Derivative $d/d\gamma_3$ along the space–time ray then coincides with derivative d/dy_4 with respect to time y_4, and space–time ray tracing equations (141) reduce to the spatial ray tracing equations (115).

5.2.2. Paraxial Gaussian Packet

We shall refer to the approximation of a Gaussian packet, obtained by substituting the exact phase function τ by its second-order Taylor expansion, as the paraxial approximation of the Gaussian packet or, briefly, as the *paraxial Gaussian packet*. The paraxial approximation of the Gaussian packet in space is an analogy of the paraxial approximation of the Gaussian beam in the plane perpendicular to the ray. We shall refer to the Gaussian packet (138) with the exact phase function τ satisfying eikonal equation (139) as the *strict Gaussian packet*.

The space–time paraxial approximation of a Gaussian packet centred at point $y_\alpha(\gamma_3)$, is

$$\mathbf{u}^{PGP}(x_\kappa) = \mathbf{U}\exp\{i\omega[N_\alpha r_\alpha + \tfrac{1}{2}N_{\alpha\beta}r_\alpha r_\beta]\}, \tag{145}$$

where

$$r_i = x_i - y_i(\gamma_3), \tag{146}$$

$$r_4 = x_4 - y_4(\gamma_3) = x_4 - \gamma_3 \tag{147}$$

is the difference of the space–time coordinates from the central point y_α. N_α defined by (140) is the space–time slowness vector of the Gaussian packet, and

$$N_{\alpha\beta} = \tau_{,\alpha\beta} \tag{148}$$

are the second derivatives of the phase function at the central point. Quantities $\mathbf{U}$, N_α and $N_{\alpha\beta}$ on the right-hand side of Eq. (145) are taken at the central point $y_\alpha(\gamma_3)$ of the Gaussian packet.

Since we have selected $N_4 = -1$, see (144),

$$N_i = p_i^R \qquad (149)$$

is the ray-theory slowness vector at the central point. The first derivatives N_α of the phase function at the central point of the Gaussian packet are thus determined by standard ray tracing and are real-valued.

The complex-valued second partial derivatives $N_{\alpha\beta}$ of the phase function with respect to Cartesian coordinates x_i at the central point y_α determine the shape of the Gaussian packet. The imaginary part of the 3×3 spatial submatrix N_{kl} of the 4×4 matrix $N_{\alpha\beta}$ should be chosen so that it is positive-definite. The imaginary part of the whole 4×4 matrix $N_{\alpha\beta}$ is singular. The evolution of the complex-valued second derivatives $N_{\alpha\beta}$ of the phase function along the spatial central ray is determined by quantities calculated by dynamic ray tracing. Dynamic ray tracing is also required for determining the complex-valued amplitude of the Gaussian packet.

The spatial effective region of the Gaussian packet may be sufficiently small for sufficiently high frequencies. Then, at any time $x_4 = y_4(\gamma_3)$, the phase function τ of the strict Gaussian packet can be replaced, in the spatial effective region, by its Taylor expansion up to the second order with respect to the spatial coordinates,

$$\mathbf{u}^{PGP}(x_\kappa) = \mathbf{U} \exp\{i\omega[N_i r_i + \tfrac{1}{2} N_{ij} r_i r_j]\}, \qquad (150)$$

where N_{ij} is the spatial part of the 4×4 matrix (148) of the second space–time derivatives of the phase function.

The difference between the strict Gaussian packet and its spatial paraxial approximation is of the asymptotic order $\omega^{-\frac{1}{2}}$. A single paraxial Gaussian packet is then a worse solution to the equations of motion than the strict Gaussian packet by the asymptotic order $\omega^{\frac{1}{2}}$. However, the differences of order $\omega^{-\frac{1}{2}}$ between the paraxial and strict Gaussian packets are odd functions of the coordinates, which therefore vanish in the integration over the space in the asymptotic summation of Gaussian packets. The error of the 3-parametric (2-parametric in 2-D models) time-harmonic superposition of paraxial Gaussian packets is thus of the asymptotic order ω^{-1}, i.e. of the same order as the error of the superposition of the strict Gaussian packets.

As the phase function, the amplitude can also be replaced by its Taylor expansion. The restriction of the amplitude up to the nth order introduces an error of the asymptotic order $\omega^{-\frac{n+1}{2}}$. Since the error of the second-order expansion of the phase function is of the order $\omega^{-\frac{1}{2}}$, the amplitude can be replaced by the constant in the paraxial approximation.

5.2.3. Equations for the Second Derivatives of Phase Function in Cartesian Coordinates

The Riccati equation for the 4×4 space–time matrix (148) may be obtained by differentiating eikonal equation (139) twice with respect to the space–time coordinates x_α, see Ralston (1983, Eq. (2.4)). The corresponding linear dynamic ray tracing system may be obtained by differentiating the ray tracing system with respect to the space–time ray parameters, see Ralston (1983, Eq. (2.6)). The Riccati equation for the 4×4 space–time matrix (148) may be reduced to the Riccati equation for its 3×3 spatial submatrix N_{ij}, see Kachalov (1984, Eq. (9)), Babich et al. (1985, Eq. (2.15)), Norris et al. (1987, Eq. (2.27a)). For the corresponding linear dynamic ray tracing system refer to Kachalov (1984, Eq. (12)), Babich et al. (1985, Eq. (2.18)), Norris et al. (1987, Eq. (2.25ab)). Here we shall directly derive the Riccati equation for the 3×3 spatial submatrix of matrix (148) and then switch to ray-centred coordinates in order to decouple it.

Differentiating eikonal equation (139) with respect to space–time coordinates and considering (144), we obtain relation

$$N_{4\alpha} = -V^{-1} V_{,\alpha} - V^2 N_i N_{i\alpha} \qquad (151)$$

for calculating N_{4i} from N_{ij}, and for calculating N_{44} from N_{4i}.

Differentiating eikonal equation (139) twice with respect to spatial coordinates, and inserting it together with space–time ray tracing equations (141) and with Eq. (151) into equation $d\tau_{,jk}/d\gamma_3 = \tau_{,ij\alpha} dy_\alpha/d\gamma_3$, we arrive at the Riccati equation

$$\frac{dN_{jk}}{d\gamma_3} = -V^2 N_{ij} N_{ik} - V^{-1} V_{,jk} - V V_{,j} N_{ik} N_i$$
$$- V V_{,k} N_{ij} N_i + V^4 N_i N_{ij} N_{lk} N_l \qquad (152)$$

for symmetric 3×3 matrix N_{ij}. Equations (151) and (152) describe the evolution of the shape of a Gaussian packet in general Cartesian coordinates.

5.2.4. Equations for the Second Derivatives of Phase Function in Ray-Centred Coordinates

Covariant derivatives M_{mn} of the phase function in ray-centred coordinates are related to the partial derivatives N_{mn} in Cartesian coordinates through equations

$$N_{jk} = H_{jm} H_{kn} M_{mn}, \qquad (153)$$
$$N_{j4} = H_{jm} M_{m4}, \qquad (154)$$
$$N_{44} = M_{44}. \qquad (155)$$

Here H_{im} is the ith component of mth basis vector of the ray-centred coordinate system, see Section 2.3.2.

Riccati equation (152) for matrix N_{ij} is equivalent to the Riccati equation for the 3×3 matrix M_{mn}, see Norris et al. (1987, Eq. (4.8d)), with the corresponding linear dynamic ray tracing system (Norris et al., 1987, Eq. (4.8ab)). An analogous Riccati equation for the 3×3 matrix of the second partial derivatives in ray-centred coordinates was derived by Babich and Ulin (1981, Eqs. (2.7)–(2.9)). The evolution equations for the 2×2 submatrix M_{MN}, corresponding to the plane perpendicular to the ray, are decoupled, i.e. they do not depend upon the outer elements M_{m3}. These equations are identical with the equations for Gaussian beams and can be treated in the same way. The equations for M_{m3} can be solved analytically (Babich and Ulin, 1981). Here we shall stick to the equations for the 3×3 matrix of the second covariant derivatives M_{mn} in ray-centred coordinates.

Equation (151) yields

$$M_{i4} = -VM_{i3} - V^{-1}V_i^{(q)} \tag{156}$$

and

$$M_{44} = V^2 M_{33} + V_3^{(q)}, \tag{157}$$

where velocity gradient $V_m^{(q)}$ in ray-centred coordinates is defined by Eq. (126).

Equation (152), transformed using (153) into ray-centred coordinates, reads

$$\frac{dM_{mn}}{d\gamma_3} = -M_{jn}\frac{dH_{ij}}{d\gamma_3}H_{im} - M_{mk}\frac{dH_{ik}}{d\gamma_3}H_{in} - V^2 M_{Km}M_{Kn}$$
$$- V^{-1}V_{mn}^{(q)} - V_m^{(q)}M_{3n} - V_n^{(q)}M_{3m}, \tag{158}$$

where covariant velocity derivatives $V_{mn}^{(q)}$ in ray-centred coordinates are defined by Eq. (129).

Inserting (125) for the derivatives of the basis vectors, Eq. (158) may be expanded into three equations

$$\frac{dM_{MN}}{d\gamma_3} = -V^2 M_{MK}M_{KN} - V^{-1}V_{MN}^{(q)}, \tag{159}$$

$$\frac{dM_{M3}}{d\gamma_3} = -M_{Mi}V_i^{(q)} - V^2 M_{MI}M_{I3} - V^{-1}V_{M3}^{(q)}, \tag{160}$$

$$\frac{dM_{33}}{d\gamma_3} = -2M_{i3}V_i^{(q)} - V^2 M_{3I}M_{I3} - V^{-1}V_{33}^{(q)}. \tag{161}$$

Equations (159) to (161) for the second covariant derivatives of the phase function in ray-centred coordinates are equivalent to the equations by Babich and Ulin (1981, Eqs. (2.7)–(2.9)) for the second partial derivatives of the phase function in ray-centred coordinates.

Riccati equation (159) is *identical with the equation for the corresponding Gaussian beam* and its solution is given by Eq. (135). Equations (160) and (161) extend the Gaussian-beam solution to a Gaussian-packet solution.

5.2.5. Solving the Equations for a Gaussian Packet

Denoting

$$\mathbf{M}_3 = \begin{pmatrix} M_{13} \\ M_{23} \end{pmatrix}, \qquad \mathbf{M}_4 = \begin{pmatrix} M_{14} \\ M_{24} \end{pmatrix}, \qquad \mathbf{v}^{(q)} = \begin{pmatrix} V_1^{(q)} \\ V_2^{(q)} \end{pmatrix}, \qquad (162)$$

we may express the solutions of Eqs. (156), (157), (160) and (161) as

$$\mathbf{M}_3 = -V^{-1}\mathbf{M}_4 - V^{-2}\mathbf{v}^{(q)}, \qquad (163)$$

$$M_{33} = V^{-2}\left[M_{44} - V_3^{(q)}\right], \qquad (164)$$

$$M_{34} = -V^{-1}M_{44}, \qquad (165)$$

$$\mathbf{M}_4 = \left[(\mathbf{Q}_1 + \mathbf{Q}_2\mathbf{M}^0)^{-1}\right]^{\mathrm{T}}\mathbf{M}_4^0, \qquad (166)$$

$$M_{44} = M_{44}^0 - \mathbf{M}_4^{0\mathrm{T}}(\mathbf{Q}_1 + \mathbf{Q}_2\mathbf{M}^0)^{-1}\mathbf{Q}_2\mathbf{M}_4^0 \qquad (167)$$

(Klimeš, 2004b). Equation (166) is convenient for calculating vector $\mathbf{M}_4$ along the central spatial ray. To calculate M_{44}, we may use Eq. (167). Quantities $\mathbf{M}_3$, M_{33} and M_{34} can then be obtained through simple relations (163) to (165).

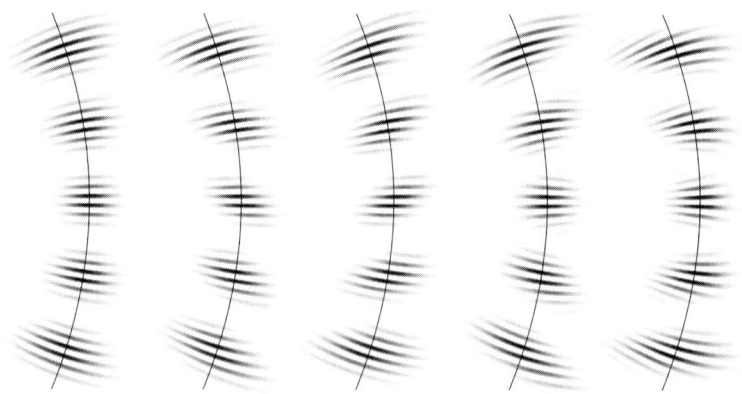

FIG. 23. Symmetric and asymmetric Gaussian packets propagating along a spatial ray. All five packets correspond to Gaussian beams of equal shapes. The first Gaussian packet is axially symmetric with respect to its central ray. The second and third packets have envelopes oblique with respect to their central rays. The frequencies of the second and third packets at the waists of the corresponding Gaussian beams are symmetric with respect to the central rays. The fourth and fifth packets have different frequencies to the left and to the right of the central rays. We may call them "dispersive". The envelopes of the fourth and fifth packets at the waists of the corresponding Gaussian beams are symmetric with respect to the central rays. During the evolution of Gaussian packets, the dispersion of the wavefronts generates obliquity of the envelopes and vice versa. The asymmetric Gaussian packets are thus both oblique and dispersive in general.

Non-vanishing vector (166) implies that the Gaussian packet is asymmetric with respect to the central ray. The evolution of the shape of a Gaussian packet along the central ray is illustrated in Fig. 23.

5.2.6. Amplitude of a Gaussian Packet

The equations for the vectorial amplitude of the Gaussian packet are identical with Eqs. (136) and (137) for the corresponding Gaussian beam.

5.3. Optimization of the Shape of Gaussian Beams or Packets

Gaussian beams and packets are high-frequency approximate solutions of the elastodynamic equation. The accuracy of these approximate solutions depends on their shape, especially on their width. The "optimum" shape depends on the distance from the source (Klimeš, 1989a).

The optimum shape of Gaussian beams depends on the error of the Gaussian-beam solution of the elastodynamic equation, which is, unfortunately, unknown for general media. However, in considerably heterogeneous media, the optimum Gaussian beams are very close to the narrowest Gaussian beams.

The root-mean-square width of each individual Gaussian beam between the source and receiver can be minimized using the algorithm by Klimeš (1989a), see Fig. 24.

The same algorithm can be used to minimize the width of a Gaussian packet, measured perpendicularly to the central ray. The optimum Gaussian packet is *sym-*

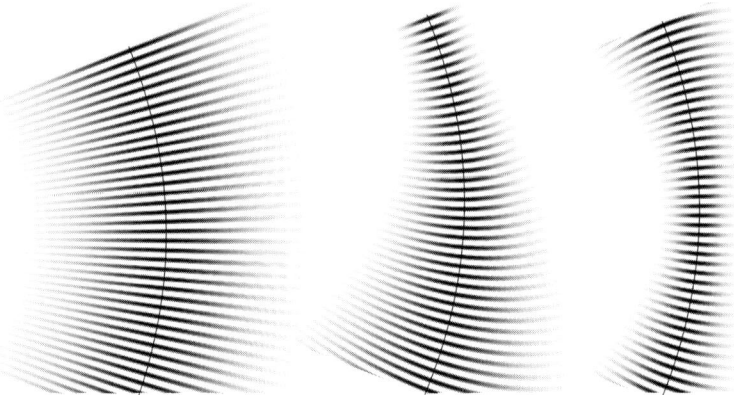

FIG. 24. Optimization of the shape of Gaussian beams. The left Gaussian beam is too wide. The middle Gaussian beam has been selected too narrow at its top end, which results in its extensive spreading downwards. The narrowest Gaussian beam with respect to the mean square width is displayed on the right.

metric with respect to its central ray, see Fig. 23. The error of the Gaussian-packet solution of the elastodynamic equation is smaller for longer Gaussian packets, but the length of Gaussian packets is limited by the accuracy of their paraxial approximation. This problem with long Gaussian packets may be overcome by decomposing each long Gaussian packet into the summation of shorter ones, similarly as composing a Gaussian beam of Gaussian packets in the asymptotic summation of Gaussian packets. Each long Gaussian packet is then numerically calculated as a combination of the paraxial approximations of the Gaussian packet from different reference points (Žáček, 2004).

To preserve the accuracy of the approximate decomposition of a general wavefield into Gaussian beams or packets (Section 5.5), the optimization of the shape of the individual beams or packets should be supplemented with smoothing the dependence of the shape on the summation parameters (Žáček, 2006a).

A possibility to optimize clearly the shape of Gaussian beams or packets in the asymptotic superposition integral (Sections 5.4.1 and 5.4.2) is probably the main difference of the summation of Gaussian beams and packets from the integral representations formally derived using the Maslov method, coherent-state method, or their various generalizations, notwithstanding formally equal integral representations.

5.4. Asymptotic Summation of Gaussian Beams and Packets

5.4.1. Asymptotic Decomposition into Gaussian Beams

A general time-harmonic wavefield given along a surface, decomposed into individual elementary waves and expressed in terms of the vectorial amplitude and travel time, can be asymptotically expressed as a two-parametric integral superposition of Gaussian beams (Klimeš, 1984a).

The two-parametric integral superposition of Gaussian beams, corresponding to time-harmonic ray-theory wavefield

$$\mathbf{u}^R(x_i, t) = \mathbf{U}^R(x_i) \exp\{i\omega[\tau^R(x_i) - t]\}, \tag{168}$$

may be expressed as (Klimeš, 1984a, Eq. (77))

$$\mathbf{u}(x_i, t) = \frac{\omega}{2\pi} \iint d\gamma_1 \, d\gamma_2 \, \mathbf{U}^R |\det(\mathbf{Q}^{(q)R})| \sqrt[2]{\det}[i(\mathbf{M}^R - \mathbf{M})]$$
$$\times \exp\{i\omega[\tau^R + p_k^R(x_k - y_k)$$
$$+ \frac{1}{2}(x_k - y_k)N_{kl}(x_l - y_l) - t]\}, \tag{169}$$

where the function $\sqrt[2]{\det}$ of a symmetric complex-valued matrix is defined by the following two conditions: (a) $\sqrt[2]{\det}(\mathbf{B})$ is the product of the square roots of the eigenvalues of matrix $\mathbf{B}$, (b) the square roots with the greatest real parts are

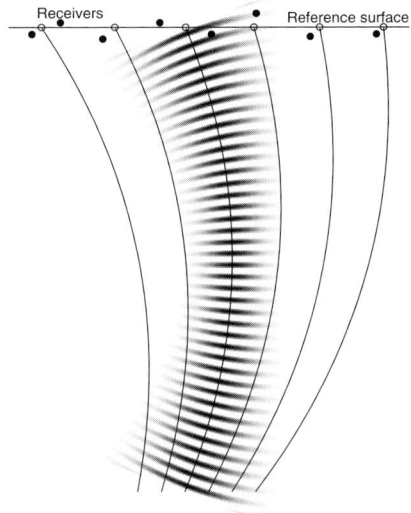

FIG. 25. Asymptotic summation of Gaussian beams. Receivers (bullets) are situated in the vicinity of the reference surface. The points of intersection of rays with the reference surface are selected as the reference points (empty circles) for the paraxial approximation of Gaussian beams.

selected. All the quantities on the right-hand side (except x_i) are taken at the reference point $y_i = y_i(\gamma_1, \gamma_2)$ selected for the paraxial approximation of a Gaussian beam concentrated to the ray with ray parameters γ_1, γ_2. The reference points are usually determined as the points of intersection of rays with a given reference surface, see Fig. 25. Receivers should be situated in the vicinity of the reference surface.

Here τ^R and p_k^R are the ray-theory travel time and ray-theory slowness vector, taken at point y_i. Similarly $\mathbf{U}^R$ is the complex-valued vectorial ray-theory amplitude at point y_i. Matrix $\mathbf{M}^R$ is the 2×2 matrix of the second derivatives of the ray-theory travel time with respect to ray-centred coordinates q_1, q_2 perpendicular to the central ray. Matrix $\mathbf{M}$ is the 2×2 matrix of the second derivatives of the complex-valued travel time of the Gaussian beam with respect to ray-centred coordinates q_1, q_2. Matrix $\mathbf{M}$ controls the shape of Gaussian beams and should be selected with the positive-definite imaginary part. Matrix N_{kl} is the 3×3 matrix of the second derivatives of the complex-valued travel time of the Gaussian beam with respect to Cartesian coordinates x_i, and is completely determined by the 2×2 matrix $\mathbf{M}$ (Klimeš, 1984a, Eqs. (77) and (78)). The transformation Jacobian from ray parameters γ_1, γ_1 to ray-centred coordinates q_1, q_2 is $|\det(\mathbf{Q}^{(q)R})|$, where $\mathbf{Q}^{(q)R}$ is the 2×2 matrix of geometrical spreading in ray-centred coordinates defined by (65).

The asymptotic integral superposition (169) is independent of the selection of the surface for decomposition of the ray-theory wavefield into Gaussian beams (Klimeš, 1984a). That is why the ray-theory wavefield may asymptotically be decomposed into Gaussian beams locally, even in its singular regions. The accuracy of the integral superposition of Gaussian beams thus depends on the optimization of the shape of beams only, not on the surface for decomposition. The variation of the optimum shape of beams with distance from the source is quite different from the evolution of the individual Gaussian beams along the same central ray. In other words, we decompose the wavefield into different Gaussian beams for different propagation distances.

5.4.2. Asymptotic Decomposition into Gaussian Packets

A time-harmonic Gaussian beam may be expressed as a one-parametric integral superposition of space–time Gaussian packets. A general time-harmonic wavefield, specified in terms of the amplitude and travel time, can be asymptotically expressed as a three-parametric integral superposition of space–time Gaussian packets (Klimeš, 1984b, 1989b).

The three-parametric integral superposition of Gaussian packets corresponding to the time-harmonic ray-theory wavefield (168) may be expressed as (Klimeš, 1984b, Eq. (51), 1989b, Eq. (31))

$$\mathbf{u}(x_i, t) = \left(\frac{\omega}{2\pi}\right)^{3/2}$$
$$\times \iiint d\gamma_1 \, d\gamma_2 \, d\gamma_3 \, \mathbf{U}^R V \left|\det(\mathbf{Q}^{(q)R})\right| \sqrt[2]{\det}[i(\mathbf{N}^R - \mathbf{N})]$$
$$\times \exp\{i\omega[\tau^R + p_k^R(x_k - y_k) + \tfrac{1}{2}(x_k - y_k)N_{kl}(x_l - y_l) - t]\}, \tag{170}$$

where all the quantities on the right-hand side (except x_k) are taken at the central point $y_i = y_i(\gamma_1, \gamma_2, \gamma_3)$ of a Gaussian packet. Points $y_i = y_i(\gamma_1, \gamma_2, \gamma_3)$ are the points of spatial rays corresponding to ray-theory wavefield $\mathbf{u}^R$. The quadrature is performed over ray parameters γ_1, γ_2 and over travel time γ_3 along the rays, see Figs. 26 and 27.

Here τ^R, p_k^R and N_{kl}^R are the ray-theory travel time, ray-theory slowness vector and the matrix of second derivatives of the ray-theory travel time in Cartesian coordinates x_i, taken at point y_i. Similarly $\mathbf{U}^R$ is the matrix of the complex-valued vectorial ray-theory amplitude at point y_i. The transformation Jacobian from ray parameters γ_l to Cartesian coordinates x_i is $V|\det(\mathbf{Q}^{(q)R})|$, where V is the wave-propagation velocity and $\mathbf{Q}^{(q)R}$ is the 2×2 matrix of geometrical spreading in ray-centred coordinates defined by (65).

Matrix N_{kl} is the 3×3 matrix of second derivatives of the complex-valued phase function of the Gaussian packet with respect to Cartesian coordinates x_i.

SEISMIC RAY METHOD: RECENT DEVELOPMENTS

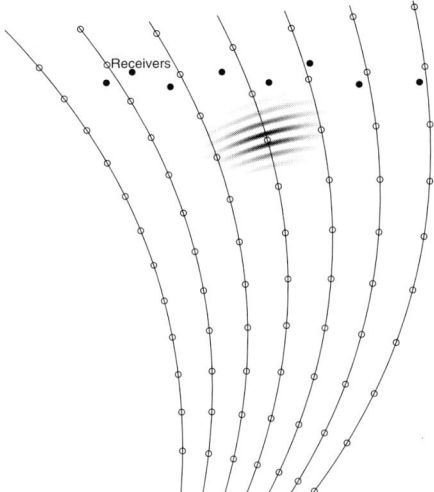

Fig. 26. Asymptotic summation of Gaussian packets. The central points (empty circles) for the paraxial approximation of Gaussian packets are regularly distributed along rays. The density of rays and of central points depends on the shape of the Gaussian packets. Receivers (bullets) may be situated arbitrarily, no reference surfaces are required.

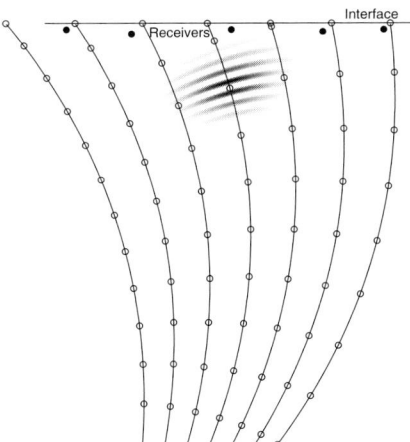

Fig. 27. Summation of Gaussian packets at an interface. Paraxial approximation cannot be applied across the interface. The points of intersection of rays with the interface serve as the reference points (empty circles) for the paraxial approximation of Gaussian packets centred beyond the interface. Receivers (bullets) may be situated arbitrarily.

Matrix N_{kl} controls the shape of Gaussian packets and should be selected with the positive-definite imaginary part.

The properties of the integral superposition of space–time Gaussian packets follow from the properties of the integral superposition of Gaussian beams.

5.4.3. Discretization Error

In numerical algorithms, the integral superposition (169) of Gaussian beams is discretized into the summation over rays. These rays are calculated by controlled initial-value ray tracing. We thus have to divide the ray-parameter domain into elements corresponding to individual rays used in the summation. We first divide the ray-parameter domain into areas corresponding to the individual ray histories. We then divide each area into elements corresponding to the individual rays. We choose each element as the Voronoi polygon of the corresponding ray, i.e. as the set of points, which are closer to the corresponding ray than to other traced rays of equal ray history. Since the discretization error is approximately controlled by contravariant ray-parameter metric tensor

$$\mathbf{G}_{(\gamma)} = \left(\mathbf{Q}^{(q)R}\right)^T (\mathbf{M} - \mathbf{M}^R)^* [\mathrm{Im}(\mathbf{M})]^{-1} (\mathbf{M} - \mathbf{M}^R) \mathbf{Q}^{(q)R} \qquad (171)$$

(Klimeš, 1986), the Voronoi polygons are defined with respect to this metric tensor. Here the 2×2 matrices $\mathbf{Q}^{(q)R}$, $\mathbf{M}^R$ and $\mathbf{M}$ are expressed in ray-centred coordinates. Matrix $\mathbf{M}$ controls the shape of the Gaussian beam concentrated at the ray, and should be determined by optimization of the shape of the Gaussian beam. Matrix $\mathbf{Q}^{(q)R}$ of geometrical spreading and matrix $\mathbf{M}^R$ of the second derivatives of travel time describe the orthonomic system of rays, at which the Gaussian beams are concentrated. Contravariant ray-parameter metric tensor (171) varies smoothly within each ray history, but is often discontinuous at the boundary between different ray histories. Metric tensor (171) thus serves as the *secondary ray-parameter metric tensor* during controlled initial-value ray tracing described in Section 2.2.2. Since we know the values of the metric tensor only at the ray-parameter points corresponding to the calculated rays, we can only determine the elements as an approximation to the Voronoi polygons.

The error due to the discretization depends on frequency and on the shape of the Gaussian beams and can be controlled by the selection of the maximum distance between rays with respect to ray-parameter metric tensor (171). This maximum distance between rays influences the sizes and shapes of the Voronoi polygons. The size and shape of a Voronoi polygon can be quantified by its moment with respect to the corresponding ray-parameter point, which represents the central ray of a Gaussian beam. Unfortunately, the exact equations for the discretization into an irregular system of central rays of Gaussian beams are not known yet.

The discretization error applicable to a system of central rays regular with respect to metric tensor (171), with smooth variation of the weights of Gaussian beams in the decomposition, was derived by Klimeš (1986), for 2-D also by Hill

(1990). This "regular" error represents the lower estimate of the discretization error. Since the upper estimate based on the discretization of the integral over a single Voronoi polygon into a single Gaussian beam (Klimeš, 1985) is too conservative, the "regular" error is now used as the "optimistic" approximation of the discretization error also for an irregular system of central rays of Gaussian beams. The assumption of smooth variation of the weights of Gaussian beams corresponds to the conditions for the integral superposition of Gaussian beams. This smooth variation includes wavefields generated by various point sources, or wavefields with smooth variation of the amplitude and travel time along an initial surface.

The error due to the discretization of the integral superposition of Gaussian packets can be controlled analogously to the error due to the discretization of the integral superposition of Gaussian beams. Refer to Klimeš (1989b) for more details.

Note that it is even more difficult to estimate the discretization error for the decomposition of a general 3-D wavefield into Gaussian packets, described in Section 5.5.

5.4.4. Linear Canonical Transforms

Various superposition integrals similar or equivalent to the asymptotic summation of Gaussian beams or to the asymptotic summation of Gaussian packets have been derived by many authors using the Maslov method, its various generalizations, or coherent-state transforms. The generalizations of the Maslov method are based on the linear canonical transform or on its special cases such as the fractional Fourier transform. Before discussing these integral transform methods, we briefly recall the definition of the linear canonical transform and mention their relation to the coherent-state transform.

Spatial coordinate x_i in phase space corresponds to operator $\hat{x}_i$ of the multiplication by x_i. The momentum coordinate p_i^x in phase space corresponds to the differential operator

$$\hat{p}_i = \frac{1}{i\omega} \frac{\partial}{\partial x_i}. \tag{172}$$

Let us consider a linear coordinate transform in phase space,

$$\begin{pmatrix} \mathbf{x} \\ \mathbf{p}^x \end{pmatrix} = \mathcal{M} \begin{pmatrix} \mathbf{x}' \\ \mathbf{p}^{x'} \end{pmatrix}, \tag{173}$$

where the complex-valued 6×6 matrix $\mathcal{M}$ may be composed of four 3×3 matrices,

$$\mathcal{M} = \begin{pmatrix} \tilde{\mathbf{Q}} & \mathbf{Q} \\ \tilde{\mathbf{P}} & \mathbf{P} \end{pmatrix}. \tag{174}$$

We shall see later that matrix $\mathcal{M}$ should be symplectic. The linear coordinate transform corresponds to the linear transform

$$\begin{pmatrix} \hat{\mathbf{x}} \\ \hat{\mathbf{p}}^x \end{pmatrix} = \mathcal{M} \begin{pmatrix} \hat{\mathbf{x}}' \\ \hat{\mathbf{p}}^{x'} \end{pmatrix} \tag{175}$$

of the operators. The *linear canonical transform* is the linear transform

$$f_{\mathcal{M}}(\mathbf{x}) = \int d^3\mathbf{x}'\, C_{\mathcal{M}}(\mathbf{x}, \mathbf{x}') f(\mathbf{x}') \tag{176}$$

of functions, with integration kernel of the form

$$C_{\mathcal{M}}(\mathbf{x}, \mathbf{x}') = c_{\mathcal{M}} \exp\left[\frac{i\omega}{2} \begin{pmatrix} \mathbf{x} \\ \mathbf{x}' \end{pmatrix}^T \begin{pmatrix} \mathbf{N}_{\mathcal{M}}^{xx} & \mathbf{N}_{\mathcal{M}}^{xx'} \\ \mathbf{N}_{\mathcal{M}}^{x'x} & \mathbf{N}_{\mathcal{M}}^{x'x'} \end{pmatrix} \begin{pmatrix} \mathbf{x} \\ \mathbf{x}' \end{pmatrix} \right], \tag{177}$$

which corresponds to transform (175) of operators. Applying both sides of operator transform (175) with (172) and (174) to the right-hand side of Eq. (176) and assuming that the integrand vanishes at the boundary of the integration volume, we can prove that matrix (174) must be *symplectic* and that

$$\mathbf{N}_{\mathcal{M}}^{x'x} = \left(\mathbf{N}_{\mathcal{M}}^{xx'}\right)^T = -\mathbf{Q}^{-1}, \tag{178}$$

$$\mathbf{N}_{\mathcal{M}}^{x'x'} = \mathbf{Q}^{-1}\tilde{\mathbf{Q}}, \tag{179}$$

$$\mathbf{N}_{\mathcal{M}}^{xx} = \mathbf{P}\mathbf{Q}^{-1}. \tag{180}$$

The relation of symplectic matrix (174) and integration kernel (177) resembles the relation of the paraxial ray propagator matrix and the paraxial two-point eikonal (Arnaud, 1972, Eq. (11.1); Červený et al., 1984, Eq. (22)).

The inverse transforms to (173) and (175) are

$$\begin{pmatrix} \mathbf{x}' \\ \mathbf{p}^{x'} \end{pmatrix} = \mathcal{M}^{-1} \begin{pmatrix} \mathbf{x} \\ \mathbf{p}^x \end{pmatrix} \tag{181}$$

and

$$\begin{pmatrix} \hat{\mathbf{x}}' \\ \hat{\mathbf{p}}^{x'} \end{pmatrix} = \mathcal{M}^{-1} \begin{pmatrix} \hat{\mathbf{x}} \\ \hat{\mathbf{p}}^x \end{pmatrix}, \tag{182}$$

where the inverse matrix to symplectic matrix $\mathcal{M}$ is

$$\mathcal{M}^{-1} = \begin{pmatrix} \mathbf{P}^T & -\mathbf{Q}^T \\ -\tilde{\mathbf{P}}^T & \tilde{\mathbf{Q}}^T \end{pmatrix}. \tag{183}$$

The inverse transform to (176) is

$$f_{\mathcal{M}^{-1}}(\mathbf{x}') = \int d^3\mathbf{x}\, C_{\mathcal{M}^{-1}}(\mathbf{x}', \mathbf{x}) f(\mathbf{x}). \tag{184}$$

Integration kernel $C_{\mathcal{M}^{-1}}(\mathbf{x}', \mathbf{x})$ is obtained from symplectic matrix $\mathcal{M}^{-1}$ by equations analogous to (177) and (178) to (180).

Constants $c_\mathcal{M}$ and $c_{\mathcal{M}^{-1}}$ must be selected so that the composition of transforms (176) and (184) forms the identity transform,

$$c_\mathcal{M} = c_{\mathcal{M}^{-1}} = \left[\sqrt[2]{\det}\left(\frac{2\pi}{\omega}\mathbf{Q}\right)\right]^{-1}. \tag{185}$$

Here the matrix function $\sqrt[2]{\det}$ is defined by the following two conditions: (a) $\sqrt[2]{\det}(\mathbf{A})$ is the product of the fourth roots of the eigenvalues of matrix $\mathbf{A}^T\mathbf{A}$, (b) the roots with the greatest real parts are selected. For more details refer to Wolf (1974, 1979) or Ozaktas *et al.* (2001).

Interesting special cases of 3-D linear canonical transform (176) are 3-D Fourier transform

$$\mathcal{M} = \begin{pmatrix} \mathbf{0} & -\mathbf{I} \\ \mathbf{I} & \mathbf{0} \end{pmatrix}, \tag{186}$$

and 3-D separable fractional Fourier transform (Condon, 1937; Ozaktas *et al.*, 2001)

$$\mathcal{M} = \begin{pmatrix} \cos(\alpha_1) & 0 & 0 & -\sin(\alpha_1) & 0 & 0 \\ 0 & \cos(\alpha_2) & 0 & 0 & -\sin(\alpha_2) & 0 \\ 0 & 0 & \cos(\alpha_3) & 0 & 0 & -\sin(\alpha_3) \\ \sin(\alpha_1) & 0 & 0 & \cos(\alpha_1) & 0 & 0 \\ 0 & \sin(\alpha_2) & 0 & 0 & \cos(\alpha_2) & 0 \\ 0 & 0 & \sin(\alpha_3) & 0 & 0 & \cos(\alpha_3) \end{pmatrix}. \tag{187}$$

For real-valued, imaginary-valued or complex-valued α_i, (187) defines real-ordered, imaginary-ordered or complex-ordered fractional Fourier transforms, respectively.

5.4.5. Coherent-State Transforms

The paraxial approximation of a Gaussian packet (coherent state) in the vicinity of its central point $\mathbf{y}$ is performed in local Cartesian coordinates

$$\mathbf{r} = \mathbf{x} - \mathbf{y}. \tag{188}$$

It is thus also reasonable to consider linear canonical transforms (176) and (184) with respect to local Cartesian coordinates $\mathbf{r}$. For a given function f, we define new function

$$\phi(\mathbf{r}, \mathbf{y}) = f(\mathbf{r} + \mathbf{y}). \tag{189}$$

The local inverse linear canonical transform

$$\psi(\mathbf{r}', \mathbf{y}) = \phi_{\mathcal{M}^{-1}}(\mathbf{r}', \mathbf{y}) \tag{190}$$

of function $\phi(\mathbf{r}, \mathbf{y})$ with respect to the first argument is called the *coherent-state transform* of function $f(\mathbf{x})$ (Klauder, 1987, Eq. (3); Foster and Huang, 1991,

Eq. (39); Thomson, 2001, Eq. (2.4)). Function $\phi(\mathbf{r}, \mathbf{y})$ may be obtained from the coherent-state transform $\psi(\mathbf{r}', \mathbf{y})$ by local linear canonical transform

$$\phi(\mathbf{r}, \mathbf{y}) = \psi_{\mathcal{M}}(\mathbf{r}, \mathbf{y}). \qquad (191)$$

Since transform (189) projects functions $f(\mathbf{x})$ onto only a small subset of functions $\phi(\mathbf{r}, \mathbf{y})$, Eq. (189) does not define a transform from $\phi(\mathbf{r}, \mathbf{y})$ to $f(\mathbf{x})$ uniquely. Function $f(\mathbf{x})$ may thus be obtained from $\phi(\mathbf{r}, \mathbf{y})$ in many ways, yielding various linear transforms of operators $\hat{\mathbf{x}}$ and $\hat{\mathbf{p}}^x$ to operators $\hat{\mathbf{r}}'$, $\hat{\mathbf{p}}^{r'}$, $\hat{\mathbf{y}}'$ and $\hat{\mathbf{p}}^{y'}$.

In application of the coherent-state method to the wave equation, we select $\mathcal{M}$ in (191) and select the transform from $\phi(\mathbf{r}, \mathbf{y})$ to $f(\mathbf{x})$. For the selected transform from $\phi(\mathbf{r}, \mathbf{y})$ to $f(\mathbf{x})$, we determine the corresponding transforms of operators $\hat{\mathbf{x}}$ and $\hat{\mathbf{p}}^x$ to operators $\hat{\mathbf{r}}'$, $\hat{\mathbf{p}}^{r'}$, $\hat{\mathbf{y}}$ and $\hat{\mathbf{p}}^y$, and obtain the wave equation in the new representation. We then find ray-theory approximate solution $\psi(\mathbf{r}', \mathbf{y})$ of the wave equation in the new representation, and transform it to high-frequency approximate solution $f(\mathbf{x})$ using transform (191) and the selected transform from $\phi(\mathbf{r}, \mathbf{y})$ to $f(\mathbf{x})$. We now present four simple examples of a transform from $\phi(\mathbf{r}, \mathbf{y})$ to $f(\mathbf{x})$.

(a) Functional transform

$$f(\mathbf{x}) = \phi(\mathbf{0}, \mathbf{x}) \qquad (192)$$

(Klauder, 1987, Eq. (4); Foster and Huang, 1991, Eq. (40); Thomson, 2001, Eq. (2.6)) yields operator transform

$$\begin{pmatrix} \hat{\mathbf{x}} \\ \hat{\mathbf{p}}^x \end{pmatrix} = \begin{pmatrix} \hat{\mathbf{y}} \\ \hat{\mathbf{p}}^y \end{pmatrix}. \qquad (193)$$

The coherent-state method with functional transform (192) thus does not change the form of the wave equation, and yields the standard ray-theory solution.

(b) Functional transform

$$f(\mathbf{x}, \mathbf{r}) = \phi(\mathbf{r}, \mathbf{x} - \mathbf{r}) \qquad (194)$$

transforms $\phi(\mathbf{r}, \mathbf{y})$ to the family $f(\mathbf{x}, \mathbf{r})$ of high-frequency approximate solutions $f(\mathbf{x})$ parametrized by $\mathbf{r}$. Functional transform (194) yields operator transform

$$\begin{pmatrix} \hat{\mathbf{x}} \\ \hat{\mathbf{p}}^x \end{pmatrix} = \begin{pmatrix} \hat{\mathbf{y}} + \hat{\mathbf{r}} \\ \hat{\mathbf{p}}^y \end{pmatrix} = \begin{pmatrix} \hat{\mathbf{y}} + \tilde{\mathbf{Q}}\hat{\mathbf{r}}' + \mathbf{Q}\hat{\mathbf{p}}^{r'} \\ \hat{\mathbf{p}}^y \end{pmatrix} \qquad (195)$$

actually used by Foster and Huang (1991, Eq. (46)) and Thomson (2001, Eqs. (4.1) and (4.2)), who then chose $\mathbf{r} = \mathbf{0}$ and transformed their high-frequency approximate solution using (192). Note that operator transform (195) with functional transform (194) can generate the standard ray-theory solution for any choice of matrix (174) in (191).

(c) Functional transform

$$f(\mathbf{x}, \mathbf{y}) = \phi(\mathbf{x} - \mathbf{y}, \mathbf{y}), \tag{196}$$

naturally corresponding to Eqs. (188) and (189), transforms $\phi(\mathbf{r}, \mathbf{y})$ to the family $f(\mathbf{x}, \mathbf{y})$ of high-frequency approximate solutions $f(\mathbf{x})$ parametrized by $\mathbf{y}$. Functional transform (196) yields operator transform

$$\begin{pmatrix} \hat{\mathbf{x}} \\ \hat{\mathbf{p}}^x \end{pmatrix} = \begin{pmatrix} \hat{\mathbf{y}} + \hat{\mathbf{r}} \\ \hat{\mathbf{p}}^r \end{pmatrix} = \begin{pmatrix} \hat{\mathbf{y}} \\ \mathbf{0} \end{pmatrix} + \mathcal{M} \begin{pmatrix} \hat{\mathbf{r}}' \\ \hat{\mathbf{p}}^{r'} \end{pmatrix}. \tag{197}$$

The coherent-state method with functional transform (196) thus generates the family of high-frequency approximate solutions obtained by application of the linear canonical transform in differently shifted coordinates.

(d) Functional transform

$$f(\mathbf{x}) = \phi(\mathbf{x}, \mathbf{0}) \tag{198}$$

reduces the coherent-state transform to the linear canonical transform.

5.4.6. Maslov Methods

The standard ray theory is derived and expressed in the "coordinate representation", i.e. with respect to the spatial coordinates. In order to obtain various special cases of the summation of Gaussian beams and packets, the high-frequency approximation may be developed with respect to 3 appreciably general "representation coordinates" $\mathbf{x}'$ chosen in 6-D phase space, see (181), and then transformed to the coordinate representation.

The original Maslov method (Maslov, 1965; Chapman and Drummond, 1982) consists in weighted combination of the standard ray-theory approximation with the Maslov methods of the first, second and third order. The Maslov method of the first order corresponds to one spatial coordinate replaced by the respective momentum coordinate. The Maslov method of the second or third order corresponds to two or three spatial coordinates replaced by the respective momentum coordinates, see (186). We obtain the generalized eikonal and transport equations by the high-frequency approximation in this representation, and solve them. The travel time in this representation is the Legendre transform of the ray-theory travel time with respect to 1, 2 or 3 coordinates. The high-frequency asymptotic approximation of the wavefield is then transformed to the coordinate representation by the 1-D, 2-D or 3-D Fourier transform, respectively. This Fourier transform forms the superposition integral. The Maslov method of the first order represents the one-parametric superposition of infinitely broad "ribbon" Gaussian beams with second-order derivatives of travel time vanishing along the summation lines. The Maslov method of the second order represents the two-parametric superposition of infinitely broad Gaussian beams with second-order derivatives of travel time vanishing along the summation surfaces. The Maslov method of the third order represents the superposition of infinitely broad Gaussian packets with vanishing second-order spatial derivatives of the phase function.

Alonso and Forbes (1998) selected each representation coordinate x'_i as a real-valued linear combination of a spatial coordinate x_i and the corresponding momentum coordinate p^x_i, see (187). They then solved the generalized eikonal and transport equations obtained by the high-frequency approximation in this representation. The travel time in this representation is the separable real-ordered fractional Legendre transform of the ray-theory travel time (Alonso and Forbes, 1995). The high-frequency asymptotic approximation of the wavefield is then transformed to the coordinate representation by the 1-D, 2-D or 3-D separable real-ordered fractional Fourier transform (Condon, 1937; Ozaktas et al., 2001). This real-ordered fractional Fourier transform forms the superposition integral. The resulting approximation represents the superposition of infinitely broad Gaussian beams (1-D, 2-D) or packets (3-D). Equivalent results have been achieved by application of the Maslov method in local curvilinear coordinates or with respect to the "reference travel time" (Kendall and Thomson, 1993). This approximation may artificially be supplemented with Gaussian windowing through the imaginary-ordered fractional Fourier transform, sometimes also called the Gaussian-windowed Fourier transform (Alonso and Forbes, 1998; Forbes and Alonso, 1998; Kravtsov and Orlov, 1999). Analogous Gaussian windowing may be introduced using the coherent-state transform (Foster and Huang, 1991).

The Maslov method yields general superpositions of Gaussian beams or packets if the representation coordinates $\mathbf{x}'$ are sufficiently general complex-valued linear combinations of phase-space coordinates $\mathbf{x}$ and $\mathbf{p}^x$ (Klimeš, 1984b, Eq. (24)). We then solve the generalized eikonal and transport equations obtained by the high-frequency approximation in the new representation. The travel time in the new representation may be obtained by the generalized Legendre transform of the ray-theory travel time (Klimeš, 1984b, Eq. (39)). The high-frequency asymptotic approximation of the wavefield is then transformed to the coordinate representation by the 3-D complex linear canonical transform (Klimeš, 1984b, Eq. (27)). Since the linear canonical transform is a special case of the coherent-state transform, see (198), analogous superpositions may also be obtained by means of the coherent-state transform.

To obtain a general superposition of Gaussian packets by the Maslov method, we transform the elastodynamic equation to phase-space coordinates $\mathbf{x}'$, $\mathbf{p}^{x'}$, given by general complex-valued symplectic transform (181), by inserting expressions (175) for operators $\hat{\mathbf{x}}$, $\hat{\mathbf{p}}^x$. The high-frequency approximation then yields the generalized eikonal and transport equations. The solutions of these equations may be expressed in terms of the solutions of the standard eikonal and transport equations. The obtained approximate high-frequency solution can then be transformed back to coordinates $\mathbf{x}'$ by 3-D complex linear canonical transform (176), which is a generalization of both the real-ordered fractional Fourier transform and imaginary-ordered fractional Fourier transform (Gaussian-windowed Fourier transform). The resulting approximate high-frequency solution is identical to general 3-parametric superposition (170) of Gaussian packets, including general 2-parametric superposition (169) of Gaussian beams as a special case. Matrix $\mathbf{N}$ controlling the shape of Gaussian packets in (170) is identical to matrix (180),

$$\mathbf{N} \equiv \mathbf{N}_{\mathcal{M}}^{xx}. \tag{199}$$

The approximate high-frequency solution (170) derived in this way is thus independent of the choice of $\mathbf{Q}$ in symplectic matrix (174), provided that we set $\mathbf{P} = \mathbf{NQ}$. Solution (170) is also independent of matrices $\tilde{\mathbf{Q}}$ and $\tilde{\mathbf{P}}$ forming with $\mathbf{Q}$ and $\mathbf{P}$ symplectic matrix $\mathcal{M}$. Klimeš (1984b) thus performed the above described derivation with matrix (174) selected in the special form

$$\mathcal{M} = \begin{pmatrix} \mathbf{0} & -\mathbf{I} \\ \mathbf{I} & -\mathbf{N} \end{pmatrix}, \tag{200}$$

which corresponds to the form

$$\mathcal{M}^{-1} = \begin{pmatrix} -\mathbf{N} & \mathbf{I} \\ -\mathbf{I} & \mathbf{0} \end{pmatrix} \tag{201}$$

of matrix (183).

Note that Foster and Huang (1991) and Thomson (2001, 2004) selected matrix (183) for the coherent state-transform (190) in the separable form

$$\mathcal{M}^{-1} = \begin{pmatrix} i\Omega & 0 & 0 & 1 & 0 & 0 \\ 0 & i\Omega & 0 & 0 & 1 & 0 \\ 0 & 0 & 1 & 0 & 0 & 0 \\ -1 & 0 & 0 & 0 & 0 & 0 \\ 0 & -1 & 0 & 0 & 0 & 0 \\ 0 & 0 & 0 & 0 & 0 & 1 \end{pmatrix}. \tag{202}$$

Operators $\hat{\mathbf{r}}$ and $\hat{\mathbf{p}}^r$ are then transformed to operators $\hat{\mathbf{r}}'$ and $\hat{\mathbf{p}}^{r'}$ using Eq. (175) with matrix

$$\mathcal{M} = \begin{pmatrix} 0 & 0 & 0 & -1 & 0 & 0 \\ 0 & 0 & 0 & 0 & -1 & 0 \\ 0 & 0 & 1 & 0 & 0 & 0 \\ 1 & 0 & 0 & i\Omega & 0 & 0 \\ 0 & 1 & 0 & 0 & i\Omega & 0 \\ 0 & 0 & 0 & 0 & 0 & 1 \end{pmatrix}. \tag{203}$$

Compared with Eqs. (199) to (201), where matrix $\mathbf{N}$ has the positive-definite imaginary part, Foster and Huang (1991) and Thomson (2001, 2004) selected the analogous 2×2 matrix with the negative-definite imaginary part.

Chapman and Keers (2002) and Chapman (2004) successfully applied the Maslov method corresponding to matrices (202) and (203) with $\Omega = 0$ in the time domain.

Note that the mathematical formalism of Maslov methods obscures the evolution of Gaussian beams or Gaussian packets along rays and makes the optimization of their shapes more difficult.

5.5. Decomposition of a General Wavefield into Gaussian Packets or Beams

Assume a time-dependent wavefield specified along a given surface, and call it the "time section". The trace of a Gaussian packet in the time section is approximately a Gabor function. The widths of the envelopes of Gabor functions are inversely proportional to the square root of frequency, not constant as for the Gabor transforms (discrete, integral) nor inversely proportional to frequency as for the wavelet transforms. Moreover, the shape of the packets has to be optimized to some extent with respect to the elastodynamic equation, and is thus often dependent on time and on the coordinates and wavenumbers along the surface. The Gabor functions corresponding to optimized Gaussian packets have envelopes considerably dependent on frequency and on the direction of propagation, and moderately dependent on the position and time. This makes the decomposition of a general wavefield into Gaussian packets intricate. Žáček (2006b) generalized the integral Gabor transform towards the approximate integral expansion of a time section into Gabor functions of varying shape (refer to Fig. 28 for an illustration). The approximate expansion into Gaussian packets has the form of a coherent-state transform. The expansion is exact if the envelopes of Gaussian packets depend on frequency and on the angle of incidence only. The system of Gaussian packets is four-parametric in 2-D and six-parametric in 3-D.

The decomposition of a general wavefield into Gaussian packets of given, optimized envelopes is crucial for Gaussian packet true-amplitude prestack depth migrations. It may also enable to develop general hybrid methods, combining the Gaussian packet summation method with finite differences, finite elements, and other highly accurate methods, which can be applied only to small parts of large models at high frequencies.

A time-dependent wavefield specified along a given surface can also be Fourier transformed into the frequency domain, where it consists of individual time-harmonic wavefields. Each time-harmonic wavefield may then be decomposed using the oversampled Gabor transform into the four-parametric (in 3-D) system of Gabor functions, which are the traces of Gaussian beams along the surface (Hill, 1990, 2001; Lugara *et al.*, 2003; Shlivinski *et al.*, 2004).

To estimate the discretization error for the decomposition of a general 3-D wavefield into Gaussian packets is much more difficult than to estimate the discretization error of the asymptotic decomposition. The upper estimate of the discretization error for the decomposition of a general 1-D wavefield into the regular two-parametric system of Gabor functions was derived by Daubechies (1991, Eq. (3.2.4)).

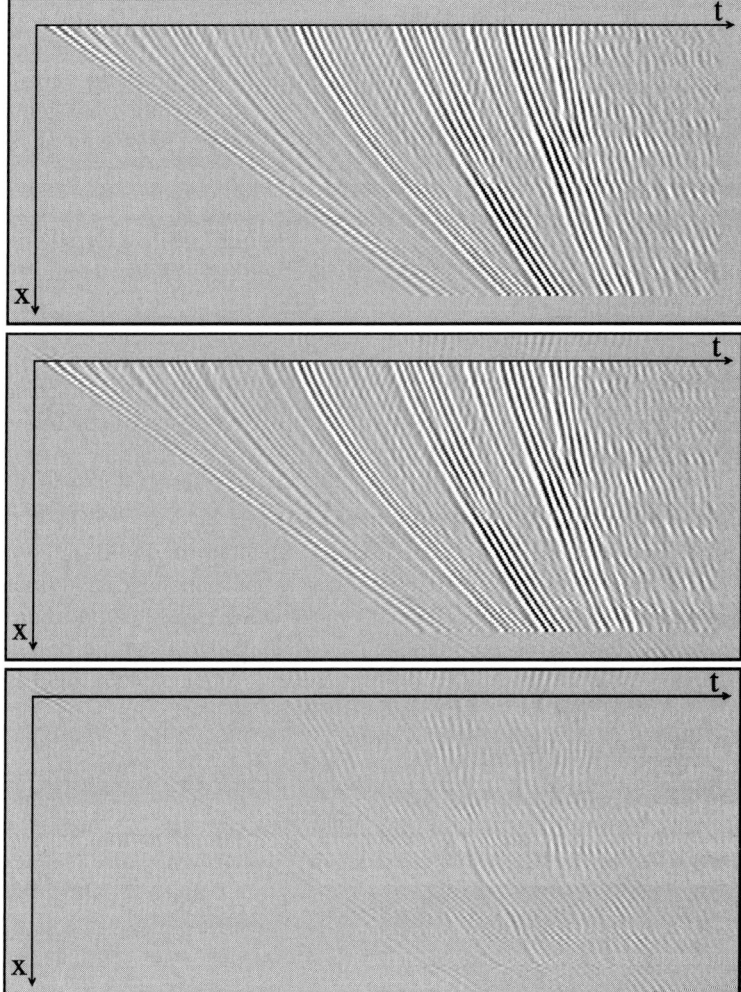

FIG. 28. Decomposition of a general wavefield into Gaussian packets. Top: Recorded wavefield. Middle: Wavefield composed of the Gabor functions corresponding to optimized Gaussian packets (six-parametric summation in 3-D). Bottom: Difference between the original wavefield and the wavefield composed of the Gabor functions.

5.6. Sensitivity of Waves to Heterogeneities

We decompose perturbations of the coefficients of the elastodynamic equation (e.g., elastic moduli and density in the elastodynamic equation) into Gabor functions

$$g(\mathbf{x}) = \exp\left[i\mathbf{k}^T(\mathbf{x} - \mathbf{y}) - \tfrac{1}{2}(\mathbf{x} - \mathbf{y})^T \mathbf{K}(\mathbf{x} - \mathbf{y}) \right] \tag{204}$$

centred at various spatial positions **y** and having various structural wavenumber vectors **k**. We consider a short-duration incident wavefield with a smooth broad–band frequency spectrum. The wavefield scattered by the perturbations is then composed of waves scattered by the individual Gabor functions. The scattered waves are estimated using the first-order Born approximation with paraxial ray approximation. Each Gabor function usually generates only a few narrow-band space–time Gaussian packets propagating in specific directions (Žáček and Klimeš, 2003). The only exceptions are broad–band forward scattering and broad–band narrow-angle scattering from the lowest structural wavenumbers, and rather rare broad–band critical scattering with mode conversion. Each scattered Gaussian packet is sensitive to just a single linear combination of the coefficients of the elastodynamic equation. This information about the Gabor function is lost if the scattered Gaussian packet does not fall into the aperture covered by the receivers and into the legible frequency band.

5.7. Migrations

A "prestack depth migration" is a simple back-projection of a wavefield, roughly approximating the inversion of wide-angle scattering. It often includes even additional rough approximations. The back-propagated wavefield is compared with the incident wavefield, forming an "image" (convolutional transform) of the gradient of a particular linear combination of the coefficients of the elastodynamic equation.

5.7.1. Gaussian Packet Migrations

The recorded wavefield (time section) is decomposed into Gaussian packets as described in Section 5.5. The individual Gaussian packets are back-propagated and compared with the incident wavefield (Žáček, 2004). The image of each back-propagated Gaussian packet is approximately formed by one or a few Gabor functions. We thus obtain not only the image of small-scale structural heterogeneities, but also the relation between the time section and the heterogeneities.

The algorithm of the true-amplitude common-source prestack depth migrations based on Gaussian packets consists of the following basic steps: optimization of the model for ray tracing and for Gaussian packets, calculation of travel times and other ray-theory quantities from the source to the dense rectangular grid of points covering the target zone, optimization of the envelopes of Gaussian packets travelling from various parts of the target zone to the receivers, decomposition of the time section into Gaussian packets, and the backprojection of the individual Gaussian packets from the time section onto the migrated image of the target zone.

5.7.2. Gaussian Beam Migrations

The recorded wavefield, Fourier transformed into the frequency domain, may be decomposed into Gaussian beams as described in Section 5.5. The individual Gaussian beams are back-propagated and compared with the incident wavefield (Hill, 1990, 2001). In the frequency domain, the image of each back-propagated Gaussian beam extends along its whole central ray.

To improve the numerical efficiency, this procedure of Gaussian beam migration can be transformed from the frequency domain to the time domain. Refer to Hill (1990, 2001) for more details.

6. RAY CHAOS, LYAPUNOV EXPONENTS, MODELS SUITABLE FOR RAY TRACING

In complex models, the behaviour of rays becomes chaotic and geometrical spreading, the number of arrivals and the density of caustic surfaces exponentially increase with travel time (Keers *et al.*, 1997). The exponential increment may be quantitatively described in terms of the Lyapunov exponents. The average twisting of ray tubes in phase space, which determines the average frequency of caustic points along rays, may be quantitatively described in terms of the rotation numbers. Note that the average Lyapunov exponent and the average rotation number are two different characteristics of ray chaos.

Lyapunov exponents enable to determine quantitative criteria on models, suitable for ray tracing, in terms of Sobolev scalar products. These criteria enable to construct the optimum models of geological structures for ray tracing by inversion of seismic data.

6.1. Lyapunov Exponents and Rotation Numbers

If heterogeneities in a model (velocity model, macro model) exceed a certain degree, average geometrical spreading exponentially increases with length of the rays and, in consequence, the average number of travel times (i.e. the average number of rays intersecting at the same point) exponentially increases with distance from the source. This behaviour of rays strictly limits the possibility of calculating two-point rays and travel times in overly complex models because:

(a) Geometrical spreading may be so large that two-point rays cannot be found within the numerical accuracy. Similarly, the ray tubes cannot be sufficiently narrow for travel-time interpolation.

(b) The number of two-point travel times at each point is so large that all travel times cannot be calculated within reasonable computational time and costs.

(c) The number of two-point travel times at each point is so large that they can hardly be useful for any application, independently of the applicability of the ray theory, which is not considered here.

It is thus of principal interest to quantify the exponential divergence of neighbouring rays with respect to the complexity of the model, and to formulate explicit criteria enabling models suitable for ray tracing to be constructed.

The exponential divergence of rays is quantified by the Lyapunov exponents (Lyapunov, 1949; Oseledec, 1968). The average frequency of caustic points along rays is quantified by the rotation numbers (Johnson, 1986). The Lyapunov exponents and the rotation numbers are two different characteristics of ray chaos.

Wolfson and Tappert (2000, Eq. (36)) estimated the Lyapunov exponent for 2-D ray tracing in random models with infinitely weak heterogeneities. Their method, designed for underwater acoustics, is thus not applicable to strong heterogeneities present in the solid Earth. Here we present the method for the quantitative estimation of the Lyapunov exponent, mean Lyapunov exponent, rotation number and mean rotation number for rays in 2-D models without interfaces according to Klimeš (2002b). The presented method is designed especially for deterministic models (velocity models, macro models) with strong heterogeneities. The method is thus not applicable to weak heterogeneities. It should be further generalized to 3-D models and to models with interfaces.

We define the Lyapunov exponents for general ray tracing, and the rotation number for 2-D ray tracing. We then introduce and estimate the *average Lyapunov exponent*, describing the average spreading of ray tubes and average number of travel times, in smooth 2-D models without interfaces. The equations allow the average exponential divergence of rays and exponential growth of the number of travel-time branches in the model to be estimated prior to ray tracing. The test of the equations for estimating the average Lyapunov exponents in a given smooth 2-D model without interfaces is illustrated on a numerical example.

6.1.1. Lyapunov Exponents for Ray Tracing

Lyapunov exponents may be defined in several ways (Lyapunov, 1949; Oseledec, 1968; Katok, 1980). Some definitions rely on an unspecified norm in phase space, which may be chosen arbitrarily. Although the phase-space norm does not affect the values of the Lyapunov exponents defined asymptotically for infinitely long rays, it may considerably affect the estimated values of the Lyapunov exponents along finite rays in finite models.

The estimates of the Lyapunov exponents based on the *characteristic values* of the DRT propagator matrix are not affected by the free parameters corresponding to the norm in phase space. On the other hand, the characteristic values oscillate along rays which makes the estimation of the Lyapunov exponents difficult. The oscillations of the characteristic values are caused by the rotation of ray tubes in phase space.

Let us denote by $\mu_1, \mu_2, \ldots, \mu_{2N}$ the *characteristic values* of the $2N \times 2N$ (in N-dimensional space) DRT propagator matrix $\mathbf{\Pi}$ in Cartesian coordinates, i.e. the solutions of the characteristic equation

$$\det[\mathbf{\Pi}(\tau, \tau_0) - \mu(\tau, \tau_0)\mathbf{I}] = 0, \tag{205}$$

sorted according to their absolute values,

$$|\mu_1| \geq |\mu_2| \geq \cdots \geq |\mu_{2N}|. \tag{206}$$

The complex-valued characteristic values, with the same absolute value, are assumed to be sorted according to their argument $-\pi < \arg\mu \leq \pi$.

The *positive Lyapunov exponents* along a ray parametrized by travel time τ may be defined as

$$\lambda_k = \limsup_{\tau \to +\infty} \frac{\ln|\mu_k(\tau, \tau_0)|}{\tau - \tau_0}, \quad k = 1, 2, \ldots, N-1. \tag{207}$$

The Lyapunov exponents are thus defined here with respect to travel time. Note that the Lyapunov exponents may also be defined with respect to another monotonic parameter along the ray, which may or may not differ from the parameter determined by the form of the Hamiltonian.

Because of the symplectic property of the DRT propagator matrix, its inverse $\mathbf{\Pi}^{-1}$ has the same set of characteristic values as $\mathbf{\Pi}$. That is why the characteristic values of all Hamiltonian systems form reciprocal pairs $\mu_1\mu_{2N} = 1$, $\mu_2\mu_{2N-1} = 1, \ldots, \mu_N\mu_{N+1} = 1$. Each positive Lyapunov exponent of ray tracing (as of other Hamiltonian systems) is thus accompanied by a negative Lyapunov exponent of the same absolute value. It is thus sufficient to study the positive Lyapunov exponents for ray tracing.

Two Lyapunov exponents λ_N and λ_{N+1} of ray tracing are always zero. The remaining $N-1$ pairs of positive and negative Lyapunov exponents correspond to the $(2N-2) \times (2N-2)$ (in N-dimensional space) DRT propagator matrix $\mathbf{\Pi}^{(q)}$ in ray-centred coordinates. It is thus often convenient to replace the DRT propagator matrix in Cartesian coordinates in Eq. (205) by the DRT propagator matrix in ray-centred coordinates.

6.1.2. Rotation Number for 2-D Ray Tracing

In isotropic 2-D models, the dynamic ray tracing equation (55) in ray-centred coordinates simplifies to equation

$$\frac{d\mathbf{\Pi}^{(q)}}{d\tau} = \begin{pmatrix} 0 & B^{(q)} \\ -C^{(q)} & 0 \end{pmatrix} \mathbf{\Pi}^{(q)}, \tag{208}$$

where the second phase-space derivatives of the Hamiltonian corresponding to the in-plane ray-centred coordinate q are

$$C^{(q)} = V^{-1} \frac{\partial^2 V}{\partial q \, \partial q} \tag{209}$$

and

$$B^{(q)} = V^2. \tag{210}$$

The solutions of dynamic ray tracing equation (208) rotate in phase space. Denote the angle of rotation of solution of Eq. (208) between travel times τ_0 and τ by $\vartheta(\tau, \tau_0)$. The exact value of angle $\vartheta(\tau, \tau_0)$ depends on the phase-space metric and is not unique. Fortunately, each increment of $\vartheta(\tau, \tau_0)$ by an integer multiple of π can uniquely be defined, at least in isotropic media. The *rotation number* φ is then defined as

$$\varphi = \lim_{\tau \to +\infty} \frac{\vartheta(\tau, \tau_0)}{\tau - \tau_0}. \tag{211}$$

The number of travel times and the number of caustics are not related in a simple way, because the number of travel times remains unchanged when the ray touches a caustic, whereas the number of travel times may be either increased by 2 or decreased by 2 when the ray crosses a caustic. The average Lyapunov exponent, characterizing the exponential increment of the number of travel times in dependence on the length of the rays, and the rotation number, characterizing the average frequency of caustic points along rays, are thus two different characteristics of ray chaos. The caustic point is understood to be the point where the ray touches the caustic.

6.1.3. *Approximation of the Positive Lyapunov Exponent in 2-D Models Without Interfaces*

We introduce parameters

$$\Lambda = \int \sqrt{\mathrm{neg}(C^{(q)} B^{(q)})} \, d\tau = \int \sqrt{\mathrm{neg}\left(V \frac{\partial^2 V}{\partial q \, \partial q}\right)} \, d\tau \tag{212}$$

and

$$\Phi = \int \sqrt{\mathrm{pos}(C^{(q)} B^{(q)})} \, d\tau = \int \sqrt{\mathrm{pos}\left(V \frac{\partial^2 V}{\partial q \, \partial q}\right)} \, d\tau \tag{213}$$

calculated along the ray. Here

$$\mathrm{pos}(x) = \max(x, 0), \quad \mathrm{neg}(x) = -\min(x, 0) \tag{214}$$

denote the positive and negative parts of real number x, respectively. Parameter Λ accounts for the exponential spreading of the ray tube in phase space, and parameter Φ describes how the ray tube is twisted in phase space. Parameter Λ increases in the *defocussing zones*, where the second velocity derivative perpendicularly to the ray is negative. Parameter Φ increases in the *focusing zones*, where the second velocity derivative perpendicularly to the ray is positive. Integral (212) is thus the sum of integrals $\Delta\Lambda_n$ over individual ray segments corresponding to defocussing zones, and integral (213) is the sum of integrals $\Delta\Phi_n$ over the individual ray segments corresponding to focusing zones.

We approximate the positive Lyapunov exponent (207) by the limit

$$\lambda_1 \approx \lim_{\tau \to +\infty} \lambda(\tau, \tau_0) \tag{215}$$

of function

$$\lambda(\tau, \tau_0) = \frac{L(\tau, \tau_0)}{\tau - \tau_0}. \tag{216}$$

The estimate of function L in Eq. (216) is (Klimeš, 2002b)

$$L(\tau, \tau_0) = \sum_{n=1}^{N} \Delta\Lambda_n + \sum_{n=1}^{N-1} \ln\left|\cos\left[\min\left(\Delta\Phi_n, \frac{\pi}{3}\right)\right]\right|$$
$$+ \frac{1}{2}\ln\left\{1 - \sin\left[2\min\left(\Delta\Phi_0, \frac{\pi}{6}\right)\right]\sin\left[2\min\left(\Delta\Phi_N, \frac{\pi}{6}\right)\right]\right\}, \tag{217}$$

where $\Delta\Lambda_n$, $n = 1, 2, \ldots, N$, correspond to the defocussing ray segments ($C^{(q)}B^{(q)} < 0$) and $\Delta\Phi_n$, $n = 0, 1, 2, \ldots, N$, correspond to the focusing ray segments ($C^{(q)}B^{(q)} > 0$), $\Delta\Phi_0$ and $\Delta\Phi_N$ may become zero if the ray starts or terminates in a defocussing region. Terms $\Delta\Lambda_n$ account for exponential deformation of the ray tube in phase space due to defocussing. After phase-space rotation of ray tubes in a focusing region, the exponential deformation does not continue in the previous deformation, and the deformations may partly cancel out. This is taken into account by the terms with $\Delta\Phi_n$ in (217). These terms are deterministic for small rotations $\Delta\Phi_n$ (the first argument of the minimum functions in (217)), and statistically averaged for greater rotations $\Delta\Phi_n$ (the second argument of the minimum functions). A rough upper estimate $L_{\max}(\tau, \tau_0) = \sum_{n=1}^{N} \Delta\Lambda_n$ of (217) may be used especially for very strong and large heterogeneities $\Lambda_n \gg \frac{\pi}{3}$.

Functions (216) and (217) are defined and have reasonable values also for rays of finite lengths. Quantity $\lambda(\tau, \tau_0)$ is thus our estimate of the Lyapunov exponent for a finite ray.

6.1.4. Approximation of the Rotation Number in 2-D Models Without Interfaces

We analogously approximate rotation number (211) by the limit

$$\varphi \approx \lim_{\tau \to +\infty} \varphi(\tau, \tau_0) \tag{218}$$

of function

$$\varphi(\tau, \tau_0) = \frac{\Phi(\tau, \tau_0)}{\tau - \tau_0}, \tag{219}$$

where

$$\Phi(\tau, \tau_0) = \sum_{n=0}^{N} \Delta \Phi_n. \tag{220}$$

This estimate is in accordance with the example of White *et al.* (1988) but has not been further numerically tested yet.

6.1.5. Lyapunov Exponent and Rotation Number for a System of Finite Rays

Equation (217) allows the average Lyapunov exponent

$$\bar{\lambda} = \frac{\sum_{\text{ray}} L_{\text{ray}}(\tau_{\text{ray}}, \tau_{\text{ray}0})}{\sum_{\text{ray}} [\tau_{\text{ray}} - \tau_{\text{ray}0}]} \tag{221}$$

over all rays to be introduced. This average Lyapunov exponent expresses the global properties of the model. However, for a single source, it is still dependent on the source geometry and position with respect to the model boundaries. The average Lyapunov exponent over various sources should describe the global properties of the model. The average Lyapunov exponent over various sources may depend on the geometry of model boundaries, but only if the statistical properties of the model are anisotropic.

Analogously, the average rotation number $\bar{\varphi}$ for the system of finite rays may be approximated by

$$\bar{\varphi} \approx \frac{\sum_{\text{ray}} \Phi_{\text{ray}}(\tau_{\text{ray}}, \tau_{\text{ray}0})}{\sum_{\text{ray}} [\tau_{\text{ray}} - \tau_{\text{ray}0}]}, \tag{222}$$

where parameter Φ along the individual rays is defined by (220).

The average distance between two consecutive caustic points along a ray, measured in terms of travel time τ, is

$$\bar{\tau}_{\text{caustic}} = \frac{\pi}{\bar{\varphi}}. \tag{223}$$

At this distance, the average number of travel times increases by the factor

$$\bar{N}_{\text{caustic}} = \exp\left(\frac{\bar{\lambda}}{\bar{\varphi}} \pi\right). \tag{224}$$

We can conjecture from Eqs. (217), (221) and (222) that

$$0 \leq \bar{\lambda} < \bar{\varphi}, \tag{225}$$

depending on the kind of heterogeneities in the model. In consequence, the average increment $\bar{N}_{\text{caustic}}$ of travel times between two consecutive caustic points depends on the kind of heterogeneities in the model, varying within interval

$$1 \leq \bar{N}_{\text{caustic}} < \exp(\pi) \simeq 23.141. \tag{226}$$

6.1.6. Average Lyapunov Exponent for the Model and Average Rotation Number for the Model

Our estimate of the Lyapunov exponent of a single finite ray depends on the position and direction of the ray. Let us now average the Lyapunov exponent over the whole model volume.

We cover the model by a dense system of parallel straight lines. For each direction of the lines, we calculate the average Lyapunov exponent (221) using Eq. (217), similarly as for a system of rays, and call it the *directional Lyapunov exponent*. We then average the calculated directional Lyapunov exponent over the directions, applying a selected directional weighting function. For instance, the directional weighting function may represent the shape of a model box, with the origin at the centre of the box or at the mean position of the intended point sources. The directional weighting function may also correspond to the probability of the ray directions estimated for ray tracing in the model. For example, if we are going to trace nearly vertical rays only and wish to estimate the average Lyapunov exponent for these rays, the directional weighting function should reflect the probability of the ray directions.

6.1.7. Numerical Example

The 2-D model designed by Jean-David Benamou is formed by the homogeneous background of velocity 1.0 km s^{-1}, perturbed by a stretched bi-sine egg-box of amplitude 0.2 km s^{-1}, see Fig. 29. The horizontal dimension of the model box is 3 km, vertical 6 km. There are focusing low-velocity regions close to the four corners. The average travel times in seconds closely correspond to the average lengths in kilometers in this model.

Some rays, shot from the point source situated at the bottom of the model box, 1.55 km from the left-hand corner, are shown in Fig. 30. Figure 31 shows the rays shot from the point source shifted to 1.85 km from the left-hand corner. Although the model is highly regular and periodic, the ray paths are quite irregular. If we divide the model into the smallest equal cells, rays enter and leave the individual cells at and in quite different positions and directions, 0% of rays being periodic.

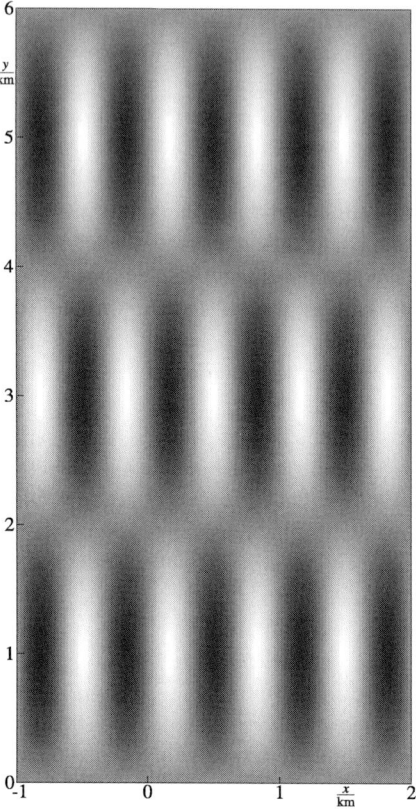

FIG. 29. The model is formed by a homogeneous background of velocity 1.0 km s^{-1}, perturbed by a stretched bi-sine egg-box of amplitude 0.2 km s^{-1}. The dark zones situated close to the four corners are focusing low-velocity regions.

The rays traced in this model can thus serve as an example of the chaotic behaviour of rays in general heterogeneous 2-D models, keeping in mind that the statistical properties of this model are strongly anisotropic.

Figure 32 displays the numbers of travel times corresponding to the point source situated at the bottom of the model box, 1.55 km from the left-hand corner. The travel times are calculated in a grid of 121×241 points covering the model box by means of interpolation within the ray cells (Bulant, 1999; Bulant and Klimeš, 1999). The maximum number of travel times found is 49. Fig. 33 shows the numbers of travel times for the point source shifted to 1.85 km from the left-hand corner. The maximum number of travel times found is 59.

Ninety directions with an angular increment of 2 degrees have been chosen to estimate the average Lyapunov exponent for the model. For each direction, the model has been covered by 45 equally spaced straight lines. The directional

FIG. 30. Rays in the model of Fig. 29, corresponding to the point source situated at the bottom of the model box, 1.55 km from the left-hand corner.

Lyapunov exponent according to Eqs. (221) and (217) has been numerically calculated along the straight lines, with a step corresponding to 45 steps along the longest line for the direction. Since the statistical properties of the model are strongly anisotropic, the directional Lyapunov exponents vary between 0.170 s^{-1} and 1.019 s^{-1}. The selected directional weighting function corresponds to the model box with the origin at the centre of the bottom edge. This directional weighting function is suitable for the point sources situated at the bottom or at the top of the model box. The average Lyapunov exponent for the model, calculated with this directional weighting function, is

$$\bar{\lambda}_{\text{model}} = 0.698 \text{ s}^{-1}. \tag{227}$$

The average Lyapunov exponent does not noticeably vary with the horizontal translation of the origin of the directional weighting function within the middle third of the horizontal model dimension.

FIG. 31. Rays in the model of Fig. 29, corresponding to the point source situated at the bottom of the model box, 1.85 km from the left-hand corner.

Figure 34 displays the natural logarithms of the average and maximum numbers of travel times along the individual horizontal grid lines of Figs. 32 and 33. The horizontal axis is the distance of the grid line from the bottom of the model box in kilometers, and serves as a rough approximation of the travel time in seconds. The slope of the straight solid lines is given by the average Lyapunov exponent (227) for the model.

The numerical example demonstrates the good correspondence between the average logarithms of the numbers of ray-theory travel times and the estimate of the average Lyapunov exponent for the model.

6.2. Models Suitable for Ray Tracing

Since the average Lyapunov exponent for the model may be approximated in terms of the Sobolev norm composed of the second velocity derivatives (Klimeš,

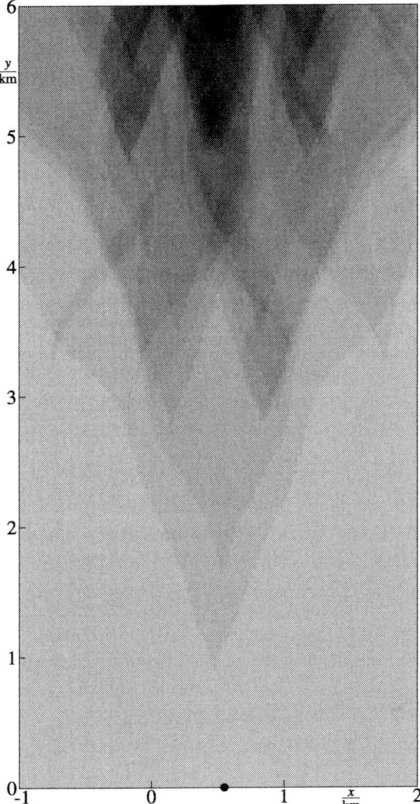

FIG. 32. The numbers of travel times corresponding to the point source situated at the bottom of the model box, 1.55 km from the left-hand corner. The maximum number of travel times found is 49.

2000b), the average Lyapunov exponent and the Sobolev norm may be used in the construction of models optimized for the calculation of ray-theory Green functions (Bulant, 2002; Žáček, 2002).

The Sobolev scalar product of two functions is a linear combination of the L_2 Lebesgue scalar products of the zero, first, second or higher derivatives of the functions. There are two main reasons for including the Sobolev norm in the objective function during the model smoothing (Žáček, 2002) or during the inversion of seismic data (Klimeš, 2002c). Firstly, the Sobolev norm regularizes ill-conditioned inversions, which enables us to control the behaviour of the model in regions not illuminated by the data. A typical example of such application is the fitting of interfaces for which the data are available only in some parts of the model (Bulant, 2002). Secondly, the minimization of the corresponding Sobolev norm during the inversion of the given data allows us to construct a model opti-

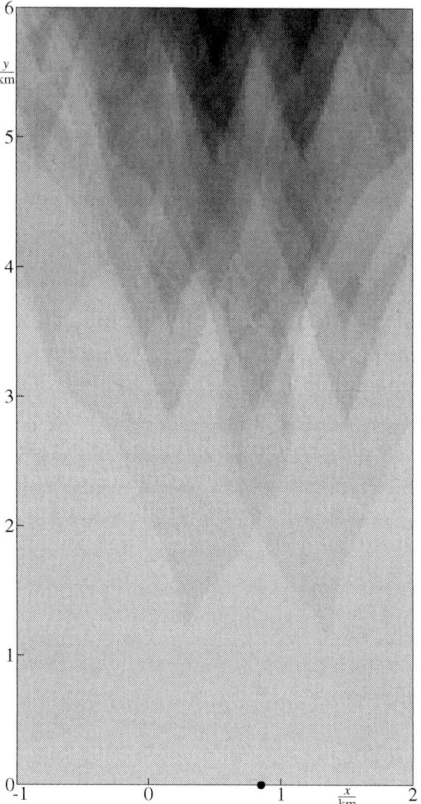

FIG. 33. The numbers of travel times corresponding to the point source situated at the bottom of the model box, 1.85 km from the left-hand corner. The maximum number of travel times found is 59.

mal for the selected computational method, i.e. for ray tracing. Note that we still have no quantitative criteria of applicability and accuracy of ray methods or their extensions. Instead, we use the criterion of the numerical efficiency of ray tracing, based on the numbers of arrivals calculated in the model.

6.2.1. Application of Sobolev Scalar Products to Smoothing Models

The Lyapunov exponent describes the complexity of ray and travel-time fields with increasing travel time. We should thus establish the maximum average Lyapunov exponent for the model. This maximum Lyapunov exponent reflects the smoothness of the model required by future applications of the model.

Since the Lyapunov exponent depends on the second velocity derivatives, we should minimize the *second velocity derivatives* during the model smoothing or

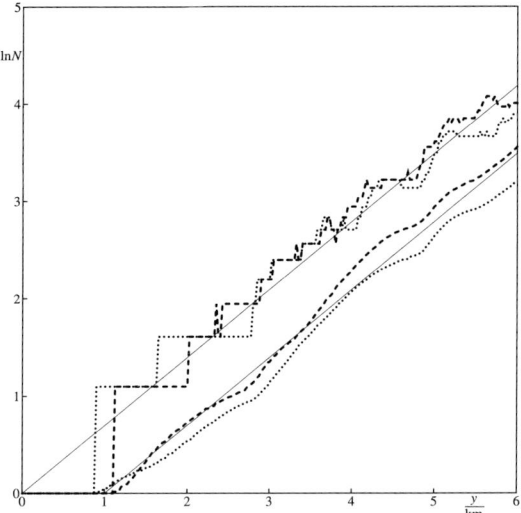

FIG. 34. The natural logarithms of the average and maximum numbers of travel times along the individual horizontal grid lines of Fig. 32 (*bold dotted lines*) and Fig. 33 (*bold dashed lines*). The horizontal axis represents the distance of the grid line from the bottom of the model box in kilometers, and serves as a rough approximation of the travel time in seconds. The slope of the straight lines *thin solid lines* is given by the average Lyapunov exponent (227) for the model.

during the inversion of seismic data. The minimization of the second velocity derivatives can be achieved by including the square of the relevant Sobolev norm of the model into the objective function. The problem now is the choice of the weighting factor of the Sobolev norm in the objective function.

We may roughly estimate the maximum Sobolev norm of the model corresponding to the maximum Lyapunov exponent (Klimeš, 2000b). Starting with zero or minimum (with respect to the stability of inversion) weighting factor, we may iteratively adjust the weighting factor so as to obtain the model with the Sobolev norm roughly equal to the estimated maximum Sobolev norm.

For this model, we calculate the average Lyapunov exponent. If the average Lyapunov exponent for the model does not accord with the maximum Lyapunov exponent, we should adjust our estimation of the maximum Sobolev norm and of the weighting factor in the objective function. If the average Lyapunov exponent for the model accords with the maximum Lyapunov exponent, we should perform ray tracing and other tests of suitability of the model for our applications. Based on these tests, we may possibly wish to adjust the maximum Lyapunov exponent, and consequently the maximum Sobolev norm and the weighting factor in the objective function.

Note that the selection of the Sobolev norm may reflect the application of the model. For example, if the model is designed for tracing only nearly vertical rays, vertical smoothing of the second velocity derivatives should be slighter than horizontal smoothing, according to the estimated deviation of rays from the vertical direction. The resulting model, optimized for nearly vertical rays, may then be much more accurate than the model smoothed for all directions of rays, including refracted ones.

7. OTHER TOPICS RELATED TO THE RAY METHOD

There are many useful topics, which are related to the ray method. Some of them are described here. The following sections are devoted to higher-order approximation of the ray method (Section 7.1), to the direct computation of first-arrival travel times (Section 7.2), to the ray method with a complex eikonal (Section 7.3), and to the hybrid methods, in which the zero-order ray approximation is combined with some other methods (Section 7.4). This text does not offer enough space to discuss in detail some other topics, which have been described in the seismological literature. Section 7.5 gives a brief list of some of them.

7.1. Higher-Order Ray Approximations

As already mentioned in Section 2, the zero-order approximation of the ray method (2) is often sufficient for solving various wave-propagation problems. In regular regions, the higher-order terms are usually small and thus it is not necessary to compute them. In singular regions, the zero-order approximation fails to describe the wavefield properly and the higher-order approximation cannot fix this anyway.

There are, however, situations, in which higher-order ray approximations play an important role. For example in situations, in which the zero-order ray amplitude or some of its components (if we are dealing with a vectorial wavefield) are zero. This occurs, for example, in the vicinity of the Brewster angle, for which R/T coefficients are zero. Another example are nodal lines of seismic sources, in whose directions the radiation function is zero. Higher-order approximations are also important in the study of the problem of R/T at interfaces of orders higher than first (at an interface of the first order, elastic moduli and density change discontinuously; at an interface of the Nth order, the $(N-1)$st derivatives of elastic moduli or density change discontinuously). A zero-order wave incident at an interface of the second order generates a zero-order transmitted wave but first-order reflected wave, for whose evaluation, the first-order approximation of the ray theory is necessary. To a certain extent, the situation with the zero amplitude of the zero-order wave also concerns head waves, which are of the first order and are generated in supercritical regions by zero-order incident waves. For more details refer to Červený (2001, Section 5.6) and to references therein.

Among the higher-order terms of the ray series, additional components, specifically the first-order additional components, play a special role. The higher-order vectorial amplitude coefficients of ray series (1) can be separated into *principal components*, whose directions coincide with the direction of the zero-order vectorial amplitude coefficient, and into *additional components*, whose directions are perpendicular to the direction of the zero-order vectorial amplitude coefficient. The principal components are obtained by integrating along the corresponding rays, and require the knowledge of high-order spatial derivatives of functions describing the model. In the case of the first-order principal components, fourth-order spatial derivatives are required. In contrast to this, the additional components are local quantities and can be evaluated relatively easily in arbitrary varying media. First-order spatial derivatives of the functions describing the medium and of the zero-order amplitude coefficients are sufficient for the evaluation of the first-order additional components. Note that the zero-order amplitude coefficients already depend on second-order spatial derivatives of the functions describing the medium so that the first-order additional components depend on third-order spatial derivatives of the functions describing the medium. The procedure described by Eisner and Pšenčík (1996), based on the substitution of the derivatives by the differences of values calculated on neighbouring rays, can be used to evaluate the derivatives of the amplitude coefficients.

Whereas the first-order principal component represents only a certain correction of the zero-order amplitude coefficient, the first-order additional components represent first non-zero terms in the directions perpendicular to the zero-order coefficient. The combination of the zero-order coefficient and of the first-order additional components thus represents a vectorial leading term of the ray series, see Eisner and Pšenčík (1996). It is called the two-component ray representation by Fradkin and Kiselev (1997), where many additional references can be found. See also Goldin (1989).

Eisner and Pšenčík (1996) showed that the use of the first-order additional components in the expression for the ray Green function in a homogeneous isotropic medium improves the ray-theory Green function much more than the addition of any other higher-order component (note that the ray series corresponding to the exact Green function in a homogeneous isotropic medium consists of three terms). This indicates that the first-order additional components could be used routinely. Their most obvious effects are deviations of the polarizations of waves from the directions specified by their zero-order ray approximation. The polarization of a P wave may be close to transverse and elliptical, the polarization of an S wave, on the contrary, can be close to longitudinal. The first-order additional components can also be considered in anisotropic media. In weakly anisotropic media, they reduce to their "isotropic" versions, see Pšenčík (1998).

In homogeneous media, it is possible, theoretically without limitation, to generate and evaluate formulae for higher-order terms with the use of symbolic manipulation software. Vavryčuk (1999) demonstrated on a numerical example

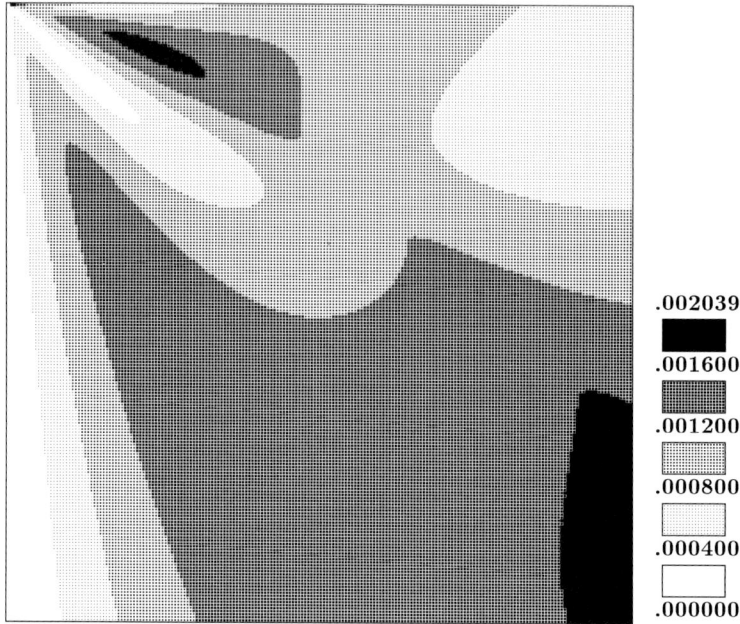

FIG. 35. Constant vertical velocity gradient model, velocity contrast 1 : 3. Travel-time errors of the first-order grid travel-time tracing method proposed by Podvin and Lecomte (1991). The absolute errors of the first-arrival travel time are scaled in seconds.

that this approach can remove the failure of the zero-order ray approximation in the proper description of the problem of coupling of two S waves propagating vertically in a weakly transversely isotropic medium with the axis of symmetry rotating in the horizontal plane. Unfortunately, in heterogeneous media this procedure cannot be used, and we have to resort to the coupling ray theory, see Section 4.

7.2. Direct Computation of First-Arrival Travel Times

Recently, considerable attention has been devoted to the direct computation of travel times $\tau(x_i)$ on 2-D and 3-D grids without using ray tracing. These methods, however, do not yield the ray-theory travel times, but the first-arrival travel times. The first-arrival travel times are not related to the propagation of energy and should strictly be distinguished from the ray-theory travel times related to energetic arrivals. In the interpretational procedures of seismic exploration and seismology (tomography, migration), the energetic arrivals should mostly be considered, first-arrival travel times playing a smaller role. For this reason, we shall discuss the methods for the computation of first-arrival travel times only

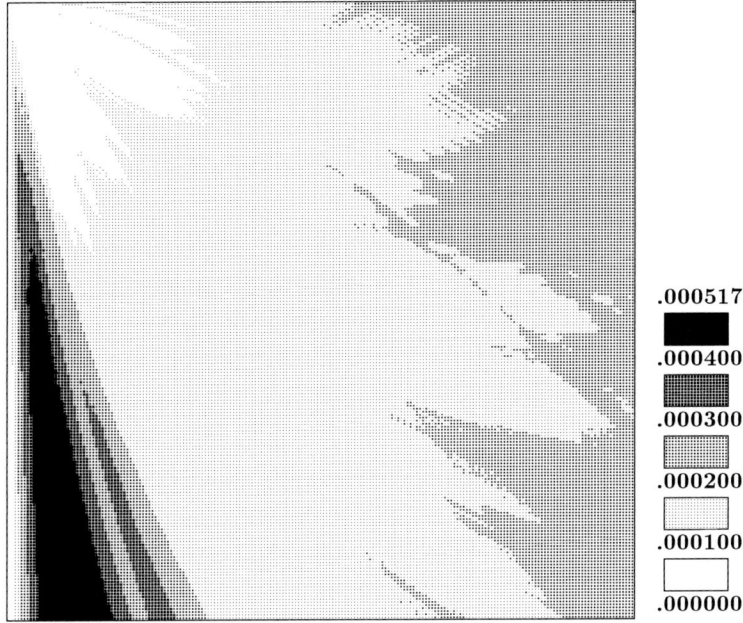

FIG. 36. Constant vertical velocity gradient model, velocity contrast 1 : 3. Travel-time errors of the network shortest-path ray tracing according to Klimeš and Kvasnička (1994). The absolute errors of the travel time are scaled in seconds.

very briefly. Two basic methods in the computation of first-arrival travel times are network shortest-path ray tracing, and "finite-difference" travel time tracing. Refer to Figs. 35–39 for simple numerical examples. The network shortest-path ray tracing is based on the theory of graphs. The trajectory corresponding to the minimum time is usually called the shortest path, where "shortest" means the minimum travel time. For more details and other references see Moser (1991), Klimeš and Kvasnička (1994). The "finite-difference" methods do not usually consist in the direct finite-difference solutions of the eikonal equation (remember that the eikonal equation is nonlinear), but only in certain finite-difference approximations for the first-arrival travel-time continuation. Many such "finite-difference" approximations have been proposed. See Vidale (1990), Klimeš (1996a), and for many other references Červený (2001, p. 187). An extension of the "finite-difference" method, which also takes into account later arrivals, was proposed by Abgrall and Benamou (1999).

It should be emphasized that the ray-theory travel times, related to energy transport, are safely obtained by shooting methods with standard ray tracing, or by the wavefront construction method, see Section 2.2.

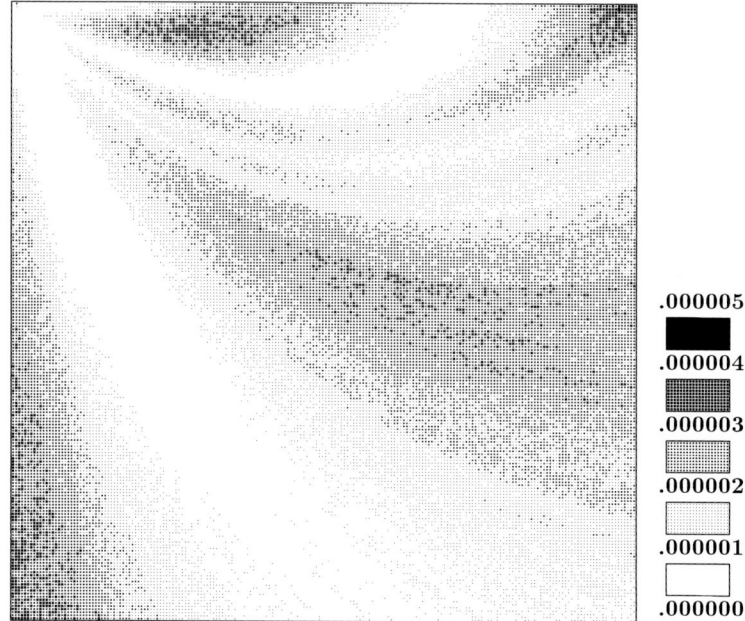

FIG. 37. Constant vertical velocity gradient model, velocity contrast 1 : 3. Travel-time errors of the second-order grid travel-time tracing according to Klimeš (1996a). The absolute errors of the travel time are scaled in seconds.

7.3. Ray Method with Complex Eikonal

In the standard zero-order approximation of the ray method, travel time (eikonal) τ and slowness vector $\mathbf{p}$ are real-valued. The ray method can, however, be extended to complex-valued τ and $\mathbf{p}$. This generalization is then called *the ray method with complex eikonal* and the rays in it are complex-valued. Note that the elementary waves along complex rays with Re $\mathbf{p}$ and Im $\mathbf{p}$ parallel are usually called *homogeneous* and with Re $\mathbf{p}$ and Im $\mathbf{p}$ non-parallel are called *inhomogeneous*.

Homogeneous and inhomogeneous waves play an important role in many wave propagation problems including perfectly elastic media. As an example of inhomogeneous waves in perfectly elastic media let us mention the waves penetrating the caustic shadow, supercritically transmitted waves, and the inhomogeneous waves generated by a point source. The literature devoted to theoretical problems of inhomogeneous waves propagating in perfectly elastic media is broad, see for example recent publications by Chapman *et al.* (1999) or Kravtsov *et al.* (1999), where many additional references can be found. However, as far as we know, the calculation of complex rays has not yet been implemented into ray tracing program packages for heterogeneous perfectly elastic

FIG. 38. A 2-D realization of a random velocity distribution.

media. Only the amplitudes of the complex rays at the point of their generation at an interface are considered when supercritically reflected waves are calculated.

In *dissipative media*, the eikonal τ is intrinsically complex-valued, and the absorption is always connected with velocity dispersion (causal absorption). Consequently, the rays are also complex-valued, see Thomson (1997). In the seismic ray theory the study of complex rays in dissipative media has not been completed yet. In most cases, it is assumed that dissipation is weak and it is taken into account only as a first-order perturbation of a perfectly elastic medium, see Fig. 40. Refer, for example, to Moczo *et al.* (1987) for isotropic dissipative media, and to Gajewski and Pšenčík (1992) for anisotropic dissipative media. Such a simple approximation is at present implemented in most ray-based program packages. As the dissipation within the Earth's interior is usually weak, the first-order perturbation method yields sufficiently accurate amplitude decay due to absorption and frequency-dependent time shift due to the relevant dispersion.

FIG. 39. First-arrival travel times calculated in the 2-D model shown in Fig. 38 using grid travel-time tracing according to Klimeš (1996a).

7.4. Hybrid Methods

In hybrid methods, the ray method is locally combined with some other method. There are several such existing combinations. We shall now discuss briefly three of them: (a) the hybrid ray-finite-difference method; (b) the hybrid ray-mode method; (c) the hybrid ray-matrix method (sometimes called the hybrid ray-reflectivity method).

In the hybrid ray-finite-difference method, the wavefield is computed in a globally smooth model, with a more complicated structure localized in a small part of the model. The ray method is used in the smooth part of the model, finite differences in the small complicated region. The basic problem is to match the solutions of both methods at the boundary of the complicated region. See, e.g., Lecomte (1996), Gjøystdal *et al.* (2002), Opršal *et al.* (2002).

In the hybrid ray-mode method, the modal and ray approaches are combined. The aim is to simulate the complete wavefield by several ray contributions and by several modes. See, e.g., Kapoor and Felsen (1995), or Zhao and Dahlen (1996).

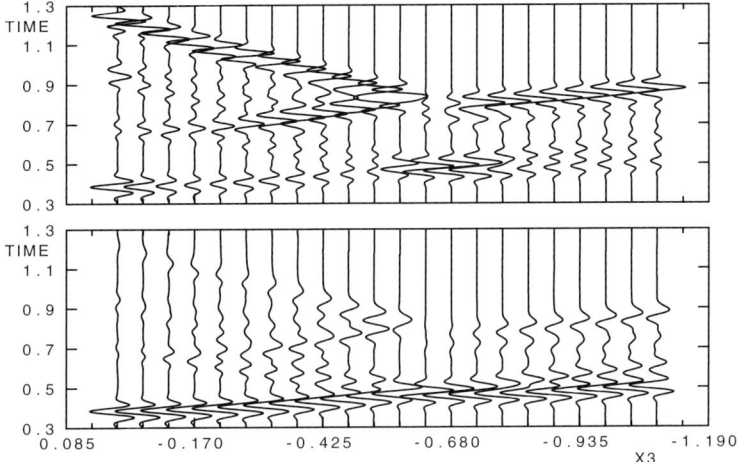

FIG. 40. Trace-normalized ray-theory seismograms without attenuation (top) and with causal attenuation (bottom) in 2-D model UNCONFORMITY by Cormier and Mellen (1984). Complex-valued travel times are approximated using the ray perturbation method. Refer to Klimeš (1996b) for more information.

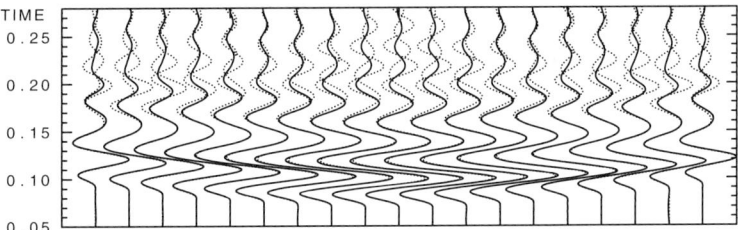

FIG. 41. Vertical component of synthetic seismograms at the top of a sedimentary layer. Solid line: elastic seismograms calculated by the ray-matrix method are quite accurate in this case. The standard ray method would yield only the onset of the wave-train. Two dotted lines: finite-difference elastic seismograms are deteriorated by reflections due to imperfect non-reflecting boundary conditions. Refer to Klimeš (2000a) for more information.

The hybrid ray-matrix method has been used in smooth models containing transition layers thin with respect to the prevailing wavelength. The ray method is again used in smooth parts of the model, the matrix method is used to compute R/T coefficients of thin layers, see Fig. 41. Červený (1989) applied this method to the Earth's crust model with the MOHO discontinuity approximated by a thin layer with varying internal structure (transition layer, laminated layer). The method was also applied to the study of tunnelling of seismic waves through a thin high-velocity layer (Červený and Aranha, 1992), and to the study of the effects of a near-surface sedimentary layer on the wavefield measured on the Earth's surface (Červený and Andrade, 1992). The method is also a part of the CRT program package (Klimeš, 2000a).

7.5. Several Other Extensions of the Ray Method

There are many other useful extensions of the ray method. We briefly mention several of them and supplement each of them by at least one reference, where more details and additional references can be found.

(a) Space–time ray method (Babich *et al.*, 1985).
(b) Surface-wave ray tracing and surface-wave Gaussian beams (Woodhouse, 1974; Yomogida, 1985, 1987; Martin and Thomson, 1997; Dahlen and Tromp, 1998; Červený, 2001, Section 3.12).
(c) Asymptotic diffraction theory including geometric theory of diffraction (Klem-Musatov and Aizenberg, 1984; Klem-Musatov, 1994; Hanyga, 1995; Hanyga *et al.*, 2001).
(d) Ray methods in directional (one way) propagation (Thomson, 1999).
(e) Ray method in Born scattering of seismic waves (Ursin and Tygel, 1997; Chapman, 2004).
(f) Kirchhoff–Helmholtz integrals (Tygel *et al.*, 1994; Schleicher *et al.*, 2001; Chapman, 2004).

ACKNOWLEDGEMENTS

Petr Bulant prepared many figures used in this chapter. Figure 28 was prepared by Karel Žáček. The authors are grateful to Petr Bulant, Veronique Farra, Einar Iversen and Luis Rivera whose comments enabled the improvement of this chapter. The research has been supported by the Grant Agency of the Czech Republic under Contracts 205/01/0927, 205/01/D097, 205/04/1104 and 205/05/2182, by the Grant Agency of the Charles University under Contract 375/2004/B-GEO/MFF, by the Grant Agency of the Academy of Sciences of the Czech Republic under Contract A3012309, by the Ministry of Education of the Czech Republic within Research Project MSM113200004, and by the members of the consortium "Seismic Waves in Complex 3-D Structures" (see "http://sw3d.mff.cuni.cz").

REFERENCES

Abgrall, R., Benamou, J.-D. (1999). Big ray-tracing and eikonal solver on unstructured grids: Application to the computation of a multivalued traveltime field in the Marmousi model. *Geophysics* **64**, 230–239.
Aki, K., Richards, P. (1980). Quantitative Seismology (2 vols.). Freeman, San Francisco.
Alonso, M.A., Forbes, G.W. (1995). Fractional Legendre transformation. *J. Phys. A* **28**, 5509–5527.

Alonso, M.A., Forbes, G.W. (1998). Asymptotic estimation of the optical wave propagator. I. Derivation of a new method. *J. Opt. Soc. Am. A* **15**, 1329–1340.
Arnaud, J.A. (1972). Modes in helical gas lenses. *Appl. Opt.* **11**, 2514–2521.
Babich, V.M. (1956). Ray method of the computation of the intensity of wave fronts. *Dokl. Akad. Nauk SSSR* **110**, 355–357 (in Russian).
Babich, V.M. (1961). Ray method of calculating the intensity of wavefronts in the case of a heterogeneous, anisotropic, elastic medium. In: Petrashen, G.I. (Ed.), *Problems of the Dynamic Theory of Propagation of Seismic Waves*, vol. 5. Leningrad Univ. Press, Leningrad, pp. 36–46 (in Russian), *Geophys. J. Int.* **118** (1994) 379–383 (English translation).
Babich, V.M., Buldyrev, V.S. (1972). Asymptotic Methods in Problems of Diffraction of Short Waves. Nauka, Moscow, (in Russian). Translated to English by Springer, Berlin, 1991, under the title Short-Wavelength Diffraction Theory.
Babich, V.M., Buldyrev, V.S., Molotkov, I.A. (1985). Space–time Ray Method. Linear and Non-linear Waves. Leningrad Univ. Press, Leningrad (in Russian).
Babich, V.M., Ulin, V.V. (1981). Complex space–time ray method and "quasiphotons". In: Babich, V.M. (Ed.), *Mathematical Problems of the Theory of Propagation of Waves*, vol. 12. Nauka, Leningrad, pp. 5–12 (in Russian), *J. Sov. Math.* **24** (1984) 269–273 (English translation).
Bakker, P.M. (1998). Phase shift at caustics along rays in anisotropic media. *Geophys. J. Int.* **134**, 515–518.
Bakker, P.M. (2002). Coupled anisotropic shear wave raytracing in situations where associated slowness sheets are almost tangent. *Pure Appl. Geophys.* **159**, 1403–1417.
Bleistein, N. (1984). Mathematical Methods for Wave Phenomena. Academic Press, New York.
Bleistein, N., Cohen, J.K., Stockwell Jr., J.W. (2001). Mathematics of Multidimensional Seismic Imaging, Migration, and Inversion. Springer, Berlin.
Bulant, P. (1996). Two-point ray tracing in 3-D. *Pure Appl. Geophys.* **148**, 421–447.
Bulant, P. (1999). Two-point ray-tracing and controlled initial-value ray-tracing in 3-D heterogeneous block structures. *J. Seismol. Exp.* **8**, 57–75.
Bulant, P. (2002). Sobolev scalar products in the construction of velocity models—application to model Hess and to SEG/EAGE Salt model. *Pure Appl. Geophys.* **159**, 1487–1506.
Bulant, P., Klimeš, L. (1999). Interpolation of ray theory traveltimes within ray cells. *Geophys. J. Int.* **139**, 273–282.
Bulant, P., Klimeš, L. (2002). Numerical algorithm of the coupling ray theory in weakly anisotropic media. *Pure Appl. Geophys.* **159**, 1419–1435.
Bulant, P., Klimeš, L. (2004). Comparison of quasi-isotropic approximations of the coupling ray theory with the exact solution in the 1-D anisotropic "oblique twisted crystal" model. *Stud. Geophys. Geod.* **48**, 97–116.
Bulant, P., Klimeš, L. (2006). Numerical comparison of the isotropic-common-ray and anisotropic-common-ray approximations of the coupling ray theory. In: *Seismic Waves in Complex 3-D Structures, Report 16*, pp. 155–178. Department of Geophysics, Charles University, Prague, online at http://sw3d.mff.cuni.cz.
Bulant, P., Klimeš, L., Pšenčík, I., Vavryčuk, V. (2004). Comparison of ray methods with the exact solution in the 1-D anisotropic "simplified twisted crystal" model. *Stud. Geophys. Geod.* **48**, 675–688.
Červený, V. (1972). Seismic rays and ray intensities in inhomogeneous anisotropic media. *Geophys. J. R. Astron. Soc.* **29**, 1–13.
Červený, V. (1982). Direct and inverse kinematic problems for inhomogeneous aniso tropic media— linearization approach. *Contrib. Geophys. Inst. Slov. Acad. Sci.* **13**, 127–133.
Červený, V. (1985). Gaussian beam synthetic seismograms. *J. Geophys.* **58**, 44–72.
Červený, V. (1989). Synthetic body wave seismograms for laterally varying media containing thin transmission layers. *Geophys. J. Int.* **99**, 331–349.
Červený, V. (2001). Seismic Ray Theory. Cambridge Univ. Press, Cambridge.

Červený, V., Andrade, F.C.M. (1992). Influence of a near-surface structure on seismic wave fields recorded at the Earth's surface. *J. Seismol. Exp.* **1**, 107–116.
Červený, V., Aranha, P.R.A. (1992). Tuneling of seismic body waves through thin high-velocity layers in complex structures. *Stud. Geophys. Geod.* **36**, 115–138.
Červený, V., Jech, J. (1982). Linearized solutions of kinematic problems of seismic body waves in inhomogeneous slightly anisotropic media. *J. Geophys.* **51**, 96–104.
Červený, V., Klimeš, L., Pšenčík, I. (1984). Paraxial ray approximation in the computation of seismic wavefields in inhomogeneous media. *Geophys. J. R. Astron. Soc.* **79**, 89–104.
Červený, V., Klimeš, L., Pšenčík, I. (1988). Complete seismic-ray tracing in three-dimensional structures. In: Doornbos, D.J. (Ed.), *Seismological Algorithms*. Academic Press, New York, pp. 89–168.
Červený, V., Molotkov, I.A., Pšenčík, I. (1977). *Ray Method in Seismology*. Charles Univ. Press, Praha.
Červený, V., Popov, M.M., Pšenčík, I. (1982). Computation of wave fields in inhomogeneous media—Gaussian beam approach. *Geophys. J. R. Astron. Soc.* **70**, 109–128.
Červený, V., Pšenčík, I. (1984). SEIS83—Numerical modelling of seismic wave fields in 2-D laterally varying layered structures by the ray method. In: Engdahl, E.R. (Ed.), *Documentation of Earthquake Algorithms, Report SE-35*. World Data Center A for Solid Earth Geophysics, Boulder, pp. 36–40.
Červený, V., Soares, J.E.P. (1992). Fresnel volume ray tracing. *Geophysics* **57**, 902–915.
Chapman, C.H. (1994). Reflection/transmission coefficient reciprocities in anisotropic media. *Geophys. J. Int.* **116**, 498–501.
Chapman, C.H. (2002). Seismic ray theory and finite frequency extensions. In: Lee, W.H.K., Kanamori, H., Jennings, P.C. (Eds.), *International Handbook of Earthquake and Engineering Seismology, Part A*. Academic Press, New York, pp. 103–123.
Chapman, C.H. (2004). *Fundamentals of Seismic Wave Propagation*. Cambridge Univ. Press, Cambridge.
Chapman, C.H., Drummond, R. (1982). Body-wave seismograms in inhomogeneous media using Maslov asymptotic theory. *Bull. Seismol. Soc. Am.* **72**, S277–S317.
Chapman, C.H., Keers, H. (2002). Application of the Maslov seismogram method in three dimensions. *Stud. Geophys. Geod.* **46**, 615–649.
Chapman, C.H., Pratt, R.G. (1992). Traveltime tomography in anisotropic media-I. Theory. *Geophys. J. Int.* **109**, 1–19.
Chapman, S.J., Lawry, J.M.H., Ockendon, J.R., Tew, R.H. (1999). On the theory of complex rays. *SIAM Rev.* **41**, 417–509.
Coates, R.T., Chapman, C.H. (1990). Quasi-shear wave coupling in weakly anisotropic 3-D media. *Geophys. J. Int.* **103**, 301–320.
Coates, R.T., Chapman, C.H. (1991). Generalized Born scattering of elastic waves in 3-D media. *Geophys. J. Int.* **107**, 231–263.
Combescure, M., Ralston, J., Robert, D. (1999). A proof of the Gutzwiller semiclassical trace formula using coherent state decomposition. *Commun. Math. Phys.* **202**, 463–480.
Condon, E.U. (1937). Immersion of the Fourier transform in a continuous group of functional transformations. *Proc. Natl. Acad. Sci.* **23**, 158–164.
Cormier, V.F., Mellen, M.H. (1984). Application of asymptotic ray theory to vertical seismic profiling. In: Toksöz, M.N., Stewart, R.R. (Eds.), *Vertical Seismic Profiling: Advanced Concepts*. Geophysical Press, London, pp. 28–44.
Dahlen, F.A., Tromp, J. (1998). *Theoretical Global Seismology*. Princeton Univ. Press, Princeton.
Daubechies, I. (1991). *Ten Lectures on Wavelets*. Society for Industrial and Applied Mathematics, Pennsylvania.
Eisner, L., Pšenčík, I. (1996). Computation of additional components of the first-order ray approximation in isotropic media. *Pure Appl. Geophys.* **148**, 227–253.
Farra, V. (1989). Ray perturbation theory for heterogeneous hexagonal anisotropic medium. *Geophys. J. Int.* **99**, 723–737.

Farra, V. (1999). Computation of second-order traveltime perturbation by Hamiltonian ray theory. *Geophys. J. Int.* **136**, 205–217.
Farra, V. (2005). First-order ray tracing for qS waves in inhomogeneous weakly anisotropic media. *Geophys. J. Int.* **161**, 309–324.
Farra, V., Le Bégat, S. (1995). Sensitivity of qP-wave traveltimes and polarization vectors to heterogeneity, anisotropy, and interfaces. *Geophys. J. Int.* **121**, 371–384.
Farra, V., Madariaga, R. (1987). Seismic waveform modeling in heterogeneous media by ray perturbation theory. *J. Geophys. Res. B* **92**, 2697–2712.
Farra, V., Pšenčík, I. (2003). Properties of the zero-, first- and higher-order approximations of attributes of elastic waves in weakly anisotropic media. *J. Acoust. Soc. Am.* **114**, 1366–1378.
Felsen, L.B., Markuvitz, N. (1973). Radiation and Scattering of Waves. Prentice Hall, Englewood Cliffs.
Forbes, G.W., Alonso, M.A. (1998). Asymptotic estimation of the optical wave propagator. II. Relative validity. *J. Opt. Soc. Am. A* **15**, 1341–1354.
Foster, D.J., Huang, J.-I. (1991). Global asymptotic solutions of the wave equation. *Geophys. J. Int.* **105**, 163–171.
Fradkin, L.Yu., Kiselev, A.P. (1997). The two-component representation of time-harmonic elastic body waves in the high- and intermediate-frequency regimes. *J. Acoust. Soc. Am.* **101**, 52–65.
Gajewski, D., Pšenčík, I. (1987). Computation of high-frequency seismic wavefields in 3-D laterally inhomogeneous anisotropic media. *Geophys. J. R. Astron. Soc.* **91**, 383–411.
Gajewski, D., Pšenčík, I. (1990). Vertical seismic profile synthetics by dynamic ray tracing in laterally varying layered anisotropic structures. *J. Geophys. Res.* **95**, 11301–11315.
Gajewski, D., Pšenčík, I. (1992). Vector wavefields for weakly attenuating anisotropic media by the ray method. *Geophysics* **57**, 27–38.
Garmany, J. (2000). Phase shifts at caustics in anisotropic media. In: *Anisotropy 2000: Fractures, Converted Waves and Case Studies*. Soc. Exp. Geophysicists, Tulsa, pp. 419–425.
Gjøystdal, H., Iversen, E., Laurain, R., Lecomte, I., Vinje, V., Åstebøl, K. (2002). Review of ray theory applications in modelling and imaging of seismic data. *Stud. Geophys. Geod.* **46**, 113–164.
Gjøystdal, H., Reinhardsen, J.E., Åstebøl, K. (1985). Computer representation of complex 3-D geological structures using a new "solid modeling" technique. *Geophys. Prosp.* **33**, 1195–1211.
Goldin, S.V. (1989). Physical analysis of the additional components of seismic waves in the first approximation of ray series. *Geol. Geophys.* **30**, 128–132.
Hanyga, A. (1982). The kinematic inverse problem for weakly laterally inhomogeneous anisotropic media. *Tectonophysics* **90**, 253–262.
Hanyga, A. (1995). Asymptotic edge-and-vertex diffraction theory. *Geophys. J. Int.* **123**, 227–290.
Hanyga, A. (1996). Point-to-curve ray tracing. *Pure Appl. Geophys.* **148**, 387–420.
Hanyga, A., Druzhinin, A.B., Dzhafarov, A.D., Frøyland, L. (2001). A Hamiltonian approach to asymptotic seismic reflection and diffraction modeling. *Geophys. Prosp.* **49**, 213–227.
Hanyga, A., Lenartowicz, E., Pajchel, J. (1984). Seismic Wave Propagation in the Earth. Elsevier, Amsterdam.
Hanyga, A., Pajchel, J. (1995). Point-to-curve ray tracing in complicated geological models. *Geophys. Prosp.* **43**, 859–872.
Hill, N.R. (1990). Gaussian beam migration. *Geophysics* **55**, 1416–1428.
Hill, N.R. (2001). Prestack Gaussian-beam depth migration. *Geophysics* **66**, 1240–1250.
Hubral, P., Schleicher, J., Tygel, M. (1992a). Three-dimensional paraxial ray properties: Part I. Basic relations. *J. Seismol. Exp.* **1**, 265–279.
Hubral, P., Schleicher, J., Tygel, M. (1992b). Three-dimensional paraxial ray properties: Part II. Applications. *J. Seismol. Exp.* **1**, 347–362.
Iversen, E. (1996). Derivatives of reflection point coordinates with respect to model parameters. *Pure Appl. Geophys.* **148**, 287–317.
Iversen, E. (2001a). First-order perturbation theory for seismic isochrons. *Stud. Geophys. Geod.* **45**, 395–444.

Iversen, E. (2001b). Ray systems for propagation of seismic isochrons. Part I: Isochron rays. In: *Expanded Abstracts of 7th International Congress Braz. Geophys. Soc.* Braz. Geophys. Soc., Rio de Janeiro, pp. 1158–1161.

Iversen, E. (2001c). Ray systems for propagation of seismic isochrons. Part II: Velocity rays. In: *Expanded Abstracts of 7th International Congress Braz. Geophys. Soc.* Braz. Geophys. Soc., Rio de Janeiro, pp. 1162–1165.

Iversen, E. (2004). The isochron ray in seismic modeling and imaging. *Geophysics* **69**, 1053–1070.

Jech, J., Pšenčík, I. (1989). First-order perturbation method for anisotropic media. *Geophys. J. Int.* **99**, 369–376.

Johnson, R.A. (1986). An example concerning the geometrical significance of the rotation number—integrated density of states. In: Arnold, L., Wihstutz, V. (Eds.), *Lyapunov Exponents*. Springer, Berlin, pp. 216–226.

Kachalov, A.P. (1984). A coordinate system for describing the "quasiphoton". In: Babich, V.M. (Ed.), *Mathematical Problems of the Theory of Propagation of Waves*, vol. 14. Nauka, Leningrad, pp. 73–76 (in Russian), *J. Sov. Math.* **32** (1986) 151–153 (English translation).

Kapoor, T.K., Felsen, L.B. (1995). Hybrid ray-mode analysis of acoustic scattering from a finite, fluid loaded plate. *Wave Motion* **22**, 109–131.

Karal, F.C., Keller, J.B. (1959). Elastic wave propagation in homogeneous and inhomogeneous media. *J. Acoust. Soc. Am.* **31**, 694–705.

Katok, S.R. (1980). The estimation from above for the topological entropy of a diffeomorphism. In: Netecki, Z., Robinson, C. (Eds.), *Global Theory of Dynamical Systems*. In: *Lecture Notes in Mathematics*, vol. 819. Springer, Berlin, pp. 258–264.

Keers, H., Dahlen, F.A., Nolet, G. (1997). Chaotic ray behaviour in regional seismology. *Geophys. J. Int.* **131**, 361–380.

Kendall, J.-M., Guest, W.S., Thomson, C.J. (1992). Ray-theory Green's function reciprocity and ray-centred coordinates in anisotropic media. *Geophys. J. Int.* **108**, 364–371.

Kendall, J.-M., Thomson, C.J. (1993). Maslov ray summation, pseudo-caustics, Lagrangian equivalence and transient seismic waveforms. *Geophys. J. Int.* **113**, 186–214.

Klauder, J.R. (1987). Semiclassical quantization of classically chaotic systems. *Phys. Rev. Lett.* **59**, 748–750.

Klem-Musatov, K.D. (1994). Theory of Seismic Diffractions. Society of Exp. Geophysicists, Tulsa.

Klem-Musatov, K.D., Aizenberg, A.M. (1984). The ray method of the theory of edge waves. *Geophys. J. R. Astron. Soc.* **79**, 35–50.

Klimeš, L. (1984a). Expansion of a high-frequency time-harmonic wavefield given on an initial surface into Gaussian beams. *Geophys. J. R. Astron. Soc.* **79**, 105–118.

Klimeš, L. (1984b). The relation between Gaussian beams and Maslov asymptotic theory. *Stud. Geophys. Geod.* **28**, 237–247.

Klimeš, L. (1985). Computation of seismic wavefields in 3-D media by the Gaussian beam method. Program package SW84. Research Report No. 68 for Geofyzika Brno, Institute of Geophysics, Charles University, Prague.

Klimeš, L. (1986). Discretization error for the superposition of Gaussian beams. *Geophys. J. R. Astron. Soc.* **86**, 531–551.

Klimeš, L. (1989a). Optimization of the shape of Gaussian beams of a fixed length. *Stud. Geophys. Geod.* **33**, 146–163.

Klimeš, L. (1989b). Gaussian packets in the computation of seismic wavefields. *Geophys. J. Int.* **99**, 421–433.

Klimeš, L. (1994). Transformations for dynamic ray tracing in anisotropic media. *Wave Motion* **20**, 261–272.

Klimeš, L. (1996a). Grid travel-time tracing: Second-order method for the first arrivals in smooth media. *Pure Appl. Geophys.* **148**, 539–563.

Klimeš, L. (1996b). Synthetic seismograms in 2-D model UNCONFORMITY. In: *Seismic Waves in Complex 3-D Structures, Report 4*, pp. 77–89. Department of Geophysics, Charles University, Prague, online at http://sw3d.mff.cuni.cz.
Klimeš, L. (1997). Phase shift of the Green function due to caustics in anisotropic media. In: *Seismic Waves in Complex 3-D Structures, Report 6*, pp. 167–173. Department of Geophysics, Charles University, Prague, online at http://sw3d.mff.cuni.cz.
Klimeš, L. (2000a). Comparison of ray-matrix and finite-difference methods in a simple 1-D model. In: *Expanded Abstracts of 70th Annual Meeting (Calgary)*. Soc. Exp. Geophysicists, Tulsa, pp. 2325–2328, online at http://sw3d.mff.cuni.cz.
Klimeš, L. (2000b). Sobolev scalar products in the construction of velocity models. In: *Seismic Waves in Complex 3-D Structures, Report 10*, pp. 15–40. Department of Geophysics, Charles University, Prague, online at http://sw3d.mff.cuni.cz.
Klimeš, L. (2002a). Second-order and higher-order perturbations of travel time in isotropic and anisotropic media. *Stud. Geophys. Geod.* **46**, 213–248.
Klimeš, L. (2002b). Lyapunov exponents for 2-D ray tracing without interfaces. *Pure Appl. Geophys.* **159**, 1465–1485.
Klimeš, L. (2002c). Application of the medium covariance functions to travel-time tomography. *Pure Appl. Geophys.* **159**, 1791–1810.
Klimeš, L. (2004a). Analytical one-way plane-wave solution in the 1-D anisotropic "simplified twisted crystal" model. *Stud. Geophys. Geod.* **48**, 75–96.
Klimeš, L. (2004b). Gaussian packets in smooth isotropic media. In: *Seismic Waves in Complex 3-D Structures, Report 14*, pp. 43–54. Department of Geophysics, Charles University, Prague, online at http://sw3d.mff.cuni.cz.
Klimeš, L. (2006a). Spatial derivatives and perturbation derivatives of amplitude in isotropic and anisotropic media. *Stud. Geophys. Geod.* **50**, 417–430.
Klimeš, L. (2006b). Common-ray tracing and dynamic ray tracing for S waves in a smooth elastic anisotropic medium. *Stud. Geophys. Geod.* **50**, 449–461.
Klimeš, L. (2006c). Ray-centred coordinate systems in anisotropic media. *Stud. Geophys. Geod.* **50**, 431–447.
Klimeš, L. (2006d). Phase shift of the Green function due to caustics in anisotropic media. *Stud. Geophys. Geod.*, submitted for publication.
Klimeš, L., Bulant, P. (2004). Errors due to the common ray approximations of the coupling ray theory. *Stud. Geophys. Geod.* **48**, 117–142.
Klimeš, L., Bulant, P. (2006). Errors due to the anisotropic-common-ray approximation of the coupling ray theory. *Stud. Geophys. Geod.* **50**, 463–477.
Klimeš, L., Kvasnička, M. (1994). 3-D network ray tracing. *Geophys. J. Int.* **116**, 726–738.
Kravtsov, Yu.A., Forbes, G.W., Asatryan, A.A. (1999). Theory and applications of complex rays. In: Wolf, E. (Ed.), *Progress in Optics*, vol. 39. Elsevier, Amsterdam, pp. 1–62.
Kravtsov, Yu.A., Orlov, Yu.I. (1980). Geometrical Optics of Inhomogeneous Media (in Russian). Nauka, Moscow. Translation to English by Springer, Berlin, 1990.
Kravtsov, Yu.A., Orlov, Yu.I. (1999). Caustics, Catastrophes and Wave Fields. Springer, Heidelberg.
Lambaré, G., Lucio, P.S., Hanyga, A. (1996). Two-dimensional multivalued traveltime and amplitude maps by uniform sampling of ray field. *Geophys. J. Int.* **125**, 584–598.
Lecomte, I. (1996). Hybrid modeling with ray tracing and finite difference. In: *Expanded Abstracts of 66th SEG Annual Meeting*. Soc. Exp. Geophysicists, Tulsa, pp. 699–702.
Lewis, R.M. (1965). Asymptotic theory of wave-propagation. *Arch. Ration. Mech. Anal.* **20**, 191–250.
Lugara, D., Letrou, C., Shlivinski, A., Heyman, E., Boag, A. (2003). Frame-based Gaussian beam summation method: Theory and applications. *Radio Sci.* **38**, VIC-27-1–VIC-27-15.
Lyapunov, A.M. (1949). Problème Général de la Stabilité du Mouvement. In: *Annals of Mathematical Studies*, vol. 17. Princeton Univ. Press, Princeton.
Martin, B.E., Thomson, C.J. (1997). Modelling surface waves in anisotropic structures. II. Examples. *Phys. Earth Planet. Int.* **103**, 253–279.

Maslov, V.P. (1965). Theory of Perturbations and Asymptotic Methods. Izd. MGU, Moscow (in Russian).

Moczo, P., Bard, P.Y., Pšenčík, I. (1987). Seismic response of 2-D absorbing structure by the ray method. *J. Geophys.* **62**, 38–49.

Moser, T.-J. (1991). Shortest path calculation of seismic rays. *Geophysics* **56**, 59–67.

Moser, T.-J. (2004). Review of the anisotropic interface ray propagator: Symplecticity, eigenvalues, invariants and applications. *Stud. Geophys. Geod.* **48**, 47–73.

Norris, A., White, B.S., Schreiffer, J.R. (1987). Gaussian wave packets in inhomogeneous media with curved interfaces. *Proc. R. Soc. London A* **412**, 93–123.

Nowack, R.L. (2003). Calculation of synthetic seismograms with Gaussian beams. *Pure Appl. Geophys.* **160**, 487–507.

Nowack, R.L., Pšenčík, I. (1991). Travel-time and ray path perturbation from isotropic to anisotropic media. *Geophys. J. Int.* **106**, 1–10.

Opršal, I., Brokešová, J., Fäh, D., Girardini, D. (2002). 3-D hybrid ray-FD and DWN-FD seismic modeling for simple models containing complex local structures. *Stud. Geophys. Geod.* **46**, 711–730.

Oseledec, V.I. (1968). A multiplicative ergodic theorem: Lyapunov characteristic numbers for dynamical systems. *Trans. Moscow Math. Soc.* **19**, 197–231.

Ozaktas, H.M., Zalevsky, Z., Kutay, M.A. (2001). The Fractional Fourier Transform with Applications in Optics and Signal Processing. John Wiley and Sons, Chichester.

Podvin, P., Lecomte, I. (1991). Finite difference computation of traveltimes in very contrasted velocity models: A massively parallel approach and its associated tools. *Geophys. J. Int.* **105**, 271–284.

Popov, M.M. (1982). A new method of computation of wave fields using Gaussian beams. *Wave Motion* **4**, 85–97.

Popov, M.M., Pšenčík, I. (1978a). Ray amplitudes in inhomogeneous media with curved interfaces. *Travaux Instit. Géophys. Acad. Tchécosl. Sci. No. 454.* In: *Geofys. Sborník*, vol. 24. Academia, Praha, pp. 111–129.

Popov, M.M., Pšenčík, I. (1978b). Computation of ray amplitudes in inhomogeneous media with curved interfaces. *Stud. Geophys. Geod.* **22**, 248–258.

Pšenčík, I. (1998). Green's functions for inhomogeneous weakly anisotropic media. *Geophys. J. Int.* **135**, 279–288.

Pšenčík, I., Dellinger, J. (2001). Quasi-shear waves in inhomogeneous weakly anisotropic media by the quasi-isotropic approach: A model study. *Geophysics* **66**, 308–319.

Pšenčík, I., Farra, V. (2005). First-order ray tracing for qP waves in inhomogeneous weakly anisotropic media. *Geophysics* **70**, D65–D75.

Pšenčík, I., Gajewski, D. (1998). Polarization, phase velocity and NMO velocity of qP waves in arbitrary weakly anisotropic media. *Geophysics* **63**, 1754–1766.

Pšenčík, I., Teles, T.N. (1996). Point-source radiation in inhomogeneous anisotropic structures. *Pure Appl. Geophys.* **148**, 591–623.

Ralston, J. (1983). Gaussian beams and the propagation of singularities. In: Littman, W. (Ed.), *Studies in Partial Differential Equations*. In: *MAA Studies in Mathematics*, vol. 23. Math. Assoc. Amer., Washington, DC, pp. 206–248.

Schleicher, J., Tygel, M., Hubral, P. (1993). 3-D true-amplitude finite-offset migration. *Geophysics* **58**, 1112–1126.

Schleicher, J., Tygel, M., Ursin, B., Bleistein, N. (2001). The Kirchhoff–Helmholtz integral for anisotropic elastic media. *Wave Motion* **34**, 353–364.

Shlivinski, A., Heyman, E., Boag, A., Letrou, C. (2004). Phase-space beam summation formulations for ultra wideband (UWB) radiation. In: *Proceedings of URSI International Symposium on Electromagnetic Theory (Pisa)*. Edizioni PLUS, Università di Pisa, pp. 936–938.

Sun, J. (2004). True-amplitude weight functions in 3-D limited aperture migration revisited. *Geophysics* **60**, 1025–1036.

Tarantola, A. (1987). Inverse Problem Theory. Elsevier, Amsterdam.
Tessmer, E. (1995). 3-D seismic modelling of general material anisotropy in the presence of the free surface by a Chebyshev spectral method. *Geophys. J. Int.* **121**, 557–575.
Thomsen, L. (1986). Weak elastic anisotropy. *Geophysics* **51**, 1954–1966.
Thomson, C.J. (1997). Complex rays and wave packets for decaying signals in inhomogeneous, anisotropic and anelastic media. *Stud. Geophys. Geod.* **41**, 345–381.
Thomson, C.J. (1999). The 'gap' between seismic ray theory and 'full' wavefield extrapolation. *Geophys. J. Int.* **137**, 364–380.
Thomson, C.J. (2001). Seismic coherent states and ray geometrical spreading. *Geophys. J. Int.* **144**, 320–342.
Thomson, C.J. (2004). Coherent-state analysis of the seismic head wave problem: An overcomplete representation and its relation to rays and beams. *Geophys. J. Int.* **157**, 1189–1205.
Thomson, C.J., Kendall, J.-M., Guest, W.S. (1992). Geometrical theory of shear-wave splitting: Corrections to ray theory for interference in isotropic/anisotropic transitions. *Geophys. J. Int.* **108**, 339–363.
Tygel, M., Schleicher, J., Hubral, P. (1994). Kirchhoff–Helmholtz theory in modelling and migration. *J. Seismol. Exp.* **3**, 203–214.
Ursin, B., Tygel, M. (1997). Reciprocal volume and surface scattering integrals for anisotropic elastic media. *Wave Motion* **26**, 31–42.
Vavryčuk, V. (1999). Applicability of higher-order ray theory for S wave propagation in inhomogeneous weakly anisotropic elastic media. *J. Geophys. Res. B* **104**, 28829–28840.
Vavryčuk, V. (2001). Ray tracing in anisotropic media with singularities. *Geophys. J. Int.* **145**, 265–276.
Vidale, J.E. (1990). Finite-difference calculation of traveltimes in three dimensions. *Geophysics* **55**, 521–526.
Vinje, V. (1997). A new interpolation criterion for controlling accuracy in wavefront construction. In: *Expanded Abstracts of 67th SEG Annual Meeting (Dallas)*. Soc. Exp. Geophysicists, Tulsa, pp. 1723–1726.
Vinje, V., Iversen, E., Åstebøl, K., Gjøystdal, H. (1996a). Estimation of multivalued arrivals in 3-D models using wavefront construction, Part I. *Geophys. Prosp.* **44**, 819–842.
Vinje, V., Iversen, E., Åstebøl, K., Gjøystdal, H. (1996b). Tracing and interpolation, Part II. *Geophys. Prosp.* **44**, 843–858.
Vinje, V., Iversen, E., Gjøystdal, H. (1993a). Traveltime and amplitude estimation using wavefront construction. *Geophysics* **58**, 1157–1166.
Vinje, V., Iversen, E., Gjøystdal, H., Åstebøl, K. (1993b). Estimation of multivalued arrivals in 3-D models using wavefront construction. In: *Extended Abstracts of 55th Annual Meeting (Stavanger)*. Eur. Assoc. Exp. Geophysicists, Zeist, p. B019.
Virieux, J. (1996). Seismic ray tracing. In: Boschi, E., Ekström, G., Morelli, A. (Eds.), *Seismic Modelling of the Earth Structure*. Instituto Nazionale di Geofisica, Roma, pp. 223–304.
White, B.S., Nair, B., Bayliss, A. (1988). Random rays and seismic amplitude anomalies. *Geophysics* **53**, 903–907.
Wolf, K.B. (1974). Canonical Transforms. I. Complex linear transforms. *J. Math. Phys.* **15**, 1295–1301.
Wolf, K.B. (1979). Integral Transforms in Science and Engineering. Plenum Press, New York.
Wolfson, M.A., Tappert, F.D. (2000). Study of horizontal multipaths and ray chaos due to ocean mezoscale structure. *J. Acoust. Soc. Am.* **107**, 154–162.
Woodhouse, J.H. (1974). Surface waves in laterally varying layered structure. *Geophys. J. R. Astron. Soc.* **37**, 461–490.
Yomogida, K. (1985). Gaussian beams for surface waves in laterally slowly-varying media. *Geophys. J. R. Astron. Soc.* **82**, 511–533.
Yomogida, K. (1987). Gaussian beams for surface waves in transversely isotropic media. *Geophys. J. R. Astron. Soc.* **88**, 297–304.

Žáček, K. (2002). Smoothing the Marmousi model. *Pure Appl. Geophys.* **159**, 1507–1526.
Žáček, K. (2004). Gaussian-packet prestack depth migration. In: *Seismic Waves in Complex 3-D Structures, Report 14*, pp. 17–25. Department of Geophysics, Charles University, Prague, online at http://sw3d.mff.cuni.cz.
Žáček, K. (2006a). Optimization of the shape of Gaussian beams. *Stud. Geophys. Geod.* **50**, 349–366.
Žáček, K. (2006b). Decomposition of the wave field into optimized Gaussian packets. *Stud. Geophys. Geod.* **50**, 367–380.
Žáček, K., Klimeš, L. (2003). Sensitivity of seismic waves to the structure. In: *Expanded Abstracts of 73rd SEG Annual Meeting (Dallas)*. Soc. Exp. Geophysicists, Tulsa, pp. 1857–1860.
Zhao, L., Dahlen, F.A. (1996). Mode-sum to ray-sum transformation in a spherical and an aspherical earth. *Geophys. J. Int.* **126**, 389–412.

INTRODUCTION TO MODE COUPLING METHODS FOR SURFACE WAVES

VALÉRIE MAUPIN

Department of Geosciences, University of Oslo, PO Box 1047 Blindern, 0316 Oslo, Norway

ABSTRACT

In laterally homogeneous models, free oscillations and surface waves are composed of modes propagating independently of each other. In laterally heterogeneous structures, the seismic wavefield is more complex, but can still be expressed up to a certain degree of approximation by modes, which are then coupled to each other by the lateral heterogeneities. We review here the different types of mode coupling methods which have been developed to model surface waves in laterally heterogeneous structures.

In the surface wave formalism, that is for a single wavetrain propagating in a flat structure, two classes of mode coupling methods exist. Those based on coupling of local modes are most appropriate for analysing the gradual evolution of the surface wavefield from one type of structure to another, and are related to surface wave raytracing. For models with localised heterogeneities, the modelling methods are based on coupling of modes calculated in a laterally homogeneous reference structure. In that case, the modes are coupled by the heterogeneities, which act as scatterers. Single- as well as multiple-scattering algorithms have been developed. We review here principally the methods for direct problems, but mention also their applicability in inverse problems.

Keywords: Surface waves, Mode coupling, Scattering, Local modes, Adiabatic modes 4

1. INTRODUCTION

The seismic waves in the gravest part of the Earth's spectrum are modelled very efficiently using mode methods. When the propagation involves the whole spherical Earth, and possibly includes wavetrains having propagated around the Earth several times, the free oscillation formalism is most appropriate. When propagation involves only a limited region of the Earth's upper structure, with limited influence of the Earth's sphericity, and when the wavefield only consists of wavetrains propagating away from the source, the surface wave mode formalism is more adequate. In that case, one uses flat models infinite both laterally and with depth. This formalism is used at a large number of scales, from the propagation of surface waves in the lithosphere, to guided waves in sedimentary layers, and to medical and material sciences applications at centimeter scale. It is the coupling methods developed in this formalism that I shall principally review here.

The propagation of surface waves and free oscillations in laterally homogeneous structures has been well understood for several decades, at least in isotropic structures. Current efforts are devoted to study wave propagation in laterally heterogeneous structures and to develop efficient data analysis tools for tomography. In this chapter, I will introduce the concepts and domains of application of the different kinds of methods which have been developed and used in the last two decades or so to model surface wave propagation in laterally heterogeneous structures in seismology. I will not make an extensive review of existing works, but rather provide keys to enter the field and gain insight into the relations between the different methods and their domains of application. I will focus on surface waves at relatively short periods, which are best represented as propagating wavetrains, and for which the free oscillation formalism is not well adapted. I will also concentrate on methods developed specifically for surface waves and will not discuss numerical methods, like finite-differences or spectral elements, which can be used to model surface waves but are not specific to them.

2. Modes and Mode Coupling

In isotropic laterally homogeneous structures, it is well known that elastic surface waves propagate as independent Rayleigh and Love wave modes. As opposed to the free oscillation formalism, where time-domain solutions as well as Laplace transforms are used, frequency-domain analysis is ubiquitous in the surface wave formalism. In two-dimensional geometry, where z is the vertical coordinate and x the horizontal one, the displacement $\vec{v}$ related to a surface wavetrain can be expressed at frequency ω by the following modal sum:

$$\vec{v}(x, z, \omega) = \sum_m a_m(\omega) \vec{f}_m(z, \omega) \exp(i\omega x / c_m(\omega)), \qquad (1)$$

where m is the mode index, a_m is the amplitude of mode m, a complex number which also contains the source phase, and does not depend on x and z, $\vec{f}_m$ is the depth dependence of the eigenfunction of mode m and c_m its phase velocity. The eigenfunctions with depth are vectorial functions which are polarized perpendicularly to the propagation plane for Love waves, in the propagation plane for Rayleigh waves, and in three directions for generalized modes in anisotropic structures. An example of such functions is given in Fig. 1, where we show the depth dependence of the eigenfunctions of the Rayleigh and Love waves fundamental modes and first three overtones at 30 s period in the isotropic PREM model (Dziewonski and Anderson, 1981).

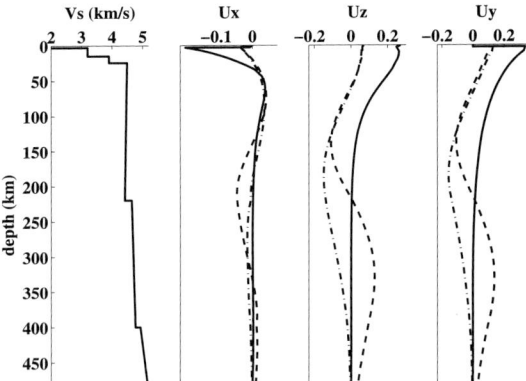

FIG. 1. Example of Love and Rayleigh wave eigenfunctions. The S-wave velocity in the PREM model is shown in the leftmost panel. The two next panels show respectively the horizontal and vertical displacement with depth of the fundamental mode and two first overtones of the Rayleigh wave in the PREM model at a period of 30 seconds. The rightmost panel shows the displacement with depth of the fundamental mode and two first overtones of the Love wave, also in the PREM model and at a period of 30 seconds. The fundamental modes are shown with solid lines, the first overtones with dashed-dotted lines, and the second overtones with dashed lines.

Here we have used a far-field expression for two-dimensional geometry to keep the equations as simple as possible, but expressions for propagation in three-dimensional geometry and those which include the near-field have a similar form, with an additional geometric attenuation in particular (see Friederich *et al.* (1993) or Snieder (1986) for example).

A remarkable characteristic of Eq. (1) and similar ones in three-dimensions is the separation of variables in the modes eigenfunctions: they combine a sinusoidal function in the horizontal direction $\exp(i\omega x/c_m(\omega))$ with a more general function $\vec{f}_m(z, \omega)$ in the depth direction. This separation is made possible by the lateral homogeneity of the structure. More general types of structure, and in particularly laterally heterogeneous structures, also exhibit a set of eigenvibrations which, by nature, propagate independently of each other. But these eigenvibrations, or true modes, may have very complicated forms. An example of true Love modes propagating in a laterally heterogeneous basin is shown in the chapter by Chen (2006) in the present book. Generally, their eigenfunctions cannot be put into a form where variables are separated. In seismology, although it would be possible to calculate the true modes of the Earth by using heavy numerical codes, we shall see that it is usually not necessary, and that other methods are more appropriate.

The Earth is dominantly stratified and lateral variations are almost always a second order feature compared to the horizontal layering. As a consequence, surface waves travel almost like a set of modes having decoupled vertical and horizontal behaviour. It is therefore less adequate to use the true modes of the structure

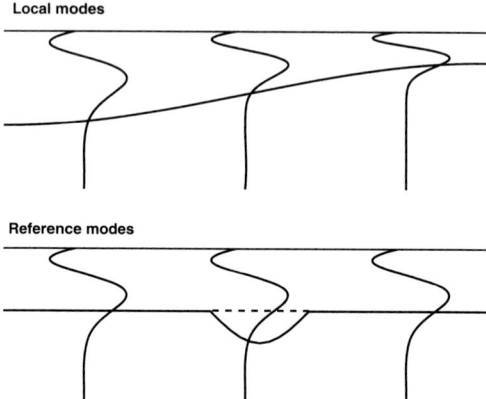

FIG. 2. Vertical cross-sections showing the difference between the local modes approach and the reference modes approach. In both panels, the model is represented schematically by a laterally varying layer over a halfspace, and the modal eigenfunctions are shown at three locations in the model. The upper panel shows the principle of the local modes approach, for which the eigenfunctions vary laterally, following the lateral evolution of the structure. In the lower panel, in addition to the laterally varying model (solid line), we define a reference model (dashed line). The eigenfunctions which are used are those calculated in the reference model only, and do not vary laterally in the structure.

than an approximate expression which has some form of separation of variable between the horizontal and vertical directions. This form can be achieved by using the modes calculated in laterally homogeneous structures connected to the laterally heterogeneous one. Since these modes are not anymore the true modes of the structure, but simply a convenient basis for representing the wavefield, they do not propagate independently of each other anymore and are said to be coupled.

As illustrated in Fig. 2, two types of modes are used in the most classical mode coupling methods: the local modes and the reference modes. At any given position, the local modes are the modes which are calculated in laterally homogeneous structure having the same profile with depth as the laterally heterogeneous structure at the given position. They evolve therefore from one place to the other, adapting to the variations of the heterogeneous structure. The reference modes on the other hand are calculated in a single reference structure close to the laterally heterogeneous structure, and are fixed for the whole structure. Depending on the problem at hand, one or the other method is best suited. Large lateral variations are usually best tackled by local mode coupling methods, whereas reference modes methods are well-suited to work in structures with more localised heterogeneities.

Whether with local or with reference modes, "mode coupling" in the surface wave formalism is a term restricted to the coupling of modes located on different dispersion branches, the so-called "cross-branch" coupling in the free oscillation literature. This is because, working at a given frequency, the coupling we consider

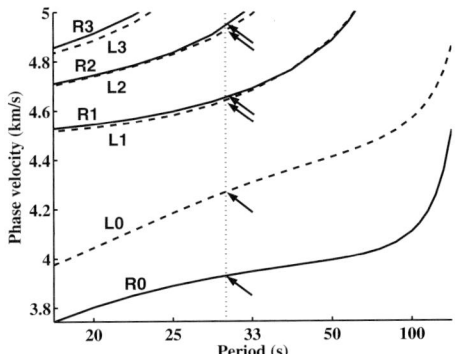

FIG. 3. Phase velocity as a function of period for the fundamental modes and the three first overtones of the Rayleigh (solid lines) and Love (dashed lines) waves in the PREM model (shown in Fig. 1). R0, L0, R1 to R3 and L1 to L3 label respectively the dispersion curves of the Rayleigh wave fundamental mode, Love wave fundamental mode, Rayleigh wave first to third overtone and Love wave first to third overtone. The vertical dashed line at 30 s period and the associated arrows show an example of which modes couple together in the surface wave formalism (the so-called cross-branch coupling in the free oscillation literature).

is the one between modes having the same frequency and belonging to different dispersion branches. This is illustrated in Fig. 3, where arrows point to a set of modes in the frequency-phase velocity plane which have to be considered together in a mode coupling analysis. There is no "along-branch" coupling in the surface wave formalism, apart from the coupling of a mode to itself. The variations of the modes phase velocities and eigenfunctions at a single frequency are not expressed in terms of along-branch coupling.

Starting with a very simple model of two welded quarter-spaces, we will first introduce local mode methods. We will show that these methods reduce to surface wave ray-tracing in structures with only long-wavelength variations. Second, we will develop reference mode methods, going from the more general multiple-scattering methods to the single scattering ones.

3. Local Modes

3.1. A Simple Example of Local Mode Coupling: Welded Spaces

The simplest model in which local mode coupling methods can be illustrated is a structure made of two quarter-spaces welded together at a vertical boundary. This problem was one of the first problems to be addressed in surface wave propagation in laterally heterogeneous structures and is extensively discussed in Levshin *et al.* (1989) and Malischewsky (1987) for example. In each quarter-space, one can define a set of Rayleigh and Love modes. These modes are called local modes because each set is valid only in its quarter-space, and not over the whole structure. When one mode, for example, a Rayleigh wave fundamental mode, is incident on the boundary from one side, it will be partly transmitted, partly reflected, and partly converted into reflected and transmitted Rayleigh overtones and Love modes, as shown in Fig. 4. The coupling between the different types of modes and between forward and backward propagating modes is controlled by the boundary condition at the vertical interface between the two quarter-spaces. This process is very reminiscent of how body waves are reflected, transmitted and converted at a plane boundary. If the boundary is perpendicular to the propagation direction of the surface waves, there is no coupling between Rayleigh and Love waves. And, similarly to body waves, for a propagation direction not normal to the boundary, the modes follow a form of Snell's law involving their phase velocities: p, the component of the horizontal slowness parallel to the strike of the vertical boundary, is invariant. This relation determines i_m, the incidence angle of the propagation in the horizontal plane, for each mode m according to the relation:

$$p = \sin(i_m)/c_m \qquad (2)$$

as illustrated in Fig. 4. The set of local modes one has to consider in case of non-normal incidence is therefore a set of modes propagating forwards and backwards in each quarter-space, but at slightly different angles i_m, in order to ensure that they all have the same p-value.

3.1.1. The Orthogonality Relation

The amplitudes of the transmission and reflection coefficients of the different modes can be determined by considering the boundary condition at the interface. For two welded quarter-spaces, the boundary condition is the continuity of the displacement and of the traction on the vertical plane between the two quarter-spaces.

It is therefore convenient to define for each mode the following displacement-stress vector:

$$\vec{u}_m = \begin{pmatrix} \vec{f}_m \\ \vec{t}_m \end{pmatrix}, \qquad (3)$$

where $\vec{f}_m$ is the displacement with depth defined in Eq. (1) and $\vec{t}_m$ is the traction with depth for mode m on planes parallel to the vertical boundary between the two quarter-spaces. (Note the difference with the traction on horizontal planes used in reflectivity methods; see Kennett (1983) for example.) In addition to the fact that the boundary condition is easily expressed in terms of these displacement-stress vectors, they provide an orthogonality relation for the modes which proves useful well beyond the case of two welded spaces. We can define the cross-product

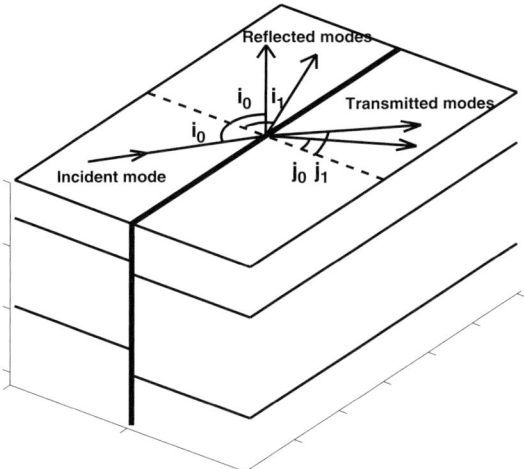

FIG. 4. Reflection and transmission of surface waves across a vertical boundary. Three-dimensional plot of two layered quarter-spaces welded together at a vertical boundary. In the horizontal upper plane, we show the propagation direction of a mode incident on the vertical boundary with an incidence angle i_0. This mode is reflected by the boundary and travels back into the quarter-space with an incidence angle i_0. It is also converted both in reflection and in transmission. The converted reflected mode has an incidence angle i_1. The transmitted modes propagate with incidence angles j_0 and j_1.

between two modes as:

$$\langle \vec{u}_m, \vec{u}_n \rangle = i \int_z \vec{f}_m^*(z,\omega) \vec{t}_n(z,\omega) - \vec{t}_m^*(z,\omega) \vec{f}_n(z,\omega)\, dz, \qquad (4)$$

where $*$ stands for complex conjugation. Note that

$$\langle \vec{u}_m, \vec{u}_n \rangle = \langle \vec{u}_n, \vec{u}_m \rangle^*. \qquad (5)$$

If modes m and n belong to the same set of local modes, it can be shown using Betti's theorem that this product is zero for $m \neq n$. For $m = n$, it is directly related to the energy flux carried by the mode across the vertical interface. Normalizing the eigenfunctions such that each mode carries the same unit energy flux,

$$\langle \vec{u}_m, \vec{u}_n \rangle = \pm \delta_{mn} \qquad (6)$$

with a '+' sign for the normalisation of the modes propagating forwards (that is positively along the horizontal coordinate axis perpendicular to the vertical interface) and a '−' sign for those propagating backwards, the set of local modes can be seen as a basis of orthonormal functions with respect to the scalar product defined in Eq. (4). The orthogonality also applies between any mode and the body waves which are not trapped in the structure.

The orthonormality (6) ensures that the amplitude of mode m in any total wavefield specified at a given position x can be simply calculated by projecting $\vec{u}$ on mode m:

$$a_m = \pm \langle \vec{u}_m, \vec{u} \rangle. \tag{7}$$

3.1.2. Transmission and Reflection Coefficients

Calculating transmission and reflection coefficients for the different modes can be seen as projecting the continuous displacement-stress at the vertical interface on the local modes belonging to the two quarter-spaces on both sides of the interface. We can write the displacement-stress on both sides as a sum of modes plus a residual wavefield associated with body waves and possibly very high modes neglected in the modal sum. The boundary condition is then:

$$\vec{u}_i + \sum_j \vec{u}_j^R a_j + \vec{u}_b^R = \sum_j \vec{u}_j^T b_j + \vec{u}_b^T, \tag{8}$$

where $\vec{u}_i$ is the incident mode, $\vec{u}_j^R$ are reflected modes with amplitudes a_j, $\vec{u}_j^T$ are transmitted modes with amplitudes b_j, and $\vec{u}_b^R$ and $\vec{u}_b^T$ are residual wavefields.

Taking the scalar product of Eq. (8) with reflected mode $\vec{u}_n^R$, the orthonormality relation (6) ensures that all terms except $-a_n$ disappear on the l.h.s. On the r.h.s., we have the projection on reflected mode n of all the forward propagating modes on the transmission side of the interface. We have therefore:

$$-a_n = \langle \vec{u}_n^R, \vec{u}_j^T \rangle b_j + \langle \vec{u}_n^R, \vec{u}_b^T \rangle. \tag{9}$$

Here we omit the $\sum$ and use Einstein summation convention. Taking now the scalar product of Eq. (8) with transmitted mode $\vec{u}_n^T$, we get similarly an equation where b_n, on the r.h.s., is singled out:

$$\langle \vec{u}_n^T, \vec{u}_i \rangle + \langle \vec{u}_n^T, \vec{u}_j^R \rangle a_j + \langle \vec{u}_n^T, \vec{u}_b^R \rangle = b_n. \tag{10}$$

An alternative to Eq. (10) is to take the scalar product with the modes propagating forwards on the incident side of the interface. One obtains then:

$$\delta_{ni} = \langle \vec{u}_n, \vec{u}_j^T \rangle b_j + \langle \vec{u}_n, \vec{u}_b^T \rangle. \tag{11}$$

If the residual terms $\vec{u}_b^R$ and $\vec{u}_b^T$ are zero or can be neglected, Eqs. (9) and (10) or Eqs. (9) and (11) can be combined to provide a linear system of equations which yield the reflection and transmission coefficients a_i and b_i. If the modal set is complete, that is if we can express the total wavefield as a sum of modes, there are no residual terms, and it can be shown that the two combinations provide exactly the same coefficients. It is possible to define a complete modal set for a vertically varying infinite halfspace by using modes with complex wavenumbers (Stange and Friederich, 1992) or radiation modes (Maupin, 1996). But in practice, we usually have a truncated mode set and not a complete one, and contributions from very high modes and body waves generated at the vertical interface are neglected. In that case, the displacement-stress on the interface is not known exactly and the reflection and transmission coefficients which are obtained are not exact. They vary depending on the number of modes included in the modal sums and depending on which combination of Eqs. (9) and (10) or (9) and (11) is used. Its and Yanovskaya (1987) and Meier et al. (1997b) use the first combination, which gives exactly the same results as if one calculates the reflection and transmission coefficients by minimizing the flux of energy in the residual terms (Malischewsky, 1987). Bostock (1991) uses the second combination in three-dimensional models. Romanelli et al. (1996) have developed a third alternative, using transmission and reflection coefficients of body waves in the individual layers before calculating the contributions of the total wavefield to the individual modes.

These different variants of the same method, denoted Green's function or mode-matching methods, lead to similar coefficients if the basis of modes which is used is appropriate. In all cases, we can note that a mode which has an eigenfunction which does not vary much across the boundary will have $\langle u_i^T, u_i \rangle$ close to 1 and all other scalar products in Eqs. (9) to (11) close to zero, leading to a high transmission coefficient, low reflection coefficient and little coupling to other modes. On the other hand, coupling is important for modes which eigenfunction is significantly modified by crossing the vertical boundary. In that case, the energy can be strongly redistributed with depth, leading to significant amplitude variations at the surface. Levshin (1985) noted that the redistribution with depth of the energy is the most important factor affecting the surface amplitude variations of surface waves during propagation through a laterally heterogeneous region, and was a more important factor than the value of the transmission coefficients themselves.

An example of the variation one can expect in eigenfunctions is given in Fig. 5. The horizontal and vertical components of the Rayleigh wave fundamental mode and the transverse component of the Love wave fundamental mode are shown in two different structures at a period of 40 seconds. One structure, derived from the surface wave tomography of the Indian Ocean by Debayle and Lévêque (1997), is a typical oceanic one, with a thin crust and a low-velocity zone in the upper mantle. The other model is the Australian model of Gaherty et al. (1999), with a thicker crust and smaller velocity variations in the upper mantle. Except for small variations in the crust, the eigenfunctions of the Rayleigh wave fundamental modes are very similar in the two models. The eigenfunctions of the Love wave fundamental mode on the other hand are very different. A significant part of the energy is trapped in the low-velocity zone in the oceanic model, whereas energy is more concentrated in the upper region in the continental model. If these two models were adjacent at a rather sharp continental margin, 40 s Rayleigh waves would therefore likely cross the margin without noticeable coupling to other modes, whereas Love waves would likely couple to other modes, in par-

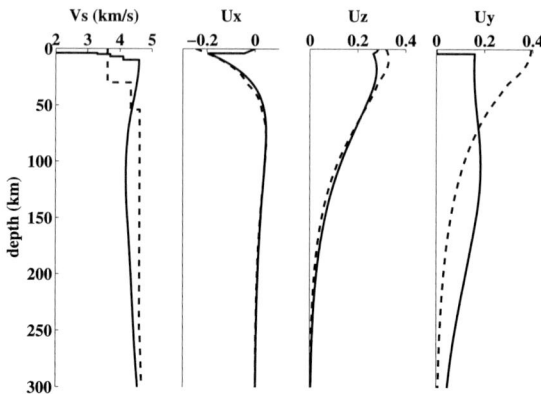

FIG. 5. Example of Love and Rayleigh wave eigenfunctions in two different structures. The leftmost panel shows the S-wave velocity with depth in a typical oceanic model (solid line) and a typical continental one (dashed line). The two next panels show, with the same line type convention, respectively the horizontal and vertical displacement with depth of the fundamental mode of the Rayleigh wave at a period of 40 seconds these two models. The rightmost panel shows the displacement with depth of the Love wave fundamental mode in the two models at the same period.

ticular to their overtones, in order to cope with the difference in propagation style on both sides of the margin.

Recent results concerning reflection and transmission coefficients at a continental margin are presented in Meier and Malischewsky (2000). Reflection coefficients have also been used for surface wave reflection tomography in Europe using coda waves (Meier et al., 1997b). An important domain of application of mode-coupling methods in welded spaces is seismic risk analysis. As described in Panza et al. (2000), synthetic seismograms can be calculated at the scale of a city or a small region to analyse seismic risk by applying mode-matching techniques in a series of welded laterally homogeneous models. The same method has been used by Du (2002) at a somewhat larger scale, to analyse and invert wave propagation across the Iberian peninsula and part of the Atlantic Ocean. Mode-matching technics have been extended to three-dimensional models with possibly non-vertical boundaries by Bostock (1991) and Bostock and Kennett (1992).

3.2. Local Mode Coupling: The General Case

The geometry of the two welded quarter-spaces is appropriate when the width of the transition zone between the two end-structures is small compared to the wavelengths of the waves. In many cases, the wavelengths involved in the wavefields and in the models are such that models with continuous variations are more suitable than models build with sharp boundaries. In that case, the wavefield can be calculated by a local mode coupling method which is a continuous variant of the welded-spaces methods described above. The wavefield is expressed as a sum of local modes $\vec{u}_m(\vec{x}, \omega)$ with laterally varying phase velocity $c_m(\omega, \vec{x})$ and laterally varying amplitudes $a_m(\omega, \vec{x})$. The local modes and their phase velocities at a location x of the horizontal plane are defined as those which would exist in a laterally homogeneous structure having the profile with depth of the laterally heterogeneous structure at that location. The total wavefield is thus:

$$\vec{u}(\omega, \vec{x}) = \sum_m a_m(\omega, x) \vec{u}_m(z, \omega, x) \exp\left(i\omega \int_0^x \frac{1}{c_m(\omega, \xi)} d\xi\right). \tag{12}$$

Here we also choose to show expressions in two-dimensions to keep them as simple as possible and highlight the principles. Note the similarity with Eq. (1). The modal sum includes modes propagating forwards and backwards, possibly at different angles but in such a way that Snell's law (Eq. (2)) is satisfied if the propagation is not perpendicular to the axis of the two-dimensional structure. Inserting this expression into the wave equation, taking the scalar product with forward and backward propagating modes, and taking advantage of the modes orthonormality (Eq. (6)), one obtains a differential equation for the amplitudes:

$$\frac{da_n}{dx} = \sum_{m \neq n} \pm \left\langle u_n, \frac{d\vec{u}_m}{dx} \right\rangle \exp\left(i\omega \int_0^x \left(\frac{1}{c_m(\omega, \xi)} - \frac{1}{c_n(\omega, \xi)}\right) d\xi\right) a_m. \tag{13}$$

This equation expresses how energy is transferred from mode m to mode n at the location x. Similar to the welded-spaces case, we see that the coupling is controlled by the lateral variation of the eigenfunctions. How much energy is transferred from mode m to mode n is controlled by the projection of the lateral derivative of the eigenfunction of mode m on the basis of functions u_n. Here also, the modes are normalized with respect to their flux of energy.

Equation (13) can be recast into a form which does not involve lateral derivatives of the eigenfunctions, but lateral derivatives of the model parameters, which are usually more easily accessible. Noting k_n and k_m the horizontal wavenumbers of modes n and m, it becomes:

$$\frac{da_n}{dx} = \sum_{m \neq n} \frac{1}{k_n - k_m} \left\langle \vec{u}_n, \frac{dC_{ijkl}}{dx}, \vec{u}_m \right\rangle$$

$$\times \exp\left(i\omega \int_0^x \left(\frac{1}{c_m(\omega, \xi)} - \frac{1}{c_n(\omega, \xi)}\right) d\xi\right) a_m, \tag{14}$$

where we refer to Maupin (1988) for the detailed expression of $\langle , , \rangle$. Let us just mention that it is an integral with depth of a linear function of the modal eigenfunctions, the layer interface slopes, and the lateral derivatives of the elastic coefficients and density. Under this form, it becomes apparent that the mode coupling is related to the horizontal gradients of the structure and that coupling is inversely proportional to wavenumber difference between the two modes. As a result, since the Rayleigh wave fundamental mode has wavenumbers markedly different from the other modes in a large domain of frequency, it couples generally weakly to higher modes or Love modes.

Although some general expressions have been derived by Tromp (1994) and by Kennett (1998), the local mode coupling problem in three-dimensional structures is not fully solved. As we will see in the next section, the different modes follow rays in smoothly varying three-dimensional structure, and obtaining practically useful expressions for mode coupling in three-dimensions can probably not be done independently of the ray-tracing equations. Making the assumption that the derivatives transverse to the rays are negligible, and that consequently all modes follow the same ray and coupling occurs only in the forward direction, Tromp (1994) derived an equation for the amplitude of a mode along its ray:

$$2k_n \frac{da_n}{ds} = \sum_{m \neq n} \frac{1}{k_n - k_m} \left\langle \vec{u}_n, \frac{dC_{ijkl}}{ds}, \vec{u}_m \right\rangle$$
$$\times \exp(i(\psi_m - \psi_n)) a_m - a_n \vec{\nabla} \cdot \vec{k}_n, \qquad (15)$$

where s is the length along the ray, ψ_m the phase of mode m, and $\langle , , \rangle$ is the same integral as in Eq. (14). The first term on the r.h.s. expresses how the amplitude varies due to transfer of energy from mode m to mode n, and the second one expresses how the amplitude of mode n varies due to focusing/defocusing of its rayfield. Note, as in previous sections, that the modes must be normalized with respect to a constant energy flux. The surface amplitude is therefore affected by a third factor, which is the redistribution with depth of the energy.

Applications of local mode coupling in seismology are limited to two-dimensional and 2.5-dimensional problems. One example of a 2.5-dimensional application is the analysis of Rayleigh waves propagating along a model of the Hawaiian swell (Maupin, 1992), from which Fig. 6 is derived. The model is based on the hypothesis that the lithosphere has been thinned along the hotspot track due to reheating, and that the thinner lithosphere in the middle of the swell connects continuously to a normal oceanic lithosphere 600 km away from the swell axis. This forms a low-velocity channel in which surface waves get trapped when they propagate in a direction close to the axis of the swell. Fig. 6 shows the polarization in the horizontal plane of the Rayleigh fundamental mode generated by a plane pure Rayleigh wave incident on the structure from the left, and recorded at different positions in the swell. At the dominant period of the wavetrain, that is 40 seconds, there is only 2% difference in phase velocity between the middle of the swell and the outskirts. For the stations 1000 km from the opening of the swell, on the l.h.s., the effect of the swell is to slightly rotate the polarization of the Rayleigh wave away from its initial polarization direction. Further away, wavetrains having propagated in different parts of the swell interfere together.

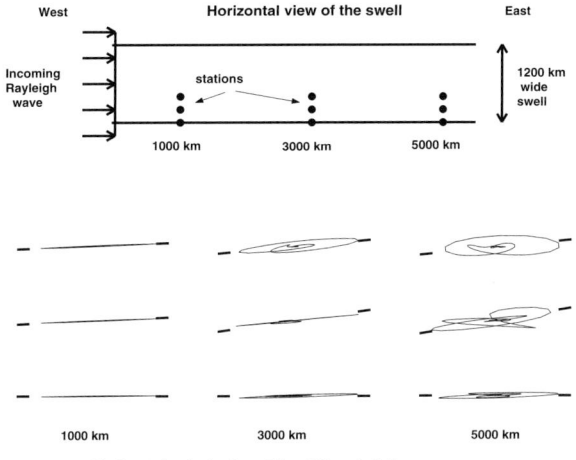

FIG. 6. Polarization in the horizontal plane of the Rayleigh wave fundamental mode at different locations in a low-velocity vertical channel. The geometry of the structure is shown in the horizontal plane in the upper part of the figure. The low-velocity channel, which represents here the Hawaiian swell, is bounded northwards and southwards by higher velocity structures. A Rayleigh wave propagates into the swell by the western side. Nine stations are distributed at different locations in the channel and the polarization of the Rayleigh wave fundamental mode at these stations, calculated with a local mode coupling method, are shown in the bottom part of the figure. The Rayleigh waves have a period range from 20 to 150 s. The deviation predicted by ray theory at the dominant period of the signal, 40 seconds, is also shown as two small thick lines close to the polarization maxima. Notice that this prediction fits quite well, except at the eastern stations, where the wavefield has propagated for long enough into the heterogeneous structure to have developed more complex features related to multipathing.

The polarization becomes quite complex despite the small amount of heterogeneity present in this model. The theoretical directions of arrival calculated by surface wave ray theory, detailed in the next section, are also shown in the polarization plots. In most cases, ray-theory predicts well the polarization of the waves, but at the stations where we have an interference of several wavetrains, leading the complex polarization pattern, simple first-arrival ray theory is of course not able to reproduce the pattern.

3.3. Surface Wave Ray Tracing

When the lateral gradients in the structure are small, that is when the variations in the model occur at wavelengths much larger than the wavelengths in the wavefield, the terms containing lateral derivatives of the model in Eqs. (14) and (15) become small and mode coupling becomes negligible. The different modes propagate independently of each other.

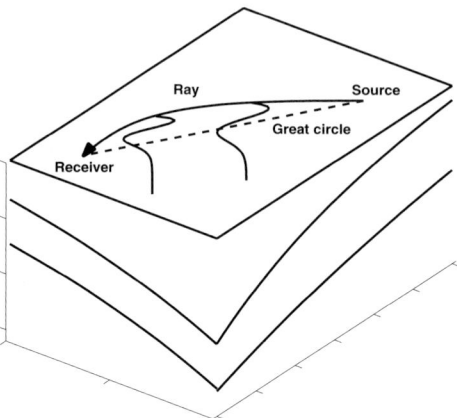

FIG. 7. Principle of surface wave ray-tracing. Three-dimensional plot of laterally varying layered halfspace. The location of a source, of a receiver, the great-circle connecting them (dashed line) and the ray followed by a mode at a given period are shown on the upper surface of the model. The mode eigenfunction, which adapts to the local structure along the ray, is also shown below the ray.

In models having three-dimensional variations, one can show that modes follow rays which can be traced in the two-dimensional horizontal plane using local phase velocity maps (Woodhouse, 1974). Since phase velocity is frequency-dependent, it is important to note that the ray trajectories are frequency-dependent. The amplitude of the displacement at the surface depends on two factors: the focusing or defocusing of the ray field, and the redistribution of energy with depth related to the deformation of the local eigenfunction along the ray. Figure 7 illustrates the propagation of a mode along a ray traced in the horizontal plane of a model with smooth heterogeneities. Notice that the eigenfunction changes along the ray, adapting to the local structure. This ray theory for surface waves assures conservation of energy. It is often referred to as adiabatic mode theory in the acoustic wave literature and to WKBJ (also noted WKB or JWKB) in seismology. A complete and rigorous treatment of this theory can be found in Tromp and Dahlen (1992).

The fact that surface waves follow rays in slowly varying heterogeneous structures has two important consequences on surface wave data: the surface waves do not necessarily arrive at a station along the great-circle path, and amplitudes need to be corrected for path effects before they can be used to analyse attenuation or sources.

In order to calculate deviations and amplifications in a given model, it is of course possible to trace the rays. But ray-tracing is time-consuming and not well-suited to implement in inverse methods. A very useful result is the fact that a number of ray attributes can be calculated at first-order as a function of the phase velocity and its derivatives along the great-circle path (Woodhouse and Wong,

1986). The most classic of these quantities is of course the travel time anomaly:

$$\delta t_m(\Delta) = -\frac{a}{c_m} \int_0^\Delta \frac{\delta c_m}{c_m} d\phi, \tag{16}$$

where we have kept the expressions of Woodhouse and Wong (1986) in spherical symmetry, with radius of the earth a, ϕ is the angular coordinate along the great-circle path and Δ the epicentral distance, c_m is the phase velocity of mode m in the reference structure and δc_m its perturbation along the great-circle. Let us note that Eq. (16) is at the base of the vast majority of the present tomographic methods for surface waves. It simply states that in a first approximation the phase velocity of a mode propagating between two locations is a linear mean of the phase velocity along the great circle path connecting these two locations.

The deviation at the receiver can be approximated similarly as a linear function of the model along the great-circle path. In this case, the variable which determines the data is the phase velocity derivative in the direction θ transverse to the great-circle:

$$\nu(\Delta) = -\int_0^\Delta \frac{\sin\phi}{\sin\Delta} \frac{1}{c_m} \frac{\partial \delta c_m}{\partial \theta} d\phi. \tag{17}$$

Note that, due to the $\frac{\sin\phi}{\sin\Delta}$ term, the transverse derivative close to the station influences the total deviation at the station more than the transverse derivative in the vicinity of the source. There is a similar expression for the deviation of the take-off angle at the source.

The variation in amplitude due to focusing or defocusing is, at first order:

$$\ln A = \frac{1}{2} \int_0^\Delta \frac{\sin(\Delta-\phi)}{\sin\Delta} \frac{1}{c_m} \left(\sin\phi \frac{\partial^2}{\partial\theta^2} - \cos\phi \frac{\partial}{\partial\phi} \right) \delta c_m \, d\phi. \tag{18}$$

In addition, the registered amplitude is affected by two elements. First, by the fact that the take-off angle at the source deviates from the great-circle path, necessitating a correction in the evaluation of the amplitude at the source according to the source radiation pattern. Secondly, following the variations of the eigenfunctions, the surface amplitude is affected by the fact that the energy is redistributed with depth along the ray. This factor can be very important. As an example, a Love wave fundamental mode at 40 s period, propagating in a smooth model from the oceanic structure shown in Fig. 5 to the continental one of the same figure, would see its amplitude increase by more than a factor of 2. This is due to the fact that, as shown in the right panel of Fig. 5, at constant energy flux, the surface amplitude is 0.4 in the continental mode, whereas it is smaller than 0.2 in the oceanic one.

Under the forms given above, arrival deviation and amplification are well-suited to implement in inverse methods. Comparison of exact ray-tracing and first-order approximations have been performed in realistic models of the Earth by Wang and Dahlen (1994) for isotropic models and by Larson et al. (1998) for anisotropic models. Wang and Dahlen (1994) analysed the rays of long-period (165 s) Rayleigh and Love waves first-arriving wavetrains as well as higher orbit wavetrains. For first-arriving wavetrains, they find that the phase anomalies are usually well-predicted by the first-order approximations. Take-off and arrival angle deviations reach up to 10 degrees in their models, but only those smaller than 5 to 6 degrees are well predicted by first-order evaluations. The same order of magnitude for the upper limit of deviations which are well modelled by first-order deviations is found by Larson et al. (1998) for higher frequency Rayleigh waves (35 s) in anisotropic models. In that case, the maximum deviations reach 20 degrees. Concerning amplitudes, note that the validity of the amplitude variations related to focusing/defocusing and calculated by ray theory, deteriorate rapidly when the paths get longer than a few tens of wavelengths (Wang and Dahlen, 1995). Compared to more exact methods, amplitude variations tend to be overestimated by ray theory due to the fact that smoothing finite-frequency effects, which leads to wavefront healing, cannot be taken into account. For shorter paths, Wang and Dahlen (1994) note that the effect of the redistribution with depth of the energy due to variation of the local eigenfunctions is an important factor which can account for 30% of the total amplitude anomalies for direct wavetrains. This is similar to the conclusions made by Levshin (1985) at a more regional scale.

Laske et al. (1994) and Laske and Masters (1998) have used the azimuthal deviation at the receiver to complement traveltime information in constraining tomographic models. Since deviations are linearly related to the gradients on the phase velocity maps, they are more sensitive than traveltimes to the short-wavelengths in the models and are therefore a very useful complementary dataset. In theory, one would need an array of stations to measure the direction of arrival of a wavetrain. Since arrays of long-period seismometers are very sparse, this deviation is measured in practice at single stations by measuring the polarization of the waves, and making the assumption that Rayleigh waves are polarized in the vertical propagation plane and the Love waves perpendicular to it. This assumption is valid in isotropic media and away from sharp heterogeneities. Refined methods are currently developed to take better into account the effect of the lateral heterogeneities in the interpretation of the polarization (Zhou et al., 2004, detailed in the next section) but the trade-off between the effect of isotropic lateral heterogeneities and the influence of anisotropy will probably be difficult to remove using single station recordings only.

Like body wave ray tracing, surface wave ray tracing can be extended beyond kinematic ray tracing to dynamic ray tracing or paraxial ray theory. The tools provided by paraxial ray theory to analyse the ray field in the vicinity of a central ray have been used by Yoshizawa and Kennett (2002) to analyse the width of the region around a surface wave ray where the phase of the wavefield is stationary. This provides an efficient way of calculating widths of rays in the general context of curved rays. Yoshizawa and Kennett (2002) argue that the effective width of rays using the stationary phase concept is about one third of the width of the first Fresnel zone, an evaluation which is still under debate. They provide examples of ray tracing and widths of rays at periods and epicentral distances appropriate for regional surface wave tomography.

4. Reference Modes

Another class of methods consists of using modes calculated in a reference structure to represent the wavefield. These methods are all based on a separation of the structure into a reference structure, which is laterally homogeneous, and superimposed heterogeneities, illustrated here with the decomposition of the elastic tensor:

$$C(\vec{x}) = C_0(z) + \epsilon \Delta C(\vec{x}), \tag{19}$$

where ϵ is used only to indicate which elements are related to the perturbation. The lateral heterogeneities act as secondary sources which couple the modes together and scatter energy away from their original propagation direction. This is illustrated in Fig. 8, which shows an example of surface wave scattering. An incident field, represented by its eigenfunctions with depth, is incident from the left in the structure. The interaction of the incident wavefield with the heterogeneity located in the model leads to different modes being diffracted from the heterogeneity. These can be Rayleigh as well as Love wave modes, fundamental

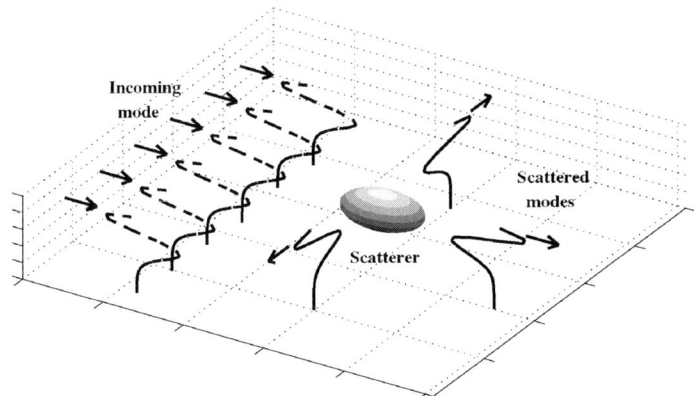

FIG. 8. Principle of scattering methods. Three-dimensional plot showing the incident wavefield (dashed eigenfunctions and associated arrows), the heterogeneity, and the modes scattered by the heterogeneity. Note that among the scattered modes, we have both converted (cross-branch coupling) and non-converted (self-coupling) modes.

and higher modes. As opposed to the local modes, the reference modes do not evolve with the structure. They constitute a basis of functions on which the total wavefield is represented. If a complete basis is used, some of the reference mode methods can handle both strong and extended heterogeneities. One should however be aware of the fact that coupling of reference modes has a different physical signification than coupling of local modes. Whereas coupling of local modes represents a real transfer of energy between modes, part of the coupling of reference modes is only a mathematical representation of the deformation of the eigenfunctions. Let us give an example to illustrate this.

Suppose that we have a Rayleigh wave fundamental mode propagating in a structure similar to the one in the lower panel of Fig. 2, with two identical homogeneous regions on the borders and an heterogeneous region in between. Let us suppose also that the eigenfunction of the Rayleigh wave fundamental mode varies significantly across the model, but that the transitions in between the different regions are very smooth. The smoothness ensures that the mode does not couple to other modes, and the coupling will be zero in the local mode coupling sense. If we now use reference modes in this context, it is natural to use the modes in the structure at the outskirts as reference modes. Due to the variation of the eigenfunction across the structure, the eigenfunction of the Rayleigh wave fundamental reference mode is not sufficient to represent properly the eigenfunction of the Rayleigh wave fundamental mode in the heterogeneous part of the model, and other modes need to be introduced there, leading to so-called coupling of the reference modes. The (reference) modal content in the heterogeneous structure does not represent the real modal content of the wavefield, but only how it can best be represented by the reference modes. When the wavetrain propagates out of the heterogeneous part of the structure, its eigenfunction returns to the reference one, and in case of smooth transition, the coupling (in the reference mode sense) to the other modes should vanish. This example illustrates the difference between the real mode coupling which occurs between local modes and the artificial one which arises using reference modes to represent a wavefield at any place. Consequently, reference mode methods need in general a larger number of modes than local mode methods in structures with similar degree of heterogeneity, and are more subject to numerical problems since the evolution of each eigenfunction needs to be followed numerically inside the structure, instead of analytically. This disadvantage is of course compensated by the fact that modes need to be calculated only in one structure and that reference mode methods are well-suited to develop inverse methods, as we will see.

4.1. Two-Dimensional Method of Kennett

The first general method developed to calculate the coupling of reference modes is due to Kennett (1984). This method for two-dimensional structures is reminiscent in many ways of the welded-spaces and local mode coupling methods of Sections 3.1 and 3.2. It is formulated with the same displacement-traction vectors, defined in Eq. (3), and uses the modal orthogonality (Eq. (6)) to derive the coupling equations. The main fundamental difference is that the wavefield is expressed on a fixed set of reference modes, calculated in a reference structure. Consequently, in order to keep a finite value to the coupling coefficients and a physical meaning to the mode amplitudes, the model should be equal to the reference structure at both ends of the two-dimensional structure, like in the lower panel of Fig. 2.

The wavefield is expressed as a sum of modes propagating forwards and backwards and having laterally varying amplitude coefficients:

$$u(\vec{x}, \omega) = \sum_m a_m(\omega, x) \vec{u}_m(z, \omega) \exp\left(i\omega \frac{x}{c_m(\omega)}\right). \tag{20}$$

Inserting into the equation of motion and using the modal orthogonality, one gets a first-order differential equation for the amplitudes coefficients:

$$\frac{da_n}{dx} = \sum_m [\vec{u}_n, \Delta C_{ijkl}, \vec{u}_m] \exp\left(i\omega\left(\frac{x}{c_m(\omega)} - \frac{x}{c_n(\omega)}\right)\right) a_m, \tag{21}$$

[, ,] is an integral with depth of a linear function of the eigenfunctions of modes m and n and of the deviation of the elastic coefficients, density and layer interfaces with respect to the reference structure. Note the similarity with Eq. (14) for local mode coupling. The main difference is the dependence of coupling on model deviations instead of model lateral derivatives and the absence of factor depending on wavenumber difference between the two modes.

These equations were recast by Kennett (1984) into differential equations for generalized transmission and reflection coefficients, a technique for solving differential equations called invariant embedding technique. These coefficients, a generalization of the reflection and transmission coefficients across a vertical boundary, express the relation between the modal content of the incident wavefield, the modal content of the wavefield which has propagated through the heterogeneous region, and the modal content of the wavefield reflected from the structure. As discussed above, the modal content inside the heterogeneous structure has no physical signification.

Let us note that this method includes multiple reflections inside the structure, is not limited to small heterogeneities, and gives an exact solution if the modal set is complete. In practice, approximations and limitations arise because one has to work with a truncated modal set, as discussed in Maupin and Kennett (1987), limiting how much the wavefield may vary within the model. One could consider combining this method with a welded quarter-spaces approach at one end of the two-dimensional model to increase the range of models it can be applied to.

4.2. Multiple-Scattering

The most powerful methods for mode coupling in three-dimensional structures today use reference modes and are based on scattering theory. In these methods, the modal orthogonality is not used explicitly, but is implicitly introduced by exploiting the dyadic form of the surface wave Green's function, where mode eigenfunctions appear twice in the expression. Its far-field form, given in Snieder (1986), is:

$$G(\vec{x}, \vec{\xi}, \omega) = \sum_m \vec{f}_m(x_3) \frac{\exp(ik_m r + i\pi/4)}{\sqrt{(\pi/2)k_m r}} \vec{f}_m^{*T}(\xi_3), \qquad (22)$$

where $\vec{f}_m$ is the displacement with depth of mode m, already defined in Eq. (1), and r is the distance between source and station. Here also, as in the expressions for two-dimensional geometry presented previously, the modes are normalized with respect to the flux of energy. An equivalent expression valid in the near-field is given in Friederich et al. (1993). Using the decomposition of the structure given in Eq. (19), the equation of elastic motion:

$$\mathcal{L}(\vec{v}) = \vec{F}, \qquad (23)$$

where $\vec{F}$ is the source factor, can be recast in a form which involves on the l.h.s. the equation of motion in the reference structure:

$$\mathcal{L}_0(\vec{v}) = \vec{F} - \Delta\mathcal{L}(\vec{v}). \qquad (24)$$

Using a decomposition of the wavefield into a Neumann series, that is a reference wavefield and perturbation terms at increasing orders of scattering:

$$\vec{v}(\vec{x}) = \vec{v}^{(0)}(\vec{x}) + \sum_{p=1}^N \epsilon^p \vec{v}^{(p)}(\vec{x}) \qquad (25)$$

Eq. (24) at power p in ϵ becomes:

$$\mathcal{L}_0(\vec{v}^{(p)}) = -\Delta\mathcal{L}(\vec{v}^{(p-1)}). \qquad (26)$$

This recurrence equation can be used to calculate $v^{(p)}$, the scattered field at order p from its value at order $(p-1)$ using the Green's function in the reference structure:

$$\vec{v}^{(p)} = -\int G_0 \Delta\mathcal{L}(\vec{v}^{(p-1)}) \, d\vec{x}. \qquad (27)$$

This expresses simply that the wavefield scattered p times arises from the interaction with the heterogeneities of the wavefield scattered $(p-1)$ times. Applied to surface waves, with wavefields and Green's function decomposed on reference modes, the very general volume integral of Eq. (27) decomposes into products of surface integrals and depth integrals. Defining $\Phi_m^{(p)}(x, y)$, which expresses the variation in the horizontal plane of the amplitude and phase of mode m at order p, Eq. (27) can be decomposed into equations for the amplitude and phase of the different modes m at order p:

$$\Phi_m^{(p)}(x, y) = \sum_n \int \mathcal{K}_{mn}(x, y, x', y') \Phi_n^{(p-1)}(x', y') \, dx' \, dy' \qquad (28)$$

which expresses that the amplitude of mode m at order p depends on the amplitudes of all the modes n at order $(p-1)$ and how they are coupled to mode m by the heterogeneities. The coupling between modes n and m depends on the coupling kernel:

$$\mathcal{K}_{mn}(x, y, x', y') = f\left(H_\nu(k_m r), \theta, \int [\vec{f}_m, \Delta C, \vec{f}_n] \, dz\right), \qquad (29)$$

where r and θ are the horizontal distance and azimuth of the receiver (x, y) as seen from the horizontal location (x', y') of the heterogeneity. $\mathcal{K}_{mn}(x, y, x', y')$ expresses how the heterogeneity located at (x', y') transfers energy from mode n to mode m, and how this energy is seen at (x, y). It is a function of three main terms: Hankel functions H_ν at orders 0 to 2 which express the propagation from (x', y') to (x, y), a term expressing the azimuthal dependence of the coupling, and integrals with depth of linear functions of the heterogeneity and eigenfunctions of modes n and m at position (x', y'). These last integrals are a three-dimensional generalisation of a similar integral introduced in the invariant embedding technique in Eq. (21). The coupling terms have an azimuthal variation in θ and 2θ in addition to a small isotropic one for the Rayleigh–Rayleigh coupling. Rayleigh–Love coupling has a node in the forward direction for isotropic heterogeneities, but not for anisotropic ones.

Multiple-scattering, in particular the Love–Rayleigh coupling which occurs in combination with anisotropic heterogeneities, has been studied by Maupin (2001). Although the disadvantage of the Neumann series is that phase shifts have to be accounted for by several orders of multiple scattering, the method proves efficient in calculating wavefields in rather strongly laterally heterogeneous models. Let us note however that, as opposed to the local mode methods described in the previous section, there is nothing in the expressions which guarantees that the total energy is conserved in scattering methods.

Scattering of elastic waves is not isotropic and has a tendency to be largest in the forward direction. Taking advantage of that, Friederich et al. (1993) and Marquering and Snieder (1995) have developed efficient forward modelling schemes which take into account multiple scattering in the forward direction, but only single scattering backwards. The calculation of the wavefield is done iteratively following the advance of the main wavefront, which is updated at each propagation step by addition of the forward scattered field. Compared to full multiple-scattering methods, where the wavefield has to be computed in the whole structure as many times as one needs to include of scattering orders, forward multiple-scattering requires the calculation of the wavefield only once in each part of the model and is therefore much more efficient from a numerical point of view. Building on such a forward multiple-scattering scheme, Friederich and Wielandt (1995) and Friederich (1998) have developed a tool to invert the vertical component of the Rayleigh wave fundamental mode under a regional array of stations. An important element in that inversion is the fact that the incident wavefield is not assumed plane, but its characteristics are inverted together with the characteristics of the structure. The effect of heterogeneities located between the source and the array of stations, like the deviation from great-circle path, amplitude variations and the non-plane nature of the incoming wavefield, can thereby be accounted for and do not bias the model under the array. The inversion scheme takes into account the effect of the structural heterogeneities on the wavefield via multiple-scattering of the fundamental mode onto itself, but, as opposed to the forward modelling scheme mentioned above, cannot account for coupling with higher modes or Love waves. It is therefore restricted to inversion of vertical components of Rayleigh fundamental modes, which are shown not to be severely contaminated by mode coupling in current situations (Friederich et al., 1993).

4.3. Single-Scattering Approximation

A number of studies, especially those related to inversion, rely on the single-scattering approximation. The expression for single-scattering of surface waves in three-dimensional structures and in the far-field approximation were given by Snieder (1986) and reviewed in Snieder (2002). Although they are a subset of Eq. (28) given above, they are more easy to analyse due the simplification related to the single-scattering and the far-field formulation. For scatterers distributed with depth but located at a single horizontal position (x', y'), the total field at $\vec{x}$ can be written as:

$$v(\vec{x}, \omega) = \sum_m \vec{f}_m(x_3) \frac{\exp(ik_m r + i\pi/4)}{\sqrt{(\pi/2)k_m r}}$$
$$\times \sum_n V_{mn}(x', y') \frac{\exp(ik_n r' + i\pi/4)}{\sqrt{(\pi/2)k_n r'}} \vec{f}_n^{*T}(\xi_3) \cdot \vec{F}, \qquad (30)$$

where

$$V_{mn}(x', y') = f\left(\theta, \int [\vec{f}_m, \Delta C, \vec{f}_n] dz\right) \qquad (31)$$

r is the horizontal distance from the scatterer to the receiver and r' the horizontal distance from the source to the scatterer. $V_{mn}(x', y')$ expresses how the scatterer couples energy from mode n to mode m. Compared with the coupling term $\mathcal{K}_{mn}$ in Eq. (29), it contains the same integrals with depth [,] and the same azimuthal dependence of the scattering, which is mainly in θ and 2θ, with an additional small isotropic component. But in this case, due to the far-field approximation, the propagation can be singled out in exponential functions. Reading from left to right, Eq. (30) expresses first the excitation of mode n by the point force $\vec{F}$, its propagation from the source to the scatterer, its coupling V to mode m at the scatterer, and then propagation of mode m from the scatterer to the receiver. For several or larger scatterers, this expression must contain in addition a sum or an integral over the horizontal location of the heterogeneities.

Since single-scattering theory provides a linear relation between heterogeneities and wavefield perturbation, it is well suited to develop inverse methods. Snieder (1986) attempted a surface wave tomography of the European lithosphere based on the single-scattering theory described above. More recent studies have refined the way surface wave scattering can be used to improve surface wave tomographic procedures.

4.3.1. Coupling of One Mode to Itself

The first class of single-scattering methods neglects coupling between modes and considers only coupling of one mode onto itself. This coupling can be shown to be directly related to the phase shifts or, equivalently, to the variations in phase velocity of the mode. Using the variational principle to calculate at first order in heterogeneity the variation δc in the local phase velocity of a mode (Takeuchi and Saito, 1972), one can show that the coupling of one mode onto itself at a given location and in the forward direction is directly related to the phase velocity variation:

$$V_{mm}(x', y') = \frac{-\delta c_m(x', y')}{c_m} \frac{k_m^2}{2} \qquad (32)$$

provided the modal eigenfunctions are normalized as in Eq. (22).

Another way to show the relation between the forward scattered field and the phase velocity perturbation is to express, as in Snieder (1986), the field scattered by a band-like heterogeneity of width D located perpendicularly to the wave propagation path. Taking into account only the coupling term of mode m onto itself, the first order scattered field is:

$$u_1 = i \delta k_m D u_0 \qquad (33)$$

which is equivalent to the first-order term in the expansion of the phase variation $\exp(i \delta k_m D)$.

In several recent studies, in particular in Spetzler *et al.* (2002) and Ritzwoller *et al.* (2002), these relations have been used to analyse the width of the region which influences the phase of a surface wavetrain. Using scattering theory, it is possible to calculate the phase shift introduced on the main wavetrain by the presence of an heterogeneity located on or partly away from the great circle path. The influence of heterogeneities located at different places laterally and at depth can be used to define a three-dimensional sensibility function called kernel. It is thereby possible to map the zone around the great circle path which effectively affects the phase measurement for a given source-station geometry, thus generalising the concept of Fresnel zone. The widths of these zones increase with period and with path length. In an application to fundamental Love waves at 150 and 40 s period, Spetzler *et al.* (2002) show how this concept can be used to improve the resolution of the models obtained with surface wave tomography. The models obtained by Ritzwoller *et al.* (2002) with a simplified version of this method applied to Rayleigh waves down to 20 s period show higher contrasts in the short wavelength band than those obtained with classical tomographic methods based on ray theory. Levshin *et al.* (2005) also use the zone of sensitivity derived in Spetzler *et al.* (2002) to analyse finite-frequency effects on major arc Rayleigh waves at periods of 50 and 100 s. The interaction of major arc surface waves with the Earth's structure is more extensive than that of minor arc surface waves. This reflects in the more complicated structure of the wavetrains, which are usually considered as more noisy, although this noise is related to the structure and should of course ideally be considered as an interesting source of information. Levshin *et al.* (2005) show that, using in the inversion the zones of sensitivity described above, major arc surface waves bring data which are precise enough to contribute to increase the resolution in regions where coverage by minor arc data is poor.

4.3.2. Coupling Between Different Modes

In addition to taking into account the coupling of one mode onto itself as in the examples above, it is possible to take into account the coupling between modes. The coupling between modes is generally speaking less important for the Rayleigh wave fundamental mode than for the other modes, as mentioned earlier. Coupling to Love waves is however not completely negligible even for Rayleigh wave fundamental modes at long periods. Zhou *et al.* (2004) show that the kernels of the phase velocities of the Rayleigh wave fundamental modes calculated using single-scattering for finite-frequency bands around 166 s period are affected by coupling to Love waves. The coupling is not important for the gross features of the kernels and can therefore be neglected if one wishes to retrieve images with the resolution of current global studies. The details, which are affected by the coupling, will however become important when we will start to try to retrieve smaller scale features, like subduction zones, from the phase of the surface waves.

An important point is although that it is only by taking into account inter-mode coupling that one can reproduce the fact that the sensitivity kernels of surface wave phase velocities have, in some cases, significant lateral variations along the propagation path. How inter-mode coupling renders the sensitivity kernels laterally dependent is illustrated in Fig. 9. The positions, at a given time, of two scattered modes are shown for two different locations of an heterogeneity along the propagation path. The location of the scattered mode m, which results from

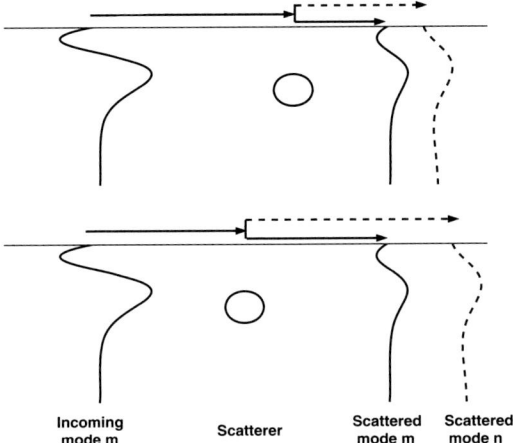

FIG. 9. Vertical cross-sections showing the influence of the horizontal location of a scatterer on the scattered wavefield. The incoming mode m is shown at the position it has at a given time t_0. The positions of the scattered modes are shown at a common later time, after interaction of the incoming mode with the scatterer, shown as a small circle in the middle of the figure. The scattered mode originating from self-coupling of the incoming mode is shown as a solid line, whereas the scattered mode which originates from cross-branch coupling is shown as a dashed line. Note that the location of the scattered mode m does not depend on the location of the heterogeneity, whereas the location of mode n does.

self-coupling of incident mode m when it interacts with the heterogeneity, does not depend on the horizontal location of the heterogeneity. This is because the energy in scattered mode m has propagated with the same velocity before and after interaction with the heterogeneity. It may have a phase shift with the main wavetrain, but this will only be due to the fact that the interaction itself can produce a phase shift (V are complex matrices, implying that mode coupling can introduce phase shifts), and will not be related to the propagation itself. On the other hand, the position of scattered mode n depends on the horizontal location of the heterogeneity. In the present case, mode n has a higher velocity than mode m and will arrive at the station earlier for heterogeneities located farther from the station than for those located close to the station. Considering only one mode and self-coupling, the lateral position of the heterogeneity along the source-receiver path does not influence how much this heterogeneity modifies the phase of the total wavefield, and sensitivity kernels are laterally homogeneous. On the other hand, if coupling between different modes is accounted for, the sensitivity kernels become laterally dependent. When higher modes are used to model body waves like S or SS waves, the kernels calculated taking mode coupling into account get a laterally varying form which actually follows the rays of the body-wave phases formed by the modal sum. This proves to be crucial to develop adequate sensitivity kernels when inverting long period S or SS waves with mode methods, as shown in Li and Tanimoto (1993), Marquering *et al.* (1998) and Zhao and Jordan (1998). The importance of coupling between different modes depends strongly on which part of the wavetrain one wants to invert.

Coupling is also more important at short periods than at longer periods. This has been shown quantitatively by Meier *et al.* (1997a) who have developed a very interesting approach to surface wave tomography, combining a first inversion based on simple ray theory or WKBJ approximation (Eq. (16)) for the mean model along each source-station path and a single scattering approach to invert the residual field. They also show that the long periods are well modelled by the first part of the inversion whereas the second inversion serves to improve the fit of the shorter periods. Combining WKBJ, which takes properly into account the phase of each mode, with single-scattering is an interesting way of avoiding the problems mentioned in Section 4.2, which is that accounting for large phase-shifts with purely reference mode methods necessitates several orders of scattering. Combinations of local modes approaches and scattering are probably the best approach to develop efficient direct and inverse methods for surface waves in three-dimensional models.

In addition to the kernels attached to the phases, scattering theory can be used to calculate other kernels, like those which relate the heterogeneities to the deviation of the polarization at the station. Zhou *et al.* (2004) show that these kernels are more affected by coupling than the phase kernels. The kernel they show for the polarization of a long-period Rayleigh wave has strong secondary lobes close to the station when coupling to Love modes is taken into account. This is not surprising since coupling from Rayleigh to Love waves at the vicinity of the receiver will introduce a transverse component arriving close in time to the main Rayleigh wavetrain, and that the polarization of the total recorded Rayleigh wave will be affected by its deviation from the great circle and its interference with the induced Love wave. This kind of interference cannot be accounted for with ray theory, used up to now to interpret the polarization deviations, and it would certainly be important to take into account inter-mode coupling in deviations analysis.

5. Conclusion and Perspectives

We have only presented the coupling methods developed in the surface wave formalism. For the longest periods in the Earth, or for problems where it is important to take into account the Earth's sphericity, mode coupling in the free oscillation formalism is better suited. The methods developed in this formalism have a number of similarities with those exposed in this chapter, but there are also a number of significant differences. To our knowledge, free oscillation mode coupling is only developed in the framework of reference modes. In general, the methods in which the wavefield is expressed in the frequency domain

(Friederich, 1999, for example) are closer to those developed in the surface wave formalism than those which are developed in the time domain or using Laplace transforms. In those cases, some approximate results can be derived for "short" times, or first-arriving wavetrains. The mode coupling discussed in this chapter is called "cross-branch" coupling in the free oscillation formalism. Since free oscillations have discrete eigenfrequencies along the dispersion branches, coupling in this framework can also signify the coupling of modes along the same dispersion branch. This along-branch coupling is another way to express the self-coupling of the modes in the surface wave formalism. Reduced to its simpler expression, it gives the "location-parameter" defined by Jordan (1978). A recent review of mode coupling methods in the framework of free oscillation theory is given in Lognonné and Clévédé (2002).

Considerable progress has been done concerning the modelisation of surface waves in laterally heterogeneous structures in the last two decades. Most methods use either local modes or reference modes. Local modes adapt to the structure and should therefore lead to more robust modelisation methods. However, we do not have yet a complete and numerically tractable expression for the coupling of the local modes in three-dimensional structures. The fact that coupling of local modes reduce to surface wave ray tracing in slowly laterally varying structures give us some clues as to the direction we should take: combining ray tracing with mode coupling is probably the way to go to develop a complete local mode coupling method in three-dimensions.

It is easier to tackle three-dimensional problems with reference modes. Forward multiple-scattering and single-scattering, despite their limited validity in strongly heterogeneous models, are very efficient from a numerical point of view and can be used in inverse methods. They require however that the modes eigenfunctions do not vary too much across the study area, and this will probably limit their application at large scales or at higher frequencies. A combination of local modes, or simply ray-tracing, with scattering by smaller heterogeneities, as proposed by Meier *et al.* (1997a), seems a good pragmatic approach to inverse problems in three-dimensions. Going to higher frequencies, taking into account finite-frequency effects, using the information contained in the wave amplitudes and in their polarizations, using higher modes, are some of the challenges than surface wave tomography faces today. All of them require taking into account mode coupling in the inversion procedures and a better understanding of the effect it has on the observed surface wavefields.

REFERENCES

Bostock, M.G. (1991). Surface wave scattering from 3-D obstacles. *Geophys. J. Int.* **104**, 351–370.
Bostock, M.G., Kennett, B.L.N. (1992). Multiple scattering of surface wave from discrete obstacles. *Geophys. J. Int.* **108**, 52–70.

Debayle, E., Lévêque, J.-J. (1997). Upper mantle heterogeneities in the Indian Ocean from waveform inversion. *Geophys. Res. Lett.* **24**, 245–248.
Chen, X. (2006). Generation and propagation of seismic SH waves in multi-layered media with irregular interfaces. In: *Advances in Wave Propagation in Heterogeneous Media. In Advances in Geophysics*, vol. 48, p. 191 (this book).
Du, Z. (2002). Waveform inversion for lateral heterogeneities using multimode surface waves. *Geophys. J. Int.* **149**, 300–312.
Dziewonski, A.M., Anderson, D.L. (1981). Preliminary reference Earth model. *Phys. Earth Planet. Int.* **184**, 297–356.
Friederich, W. (1998). Wave-theoretical inversion of teleseismic surface waves in a regional network: Phase-velocity maps and a three-dimensional upper-mantle shear-wave-velocity model for southern Germany. *Geophys. J. Int.* **132**, 203–225.
Friederich, W. (1999). Propagation of shear wave and surface waves in a laterally heterogeneous mantle by multiple forward scattering. *Geophys. J. Int.* **136**, 180–204.
Friederich, W., Wielandt, E. (1995). Interpretation of seismic surface waves in regional networks: Joint estimation of wavefield geometry and local phase velocity. *Geophys. J. Int.* **120**, 731–744.
Friederich, W., Wielandt, E., Stange, S. (1993). Multiple forward scattering of surface waves: Comparison with an exact solution Born single-scattering methods. *Geophys. J. Int.* **112**, 264–275.
Gaherty, J.B., Kato, M., Jordan, T.H. (1999). Seismological structure of the upper mantle: A regional comparison of seismic layering. *Phys. Earth Planet. Int.* **110**, 21–41.
Its, E., Yanovskaya, T.B. (1987). Propagation of surface waves in a half-space with vertical, inclined or curved interfaces. *Wave Motion* **7**, 79–84.
Jordan, T.H. (1978). A procedure for estimating lateral variations for low-frequency eigenspectra data. *Geophys. J. R. Astron. Soc.* **52**, 441–465.
Kennett, B.L.N. (1983). Seismic Wave Propagation in Stratified Media. Cambridge Univ. Press, Cambridge.
Kennett, B.L.N. (1984). Guided wave propagation in laterally varying media-I. Theoretical development. *Geophys. J. R. Astron. Soc.* **79**, 235–255.
Kennett, B.L.N. (1998). Guided waves in three-dimensional structures. *Geophys. J. Int.* **133**, 159–174.
Laske, G., Masters, G. (1998). Surface-waves polarization data and global anisotropic structure. *Geophys. J. Int.* **132**, 508–520.
Laske, G., Masters, G., Zürn, W. (1994). Frequency-dependent polarization measurements of long-period surface waves and their implications for global phase velocity maps. *Phys. Earth Planet. Int.* **84**, 111–137.
Larson, E.W.F., Tromp, J., Ekström, G. (1998). Effects of slight anisotropy on surface waves. *Geophys. J. Int.* **132**, 654–666.
Levshin, A.L. (1985). Effects of lateral inhomogeneities on surface wave amplitude measurements. *Ann. Geophys.* **3**, 511–518.
Levshin, A.L., Barmin, M., Ritzwoller, M., Trampert, J. (2005). Minor-arc and major-arc global surface wave diffraction tomography. *Phys. Earth Planet. Int.* **149**, 205–223.
Levshin, A.L., Yanovskaya, T.B., Lander, A.L., Bukchin, B.G., Barmin, M.P., Ratnikova, L.P., Its, E. (1989). Seismic Surface Waves in a Laterally Inhomogeneous Earth. In: Keilis-Borok, V.I. (Ed.), Kluwer, Dordrecht.
Li, X.D., Tanimoto, T. (1993). Waveforms of long-period body waves in a slightly aspherical Earth model. *Geophys. J. Int.* **112**, 92–102.
Lognonné, P., Clévédé, E. (2002). Normal modes of the Earth and planets. In: *International Handbook of Earthquake and Engineering Seismology*, vol. 81A, pp. 125–147. Int'l Assoc. Seismo. and Phys. Earth's Interior, committee on education.
Malischewsky, P. (1987). Surface Waves and Discontinuites. Akademie Verlag, Berlin.
Marquering, H., Snieder, R. (1995). Surface-wave mode coupling for efficient modelling and inversion of body-wave phases. *Geophys. J. Int.* **120**, 186–208.

Marquering, H., Nolet, G., Dahlen, F.A. (1998). Three-dimensional waveform sensitivity kernels. *Geophys. J. Int.* **132**, 521–534.

Maupin, V. (1988). Surface waves across 2-D structures: a method based on coupled local modes. *Geophys. J.* **93**, 173–185.

Maupin, V. (1992). Modelling of laterally trapped surface waves with application to Rayleigh waves in the Hawaiian swell. *Geophys. J. Int.* **110**, 553–570.

Maupin, V. (1996). The radiation modes of a vertically varying half-space: a new representation of the complete Green's function in terms of modes. *Geophys. J. Int.* **126**, 762–780.

Maupin, V. (2001). A multiple-scattering scheme for modelling surface wave propagation in isotropic and anisotropic three-dimensional structures. *Geophys. J. Int.* **146**, 332–348.

Maupin, V., Kennett, B.L.N. (1987). On the use of truncated modal expansions in laterally varying media. *Geophys. J. R. Astron. Soc.* **91**, 837–851.

Meier, T., Malischewsky, P.G. (2000). Approximation of surface wave mode conversion at a passive continental margin by a mode-matching technique. *Geophys. J. Int.* **141**, 12–24.

Meier, T., Lebedev, S., Nolet, G., Dahlen, F.A. (1997a). Diffraction tomography using multimode surface waves. *J. Geophys. Res.* **102**, 8255–8267.

Meier, T., Malischewsky, P.G., Neunhofer, H. (1997b). Reflection and transmission coefficients of surface waves at a vertical dicontinuity and imaging of lateral heterogeneity using reflected fundamental Rayleigh waves. *Bull. Seismol. Soc. Am.* **87**, 1648–1661.

Panza, G.F., Romanelli, F., Vaccari, F. (2000). Realistic modelling of waveforms in laterally heterogeneous anelastic media by modal summation. *Geophys. J. Int.* **143**, 340–352.

Ritzwoller, M.H., Shapiro, N.M., Barmin, M.P., Levshin, A.L. (2002). Global surface wave diffraction tomography. *J. Geophys. Res.* **107**, 2235, doi:10.1029/2002JB001777.

Romanelli, F., Bing, Z., Vaccari, F., Panza, G.F. (1996). Analytical computation of reflection and transmission coupling coefficients for Love waves. *Geophys. J. Int.* **125**, 132–138.

Snieder, R. (1986). 3-D linearized scattering of surface waves and a formalism for surface wave holography. *Geophys. J. R. Astron. Soc.* **84**, 581–605.

Snieder, R. (2002). Scattering of surface waves. In: Pike, R., Sabatier, P. (Eds.), *Scattering and Inverse Scattering in Pure and Applied Science*. Academic Press, San Diego, pp. 528–542.

Spetzler, J., Trampert, J., Snieder, R. (2002). The effect of scattering in surface wave tomography. *Geophys. J. Int.* **149**, 755–767.

Stange, S., Friederich, W. (1992). Guided wave propagation across sharp lateral heterogeneities: the complete wavefield at plane vertical discontinuities. *Geophys. J. Int.* **109**, 183–190.

Takeuchi, H., Saito, M. (1972). Seismic surface waves. In: *Methods in Computational Physics*, vol. 11. Academic Press, New York, pp. 217–295.

Tromp, J. (1994). A coupled local-mode analysis of surface-wave propagation in a laterally heterogeneous waveguide. *Geophys. J. Int.* **117**, 153–161.

Tromp, J., Dahlen, F.A. (1992). Variational principles for surface wave propagation on a laterally heterogeneous Earth-II. Frequency-domain JWKB theory. *Geophys. J. Int.* **109**, 599–619.

Wang, Z., Dahlen, F.A. (1994). JWKB surface-waves seismograms on a laterally heterogeneous Earth. *Geophys. J. Int.* **119**, 381–401.

Wang, Z., Dahlen, F.A. (1995). Validity of surface wave ray theory on a laterally heterogeneous Earth. *Geophys. J. Int.* **123**, 757–773.

Woodhouse, J.H. (1974). Surface waves in a laterally varying layered structure. *Geophys. J. R. Astron. Soc.* **37**, 461–490.

Woodhouse, J.H., Wong, Y.K. (1986). Amplitude, phase and path anomalies of mantle waves. *Geophys. J. R. Astron. Soc.* **87**, 753–773.

Yoshizawa, K., Kennett, B.L.N. (2002). Determination of the influence zone for surface wave paths. *Geophys. J. Int.* **149**, 440–453.

Zhao, L., Jordan, T.H. (1998). Sensitivity of frequency-dependent traveltimesto laterally heterogeneous, anisotropic Earth structure. *Geophys. J. Int.* **133**, 683–704.

Zhou, Y., Dahlen, F., Nolet, G. (2004). Three-dimensional sensitivity kernels for surface wave observables. *Geophys. J. Int.* **158**, 142–168.

BOUNDARY INTEGRAL EQUATIONS AND BOUNDARY ELEMENTS METHODS IN ELASTODYNAMICS

MICHEL BOUCHON[1] AND FRANCISCO J. SÁNCHEZ-SESMA[2]

[1] Université Joseph Fourier, IRIGM, BP 53, F-38041 Grenoble, France
[2] Institute of Geophysics, Mexico National Autonomous University (UNAM), Mexico

ABSTRACT

We review the application of boundary integral equation (BIE) methods to elastic wave propagation problems. BIE methods express the wavefield as an integral equation defined over the boundary of the domain studied. They can be grouped into two families. "Direct" BIE relate the wavefield (generally the displacements and tractions) to the values of the wavefield at the boundary of the domain, while "Indirect" BIE rely on an intermediate unknown, which is usually a distribution of fictitious sources along the boundary. We present the mathematical bases of the methods and discuss and review their various numerical implementations. We illustrate some of the applications for seismic wave propagation.

1. INTRODUCTION

In several areas of physics, like solid mechanics, electromagnetism, fluid dynamics, or thermodynamics, the resolution of a problem can be reduced to an integral equation defined over the boundary of the domain studied. This expresses the fact that the totality of the information pertaining to the domain is carried by the boundary. Apparently this remarkable fact was discovered by G. Green although some attribute the finding to C.F. Gauss. In any event, the Boundary Integral Equations (BIE) formulate the solution of a problem in terms of values at the domain's boundary. BIE methods arise from potential theory and can be grouped into two families. "Direct" BIE relate the physical variables (the wavefield of displacements and tractions in elastodynamics) in a given domain to the values taken by these variables at the boundary. "Indirect" BIE, on the other hand, rely on an intermediate unknown, which is usually a distribution of fictitious (or apparent) sources along the boundary.

The resolution of the direct or indirect BIE must be done numerically. The first step consists in discretizing the boundary into a set of points or into a set of elements. In the latter case, the BIE method is often referred to as a Boundary

Element Method (BEM) while the indirect BIE may be named as indirect BEM (IBEM). In elastodynamics, which is our domain of interest here, the second step is the evaluation of the medium Green's functions. In doing so, one needs to avoid the inherent singularities of having to evaluate the wavefield at the source location itself. To this effect, various formulations of the BIE have been developed which are un-conditionally stable. The third and final step is to solve the resulting linear system of equations.

Compared to domain methods, like finite-differences or finite-elements, BIE methods have an interesting conceptual advantage which is the reduction of one space dimension for both discretization and handling of the unknowns. For instance, two-dimensional (2-D) problems are reduced to integral representations along lines (thus, this is equivalent to 1-D problems). This advantage is enhanced because the domain effectively considered can be much larger than in domain methods which require to make explicit such size. Moreover, BIE methods (in frequency) do not need absorbing boundaries. Indeed, we have to mention that BIE are more accurate than domain methods. They match more easily the boundary conditions and do not suffer from grid dispersion. They easily fulfill radiation conditions at infinity and can handle complex boundary geometries for which domain methods would require complicated grid setting. On the other hand, the benefits are counter-balanced by the global character of boundary approaches, each equation involves the full wavefield (at least in the considered sub domain), while in domain methods, only the local field values are related by the governing equations. In the BIE the field at each point of the boundary is linked to all boundary field values and, as a consequence, the linear system that needs to be solved is dense. For this reason, sometimes they are more demanding in computer time and memory space, because they require solving non-sparse linear systems.

As the BIE methods directly express the physics of the problem, they also provide potentially valuable insight into the problems they tackle. They cannot be easily designed to handle non-linear problems. They are relatively well adapted to study acoustic or seismic wave propagation in the shallow Earth which can often be described as consisting of relatively homogeneous layers separated by sharp boundaries.

After describing the direct and indirect BIE methods, we will review and discuss some salient numerical aspects of these techniques. We will also present and discuss other boundary methods, in which approximate expressions of the wavefield, such as plane wave decompositions or basis function expansions are used to set up boundary equations from which relatively fast, approximate solutions can be obtained.

Excellent reviews of the available literature on BEM in elastodynamics up to the early 90s are those by Manolis and Beskos (1988), Brebbia and Domínguez (1992), and Domínguez (1993). These researchers focused on various structural engineering applications. Herein the emphasis is laid on seismological aspects.

2. The Direct BIE

The formulation of the direct BIE in elastodynamics can be traced back to the pioneering work of Somigliana (1886), more than a century ago. It stems from the theory of elastic potentials and states that field values of a problem of linear elasticity in a given domain are fully determined from the displacements and tractions along the domain's boundary. It can be seen as an application of the Maxwell–Betti reciprocity theorem. Somigliana's work is based upon the classical work by Stokes (1849) who found the fundamental response (or Green's function) for a homogeneous, isotropic, elastic medium.

Considering a volume of elastic material V bounded by a surface S (see Fig. 1), the displacement field u_i at a point $\mathbf{x}$ and time t can be expressed as a function of the values of the displacement and traction t_j along the boundary through Somigliana's representation theorem:

$$cu_i(\mathbf{x},t) = \int_0^t d\tau \int_S \bigl[t_j(\xi,\tau)G_{ji}(\xi,t,\mathbf{x},\tau) - u_j(\xi,\tau)\Sigma_{jik}(\xi,t,\mathbf{x},\tau)n_k(\xi)\bigr] dS(\xi) \tag{1}$$

where $G_{ji}(\xi,t,\mathbf{x},\tau)$ and $\Sigma_{jik}(\xi,t,\mathbf{x},\tau)$ are the responses, in terms of displacement and stress, at time t and point ξ of an infinite homogeneous medium to a unit force impulse, applied at time τ at point $\mathbf{x}$ in the direction i. $n_k(\xi)$ is the normal to the boundary pointing outside V. The volumetric body sources are assumed to be null, but, if present, their contribution can easily be added. c is a constant that takes the values 1, 0.5 or 0 for $\mathbf{x}$, respectively, inside V, on S, or outside V, assuming that S has smooth boundaries. These values for c appear to be somewhat obscure, but they come from the volume integration of the product $\delta(\mathbf{x}-\xi, \mathbf{t}-\tau)\mathbf{u}_i(\xi,\tau)$ and the value accounts for the integration of the Dirac's delta and its position.

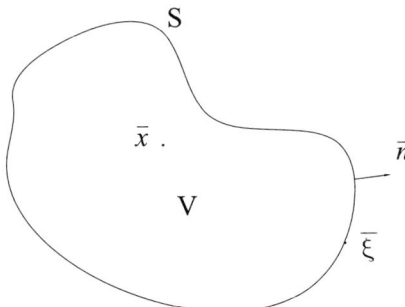

FIG. 1. Elastic space in which a closed surface S is defined. The vector ξ defines position along this surface. Domain V is the *internal* region. Position vector $\mathbf{x}$ within this region is depicted. Note that the normal vector $\mathbf{n}$ points toward domain V' which is the complement of V, or *external* region.

The Green's function for the homogeneous, isotropic, elastic medium can be written as (see e.g. Aki and Richards, 1980).

$$G_{ij}(\mathbf{x}, t, \xi, \tau) = \frac{1}{4\pi \mu r}\left(f_2 \delta_{ij} + (f_1 - f_2)\gamma_i \gamma_j\right) \qquad (2)$$

where, $\delta_{ii} = 1$ and $\delta_{ij} = 0$ if $i \neq j$, $\gamma_i = (x_i - \xi_i)/r$ is the unit vector from ξ to $\mathbf{x}$, and r is the distance between these points. The functions f_1 and f_2 are given by

$$f_1(r, t) = \left[\frac{\beta}{\alpha}\right]^2 \delta\left(t - \frac{r}{\alpha}\right) + \frac{2}{(r/\beta)^2} \int_{r/\alpha}^{r/\beta} \delta(t - \tau)\tau \, d\tau \qquad (3)$$

and

$$f_2(r, t) = \delta\left(t - \frac{r}{\beta}\right) - \frac{1}{(r/\beta)^2} \int_{r/\alpha}^{r/\beta} \delta(t - \tau)\tau \, d\tau. \qquad (4)$$

where α and β are the P and S wave velocities and μ is the shear modulus.

These functions clearly show in the first term of the left-hand side the far-source field terms which in 3-D have the same waveform as the applied force (in this case an impulse). f_1 is proportional to the radial displacement while f_2 is proportional to the transverse one. The integral in these expressions samples the slowness space between those of P and S waves.

If the load is applied permanently we have to integrate in time from zero to infinity and we have then $f_1 = 1$ and $f_2 = 1/2(1 + \beta^2/\alpha^2)$, which correspond to Kelvin's solution for the static Green function.

In the frequency domain the Fourier transform allows to find the corresponding expressions for f_1 and f_2. The functions thus obtained are regular and can be expressed using complex exponentials as described in Sánchez-Sesma and Luzón (1995). The fact of significance here is the singularity at the point of application of the load. It is of the type $1/r$, precisely the one for the static Kelvin problem.

The solution for the 2-D case is somewhat similar. We can express the corresponding solution in terms of radicals when in the time domain according to the solution by Eason et al. (1956). In the frequency domain the 2-D Green's function is expressed in terms of cylindrical Hankel functions. In 2-D the singularity at the point of application of the load is logarithmic. The reader is referred to Sánchez-Sesma and Campillo (1991) for details.

Noticing that the traction associated to the Green's function is given by $T_{ji}(\xi, t, \mathbf{x}, \tau) = \Sigma_{jik}(\xi, t, \mathbf{x}, \tau) n_k(\xi)$ then we can write:

$$cu_i(\mathbf{x}, t) = \int_0^t d\tau \int_S \left[t_j(\xi, \tau) G_{ji}(\xi, t, \mathbf{x}, \tau) - u_j(\xi, \tau) T_{ji}(\xi, t, \mathbf{x}, \tau)\right] dS(\xi). \qquad (5)$$

This equation is the basis for BIE. Discretizing the integral, the boundary integral equations can be transformed into a system of linear equations with the displacements and tractions at the boundaries as unknowns. Provided that sufficient boundary conditions are specified, this system can be solved. An important point here is that this procedure implies that the Green's functions have to be evaluated at the very location of the sources and that the singularity of the Green's functions at the sources have to be dealt with properly. For homogeneous medium problems, G_{ji} and T_{ji} are well known and their singularities can be handled with relative ease. For Green's displacements the singularity is weak and can be dealt with analytically. On the other hand, the singularity associated to Green's tractions should be considered in the Cauchy's principal value sense. Since integration in the sense of Riemann could diverge for this kind of function, it is necessary to implement a numerical approximation technique, in order to obtain the desired system of equations.

Applying Hooke's law to Eqs. (1) or (5), the stress field can be evaluated within the domain. However, if the stresses are required on S, like in a dynamic rupture problem or in the scattering of waves by a crack, special care is needed because the resulting kernel becomes hyper-singular. For clarity, if r is the distance between any point in the solid $\mathbf{x}$ and the point ξ, in which the unit load is applied, the different singularity levels are denoted as: the $(1/r)$ type of singularity "weakly-singular", the $(1/r^2)$ type of singularity "strongly-singular", and $(1/r^3)$ "hyper-singular".

3. THE INDIRECT BIE

The indirect formulation of the elastodynamic problem expresses the wavefield as an integral over the boundary of elementary source radiations:

$$u_i(\mathbf{x}, t) = \int_0^t d\tau \int_S \phi_j(\xi, \tau) G_{ij}(\mathbf{x}, t, \xi, \tau) \, dS(\xi) \tag{6}$$

where $\phi_j(\xi, \tau)$ denotes a force density distribution applied at time τ at point ξ of the surface S. In this equation we have again assumed that the volumic body sources are null. If present, their contribution simply needs to be added to the integral term.

An example of this formulation is given in Vai et al. (1999); they studied wave propagation in two-dimensional irregularity layered elastic media for internal linear sources. In Figs. 2 and 3, two different layered media considered in their article are presented with the corresponding synthetic seismograms. Results are compared with solutions obtained with the spectral element method giving a magnificent agreement.

Equation (6) can be seen as the mathematical transcription of Huygens' principle, which states that each point of a diffracting boundary acts as a secondary source of wave radiation. The principle applies to any wavefront as well. In Fig. 4

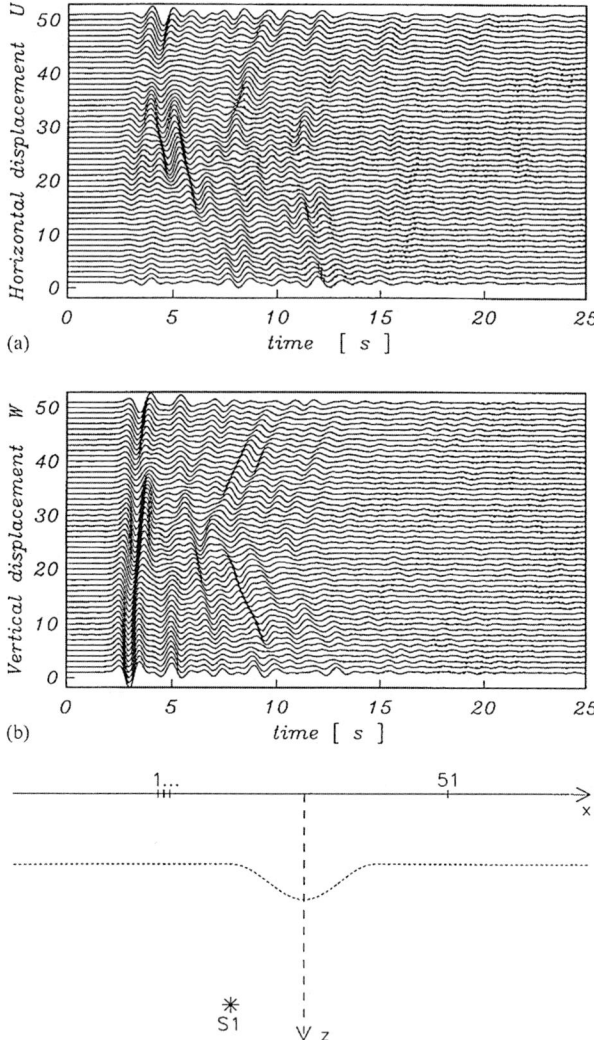

FIG. 2. Comparison between two methods: indirect boundary element method (continuous line) and spectral element method (dashed line). The problem configuration is shown at the bottom of the figure. The agreement between the traces is impressive (after Vai *et al.* (1999)).

a wavefront and its elementary sources' radiation are depicted (note that Huygens' principle implies the reconstruction of the forward wavefront and the cancellation of backward waves). Equation (6) is also the mathematical expression of the exploding reflector concept widely used in seismic exploration (e.g. Claerbout, 1985). The fictitious force distribution ϕ_j is an intermediate variable which needs

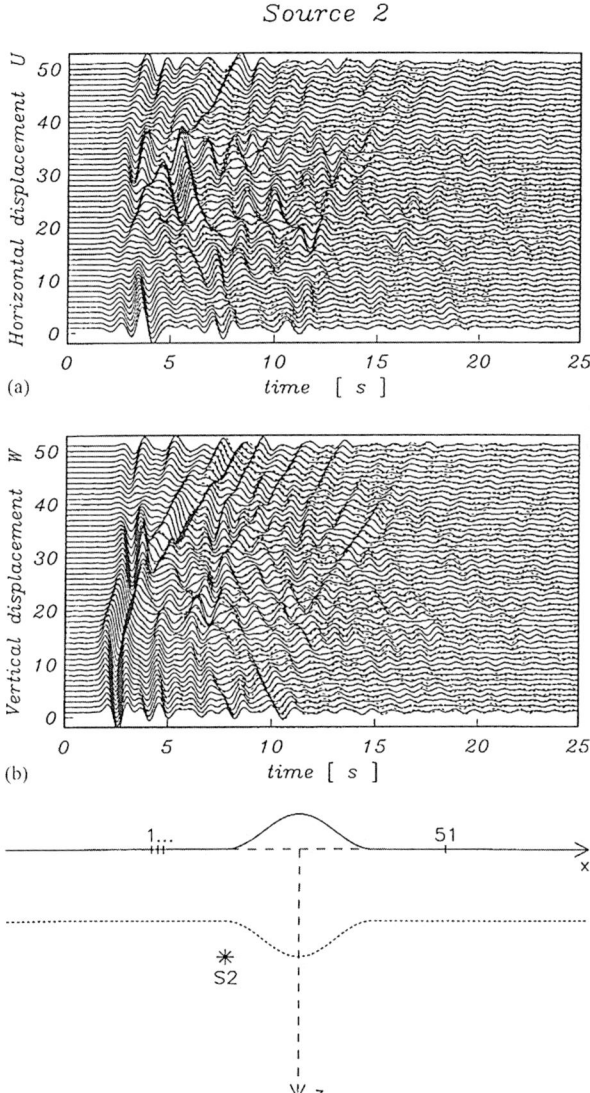

FIG. 3. (Bottom) Problem configuration and (top) corresponding synthetic seismograms. Results obtained with the indirect boundary element method (continuous line) are compared with those obtained with the spectral element method (dashed line). Edge effects introduce noise in the latter method, therefore, late phases and amplitudes are not precisely reproduced (after Vai *et al.* (1999)).

to be solved in the boundary integral equations before the wavefield can be computed.

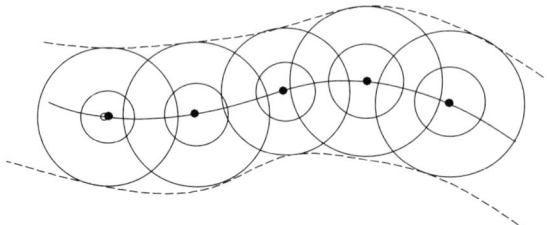

FIG. 4. Huygens' principle.

We now show that Eq. (6) can be derived from the representation theorem expressed by Eq. (1). In order to do this, let us first remove the time dependence and integration, which become implicit, for simplicity of the notations. With reference to Fig. 1, let us assume that $u'_j(\xi)$ is a solution of the exterior problem with boundary traction $t'_j(\xi)$. Let us assume also that the same elastic material occupies the interior and exterior regions. It is clear then that both interior and exterior regions have the same Green's function, so we can write

$$c' u'_i(\mathbf{x}) = -\int_S \left[t'_j(\xi) G_{ji}(\xi, \mathbf{x}) - u'_j(\xi) \Sigma_{jik}(\xi, \mathbf{x}) n_k(\xi) \right] dS(\xi), \qquad (7)$$

where c' is a complementary constant that takes the value $0, 0.5$ or 1 for $\mathbf{x}$, respectively, inside V, on S, or outside V. The negative sign of the equation is due to the fact that we keep the orientation of the normal to the boundary assumed earlier for the interior problem.

Summing up Eqs. (1) and (7), we have

$$cu_i(\mathbf{x}) + c' u'_i(\mathbf{x})$$
$$= \int_S \left[(t_j(\xi) - t'_j(\xi)) G_{ji}(\xi, \mathbf{x}) \right.$$
$$\left. - (u_j(\xi) - u'_j(\xi)) \Sigma_{jik}(\xi, \mathbf{x}) n_k(\xi) \right] dS(\xi). \qquad (8)$$

If we now impose that at the boundary $u_j(\xi) = u'_j(\xi)$, Eq. (5) becomes

$$cu_i(\mathbf{x}) + c' u'_i(\mathbf{x}) = \int_S \left[t_j(\xi) - t'_j(\xi) \right] G_{ji}(\xi, \mathbf{x}) \, dS(\xi). \qquad (9)$$

Taking into account that $c + c' = 1$, renaming $u'_i(\mathbf{x})$ as $u_i(\mathbf{x})$ in the exterior domain and using the symmetry of the Green's tensor, this can be expressed as:

$$u_i(\mathbf{x}) = \int_S \phi_j(\xi) G_{ij}(\mathbf{x}, \xi) \, dS(\xi) \qquad (10)$$

where $\phi_j(\xi) = t_j(\xi) - t'_j(\xi)$.

This last integral is now similar to Eq. (6), except for the implicitly assumed time dependence. This single-layer integral representation was studied in detail by Kupradze (1963). He showed that the displacement field is continuous *across* S if $\phi_j(\xi)$ is continuous *along* S. This is in agreement with our choice of $u_j(\xi) = u'_j(\xi)$. In fact, one can also impose instead continuous traction: $t_j(\xi) = t'_j(\xi)$, then $u_j(\xi) - u'_j(\xi)$ would be unknown and the displacement field would be expressed in terms of the stress Green's tensor. This is a usual approach in dealing with certain crack problems (e.g. Bonnet, 1989; Coutant, 1989). With our derivation, that closely follows that of Bonnet (1986), we have shown that the single-layer integral stems from Somigliana's representation theorem.

This integral representation allows computation of stresses and tractions by direct application of Hooke's law. However, when **x** is on the boundary, the operation requires particular care. From a limiting process based on equilibrium considerations around an internal neighborhood of the boundary, we can write, for **x** on S:

$$t_i(\mathbf{x}) = \pm \frac{1}{2}\phi_i(\mathbf{x}) + \int_S \phi_j(\xi) T_{ij}(\mathbf{x}, \xi) \, dS(\xi). \tag{11}$$

The first term of the right-hand side must be dropped if **x** is inside V. The signs $(+, -)$ are applied if **x** tends to S from (*inside, outside*). In Fig. 5 a free body diagram around an antiplane unit line load is depicted. We show the "interior" region as well. For symmetry reasons the boundary traction is one half of the antiplane load. Had the depicted regions been welded (including the load) the traction at the "boundary" would be minus one half the unit load. This result was found by Kupradze (1963) and for potential problems by Fredholm as early as 1900 (see Webster, 1955).

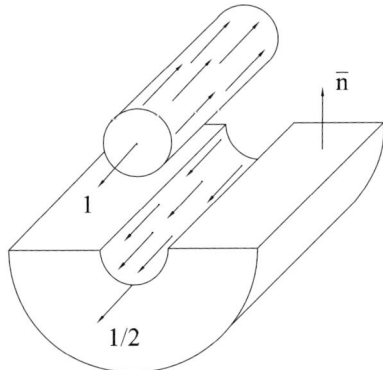

FIG. 5. Free body diagram around an antiplane unit load. The "interior" region will have a "boundary" traction of $+1/2$.

4. THE CALCULATION OF THE GREEN'S FUNCTIONS

4.1. Approximate Representations

Early applications of BIE methods in elastodynamics have dealt with the inherent singularities of the integrand kernels, as the Green's functions have to be evaluated at the very location of the sources. The pioneering works by Wong and Jennings (1975), and Sills (1978) are good examples of these efforts. However, in several practical instances a simplified approach was needed. The chosen strategy was to avoid these singularities and reliable methods have been developed where the sources are distributed not on the boundary itself, but some distance away. A number of sources, located in the volume itself or outside it were considered, and the boundary conditions were matched at a number of collocation points on the boundary, usually in a least-squares sense. These simplified formulations have been introduced and developed for two-dimensional problems by Sánchez-Sesma (1978), Sánchez-Sesma and Rosenblueth (1979), Sánchez-Sesma and Esquivel (1979), Wong (1982), Dravinski (1982, 1983), Mossessian and Dravinski (1987), Bravo *et al.* (1988). Figure 6 shows sources located outside the irregular exterior region. For the antiplane case it suffices to include image sources to satisfy the half-space free boundary conditions. The collocation points along the irregular boundary are depicted as well.

An alternate, but related approach consists in representing the diffracted field by a linear combination of a complete family of wave functions which are solutions of the wave equation (Herrera, 1980). These families of functions can be constructed with single or multi-polar sources having their singularities outside the region of interest and fulfilling the appropriate radiation conditions. An irregular topography is illustrated in Fig. 7. A multipole source is shown with its associated radiation pattern. The coefficients of the linear forms thus constructed, are obtained from a least-square matching of boundary conditions at a number of collocation points, larger than the order of the expansion. Results have been obtained for two-dimensional topographies by Sánchez-Sesma *et al.* (1982) and

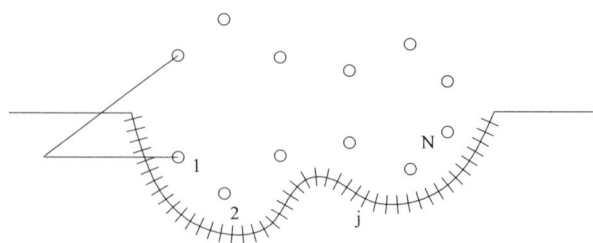

FIG. 6. Sources (1 to N) considered with their respective images to account for the half-space free boundary conditions. The collocation points are shown.

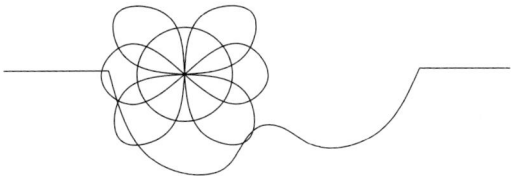

FIG. 7. Multipole source.

for three-dimensional problems by Sánchez-Sesma (1983). Of course, more than one multipole can be used (see for instance Imhof (1996)).

4.2. 2-D Green's Functions

A number of approaches have been developed which completely remove the singularities from the boundary integral formulation. One such way, within an indirect BEM formulation, was introduced by Sánchez-Sesma and Campillo (1991) who expressed the Green's functions in terms of Hankel functions. Then, using the corresponding series expansions of these functions, they obtained analytical expressions for the integral of the displacement Green's functions calculated on the source element itself. Moreover, they showed that the contribution there was simply $\pm\frac{1}{2}\phi_i(\mathbf{x})$ because the integral of the traction Green's functions at the source location has zero Cauchy's principal value. To compute the distant influence of the constant-value elements, they performed the integrations using Gaussian quadrature. This approach has been further developed by Sánchez-Sesma et al. (1993) and Luzón et al. (1995).

Kawase (1988) and Kawase and Aki (1989, 1990) used the discrete wavenumber representation of the Green's functions, in a direct BEM formulation. They showed that the analytical integration of these expressions in each of the wavenumbers over a constant-value element removes any singularity. The canonical example of a circular canyon under incidence of SV waves is shown in Fig. 8.

Bouchon (1985, 1987), Campillo and Bouchon (1985), Campillo (1987a, 1987b), Coutant (1989), Bouchon et al. (1989), Gaffet and Bouchon (1989, 1991), Axilrod and Ferguson (1990), Campillo et al. (1993), Gaffet (1995), and Karabulut and Ferguson (1996) also used the discrete wavenumber representation of the Green's functions. Then, they discretized the boundaries at equal spatial interval, which results in a periodicity in the wavenumber domain and leads to Green's function expressions in the form of finite summations over discrete wavenumbers, which do not contain singularities. An example of seismograms calculated in this way for a buried dislocation line in the earth crust is presented in Fig. 9. When the boundaries cannot be easily discretized at equal spatial interval, the Green's function expressions are integrated analytically over constant-value elements, again yielding non-singular expressions (Bouchon and Coutant, 1994).

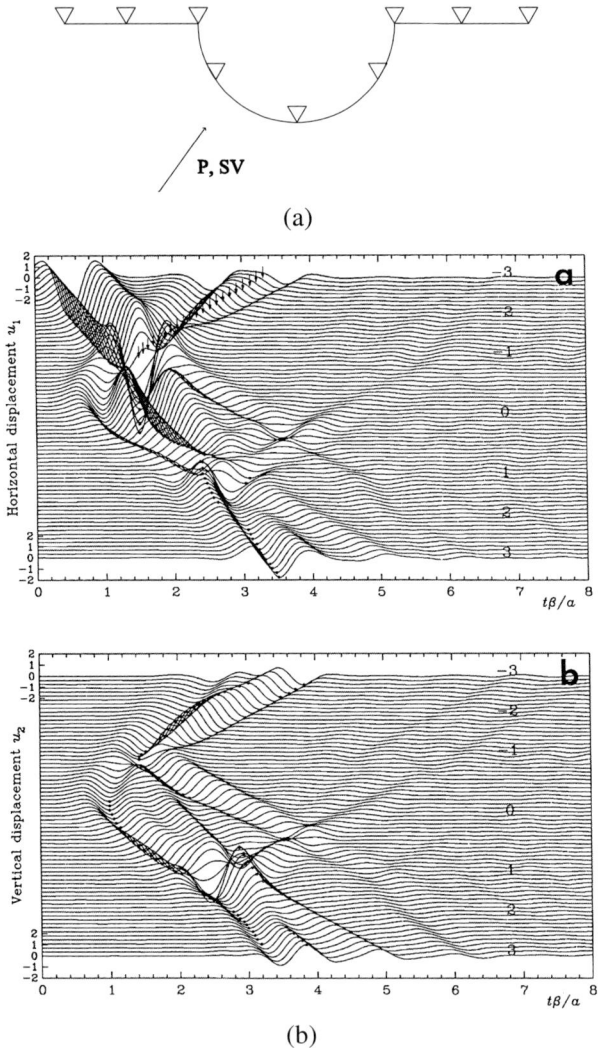

FIG. 8. (a) Geometry considered by Kawase (1988). (b) Synthetic seismograms along the surface in and around a semi-circular canyon due to an incident SV wave with a 30° incident angle.

4.3. 2.5-D Green's Functions

An interesting and important problem in elastodynamics, when dealing with the shallow earth, easier to solve than the full 3-D problem, is the one of calculating the 3-D wave field in a 2-D varying medium. When the elastic excitation is a plane wave propagating at some angle from the boundary axis, Luco *et al.* (1990) and Pedersen *et al.* (1995, 1994) proposed a formulation where the Green's func-

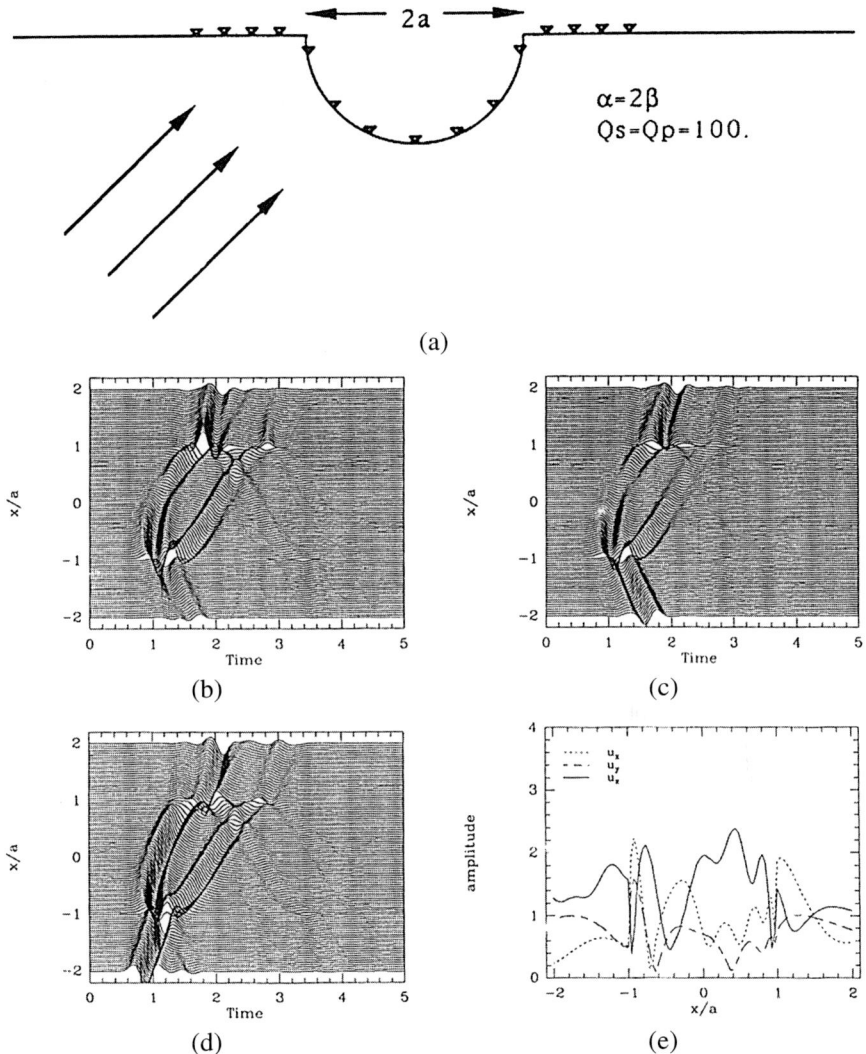

FIG. 9. (a) Semi-circular canyon used in the simulations. Displacement amplitudes across the canyon for an incident plane SV wave, with incidence angle of 45° and azimuth $\phi = 30°$. Synthetic seismograms, (b) u_x, (c) u_y, (d) u_z, and (e) spectral amplitudes along the profile (after Pedersen et al. (1994)).

tions are those of a harmonic point force moving along the boundary axis at the phase velocity of the incoming field. They then derived analytical expressions of the Green's functions in terms of Hankel functions that they integrated in the same way as Sánchez-Sesma and Campillo (1991). In Fig. 10a, a semi-circular canyon

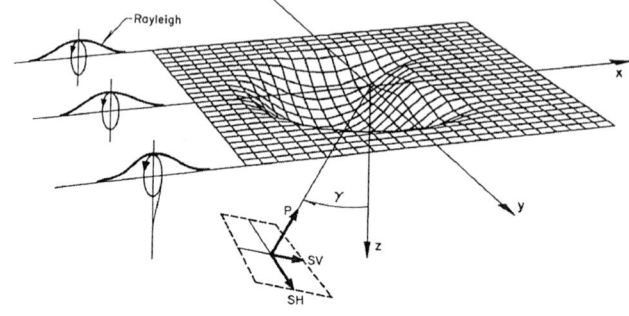

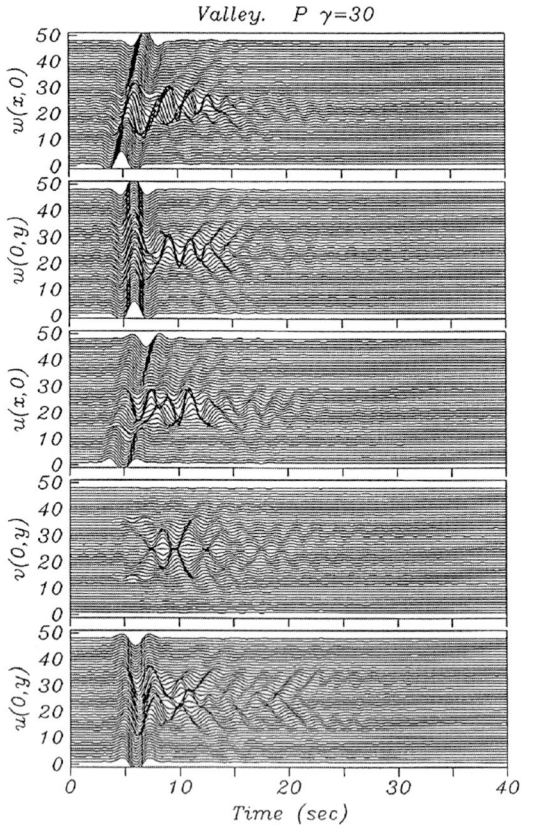

FIG. 10. Perspective view of the irregular Valley's basement. Incidence of plane P, SH, SV, and Rayleigh waves. Synthetic seismograms for u, v and w at 48 receivers equally spaced along the x or y axes. The range of x or y is between $-1.82a$ and $1.74a$, where $a = 4$ km. Irregular valley under incidence of P waves with $\gamma = 30°$ and $\phi = 0°$. The incident waveform is a Ricker wavelet with characteristic period of $t_P = 3$ s (after Sánchez-Sesma and Luzón (1995)).

used in the simulations is depicted with its receptors across the canyon to obtain the displacement amplitudes for an incident plane SV wave. Synthetic seismograms and spectral amplitudes are also shown (Fig. 10b–e) (*after* Pedersen *et al.*, 1994).

Similar expressions for the 2.5-D Green's functions are due to Tadeu and Kausel (2000). Papageorgiou and Pei (1998) used, for the same type of problem, the discrete wavenumber representation of the Green's functions, which they integrated over constant-value elements, in a direct BE formulation.

Takenaka *et al.* (1996a) extended the boundary integral equation—discrete wavenumber method to the case of a point source in a 2-D medium. By applying a spatial Fourier transform to the 3-D integral equations, they decomposed the 3-D problem into a set of independent 2-D problems. Fujiwara (1996a) developed for the same problem a direct BE approach.

4.4. 3-D Green's Functions

In three dimensions, the singularities of the displacement and stress Green's functions are of the form $1/r$ and $1/r^2$, respectively, where r is the source-receiver distance. Like in two dimensions, the presence of these singularities excludes a purely numerical evaluation of the integrals. The calculation of the Green's function integral at the location of the source has to be performed analytically. Several schemes have been proposed to this effect.

Kim and Papageorgiou (1993) extended to 3-D the formulation of Kawase (1988). Sánchez-Sesma and Luzón (1995) and Luzón *et al.* (1997) discretized the boundary into planar circular elements, on which the 3-D infinite medium Green's function can be integrated analytically. Using an irregular interface the response of a 3-D alluvial valley has been computed for incident plane P, SH, SV and Rayleigh waves. In Fig. 11 a view of the irregular basement is shown. Synthetic seismograms for u, v and w at 48 surface receivers along the x or y axes are given for incident P waves. The incident waveform is a Ricker wavelet with characteristic period of 3 s. Bouchon *et al.* (1996), and Durand *et al.* (1999) used the discrete wavenumber representation of the Green's function and discretized the boundary at equal spatial intervals along the two horizontal directions. This transforms the boundary integral equations into finite discrete summations over wavenumbers, which are free from singularity.

4.5. Non-Homogeneous Media Green's Functions

The Green's functions used in BIE/BE formulations are usually infinite homogeneous media Green's functions which require the homogeneity of the sub-domains of interest. When one of the domains is a flat layered medium Green's functions appropriate for such medium can be used (e.g. Bouchon, 1993;

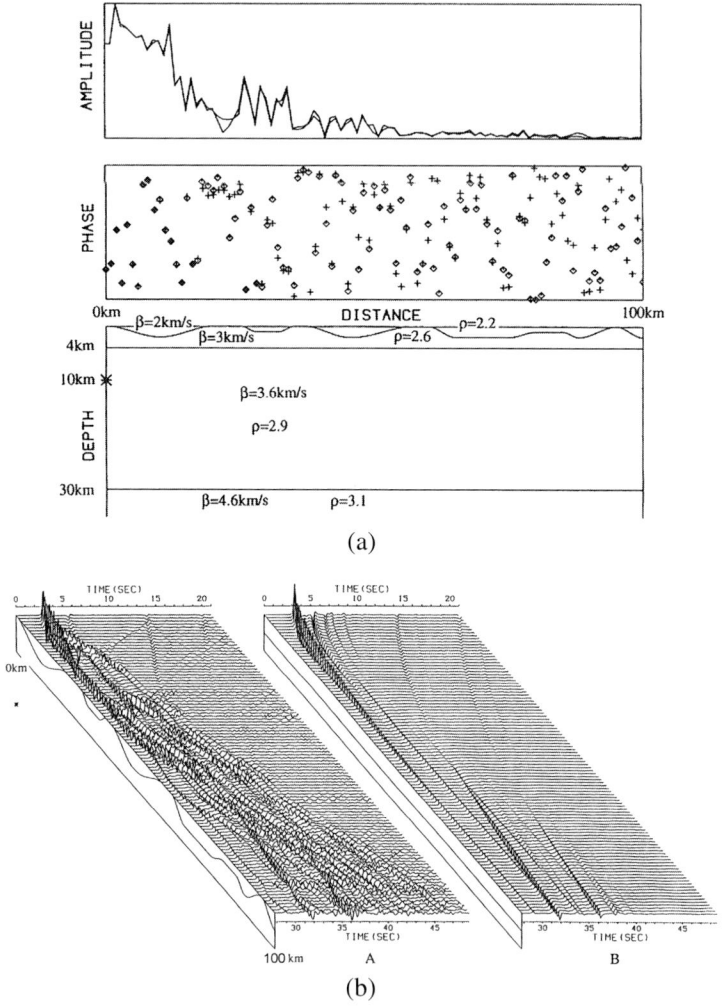

FIG. 11. (a) Source-medium configuration considered and corresponding amplitude and phase of the surface displacement at 2 Hz. The source location is indicated by the star. (b) A. Ground horizontal velocity produced by an antiplane dislocation for the crustal model of panel (a). The epicentral distance range extends from 0 to 100 km. A reduced velocity of 3.6 km/s is applied to the traces. (b) B. Same as (b) A, when the sediment bed rock interface is flat (after Bouchon et al. (1996)).

Hisada et al., 1993; Fujiwara and Takenaka, 1994). For a constant-gradient elastic isotropic medium, i.e. a medium with linear variation of wave velocity, analytical approximations of 2-D Green's functions have been developed (Janod and Coutant, 2000; Sánchez-Sesma et al., 2001) and applied for inhomogeneous allu-

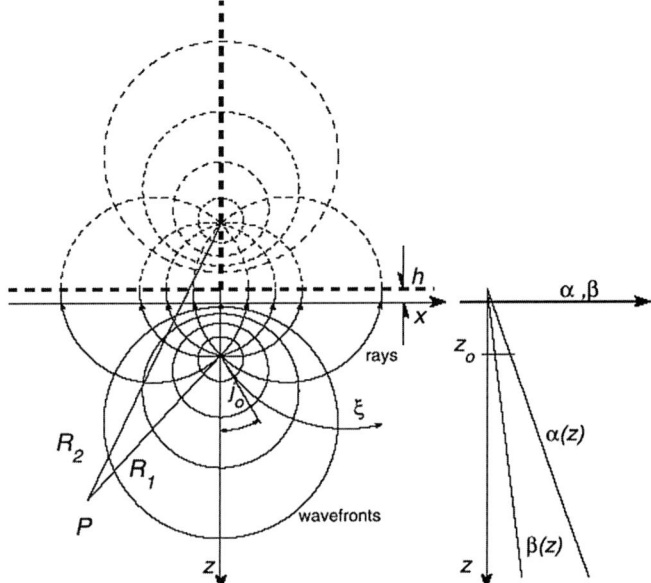

FIG. 12. Schematic illustration of the bipolar coordinate system. In a medium with linear variation of wave propagation velocity (see insert at right) both rays and wave fronts are circular. Rays depend upon take off angle j_0. Wave fronts depend on traveltime $\tau = \xi t_0$, where $t_0 = h/\beta(0)$ and ξ is one of the bipolar coordinates (the other is j_0). The distance R_1 and R_2 from the poles to point P are displayed as well. The plane for which $z = -h$ corresponds to null propagation velocities.

vial basins (Luzón et al., 2003; 2004). In Fig. 12, a 2-D wave propagation scheme for an inhomogeneous medium with constant-gradient velocity is shown. In this medium both wave fronts and rays are circular. In the figure the orthogonal bipolar coordinates are depicted. They are useful to represent wave propagation.

5. Discretization and Inversion

The discretization of the boundaries leads to the discretization of the BIE and the application of the boundary conditions transforms the BIE into a system of linear equations which, in general, is not symmetric. This fact is sometimes undesirable, particularly for large size matrices. The resolution of this system is usually done implicitly, and in the frequency domain. The discretization of the boundaries is generally done by approximating the surfaces by a set of elements over which the field values or the source densities are held constant. Non-constant value elements may also be constructed with interpolation functions similar to those used in finite elements methods. The geometry of the elements depends on the configuration studied and on the way the method is formulated. Discretization may also be achieved by representing the boundary source density functions by distributions. The element size is usually not frozen; it varies with frequency. A minimum number of about three elements per shortest wavelength is usually adopted.

The resolution of the resulting linear system of equations is generally performed by Gaussian elimination when the size of the system is not too large, and by iterative methods like the conjugate gradient, when the size of the system is large.

Different schemes have been proposed to reduce the size of the linear system to invert. Bouchon *et al.* (1996) and Ortiz-Alemán *et al.* (1998) applied an amplitude threshold (fraction of the maximum value) below which the matrix elements of the system are not considered, thus resulting in a sparse system. This approach, which only retains the preponderant terms, may still lead to accurate solutions after eliminating as much as 90% of the matrix terms. In Fig. 13a a 3D mountain of irregular shape on the surface of an elastic half-space is presented, while in Fig. 13b, the seismic response for SH waves incidence is depicted. Displacement amplitudes were computed for different thresholds.

Takenaka *et al.* (1996b) forced the matrix into a band structure by restricting the number of basis functions allowed to couple with each other, thus discarding elements beyond a certain distance from the diagonal.

6. Time Domain Implementation

In order to overcome memory limitations, which arise from solving very large linear systems, BIE methods may be implemented in the time domain. The BIE then gives the expression of the displacement field $\mathbf{u}$ at time t as a function of the values of $\mathbf{u}$ and $\mathbf{t}$ at times $\tau < t$. Its discretization leads to an implicit formulation which can be solved for each time step (Mansur and Brebbia, 1982, 1985; Rice and Sadd, 1984; Banerjee *et al.*, 1986).

It is, however, often more interesting to render this formulation explicit, which can be done by using small time steps and interpolation functions. This explicit scheme is then solved in a time-stepping algorithm, which requires little memory and bears much resemblance to the implementation of domains methods like finite differences or finite elements (Cole *et al.*, 1978; Peirce and Siebrits, 1997; Janod and Coutant, 2000). It also requires the introduction of absorbing boundaries. Figure 14a shows the topography considered by Janod and Coutant (2000), consisting of a hemispherical canyon, overlying a homogeneous half space. Corresponding seismograms are displayed in Fig. 14b.

In this explicit formulation, linear time interpolation may be used to calculate time integrals while Gaussian quadrature based on polynomial interpolation is used for space integration (Janod and Coutant, 2000). However, time implementations of BE are extremely sensitive to numerical noise (Cole *et al.*, 1978;

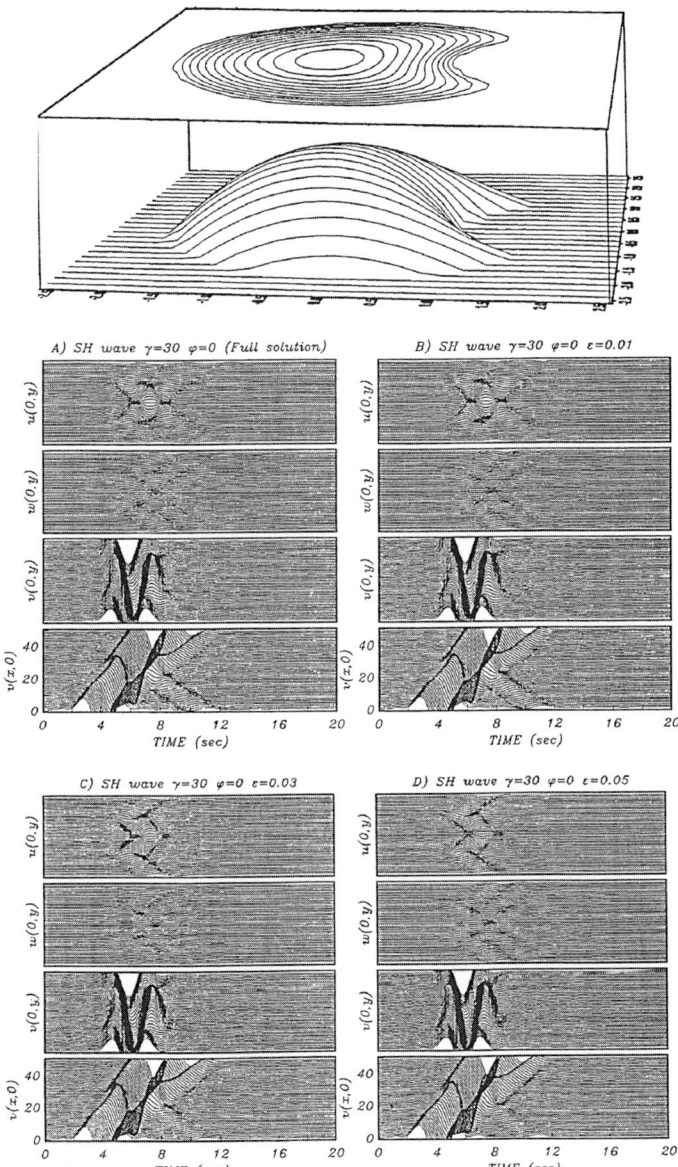

FIG. 13. (Top) Plan and cross section of the three-dimensional mountain of irregular shape on the surface of an elastic half-space. (a)–(d) Synthetic seismograms for incident SH waves, with azimuth $\phi = 0$, and incidence angle $\gamma = 30°$. A Ricker wavelet is considered as incoming wave, with a characteristic period $t_p = 3$ s. Displacements are computed for different thresholds (ε = fraction of the maximum coefficient value).

Janod and Coutant, 2000), particularly through the calculation of time derivatives. This noise level increases with the number of time iterations.

7. OTHER BOUNDARY METHODS

One of the most successful boundary method in elastodynamics is the Aki–Larner method. This method, developed by Aki and Larner (1970), represents the harmonic wave field diffracted by a surface or interface as a superposition of plane waves with unknown coefficients propagating away from the diffracting boundary. Inhomogeneous waves are allowed. This discrete summation results from an assumed spatial periodicity of the medium. The equations, in displacement and stress, expressing the boundary conditions are Fourier transformed in space, which results in a linear system of equations. The solution of this system yields the plane wave coefficients. One of the unique feature of the method is its use of complex frequency. This method provides an approximate solution, which is very accurate when the geometry of the boundary is smooth. For this reason it has been applied extensively to scattering problems in two and three dimensions (Bouchon, 1973; Bard and Bouchon, 1980a, 1980b,

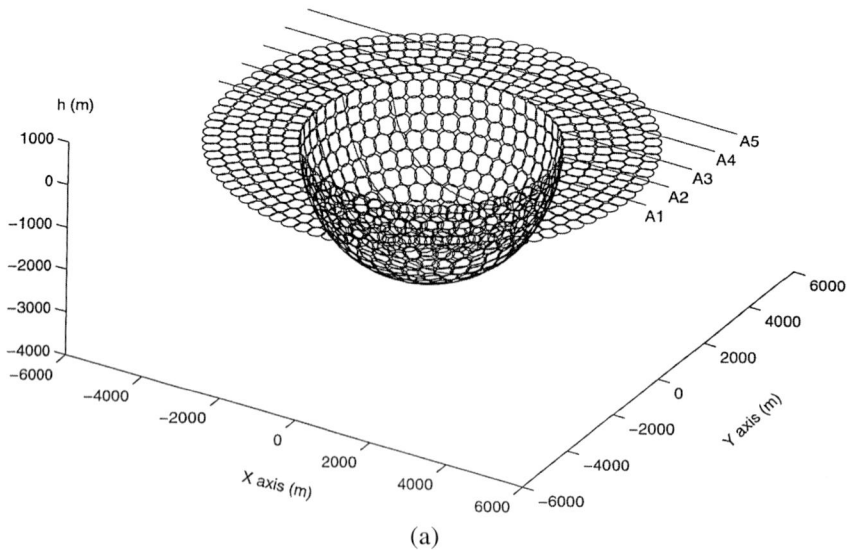

FIG. 14. (a) Topography considered with sharp edges, consisting of a hemispherical canyon of radius $a = 3000$ m, overlaying a homogeneous half-space. A vertical plane S wave polarized along the x-axis is diffracted by canyon. (b) Corresponding horizontal and vertical seismograms (after Janod and Coutant (2000)).

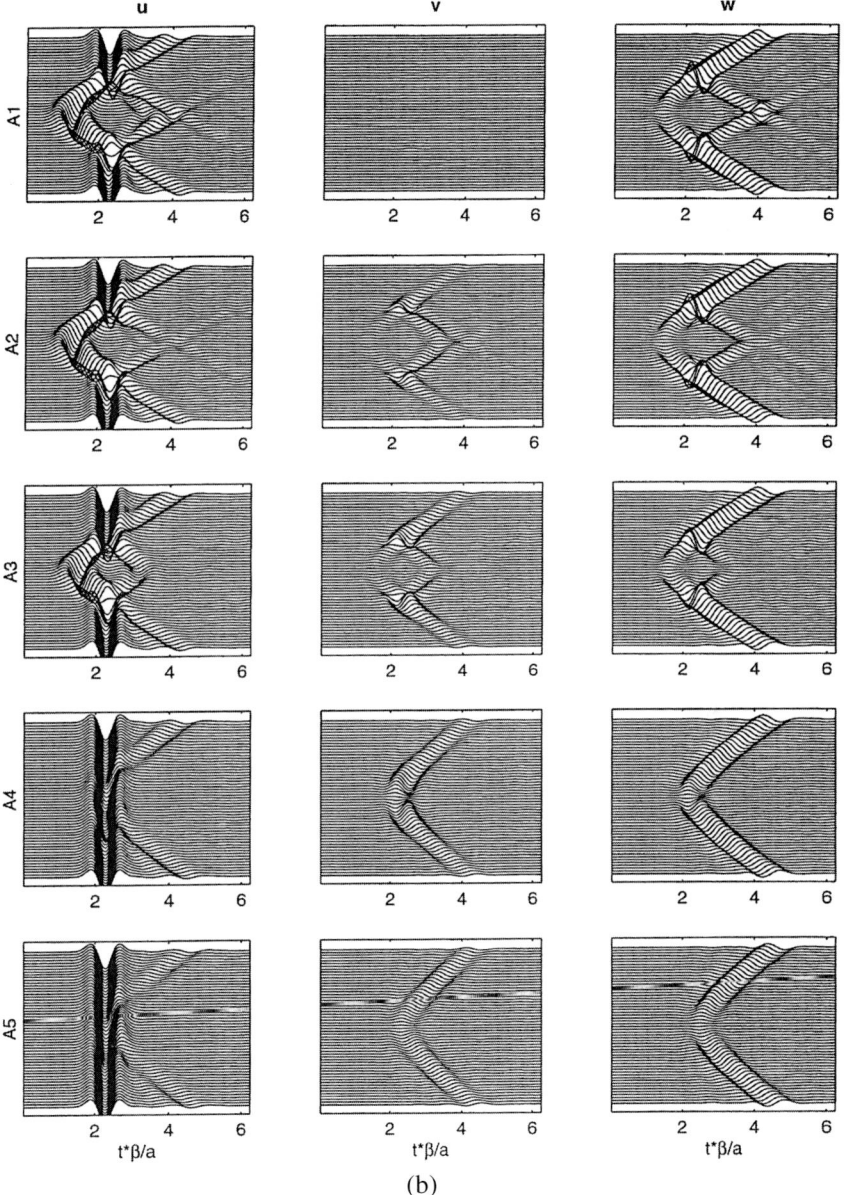

FIG. 14. Continued.

1985; Bard, 1982; Géli *et al.*, 1988; Horike *et al.*, 1990; Ohori *et al.*, 1992; Uebayashi *et al.*, 1992). In Fig. 15 (Bard and Bouchon, 1980b), the computed valley response for a P incident Ricker wavelet is shown. This excitation has a

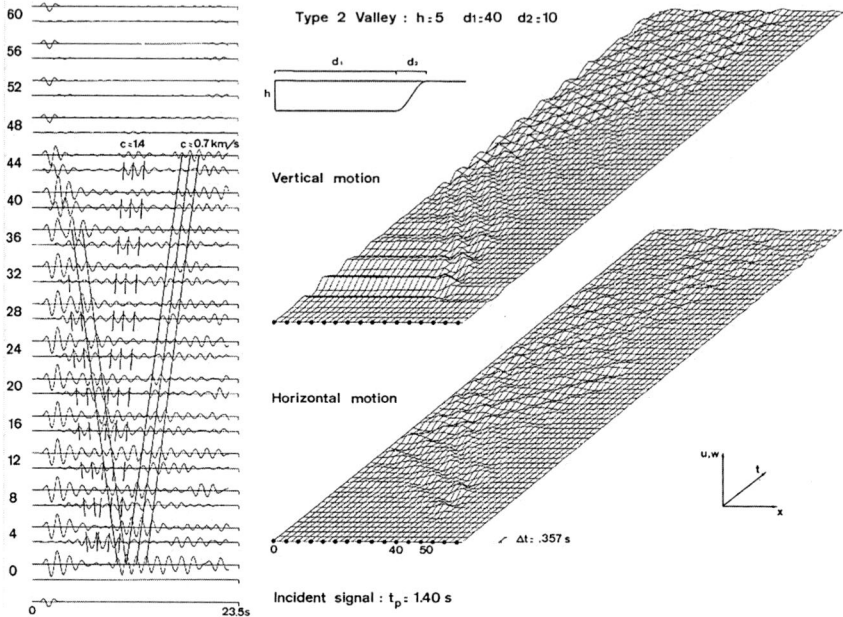

FIG. 15. Response of a high-contrast type 2 (shallow) valley with maximum depth $h = 500$ m, half-width $D = 5$ km, to a vertically incident P Ricker wavelet of characteristic period $t_p = 1.40$ s. Left: The traces represent the horizontal (bottom u) and vertical (upper, w) displacement signal within the valley. Right: Diagrams showing the spatial (x) and temporal (t) evolution of the surface displacement components in the valley and in its immediate proximity. The dots indicate the location of the sites where the seismograms (left part) are computed. In both figures, the length unit is 100 m, the time unit is 1 s, and only one side of the valley is represented because of the symmetry of the problem. The vertical scale is the same for the two components.

characteristic period of 1.4 s, which is near the P fundamental resonance period of the layer. Besides the reflections at the surface and at the interface, which are quite clear over the plane part of the valley, one can see some laterally propagating disturbances, which are generated at the edges of the basin, and subsequently travel across the valley. The vertical displacement is characterized by the presence of only one such disturbance, but the horizontal displacement diagram reveals the existence of two distinct phases, which are almost simultaneously generated and then separated because of different phase velocities.

Other related approaches, formulated in terms of propagator matrices or generalized reflection/transmission matrices have been developed by Kohketsu (1987); Chen (1990, 1996), and Koketsu et al. (1991).

A strategy aiming at reducing the size of the linear system consists in expanding the wavefield into normal modes (Hisada et al., 1993; Fujiwara and Takenaka, 1994; Hatayama and Fujiwara, 1998).

One of the drawbacks of BIE methods is that matrices are not symmetric. This may restrict to some degree the efficiency of solvers. To mitigate this the Galerkin method has been implemented and this leads to symmetric systems of equations. This variational approach may provide advantages despite the fact that double integrations have to be carried along all the elements (Bonnet *et al.*, 1998).

8. Hybrid Methods

In many ways, boundary and domain methods can be viewed as complementary techniques, and their combination may lead to more efficient solutions than relying on one technique only. Formulations combining BE and FE have been proposed by Mossessian and Dravinski (1987), Regan and Harkrider (1989), Fujiwara (1996b), Zhang *et al.* (1998), and Fu (2002). This combination is particularly relevant for media containing both irregular boundaries and volume heterogeneities (Fu, 2002). Figure 16, shows the 2-D basin model considered by Fujiwara (1996b), and the wavefield response of such sedimentary structure when the excitation is a point source. The figure depicts the frequency response for 0.125 Hz. In Fujiwara (1996b), a set of synthetic seismograms is shown and the emergence of basin-generated surface Love and Rayleigh waves is pointed out.

Other hybrid methods have been proposed which combine BE with Born series approximation (Schuster, 1985), Gaussian beam (Benites and Aki, 1989), and generalized screen propagator (Fu and Wu, 2001).

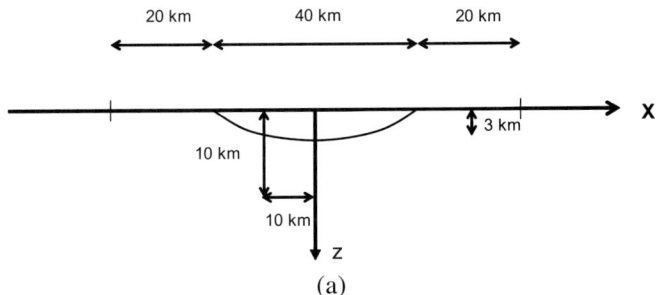

FIG. 16. (a) Two-dimensional basin model used for calculation of response from a point source. The P- and S-wave velocities and the density for the medium in the basin are $\alpha_1 = 1.8$ km/s, $\beta_1 = 0.8$ km/s, and $\rho_1 = 1.8$ g/cm^3, respectively. Those values for the half-space are $\alpha_2 = 4.5$ km/s, $\beta_2 = 2.5$ km/s, and $\rho_2 = 2.5$ g/cm^3. The origin of coordinates is at the center of the basin and the y-axis is taken in the homogeneous direction. The shape of the basin-basement boundary is given by $z(x) = 3\sin(x/40)$. The source location is $(0, 0, -10)$ km. (b) Responses of the free surface in the frequency domain (period of 8 s). The range shown in the figures is from -40 to 40 km along the x-axis and from -80 to 80 along the y-axis.

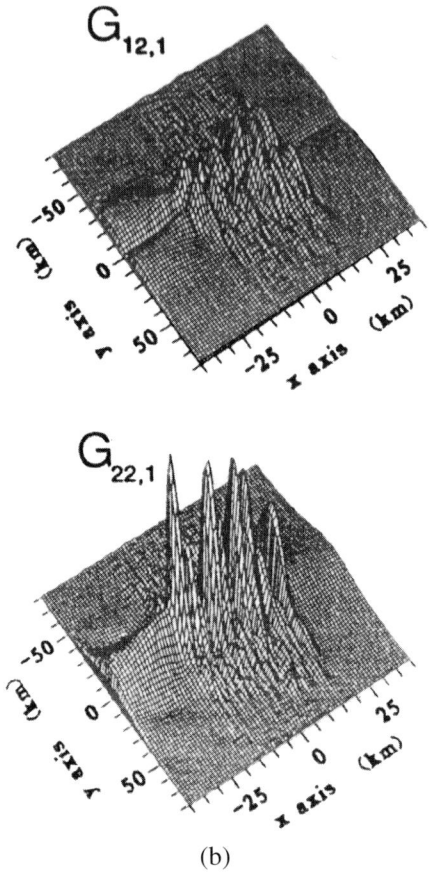

(b)

Fig. 16. Continued.

9. Domains of Application

BIE methods have been extensively applied to model or simulate seismic wave propagation in the shallow earth. They have helped understand and quantify the effect of irregular topography, like hills, mountains, or valleys on earthquake ground motion (Bouchon, 1973, 1985; Wong and Jennings, 1975; Sánchez-Sesma, 1978, 1983; Sills, 1978; Sánchez-Sesma and Rosenblueth, 1979; Bard, 1982; Wong, 1982; Sánchez-Sesma et al., 1982; Géli et al., 1988; Kawase, 1988; Gaffet and Bouchon, 1989; Axilrod and Ferguson, 1990; Sánchez-Sesma and Campillo, 1991; Pedersen et al., 1994; Yokoi and Takenaka, 1995; Bouchon et al., 1996; Takenaka et al., 1996a; Luzón et al., 1997; Ortiz-Alemán et al., 1998; Yokoi and Sánchez-Sesma, 1998; Durand et al., 1999; Janod and Coutant, 2000).

They have been extremely useful to investigate the seismic response of sedimentary basins and alluvial valleys, two widely encountered geological structures where much of the world population is concentrated (Aki and Larner, 1970; Sánchez-Sesma and Esquivel, 1979; Bard and Bouchon, 1980a, 1980b, 1985; Dravinski, 1982, 1983; Mossessian and Dravinski, 1987; Bravo *et al.*, 1988; Campillo *et al.*, 1989; Chávez-García and Bard, 1989; Kawase and Aki, 1989, 1990; Horike *et al.*, 1990; Gaffet and Bouchon, 1991; Papageorgiou and Kim, 1991; Ohori *et al.*, 1992; Jongmans and Campillo, 1993; Fujiwara and Takenaka, 1994; Hatayama *et al.*, 1995; Pedersen *et al.*, 1995; Sánchez-Sesma and Luzón, 1995; Fujiwara, 1996a; Takenaka *et al.*, 1996b; Zhang and Papageorgiou, 1996; Hatayama and Fujiwara, 1998; Riepl *et al.*, 2000).

BIE methods have also been successful in modelling seismic wave propagation at regional distances in laterally-varying crustal structures, in particular to study Lg wave and surface wave propagation and attenuation (Campillo, 1987b; Campillo *et al.*, 1993; Bouchon and Coutant, 1994; Gibson and Campillo, 1994; Paul *et al.*, 1996; Pedersen *et al.*, 1996; Fu and Wu, 2001; Fu *et al.*, 2002). A example is shown in Fig. 17. The snapshots of *SH* waves traveling through a homogeneous wave guide with randomly-rough free surface are depicted in the figure (from Fu *et al.*, 2002). The point source is located on the left boundary at 2.0 km of depth. The snapshots of the figure, demonstrate the development of both mantle and head waves. The formation of crustal guided waves as repetitive reflections at the free surface and the Moho are illustrated as well.

BIE methods have been helpful to study problems in seismic exploration, either in modelling surface seismic profiles (Campillo, 1987a; Paul and Campillo, 1988; Bouchon *et al.*, 1989) or borehole acoustic logging (Bouchon and Schmitt, 1989; Bouchon, 1993; Dong *et al.*, 1995).

One domain for which BIE methods are extremely well adapted is wave propagation in fractured media or in media containing small scale heterogeneities. They allow a very precise application of stress boundary conditions along the crack or inclusion surfaces and they accurately handle the differences in scale between the wavelengths and the diffracting objects. These simulations have helped understand the importance of multiple scattering (Benites *et al.*, 1992) and the anisotropy and attenuation of seismic waves in fractured media (Coutant, 1989; Benites *et al.*, 1992; Yomogida and Benites, 1995; Kelner *et al.*, 1999a, 1999b).

BIE methods have also been extensively applied to problems in fracture dynamics. They have played an important role in our understanding of how cracks propagate and how earthquake faults rupture (Burridge, 1969; Das and Aki, 1977; Das, 1980; Andrews, 1985; Das and Kostrov, 1987; Koller *et al.*, 1992; Cochard and Madariaga, 1994, 1996; Fukuyama and Madariaga, 1995, 1998; Yamashita and Fukuyama, 1996; Bouchon and Streiff, 1997; Kame and Yamashita, 1999; Aochi *et al.*, 2000; Tada *et al.*, 2000; Aochi and Madariaga, 2003). BIE formulation in this domain is closely related to the one developed for elastic wave

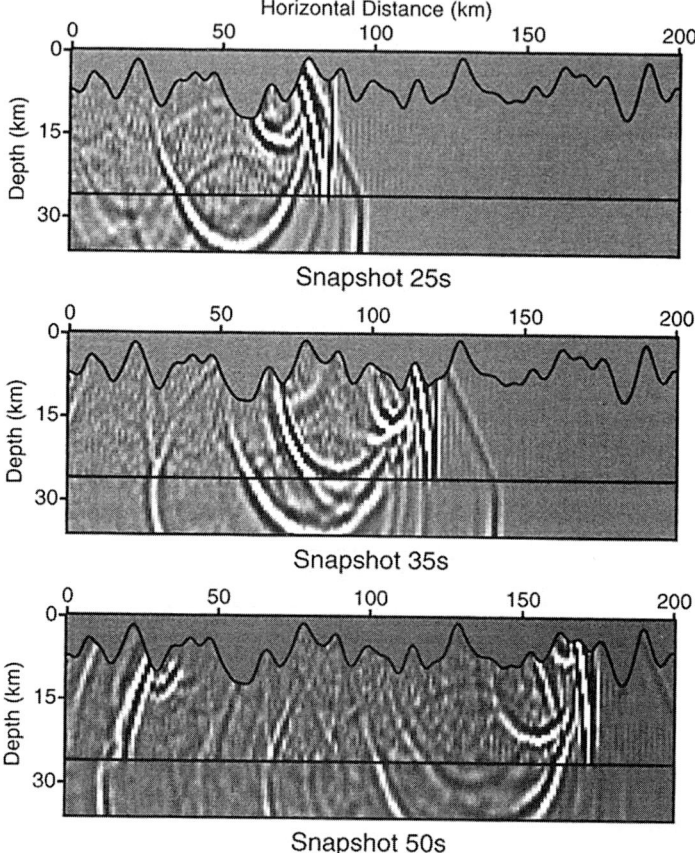

FIG. 17. Snapshots of SH waves traveling through a homogeneous wave guide with a randomly rough free surface (the rms height $\sigma = 2.5$ km). The source depth is 2.0 km, and the wave-guide thickness is 20 km. Seismograms are computed in the frequency range of 0–2 Hz. The snapshots of the figure clearly demonstrate the development of both mantle and head waves. Crustal guided waves, formed as multiple reflections at the free surface and the Moho are also depicted (after Fu *et al.* (2002).

propagation as the problem is mathematically almost identical to the diffraction of elastic waves by a crack whose size varies in time.

10. Concluding Remarks

This rapid passage over the BIE in elastodynamics essentially covers certain developments in the Earth sciences and seismology of the last three decades. There is a number of problems that may benefit from advances in other areas.

Variational, domain decomposition and subdomain hybrid formulations hold the promise to give the power to deal with more complex configurations...

ACKNOWLEDGEMENTS

The critical comments of V. Maupin and R.-S. Wu helped to sharpen the focus of this work. Thanks are given to M. Campillo, O. Coutant for their comments and suggestions. The multifarious help of E. Flores was crucial to finish this work. Our research benefited from various sources of support along the years. We want to thank the French CNRS, and the Mexican CONACYT and DGAPA-UNAM.

REFERENCES

Aki, K., Larner, K. (1970). Surface motion of a layered medium having an irregular interface due to incident plane SH waves. *J. Geophys. Res.* **75**, 933–954.
Aki, K., Richards, P.G. (1980). Quantitative Seismology, vol. 1. Freeman and Co, San Francisco.
Andrews, D.J. (1985). Dynamic plane-strain shear rupture with a slip-weakening friction law calculated by a boundary integral method. *Bull. Seismol. Soc. Am.* **75**, 1–21.
Aochi, H., Fukuyama, E., Matsu'ura, M. (2000). Spontaneous rupture propagation on a non-planar fault in 3-D elastic medium. *Pure Appl. Geophys.* **157**, 2003–2027.
Aochi, H., Madariaga, R. (2003). The 1999 Izmit, Turkey, earthquake: Non-planar fault structure, dynamic rupture process, and strong ground motion. *Bull. Seismol. Soc. Am.* **93**, 1249–1266.
Axilrod, H.D., Ferguson, J.F. (1990). SH-wave scattering from a sinusoidal grating: An evaluation of four discrete wavenumber methods. *Bull. Seismol. Soc. Am.* **80**, 643–655.
Banerjee, P.K., Ahmad, S., Manolis, G.D. (1986). Transient elastodynamic analysis of three-dimensional problems by Boundary Element Methods. *Earthquake Engrg. Struct. Dyn.* **14**, 933–949.
Bard, P.Y. (1982). Diffracted waves and displacement field over two-dimensional topographies. *Geophys. J. R. Astron. Soc.* **71**, 731–760.
Bard, P.Y., Bouchon, M. (1980a). The seismic response of sediment-filled valleys. Part 1: The case of incident SH waves. *Bull. Seismol. Soc. Am.* **70**, 1263–1286.
Bard, P.Y., Bouchon, M. (1980b). The seismic response of sediment-filled valleys. Part 2: The case of incident P and SV waves. *Bull. Seismol. Soc. Am.* **70**, 1921–1941.
Bard, P.Y., Bouchon, M. (1985). The two-dimensional resonance of sediment-filled valleys. *Bull. Seismol. Soc. Am.* **75**, 519–541.
Benites, R., Aki, K. (1989). Boundary integral—Gaussian beam method for seismic wave scattering: SH waves in two-dimensional media. *J. Acoust. Soc. Am.* **86**, 375–386.
Benites, R., Aki, K., Yomogida, K. (1992). Multiple scattering of SH waves in 2-D media with many cavities. *Pure Appl. Geophys.* **138**.
Bonnet, M. (1986). Méthode des équations intégrales régularisées en élastodynamique. Doctorate Thesis. Ecole Nationale des Ponts-et-Chaussées, Paris.
Bonnet, M. (1989). Regular boundary integral equations for three-dimensional finite or infinite bodies with and without curved cracks in elastodynamics. In: Brebbia, C.A., Zamani, N.G. (Eds.), *Boundary Element Techniques: Applications in Engineering*. Computational Mechanics Publications, Southampton.

Bonnet, M., Maier, G., Polizzotto, G. (1998). Symmetric Galerkin boundary element methods. *Appl. Mech. Rev.* **51**, 669–704.
Bouchon, M. (1973). Effect of topography on surface motion. *Bull. Seismol. Soc. Am.* **63**, 715–732.
Bouchon, M. (1985). A simple complete numerical solution to the problem of diffraction of SH waves by an irregular surface. *J. Acoust. Soc. Am.* **77**, 1–5.
Bouchon, M. (1987). Diffraction of elastic waves by cracks or cavities using the discrete wavenumber method. *J. Acoust. Soc. Am.* **81**, 1671–1676.
Bouchon, M. (1993). A numerical simulation of the acoustic and elastic wavefields radiated by a source in a fluid-filled borehole embedded in a layered medium. *Geophysics* **58**, 475–481.
Bouchon, M., Schmitt, D.P. (1989). Full-wave acoustic logging in an irregular borehole. *Geophysics* **54**, 758–765.
Bouchon, M., Campillo, M., Gaffet, S. (1989). A boundary integral equation—discrete wavenumber representation method to study wave propagation in multilayered media having irregular interfaces. *Geophysics* **54**, 1134–1140.
Bouchon, M., Coutant, O. (1994). Calculation of synthetic seismograms in a laterally-varying medium by the boundary element—discrete wavenumber method. *Bull. Seismol. Soc. Am.* **84**, 1869–1881.
Bouchon, M., Schultz, C.A., Toksöz, M.N. (1996). Effect of 3-D topography on seismic motion. *J. Geophys. Res.* **101**, 5835–5846.
Bouchon, M., Streiff, D. (1997). Propagation of a shear crack on a non-planar fault: A method of calculation. *Bull. Seismol. Soc. Am.* **87**, 61–66.
Bravo, M.A., Sánchez-Sesma, F.J., Chávez-García, F.J. (1988). Ground motion on stratified alluvial deposits for incident SH waves. *Bull. Seismol. Soc. Am.* **78**, 436–450.
Brebbia, C.A., Domínguez, J. (1992). Boundary Elements: An Introductory Course, second ed. Computation Mechanics Publications, Southampton/McGraw–Hill, New York.
Burridge, R. (1969). The numerical solution of certain integral equations with non-integrable kernels arising in the theory of crack propagation and elastic wave diffraction. *Philos. Trans. R. Soc. London A* **265**, 353–381.
Campillo, M. (1987a). Modeling of SH-wave propagation in an irregularly layered medium—Application to seismic profiles near a dome. *Geophys. Prosp.* **35**, 236–249.
Campillo, M. (1987b). Lg wave propagation in a laterally varying crust and the distribution of the apparent quality factor in central France. *J. Geophys. Res.* **92**, 12604–12614.
Campillo, M., Bouchon, M. (1985). Synthetic seismograms in a laterally varying medium by the discrete wavenumber method. *Geophys. J. R. Astron. Soc.* **83**, 307–317.
Campillo, M., Gariel, J.C., Aki, K., Sánchez-Sesma, F.J. (1989). Destructive strong ground motion in Mexico City: Source, path, and site effects during great 1985 Mechoacan earthquake. *Bull. Seismol. Soc. Am.* **79**, 1718–1735.
Campillo, M., Feignier, B., Bouchon, M., Béthoux, N. (1993). Attenuation of crustal waves across the Alpine range. *J. Geophys. Res.* **98**, 1987–1996.
Chávez-García, F.J., Bard, P.Y. (1989). Effect of random thickness variations on the seismic response of a soft soil layer: Applications to Mexico City. In: Cakmak, A.S., Herrera, I. (Eds.), *Engineering Seismology and Site Response*. Computational Mechanics Publications, Southampton, pp. 247–261.
Chen, X. (1990). Seismogram synthesis for multi-layered media with irregular interfaces by global generalized reflection/transmission matrices method. 1. Theory of two-dimensional SH case. *Bull. Seismol. Soc. Am.* **80**, 1696–1724.
Chen, X. (1996). Seismogram synthesis for multi-layered media with irregular interfaces by global generalized reflection/transmission matrices method. 3. Theory of 2-D P-SV case. *Bull. Seismol. Soc. Am.* **86**, 389–405.
Claerbout, J.F. (1985). Imaging the Earth's Interior. Blackwell Scientific Publishing.
Cochard, A., Madariaga, R. (1994). Dynamic faulting under rate-dependent friction. *Pure Appl. Geophys.* **142**, 419–445.

Cochard, A., Madariaga, R. (1996). Complexity of seismicity due to highly rate-dependent friction. *J. Geophys. Res.* **101**, 25321–25336.
Cole, D.M., Kosloff, D.D., Minster, J.B. (1978). A numerical Boundary Integral Equation Method for elastodynamics, I. *Bull. Seismol. Soc. Am.* **68**, 1331–1357.
Coutant, O. (1989). Numerical study of the diffraction of elastic waves by fluid-filled cracks. *J. Geophys. Res.* **94**, 17805–17818.
Das, S. (1980). A numerical method for determination of source time functions for general three-dimensional rupture propagation. *Geophys. J. R. Astron. Soc.* **622**, 591–604.
Das, S., Aki, K. (1977). A numerical study of two-dimensional spontaneous rupture propagation. *Geophys. J. R. Astron. Soc.* **50**, 643–668.
Das, S., Kostrov, B.V. (1987). On the numerical boundary integral equation method for three-dimensional dynamic shear crack problems. *J. Appl. Mech.* **54**, 99–104.
Domínguez, J. (1993). Boundary Elements in Dynamics. Computational Mechanics Publications, Southampton.
Dong, W., Bouchon, M., Toksöz, M.N. (1995). Borehole seismic-source radiation in layered isotropic and anisotropic media: boundary element modeling. *Geophysics* **60**, 735–747.
Dravinski, M. (1982). Influence of interface depth upon strong ground motion. *Bull. Seismol. Soc. Am.* **72**, 597–614.
Dravinski, M. (1983). Scattering of plane harmonic SH wave by dipping layers of arbitrary shape. *Bull. Seismol. Soc. Am.* **73**, 1303–1319.
Durand, S., Gaffet, S., Virieux, J. (1999). Seismic diffracted waves from topography using 3-D discrete wavenumber-boundary integral equation simulation. *Geophysics* **64**, 572–578.
Eason, G., Fulton, I., Sneddon, I.N. (1956). The generation of waves in an infinite elastic solid by variable body. *Philos. Trans. R. Soc. London* **248**, 575–607.
Fu, L.Y. (2002). Seismogram synthesis for piecewise heterogeneous media. *Geophys. J. Int.* **150**, 800–808.
Fu, L.Y., Wu, R.S. (2001). A hybrid BE-GS method for modeling regional wave propagation. *Pure Appl. Geophys.* **158**, 1251–1277.
Fu, L.Y., Wu, R.S., Campillo, M. (2002). Energy partition and attenuation of regional phases by random free surface. *Bull. Seismol. Soc. Am.* **92**, 1992–2007.
Fujiwara, H. (1996a). Three-dimensional wavefield in a two-dimensional basin structure due to point source. *J. Phys. Earth* **44**, 1–22.
Fujiwara, H. (1996b). Seismic wavefields in multilayered media calculated by hybrid combination of boundary element method and thin layer finite element method—The case of SH-wavefields. *J. Phys. Earth* **44**, 61–77.
Fujiwara, H., Takenaka, H. (1994). Calculation of surface waves for a thin basin structure using a direct boundary element method with normal modes. *Geophys. J. Int.* **117**, 69–91.
Fukuyama, E., Madariaga, R. (1995). Integral equation method for planar crack with arbitrary shape in 3-D elastic medium. *Bull. Seismol. Soc. Am.* **85**, 614–628.
Fukuyama, E., Madariaga, R. (1998). Rupture dynamics of a planar fault in a 3-D elastic medium: Rate- and slip-weakening friction. *Bull. Seismol. Soc. Am.* **88**, 1–17.
Gaffet, S. (1995). Teleseismic waveform modeling including geometrical effects of superficial geological structures near to seismic sources. *Bull. Seismol. Soc. Am.* **85**, 1068–1079.
Gaffet, S., Bouchon, M. (1989). Effects of two-dimensional topographies using the discrete wavenumber—boundary integral equation method in P-SV cases. *J. Acoust. Soc. Am.* **85**, 2277–2283.
Gaffet, S., Bouchon, M. (1991). Source location and valley shape effects on the P-SV displacement field using a boundary integral equation—discrete wavenumber representation method. *Geophys. J. Int.* **106**, 341–355.
Géli, L., Bard, P.Y., Jullien, B. (1988). The effect of topography on earthquake ground motion: A review and new results. *Bull. Seismol. Soc. Am.* **78**, 42–63.

Gibson, R.L., Campillo, M. (1994). Numerical simulation of high- and low-frequency Lg-wave propagation. *Geophys. J. Int.* **118**, 47–56.

Hatayama, K., Matsunami, K., Iwata, T., Irikura, K. (1995). Basin induced Love waves in the eastern part of the Osaka basin. *J. Phys. Earth* **43**, 131–155.

Hatayama, K., Fujiwara, H. (1998). Excitation of secondary Love and Rayleigh waves in a three-dimensional sedimentary basin evaluated by a direct boundary element method with normal modes. *Geophys. J. Int.* **133**, 260–278.

Herrera, I. (1980). Boundary methods: A criterion for completeness. *Proc. Natl. Acad. Sci.* **77**, 4395–4398.

Hisada, Y., Aki, K., Teng, T. (1993). 3-D simulation of surface wave propagation in the Kanto sedimentary basin, Part 2: Application of the surface wave BEM. *Bull. Seismol. Soc. Am.* **83**, 1700–1720.

Horike, M., Uebayashi, H., Takeuchi, T. (1990). Seismic response in three-dimensional sedimentary basin due to plane S wave incidence. *J. Phys. Earth* **38**, 261–284.

Janod, F., Coutant, O. (2000). Seismic response of three-dimensional topographies using a time-domain boundary element method. *Geophys. J. Int.* **142**, 603–614.

Jongmans, D., Campillo, M. (1993). The response of the Ubaye Valley (France) for incident SH and SV waves: Comparison between measurements and modeling. *Bull. Seismol. Soc. Am.* **83**, 907–924.

Imhof, M. (1996). Scattering of elastic waves using non-orthogonal expansions. Doctorate Thesis. Massachusetts Institute of Technology, Massachusetts.

Kame, N., Yamashita, T. (1999). Simulation of the spontaneous growth of a dynamic crack without constraints on the crack tip path. *Geophys. J. Int.* **139**, 345–358.

Karabulut, H., Ferguson, J.F. (1996). SH wave propagation by discrete wavenumber boundary integral modeling in transversely isotropic medium. *Bull. Seismol. Soc. Am.* **86**, 524–529.

Kawase, H. (1988). Time-domain response of a semicircular canyon for incident SV, P, and Rayleigh waves calculated by the discrete wavenumber boundary element method. *Bull. Seismol. Soc. Am.* **78**, 1415–1437.

Kawase, H., Aki, K. (1989). A study on the response of a soft basin for incident S, P, and Rayleigh waves with special reference to the long duration observed in Mexico City. *Bull. Seismol. Soc. Am.* **79**, 1361–1382.

Kawase, H., Aki, K. (1990). Topography effect at the critical SV wave incidence: Possible explanation of damage pattern by the Whittier Narrows, California, earthquake of 1 October 1987. *Bull. Seismol. Soc. Am.* **80**, 1–22.

Kelner, S., Bouchon, M., Coutant, O. (1999a). Numerical simulation of the propagation of P-waves in fractured media. *Geophys. J. Int.* **137**, 197–206.

Kelner, S., Bouchon, M., Coutant, O. (1999b). Characterization of fractures in shallow granite from the modeling of the anisotropy and attenuation of seismic waves. *Bull. Seismol. Soc. Am.* **89**, 706–717.

Kim, J., Papageorgiou, A.S. (1993). Discrete wavenumber boundary-element method for 3-D scattering problems. *J. Engrg. Mech. ASCE* **119**, 603–624.

Kohketsu, K. (1987). 2-D reflectivity method and synthetic seismograms for irregularly layered structures-I. SH-wave generation. *Geophys. J. Int.* **89**, 821–838.

Koketsu, K., Kennett, B.L.N., Takeneka, H. (1991). 2-D reflectivity method and synthetic seismograms for irregularly layered structures-II. Invariant embedding approach. *Geophys. J. Int.* **105**, 119–130.

Koller, M.G., Bonnet, M., Madariaga, R. (1992). Modelling of dynamic crack propagation using time-domain boundary integral equations. *Wave Motion* **16**, 339–366.

Kupradze, V.D. (1963). Dynamical problems in elasticity. In: Sneddon, I.N., Hill, R. (Eds.), *Progress in Solid Mechanics,* vol. III. North-Holland, Amsterdam.

Luco, J.E., Wong, H.L., De Barros, F.C.P. (1990). Three-dimensional response of a cylindrical canyon in a layered half-space. *Earthquake Engrg. Struct. Dyn.* **19**, 799–817.

Luzón, F., Aoi, S., Fäh, D., Sánchez-Sesma, F.J. (1995). Simulation of the seismic response of a 2-D sedimentary basin: A comparison between the indirect boundary element method and a hybrid technique. *Bull. Seismol. Soc. Am.* **85**, 1501–1506.

Luzón, F., Sánchez-Sesma, F.J., Rodríguez-Zúniga, J.L., Posadas, A.M., García, J.M., Martín, J., Romacho, M.D., Navarro, M. (1997). Diffraction of P, S and Rayleigh waves by three-dimensional topographies. *Geophys. J. Int.* **129**, 571–578.

Luzón, F., Ramírez-Guzmán, L., Sánchez-Sesma, F.J., Posadas, A. (2003). Propagation of SH elastic waves in deep sedimentary basins with an oblique velocity gradient. *Wave motion* **38**, 11–23.

Luzón, F., Ramírez-Guzmán, L., Sánchez-Sesma, F.J., Posadas, A. (2004). Simulation of the seismic response of sedimentary basins with constant-gradient of velocity. *Pure Appl. Geophys.* **161**, 1533–1547.

Manolis, G.D., Beskos, D.E. (1988). Boundary Element Methods in Elastodynamics. Unwin Hyman, London.

Mansur, W.J., Brebbia, C.A. (1982). Numerical implementation of the boundary element method for two-dimensional transient scalar wave propagation problems. *Appl. Math. Mod.* **6**, 299–306.

Mansur, W.J., Brebbia, C.A. (1985). Further developments on the solution of the transient scalar wave equation. In: Brebbia, C.A. (Ed.), *Topics in Boundary Element Research, Vol. 2*, pp. 88–123.

Mossessian, T.K., Dravinski, M. (1987). Application of a hybrid method for scattering of P, SV and Rayleigh waves by near-surface irregularities. *Bull. Seismol. Soc. Am.* **77**, 1784–1803.

Ohori, M., Koketsu, K., Minami, T. (1992). Seismic response of three-dimensional sediment-filled valleys due to incident plane waves. *J. Phys. Earth* **40**, 209–222.

Ortiz-Alemán, C., Sánchez-Sesma, F.J., Rodríguez-Zúñiga, J.L., Luzón, F. (1998). Computing topographical 3-D site effects using a fast IBEM/Conjugate gradient approach. *Bull. Seismol. Soc. Am.* **88**, 393–399.

Papageorgiou, A.S., Kim, J. (1991). Study of the propagation and amplification of seismic waves in Caracas valley with reference to the 29 July 1967 earthquake: SH waves. *Bull. Seismol. Soc. Am.* **81**, 2214–2233.

Papageorgiou, A.S., Pei, D. (1998). A discrete wavenumber boundary element method for 2.5-D elastodynamic scattering problems. *Earthq. Engrg. Struct. Dyn.* **27**, 619–638.

Paul, A., Campillo, M. (1988). Diffraction and conversion of elastic waves at a corrugated interface. *Geophysics* **53**, 1415–1424.

Paul, A., Jongmans, D., Campillo, M., Malin, P., Baumont, D. (1996). Amplitudes of regional seismic phases in relation to the crustal structure of the Sierra Nevada, California. *J. Geophys. Res.* **101**, 25243–25254.

Pedersen, H.A., Sánchez-Sesma, F.J., Campillo, M. (1994). Three-dimensional scattering by two-dimensional topographies. *Bull. Seismol. Soc. Am.* **84**, 1169–1183.

Pedersen, H.A., Campillo, M., Sánchez-Sesma, F.J. (1995). Azimuth dependent wave amplification in alluvial valleys. *Soil Dyn. Earthquake Engrg.* **14**, 289–300.

Pedersen, H.A., Maupin, V., Campillo, M. (1996). Wave diffraction in multilayered media with the Indirect Boundary Element Method: Application to 3-D diffraction of long-period surface waves by 2-D lithospheric structures. *Geophys. J. Int.* **125**, 545–558.

Peirce, A., Siebrits, E. (1997). Stability analysis and design of time-stepping schemes for general elastodynamics boundary element models. *Int. J. Num. Methods Engrg.* **40**, 319–342.

Regan, J., Harkrider, D.G. (1989). Seismic representation theorem coupling: synthetic SH mode sum seismograms for non-homogeneous paths. *Geophys. J. Int.* **98**, 429–446.

Riepl, J., Zahradnik, J., Plicka, V., Bard, P.Y. (2000). About the efficiency of numerical 1-D and 2-D modelling of site effects in basin structures. *Pure Appl. Geophys.* **157**, 319–342.

Rice, J.M., Sadd, M.H. (1984). Propagation and scattering of SH-waves in semi-infinite domains using a time-dependent boundary element method. *J. Appl. Mech.* **51**, 641–645.

Sánchez-Sesma, F.J. (1978). Ground motion amplification due to canyons of arbitrary shape. In: *Proceedings of the 2nd International Conference on Microzonation*. San Francisco, pp. 729–738.

Sánchez-Sesma, F.J. (1983). Diffraction of elastic waves by three-dimensional surface irregularities. *Bull. Seismol. Soc. Am.* **73**, 1621–1636.
Sánchez-Sesma, F.J., Rosenblueth, E. (1979). Ground motion at canyons of arbitrary shape under incident SH waves. *Earthquake Engrg. Struct. Dyn.* **7**, 441–450.
Sánchez-Sesma, F.J., Esquivel, J. (1979). Ground motion on alluvial valleys under incident plane SH waves. *Bull. Seismol. Soc. Am.* **69**, 1107–1120.
Sánchez-Sesma, F.J., Herrera, I., Avilés, J. (1982). A boundary method for elastic wave diffraction: Application to scattering of SH waves by surface irregularities. *Bull. Seismol. Soc. Am.* **72**, 473–490.
Sánchez-Sesma, F.J., Campillo, M. (1991). Diffraction of P, SV, and Rayleigh waves by topographic features: a boundary integral formulation. *Bull. Seismol. Soc. Am.* **81**, 2234–2253.
Sánchez-Sesma, F.J., Ramos-Martínez, J., Campillo, M. (1993). An indirect boundary element method applied to simulate the seismic response of alluvial valleys for incident P, S and Rayleigh waves. *Earthquake Engrg. Struct. Dyn.* **22**, 279–295.
Sánchez-Sesma, F.J., Luzón, F. (1995). Seismic response of three-dimensional alluvial valleys for incident P, S and Rayleigh waves. *Bull. Seismol. Soc. Am.* **85**, 269–284.
Sánchez-Sesma, F.J., Madariaga, R., Irikura, K. (2001). An approximate elastic two-dimensional Green's function for a constant-gradient medium. *Geophys. J. Int.* **146**, 237–248.
Schuster, G.T. (1985). A hybrid BIE + Born series modeling scheme: Generalized Born series. *J. Acoust. Soc. Am.* **77**, 865–879.
Sills, L.B. (1978). Scattering of horizontally-polarized waves by surface irregularities. *Geophys. J. R. Astron. Soc.* **54**, 319–348.
Somigliana, C. (1886). Sopra l'equilibrio di un corpo elastico isotrope. *Nuovo Cimento* **20**, 181–185.
Stokes, G.G. (1849). On the dynamical theory of diffraction. *Cambridge Trans. Phil. Soc. Vol. IX.*
Tadeu, A., Kausel, E. (2000). Green's functions for two-and-a-half dimensional elastodynamic problems. *J. Engrg. Mech. ASCE* **126**, 1093–1097.
Tada, T., Fukuyama, E., Madariaga, R. (2000). Non-hypersingular boundary integral equations for 3-D non-planar crack dynamics. *Comput. Mech.* **25**, 613–626.
Takenaka, H., Kennett, B.L.N., Fujiwara, H. (1996a). Effect of 2-D topography on the 3-D seismic wavefield using a 2.5-D discrete wavenumber—boundary integral equation method. *Geophys. J. Int.* **124**, 741–755.
Takenaka, H., Ohori, M., Koketsu, K., Kennett, B.L.N. (1996b). An efficient approach to the seismogram synthesis for a basin structure using propagation invariants. *Bull. Seismol. Soc. Am.* **86**, 379–388.
Uebayashi, H., Horike, M., Takeuchi, Y. (1992). Seismic motion in a three-dimensional arbitrarily-shaped sedimentary basin due to a rectangular dislocation source. *J. Phys. Earth* **40**, 223–240.
Vai, R., Castillo-Covarrubias, J.M., Sánchez-Sesma, F.J., Komatitsch, D., Vilotte, J.P. (1999). Elastic wave propagation in an irregularly layered medium. *Soil Dyn. Earthquake Engrg.* **18**, 11–18.
Webster, A.G. (1955). Partial Differential Equations in Mathematical Physics. Dover, New York.
Wong, H.L. (1982). Effect of surface topography on the diffraction of P, SV and Rayleigh waves. *Bull. Seismol. Soc. Am.* **72**, 1167–1183.
Wong, H.L., Jennings, P.C. (1975). Effects of canyon topography on strong ground motion. *Bull. Seismol. Soc. Am.* **65**, 1239–1257.
Yamashita, T., Fukuyama, E. (1996). Apparent critical slip displacement caused by the existence of a fault zone. *Geophys. J. Int.* **125**, 459–472.
Yokoi, T., Takenaka, H. (1995). Treatment of an infinitely extended free surface for indirect formulation of the boundary element method. *J. Phys. Earth* **43**, 79–103.
Yokoi, T., Sánchez-Sesma, F.J. (1998). A hybrid calculation technique of the Indirect Boundary Element Method and the analytical solutions for three-dimensional problems of topography. *Geophys. J. Int.* **133**, 121–139.
Yomogida, K., Benites, R. (1995). Relation between direct wave Q and coda Q: A numerical approach. *Geophys. J. Int.* **123**, 471–483.

Zhang, B., Papageorgiou, A.S. (1996). Simulation of the response of the Marina district basin, San Francisco, California, to the 1989 Loma Prieta earthquake. *Bull. Seismol. Soc. Am.* **86**, 1382–1400.

Zhang, B., Papageorgiou, A.S., Tassoulas, J.L. (1998). A hybrid numerical technique, combining the finite-element and boundary-element methods, for modeling the 3-D response. *Bull. Seismol. Soc. Am.* **88**, 1036–1050.

GENERATION AND PROPAGATION OF SEISMIC SH WAVES IN MULTI-LAYERED MEDIA WITH IRREGULAR INTERFACES

XIAO-FEI CHEN

*Laboratory of Computational Geodynamics, Department of Geophysics,
School of Earth and Space Sciences, Peking University, Beijing 100871, China*

ABSTRACT

This chapter is devoted to summarize the global generalized reflection/transmission matrices method developed by the author earlier for solving the generation and propagation of seismic SH wave in multi-layered media with irregular interfaces. This method is an extension or generalization of the classic generalized reflection/transmission coefficients method developed for solving the problem of seismic waves' generation and propagation in laterally homogeneous layered media. It can be directly applied to simulate the seismic SH wave field for a variety of complex structure models such as the irregular topography, basin, soft or hard inclusions, and laterally varying sub-surfaces, etc. These models are important for seismological study. Because of its analytical nature, this method is also successfully applied to establish the theoretical formulation of modes and surface wave in multi-layered media with irregular interfaces. Besides the systematic derivation of the theoretical formulation, a number of numerical examples are presented in this chapter to show the versatile applicability of this theoretical method.

Keywords: Seismic waves, Multi-layered media with irregular interfaces, Global generalized reflection/transmission matrices, Surface waves in lateral heterogeneous media

1. INTRODUCTION

In classic seismology, structure of the earth is often modelled by stratified medium model, i.e., laterally homogeneous model. In past three decades, numerous theories and effective algorithms for simulating seismic wave excitation and propagation in such kind of idealized structure model were developed, among them the reflectivity method (or generalized R/T coefficients method) (e.g., Fuchs and Muller, 1971; Kennett, 1983; Luco and Apsel, 1983; Yao and Harkrider, 1983; Chen, 1999b) and generalized ray method (e.g., Helmberger and Harkrider, 1978; Helmberger, 1983) are the most popular and widely used in seismological study.

The real structure of the earth, however, is full of heterogeneity although the lateral heterogeneity is generally weaker than the vertical heterogeneity. Thus the stratified structure model is a good approximation only for the case of relatively low frequency or some problems with local scale. For many problems, the lateral heterogeneity of the structure cannot be ignored and sometimes even plays a key role.

Since the 1970s, numerous studies on seismic wave excitation and propagation in laterally heterogeneous media have been done. Those studies can be classified into the pure numerical methods, high-frequency asymptotic methods and plane-wave decomposition methods. Although the pure numerical methods, such as finite difference methods (e.g., Boore, 1972; Levander, 1988; Moczo et al., 2004; Moczo et al., 2006), pseudo-spectral method (e.g., Kosloff and Baysal, 1982; Furumura et al., 1998; Wang et al., 2001), spectral element method (e.g., Komatitsch and Vilotte, 1998; Chaljub et al., 2006), finite element and boundary integral methods (e.g., Smith, 1974; Day, 1977; Sanchez-Sesma and Rosenbluth, 1979; Dravinski, 1982; Kawase, 1988; Bouchon and Sanchez-Sesma, 2006) usually provide satisfactory solutions, they require extensive computations. The high-frequency asymptotic methods (e.g., Cerveny et al., 1977; Hong and Helmberger, 1978; Chapman and Drummond, 1982; Frazer and Sen, 1985; Frazer, 1987; Cerveny et al., 2006) are only appropriate and efficient for computing body waves at relatively high-frequency cases, but not for surface wave's problems. The plane-wave decomposition methods (e.g., Aki and Larner, 1970; Bouchon, 1973; Bard and Bouchon, 1980a, 1980b), on the contrary, usually favor the lower frequency problems.

Recently the author developed a new method, the *global R/T matrices method*, for simulating the seismic wave excitation and propagation in an arbitrarily multi-layered medium with irregular interfaces (e.g., Chen, 1990, 1995, 1996; Cao et al., 2004), which can be regarded as an extension of the generalized R/T coefficients method for the horizontally layered case (e.g., Kennett, 1983; Chen, 1999b) to the case of multi-layered media with irregular interfaces by incorporating T-matrix approach (e.g., Waterman, 1969, 1975). This method provides an efficient scheme to synthesize seismograms in the multi-layered media with irregular interfaces due to an arbitrary source. In this section, I will summarize the basic idea and fundamental formulation of this new method, demonstrate its validity and show its applicability for synthetic seismograms in a variety of laterally heterogeneous structure.

2. Seismic SH Wave Generation and Propagation in Multi-Layered Media with Irregular Interfaces

For simplicity, in this section we will call *the multi-layered media with irregular interfaces* as *irregular multi-layered media*, or simply abbreviated to *IMLM*. Such

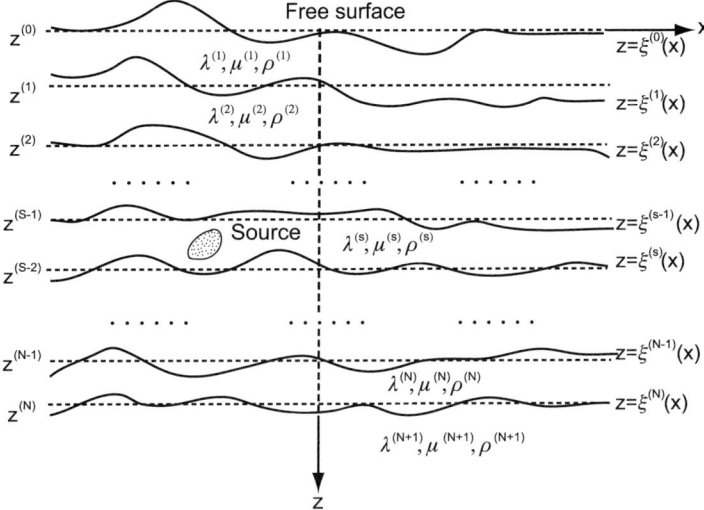

FIG. 1. Configuration of the problem considered in this study.

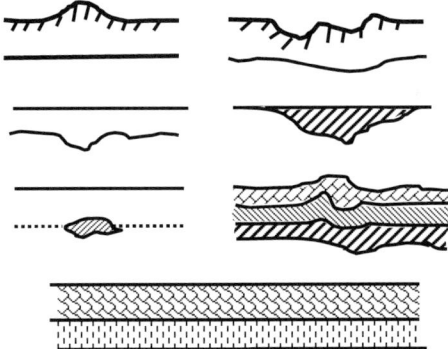

FIG. 2. Some typical structure models of seismologically interesting that can be modeled by the general model of multi-layers with irregular interfaces.

medium model is illustrated in Fig. 1, and it represents a variety of seismologically interesting complex models, such as the irregular topography, basin, soft or hard inclusions, and laterally varying sub-subsurfaces, etc., as shown in Fig. 2.

2.1. Integral Equation and its Discrete Wave Number Representation

The problem to be studied here is illustrated in Fig. 1. There are N homogeneous layers over a half-space, and each layer is bounded by irregular interfaces,

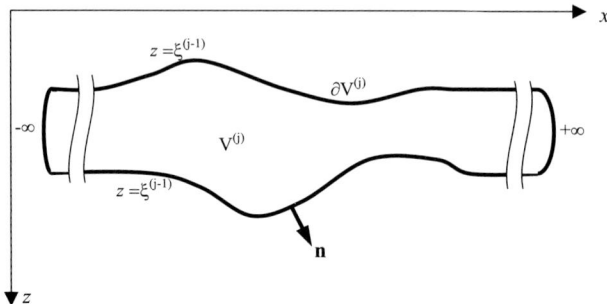

FIG. 3. Geometry of the jth layer of the general structure model considered in this study.

$z = \xi^{(j-1)}(x)$ and $z = \xi^{(j)}(x)$. The uppermost interface, $z = \xi^{(0)}(x)$, is a free surface and an arbitrary source is embedded in the sth layer. The elastic properties of each layer are described by the shear modulus $\mu^{(j)}$, and density $\rho^{(j)}$. For simplicity, we here only consider the two-dimensional SH problem. Then, the spectra of displacement $W^{(j)}(\mathbf{x}, \omega)$ in the jth layer satisfies the following equation,

$$\nabla^2 W^{(j)}(\mathbf{x}, \omega) + \left(k_0^{(j)}\right)^2 W^{(j)}(\mathbf{x}, \omega) = -\delta_{js} f(\mathbf{x}, \omega), \quad \mathbf{x} \in V^{(j)},$$
$$\text{for } j = 1, 2, \ldots, N, N+1, \tag{1}$$

where $(k_0^{(j)})^2 = \omega^2 \rho^{(j)}/\mu^{(j)}$, δ_{js} is Kronecker delta; $f(\mathbf{x}, \omega)$ is the distribution of body force (or equivalent body force); $V^{(j)}$ is the volume of the jth layer enclosed by surfaces j and $(j-1)$ and two arcs at infinity (see Fig. 3). The corresponding full-space Green's function $G^{(j)}(\mathbf{x}, \mathbf{x}')$ satisfies the following equations,

$$\nabla^2 G^{(j)}(\mathbf{x}, \mathbf{x}') + \left(k_0^{(j)}\right)^2 G^{(j)}(\mathbf{x}, \mathbf{x}') = -4\pi \delta(\mathbf{x} - \mathbf{x}'),$$
$$\mathbf{x}, \mathbf{x}' \in (-\infty, +\infty) \times (-\infty, +\infty). \tag{1a}$$

We multiply Eq. (1) by $G^{(j)}(\mathbf{x}, \mathbf{x}')$, multiply Eq. (1a) by $-W^{(j)}(x, \omega)$, then combine them together and integrate it over $V^{(j)}$, and obtain the following integral equation

$$\delta_{js} \int_{V^{(j)}} f(\mathbf{x}, \omega) G^{(j)}(\mathbf{x}, \mathbf{x}') \, dV(\mathbf{x})$$
$$+ \int_{\partial V^{(j)}} \mathbf{n} \cdot \left\{ G^{(j)}(\mathbf{x}, \mathbf{x}') \nabla W^{(j)}(\mathbf{x}, \omega) - W^{(j)}(\mathbf{x}, \omega) \nabla G^{(j)}(\mathbf{x}, \mathbf{x}') \right\} dl(\mathbf{x})$$
$$= \begin{cases} 4\pi W^{(j)}(\mathbf{x}', \omega), & \text{for } \mathbf{x}' \text{ inside } V^{(j)}, \\ 0, & \text{for } \mathbf{x}' \text{ outside } V^{(j)}, \end{cases} \quad j = 1, 2, \ldots, N, N+1. \tag{2}$$

Where, $\partial V^{(j)}$ is the boundary of $V^{(j)}$, and $\mathbf{n}$ is normal vector of $\partial V^{(j)}$. As shown in Fig. 3, $\partial V^{(j)}$ consists of interfaces $z = \xi^{(j-1)}(x)$ and $z = \xi^{(j)}(x)$, and the arcs at infinity. Considering the contributions scattered from these arcs to the observed wave field are negligible, we can perform the surface integral of Eq. (2) only along the interfaces j and $j-1$ instead of the entire $\partial V^{(j)}$.

In this study, the interface irregularity is assumed to be localized, so we can use the discrete wave number approach to represent this problem. As shown in Appendix A, the discrete wave number representation of Eq. (2) is

$$\sum_{n=-\infty}^{+\infty} \left[s_n(z', \omega) \delta_{sj} + a_n^{(j)}(z', \omega) \right] e^{ix'k_n}$$

$$= \begin{cases} W^{(j)}(\mathbf{x}', \omega), & \text{for } \xi^{(j-1)}(x') < z' < \xi^{(j)}(x'), \quad -\frac{L}{2} < x' < \frac{L}{2}, \\ 0, & \text{for } z' < \xi^{(j-1)}(x'), \text{ or } z' > \xi^{(j)}(x'), \quad -\frac{L}{2} < x' < \frac{L}{2}, \end{cases}$$
(3a)

for $j = 1, 2, \ldots, N, N+1$; where L is the length of periodicity in x coordinate, $s_n(z', \omega)$ represents the source contribution and is given by

$$s_n(z', \omega) = \frac{i}{2Lv_n^{(s)}} \int_{-L/2}^{L/2} \int_{\xi^{(j-1)}(x)}^{\xi^{(j)}(x)} f(x, z) e^{-i(k_n x - v_n^{(j)}|z-z'|)} \, dx \, dz; \quad (3b)$$

$a_n(z', \omega)$ associated with the scattered wave field is given by

$$a_n(z', \omega) = \frac{1}{2Lv_n^{(j)}} \int_{-L/2}^{L/2} dx \left\{ \frac{i\tau^{(j)}[x, \xi^{(j)}(x)]}{\mu^{(j)}} \sqrt{1 + [\dot{\xi}^{(j)}(x)]^2} \right.$$
$$+ \left[\dot{\xi}^{(j)}(x) k_n + v_n^{(j)} \operatorname{sgn}(\xi^{(j)}(x) - z') \right] W^{(j)}[x, \xi^{(j)}(x)] \right\}$$
$$\times e^{-i(k_n x - v_n^{(j)}|\xi^{(j)}(x)-z'|)}$$
$$- \frac{1}{2Lv_n^{(j)}} \int_{-L/2}^{L/2} dx \left\{ \frac{i\tau^{(j)}[x, \xi^{(j-1)}(x)]}{\mu^{(j)}} \sqrt{1 + [\dot{\xi}^{(j-1)}(x)]^2} \right.$$
$$+ \left[\dot{\xi}^{(j-1)}(x) k_n + v_n^{(j)} \operatorname{sgn}(\xi^{(j-1)}(x) - z') \right]$$
$$\times W^{(j)}[x, \xi^{(j-1)}(x)] \right\} e^{-i(k_n x - v_n^{(j)}|\xi^{(j-1)}(x)-z'|)}, \quad (3c)$$

where $k_n = 2\pi n/L$, $v_n^{(j)} = \sqrt{(k_0^{(j)})^2 - (k_n)^2}$ with $\operatorname{Im}\{v_n^{(j)}\} > 0$ ($n = 0$, $\pm 1, \pm 2, \pm 3, \ldots$), $\dot{\xi}^{(j)}(x)$ is the derivative of $\xi^{(j)}(x)$ with respect to x, and $\tau^{(j)}[x, \xi^{(j)}(x)]$ and $\tau^{(j)}[x, \xi^{(j-1)}(x)]$ are traction fields on the jth and $(j-1)$th interfaces.

Equation (3a) shows that to obtain the total wave field we need to know $s_n(z', \omega)$ and $a_n(z', \omega)$, the representations of the direct radiated wave from the source and the scattered wave filed by the surrounding interfaces, respectively. For convenience, we simply call them the source term and scattered wave term, respectively. Using Eq. (3b), we can directly calculate the source term $s_n(z', \omega)$

from the known source distribution. Involving two unknowns $W^{(j)}(\mathbf{x}, \omega)$ and $\tau^{(j)}(\mathbf{x}, \omega)$, the scattered wave term $a_n(z', \omega)$ cannot be determined directly from Eq. (3c). Thus, our next step is to evaluate the scattered wave term $a_n(z', \omega)$. We shall first convert these integral equations into a set of simultaneous matrix equations by spanning the unknown wave fields on the interfaces, and then solve these simultaneous matrix equations.

2.2. Simultaneous Matrix Equations

To convert integral equation (3) into matrix equations, we need to expand the wave field at each interface by a set of appropriate function basis. Consider an interface, $z = \xi(x)$, shown in Fig. 4, the wave fields above and below the interface can be expressed, respectively, as (e.g., Waterman, 1975),

$$w(x, z) = \sum_{m=-\infty}^{+\infty} c_m^{(-)} \psi_m^{(-)}(x, z), \quad \text{for } z < \xi_{\min},$$

$$w(x, z) = \sum_{m=-\infty}^{+\infty} c_m^{(+)} \psi_m^{(+)}(x, z), \quad \text{for } z > \xi_{\max},$$

where the bases are

$$\psi_m^{(\pm)} = e^{i(xk_m \pm zv_m)},$$

where $k_m = 2\pi m/L$, $v_m = \sqrt{(k_\beta)^2 - (k_m)^2}$ with $\text{Im}\{v_m\} > 0$, and $k_\beta = \omega/\beta$. Here we used "+" and "−" to denote the lower and upper media, respectively. We

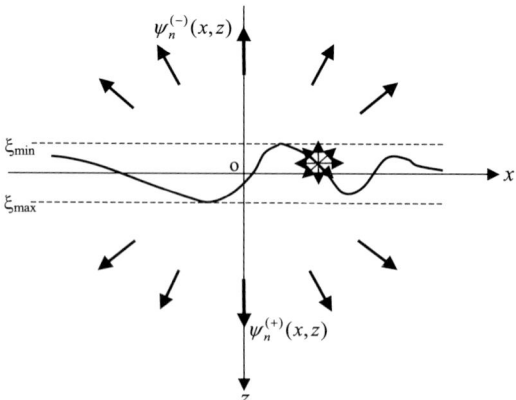

FIG. 4. Configuration of the wave fields around an irregular interface.

recognized that neither $\psi_m^{(-)}(x, z)$ nor $\psi_m^{(+)}(x, z)$ alone can correctly describe the wave field at the interface. Because the total wave field at an irregular interface should include both up-going wave $\psi_m^{(-)}(x, z)$ and down-going wave $\psi_m^{(+)}(x, z)$, rather than only one of them. To describe the wave field at the interface more accurately, we define following basis

$$\hat{u}_m^{(j)}(x) = e^{[ixk_m + iv_m^{(j+1)}|\Delta\xi^{(j)}(x)|]}, \tag{4a}$$

for displacement field at the jth interface. Where $m = 0, \pm1, \pm2, \pm3, \ldots$ and $\Delta\xi^{(j)}(x) = \xi^{(j)}(x) - z^{(j)}$ is the topographic fluctuation at jth interface (see Fig. 1 where $z^{(j)}$ and $z^{(j-1)}$ are the reference depths of the jth and $(j-1)$th interfaces). Considering the constitutive relation (e.g., Aki and Richards, 1980), we construct the basis for traction field at the jth interface as

$$\hat{t}_m^{(j)}(x) = i\mu^{(j+1)}\{n_x^{(j)}(x)k_m + n_z^{(j)}(x)v_m^{(j+1)} \operatorname{sgn}[\Delta\xi^{(j)}(x)]\}$$
$$\times e^{[ixk_m + iv_m^{(j+1)}|\Delta\xi^{(j)}(x)|]}, \tag{4b}$$

where $\mathbf{n}^{(j)}(x)$ is the normal unit vector of the jth interface. Therefore the expansions of wave fields at the arbitrary jth interface ($z = \xi^{(j)}(x)$) have the following forms

$$W^{(j)}[x, \xi^{(j)}(x); \omega] = W^{(j+1)}[x, \xi^{(j)}(x); \omega] = \sum_{m=-\infty}^{+\infty} \alpha_m^{(j)} \hat{u}_m^{(j)}(x), \tag{5a}$$

and

$$\tau^{(j)}[x, \xi^{(j)}(x); \omega] = \tau^{(j+1)}[x, \xi^{(j)}(x); \omega] = \sum_{m=-\infty}^{+\infty} \beta_m^{(j)} \hat{t}_m^{(j)}(x), \tag{5b}$$

for $j = 0, 1, 2, \ldots, N$; where $\alpha_m^{(j)}$ and $\beta_m^{(j)}$ are the expansion coefficients for the jth interface. It should be stressed that in the above expansions we have imposed the continuities of displacement and traction at each interface. Substitution of these expansions into Eq. (3c) yields

$$a_n^{(j)}(z', \omega) = \sum_{m=-\infty}^{+\infty} \{(\mathbf{Q}_\uparrow^{(j)}(z', \omega))_{nm} \alpha_m^{(j)} + (\mathbf{P}_\uparrow^{(j)}(z', \omega))_{nm} \beta_m^{(j)}\}$$
$$- \sum_{m=-\infty}^{+\infty} \{(\mathbf{Q}_\downarrow^{(j)}(z', \omega))_{nm} \alpha_m^{(j-1)} + (\mathbf{P}_\downarrow^{(j)}(z', \omega))_{nm} \beta_m^{(j-1)}\}, \tag{6}$$

where $\mathbf{Q}_\uparrow^{(j)}(z', \omega)$, $\mathbf{P}_\uparrow^{(j)}(z', \omega)$, $\mathbf{Q}_\downarrow^{(j)}(z', \omega)$ and $\mathbf{P}_\downarrow^{(j)}(z', \omega)$ are given by

$$(\mathbf{Q}_\uparrow^{(j)}(z', \omega))_{nm} = \frac{1}{2v_n^{(j)}L} \int_{-L/2}^{L/2} \{\dot{\xi}^{(j)}(x)k_n + v_n^{(j)} \operatorname{sgn}[\xi^{(j)}(x) - z']\}$$
$$\times e^{i\Phi_\uparrow^{(j)}(x,z')} dx, \tag{6a}$$

$$\left(\mathbf{P}_\uparrow^{(j)}(z',\omega)\right)_{nm}$$
$$= \frac{\mu^{(j+1)}}{2\mu^{(j)}v_n^{(j)}L} \int_{-L/2}^{L/2} \{\dot{\xi}^{(j)}(x)k_m - v_m^{(j+1)}\operatorname{sgn}[\Delta\xi^{(j)}(x)]\}e^{i\Phi_\uparrow^{(j)}(x,z')}\,dx, \tag{6b}$$

$$\left(\mathbf{Q}_\downarrow^{(j)}(z',\omega)\right)_{nm}$$
$$= \frac{1}{2v_n^{(j)}L} \int_{-L/2}^{L/2} \{\dot{\xi}^{(j-1)}(x)k_n + v_n^{(j)}\operatorname{sgn}[\xi^{(j-1)}(x) - z']\}e^{i\Phi_\downarrow^{(j)}(x,z')}\,dx, \tag{6c}$$

and

$$\left(\mathbf{P}_\downarrow^{(j)}(z',\omega)\right)_{nm}$$
$$= \frac{1}{2v_n^{(j)}L} \int_{-L/2}^{L/2} \{\dot{\xi}^{(j-1)}(x)k_m - v_m^{(j)}\operatorname{sgn}[\Delta\xi^{(j-1)}(x)]\}e^{i\Phi_\downarrow^{(j)}(x,z')}\,dx, \tag{6d}$$

where

$$\Phi_\downarrow^{(j)}(x,z') = (k_m - k_n)x + v_n^{(j)}|\xi^{(j-1)}(x) - z'| + v_m^{(j)}|\Delta\xi^{(j-1)}(x)|,$$

and

$$\Phi_\uparrow^{(j)}(x,z') = (k_m - k_n)x + v_n^{(j)}|\xi^{(j)}(x) - z'| + v_m^{(j+1)}|\Delta\xi^{(j)}(x)|.$$

In the above formulation, we used the up and down arrows to indicate the terms associated with up-going and down-going waves, respectively. In rest of this chapter, usage of up and down arrows will also follow such principle. We see that the problem of solving integral equation (3) is now reduced to the problem of seeking a set of unknown expansion coefficients. Next, we shall find the equations by which these coefficients are related. According to Eq. (3a) we have,

$$0 = \sum_{n=-\infty}^{+\infty} \{s_n(z',\omega)\delta_{sj} + a_n^{(j)}(z',\omega)\}e^{ik_n x'},$$
$$\text{for } z' < \xi_{\min}^{(j-1)} \quad \text{or} \quad z' > \xi_{\max}^{(j)}, \tag{7}$$

where $j = 0, 1, 2, \ldots, N$; and $\xi_{\min}^{(j-1)}$ and $\xi_{\max}^{(j)}$ are defined by

$$\xi_{\max}^{(j)} = \max\{\xi^{(j)}(x) \mid -L/2 < x < L/2\},$$

$$\xi_{\min}^{(j-1)} = \min\{\xi^{(j-1)}(x) \mid -L/2 < x < L/2\},$$

and we also assume that $\xi_{\min}^{(j-1)} \leq \xi_{\min}^{(j)}$ and $\xi_{\max}^{(j)} \geq \xi_{\max}^{(j-1)}$. Since $\{e^{ik_n x'}\}$ is an orthogonal function set over $-L/2 < x' < L/2$ and $\{s_n(z',\omega)\delta_{sj} + a_n^{(j)}(z',\omega)\}$

is independent of x' in the above case, thus Eq. (7) can be reduced to

$$0 = \{s_n(z', \omega)\delta_{sj} + a_n^{(j)}(z', \omega)\},$$
$$\text{for } z' < \xi_{\min}^{(j-1)} \quad \text{or} \quad z' > \xi_{\max}^{(j)}, \quad j = 1, 2, \ldots, N. \tag{8}$$

For each layer, Eq. (8) implies that for $z' < \xi_{\min}^{(j-1)}$

$$-\delta_{sj}\left(s_{0\uparrow}(\omega)\right)_n e^{iv_n^{(j)}(\xi_{\min}^{(j)}-z')}$$

$$= \sum_{m=-\infty}^{+\infty} \{(\mathbf{Q}_{\uparrow\uparrow}^{(j)}(\omega))_{nm}\alpha_m^{(j)} + (\mathbf{P}_{\uparrow\uparrow}^{(j)}(\omega))_{nm}\beta_m^{(j)}\} e^{iv_n^{(j)}(\xi_{\min}^{(j)}-z')}$$

$$- \sum_{m=-\infty}^{+\infty} \{(\mathbf{Q}_{\downarrow\uparrow}^{(j)}(\omega))_{nm}\alpha_m^{(j-1)} + (\mathbf{P}_{\downarrow\uparrow}^{(j)}(\omega))_{nm}\beta_m^{(j-1)}\} e^{iv_n^{(j)}(\xi_{\min}^{(j-1)}-z')}$$
$$\tag{8a}$$

and for $z' > \xi_{\max}^{(j)}$

$$-\delta_{sj}\left(s_{0\downarrow}(\omega)\right)_n e^{iv_n^{(j)}(z'-\xi_{\max}^{(j-1)})}$$

$$= \sum_{m=-\infty}^{+\infty} \{(\mathbf{Q}_{\uparrow\downarrow}^{(j)}(\omega))_{nm}\alpha_m^{(j)} + (\mathbf{P}_{\uparrow\downarrow}^{(j)}(\omega))_{nm}\beta_m^{(j)}\} e^{iv_n^{(j)}(z'-\xi_{\max}^{(j)})}$$

$$- \sum_{m=-\infty}^{+\infty} \{(\mathbf{Q}_{\downarrow\downarrow}^{(j)}(\omega))_{nm}\alpha_m^{(j-1)} + (\mathbf{P}_{\downarrow\downarrow}^{(j)}(\omega))_{nm}\beta_m^{(j-1)}\} e^{iv_n^{(j)}(z'-\xi_{\max}^{(j-1)})},$$
$$\tag{8b}$$

where

$$\left(s_{0\uparrow}(\omega)\right)_n = \frac{i}{2Lv_n^{(s)}} \int_{-L/2}^{L/2} dx \int_{\xi^{(s-1)}(x)}^{\xi^{(s)}(x)} dz \, f(x,z) e^{-i[k_n x - v_n^{(s)}(z-\xi_{\min}^{(s)})]}, \tag{8c}$$

$$\left(s_{0\downarrow}(\omega)\right)_n = \frac{i}{2Lv_n^{(s)}} \int_{-L/2}^{L/2} dx \int_{\xi^{(s-1)}(x)}^{\xi^{(s)}(x)} dz \, f(x,z) e^{-i[k_n x + v_n^{(s)}(z-\xi_{\max}^{(s-1)})]}, \tag{8d}$$

and

$$(\mathbf{Q}_{\uparrow\uparrow}^{(j)}(\omega))_{nm} = \frac{1}{2v_n^{(j)}L} \int_{-L/2}^{L/2} \{\dot{\xi}^{(j)}(x)k_n + v_n^{(j)}\} e^{i\Psi_{\uparrow\uparrow}^{(j)}(x)} \, dx, \tag{8e}$$

$$(\mathbf{Q}_{\uparrow\downarrow}^{(j)}(\omega))_{nm} = \frac{1}{2v_n^{(j)}L} \int_{-L/2}^{L/2} \{\dot{\xi}^{(j)}(x)k_n - v_n^{(j)}\} e^{i\Psi_{\uparrow\downarrow}^{(j)}(x)} \, dx, \tag{8f}$$

$$(\mathbf{Q}_{\downarrow\uparrow}^{(j)}(\omega))_{nm} = \frac{1}{2v_n^{(j)}L} \int_{-L/2}^{L/2} \{\dot{\xi}^{(j-1)}(x)k_n + v_n^{(j)}\} e^{i\Psi_{\downarrow\uparrow}^{(j)}(x)} \, dx, \tag{8g}$$

$$(\mathbf{Q}_{\downarrow\downarrow}^{(j)}(\omega))_{nm} = \frac{1}{2v_n^{(j)}L} \int_{-L/2}^{L/2} \{\dot{\xi}^{(j-1)}(x)k_n - v_n^{(j)}\} e^{i\Psi_{\downarrow\downarrow}^{(j)}(x)} \, dx, \tag{8h}$$

$$(\mathbf{P}_{\uparrow\uparrow}^{(j)}(\omega))_{nm}$$
$$= \frac{\mu^{(j+1)}}{2\mu^{(j)}v_n^{(j)}L} \int_{-L/2}^{L/2} \{\dot{\xi}^{(j)}(x)k_m - v_m^{(j+1)} \operatorname{sgn}[\Delta\xi^{(j)}(x)]\} e^{i\Psi_{\uparrow\uparrow}^{(j)}(x)} \, dx, \tag{8i}$$

$$(\mathbf{P}_{\uparrow\downarrow}^{(j)}(\omega))_{nm}$$
$$= \frac{\mu^{(j+1)}}{2\mu^{(j)}v_n^{(j)}L} \int_{-L/2}^{L/2} \{\dot{\xi}^{(j)}(x)k_m - v_m^{(j+1)} \operatorname{sgn}[\Delta\xi^{(j)}(x)]\} e^{i\Psi_{\uparrow\downarrow}^{(j)}(x)} \, dx, \tag{8j}$$

$$(\mathbf{P}_{\downarrow\uparrow}^{(j)}(\omega))_{nm}$$
$$= \frac{1}{2v_n^{(j)}L} \int_{-L/2}^{L/2} \{\dot{\xi}^{(j-1)}(x)k_m - v_m^{(j)} \operatorname{sgn}[\Delta\xi^{(j-1)}(x)]\} e^{i\Psi_{\downarrow\uparrow}^{(j)}(x)} \, dx, \tag{8k}$$

$$(\mathbf{P}_{\downarrow\downarrow}^{(j)}(\omega))_{nm}$$
$$= \frac{1}{2v_n^{(j)}L} \int_{-L/2}^{L/2} \{\dot{\xi}^{(j-1)}(x)k_m - v_m^{(j)} \operatorname{sgn}[\Delta\xi^{(j-1)}(x)]\} e^{i\Psi_{\downarrow\downarrow}^{(j)}(x)} \, dx, \tag{8l}$$

with

$$\Psi_{\uparrow\uparrow}^{(j)}(x) = (k_m - k_n)x + v_n^{(j)}(\xi^{(j)}(x) - \xi_{\min}^{(j)}) + v_m^{(j+1)}|\Delta\xi^{(j)}(x)|,$$

$$\Psi_{\uparrow\downarrow}^{(j)}(x) = (k_m - k_n)x + v_n^{(j)}(\xi_{\max}^{(j)} - \xi^{(j)}(x)) + v_m^{(j+1)}|\Delta\xi^{(j)}(x)|,$$

$$\Psi_{\downarrow\uparrow}^{(j)}(x) = (k_m - k_n)x + v_n^{(j)}(\xi^{(j-1)}(x) - \xi_{\min}^{(j-1)}) + v_m^{(j)}|\Delta\xi^{(j-1)}(x)|,$$

$$\Psi_{\downarrow\downarrow}^{(j)}(x) = (k_m - k_n)x + v_n^{(j)}(\xi_{\max}^{(j-1)} - \xi^{(j-1)}(x)) + v_m^{(j)}|\Delta\xi^{(j-1)}(x)|.$$

Where $\mathbf{s}_{0\uparrow}(\omega)$ and $\mathbf{s}_{0\downarrow}(\omega)$ are source terms contained seismic source information; $\mathbf{Q}_{\uparrow\uparrow}^{(j)}(\omega)$, $\mathbf{Q}_{\uparrow\downarrow}^{(j)}(\omega)$, $\mathbf{Q}_{\downarrow\uparrow}^{(j)}(\omega)$, $\mathbf{Q}_{\downarrow\downarrow}^{(j)}(\omega)$, $\mathbf{P}_{\uparrow\uparrow}^{(j)}(\omega)$, $\mathbf{P}_{\uparrow\downarrow}^{(j)}(\omega)$, $\mathbf{P}_{\downarrow\uparrow}^{(j)}(\omega)$ and $\mathbf{P}_{\downarrow\downarrow}^{(j)}(\omega)$ are constant matrices whose explicit expressions are given by integrals along the interfaces, and they are defined as *interface-matrices*.

In matrix form, Eq. (8) can be written as

$$-\delta_{sj}\mathbf{E}_{\min}^{(j)}\mathbf{s}_{0\uparrow}(\omega)$$
$$= \mathbf{E}_{\min}^{(j)}\{\mathbf{Q}_{\uparrow\uparrow}^{(j)}\underline{\alpha}^{(j)} + \mathbf{P}_{\uparrow\uparrow}^{(j)}\underline{\beta}^{(j)}\} - \{\mathbf{Q}_{\downarrow\uparrow}^{(j)}\underline{\alpha}^{(j-1)} + \mathbf{P}_{\downarrow\uparrow}^{(j)}\underline{\beta}^{(j-1)}\}, \tag{9a}$$

$$-\delta_{sj}\mathbf{E}_{max}^{(j)}\mathbf{s}_{0\downarrow}(\omega)$$
$$= \{\mathbf{Q}_{\uparrow\downarrow}^{(j)}\underline{\alpha}^{(j)} + \mathbf{P}_{\uparrow\downarrow}^{(j)}\underline{\beta}^{(j)}\} - \mathbf{E}_{max}^{(j)}\{\mathbf{Q}_{\downarrow\downarrow}^{(j)}\underline{\alpha}^{(j-1)} + \mathbf{P}_{\downarrow\downarrow}^{(j)}\underline{\beta}^{(j-1)}\}, \quad (9b)$$

for $j = 1, 2, \ldots, N$; where

$$\underline{\alpha}^{(j)} = \left(\ldots, \alpha_{-M}^{(j)}, \ldots, \alpha_{-1}^{(j)}, \alpha_0^{(j)}, \ldots, \alpha_M^{(j)}, \ldots\right)^T,$$

$$\underline{\beta}^{(j)} = \left(\ldots, \beta_{-M}^{(j)}, \ldots, \beta_{-1}^{(j)}, \beta_0^{(j)}, \ldots, \beta_M^{(j)}, \ldots\right)^T,$$

$$\mathbf{E}_{max}^{(j)} = \text{diag}\{e^{iv_n^{(j)}(\xi_{max}^{(j)} - \xi_{max}^{(j-1)})}, \ n = 0, \pm 1, \pm 2, \ldots\},$$

$$\mathbf{E}_{min}^{(j)} = \text{diag}\{e^{iv_n^{(j)}(\xi_{min}^{(j)} - \xi_{min}^{(j-1)})}, \ n = 0, \pm 1, \pm 2, \ldots\}.$$

In the half-space ($j = N+1$, the lower-most layer), we have only one equation

$$0 = -\{\mathbf{Q}_{\downarrow\uparrow}^{(N+1)}\underline{\alpha}^{(N)} + \mathbf{P}_{\downarrow\uparrow}^{(N+1)}\underline{\beta}^{(N)}\}. \quad (9c)$$

Here, we assumed that no sources exist in the half-space. For the case in which there exists a source in the half-space, we can introduce a fictitious interface to preserve the validity of Eq. (9c). The traction-free condition at the top surface leads to

$$\underline{\beta}^{(0)} = 0. \quad (9d)$$

So far, we have derived $(2N+1)$ independent matrix equations (9a) to (9d), for the same number of unknowns vectors $\{\underline{\alpha}^{(j)}, \underline{\beta}^{(j)}\}$. In principle, these unknowns can be directly determined by solving these $(2N+1)$ simultaneous matrix equations. Such a method, however, is not efficient, particularly when the number of layers becomes large. In the following sections, we shall develop an efficient method to solve the above matrix equations.

2.3. Global Generalized Reflection/Transmission Matrices

The algorithm of generalized reflection/transmission (hereafter abbreviated to R/T) coefficients has been very successful in dealing with the problem of propagation and excitation of seismic waves in horizontally homogeneous layered media (e.g., Fuchs and Muller, 1971; Kennett, 1983; Luco and Apsel, 1983; Yao and Harkrider, 1983; Chen, 1999b). We attempt to extend that effective algorithm to our current problem, that is, seismic wave's propagation and excitation in multi-layered media with irregular interfaces. To do so, we first introduce the concept of *generalized amplitudes* of the wave-fields in an arbitrary irregular layer, and then introduce the global generalized R/T matrices.

2.3.1. Generalized Amplitudes

To introduce the concept of *generalized amplitude* of wave field for the IMLM (irregular multi-layered media), let us consider a simplified IMLM case that satisfies the simplicity assumption

$$\xi_{\min}^{(j)} > \xi_{\max}^{(j-1)}, \quad j = 1, 2, \ldots, N. \tag{10}$$

Under this assumption, as shown in Fig. 5a, each layer can be divided into three sub-regions: the upper region, V_U, the middle region, V_M, and the lower region, V_L. The wave fields in V_U and V_L are complicated. In V_U, for instance, the up-going scattered wave not only comes from the lower boundary, but also from some part of the upper boundary that is lower than the observation point. A similar problem occurs in V_L. In the middle sub-region V_M, however, the wave field is similar to that in horizontally layered media and can be decomposed into up-going and down-going waves with constant amplitudes as,

$$a_n^{(j)}(z', \omega) = \left[\mathbf{a}_\downarrow^{(j)}(\omega)\right]_n e^{i v_n^{(j)}(z' - \xi_{\max}^{(j-1)})} + \left[\mathbf{a}_\uparrow^{(j)}(\omega)\right]_n e^{i v_n^{(j)}(\xi_{\min}^{(j)} - z')},$$
$$\text{for } \xi_{\max}^{(j-1)} < z' < \xi_{\min}^{(j)}, \tag{11}$$

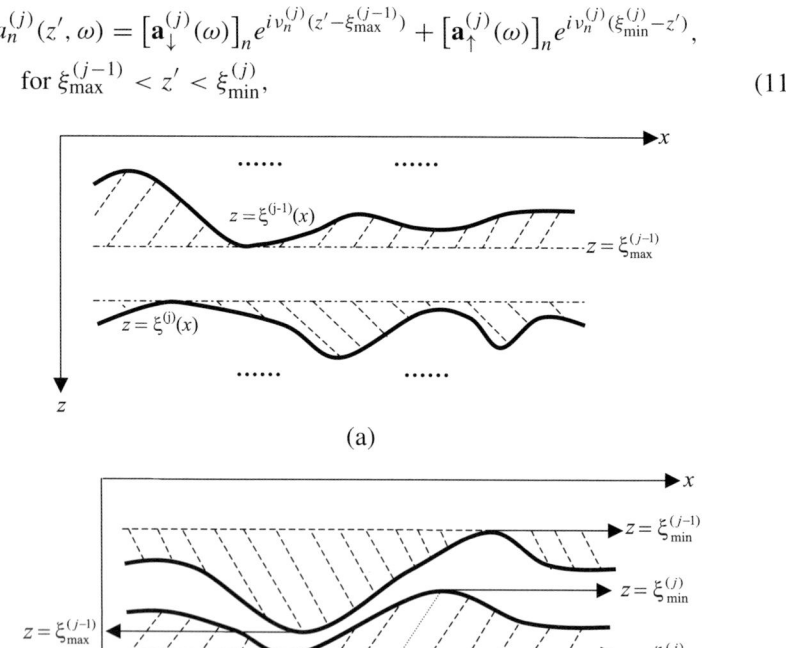

FIG. 5. (a) The structure model that satisfies the simplicity assumption; (b) the general structure model that may not satisfy the simplicity assumption.

where

$$\begin{cases} \mathbf{a}_\downarrow^{(j)}(\omega) = -\{\mathbf{Q}_{\downarrow\downarrow}^{(j)}(\omega)\underline{\alpha}^{(j-1)} + \mathbf{P}_{\downarrow\downarrow}^{(j)}(\omega)\underline{\beta}^{(j-1)}\}, \\ \mathbf{a}_\uparrow^{(j)}(\omega) = \{\mathbf{Q}_{\uparrow\uparrow}^{(j)}(\omega)\underline{\alpha}^{(j)} + \mathbf{P}_{\uparrow\uparrow}^{(j)}(\omega)\underline{\beta}^{(j)}\}, \\ \text{for } j = 1, 2, \ldots, N, N+1. \end{cases} \quad (11a)$$

We define the middle sub-region V_M *normal region*, while the sub-regions V_U and V_L *complex regions*. Note that in the simplified IMLM case, Eq. (11a) gives rise to the explicit expressions of the constant amplitudes of up- and down-going waves inside the normal region of the arbitrary jth layer. However, the right-hand side of this equation is mathematically well defined even for the general IMLM case (see Fig. 5b) in which the simplicity assumption (Eq. (10)) does not hold, and the scattered wave fields thus cannot be decomposed into the form of Eq. (11). Therefore, for the general IMLM case, we introduce the concepts of *generalized amplitudes* of the wave filed inside the jth layer as the constant quantities defined by Eq. (11a). Obviously for the simplified IMLM case, they become the actual amplitudes of up- and down-going waves within the normal region. On the other hand, the definitions of generalized amplitudes, Eq. (11a), can be viewed as a transform from the unknowns $\{\underline{\alpha}^{(j)}, \underline{\beta}^{(j)}\}$ to the intermediate quantities, i.e., *generalized amplitudes* $\{\mathbf{a}_\uparrow^{(j)}(\omega), \mathbf{a}_\downarrow^{(j)}(\omega)\}$, and the corresponding inverse transform is given by

$$\begin{bmatrix} \underline{\alpha}^{(j)} \\ \underline{\beta}^{(j)} \end{bmatrix} = \begin{bmatrix} \mathbf{Q}_{\uparrow\uparrow}^{(j)} & \mathbf{P}_{\uparrow\uparrow}^{(j)} \\ -\mathbf{Q}_{\downarrow\downarrow}^{(j+1)} & -\mathbf{P}_{\downarrow\downarrow}^{(j+1)} \end{bmatrix}^{-1} \begin{bmatrix} \mathbf{a}_\uparrow^{(j)} \\ \mathbf{a}_\downarrow^{(j+1)} \end{bmatrix}, \quad \text{for } j = 1, 2, 3, \ldots, N; \quad (12)$$

and

$$\underline{\alpha}^{(0)} = -\left(\mathbf{Q}_{\downarrow\downarrow}^{(1)}\right)^{-1}\mathbf{a}_\downarrow^{(1)}. \quad (12a)$$

Here the free surface condition $\underline{\beta}^{(0)} = 0$, has been used. These equations allow us firstly to find $\mathbf{a}_\uparrow^{(j)}$ and $\mathbf{a}_\downarrow^{(j)}$, and then determine the expansion coefficients $\underline{\alpha}^{(j)}$ and $\underline{\beta}^{(j)}$. To solve the generalized amplitudes $\mathbf{a}_\uparrow^{(j)}$ and $\mathbf{a}_\downarrow^{(j)}$, we shall extend the method of *generalized R/T coefficients* (e.g., Luco and Apsel, 1983; Chen, 1999b) developed for case of horizontally homogeneous layered media to the present problem.

2.3.2. Global Modified R/T Matrices

First, we shall extend the definition of modified R/T coefficients for the case of horizontally homogeneous layered media, which describes the reflection/transmission effect about a single interface between two homogeneous materials. In the present case, we define the *global modified R/T matrices* for each interface by the following relations:

$$\begin{cases} \mathbf{a}_\uparrow^{(j)} = \mathbf{R}_{\downarrow\uparrow}^{(j)}\big(\mathbf{a}_\downarrow^{(j)} + \delta_{j,s}\mathbf{s}_{0\downarrow}\big) + \mathbf{T}_{\uparrow\uparrow}^{(j)}\big(\mathbf{a}_\uparrow^{(j+1)} + \delta_{j+1,s}\mathbf{s}_{0\uparrow}\big), \\ \mathbf{a}_\downarrow^{(j+1)} = \mathbf{T}_{\downarrow\downarrow}^{(j)}\big(\mathbf{a}_\downarrow^{(j)} + \delta_{j,s}\mathbf{s}_{0\downarrow}\big) + \mathbf{R}_{\uparrow\downarrow}^{(j)}\big(\mathbf{a}_\uparrow^{(j+1)} + \delta_{j+1,s}\mathbf{s}_{0\uparrow}\big), \\ j = 1, 2, \ldots, N; \end{cases} \quad (13)$$

where $\mathbf{R}_{\downarrow\uparrow}^{(j)}$, $\mathbf{T}_{\uparrow\uparrow}^{(j)}$, $\mathbf{T}_{\downarrow\downarrow}^{(j)}$ and $\mathbf{R}_{\uparrow\downarrow}^{(j)}$ are the *global modified R/T matrices* for the jth interface. Although the definition of the R/T matrices in the present problem is similar to that of R/T coefficients in case of the horizontally homogeneous layered media, there is a significant difference between them. In the latter case, a single horizontal wave number k_n applies to all layers. In the present case, however, the R/T coefficients will include interactions among different wave numbers; consequently, full matrices must be used to represent such interactions of wave numbers, instead of the scalars or diagonal matrices used for the classic case of horizontally homogeneous layered media.

Secondly, we shall find the explicit expression of these R/T matrices in terms of the known *interface-matrices* given by Eqs. (8e) to (8l). In the arbitrary jth layer, substitution of Eq. (11a) into Eqs. (9a) and (9b) yields,

$$\begin{cases} \mathbf{a}_\downarrow^{(j)} + \delta_{j,s}\mathbf{s}_{0\downarrow} = -\big(\mathbf{E}_{\max}^{(j)}\big)^{-1}\big\{\mathbf{Q}_{\uparrow\downarrow}^{(j)}\underline{\alpha}^{(j)} + \mathbf{P}_{\uparrow\downarrow}^{(j)}\underline{\beta}^{(j)}\big\}, \\ \mathbf{a}_\uparrow^{(j+1)} + \delta_{j+1,s}\mathbf{s}_{0\uparrow} = \big(\mathbf{E}_{\min}^{(j+1)}\big)^{-1}\big\{\mathbf{Q}_{\downarrow\uparrow}^{(j+1)}\underline{\alpha}^{(j)} + \mathbf{P}_{\downarrow\uparrow}^{(j+1)}\underline{\beta}^{(j)}\big\}. \end{cases} \quad (14)$$

Putting Eq. (12) into (14), and comparing the result with Eq. (13), we finally obtain

$$\begin{bmatrix} \mathbf{R}_{\downarrow\uparrow}^{(j)} & \mathbf{T}_{\uparrow\uparrow}^{(j)} \\ \mathbf{T}_{\downarrow\downarrow}^{(j)} & \mathbf{R}_{\uparrow\downarrow}^{(j)} \end{bmatrix} = \begin{bmatrix} \mathbf{Q}_{\uparrow\uparrow}^{(j)} & \mathbf{P}_{\uparrow\uparrow}^{(j)} \\ -\mathbf{Q}_{\downarrow\downarrow}^{(j+1)} & -\mathbf{P}_{\downarrow\downarrow}^{(j+1)} \end{bmatrix} \begin{bmatrix} -\mathbf{Q}_{\uparrow\downarrow}^{(j)} & -\mathbf{P}_{\uparrow\downarrow}^{(j)} \\ \mathbf{Q}_{\downarrow\uparrow}^{(j+1)} & \mathbf{P}_{\downarrow\uparrow}^{(j+1)} \end{bmatrix}^{-1}$$

$$\times \begin{bmatrix} \mathbf{E}_{\max}^{(j)} & 0 \\ 0 & \mathbf{E}_{\min}^{(j+1)} \end{bmatrix}, \quad (15)$$

for $j = 1, 2, \ldots, N$.

2.3.3. Global Generalized R/T Matrices

Once the global modified R/T matrices are known, we can introduce the *global generalized R/T matrices*, the extension of the generalized R/T coefficients in horizontally layered case to the present case. The global generalized R/T matrices, $\hat{\mathbf{R}}_{\downarrow\uparrow}^{(j)}$, $\hat{\mathbf{T}}_{\uparrow\uparrow}^{(j)}$, $\hat{\mathbf{T}}_{\downarrow\downarrow}^{(j)}$ and $\hat{\mathbf{R}}_{\uparrow\downarrow}^{(j)}$, are defined by the following equations:

$$\begin{cases} \mathbf{a}_\uparrow^{(j)} = \hat{\mathbf{T}}_{\uparrow\uparrow}^{(j)}\big(\mathbf{a}_\uparrow^{(j+1)} + \delta_{j+1,s}\mathbf{s}_{0\uparrow}\big), \\ \mathbf{a}_\downarrow^{(j+1)} = \hat{\mathbf{R}}_{\uparrow\downarrow}^{(j)}\big(\mathbf{a}_\uparrow^{(j+1)} + \delta_{j+1,s}\mathbf{s}_{0\uparrow}\big), \end{cases} \quad j = 1, 2, \ldots, s-1; \quad (16a)$$

and

$$\mathbf{a}_\downarrow^{(1)} = \hat{\mathbf{R}}_{\uparrow\downarrow}^{(0)}(\mathbf{a}_\uparrow^{(1)} + \delta_{1,s}\mathbf{s}_{0\uparrow}) \tag{16b}$$

for the interfaces above source; and

$$\begin{cases} \mathbf{a}_\downarrow^{(j+1)} = \hat{\mathbf{T}}_{\downarrow\downarrow}^{(j)}(\mathbf{a}_\downarrow^{(j)} + \delta_{j,s}\mathbf{s}_{0\downarrow}), \\ \mathbf{a}_\uparrow^{(j)} = \hat{\mathbf{R}}_{\downarrow\uparrow}^{(j)}(\mathbf{a}_\downarrow^{(j)} + \delta_{j,s}\mathbf{s}_{0\downarrow}), \end{cases} \quad j = s, s+1, \ldots, N+1; \tag{16c}$$

for the interfaces below source. Substituting Eqs. (16a) and (16c) into Eq. (13), and using Eqs. (9d), (11a) and (12a), we finally obtain the following recursive relations

$$\begin{cases} \hat{\mathbf{R}}_{\uparrow\downarrow}^{(0)} = -\mathbf{Q}_{\downarrow\downarrow}^{(1)}(\mathbf{Q}_{\downarrow\uparrow}^{(1)})^{-1}\mathbf{E}_{\min}^{(1)}, \\ \hat{\mathbf{T}}_{\uparrow\uparrow}^{(j)} = (\mathbf{I} - \mathbf{R}_{\downarrow\uparrow}^{(j)}\hat{\mathbf{R}}_{\uparrow\downarrow}^{(j-1)})^{-1}\mathbf{T}_{\uparrow\uparrow}^{(j)}, \quad j = 1, 2, \ldots, s-1; \\ \hat{\mathbf{R}}_{\uparrow\downarrow}^{(j)} = \mathbf{R}_{\uparrow\downarrow}^{(j)} + \mathbf{T}_{\downarrow\downarrow}^{(j)}\hat{\mathbf{R}}_{\uparrow\downarrow}^{(j-1)}\hat{\mathbf{T}}_{\uparrow\uparrow}^{(j)}, \end{cases} \tag{17a}$$

for the interfaces above the source, and

$$\begin{cases} \hat{\mathbf{R}}_{\downarrow\uparrow}^{(N+1)} = 0, \\ \hat{\mathbf{T}}_{\downarrow\downarrow}^{(j)} = (\mathbf{I} - \mathbf{R}_{\uparrow\downarrow}^{(j)}\hat{\mathbf{R}}_{\downarrow\uparrow}^{(j+1)})^{-1}\mathbf{T}_{\downarrow\downarrow}^{(j)}, \quad j = s, s+1, \ldots, N; \\ \hat{\mathbf{R}}_{\downarrow\uparrow}^{(j)} = \mathbf{R}_{\downarrow\uparrow}^{(j)} + \mathbf{T}_{\uparrow\uparrow}^{(j)}\hat{\mathbf{R}}_{\downarrow\uparrow}^{(j+1)}\hat{\mathbf{T}}_{\downarrow\downarrow}^{(j)}, \end{cases} \tag{17b}$$

for the interfaces below the source. The explicit expression of $\hat{\mathbf{R}}_{\uparrow\downarrow}^{(0)}$ is derived by applying the free surface condition at $z = \xi^{(0)}(x)$, and $\hat{\mathbf{R}}_{\downarrow\uparrow}^{(N+1)} = 0$ is equivalent to the radiation condition at $z = +\infty$. The above recursive relations provide an efficient iterative scheme for determining the *global generalized R/T matrices* from the *global modified R/T matrices*.

2.4. Solution Synthesis

We have, so far, introduced the global generalized R/T matrices, and developed an effective recursive scheme to calculate them from the well-defined interface-matrices. We shall now synthesize the solution of our problem by using these global generalized R/T matrices.

2.4.1. General Solution

Firstly, let us determine the intermediate quantities $\mathbf{a}_\uparrow^{(j)}$ and $\mathbf{a}_\downarrow^{(j)}$, that is, the generalized amplitudes of up- and down-going scattered waves inside an arbitrary irregular layer. According to the definitions of the global generalized R/T matrices, i.e., Eqs. (16a) to (16c), we have

$$\begin{cases} \mathbf{a}_\uparrow^{(j)} = \hat{\mathbf{T}}_{\uparrow\uparrow}^{(j)} \hat{\mathbf{T}}_{\uparrow\uparrow}^{(j+1)} \cdots \hat{\mathbf{T}}_{\uparrow\uparrow}^{(s-1)} (\mathbf{a}_\uparrow^{(s)} + \mathbf{s}_{0\uparrow}), \\ \mathbf{a}_\downarrow^{(j)} = \hat{\mathbf{R}}_{\uparrow\downarrow}^{(j-1)} \mathbf{a}_\uparrow^{(j)}, \end{cases} \quad j = 1, 2, \ldots, s-1; \quad (18a)$$

for the layers above the source, and

$$\begin{cases} \mathbf{a}_\downarrow^{(j)} = \hat{\mathbf{T}}_{\downarrow\downarrow}^{(j-1)} \hat{\mathbf{T}}_{\downarrow\downarrow}^{(j-2)} \cdots \hat{\mathbf{T}}_{\downarrow\downarrow}^{(s)} (\mathbf{a}_\downarrow^{(s)} + \mathbf{s}_{0\downarrow}), \\ \mathbf{a}_\uparrow^{(j)} = \hat{\mathbf{R}}_{\downarrow\uparrow}^{(j)} \mathbf{a}_\downarrow^{(j)}, \end{cases} \quad j = s+1, s+2, \ldots, N; \quad (18b)$$

for the layers below the source. It can be seen that in order to evaluate these generalized amplitudes, we have to know the explicit expression of $\mathbf{a}_\uparrow^{(s)}$ and $\mathbf{a}_\downarrow^{(s)}$, the generalized amplitudes of scattered waves in the source layer. Inside the source layer, we have (see Eqs. (18a) and (18b))

$$\begin{cases} \mathbf{a}_\uparrow^{(s)} = \hat{\mathbf{R}}_{\downarrow\uparrow}^{(s)} (\mathbf{a}_\downarrow^{(s)} + \mathbf{s}_{0\downarrow}), \\ \mathbf{a}_\downarrow^{(s)} = \hat{\mathbf{R}}_{\uparrow\downarrow}^{(s-1)} (\mathbf{a}_\uparrow^{(s)} + \mathbf{s}_{0\uparrow}). \end{cases} \quad (19a)$$

After simple algebraic operations, we obtain

$$\begin{cases} (\mathbf{a}_\uparrow^{(s)} + \mathbf{s}_{0\uparrow}) = [\mathbf{I} - \hat{\mathbf{R}}_{\downarrow\uparrow}^{(s)} \hat{\mathbf{R}}_{\uparrow\downarrow}^{(s-1)}]^{-1} (\mathbf{s}_{0\uparrow} + \hat{\mathbf{R}}_{\downarrow\uparrow}^{(s)} \mathbf{s}_{0\downarrow}), \\ (\mathbf{a}_\downarrow^{(s)} + \mathbf{s}_{0\downarrow}) = [\mathbf{I} - \hat{\mathbf{R}}_{\uparrow\downarrow}^{(s-1)} \hat{\mathbf{R}}_{\downarrow\uparrow}^{(s)}]^{-1} (\mathbf{s}_{0\downarrow} + \hat{\mathbf{R}}_{\uparrow\downarrow}^{(s-1)} \mathbf{s}_{0\uparrow}). \end{cases} \quad (19b)$$

Substituting Eq. (19b) into Eqs. (18a) and (18b), we obtain the generalized amplitudes for an arbitrary layer as

$$\begin{cases} \mathbf{a}_\uparrow^{(j)} = \hat{\mathbf{T}}_{\uparrow\uparrow}^{(j)} \hat{\mathbf{T}}_{\uparrow\uparrow}^{(j+1)} \cdots \hat{\mathbf{T}}_{\uparrow\uparrow}^{(s-1)} [\mathbf{I} - \hat{\mathbf{R}}_{\downarrow\uparrow}^{(s)} \hat{\mathbf{R}}_{\uparrow\downarrow}^{(s-1)}]^{-1} (\mathbf{s}_{0\uparrow} + \hat{\mathbf{R}}_{\downarrow\uparrow}^{(s)} \mathbf{s}_{0\downarrow}), \\ \mathbf{a}_\downarrow^{(j)} = \hat{\mathbf{R}}_{\uparrow\downarrow}^{(j-1)} \mathbf{a}_\uparrow^{(j)}, \\ j = 1, 2, \ldots, s-1; \end{cases} \quad (20a)$$

for the layers above the source, and

$$\begin{cases} \mathbf{a}_\downarrow^{(j)} = \hat{\mathbf{T}}_{\downarrow\downarrow}^{(j-1)} \hat{\mathbf{T}}_{\downarrow\downarrow}^{(j-2)} \cdots \hat{\mathbf{T}}_{\downarrow\downarrow}^{(s)} [\mathbf{I} - \hat{\mathbf{R}}_{\uparrow\downarrow}^{(s-1)} \hat{\mathbf{R}}_{\downarrow\uparrow}^{(s)}]^{-1} (\mathbf{s}_{0\downarrow} + \hat{\mathbf{R}}_{\uparrow\downarrow}^{(s-1)} \mathbf{s}_{0\uparrow}), \\ \mathbf{a}_\uparrow^{(j)} = \hat{\mathbf{R}}_{\downarrow\uparrow}^{(j)} \mathbf{a}_\downarrow^{(j)}, \\ j = s+1, s+2, \ldots, N; \end{cases} \quad (20b)$$

for the layers below the source.

Once the generalized amplitudes $\{\mathbf{a}_\uparrow^{(j)}$ and $\mathbf{a}_\downarrow^{(j)}; j = 1, 2, \ldots, N+1\}$ are obtained, we can immediately find the expansion coefficients of wave field at interfaces $\{\underline{\alpha}^{(j)}, \underline{\beta}^{(j)}\}$ by using the inverse transform formulas (12) and (12a). Finally, substituting these solved expansion coefficients back into Eqs. (3a) and (6), we obtain the displacement spectra inside an arbitrary layer as follows:

$$W^{(j)}(\mathbf{x}, \omega)$$
$$= \sum_{m=-\infty}^{+\infty} \left[\mathbf{L}_\uparrow^{(j)}(z, \omega)\mathbf{a}_\uparrow^{(j)}(\omega) + \mathbf{L}_\downarrow^{(j)}(z, \omega)\mathbf{a}_\downarrow^{(j)}(\omega) + \delta_{sj}\mathbf{s}(z, \omega) \right]_m e^{ixk_m},$$
$$\text{for } j = 1, 2, \ldots, N, N+1; \tag{21}$$

where

$$\mathbf{L}_\uparrow^{(j)}(z, \omega) = \mathbf{Q}_\uparrow^{(j)}(z, \omega)\mathbf{A}_{11}^{(j)} + \mathbf{P}_\uparrow^{(j)}(z, \omega)\mathbf{A}_{21}^{(j)}$$
$$- \left\{ \mathbf{Q}_\downarrow^{(j)}(z, \omega)\mathbf{A}_{11}^{(j-1)} + \mathbf{P}_\downarrow^{(j)}(z, \omega)\mathbf{A}_{21}^{(j-1)} \right\} \hat{\mathbf{T}}_{\uparrow\uparrow}^{(j-1)}, \tag{21a}$$

$$\mathbf{L}_\downarrow^{(j)}(z, \omega) = \left\{ \mathbf{Q}_\uparrow^{(j)}(z, \omega)\mathbf{A}_{12}^{(j)} + \mathbf{P}_\uparrow^{(j)}(z, \omega)\mathbf{A}_{22}^{(j)} \right\} \hat{\mathbf{T}}_{\downarrow\downarrow}^{(j)}$$
$$- \mathbf{Q}_\downarrow^{(j)}(z, \omega)\mathbf{A}_{12}^{(j-1)} - \mathbf{P}_\downarrow^{(j)}(z, \omega)\mathbf{A}_{22}^{(j-1)}, \tag{21b}$$

$$\begin{bmatrix} \mathbf{A}_{11}^{(j)} & \mathbf{A}_{12}^{(j)} \\ \mathbf{A}_{21}^{(j)} & \mathbf{A}_{22}^{(j)} \end{bmatrix} = \begin{bmatrix} \mathbf{Q}_{\uparrow\uparrow}^{(j)} & \mathbf{P}_{\uparrow\uparrow}^{(j)} \\ -\mathbf{Q}_{\downarrow\downarrow}^{(j+1)} & -\mathbf{P}_{\downarrow\downarrow}^{(j+1)} \end{bmatrix}^{-1} \tag{21c}$$

and $\mathbf{s}(z, \omega)$ is given in Eq. (3b). After substituting Eqs. (20a) and (20b) into (21), solution (21) can be rewritten in the following compact form:

(a) Inside the layers above the source layer (i.e., $j = 1, 2, \ldots, s-1$), we have

$$W^{(j)}(x, z, \omega)$$
$$= \mathbf{e}^T(x)\mathbf{Y}^{(j)}(z, \omega)\hat{\mathbf{R}}_{\uparrow\downarrow}^{(j-1)}(\omega)\hat{\mathbf{T}}_{\uparrow\uparrow}^{(j)}(\omega)\hat{\mathbf{T}}_{\uparrow\uparrow}^{(j+1)}(\omega) \cdots \hat{\mathbf{T}}_{\uparrow\uparrow}^{(s-1)}(\omega)$$
$$\times \left[\mathbf{I} - \hat{\mathbf{R}}_{\downarrow\uparrow}^{(s)}(\omega)\hat{\mathbf{R}}_{\uparrow\downarrow}^{(s-1)}(\omega) \right]^{-1} \left[\mathbf{s}_{0\uparrow}(\omega) + \hat{\mathbf{R}}_{\downarrow\uparrow}^{(s)}(\omega)\mathbf{s}_{0\downarrow}(\omega) \right]. \tag{22a}$$

(b) Inside the layers below the source layer (i.e., $j = s+1, s+2, \ldots, N, N+1$), we have

$$W^{(j)}(x, z, \omega)$$
$$= \mathbf{e}^T(x)\mathbf{Y}^{(j)}(z, \omega)\hat{\mathbf{T}}_{\downarrow\downarrow}^{(j-1)}(\omega)\hat{\mathbf{T}}_{\downarrow\downarrow}^{(j-2)}(\omega) \cdots \hat{\mathbf{T}}_{\downarrow\downarrow}^{(s)}(\omega)$$
$$\times \left[\mathbf{I} - \hat{\mathbf{R}}_{\uparrow\downarrow}^{(s-1)}(\omega)\hat{\mathbf{R}}_{\downarrow\uparrow}^{(s)}(\omega) \right]^{-1} \left[\mathbf{s}_{0\downarrow}(\omega) + \hat{\mathbf{R}}_{\uparrow\downarrow}^{(s-1)}(\omega)\mathbf{s}_{0\uparrow}(\omega) \right]. \tag{22b}$$

(c) Inside the source layer ($j = s$), we have

$$W^{(j)}(x, z, \omega)$$
$$= \mathbf{e}^T(x)\left\{ \mathbf{Y}^{(s)}(z, \omega)\hat{\mathbf{R}}_{\uparrow\downarrow}^{(s-1)}(\omega)\left[\mathbf{I} - \hat{\mathbf{R}}_{\downarrow\uparrow}^{(s)}(\omega)\hat{\mathbf{R}}_{\uparrow\downarrow}^{(s-1)}(\omega) \right]^{-1} \right.$$
$$\left. \times \left[\mathbf{s}_{0\uparrow}(\omega) + \hat{\mathbf{R}}_{\downarrow\uparrow}^{(s)}(\omega)\mathbf{s}_{0\downarrow}(\omega) \right] + \mathbf{s}(z, \omega) \right\}. \tag{22c}$$

Where

$$\mathbf{e}^T(x) = \{e^{ixk_m}; m = 0, \pm 1, \pm 2, \ldots\}, \tag{22d}$$

$$\mathbf{Y}^{(j)}(z,\omega) = \mathbf{L}_\uparrow^{(j)}(z,\omega)\hat{\mathbf{R}}_{\downarrow\uparrow}^{(j)}(\omega) + \mathbf{L}_\downarrow^{(j)}(z,\omega). \tag{22e}$$

Equations (22a) to (22c) are the complete solution in frequency domain of SH wave in the multi-layered media with irregular interfaces generated by an arbitrary seismic source. Compared with the classic solution of generalized R/T coefficients method for the case of horizontally homogeneous layered media (e.g., Luco and Apsel, 1983; Yao and Harkrider, 1983; Chen, 1999b), this solution is in an identical form with that. Moreover, we can demonstrate later that this solution, Eq. (22), can be analytically reduced to the classic solution of generalized R/T coefficients method when the topography of each interface becomes flat. In this sense, the method presented here can be viewed as a generalization of the classic *generalized R/T coefficients method*, therefore it is named as *global generalized R/T matrices method* (e.g., Chen, 1990, 1995, 1996; Cao et al., 2004).

2.4.2. Solutions for Some Special Cases

In the previous sections, we have derived the general solution of our target problem. For some special cases of seismological interest, the general solution can be further simplified.

Solution for Horizontally Homogeneous Layered Media This is a trivial case in which all interfaces are horizontal, that is

$$\xi^{(j)}(x) = z^{(j)} \quad \text{and} \quad \Delta\xi^{(j)}(x) = 0, \quad j = 1, 2, \ldots, N.$$

Thus, the interface-matrices, Eqs. (8e) to (8l), can be analytically evaluated to be

$$(\mathbf{Q}_{\downarrow\uparrow}^{(j+1)})_{nm} = \frac{1}{2}\delta_{nm}, \qquad (\mathbf{Q}_{\downarrow\downarrow}^{(j+1)})_{nm} = -\frac{1}{2}\delta_{nm},$$

$$(\mathbf{Q}_{\uparrow\downarrow}^{(j)})_{nm} = -\frac{1}{2}\delta_{nm}, \qquad (\mathbf{Q}_{\uparrow\uparrow}^{(j)})_{nm} = \frac{1}{2}\delta_{nm},$$

and

$$(\mathbf{P}_{\downarrow\uparrow}^{(j+1)})_{nm} = \frac{-1}{2}\delta_{nm}, \qquad (\mathbf{P}_{\downarrow\downarrow}^{(j+1)})_{nm} = \frac{-1}{2}\delta_{nm},$$

$$(\mathbf{P}_{\uparrow\downarrow}^{(j)})_{nm} = \frac{-\mu^{(j+1)}v_n^{(j+1)}}{2\mu^{(j)}v_n^{(j)}}\delta_{nm}, \qquad (\mathbf{P}_{\uparrow\uparrow}^{(j)})_{nm} = \frac{-\mu^{(j+1)}v_n^{(j+1)}}{2\mu^{(j)}v_n^{(j)}}\delta_{nm},$$

for $j = 1, 2, \ldots, N$. Then the corresponding *global modified R/T matrices* can be found to be

$$(\mathbf{R}_{\downarrow\uparrow}^{(j)})_{nm} = \frac{\mu^{(j)} v_n^{(j)} - \mu^{(j+1)} v_n^{(j+1)}}{\mu^{(j)} v_n^{(j)} + \mu^{(j+1)} v_n^{(j+1)}} e^{iv_n^{(j)}(z^{(j)} - z^{(j-1)})} \delta_{nm}, \tag{23a}$$

$$(\mathbf{R}_{\uparrow\downarrow}^{(j)})_{nm} = \frac{\mu^{(j+1)} v_n^{(j+1)} - \mu^{(j)} v_n^{(j)}}{\mu^{(j)} v_n^{(j)} + \mu^{(j+1)} v_n^{(j+1)}} e^{iv_n^{(j+1)}(z^{(j+1)} - z^{(j)})} \delta_{nm}, \tag{23b}$$

$$(\mathbf{T}_{\uparrow\uparrow}^{(j)})_{nm} = \frac{2\mu^{(j+1)} v_n^{(j+1)}}{\mu^{(j)} v_n^{(j)} + \mu^{(j+1)} v_n^{(j+1)}} e^{iv_n^{(j+1)}(z^{(j+1)} - z^{(j)})} \delta_{nm}, \tag{23c}$$

and

$$(\mathbf{T}_{\downarrow\downarrow}^{(j)})_{nm} = \frac{2\mu^{(j)} v_n^{(j)}}{\mu^{(j)} v_n^{(j)} + \mu^{(j+1)} v_n^{(j+1)}} e^{iv_n^{(j)}(z^{(j)} - z^{(j-1)})} \delta_{nm}, \tag{23d}$$

for $j = 1, 2, \ldots, N$. It can be seen that the modified global R/T matrices in this case become diagonal ones because there are no interactions among different horizontal wave numbers in this case, whereas the expressions of the diagonal elements are identical with those obtained directly for horizontally homogeneous layered media case (e.g., Chen, 1999b). Since our general solution has identical form as that of the classic solution (e.g., the recursive formulas for calculating the global generalized R/T matrices, the final solution, etc.), the rest of our solution for this trivial case will be identical to that of the classic solution (e.g., Chen, 1999b). Therefore, the general solution obtained in this study can be analytically reduced to the classic solution when applied to the horizontally homogeneous layered media case.

Ground Motion For most of seismological problems, we only interest in the ground motion. According to Eq. (5a), the expression of wave field at the free surface (i.e., $z = \xi^{(0)}(x)$) has following simple form,

$$W^{(0)}[x, \xi^{(0)}(x), \omega] = \sum_m \alpha_m^{(0)} e^{i[k_m x + v_m^{(1)} |\xi^{(0)}(x) - z^{(0)}|]}, \tag{24}$$

where the coefficient vector $\underline{\alpha}^{(0)}$ can be determined by using Eqs. (12a) and (16b) as follows:

$$\underline{\alpha}^{(0)} = \begin{cases} (\mathbf{Q}_{\downarrow\uparrow}^{(1)})^{-1} \mathbf{E}_{min}^{(1)} [\mathbf{I} - \hat{\mathbf{R}}_{\downarrow\uparrow}^{(1)} \hat{\mathbf{R}}_{\uparrow\downarrow}^{(0)}]^{-1} (\mathbf{s}_{0\uparrow} + \hat{\mathbf{R}}_{\downarrow\uparrow}^{(1)} \mathbf{s}_{0\downarrow}), & \text{for } s = 1, \\ (\mathbf{Q}_{\downarrow\uparrow}^{(1)})^{-1} \mathbf{E}_{min}^{(1)} \hat{\mathbf{T}}_{\uparrow\uparrow}^{(1)} \hat{\mathbf{T}}_{\uparrow\uparrow}^{(2)} \cdots \hat{\mathbf{T}}_{\uparrow\uparrow}^{(s-1)} [\mathbf{I} - \hat{\mathbf{R}}_{\downarrow\uparrow}^{(s)} \hat{\mathbf{R}}_{\uparrow\downarrow}^{(s-1)}]^{-1} \\ \quad \times (\mathbf{s}_{0\uparrow} + \hat{\mathbf{R}}_{\downarrow\uparrow}^{(s)} \mathbf{s}_{0\downarrow}), & \text{for } s > 1. \end{cases} \tag{24a}$$

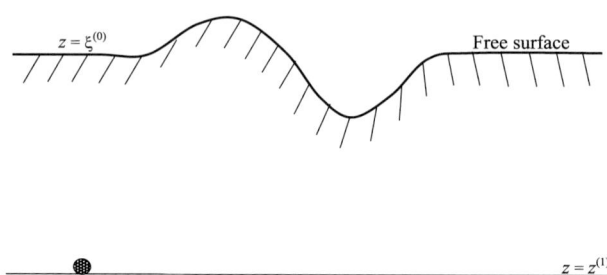

FIG. 6. Configuration of the topography problem.

Topography Problem For topography problem shown in Fig. 6, the solution can be further simplified. To directly use the general solution (22), we introduce a dummy flat interface, $z = z^{(1)}$, just below the source. According to Eqs. (23a), (23b) and (17b), we find $\hat{\mathbf{R}}_{\downarrow\uparrow}^{(1)} = 0$, putting it into Eqs. (24a) and then (24) yields,

$$W^{(0)}\left[x, \xi^{(0)}(x), \omega\right] = \sum_m \left\{\left[\mathbf{Q}_{\downarrow\uparrow}^{(1)}(\omega)\right]^{-1}\hat{\mathbf{s}}(\omega)\right\}_m e^{i[k_m x + v_m^{(1)}|\xi^{(0)}(x) - z^{(0)}|]}, \quad (25)$$

where $\hat{\mathbf{s}}(\omega) = \mathbf{E}_{\min}^{(1)}(\omega)\mathbf{s}_{0\uparrow}$ and its explicit expression is given by

$$\left[\hat{\mathbf{s}}(\omega)\right]_n = \frac{i}{2v_n^{(1)} L} \int_{-L/2}^{L/2} dx \int_{\xi^{(0)}(x)}^{+\infty} f(x, z) e^{-i[k_n x - v_n^{(1)}(z - \xi_{\min}^{(0)})]} dz. \quad (25a)$$

As we can see that although the solution obtained in this study is complete for any frequency in the sense of discrete wave number representation, all the matrices involved in the formulation are with infinite dimension. In other words, if we want to obtain exact solution, we must include the contributions from all discrete horizontal wave numbers no matter how small they are. Obviously, it is not practical. However, an approximate solution with a controllable accuracy is possible by truncating the dimension of the matrices and vectors in our formulation at a certain finite order. The truncated order N is determined by $\kappa(L/\lambda_m)$, where κ is a coefficients, L is the periodic length and λ_m is minimum wavelength. According to our experience, for 1% precision, $\kappa = 4$–10, and the lower the frequency is, the larger the κ; the larger of the topographic/interfacial slope, the lager the κ.

2.5. Numerical Validation by the Analytical Solution

We have developed an analytical solution for synthesizing seismograms in a generally irregular multi-layered media. Before applying this general formulation to practical problems, we should test its validity by comparing our results with the existing correct solutions.

Trifunac (1973) derived an analytical solution for the problem of seismic waves scattering due to a semi-cylindrical canyon as shown in Fig. 7a. This is a topography problem that could be solved by our algorithm. It is noted that the validity of our formulations depends on the choice of the basis functions for interfaces (including surface) wave-field expansion. Therefore, this semi-cylindrical canyon topography problem is a crucial test for our formulation because it is a tough topography model in which there are two sharp edges (with infinite derivatives) at $x = \pm a$.

Figure 7b shows the comparisons of the results (solid lines) directly computed by our method with the analytical solutions (dotted lines) for various dimensionless frequencies η for a vertical incidence. The dimensionless frequency is defined as $\eta = 2a/\lambda$, where λ is the wave length and the actual frequency is related to η by $f = \eta(\beta/2a)$. The comparison for 30° incidence and $\eta = 1.0$ is shown Fig. 8. In all cases, we have chosen $a = 1.0$ km, $V = 1.0$ km/s, $L = 40$ km, and $\omega_I^f = \pi V/2L$. Here ω_I^f is an imaginary frequency and its function is similar, but not exactly same, to the imaginary frequency (ω_I) used for computing time-domain solution based on discrete wave number representation (e.g., Bouchon and Aki, 1977). In our calculation, we used a complex circular frequency $\omega^* = 2\pi f + i\omega_I^f$, instead of the true real circular frequency $\omega = 2\pi f$ alone, to archive an optimal solution in frequency domain. In Appendix B, we theoretically demonstrate the choice of $\omega_I^f = \pi V/2L$ indeed is an optimal value for obtaining a best frequency domain solution.

Although these comparisons of the results in frequency domain show excellent agreements between ours with the analytical ones, we can see some minor discrepancies. This is caused by the application of discrete wave number algorithm and the introduction of periodicity of both in structure and source. In fact, the discrete wave number algorithm can ensure the accuracy of the solution in time domain but not in frequency domain (e.g., Bouchon and Aki, 1977). However, the discrepancies of the solutions in frequency domain can be decreased by increasing the periodicity length L, as shown in Fig. 9 where the results of ours (solid lines) and the analytical ones (dotted lines) for various L are compared.

We also compared the corresponded time domain solutions, i.e. the synthetic seismograms, and the results are shown in Fig. 10. In the calculation, we use Ricker wavelet (Ricker, 1977) as source-time function, that is,

$$S(t) = \left(\frac{u^2}{4} - \frac{1}{2}\right)\frac{\sqrt{\pi}}{2} \exp\left(-\frac{u^2}{4}\right), \tag{26}$$

where $u = 2\pi t f_c$, f_c is the center frequency of Ricker wavelet. Here we take $f_c = 1.0$. Figure 10a shows the ground displacement responses calculated by our algorithm for a vertical incidence, and Fig. 10b shows the differences of seismograms between ours and the analytical ones, once again an excellent agreement has been indicated.

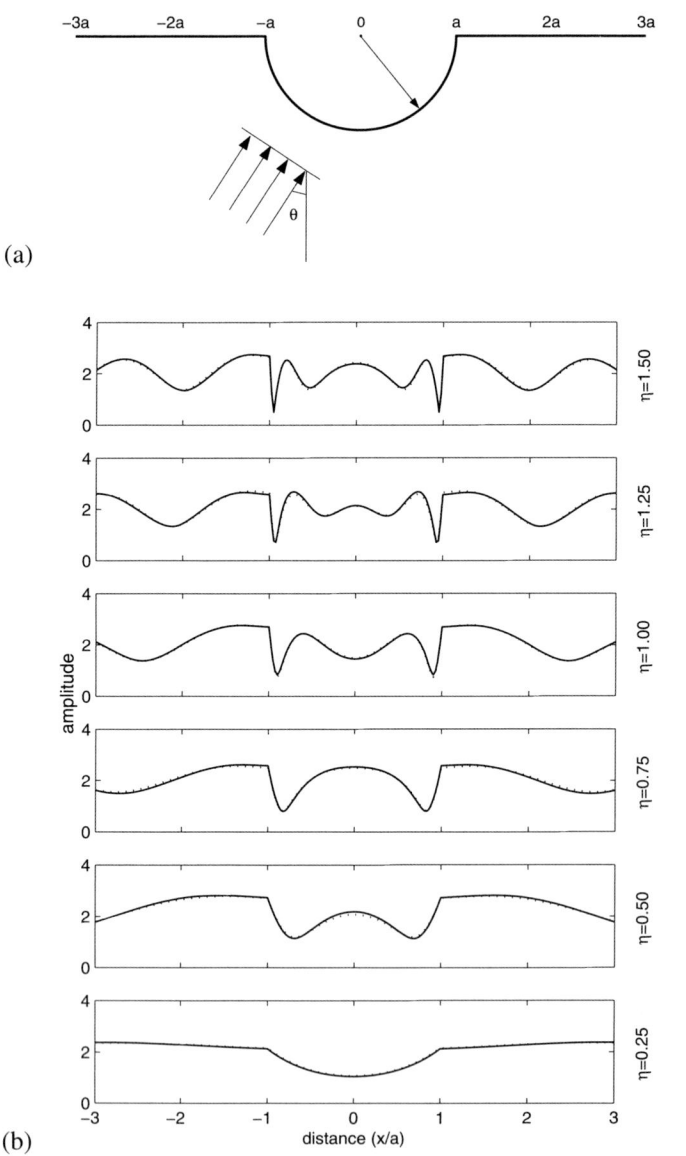

FIG. 7. Comparisons with analytical solutions by Trifunac (1973) for the scattering problem due to a semi-cylindrical canyon. (a) The configuration of semi-cylindrical canyon; (b) the frequency responses to vertical incident SH wave for various dimensionless frequencies η (see explanation in the text), where the solid lines are the results obtained by the Chen's method (1990, 1995, 1996), the dotted lines are the analytical results by Trifunac.

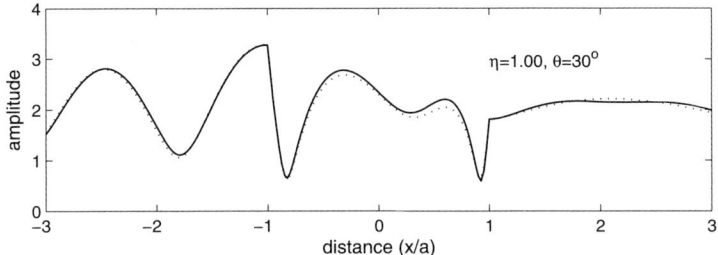

FIG. 8. Same comparison as Fig. 7, but for 30° oblique incidence, where $\eta = 1.0$.

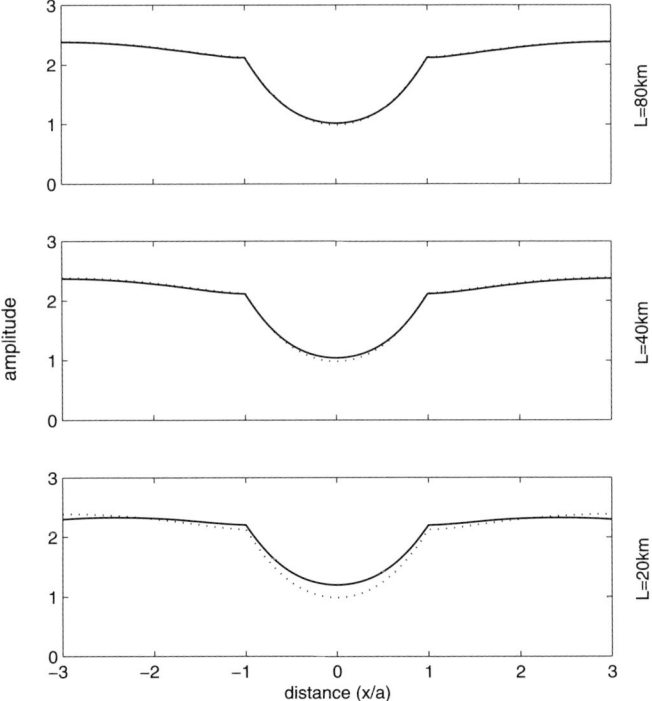

FIG. 9. Same comparison as Fig. 8 but due to vertical incidence and for various periodicity length L.

In summary, the excellent agreement between our results and the analytical solutions for the semi-cylindrical canyon case shown above confirm the appropriateness of the basis functions of wave field expansion on surface (or interface) employed in our method, and thus the validity of the derived formulations.

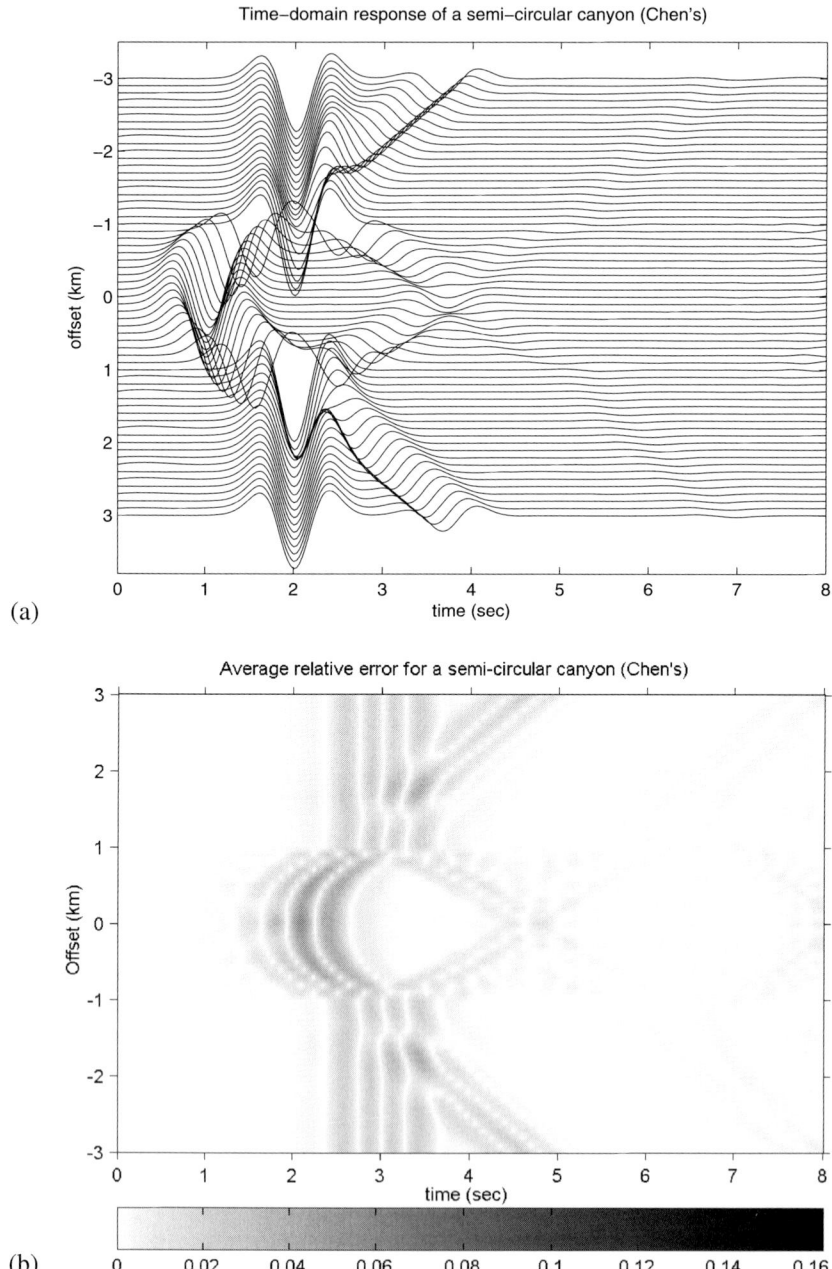

FIG. 10. (a) Time domain responses on the ground of a semi-cylindrical canyon due to a vertical incidence; (b) the differences between Chen's results and analytical solutions.

2.6. Comparisons with Other Existing Methods

Compared with pure numerical methods (such as finite difference method, finite element method, boundary element method and pseudo-spectral method, etc.), our method belongs to the semi-analytical method which is expected having a good balance between the accuracy and efficiency in solution. The Aki–Larner method (hereafter abbreviated to AL) (e.g., Aki and Larner, 1970; Bouchon, 1973; Bard and Bouchon, 1980a, 1980b) with its extension (e.g., Kohketsu, 1987; Kohketsu et al., 1991; Takenaka et al., 1996) and the Bouchon–Campillo method (e.g., Bouchon, 1985; Campillo and Bouchon, 1985; Bouchon et al., 1989) (hereafter abbreviated as BC) are other two most representative semi-analytical methods developed during the past two decades. The AL method is the pioneering work among the semi-analytical methods. One of the most important components of AL method is the Rayleigh ansatz (e.g., Aki and Larner, 1970), which works only for case with relatively smooth and small interfacial/topographic fluctuations, thus limits its application. The BC method, on the other hand, is a hybrid algorithm of the boundary integral equation and the discrete wave number approach (e.g., Bouchon, 1985), and can be applied to case with arbitrarily irregular interfaces/surface. In this section, we shall compare our method mainly with these two semi-analytical methods.

2.6.1. Comparisons for the Semi-Cylindrical Canyon Topography Problem

First, we still use the example of semi-cylindrical canyon to compare our method with the AL and BC. Figure 11 shows the comparisons between the results yielded from the AL, BC and our method (abbreviated to Chen's), with the analytical solution caused by the vertical incidence with $\eta = 0.5, 1.5, 2.5$. From this figure, we can see that the results of Chen's and BC are stable, i.e., as the truncated number of discrete wave number sum N increases, they are steadily getting more accurate; however, AL's solution is unstable. To quantitatively describe the stability and robustness of these methods, we define the average relative error by

$$E_{AR} = \frac{1}{M} \sum_{j=1}^{M} \frac{|u_j^c - u_j^a|}{|u_j^a|}, \qquad (27)$$

where u_j^c is the displacement at position j yielded from each semi-analytical method, u_j^a is the analytical displacement at position j, and M is the total number of recorders. Figure 12 shows the corresponding E_{AR} of the three methods as a function of N for $\eta = 1.5$. The abscissas of the stars in Fig. 12 denote N chosen for the three methods, respectively, in Fig. 11. We can find that the results computed by Chen's converge soon with the increase of N and the error is relatively small ($E_{AR} = 1.9\%$ when $N = 401$). The results given by BC converge a little slower with the increase of N ($E_{AR} = 7.7\%$ when $N = 801$), and the results still

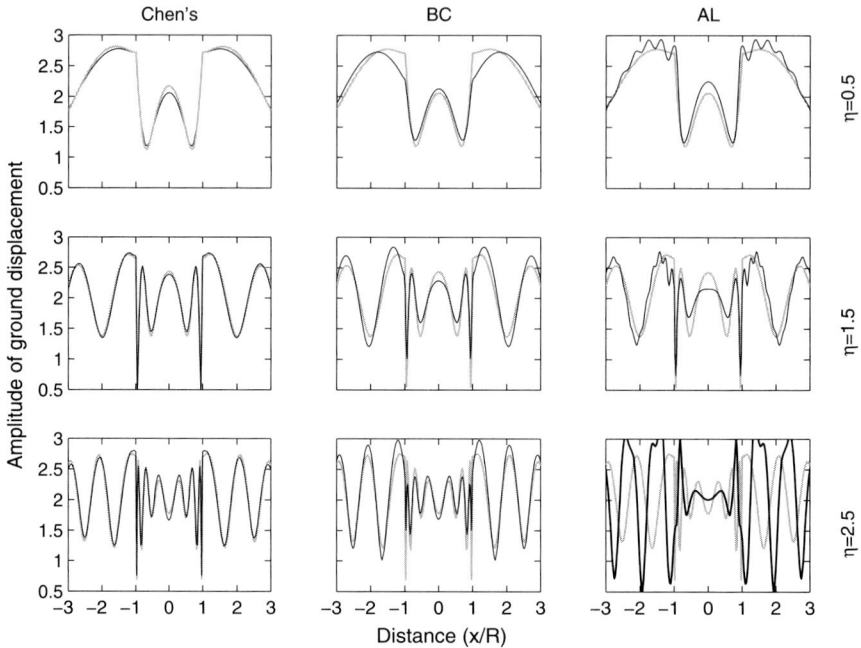

FIG. 11. The response in frequency domain on the ground of a semi-cylindrical canyon to a vertically incident plane wave for various dimensionless frequencies. The thick gray lines denote the analytical solutions, and the thin black lines denote the best results yielded from the three methods. (a) $\eta = 0.5$; (b) $\eta = 1.5$; (c) $\eta = 2.5$.

get better until $N = 1601$. However, the results computed by AL vary extremely with N and do not converge. The average relative error for $N = 200$ denoted by the star in Fig. 11 is about 9.3%. And we choose N as 401, 801, 201 and 481, 1201, 241 for Chen's, BC, AL when $\eta = 0.5$ and 2.5.

Figure 13 shows the comparison of synthetic seismograms at several selected ground points ($x = 0, 1, 2, 3$) yielded from Chen's, BC and AL methods (thin black lines) and the analytical solutions (thick gray lines) for the semi-cylindrical canyon subject to a vertical incidence. In the calculation, Ricker wavelet with $f_c = 1.0$ Hz is used as source time function. We can see that Chen's results as expected agree very well with the analytical solutions; BC's results agree relatively well with analytical solutions in the region varying smoothly (around $x = 0, 2, 3$), but do not agree well in the region varying extremely (around $x = 1$). As for AL method, except in the very smoothly varying region (around $x = 0, 3$) the direct wave agrees well with analytical solutions, it cannot provide correct solution for this problem, particularly in the relatively steep region. This is caused by the limit of Rayleigh ansatz on which AL was built up (e.g., Dravinski, 1982). In fact, Aki–Larner method was developed to compute the propagation of seismic

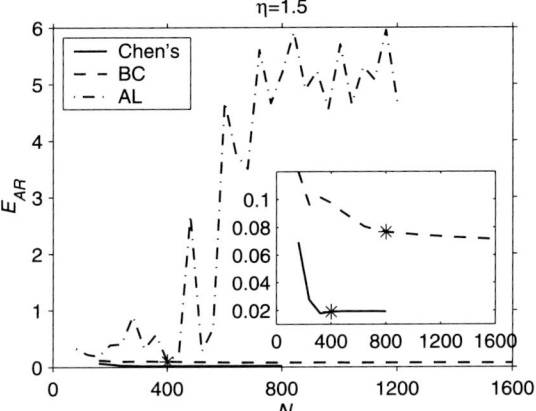

FIG. 12. The corresponding average relative error (E_{AR}) of the three methods as the function of discrete wavenumber samples N when the plane wave with $\eta = 1.5$ is vertically incident. The solid lines, the dashed lines, and the dash-dot line denote the results yielded from Chen's method, BC method, and AL method. The abscissas of the stars in this figure denote N we choose in Fig. 3 for the three methods.

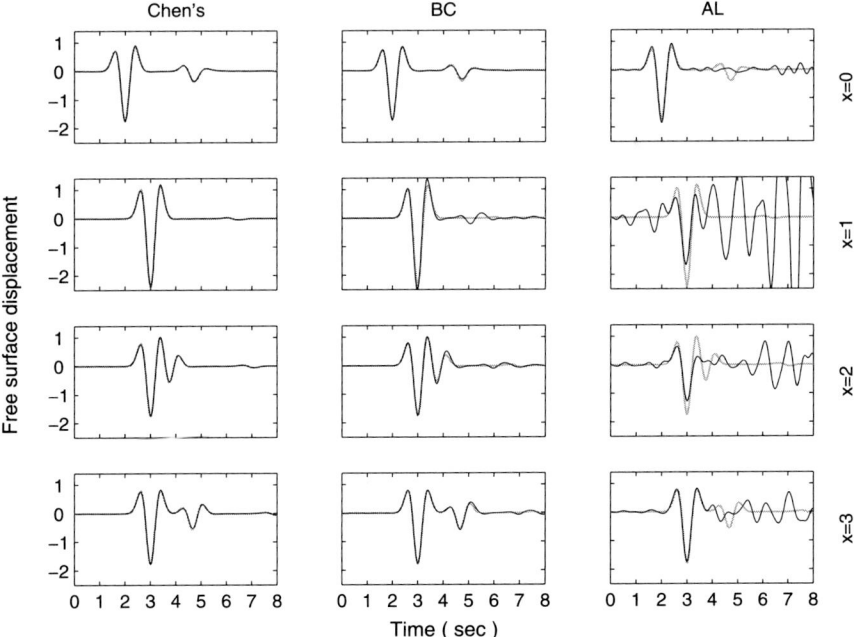

FIG. 13. The synthetic seismograms on the ground $x = 0, 1, 2, 3$ of a semi-cylindrical canyon to a vertically incident plane wave. The thick lines denote the results yielded from Chen's, BC and AL methods, and the thin lines denote the analytical solutions. We choose Ricker wavelet with characteristic frequency 1.0 as the source-time function.

waves in the multi-layered media with smooth irregular interfaces in its original purpose.

2.6.2. Comparisons for Gaussian Valley Topography Problem

We now further compare our method with AL and BC methods by using a smooth topography model, the Gaussian shape valley with a variable slope, which is defined by the function below:

$$\zeta(x) = \begin{cases} h[1 - (x/d)^2]e^{-3(x/d)^2} & (|x| \leqslant d), \\ 0 & (|x| > d). \end{cases} \quad (28)$$

Here, d and h represent the half-width and depth of the valley, respectively (see Fig. 14a), h/d is the slope of the topography. Since there is no analytical solution to be compared in this case, we introduce the *average incremental residual* E_A to measure the convergence of solutions, and thus assess their accuracy. E_A is defined as,

$$E_A = \frac{1}{M|N_2 - N_1|} \sum_{j=1}^{M} \frac{|u_j^{N_2} - u_j^{N_1}|}{|u_j^{N_2}|}, \quad (29)$$

where $u_j^{N_1}$ and $u_j^{N_2}$ represent the displacements at position j yielded from each semi-analytical method for $N = N_1$ and N_2 respectively, and M is the total number of recorders. In our calculation, the increment of wave number is taken as 5, that is $N_2 = N_1 + 5$.

Figures 14b and c shows the results for vertical incidence with $\eta = 1.0$ (here $\eta = 2d/\lambda$), and for various slopes: $h/d = 0.2, 0.4, 0.6, 0.8$. Other computation parameters are the same as the semi-cylindrical canyon case. Figure 14b shows comparisons of the variation of average incremental residuals (E_A) with N for the each method; and Fig. 14c shows the corresponded ground displacement distributions. We can see that both Chen's and BC methods are convergent, thus works very well for all slopes; the AL method, however, is survived for the relatively small slopes (0.2, 0.4), and fails for larger slopes, indicating the limitation of Rayleigh ansatz.

Figure 15 shows the synthetic seismograms on the ground yielded from Chen's (solid lines), BC (dotted lines) and AL (gray lines) methods for slope $h/d = 0.2$, 0.4, 0.6, 0.8. We see that the seismograms yielded from all these methods agree well when h/d is relatively small ($h/d = 0.2, 0.4$), but with the increase of h/d, the difference among the results yielded from these methods gradually increases, however, results yielded from Chen's and BC agree well consistently. With the increase of h/d, although the difference between the results from BC and Chen's also increases in regions where the topography varying extremely (around $x = 0$, 1), it is too small to be distinguished; the differences between the results from AL and other two methods become distinguishable, and the larger the slope, the larger

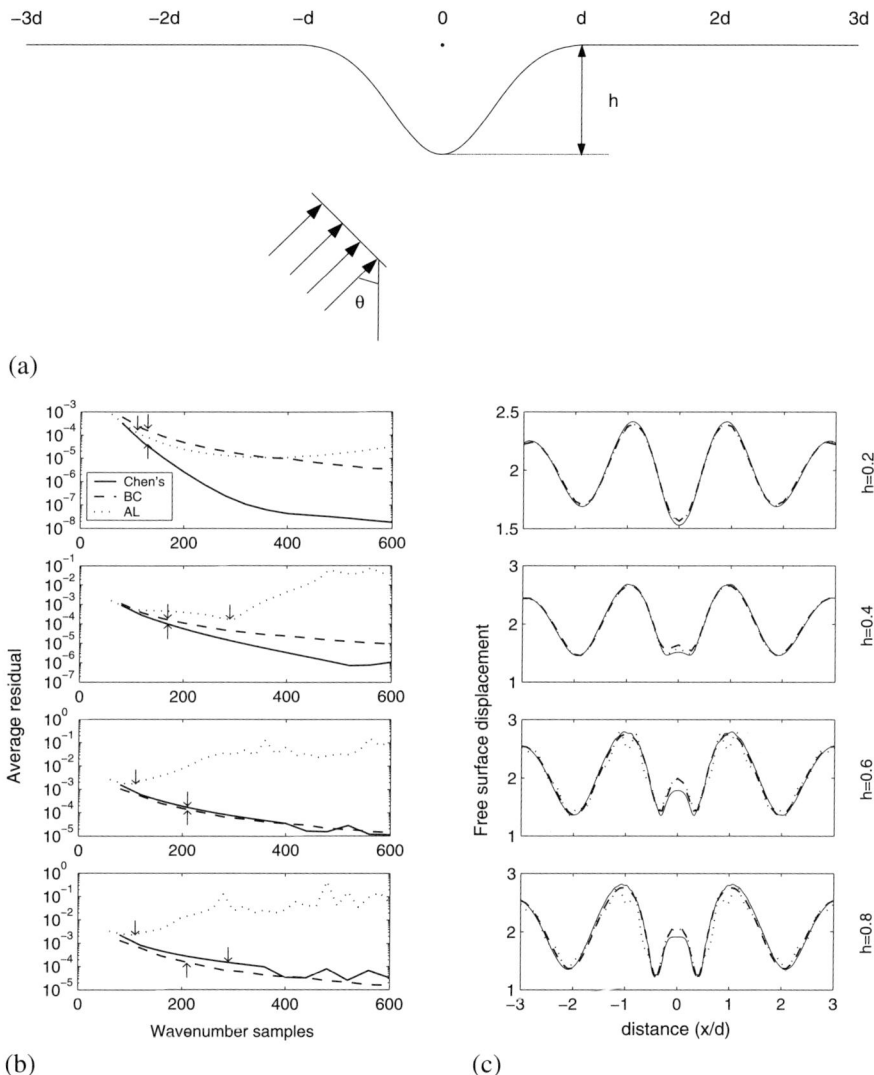

FIG. 14. The results of a Gaussian valley with various slopes ($d = 1$, $h = 0.2, 0.4, 0.6, 0.8$) to a vertically incident plane wave for dimensionless frequency 1.0. (a) The configuration of Gaussian valley, where d, h is the half-width and depth of the valley respectively, and θ is the incident angle. (b) The average residuals of response in frequency domain on the ground as the function of N; the solid lines, the dashed lines, and the dash-dot lines denote the results yielded from Chen's, BC, and AL method. (c) The displacement distribution on the ground in frequency domain, the relevant N in the computation is shown by the arrows in (b).

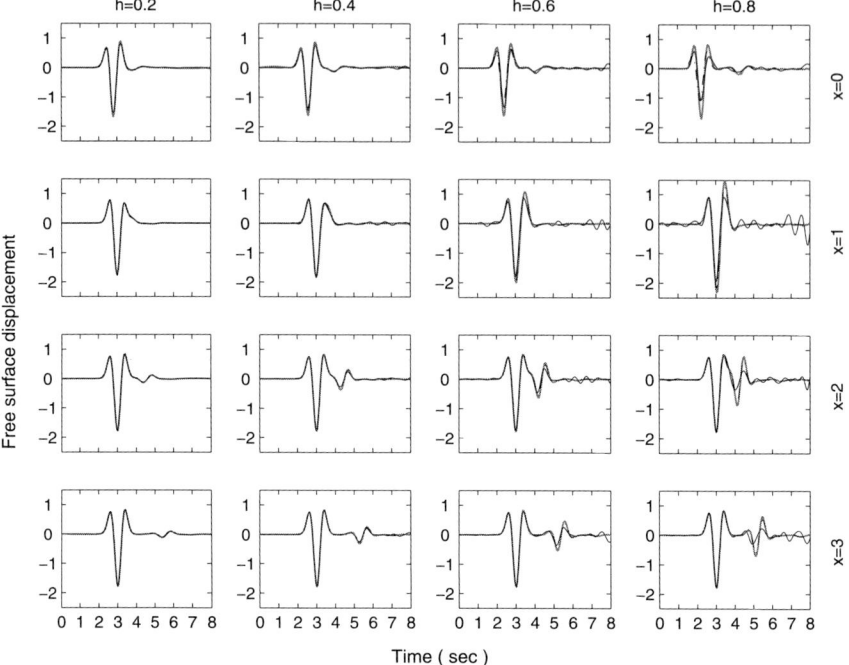

FIG. 15. The synthetic seismograms on the ground $x = 0, 1, 2, 3$ of a Gaussian valley with various slopes ($d = 1$, $h = 0.2, 0.4, 0.6, 0.8$) to a vertically incident plane wave. The thick solid lines, the dash-dotted lines, and the thin solid lines denote the results yielded from Chen's method, BC method, and AL method respectively. Here, we choose Ricker wavelet with characteristic frequency 1.0 as the source-time function.

the differences. But the whole results are much better than those in the semi-cylindrical canyon case.

2.6.3. Comparison of Computational Efficiency with Other Methods

To evaluate the computational efficiency of our method, we still take the topography problem of Gaussian shape valley as an example to compute the frequency domain response by each method for a given N, the truncated number of discrete wave number sum; and then compare the corresponding computation time consumed by each method. In the calculation, we take $d = 1.0$, $h = 0.2$ and $\eta = 1.0$, and results are shown in Fig. 16. We can see that due to the use of FFT, AL is the fastest among the three methods; BC is the slowest because of its regular sampling; and Chen's method has moderate computation efficiency. When considering that Chen's usually converges faster, it has better computa-

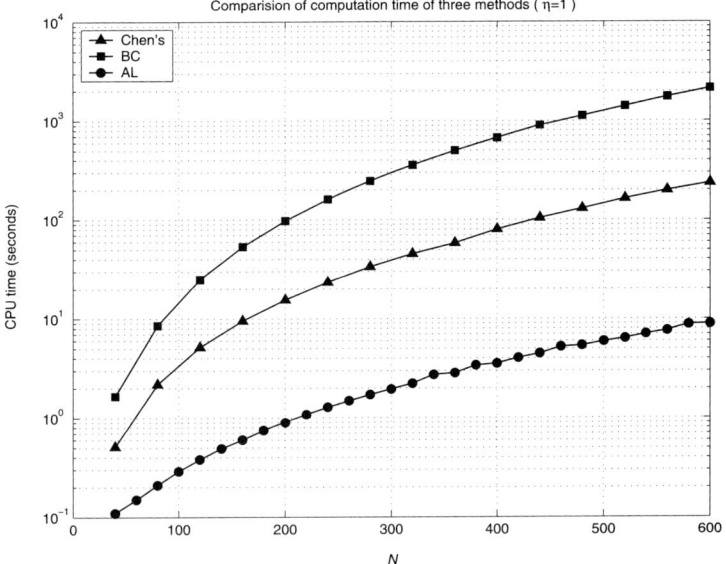

FIG. 16. The computation time on PC/AMD XP1800+ as a function of discrete wavenumber samples for a Gaussian valley with $d = 1$, $h = 0.2$ due to a vertically incident plane wave with $\eta = 1.0$. The triangle line, the rectangle line, and the circle line denote the results yielded from Chen's method, BC method, and AL method respectively.

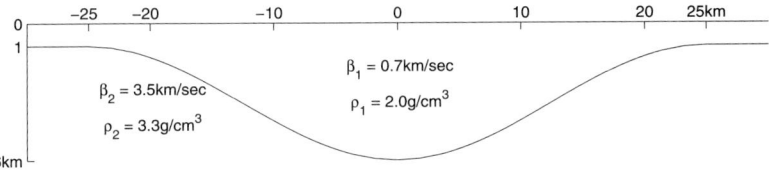

FIG. 17. Configuration of a smooth basin model.

tion efficiency in actual applications. The computation is done on a computer of PC/AMD XP1800+.

2.6.4. Comparisons with More Methods

The last example to be compared here is an open basin model shown in Fig. 17, to which our solution also can be directly applied. For this example, there was exist an extensive comparison among several classic methods such as the Asymptotic Ray Theory (ART) (e.g., Cerveny et al., 1977; Chapman and Drummond, 1982), Glorified Optics (GO) (e.g., Hong and Helmberger, 1978), Aki–Larner (AL) (e.g., Aki and Larner, 1970; Bouchon, 1973; Bard and Bouchon, 1980a, 1980b), Finite Difference (FD) (e.g., Boore, 1972; Levander, 1988; Moczo et al., 2004; Moczo et al., 2006), Finite Element (FE) (e.g., Smith, 1974; Day, 1977; Sanchez-Sesma and Rosenbluth, 1979; Dravinski, 1982; Kawase, 1988) and Kuketsu's Reflectivity method (an extension of Aki–Larner method) (e.g., Kohketsu, 1987). The basin model consists of a flat surface and an irregular interface whose lateral variation is described by following function:

$$\xi(x) = D + \frac{C}{2}\left\{1 - \cos\left[\frac{2\pi}{w}\left(x - \frac{w}{2}\right)\right]\right\}, \tag{30}$$

where $w = 50$ km, $D = 1$ km and $C = 5$ km. This is a smoothly varying interface. Inside the basin, the media parameters are $\beta_1 = 0.7$ km/s and $\rho_1 = 2.0$ g/cm^3; outside the basin, $\beta_2 = 3.5$ km/s and $\rho_2 = 3.3$ g/cm^3. The source time function is Ricker wavelet with center frequency of 1/18.3 Hz and a 30 s time shift.

The computed responses due to a vertical plane wave incidence and comparisons with results from other methods are shown in Fig. 18. We can see that the earlier arrivals of all the methods shown here are consistent; as for the later arrivals, our results are likely consistent with those of AL and RF. This is not surprising. Actually, Koketsu's Reflectivity method belongs to the Aki–Larner method, though it advanced one step with the incorporation of propagator matrix. According to the demonstration above, Aki–Larner method is applicable to smoothly varying structure, this is case of the current problem in which the lateral variation of the interface is very smooth with slope $= 0.2$.

2.7. More Numerical Examples

In the previous sections, we have numerically validated our algorithm by the comparisons with other existing valid solutions. In this section, we shall use more examples to show the versatile applicability of our algorithm to a variety of lateral heterogeneous structure models with seismological interests show in Fig. 2. The results will be shown in this section are all seismograms, that is the time domain responses of displacement on the ground. In the following calculations of synthetic seismograms, the Ricker wavelet is used as source time function for all cases.

2.7.1. Seismic Responses due to Basin Models

The first basin model considered is a relativity smooth one as shown in Fig. 19a, where all the model parameters are same as those in Fig. 17, except the basin bottom was uplifted by 1 km, making a closed smooth basin model. The synthetic seismograms due to a vertical plane wave incidence are plotted in Fig. 19b. Strong and last longer reverberations are observed on the top of the basin, which obviously are caused by the complex interferences inside the soft basin. This kind of

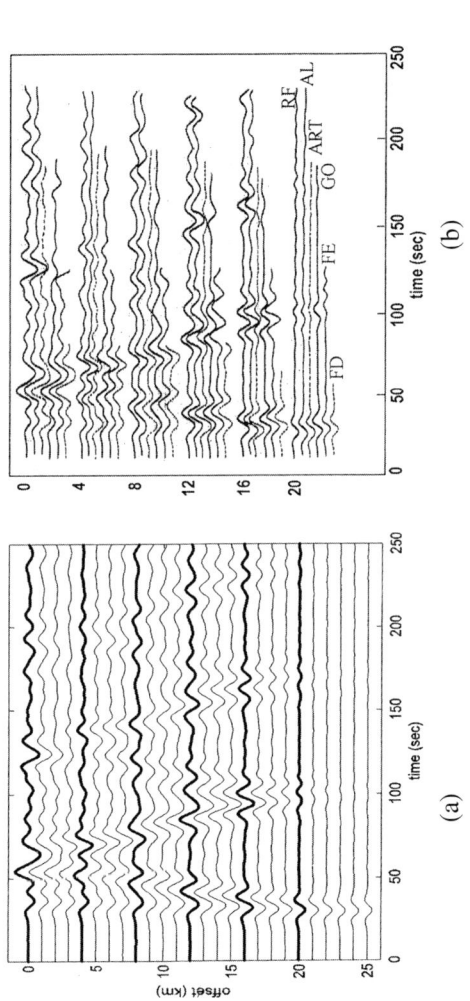

FIG. 18. The synthetic seismograms on the ground of the basin shown in Fig. 16. (a) Results from various other methods on the selected observed points ($x = 0, 4, 8, 12, 16, 20$ km). (b) Results from Chen's method, and the thick lines denote results on the same observed points as (a).

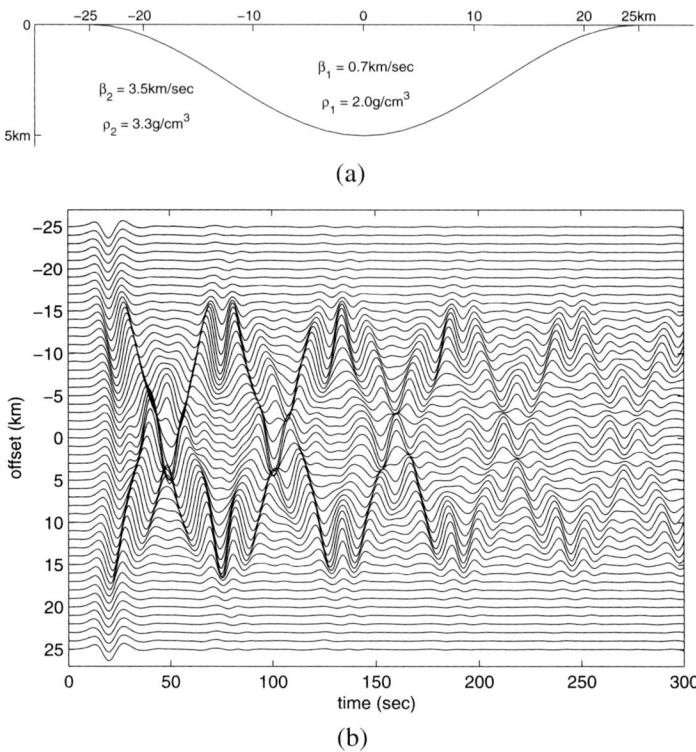

FIG. 19. (a) A closed basin model, where all parameters are same that of Fig. 16, except the basement rock is uplifted by 1 km. (b) The time domain responses on the ground, a strong and long-lasting reverberation is observed at central part of the basin's top.

basin reverberation is particularly an interesting issue in the filed of seismic strong ground motion and seismological engineering.

The second basin model considered here is illustrated in Fig. 20a, which is a steeper open basin than the previous one. The corresponding synthetic seismograms due to a vertical plane wave incidence are displayed in Fig. 20b. As a comparison, we also calculate synthetic seismograms for a closed steeper basin model whose structure parameters are same as the open model shown in Fig. 20a, except the bottom of basin was uplifted by 1 km as shown in Fig. 21a. The computed results for this closed steeper basin model are shown in Fig. 21b. In addition, we take center frequency $f_c = 1/1.83$ Hz for both cases. By comparing these results, we find that the reverberation and amplification caused by the closed basin are mainly concentrated on the basin area, little energy is scattered outward the basin. In the case of open basin, however, more energy is scattered outward the basin, besides the reverberation and amplification inside. Since more energy has

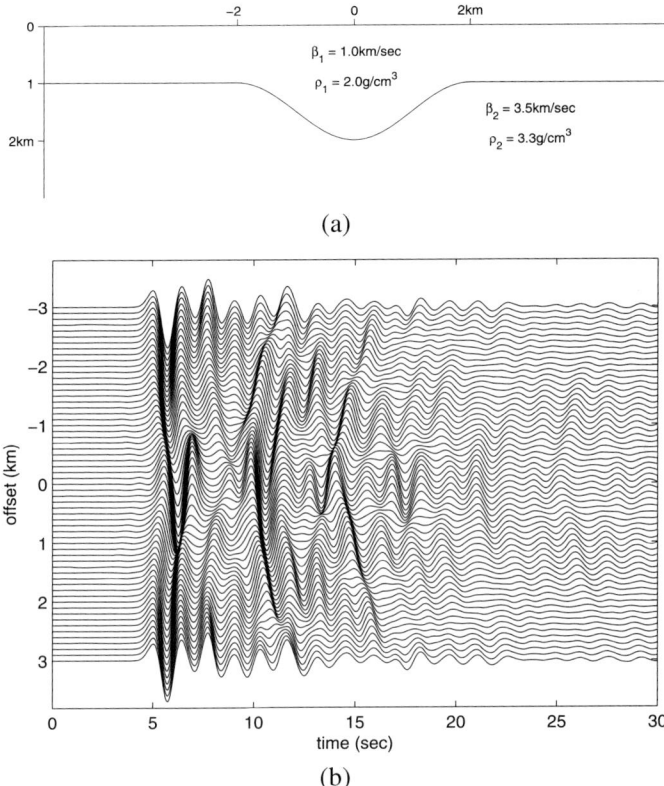

FIG. 20. (a) A steep open basin model. (b) The corresponding time domain responses on the ground, a strong and complex reverberation is observed.

been trapped in the closed basin, the corresponded reverberation is stronger and last longer than that of the open basin.

2.7.2. Seismic Scattering Problem by Elastic Inclusions

Besides the basin problem, our general algorithm can also deal with the problem of seismic scattering by elastic inclusions. Figure 22a shows a hard inclusion whose seismic wave velocity ($\beta_i = 3.5$ km/s) and density ($\rho_i = 2.8$ g/cm^3) are larger than those of background medium ($\beta_i = 2.0$ km/s, $\rho_i = 2.3$ g/cm^3), and the corresponded ground displacement responses due to a vertical plane wave incidence are plotted in Fig. 22b. As a comparison, we also calculate the ground displacement responses due to a soft inclusion (see Fig. 23) whose wave velocity and density ($\beta_i = 2.0$ km/s, $\rho_i = 2.3$ g/cm^3) are smaller than those of the background ($\beta_i = 3.5$ km/s, $\rho_i = 2.8$ g/cm^3). By comparing these two cases, we

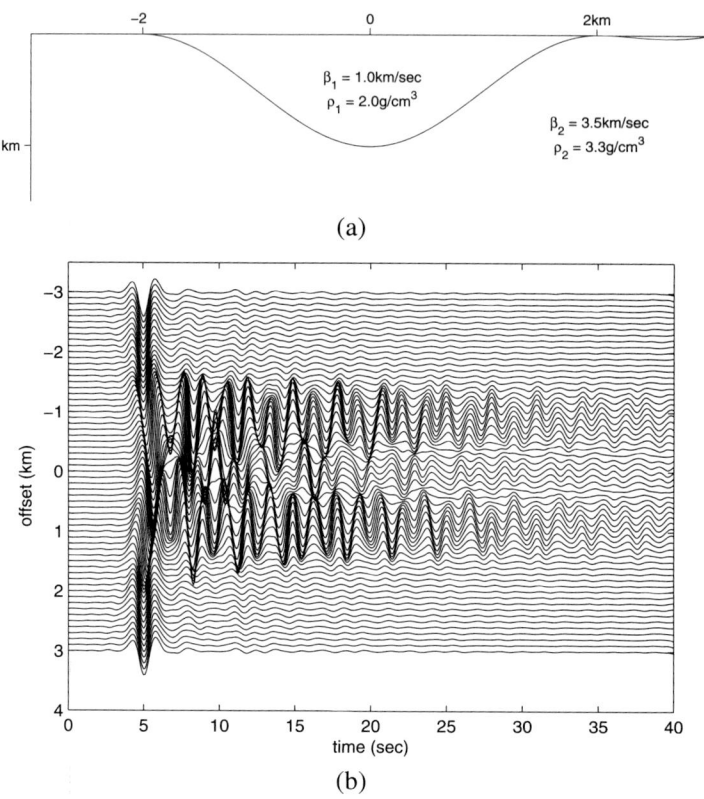

FIG. 21. (a) A closed basin model, where all parameters are same that of Fig. 20a, except the basement rock is uplifted by 1 km. (b) The time domain responses on the ground, it is observed that the strong reverberation concentrated inside the basin.

find that soft inclusion generates stronger reverberations than the hard one. Figure 24 show the scattering due to two soft inclusions, a more complicated pattern is observed.

2.7.3. Seismograms for General Multi-Layered Media with Irregular Interfaces

Simulating the excitation and propagation of seismic waves in general multi-layered media with irregular interfaces is the goal of the theory of global generalized R/T matrices presented in this chapter. Having seen a variety of applications of this algorithm above, we now turn to show its applicability to the general complex irregular layered media cases.

The first model considered is an irregular Moho model shown in Fig. 25a. We shall consider two cases for this model. First, we consider the case in which a

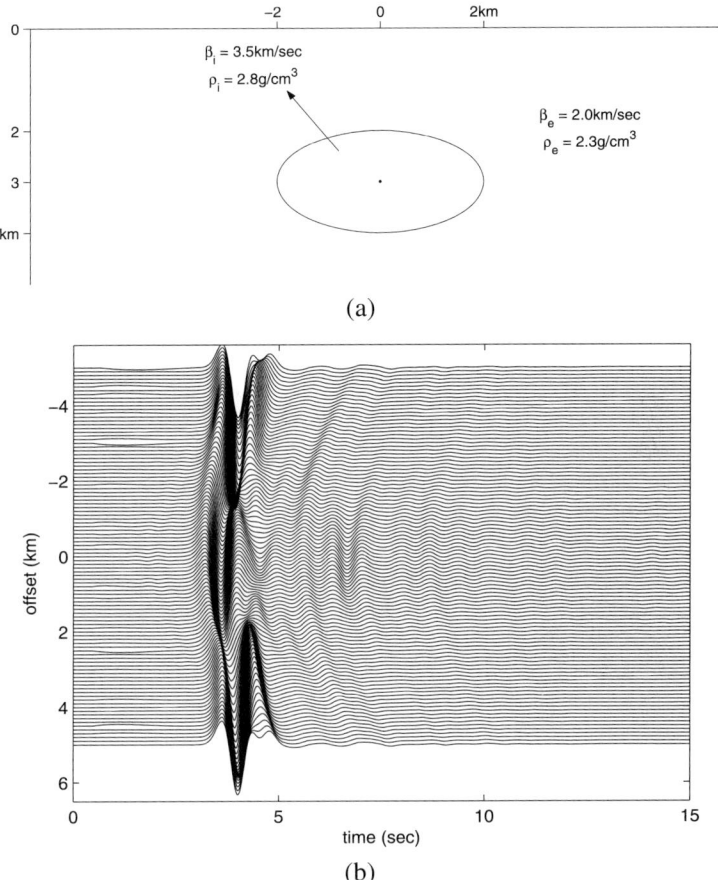

FIG. 22. Scattering problem by a hard inclusion. (a) Configuration of the model. (b) The time-domain responses on the ground.

point single-force type source is located at $x_s = -25$ km, and $z_s = 5$ km, the center frequency of source time function is 0.15 Hz, and time-shift t_s is 15 s. The corresponding synthetic seismograms recorded from $x = -20$ km to 200 km on the ground are shown in Fig. 25a. To see the effects of the lateral variation of Moho on the seismograms recorded on the surface, we consider a reference model (shown in Fig. 25b) which is same as the irregular Moho model but without the lateral variation in the Moho; and the calculated synthetic seismograms are shown in Fig. 26b. Comparing Figs. 26a and b, we find that although the Moho's irregularity is localized in the range $(-20$ km, 20 km$)$, its effects on the seismograms occur over a much wider range, particularly in the right side of the source. Another interesting feature of the comparison is that the localized Moho irregularity

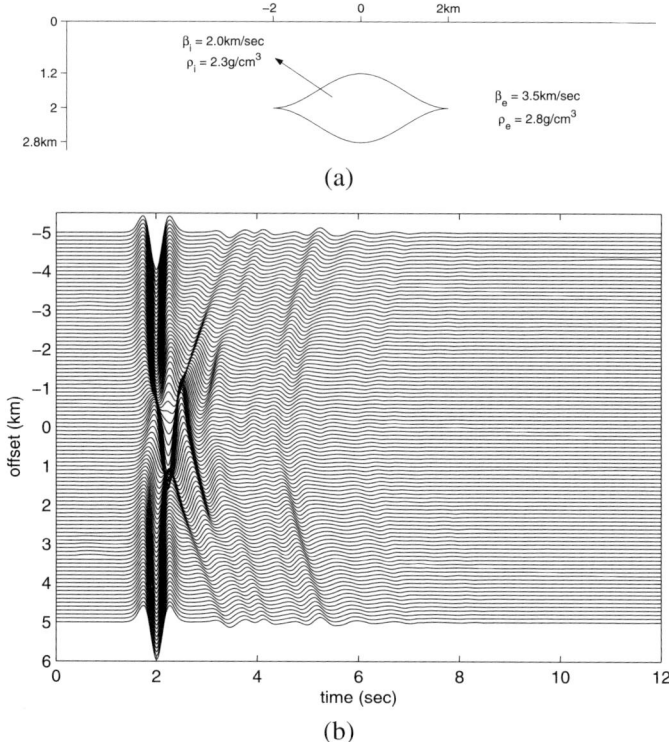

FIG. 23. Scattering problem by a soft inclusion. (a) Configuration of the model. (b) The time-domain responses on the ground.

does not only affect its nearby surrounding area (e.g., -20 km to 50 km), but also the remote area (e.g., beyond 150 km); moreover, the most affected phases around the Moho irregularity region are body waves, whereas the most affected phases in the remote area is the surface wave.

We have seen that effects of Moho irregularity on the seismograms for the case where seismic point source is located near the Moho irregularity, let us now consider the case in which seismic point source is located far away from ($x_s = 125$ km, $z_s = 5$ km) the Moho irregularity as shown in Fig. 27. The synthetic seismograms for the irregular Moho model are shown in Fig. 27a, and the differences of synthetic seismograms between the irregular model and flat Moho model are displayed in Fig. 27b. The differences of the synthetic seismograms for two models are measured by a normalized difference defined below,

$$\varepsilon(x_i, t) = \frac{|u(x_i, t) - u_0(x_i, t)|}{\max\{|u_0(x_i, t)|; 0 < t < T\}},$$

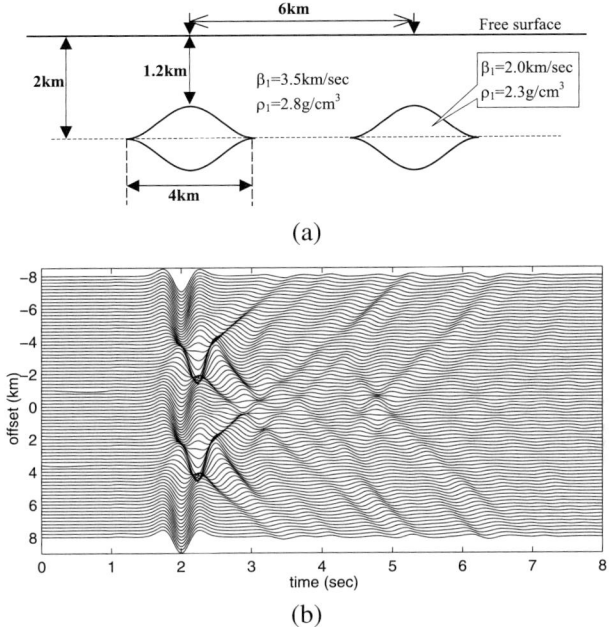

FIG. 24. Scattering problem by two soft inclusions. (a) Configuration of the model. (b) The time-domain responses on the ground.

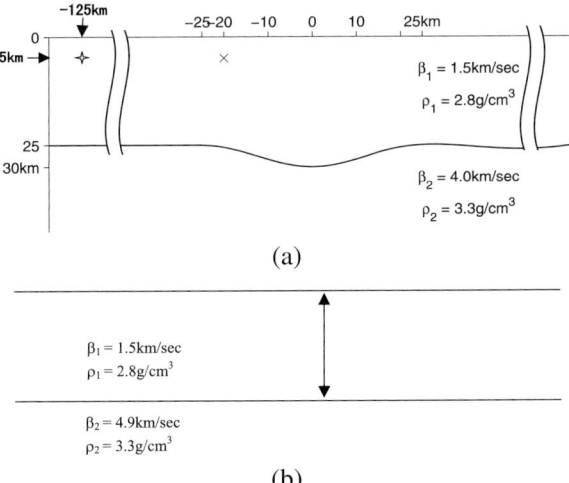

FIG. 25. Laterally varying Moho mode. (a) Configuration of lateral varying mode, where "×", and "✧" are the source locations, respectively, for the two problems considered next. (b) The referenced flat Moho model.

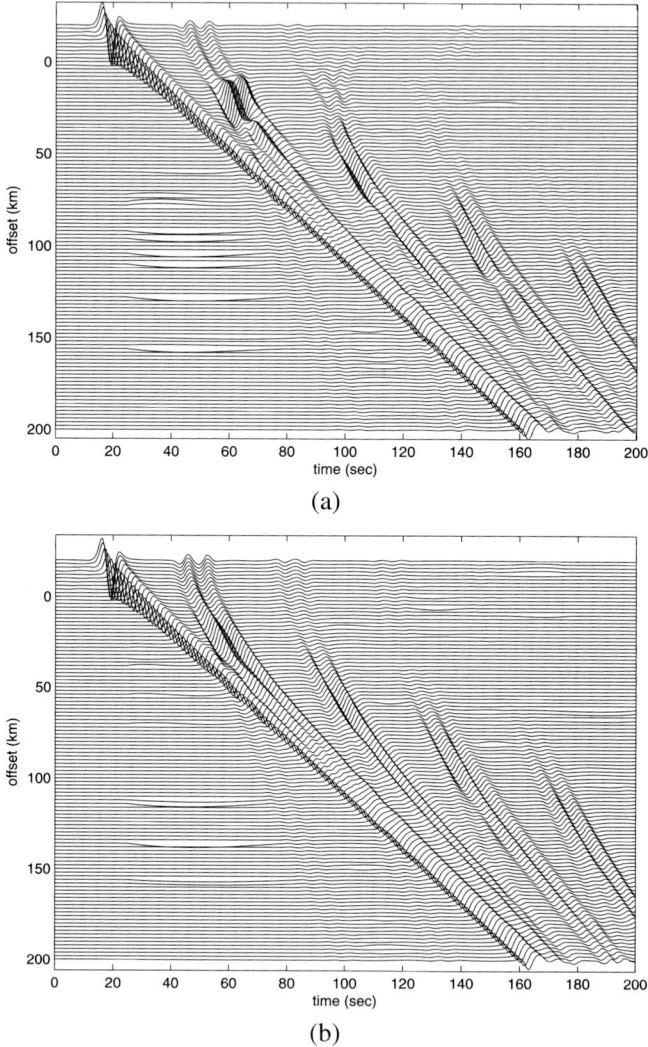

FIG. 26. The synthetic seismograms on the ground due to a point source located at ($x_S = -20$ km, $z_S = 5$ km, denoted by "×"). (a) The synthetic seismograms generated in the structure with an irregular Moho. (b) Same results for the referenced flat Moho model.

where $u(x_i, t)$ and $u_0(x_i, t)$ are the synthetic seismograms for the irregular and flat Moho models, respectively, at receiver x_i. The gray color scale denotes the value of $\varepsilon(x_i, t)$. From the comparison, we can see that the effects of Moho irregularity on the seismograms in this case mainly occur over the region $x > 0$, the right side of the Moho irregularity. And in the region $x < 0$, the effects of Moho irregularity

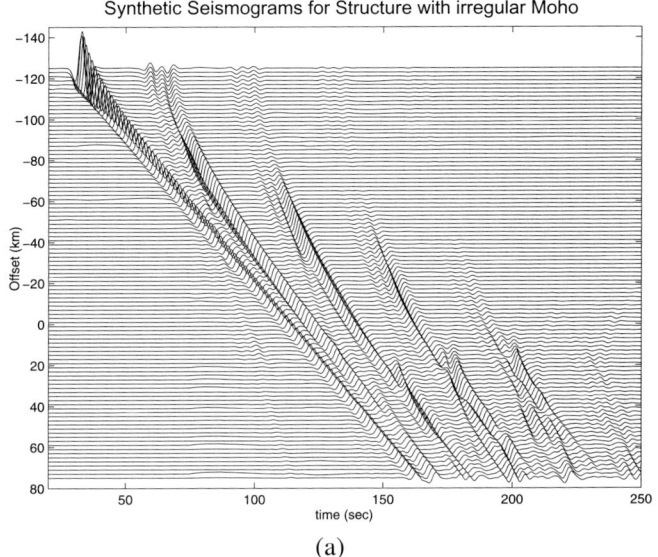

(a)

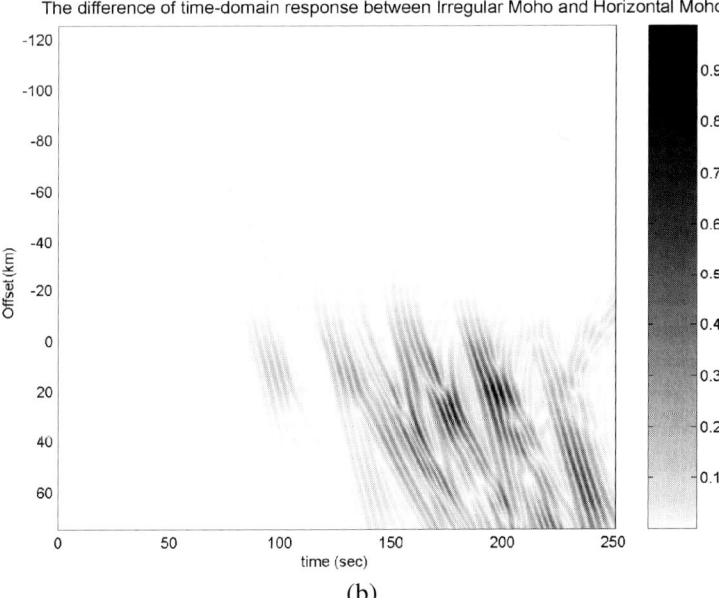

(b)

FIG. 27. The synthetic seismograms on the ground due to a point source located at ($x_S = -125$ km, $z_S = 5$ km, denoted by "◇"). (a) The synthetic seismograms generated in the structure with an irregular Moho. (b) Differences of the synthetic seismograms generated in the structures of irregular Moho and flat Moho shown in Fig. 25, where the gray color scale represents the amplitudes of normalized differences of seismograms at each receiver for the two models (see the text for detail).

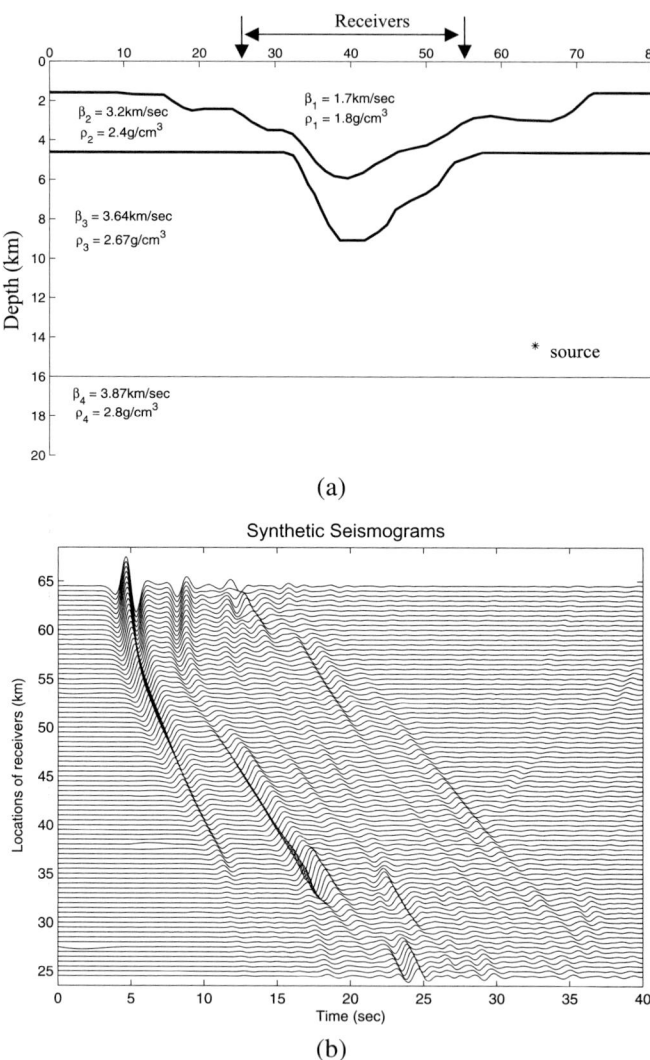

FIG. 28. (a) A cross-section profile of Los Angeles basin model. (b) The corresponding synthetic seismograms observed on the ground.

are negligible, and thus the modeling based on 1-D model is applicable here. However, in the affected region ($x > 0$), the waves (particularly the surface wave) are strongly affected by the Moho irregularity, 2-D modeling must be applied to interpret unusual wiggles in the seismograms.

The last model considered here is coined according to a realistic model, a cross section of Los Angeles basin (e.g., Chen, 1995). It is a four-layer basin model as shown in Fig. 28a, in which the upper most layer is soft sediment. We calculate the synthetic seismograms recorded top of the basin due to a point double-couple source located at position $x_s = 65$ km and $z_s = 14.5$ km, and results are shown in Fig. 28b, where Ricker wavelet with $f_c = 0.6$ Hz is used as source time function. The results show more complicated patterns:

(1) There is a strong secondary phase which is caused by the high contrast bottom of the top layer.
(2) There are two focusing areas around the points at the places $x = 35$ km and $x = 62$ km respectively, which can be attributed to the local shape of soft basin.
(3) The amplitude of predominant first arrival starts fading at place $x = 35$ km, and beyond the faded epicentral distance, seismic arrivals are complicated; these are caused by the complex shapes of the basin structure and the particularly source location as well.

It should be mentioned that the dominant computations involved in this problem are those related to the *irregular interfaces* only, the computations on the flat interfaces are 1–2 order of magnitudes lower. Because, according to our general formulation, the interface matrices and thus the global R/T matrices can be analytically reduced to diagonal ones when the interfaces become flat as demonstrated in formulas (23a) to (23d); accordingly the various matrix manipulations become scalar manipulations, thus the computations are dramatically reduced. Generally speaking, more flat layers in the considered model will not accordingly increase the computations much based on our theoretical frame, and such merit of our general algorithm will enhance its applicability to more variety seismological interesting problems.

3. LOVE WAVE IN MULTI-LAYERED MEDIA WITH IRREGULAR INTERFACES

The classic theory of seismic surface waves plays an important role in exploring the structure of crust and upper mantle, determining the earthquake source parameters, and understanding the dynamic process of seismic source. The development of classic theory of seismic surface wave started from Haskell's pioneering work (1953), underwent continuous efforts of Keilis-Borok and Yanoskaya (1962); Knopoff (1964); Harkrider and Anderson (1966); Saito (1967) and others, and was well established in the 1970s. Takeuchi and Saito (1972) and Aki and Richards (1980) systematically summarized this classic theory in their works. The classic theory of surface waves was established based on the laterally homogeneous earth model, e.g., the plane multi-layered media model used in studying the local and regional earthquake events, and the spherically stratified media model used in studying the seismological problems at global scale. However, the laterally homogeneous model is only an approximate and idealized medium model. In many actual problems, the lateral heterogeneity of media cannot be ignored, and sometimes it is even a key factor of the problem, whereas the classic theory of surface waves fails in such cases.

Since the 1970s, the classic theory of surface waves has been extended to the laterally heterogeneous case (e.g., Gjevik, 1973; Woodhouse, 1974; Yomogida, 1985; Zhu and Zeng, 1993) based on either of the following two approaches: (1) modes-coupling method, and (2) regionalization approach. The modes-coupling method (e.g., Snieder, 1987; Maupin, 2006) was based on the perturbation technique that only works for weaker lateral heterogeneity. The regionalization approach (e.g., Gjevik, 1973; Woodhouse, 1974; Yomogida, 1985; Zhu and Zeng, 1993) is built upon the hypothesis of regionalization, that is, the classic surface wave theory can be applied to each local quasi 1-D (quasi-stratified medium) structure by ignoring the interactions among different normal modes. Obviously, this hypothesis is appropriate for weak lateral variation and is questionable for strong lateral heterogeneity.

In this section, we shall summarize a new theory of Love waves in irregular multi-layered media (a typical laterally heterogeneous media model) developed by the author recently (e.g., Chen, 1999a). It will be seen that this theory of Love wave was rigorously established based on the analytical formulation of the *global generalized R/T matrix* method summarized in the previous sections of this chapter.

3.1. Modal Solution of Love Wave in an Arbitrarily Irregular Multi-Layered Medium

The modal solutions are non-trivial solutions of the free elastodynamic equation (without force term) under appropriate boundary conditions (Chen, 1993). This is a general definition for both the laterally homogeneous structure model and the laterally heterogeneous structure model. It has been successfully applied to determine the normal modes of the laterally homogeneous multi-layered media (Chen, 1993), and we will demonstrate below that this definition can also be applied to determine the modal solution of an arbitrarily irregular multi-layered medium.

3.1.1. Characteristic Frequencies and Modal Solution of Love Wave in IMLM

In the previous sections, we have successfully derived a complete formulation for computing the excitation and propagation of SH waves in arbitrarily irregular multi-layered media. According to the result, the general solution of *free elastodynamic equation* for SH waves in an arbitrarily irregular multi-layered medium have the following form:

$$W^{(j)}(x, z, \omega) = \mathbf{e}^T(x)\{\mathbf{L}_\uparrow^{(j)}(z, \omega)\mathbf{a}_\uparrow^{(j)} + \mathbf{L}_\downarrow^{(j)}(z, \omega)\mathbf{a}_\downarrow^{(j)}\}, \tag{31}$$

where, the unknown generalized amplitudes $\mathbf{a}_\uparrow^{(j)}$ and $\mathbf{a}_\downarrow^{(j)}$ satisfy

$$\begin{cases} \mathbf{a}_\downarrow^{(j)} = \hat{\mathbf{T}}_{\downarrow\downarrow}^{(j-1)}\hat{\mathbf{T}}_{\downarrow\downarrow}^{(j-2)}\cdots\hat{\mathbf{T}}_{\downarrow\downarrow}^{(1)}\mathbf{a}_\downarrow^{(1)}, \\ \mathbf{a}_\uparrow^{(j)} = \hat{\mathbf{R}}_{\downarrow\uparrow}^{(j)}\mathbf{a}_\downarrow^{(j)}, \end{cases} \quad \text{for } j = 1, 2, \ldots, N+1. \qquad (32)$$

Obviously, once the generalized amplitude in the top layer $\mathbf{a}_\downarrow^{(1)}$ is found, all others can be immediately determined. Inside the top layer, we have

$$\begin{cases} \mathbf{a}_\downarrow^{(1)} = \hat{\mathbf{R}}_{\uparrow\downarrow}^{(0)}\mathbf{a}_\uparrow^{(1)}, \\ \mathbf{a}_\uparrow^{(1)} = \hat{\mathbf{R}}_{\downarrow\uparrow}^{(1)}\mathbf{a}_\downarrow^{(1)}. \end{cases}$$

Combining the above two equations yields

$$\{\mathbf{I} - \hat{\mathbf{R}}_{\uparrow\downarrow}^{(0)}(\omega)\hat{\mathbf{R}}_{\downarrow\uparrow}^{(1)}(\omega)\}\mathbf{a}_\downarrow^{(1)}(\omega) = 0. \qquad (33)$$

Thus, the existence of a non-trivial solution leads to the following dispersion equation

$$\det\{\mathbf{I} - \hat{\mathbf{R}}_{\uparrow\downarrow}^{(0)}(\omega)\hat{\mathbf{R}}_{\downarrow\uparrow}^{(1)}(\omega)\} = 0, \qquad (34)$$

where, $\det\{\mathbf{I} - \hat{\mathbf{R}}_{\uparrow\downarrow}^{(0)}(\omega)\hat{\mathbf{R}}_{\downarrow\uparrow}^{(1)}(\omega)\}$ is defined as a secular function. It is noted that dispersion equations (33) and (34) are of the same form as those for determining Love modes in laterally homogeneous layered media case (Chen, 1993). Only certain number of discrete frequencies (ω_m; $m = 0, 1, 2, \ldots, M$) satisfy the dispersion equation (34), and they are the *characteristic frequencies* of the fundamental mode and higher modes. Substituting these frequencies back into Eq. (33), we can determine the non-zero vector $\mathbf{a}_\downarrow^{(1)}(\omega_m)$. Alternatively, Eq. (33) can be solved by using the method for solving eigen-value problem. To solve equation (33), we consider the following eigen-value equation,

$$\{\mathbf{I} - \hat{\mathbf{R}}_{\uparrow\downarrow}^{(0)}(\omega)\hat{\mathbf{R}}_{\downarrow\uparrow}^{(1)}(\omega)\}\mathbf{q}_n(\omega) = \eta_n(\omega)\mathbf{q}_n(\omega), \qquad (35)$$

where $\eta_n(\omega)$ and $\mathbf{q}_n(\omega)$ are the nth eigen-value and eigen-vectors of matrix $\mathbf{I} - \hat{\mathbf{R}}_{\uparrow\downarrow}^{(0)}\hat{\mathbf{R}}_{\downarrow\uparrow}^{(1)}$. Although there are N_E eigen-values and eigen-vectors for any frequency (N_E is the dimension of matrix $\mathbf{I} - \hat{\mathbf{R}}_{\uparrow\downarrow}^{(0)}\hat{\mathbf{R}}_{\downarrow\uparrow}^{(1)}$), obviously only those with zero eigen-values are the solutions of Eq. (33). For the nth eigen-value $\eta_n(\omega)$, there exist finite number of discrete frequencies ($\omega_{n,\nu}$) corresponding to zero eigen-value, that is

$$\eta_n(\omega_{n,\nu}) = 0, \quad \text{for } \nu = 0, 1, 2, \ldots, K, \qquad (36a)$$

where the index ν indicates the order of modal solution. The corresponding eigen-vectors are the non-zero solutions of Eq. (33), i.e.,

$$\mathbf{a}_\downarrow^{(1)}(\omega_{n,\nu}) = \mathbf{q}_n(\omega_{n,\nu}), \quad \text{for } \nu = 0, 1, 2, \ldots, K. \qquad (36b)$$

Then the corresponding eigen-displacements $\{\phi_{n,\nu}(x, z), n = 0, \pm 1, \pm 2, \ldots, N_E;$ $\nu = 0, 1, 2, \ldots, K\}$ can be determined as follows,

$$\phi_{n,\nu}^{(j)}(x, z) = \mathbf{e}^T(x)\mathbf{Y}^{(j)}(z, \omega_{n,\nu})\hat{\mathbf{T}}_{\downarrow\downarrow}^{(j-1)}\hat{\mathbf{T}}_{\downarrow\downarrow}^{(j-2)} \cdots \hat{\mathbf{T}}_{\downarrow\downarrow}^{(1)}(\omega_{n,\nu})\mathbf{a}_{\downarrow}^{(1)}(\omega_{n,\nu}),$$

for $(x, z) \in V^j$. \hfill (37)

Equation (37) gives rise to the analytic expression of modal solution of Love wave in an arbitrary irregular multi-layered medium. Unlike the horizontally homogeneous layered media case, the modal solutions of Love wave here are non-separable in (x, z) coordinates; they may only be given in (x, z)-domain simultaneously. This is the nature of laterally heterogeneous media (e.g., Rial and Ling, 1992).

3.1.2. Numerical Examples of Modal Solutions in IMLM

Let us use a particular example to illustrate the previous results. The considered structure model is shown in Fig. 29a. For comparison, we choose a horizontally homogeneous layered structure model shown in Fig. 29b as the reference model. It should be pointed out that the secular function defined in Eq. (34) is a complex value function, thus its roots correspond to the simultaneous zeros of both real and imaginary parts of the function. For simplicity, we search the roots of the module of the complex valued secular function, which is a real function and having same set of roots as the complex valued secular function. The module of secular function is plotted in Fig. 30. We find that there do exist many roots in the frequency range considered here (0.1–1.0 Hz). For example, around 0.6 Hz there exists a root at $f_1 = 0.6111$ Hz. To obtain the corresponding non-zero amplitude vector of down-going wave $\mathbf{a}_{\downarrow}^{(1)}$, we first calculate the eigen-values $\eta_n(\omega)$ and eigen-vectors $\mathbf{q}_n(\omega)$ of the dispersion matrix $\mathbf{I} - \hat{\mathbf{R}}_{\uparrow\downarrow}^{(0)}(\omega)\hat{\mathbf{R}}_{\downarrow\uparrow}^{(1)}(\omega)$ at $\omega = 2\pi f_1$,

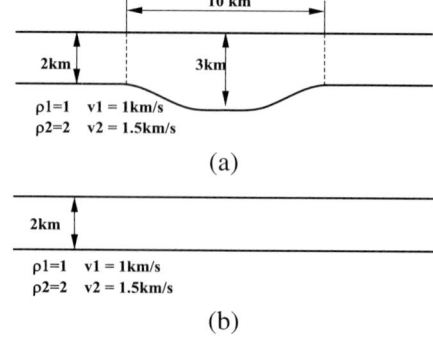

FIG. 29. (a) An irregular layered structure model; (b) and the reference model.

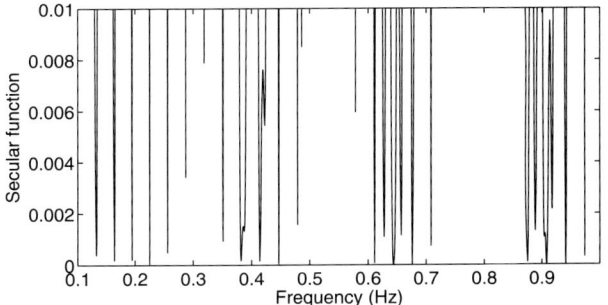

FIG. 30. Plot of the module of secular function for the structure model shown in Fig. 29a. A close look around the moderate frequency indicates a root at $f = 0.6111$ Hz.

then the eigen-vectors corresponding to zero eigen-value are the non-trivial solution of dispersion equation (34). The eigen-values at $\omega = 2\pi f_1$ are plotted in Fig. 31a, we can see the ninth and tenth eigen-values are zero within accuracy of 10^{-2}. The corresponding eigen-vectors are shown in Figs. 31b and c, and they are the non-trivial vector solutions of dispersion equation (34). Once having the non-zero solution $\mathbf{a}_\downarrow^{(1)}$, we can finally compute $\phi_\nu(x, z)$ the modal function of Love wave for this laterally varying layered media case. Without losing the generality, we use the ninth eigen-vector to compute the modal function. Figures 32a and b show, respectively, the real and imaginary parts of the modal function in 2-D space. For comparison, we display the corresponding 2-D modal function for the reference horizontally layered media mode in Figs. 32c and d. In fact, because of the horizontal homogeneity of the reference structure model, the corresponding modal function has the following simple analytic expression (e.g., Aki and Richards, 1980),

$$u_\nu(x, z) = l_\nu(z) e^{i|x|k(\omega_\nu)}, \tag{38}$$

where, $k(\omega_\nu)$ is the wave number corresponding to eigen-frequency ω_ν; $l_\nu(z)$ is the eigen-function of Love wave for the reference media mode, which is usually called as *normal mode* of Love wave. Clearly, the modal displacement $u_\nu(x, z)$ here is separable in vertical variable z and horizontal variable x: $l_\nu(z)$ gives the vertical dependence of $u_\nu(x, z)$, and $\exp\{i|x|k(\omega_\nu)\}$ governs the horizontal dependency. The vertical dependence $l_\nu(z)$ generally is a numerical solution for a given stratified medium, and it reflects the influence of vertical heterogeneity of the medium. Whereas the horizontal dependence is a simple analytic function, e.g., sine or cosine function for this 2-D case and Hankle function for 3-D case (e.g., Aki and Richards, 1980). Therefore, $l_\nu(z)$ is generally used to characterize the main feature of Love wave in the case of stratified media.

For the multi-layered media with irregular interfaces, however, the corresponded modal function of Love wave cannot be factorized into z-dependence

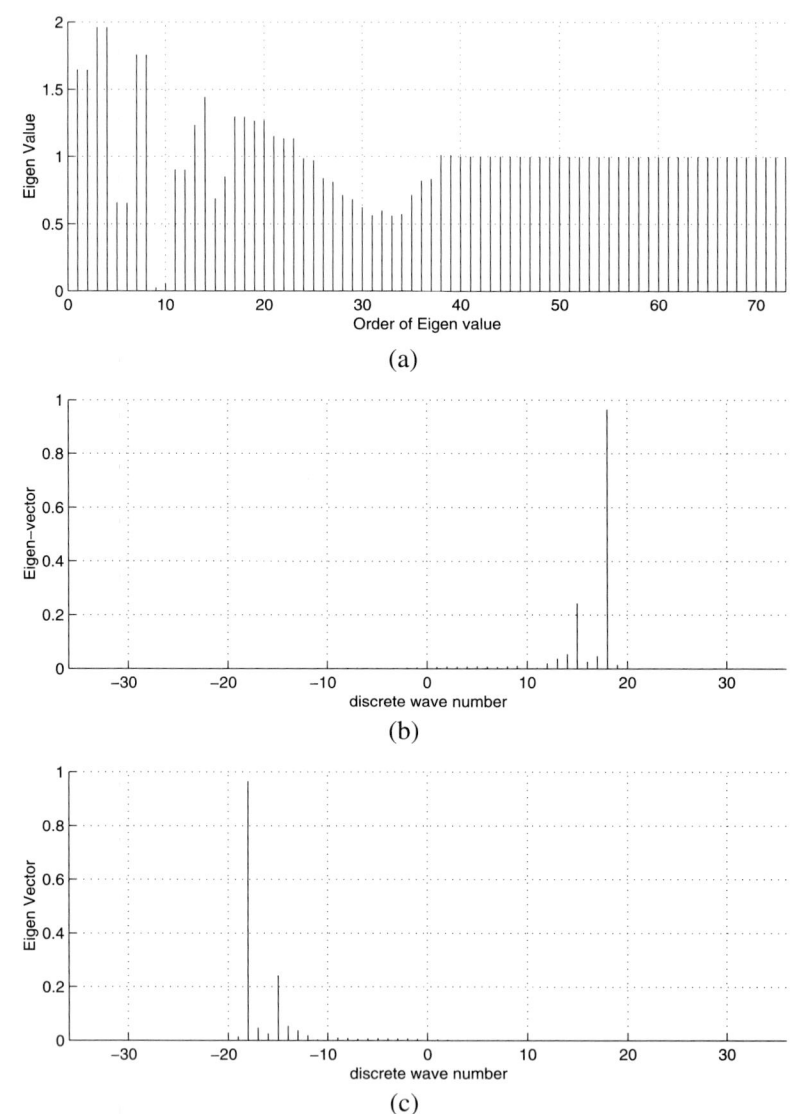

FIG. 31. (a) Distribution of eigen-values for the characteristic frequency $f_1 = 0.6111$ Hz, from which we can see the ninth and tenth eigen-states have zero eigen-values; (b) the ninth eigen-vector; (c) the tenth eigen-vector.

and x-dependence, it is non-separable function in (x, z) coordinates (e.g., Rial and Ling, 1992). Therefore, we must use the 2-D modal function $\phi_\nu(x, z)$ to characterize the Love wave in this laterally varying layered medium. As shown in Figs. 32a and 32b, the modal function of Love wave for a laterally homoge-

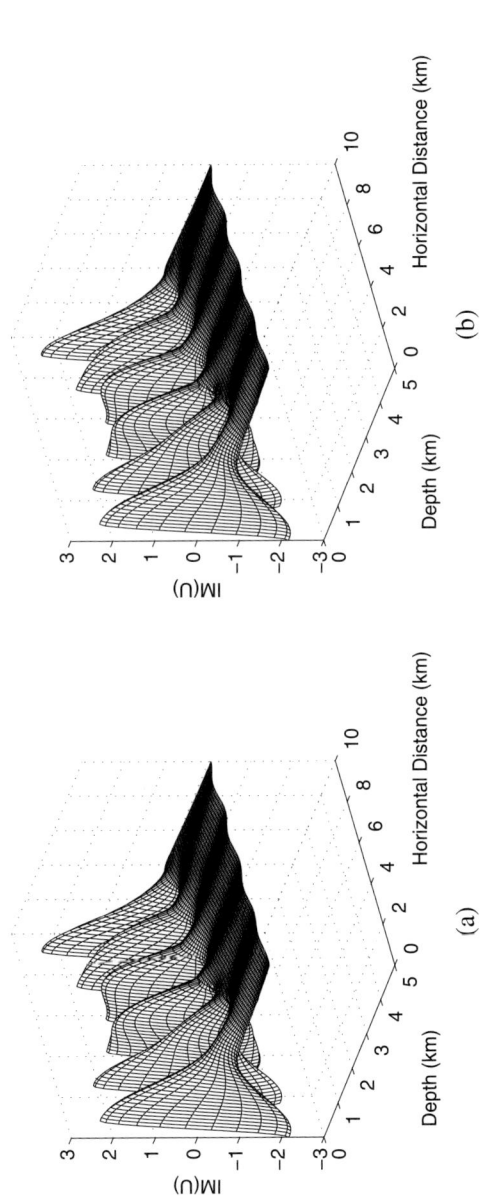

FIG. 32. (a) and (b) show the real and imaginary parts, respectively, of the modal solution corresponding to the ninth eigen-vector shown in Fig. 31b; (c) and (d) are same as (a) and (b) but for the reference structure model shown in Fig. 29b.

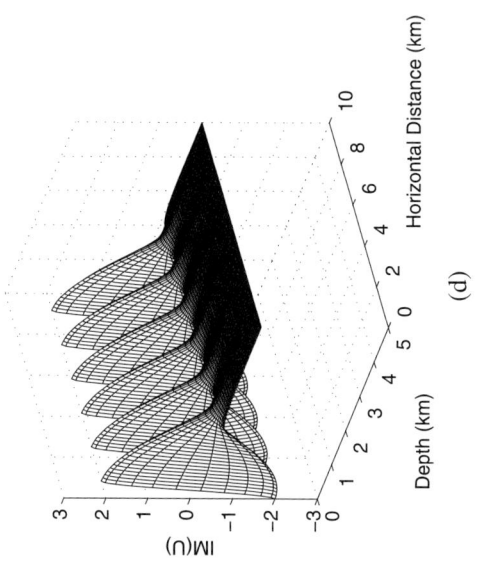

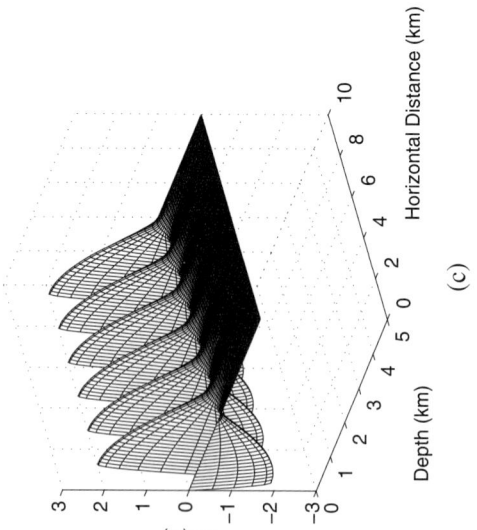

FIG. 32. Continued.

neous layered medium presents a regular shape in (x, z) space, whereas the one for IMLM presents an irregular shape that seems to be an distorted version of the formal one. This distortion is due to the coupling among different horizontal wave numbers, which is caused by lateral irregularity of the medium. We defined the 2-D modal solution $\phi_\nu(x, z)$ as *distorted mode*. Obviously, the concept of *distorted mode* is a natural generalization of the normal mode, that is, when the IMLM becomes laterally homogeneous layered media, the *distorted mode* is accordingly reduced to the *normal mode*. The modes shown in Figs. 32a and b are fundamental modes at $f = 0.6111$ Hz. Figure 33 shows the first higher distorted mode at $f = 0.6571$ Hz and the corresponding first higher normal mode for the reference model. Figures 34a and b show the fundamental distorted mode and normal model at relatively higher frequency ($f = 0.9410$ Hz) for the same problem as stated previously. In this higher frequency case, we see a mode with much distorted shape, indicating the stronger coupling among different horizontal wave numbers.

3.2. Excitation Formula of Love Wave in Irregular Multi-Layered Media

To derive the excitation formulation of Love wave in arbitrarily irregular multi-layered media, we rewrite the general solution of SH waves due to an arbitrary source (i.e., Eq. (22)) as follows:

$$W^{(j)}(x, z, \omega)$$
$$= \mathbf{e}^T(x)\mathbf{Y}^{(j)}(z, \omega)\hat{\mathbf{R}}_{\uparrow\downarrow}^{(j-1)}(\omega)\hat{\mathbf{T}}_{\uparrow\uparrow}^{(j)}(\omega)\hat{\mathbf{T}}_{\uparrow\uparrow}^{(j+1)}(\omega)\cdots\hat{\mathbf{T}}_{\uparrow\uparrow}^{(s-1)}(\omega)$$
$$\times \left[\mathbf{I} - \hat{\mathbf{R}}_{\downarrow\uparrow}^{(s)}(\omega)\hat{\mathbf{R}}_{\uparrow\downarrow}^{(s-1)}(\omega)\right]^{-1}\left[\mathbf{s}_{0\uparrow}(\omega) + \hat{\mathbf{R}}_{\downarrow\uparrow}^{(s)}(\omega)\mathbf{s}_{0\downarrow}(\omega)\right],$$
for $z < z_s$ (39a)

and

$$W^{(j)}(x, z, \omega)$$
$$= \mathbf{e}^T(x)\mathbf{Y}^{(j)}(z, \omega)\hat{\mathbf{T}}_{\downarrow\downarrow}^{(j-1)}(\omega)\hat{\mathbf{T}}_{\downarrow\downarrow}^{(j-2)}(\omega)\cdots\hat{\mathbf{T}}_{\downarrow\downarrow}^{(s)}(\omega)$$
$$\times \left[\mathbf{I} - \hat{\mathbf{R}}_{\uparrow\downarrow}^{(s-1)}(\omega)\hat{\mathbf{R}}_{\downarrow\uparrow}^{(s)}(\omega)\right]^{-1}\left[\mathbf{s}_{0\downarrow}(\omega) + \hat{\mathbf{R}}_{\uparrow\downarrow}^{(s-1)}(\omega)\mathbf{s}_{0\uparrow}(\omega)\right],$$
for $z > z_s$. (39b)

The Love wave is a special part of whole SH wave train. The complete seismogram of SH wave in this laterally heterogeneous medium can be calculated by

$$w^{(j)}(x, z, t) = \int_{-\infty}^{+\infty} W^{(j)}(x, z, \omega)e^{i\omega t}\,d\omega. \tag{40}$$

The spectrum of displacement $W^{(j)}(x, z, \omega)$, say for $z < z_s$, is given by Eq. (39a). Based on the definition of the global generalized R/T matrices, we obtain the

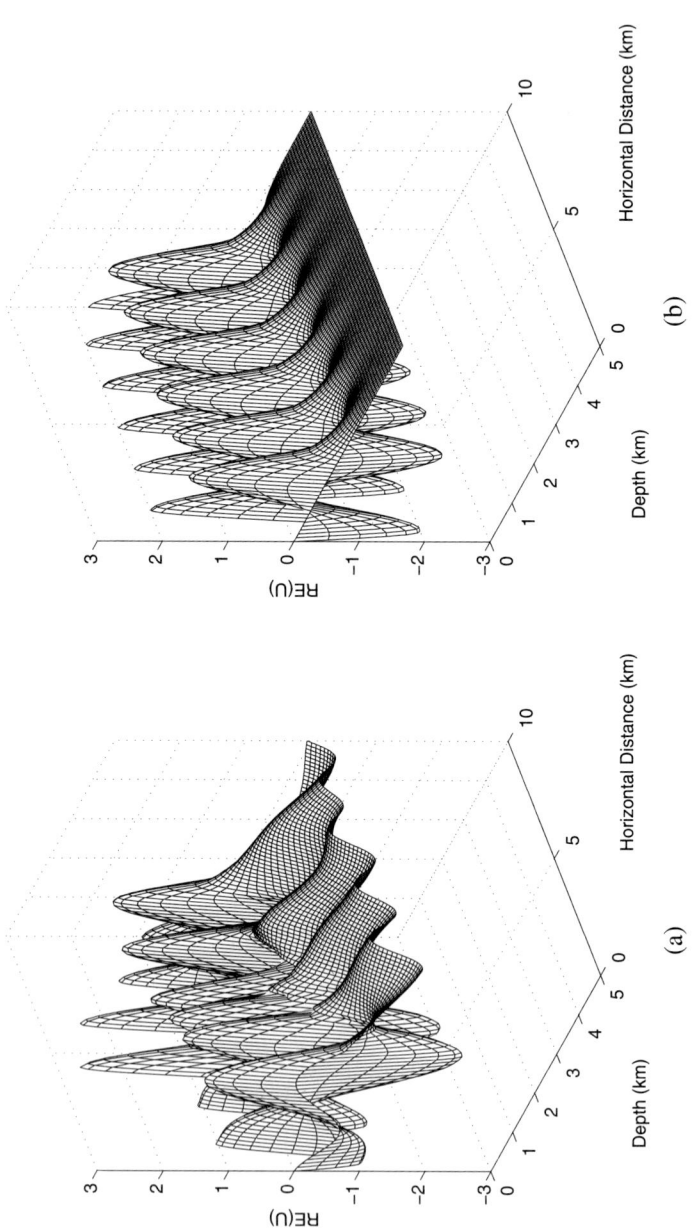

FIG. 33. Comparison between (a) the 1st higher distorted mode in the structure model shown in Fig. 29a and (b) the 1st higher normal mode in the reference structure model shown in Fig. 29b. Here frequency $f = 0.6571$ Hz.

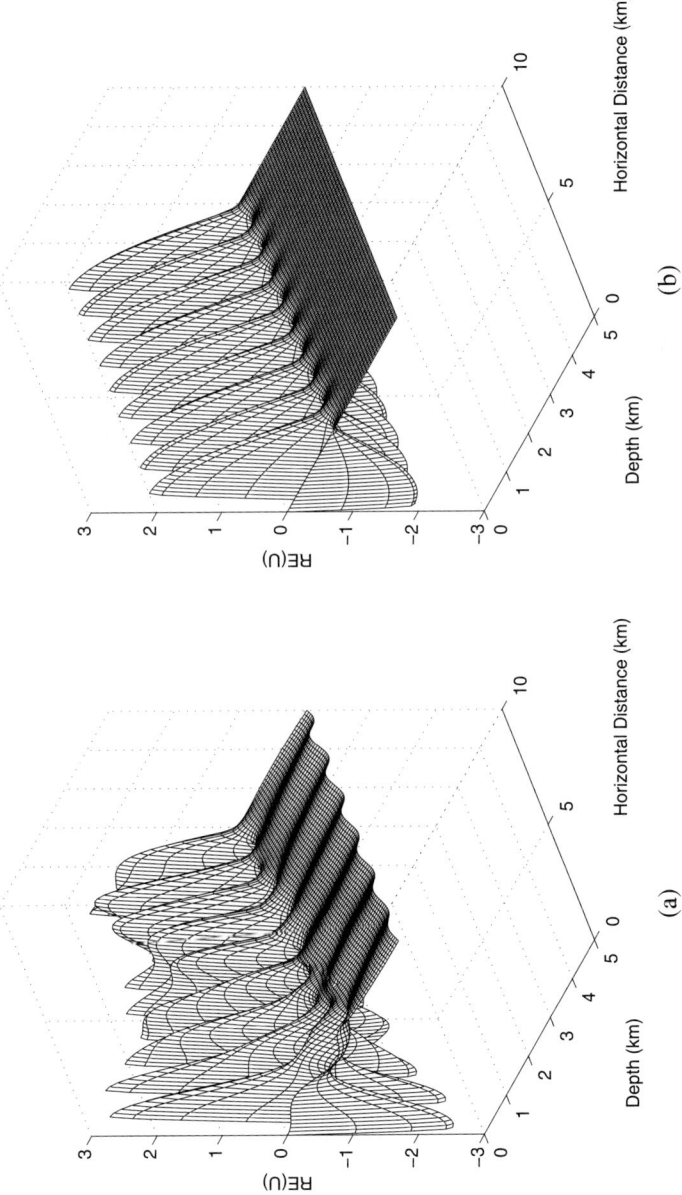

FIG. 34. Comparison between (a) distorted mode and (b) normal mode for relatively higher frequency case ($f = 0.941$ Hz).

following formula (see demonstration in Appendix C)

$$\hat{\mathbf{R}}_{\uparrow\downarrow}^{(j-1)}\hat{\mathbf{T}}_{\uparrow\uparrow}^{(j)}\hat{\mathbf{T}}_{\uparrow\uparrow}^{(j+1)}\cdots\hat{\mathbf{T}}_{\uparrow\uparrow}^{(s-1)}(\omega)\bigl[\mathbf{I} - \hat{\mathbf{R}}_{\uparrow\uparrow}^{(s)}(\omega)\hat{\mathbf{R}}_{\downarrow\uparrow}^{(s-1)}(\omega)\bigr]^{-1}$$
$$= \hat{\mathbf{T}}_{\downarrow\downarrow}^{(j)}\hat{\mathbf{T}}_{\downarrow\downarrow}^{(j-1)}\cdots\hat{\mathbf{T}}_{\downarrow\downarrow}^{(1)}(\omega)\bigl[\mathbf{I} - \hat{\mathbf{R}}_{\uparrow\downarrow}^{(0)}(\omega)\hat{\mathbf{R}}_{\downarrow\uparrow}^{(1)}(\omega)\bigr]^{-1}$$
$$\times \hat{\mathbf{R}}_{\uparrow\downarrow}^{(0)}(\omega)\hat{\mathbf{T}}_{\uparrow\uparrow}^{(1)}\hat{\mathbf{T}}_{\uparrow\uparrow}^{(2)}\cdots\hat{\mathbf{T}}_{\uparrow\uparrow}^{(s-1)}(\omega). \tag{41}$$

Moreover, for a given frequency ω, the inverse of matrix "$\mathbf{I} - \hat{\mathbf{R}}_{\uparrow\downarrow}^{(0)}(\omega)\hat{\mathbf{R}}_{\downarrow\uparrow}^{(1)}(\omega)$" can be spanned by its eigen-vector set as (see Appendix D):

$$\bigl[\mathbf{I} - \hat{\mathbf{R}}_{\uparrow\downarrow}^{(0)}(\omega)\hat{\mathbf{R}}_{\downarrow\uparrow}^{(1)}(\omega)\bigr]^{-1} = \sum_{n}\sum_{m}\frac{Z_{nm}(\omega)}{\eta_n(\omega)}\mathbf{q}_n(\omega)\bigl[\mathbf{q}_m(\omega)^*\bigr]^T, \tag{42}$$

where, matrix $\mathbf{Z}(\omega)$ is made of $\{\mathbf{q}_n(\omega); n = 0, \pm 1, \pm 2, \ldots\}$ via the following formula:

$$\bigl[\mathbf{Z}(\omega)^{-1}\bigr]_{nm} = \bigl[\mathbf{q}_n(\omega)^*\bigr]^T \mathbf{q}_m(\omega). \tag{43}$$

Substituting Eqs. (41) and (42) into Eq. (39a), we obtain

$$W^{(j)}(x, z, \omega)$$
$$= \mathbf{e}^T(x)\mathbf{Y}^{(j)}(z,\omega)\hat{\mathbf{T}}_{\downarrow\downarrow}^{(j)}\hat{\mathbf{T}}_{\downarrow\downarrow}^{(j-1)}\cdots\hat{\mathbf{T}}_{\downarrow\downarrow}^{(1)}(\omega)\bigl[\mathbf{I} - \hat{\mathbf{R}}_{\uparrow\downarrow}^{(0)}(\omega)\hat{\mathbf{R}}_{\downarrow\uparrow}^{(1)}(\omega)\bigr]^{-1}$$
$$\times \hat{\mathbf{R}}_{\uparrow\downarrow}^{(0)}(\omega)\hat{\mathbf{T}}_{\uparrow\uparrow}^{(1)}\hat{\mathbf{T}}_{\uparrow\uparrow}^{(2)}\cdots\hat{\mathbf{T}}_{\uparrow\uparrow}^{(s-1)}(\omega)\bigl[\mathbf{s}_{0\uparrow}(\omega) + \hat{\mathbf{R}}_{\downarrow\uparrow}^{(s)}(\omega)\mathbf{s}_{0\downarrow}(\omega)\bigr]$$
$$= \sum_{n}\frac{1}{\eta_n(\omega)}\bigl\{\mathbf{e}^T(x)\mathbf{Y}^{(j)}(z,\omega)\hat{\mathbf{T}}_{\downarrow\downarrow}^{(j)}\hat{\mathbf{T}}_{\downarrow\downarrow}^{(j-1)}\cdots\hat{\mathbf{T}}_{\downarrow\downarrow}^{(1)}(\omega)\mathbf{q}_n(\omega)\bigr\}p_n(\omega), \tag{44}$$

where,

$$p_n(\omega) = \sum_{m}Z_{nm}(\omega)\bigl[\mathbf{q}_m(\omega)^*\bigr]^T \hat{\mathbf{R}}_{\uparrow\downarrow}^{(0)}(\omega)\hat{\mathbf{T}}_{\uparrow\uparrow}^{(1)}\hat{\mathbf{T}}_{\uparrow\uparrow}^{(2)}\cdots\hat{\mathbf{T}}_{\uparrow\uparrow}^{(s-1)}(\omega)$$
$$\times \bigl[\mathbf{s}_{0\uparrow}(\omega) + \hat{\mathbf{R}}_{\downarrow\uparrow}^{(s)}(\omega)\mathbf{s}_{0\downarrow}(\omega)\bigr]. \tag{44a}$$

Obviously, $p_n(\omega)$ contains source information. Inserting Eq. (44) into (40) yields,

$$w^{(j)}(x, z, t)$$
$$= \sum_{n}\int_{-\infty}^{+\infty}\frac{1}{\eta_n(\omega)}\bigl\{\mathbf{e}^T(x)\mathbf{Y}^{(j)}(z,\omega)\hat{\mathbf{T}}_{\downarrow\downarrow}^{(j)}\hat{\mathbf{T}}_{\downarrow\downarrow}^{(j-1)}\cdots\hat{\mathbf{T}}_{\downarrow\downarrow}^{(1)}(\omega)\mathbf{q}_n(\omega)\bigr\}$$
$$\times p_n(\omega)e^{i\omega t}\,d\omega. \tag{45}$$

Formula (45) gives rise to a complete time-domain solution of SH wave in 2-D irregular multi-layered medium, which includes body waves and surface waves. According to Eqs. (36a) and (36b), Love wave is determined by the *characteristic frequencies* $\omega_{n,\nu}$ which makes $\eta_n(\omega_{n,\nu}) = 0$. To extract the components of Love wave excited by a seismic source, we only need to take the contributions from those characteristic frequencies (i.e., zeros of $\eta_n(\omega)$) into account as done

SEISMIC SH WAVES IN IRREGULAR MULTI-LAYERED 245

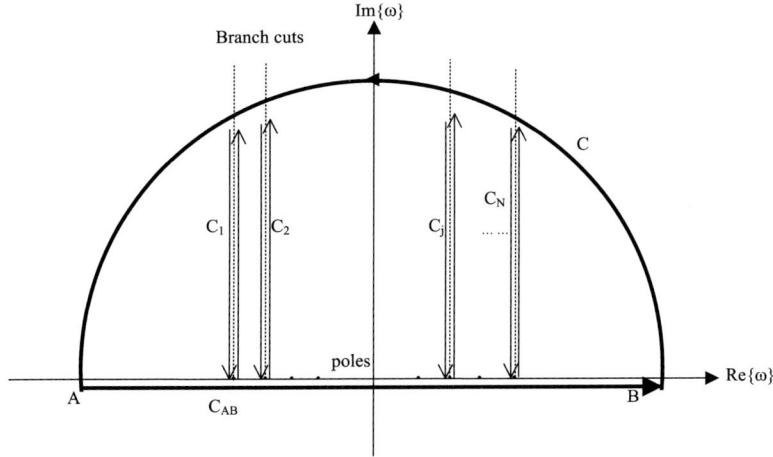

FIG. 35. A contour integration path in complex-ω plane showing the contributions from poles (Love waves) and branch cuts (body waves and leaky modes).

in deriving the excitation formula for classic Love wave (e.g., Aki and Richards, 1980). To do so, we consider the following contour integral,

$$\int_{AB} + \left\{ \int_{C_1} + \int_{C_2} + \cdots + \int_{C_N} \right\} + \int_{C} = 2\pi i \sum \text{Residues at } \omega_{n,\nu}. \quad (46)$$

As shown in Fig. 35, the integration along real frequency axis $\int_{AB}$ is the original integration (45), and integration along half-circle C, $\int_{C}$, vanishes as the radius tends to infinity, hence we have

$$w^{(j)}(x, z, t) = \int_{AB} = -\left\{ \int_{C_1} + \int_{C_2} + \cdots + \int_{C_N} \right\}$$
$$+ 2\pi i \sum \text{Residues at } \omega_{n,\nu}. \quad (47)$$

Moreover, integrals around branch-cuts $C_1, C_2, \ldots, C_N$ (i.e., $\int_{C_1} + \int_{C_2} + \cdots + \int_{C_N}$) include the body waves and leaky mode contributions (e.g., Aki and Richards, 1980), whereas the residues at $\omega_{n,\nu}$ generate the Love wave. Therefore, Love wave can be separated from the whole SH wave train as follows:

$$w_L^{(j)}(x, z, t)$$
$$= 2\pi i \sum_{n,\nu} \text{Residues at } \omega_{n,\nu} = 2\pi i \sum_{n,\nu} \frac{p_n(\omega_{n,\nu}) e^{i\omega_{n,\nu} t}}{\{\partial_\omega \eta(\omega)\}|_{\omega = \omega_{n,\nu}}}$$
$$\times \left\{ \mathbf{e}^T(x) \mathbf{Y}^{(j)}(z, \omega_{n,\nu}) \hat{\mathbf{T}}_{\downarrow\downarrow}^{(j)} \hat{\mathbf{T}}_{\downarrow\downarrow}^{(j-1)} \cdots \hat{\mathbf{T}}_{\downarrow\downarrow}^{(1)}(\omega_{n,\nu}) \mathbf{q}_n(\omega_{n,\nu}) \right\}. \quad (48)$$

Using Eq. (37), formula (48) can be simplified as

$$w_L^{(j)}(x,z,t) = \sum_n \sum_\nu \frac{2\pi p_n(\omega_{n,\nu})}{\{\partial_\omega \eta(\omega)\}|_{\omega=\omega_{n,\nu}}} \phi_{n,\nu}^{(j)}(x,z) e^{i(\omega_{n,\nu}t+\frac{\pi}{2})}. \qquad (49)$$

For $z < z_s$, we can obtain a result identical to (49).

Equation (49) is the excitation formula of Love wave in irregular multi-layered media. It can be seen that the Love wave excited by a seismic source can be represented as a superposition of *distorted modes*, $\phi_{n,\nu}^{(j)}(x,z)$. In contrast, the Love wave in laterally homogeneous layered media is represented by a superposition of *normal modes* according to the classic theory of Love wave (e.g., Aki and Richards, 1980; Keilis-Borok and Yanoskaya, 1962). In some cases, for example the relatively lower frequency case, the fundamental mode contains dominant energy of Love wave, thus in such case one can consider only the contribution from the fundamental mode by ignoring contributions from higher modes, that is,

$$w_L^{(j)}(x,z,t) \approx \sum_n \frac{2\pi p_n(\omega_{n,0})}{\{\partial_\omega \eta(\omega)\}|_{\omega=\omega_{n,0}}} \phi_{n,0}^{(j)}(x,z) e^{i(\omega_{n,0}t+\frac{\pi}{2})}. \qquad (50)$$

3.3. Comparisons with Classic Theory of Love Wave in Laterally Homogeneous Multi-Layered Media

The excitation formula of Love wave for the case of irregular multi-layered media, Eq. (49), looks quite similar to that for the case of laterally homogeneous multi-layered media (e.g., Aki and Richards, 1980). Moreover, the dispersion equation (34) also has a similar form to that for determining classic normal mode of Love wave in laterally homogeneous media (e.g., Chen, 1993). If the new formula is correct, it must be somehow a more generalized version of the classic excitation formula of Love wave for laterally homogeneous layered media. In what follows, we shall demonstrate that the theory of Love wave derived here is indeed an extension of the *alternative version of classic theory for Love wave*.

3.3.1. The Alternative Version of Classic Theory of Love Wave

To understand the relationship between our new Love wave's theory and the classic Love wave's theory, let us re-consider the classic theory of Love wave in laterally homogeneous multi-layered media. Using the method of generalized reflection/transmission coefficients, the SH wave field in a 2-D stratified layered media due to an arbitrary point source can be calculated by following formula (e.g., Chen, 1993, 1999b),

$$u^{(j)}(x, z, t)$$
$$= \int_{-\infty}^{+\infty} e^{i\omega t} \, d\omega \int_{-\infty}^{+\infty} \frac{e^{ikx}}{2\pi \Delta(\omega, k)}$$
$$\times \{Y^{(j)}(z, \omega, k) \hat{T}_d^{(j)} \hat{T}_d^{(j-1)} \cdots \hat{T}_d^{(1)}(\omega, k)\} g(\omega, k) \, dk, \quad (51)$$

where

$$Y^{(j)}(z, \omega, k) = e^{i\nu^{(j)}(k)(z - z^{(j-1)})} + \hat{R}_{du}^{(j)}(\omega, k) e^{i\nu^{(j)}(k)(z^{(j)} - z)}, \quad (51a)$$

$$\Delta(\omega, k) = 1 - \hat{R}_{ud}^{(0)}(\omega, k) \hat{R}_{du}^{(1)}(\omega, k), \quad (51b)$$

$$g(\omega, k) = \hat{R}_{ud}^{(0)}(\omega, k) \hat{T}_u^{(1)} \hat{T}_u^{(2)} \cdots \hat{T}_u^{(s-1)}(\omega, k)$$
$$\times \left[s_u(\omega, k) + \hat{R}_{du}^{(s)}(\omega, k) s_d(\omega, k) \right], \quad (51c)$$

with

$$s_u(\omega, k) = \frac{iF(\omega)}{2\nu^{(s)}(k)} \int_{z^{(s-1)}}^{z_s} f(z) e^{i\nu^{(s)}(k)(z - z^{(s)})} \, dz$$

and

$$s_d(\omega, k) = \frac{iF(\omega)}{2\nu^{(s)}(k)} \int_{z^{(s-1)}}^{z_s} f(z) e^{i\nu^{(s)}(k)(z^{(s-1)} - z)} \, dz.$$

Here, $\hat{R}(\omega, k)$ and $\hat{T}(\omega, k)$ are the generalized reflection/transmission coefficients for SH wave; $\Delta(\omega, k)$ is the dispersion function; $g(\omega, k)$ is a known function contained seismic source information. Two notes about Eq. (51) are stressed here: (1) accurately evaluating the double integration will give rise to complete time-domain solution (including body waves and surface waves) in given time window; (2) the integral order of the double integration is exchangeable, i.e., the integrations over k and ω are in equal position. Conventionally, one carries out the k-integration first and obtains the spectrum $U^{(j)}(x, z, \omega)$, and then carries out the ω-integral, and finally obtains the time-domain solution, i.e. the synthetic seismogram. In actual implementation, the outer integral, the ω-integral, usually is carried out by discrete Fourier transform (e.g., FFT),

$$u^{(j)}(x, z, t) = \sum_{n=-N}^{N} e^{i\omega_n t} U^{(j)}(x, z, \omega_n) \Delta\omega, \quad (52)$$

and

$$U^{(j)}(x, z, \omega_n)$$
$$= \int_{-\infty}^{+\infty} \frac{g(\omega_n, k)}{2\pi \Delta(\omega_n, k)} \{Y^{(j)}(z, \omega_n, k) \hat{T}_d^{(j)} \hat{T}_d^{(j-1)} \cdots \hat{T}_d^{(1)}(\omega_n, k)\} e^{ikx} \, dk.$$
$$(52a)$$

For the given ω_n, dispersion function $\Delta(\omega_n, k)$ vanishes when k takes certain discrete values: $k_{n,0}, k_{n,1}, k_{n,2}, \ldots, k_{n,M_n}$. In other words, $\{k_{n,\nu}; \nu = 0, 1, 2, \ldots, M_n\}$ are roots of $\Delta(\omega_n, k)$. Notice that $(\omega_n/k_{n,\nu})$ turns out to be the νth order phase velocity of Love wave for given ω_n (e.g., Aki and Richards, 1980). Obviously, the roots of dispersion function $\{k_{n,\nu}\}$ are poles of the integrand in Eq. (52a). The net contributions from the poles will yield the frequency-domain Love waves (e.g., Aki and Richards, 1980),

$$U_L^{(j)}(x, z, \omega_n) = \sum_{\nu=0}^{M_n} \frac{g(\omega_n, k_{n,\nu})}{\{\partial_k \Delta(\omega_n, k)\}|_{k=k_{n,\nu}}} l^{(j)}(z, \omega_n, k_{n,\nu}) e^{i(k_{n,\nu}x + \frac{\pi}{2})}, \quad (53)$$

where

$$l^{(j)}(z, \omega_n, k_{n,\nu}) = Y^{(j)}(z, \omega_n, k_{n,\nu}) \hat{T}_d^{(j)} \hat{T}_d^{(j-1)} \cdots \hat{T}_d^{(1)} (\omega_n, k_{n,\nu}), \quad (53a)$$

is the νth normal mode corresponding to $k_{n,\nu}$ for a given ω_n (e.g., Chen, 1993). It should be pointed out that Eq. (53) is derived from Eq. (52a) by applying residue theorem (e.g., Arfken and Weber, 2005). Finally, substituting Eq. (53) into (52) yields the time-domain Love waves,

$$u_L^{(j)}(x, z, t) = \sum_{n=-N}^{N} e^{i\omega_n t} \Delta\omega \sum_{\nu=0}^{M_n} \frac{g(\omega_n, k_{n,\nu})}{\{\partial_k \Delta(\omega_n, k)\}|_{k=k_{n,\nu}}}$$
$$\times l^{(j)}(z, \omega_n, k_{n,\nu}) e^{i(k_{n,\nu}x + \frac{\pi}{2})}. \quad (54)$$

This is the classic formula for synthesizing waveform of Love waves in laterally homogeneous medium (e.g., Aki and Richards, 1980), and it is derived in the conventional manner as shown above: carrying out k-integral first, then ω-integral.

Since the integral order, over k and ω, of the double integral equation (51) is exchangeable, we can equivalently evaluate the double integral equation (51) by an alternative order: carrying out ω-integral first, then k-integral. Similarly, the k-integral is also evaluated by the discrete spatial Fourier transform,

$$u^{(j)}(x, z, t) = \sum_{n=-N}^{N} e^{ik_n x} V^{(j)}(k_n, z, t) \Delta k, \quad (55)$$

where

$$V^{(j)}(k_n, z, t)$$
$$= \int_{-\infty}^{+\infty} \frac{g(\omega, k_n)}{2\pi \Delta(\omega, k_n)} \{Y^{(j)}(z, \omega, k_n) \hat{T}_d^{(j)} \hat{T}_d^{(j-1)} \cdots \hat{T}_d^{(1)}(\omega, k_n)\} e^{i\omega t} \, d\omega. \quad (55a)$$

For the given k_n, dispersion function $\Delta(\omega, k_n)$ vanishes when ω takes certain discrete values: $\omega_{n,0}, \omega_{n,1}, \omega_{n,2}, \ldots, \omega_{n,M_n}$. Similarly, $\{\omega_{n,\nu}\}$ are poles of the integrand in Eq. (55a), their sole contributions are also the Love waves but with an alternative form,

$$u_L^{(j)}(x,z,t) = \sum_{n=-N}^{N} \sum_{\nu=0}^{M_n} \frac{g(\omega_{n,\nu},k_n)\Delta k}{\{\partial_\omega \Delta(\omega,k_n)\}|_{\omega=\omega_{n,\nu}}}$$
$$\times \{l^{(j)}(z,\omega_{n,\nu},k_n)e^{ik_n x}\}e^{(i\omega_{n,\nu}t+\frac{\pi}{2})}. \tag{56}$$

This is the *alternative version* of classic excitation formula of Love wave in laterally homogeneous multi-layered structure, where $\omega_{n,\nu}$ are characteristic frequencies for the given k_n.

3.3.2. Comparisons with Classic Theory of Love Wave in Laterally Varying Multi-Layered Media

By comparing Eqs. (56) with (49), we find that the newly derived excitation formula (i.e., Eq. (49)) of Love wave in irregularly multi-layered structure is an *extension* of the *alternative version* of classic Love wave's excitation formula (Eq. (56)). In fact, when the irregular interfaces become flat, the new excitation formula (Eq. (49)) of Love wave for arbitrarily irregular multi-layered structure will naturally become identical to Eq. (56). We can demonstrate this as follows.

As given in Eq. (49), when the irregular interfaces become flat, all global generalized reflection/transmission matrices become diagonal ones whose diagonal elements are the classic generalized reflection/transmission coefficients. Accordingly, eigen-vectors $\mathbf{q}_n(\omega)$ and eigen-value $\eta_n(\omega)$ become

$$\mathbf{q}_n(\omega) = \{\delta_{n,m}\hat{\mathbf{e}}_m\} \quad \text{and} \quad \eta_n(\omega) = \Delta(\omega,k_n), \tag{57}$$

where

$$\hat{\mathbf{e}}_m = (\underbrace{\ldots,0,0,\ldots,0}_{\text{before } m\text{th element}},1,\underbrace{0,\ldots,0,0,\ldots}_{\text{after } m\text{th element}})^T. \tag{58}$$

Using these results, the distorted mode can be reduced to

$$\phi_{n,\nu}^{(j)}(x,z) = \mathbf{e}^T(x)\mathbf{Y}^{(j)}(z,\omega_{n,\nu})\hat{\mathbf{T}}_{\downarrow\downarrow}^{(j)}\hat{\mathbf{T}}_{\downarrow\downarrow}^{(j-1)}\cdots\hat{\mathbf{T}}_{\downarrow\downarrow}^{(1)}(\omega_{n,\nu})\mathbf{q}_n(\omega_{n,\nu})$$
$$= e^{ixk_n}Y^{(j)}(z,\omega_{n,\nu},k_n)\hat{T}_d^{(j)}\hat{T}_d^{(j-1)}\cdots\hat{T}_d^{(1)}(\omega_{n,\nu},k_n)$$
$$= e^{ixk_n}l^{(j)}(z,\omega_{n,\nu},k_n). \tag{59}$$

Similarly, we find

$$p_n(\omega) = g(\omega,k_n)\Delta k/2\pi. \tag{60}$$

Finally, by using formulas (56) to (60), Eq. (49) can be reduced as

$$w_L^{(j)}(x,z,t) = \sum_n \sum_\nu \frac{g(\omega_{n,\nu},k_n)\Delta k}{\{\partial_\omega \Delta(\omega,k_n)\}|_{\omega=\omega_{n,\nu}}}$$
$$\times \{l^{(j)}(z,\omega_{n,\nu},k_n)e^{ik_n x}\}e^{(i\omega_{n,\nu}t+\frac{\pi}{2})}. \tag{61}$$

TABLE 1. Comparisons between classic (alternative form) and new Love wave theories

Alternative form Classic Theory (laterally homogeneous structure)	New Theory (irregularly layered structure)
$l^{(j)}(z, \omega_{n,v}, k_n)e^{ixk_n}$	$w_{n,v}^{(j)}(x, z)$
$\Delta(\omega, k_n)$	$\eta_n(\omega)$
$g(\omega, k_n)\Delta k/2\pi$	$p_n(\omega)$
$\{\delta_{n,m}\hat{\mathbf{e}}_m\}$	$\mathbf{q}_n(\omega)$

This result is identical to Eq. (56). We have seen that when irregularly layered media become laterally homogeneous layered ones, the *distorted modes* accordingly become *normal modes* and the new excitation formula become the classic excitation formula of Love waves. Therefore, the excitation formula derived here is indeed an extension of the classic formula of Love waves.

Finally, according to the above analysis, we summarize the correspondence between the alternative form classic theory of Love waves (for laterally homogeneous layered structure) and our new theory of Love waves (for irregularly layered structure) in Table 1.

4. Concluding Remarks

In this section, the general theoretical formulation of the method of global generalized R/T matrices for SH problem has been reviewed and summarized. This method was developed by the author previously for providing an efficient algorithm to compute the seismograms in irregular multi-layered media case (e.g., Chen, 1990, 1995, 1996; Cao et al., 2004). It can be viewed as an extension of the generalized R/T coefficients method for the horizontally layered case (e.g., Kennett, 1983; Chen, 1999b) to the layered media with irregular interfaces. Because of the use of a recursive scheme in computing the generalized R/T matrices, this method is efficient, particularly for the case with a large number of irregular layers. As demonstrated in this section, our general formulation can be analytically reduced to the classic formulation of generalized R/T coefficient method when the interfaces become flat.

The validity of the general formulation presented here has been verified by comparing the numerical results with the analytical solutions for the scattering problem due to a semi-cylindrical canyon, and comparing with the results of other existing valid methods. The versatile applicability of this method has been demonstrated by calculating the synthetic seismograms for a variety of selected multi-layered media cases. All these reveal that the method of global generalized R/T matrices summarized in this chapter offers an effective means to compute the synthetic seismograms in the laterally varying media which can be modelled as irregular multi-layered media.

Based on the theoretical formulation of the global generalized R/T matrices method, we derived the fundamental formulation of Love waves in arbitrarily irregular multi-layered media. From the basic principle that the modal solutions are the non-trivial solutions of the free elastodynamic equation under appropriate boundary conditions (Chen, 1993), we derived the characteristic frequencies and the corresponding distorted modes of Love wave in irregular multi-layered media. The basic principle used here for defining modal solutions is a general one, and it is identical to that used for defining normal mode solutions in laterally homogeneous layered media (e.g., Aki and Richards, 1980; Keilis-Borok and Yanoskaya, 1962; Chen, 1993). Moreover, we derived the formulation for computing the synthetic seismograms of Love waves in irregular multi-layered media due to an arbitrary point source, that is, the excitation formulation of Love waves by using the formulation of global generalized R/T matrices method and the modal solutions. Similar to the result for laterally homogeneous layered structure, the Love waves radiated from a source in irregular multi-layered media can be expressed as a superposition of distorted modes, and the corresponding modal partition coefficients contain seismic source information. We also demonstrated that the new excitation formulation of Love waves is an extension (or generalization) of the alternative version of classic Love waves. Since the structure model used here is quite arbitrary, the new formulation of Love waves derived here can be applied to study a variety of seismological problems ranging from resonated motion in a sedimentary basin structure to Love wave's excitation in irregular multi-layered media. It offers an alternative mean to understand the nature of Love waves in laterally heterogeneous media rather than mode-coupling and regionalization approaches.

Although the presentations in this chapter is limited in two-dimensional SH case, this method can be directly extended to the two-dimensional P-SV case and three-dimensional case. Actually, the theoretical formulation for two-dimensional P-SV case had been done (Chen, 1991, 1996). Further efforts along this line would be valuable contribution to the subject of seismic wave propagation in heterogeneous earth.

Acknowledgements

I dedicate this work in memory of Professor Keiiti Aki. This work initiated during my doctoral study under Aki's supervision in University of Southern California. I also acknowledge Dr. Zengxi Ge and Mr. Jun Cao in Peking University for their assistance in preparing the manuscript, and the stimulating discussions with them during past few years. This research is supported by the National Natural Science Foundation of China (grant Nos. 40134010, 40474011, 40521002) and also by the National Basic Research Program of China (2004CB418404).

APPENDIX A

In this appendix, we shall prove that Eq. (3) is the discrete-wave number representation of Eq. (2). In order to derive Eq. (3) from Eq. (2), we assume that both the geometry of medium and the distribution of sources are periodic (see Fig. A.1), i.e.

$$\xi^{(j-1)}(x) = \xi^{(j-1)}(x + nL), \quad j = 1, 2, \ldots, N, N + 1, \quad \text{(A.1)}$$

$$f(x, z) = f(x + nL, z), \quad \text{(A.2)}$$

where $n = 0, \pm 1, \pm 2, \ldots$, L is the length of periodicity of this array. The periodicity of both the medium and source will lead to a periodic wave field,

$$\begin{cases} W^{(j-1)}(x, z) = W^{(j-1)}(x + nL, z), \\ \tau^{(j-1)}(x, z) = \tau^{(j-1)}(x + nL, z), \end{cases} \quad j = 1, 2, \ldots, N, N + 1, \quad \text{(A.3)}$$

where, $W^{(j-1)}(x, z)$ and $\tau^{(j-1)}(x, z)$ are the spectra of displacement and traction, respectively. To demonstrate the problem posed above, we will use the following integral formula of Green's function $G^{(j)}(\mathbf{x}, \mathbf{x}')$ (Morse and Feshbach, 1953),

$$G^{(j)}(\mathbf{x}, \mathbf{x}') = i \int_{-\infty}^{+\infty} \frac{\exp\{ik(x' - x) + i\nu^{(j)}|z - z'|\}}{\nu^{(j)}} \, dk, \quad \text{(A.4)}$$

where,

$$\nu^{(j)} = \sqrt{\left(k_0^{(j)}\right)^2 - k^2}, \quad \text{Im}\{\nu^{(j)}\} > 0.$$

Using this formula, we can evaluate the first integral of Eq. (2) as,

$$I_1 = \int_{V^{(j)}} \delta_{sj} f(\mathbf{x}) G^{(j)}(\mathbf{x}, \mathbf{x}') \, dV(\mathbf{x})$$

$$= i\delta_{sj} \int_{-\infty}^{+\infty} \frac{dk}{\nu^{(j)}} \sum_{n=-\infty}^{+\infty} \int_{-L/2}^{L/2} dx \int_{\xi^{(j-1)}(x+nL)}^{\xi^{(j)}(x+nL)} dz \, f(x + nL, z)$$

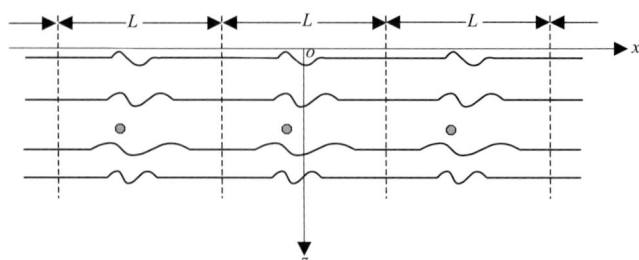

FIG. A.1. A periodic array corresponded to the discrete wave number representation.

$$\times \exp\{ik(x' - nL - x) + iv^{(j)}|z - z'|\}.$$

Here, we break up the integral over the entire periodic surface into a sum of integrals over individual periods. Considering the periodic conditions, Eqs. (A.1) and (A.2), the above expression reduces to

$$I_1 = i\delta_{sj} \int_{-\infty}^{+\infty} \frac{dk}{v^{(j)}} \left\{ \sum_{n=-\infty}^{+\infty} e^{-inkL} \right\} \int_{-L/2}^{L/2} dx \int_{\xi^{(j-1)}(x)}^{\xi^{(j)}(x)} dz\, f(x,z)$$
$$\times \exp[ik(x' - x) + iv^{(j)}|z - z'|]. \tag{A.5}$$

Substituting the delta function identity (Waterman, 1975)

$$\sum_{n=-\infty}^{+\infty} \exp(-i2\pi\alpha n) = \sum_{n=-\infty}^{+\infty} \delta(\alpha - n) \tag{A.6}$$

into Eq. (A.5), then carrying out the integral with respect to k, we finally obtain,

$$I_1 = \sum_{n=-\infty}^{+\infty} \frac{2\pi i \delta_{sj}}{L v^{(j)}} \int_{-L/2}^{L/2} dx \int_{\xi^{(j-1)}(x)}^{\xi^{(j)}(x)} f(x,z)$$
$$\times \exp[ik_n(x' - x) + iv_n^{(j)}|z - z'|]\, dz, \tag{A.7}$$

where

$$k_n = \frac{2\pi n}{L}, \quad v_n^{(j)} = \sqrt{(k_0^{(j)})^2 - (k_n)^2}, \quad \text{Im}\{v_n^{(j)}\} > 0.$$

The second integral of Eq. (2) can be treated in a similar way. Substituting the integral formula of Green's function, Eq. (A.4), into the second integral, we have

$$I_2 = \int_{\partial V^{(j)}} \mathbf{n} \cdot \left\{ G^{(j)}(\mathbf{x}, \mathbf{x}') \nabla W^{(j)}(\mathbf{x}) - W^{(j)}(\mathbf{x}) \nabla G^{(j)}(\mathbf{x}, \mathbf{x}') \right\} dl(\mathbf{x})$$

$$= i \int_{-\infty}^{+\infty} \frac{dk}{v^{(j)}} \Bigg\{ \int_{-\infty}^{+\infty} dx \Bigg\{ \frac{\tau^{(j)}[x, \xi^{(j)}(x)]}{\mu^{(j)}} \sqrt{1 + [\dot{\xi}^{(j)}(x)]^2}$$
$$- i[\dot{\xi}^{(j)}(x)k + v^{(j)} \operatorname{sgn}(\xi^{(j)}(x) - z')] W^{(j)}[x, \xi^{(j)}(x)] \Bigg\}$$
$$\times e^{ik(x'-x)+iv^{(j)}|\xi^{(j)}(x)-z'|}$$
$$- \int_{-\infty}^{+\infty} dx \Bigg\{ \frac{\tau^{(j)}[x, \xi^{(j-1)}(x)]}{\mu^{(j)}} \sqrt{1 + [\dot{\xi}^{(j-1)}(x)]^2} - i[\dot{\xi}^{(j-1)}(x)k$$
$$+ v^{(j)} \operatorname{sgn}(\xi^{(j-1)}(x) - z')] W^{(j)}[x, \xi^{(j-1)}(x)] \Bigg\}$$
$$\times e^{ik(x'-x)+iv^{(j)}|\xi^{(j-1)}(x)-z'|} \Bigg\}.$$

Here, for jth interface we used that $\tau^{(j)}(x) = \mu^{(j)}\mathbf{n}\cdot\nabla W^{(j)}(\mathbf{x})$, and

$$n_x\left(x, \xi^{(j)}(x)\right) = \frac{-\dot{\xi}^{(j)}(x)}{\sqrt{1+[\dot{\xi}^{(j)}(x)]^2}} \quad \text{and}$$

$$n_z\left(x, \xi^{(j)}(x)\right) = \frac{1}{\sqrt{1+[\dot{\xi}^{(j)}(x)]^2}};$$

for $(j-1)$th interface, we used

$$n_x\left(x, \xi^{(j-1)}(x)\right) = \frac{\dot{\xi}^{(j-1)}(x)}{\sqrt{1+[\dot{\xi}^{(j-1)}(x)]^2}} \quad \text{and}$$

$$n_z\left(x, \xi^{(j-1)}(x)\right) = \frac{-1}{\sqrt{1+[\dot{\xi}^{(j-1)}(x)]^2}}.$$

Upon using the periodic conditions, Eqs. (A.1) to (A.3), the integral I_2 can be further reduced to

$$I_2 = i\int_{-\infty}^{+\infty} \frac{dk}{v^{(j)}} \left(\sum_{n=-\infty}^{+\infty} e^{-inLk}\right)$$

$$\times \left\{\int_{-L/2}^{L/2} dx \left\{\frac{\tau^{(j)}[x, \xi^{(j)}(x)]}{\mu^{(j)}}\sqrt{1+[\dot{\xi}^{(j)}(x)]^2}\right.\right.$$

$$\left. - i[\dot{\xi}^{(j)}(x)k + v^{(j)}\,\mathrm{sgn}(\xi^{(j)}(x) - z')]W^{(j)}[x, \xi^{(j)}(x)]\right\}$$

$$\times e^{ik(x'-x)+iv^{(j)}|\xi^{(j)}(x)-z'|}$$

$$- \int_{-L/2}^{L/2} dx \left\{\frac{\tau^{(j)}[x, \xi^{(j-1)}(x)]}{\mu^{(j)}}\sqrt{1+[\dot{\xi}^{(j-1)}(x)]^2}\right.$$

$$\left. - i[\dot{\xi}^{(j-1)}(x)k + v^{(j)}\,\mathrm{sgn}(\xi^{(j-1)}(x) - z')]W^{(j)}[x, \xi^{(j-1)}(x)]\right\}$$

$$\left.\times e^{ik(x'-x)+iv^{(j)}|\xi^{(j-1)}(x)-z'|}\right\}.$$

Using the delta function identity Eq. (A.6), then carrying out the integral with respect to k, we finally find,

$$I_2 = \sum_{n=-\infty}^{+\infty} \frac{2\pi}{Lv_n^{(j)}} \left\{\int_{-L/2}^{L/2} dx \left\{\frac{i\tau^{(j)}[x, \xi^{(j)}(x)]}{\mu^{(j)}}\sqrt{1+[\dot{\xi}^{(j)}(x)]^2}\right.\right.$$

$$\left.\left. + [\dot{\xi}^{(j)}(x)k_n + v_n^{(j)}\,\mathrm{sgn}(\xi^{(j)}(x) - z')]W^{(j)}[x, \xi^{(j)}(x)]\right\}\right.$$

$$\times e^{ik_n(x'-x)+iv_n^{(j)}|\xi^{(j)}(x)-z'|}$$

$$-\int_{-L/2}^{L/2} dx \left\{ \frac{i\tau^{(j)}[x,\xi^{(j-1)}(x)]}{\mu^{(j)}} \sqrt{1+\left[\dot{\xi}^{(j-1)}(x)\right]^2} \right.$$

$$+ \left[\dot{\xi}^{(j-1)}(x)k_n + v_n^{(j)} \operatorname{sgn}(\xi^{(j-1)}(x)-z')\right] W^{(j)}\left[x,\xi^{(j-1)}(x)\right]\right\}$$

$$\left. \times e^{ik_n(x'-x)+iv_n^{(j)}|\xi^{(j-1)}(x)-z'|} \right\}.$$

Substituting this expression and Eq. (A.7) into Eq. (2), we obtain Eq. (3).

APPENDIX B

Unlike the solution in time domain, in frequency-domain one cannot obtain an accurate solution based on discrete wave number approach, but one can obtain an optimal solution for SH topography problem by properly choosing an imaginary frequency. Such a proper imaginary frequency is determined as follows.

The frequency-domain solution for the original physical problem has an integral form of

$$U(\omega) = \int_{-\infty}^{+\infty} \frac{f(k,\omega)}{v(k,\omega)} dk, \tag{B.1}$$

where $f(k,\omega)$ is a smooth and non-singular function, $v(k,\omega)$ is the vertical wave number, i.e., $v(k,\omega) = \sqrt{(k_0)^2 - k^2}$, $k_0 = \omega/V$ and $\operatorname{Im}\{v(k,\omega)\} \geq 0$. In our discrete wave number approach, on the other hand, the integral (B.1) is evaluated by a summation,

$$U(\omega) = \sum_{n=-\infty}^{+\infty} \frac{f(k_n,\omega)}{v(k_n,\omega)} \Delta k, \tag{B.2}$$

where $k_n = 2\pi n/L$ and $\Delta k = 2\pi/L$. Since $f(k,\omega)$ is a non-singular function and $1/v(k_n,\omega)$ is singular only in a small neighboring area around singular point k_0. Thus, all the terms in summation (B.1) are appropriate except the term which across over the singular point. The remedy for this singular term is introducing a small imaginary frequency ω_I^f so that for non-singular terms its effects are negligible, while for singular term it can avoid the singularity and provide an optimal value for matching the original integral over the same patch, that is

$$\frac{f(k_0,\omega)\Delta k}{\sqrt{(k_0+ik_I^f)^2 - (k_0)^2}} = \int_{k_0-\Delta k/2}^{k_0+\Delta k/2} \frac{f(k,\omega)}{\sqrt{(k_0)^2 - k^2}} dk,$$

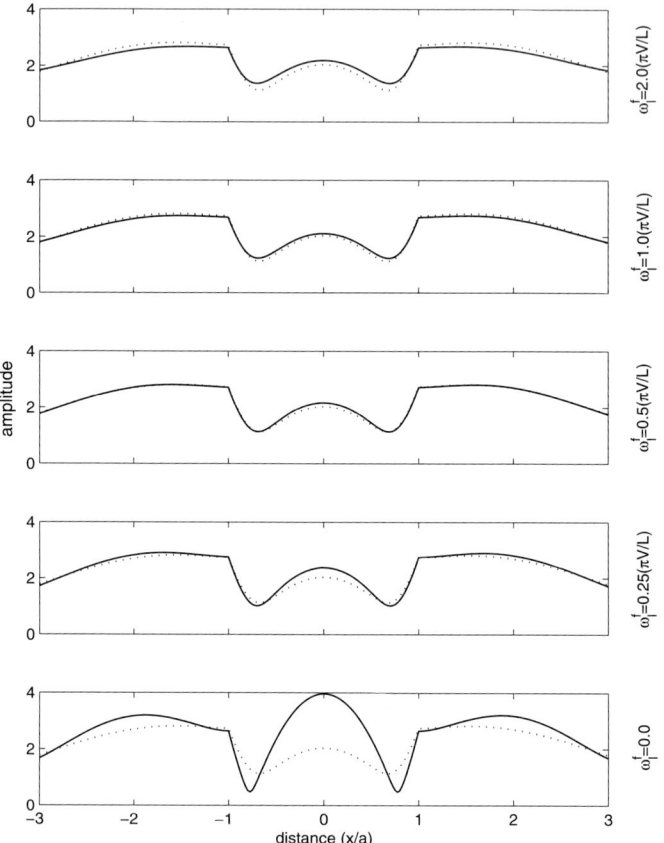

FIG. B.1. The ground responses of a semi-circular canyon to vertical incident SH waves for various values of ω_I^f for the case of $\eta = 0.5$, $a = 1$ km, $V = 2$ km/s and $L = 40$ km. The solid lines denote our results and the dotted lines denote the exact ones. We can see a best fitting when $\omega_I^f = 0.5\pi V/L$.

where $k_I^f = \omega_I^f / V$. Assuming $k_I^f \ll k_0$, the above equation can be reduced to

$$\frac{f(k_0, \omega)\Delta k}{\sqrt{2ik_0 k_I^f}} \approx \frac{2f(k_0, \omega)}{\sqrt{2k_0}} \sqrt{\frac{\Delta k}{2}} (1 - i).$$

Finally we find

$$\omega_I^f = k_I^f V \approx \frac{\pi V}{2L}. \tag{B.3}$$

This is the formula for determining the imaginary frequency by using which our discrete wave number approach can directly provide an optimal solution in fre-

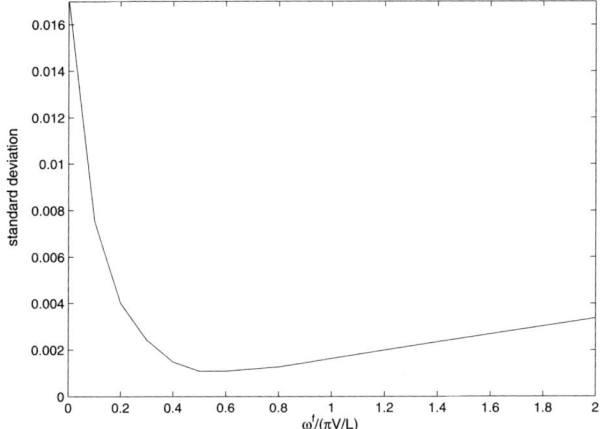

FIG. B.2. The standard deviation of our solutions from exact ones for 20 different ω_I^f. The same calculation parameters as Fig. B.1 are used. We see a minimum deviation at $\omega_I^f = 0.5\pi V/L$.

quency domain for an SH topography problem. Figures B.1 and B.2 show the effects of the choice of ω_I^f. In Fig. B.1 we directly compute our solutions (solid lines) with exact ones (dotted lines) for various values of ω_I^f for the case of $\eta = 0.5$ or $\lambda = a/4$, while in Fig. B.2 we show the standard deviation of our results from exact ones. These results confirm the validity of the formula (B.3).

APPENDIX C

In this appendix we shall proof formula (41) in the text. To do so, let us proof three lemmas first:

LEMMA 1.
$$[\mathbf{I} - \hat{\mathbf{R}}_{\downarrow\uparrow}^{(j)} \hat{\mathbf{R}}_{\uparrow\downarrow}^{(j-1)}]^{-1}$$
$$= (\hat{\mathbf{T}}_{\uparrow\uparrow}^{(1)} \hat{\mathbf{T}}_{\uparrow\uparrow}^{(2)} \cdots \hat{\mathbf{T}}_{\uparrow\uparrow}^{(j-1)})^{-1} [\mathbf{I} - \hat{\mathbf{R}}_{\downarrow\uparrow}^{(1)} \hat{\mathbf{R}}_{\uparrow\downarrow}^{(0)}]^{-1} (\hat{\mathbf{T}}_{\uparrow\uparrow}^{(1)} \hat{\mathbf{T}}_{\uparrow\uparrow}^{(2)} \cdots \hat{\mathbf{T}}_{\uparrow\uparrow}^{(j-1)});$$

LEMMA 2.
$$\hat{\mathbf{R}}_{\downarrow\uparrow}^{(j-1)} (\hat{\mathbf{T}}_{\uparrow\uparrow}^{(1)} \hat{\mathbf{T}}_{\uparrow\uparrow}^{(2)} \cdots \hat{\mathbf{T}}_{\uparrow\uparrow}^{(j-1)})^{-1} = (\hat{\mathbf{T}}_{\downarrow\downarrow}^{(j-1)} \hat{\mathbf{T}}_{\downarrow\downarrow}^{(j-2)} \cdots \hat{\mathbf{T}}_{\downarrow\downarrow}^{(1)}) \hat{\mathbf{R}}_{\uparrow\downarrow}^{(0)};$$

LEMMA 3.
$$\hat{\mathbf{R}}_{\uparrow\downarrow}^{(0)} [\mathbf{I} - \hat{\mathbf{R}}_{\downarrow\uparrow}^{(1)} \hat{\mathbf{R}}_{\uparrow\downarrow}^{(0)}]^{-1} = [\mathbf{I} - \hat{\mathbf{R}}_{\uparrow\downarrow}^{(0)} \hat{\mathbf{R}}_{\downarrow\uparrow}^{(1)}]^{-1} \hat{\mathbf{R}}_{\uparrow\downarrow}^{(0)}.$$

PROOF OF LEMMA 1. According to the definition of global R/T matrices, we can easily verify the following formula,

$$(\hat{\mathbf{T}}_{\uparrow\uparrow}^{(1)}\hat{\mathbf{T}}_{\uparrow\uparrow}^{(2)}\cdots\hat{\mathbf{T}}_{\uparrow\uparrow}^{(j-1)})[\mathbf{I}-\hat{\mathbf{R}}_{\downarrow\uparrow}^{(j)}\hat{\mathbf{R}}_{\uparrow\downarrow}^{(j-1)}]$$
$$=[\mathbf{I}-\hat{\mathbf{R}}_{\downarrow\uparrow}^{(1)}\hat{\mathbf{R}}_{\uparrow\downarrow}^{(0)}](\hat{\mathbf{T}}_{\uparrow\uparrow}^{(1)}\hat{\mathbf{T}}_{\uparrow\uparrow}^{(2)}\cdots\hat{\mathbf{T}}_{\uparrow\uparrow}^{(j-1)}).$$

Thus,

$$[\mathbf{I}-\hat{\mathbf{R}}_{\downarrow\uparrow}^{(j)}\hat{\mathbf{R}}_{\uparrow\downarrow}^{(j-1)}]$$
$$=(\hat{\mathbf{T}}_{\uparrow\uparrow}^{(1)}\hat{\mathbf{T}}_{\uparrow\uparrow}^{(2)}\cdots\hat{\mathbf{T}}_{\uparrow\uparrow}^{(j-1)})^{-1}[\mathbf{I}-\hat{\mathbf{R}}_{\downarrow\uparrow}^{(1)}\hat{\mathbf{R}}_{\uparrow\downarrow}^{(0)}](\hat{\mathbf{T}}_{\uparrow\uparrow}^{(1)}\hat{\mathbf{T}}_{\uparrow\uparrow}^{(2)}\cdots\hat{\mathbf{T}}_{\uparrow\uparrow}^{(j-1)}),$$

and its corresponding inverse is,

$$[\mathbf{I}-\hat{\mathbf{R}}_{\downarrow\uparrow}^{(j)}\hat{\mathbf{R}}_{\uparrow\downarrow}^{(j-1)}]^{-1}$$
$$=(\hat{\mathbf{T}}_{\uparrow\uparrow}^{(1)}\hat{\mathbf{T}}_{\uparrow\uparrow}^{(2)}\cdots\hat{\mathbf{T}}_{\uparrow\uparrow}^{(j-1)})^{-1}[\mathbf{I}-\hat{\mathbf{R}}_{\downarrow\uparrow}^{(1)}\hat{\mathbf{R}}_{\uparrow\downarrow}^{(0)}]^{-1}(\hat{\mathbf{T}}_{\uparrow\uparrow}^{(1)}\hat{\mathbf{T}}_{\uparrow\uparrow}^{(2)}\cdots\hat{\mathbf{T}}_{\uparrow\uparrow}^{(j-1)}). \quad \text{(C.1)}$$

□

PROOF OF LEMMA 2. According to the definition of the global generalized R/T matrices, we have

$$(\hat{\mathbf{T}}_{\uparrow\uparrow}^{(1)}\hat{\mathbf{T}}_{\uparrow\uparrow}^{(2)}\cdots\hat{\mathbf{T}}_{\uparrow\uparrow}^{(j-1)})\mathbf{a}_{\uparrow}^{(j)}=\mathbf{a}_{\uparrow}^{(1)},$$

that is,

$$\mathbf{a}_{\uparrow}^{(j)}=(\hat{\mathbf{T}}_{\uparrow\uparrow}^{(1)}\hat{\mathbf{T}}_{\uparrow\uparrow}^{(2)}\cdots\hat{\mathbf{T}}_{\uparrow\uparrow}^{(j-1)})^{-1}\mathbf{a}_{\uparrow}^{(1)}. \quad \text{(C.2)}$$

On the other hand, we have

$$\hat{\mathbf{R}}_{\uparrow\downarrow}^{(j-1)}\mathbf{a}_{\uparrow}^{(j)}=\mathbf{a}_{\downarrow}^{(j)}$$
$$=(\hat{\mathbf{T}}_{\downarrow\downarrow}^{(j-1)}\hat{\mathbf{T}}_{\downarrow\downarrow}^{(j-2)}\cdots\hat{\mathbf{T}}_{\downarrow\downarrow}^{(1)})\mathbf{a}_{\downarrow}^{(1)}$$
$$=(\hat{\mathbf{T}}_{\downarrow\downarrow}^{(j-1)}\hat{\mathbf{T}}_{\downarrow\downarrow}^{(j-2)}\cdots\hat{\mathbf{T}}_{\downarrow\downarrow}^{(1)})\hat{\mathbf{R}}_{\uparrow\downarrow}^{(0)}\mathbf{a}_{\uparrow}^{(1)}. \quad \text{(C.3)}$$

Substituting Eq. (C.2) into (A1)–(A3) yields,

$$\hat{\mathbf{R}}_{\uparrow\downarrow}^{(j-1)}(\hat{\mathbf{T}}_{\uparrow\uparrow}^{(1)}\hat{\mathbf{T}}_{\uparrow\uparrow}^{(2)}\cdots\hat{\mathbf{T}}_{\uparrow\uparrow}^{(j-1)})^{-1}=(\hat{\mathbf{T}}_{\downarrow\downarrow}^{(j-1)}\hat{\mathbf{T}}_{\downarrow\downarrow}^{(j-2)}\cdots\hat{\mathbf{T}}_{\downarrow\downarrow}^{(1)})\hat{\mathbf{R}}_{\uparrow\downarrow}^{(0)}. \quad \text{(C.4)}$$

□

PROOF OF LEMMA 3. According to the definition of function of matrix, we have

$$[\mathbf{I}-\hat{\mathbf{R}}_{\downarrow\uparrow}^{(1)}\hat{\mathbf{R}}_{\uparrow\downarrow}^{(0)}]^{-1}=\mathbf{I}+(\hat{\mathbf{R}}_{\downarrow\uparrow}^{(1)}\hat{\mathbf{R}}_{\uparrow\downarrow}^{(0)})+(\hat{\mathbf{R}}_{\downarrow\uparrow}^{(1)}\hat{\mathbf{R}}_{\uparrow\downarrow}^{(0)})^{2}+(\hat{\mathbf{R}}_{\downarrow\uparrow}^{(1)}\hat{\mathbf{R}}_{\uparrow\downarrow}^{(0)})^{3}+\cdots.$$

By multiplying $\hat{\mathbf{R}}_{\uparrow\downarrow}^{(0)}$ in the above equation, we obtain

$$\hat{\mathbf{R}}_{\uparrow\downarrow}^{(0)}[\mathbf{I} - \hat{\mathbf{R}}_{\downarrow\uparrow}^{(1)}\hat{\mathbf{R}}_{\uparrow\downarrow}^{(0)}]^{-1}$$
$$= \hat{\mathbf{R}}_{\uparrow\downarrow}^{(0)} + \hat{\mathbf{R}}_{\uparrow\downarrow}^{(0)}(\hat{\mathbf{R}}_{\downarrow\uparrow}^{(1)}\hat{\mathbf{R}}_{\uparrow\downarrow}^{(0)}) + \hat{\mathbf{R}}_{\uparrow\downarrow}^{(0)}(\hat{\mathbf{R}}_{\downarrow\uparrow}^{(1)}\hat{\mathbf{R}}_{\uparrow\downarrow}^{(0)})^{2} + \hat{\mathbf{R}}_{\uparrow\downarrow}^{(0)}(\hat{\mathbf{R}}_{\downarrow\uparrow}^{(1)}\hat{\mathbf{R}}_{\uparrow\downarrow}^{(0)})^{3} + \cdots$$
$$= \{\mathbf{I} + (\hat{\mathbf{R}}_{\uparrow\downarrow}^{(0)}\hat{\mathbf{R}}_{\downarrow\uparrow}^{(1)}) + (\hat{\mathbf{R}}_{\uparrow\downarrow}^{(0)}\hat{\mathbf{R}}_{\downarrow\uparrow}^{(1)})^{2} + (\hat{\mathbf{R}}_{\uparrow\downarrow}^{(0)}\hat{\mathbf{R}}_{\downarrow\uparrow}^{(1)})^{3} + \cdots\}\hat{\mathbf{R}}_{\uparrow\downarrow}^{(0)}$$
$$= [\mathbf{I} - \hat{\mathbf{R}}_{\uparrow\downarrow}^{(0)}\hat{\mathbf{R}}_{\downarrow\uparrow}^{(1)}]^{-1}\hat{\mathbf{R}}_{\uparrow\downarrow}^{(0)}, \qquad (C.5)$$

therefore, Lemma 3 is proved. □

PROOF OF EQ. (41) IN THE TEXT. Having the five lemmas proven above, we are now ready to derive Eq. (41) in the text. Using Lemma 1, we have

$$\hat{\mathbf{R}}_{\uparrow\downarrow}^{(j-1)}\hat{\mathbf{T}}_{\uparrow\uparrow}^{(j)}\hat{\mathbf{T}}_{\uparrow\uparrow}^{(j-1)} \cdots \hat{\mathbf{T}}_{\uparrow\uparrow}^{(s-1)}[\mathbf{I} - \hat{\mathbf{R}}_{\downarrow\uparrow}^{(s)}\hat{\mathbf{R}}_{\uparrow\uparrow}^{(s-1)}]^{-1}$$
$$= \hat{\mathbf{R}}_{\uparrow\downarrow}^{(j-1)}\hat{\mathbf{T}}_{\uparrow\uparrow}^{(j)}\hat{\mathbf{T}}_{\uparrow\uparrow}^{(j-1)} \cdots \hat{\mathbf{T}}_{\uparrow\uparrow}^{(s-1)}(\hat{\mathbf{T}}_{\uparrow\uparrow}^{(1)}\hat{\mathbf{T}}_{\uparrow\uparrow}^{(2)} \cdots \hat{\mathbf{T}}_{\uparrow\uparrow}^{(s-1)})^{-1}$$
$$\times [\mathbf{I} - \hat{\mathbf{R}}_{\downarrow\uparrow}^{(1)}\hat{\mathbf{R}}_{\uparrow\downarrow}^{(0)}]^{-1}(\hat{\mathbf{T}}_{\uparrow\uparrow}^{(1)}\hat{\mathbf{T}}_{\uparrow\uparrow}^{(2)} \cdots \hat{\mathbf{T}}_{\uparrow\uparrow}^{(s-1)})$$
$$= \hat{\mathbf{R}}_{\uparrow\downarrow}^{(j-1)}(\hat{\mathbf{T}}_{\uparrow\uparrow}^{(1)}\hat{\mathbf{T}}_{\uparrow\uparrow}^{(2)} \cdots \hat{\mathbf{T}}_{\uparrow\uparrow}^{(j-1)})^{-1}[\mathbf{I} - \hat{\mathbf{R}}_{\downarrow\uparrow}^{(1)}\hat{\mathbf{R}}_{\uparrow\downarrow}^{(0)}]^{-1}(\hat{\mathbf{T}}_{\uparrow\uparrow}^{(1)}\hat{\mathbf{T}}_{\uparrow\uparrow}^{(2)} \cdots \hat{\mathbf{T}}_{\uparrow\uparrow}^{(s-1)}).$$
$$\qquad (C.6)$$

Then, applying Lemmas 2 and 3 to the above equation, we obtain

$$\hat{\mathbf{R}}_{\uparrow\downarrow}^{(j-1)}(\hat{\mathbf{T}}_{\uparrow\uparrow}^{(1)}\hat{\mathbf{T}}_{\uparrow\uparrow}^{(2)} \cdots \hat{\mathbf{T}}_{\uparrow\uparrow}^{(j-1)})^{-1}[\mathbf{I} - \hat{\mathbf{R}}_{\downarrow\uparrow}^{(1)}\hat{\mathbf{R}}_{\uparrow\downarrow}^{(0)}]^{-1}(\hat{\mathbf{T}}_{\uparrow\uparrow}^{(1)}\hat{\mathbf{T}}_{\uparrow\uparrow}^{(2)} \cdots \hat{\mathbf{T}}_{\uparrow\uparrow}^{(s-1)})$$
$$= (\hat{\mathbf{T}}_{\downarrow\downarrow}^{(j-1)}\hat{\mathbf{T}}_{\downarrow\downarrow}^{(j-2)} \cdots \hat{\mathbf{T}}_{\downarrow\downarrow}^{(1)})\hat{\mathbf{R}}_{\uparrow\downarrow}^{(0)}[\mathbf{I} - \hat{\mathbf{R}}_{\downarrow\uparrow}^{(1)}\hat{\mathbf{R}}_{\uparrow\downarrow}^{(0)}]^{-1}(\hat{\mathbf{T}}_{\uparrow\uparrow}^{(1)}\hat{\mathbf{T}}_{\uparrow\uparrow}^{(2)} \cdots \hat{\mathbf{T}}_{\uparrow\uparrow}^{(s-1)})$$
$$= (\hat{\mathbf{T}}_{\downarrow\downarrow}^{(j-1)}\hat{\mathbf{T}}_{\downarrow\downarrow}^{(j-2)} \cdots \hat{\mathbf{T}}_{\downarrow\downarrow}^{(1)})[\mathbf{I} - \hat{\mathbf{R}}_{\uparrow\downarrow}^{(0)}\hat{\mathbf{R}}_{\downarrow\uparrow}^{(1)}]^{-1}\hat{\mathbf{R}}_{\uparrow\downarrow}^{(0)}(\hat{\mathbf{T}}_{\uparrow\uparrow}^{(1)}\hat{\mathbf{T}}_{\uparrow\uparrow}^{(2)} \cdots \hat{\mathbf{T}}_{\uparrow\uparrow}^{(s-1)}).$$
$$\qquad (C.7)$$

Finally, combining formulas (C.6) and (C.7) yields

$$\hat{\mathbf{R}}_{\uparrow\downarrow}^{(j-1)}\hat{\mathbf{T}}_{\uparrow\uparrow}^{(j)}\hat{\mathbf{T}}_{\uparrow\uparrow}^{(j-1)} \cdots \hat{\mathbf{T}}_{\uparrow\uparrow}^{(s-1)}[\mathbf{I} - \hat{\mathbf{R}}_{\downarrow\uparrow}^{(s)}\hat{\mathbf{R}}_{\uparrow\uparrow}^{(s-1)}]^{-1}$$
$$= (\hat{\mathbf{T}}_{\downarrow\downarrow}^{(j-1)}\hat{\mathbf{T}}_{\downarrow\downarrow}^{(j-2)} \cdots \hat{\mathbf{T}}_{\downarrow\downarrow}^{(1)})[\mathbf{I} - \hat{\mathbf{R}}_{\uparrow\downarrow}^{(0)}\hat{\mathbf{R}}_{\downarrow\uparrow}^{(1)}]^{-1}\hat{\mathbf{R}}_{\uparrow\downarrow}^{(0)}(\hat{\mathbf{T}}_{\uparrow\uparrow}^{(1)}\hat{\mathbf{T}}_{\uparrow\uparrow}^{(2)} \cdots \hat{\mathbf{T}}_{\uparrow\uparrow}^{(s-1)}).$$

This is Eq. (41) used in the text. □

APPENDIX D

To proof Eq. (42), we shall demonstrate following lemma first.

LEMMA. *If matrix* $[\mathbf{I} - \hat{\mathbf{R}}_{\uparrow\downarrow}^{(0)}(\omega)\hat{\mathbf{R}}_{\downarrow\uparrow}^{(1)}(\omega)]$ *is invertible, then it can be spanned by its eigen-vectors as*

$$[\mathbf{I} - \hat{\mathbf{R}}_{\uparrow\downarrow}^{(0)}(\omega)\hat{\mathbf{R}}_{\downarrow\uparrow}^{(1)}(\omega)] = \sum_{m=-M}^{M}\sum_{n=-M}^{M}\eta_m(\omega)Z_{mn}(\omega)\mathbf{q}_m(\omega)[\mathbf{q}_n(\omega)^*]^T,$$
$$\qquad (D.1)$$

where, $\mathbf{q}_n(\omega)$ and $\eta_n(\omega)$ is the nth eigen-vector and the corresponding eigen-values of this matrix, that is

$$[\mathbf{I} - \hat{\mathbf{R}}_{\uparrow\downarrow}^{(0)}(\omega)\hat{\mathbf{R}}_{\downarrow\uparrow}^{(1)}(\omega)]\mathbf{q}_n(\omega) = \eta_n(\omega)\mathbf{q}_n(\omega); \tag{D.2}$$

and its expansion coefficient matrix $\mathbf{Z}(\omega)$ is given by:

$$\{\mathbf{Z}(\omega)^{-1}\}_{mn} = [\mathbf{q}_m(\omega)^*]^T \mathbf{q}_n(\omega). \tag{D.3}$$

PROOF OF LEMMA. Since matrix $[\mathbf{I} - \hat{\mathbf{R}}_{\uparrow\downarrow}^{(0)}(\omega)\hat{\mathbf{R}}_{\downarrow\uparrow}^{(1)}(\omega)]$ is non-singular, it's eigen-vectors $\{\mathbf{q}_n(\omega)\}$ are linearly independent. Thus, they can be taken as a set of basis in the corresponding linear space $\mathfrak{R}$. Spanning by this set of basis, $[\mathbf{I} - \hat{\mathbf{R}}_{\uparrow\downarrow}^{(0)}(\omega)\hat{\mathbf{R}}_{\downarrow\uparrow}^{(1)}(\omega)]$ can be expressed as:

$$[\mathbf{I} - \hat{\mathbf{R}}_{\uparrow\downarrow}^{(0)}(\omega)\hat{\mathbf{R}}_{\downarrow\uparrow}^{(1)}(\omega)] = \sum_{m=-M}^{M}\sum_{n=-M}^{M} A_{mn}\mathbf{q}_m(\omega)[\mathbf{q}_n(\omega)^*]^T. \tag{D.4}$$

Substituting (D.4) into (D.2) yields,

$$\sum_{m=-M}^{M}\sum_{n=-M}^{M} A_{mn}\mathbf{q}_m(\omega)\{[\mathbf{q}_n(\omega)^*]^T \mathbf{q}_k(\omega)\} = \eta_k(\omega)\mathbf{q}_k(\omega). \tag{D.5}$$

Note that $\{[\mathbf{q}_n(\omega)^*]^T \mathbf{q}_k(\omega)\}$ is a scalar and is denoted as $H_{nk}(\omega)$, i.e.,

$$H_{nk}(\omega) = \{[\mathbf{q}_n(\omega)^*]^T \mathbf{q}_k(\omega)\}. \tag{D.6}$$

Taking inner product on both sides of Eq. (D.5), i.e., $[\mathbf{q}_l(\omega)^*]^T$ (D.5), we obtain

$$\sum_{m=-M}^{M}\sum_{n=-M}^{M} A_{mn}\{[\mathbf{q}_l(\omega)^*]^T \mathbf{q}_m(\omega)\}\{[\mathbf{q}_n(\omega)^*]^T \mathbf{q}_k(\omega)\}$$
$$= \eta_k\{[\mathbf{q}_l(\omega)^*]^T \mathbf{q}_k(\omega)\}. \tag{D.7}$$

In a compact form, (D.7) can be rewritten as

$$\mathbf{HAH} = \mathbf{H}\eta, \tag{D.8}$$

where, η is a diagonal matrix, i.e.,

$$\eta = \text{diag}\{\eta_{-M}, \eta_{-(M-1)}, \ldots, \eta_{-1}, \eta_0, \eta_1, \ldots, \eta_{M-1}, \eta_M\}.$$

From Eq. (D.8), we find

$$\mathbf{A} = \eta\mathbf{H}^{-1} = \eta\mathbf{Z}, \tag{D.9}$$

and

$$\mathbf{H}^{-1} = \mathbf{Z}. \tag{D.10}$$

Finally, substituting Eq. (D.9) into (D.4) yields (D.1). Therefore, lemma is proved. □

PROOF OF EQ. (42). Now, we shall derive Eq. (42) by using the lemma as follows. First, let us note that the inverse of matrix $[\mathbf{I} - \hat{\mathbf{R}}_{\uparrow\downarrow}^{(0)}(\omega)\hat{\mathbf{R}}_{\downarrow\uparrow}^{(1)}(\omega)]$ still belongs to the linear space $\mathfrak{R}$, thus it can still be spanned by the eigen-vectors $\{\mathbf{q}_n(\omega)\}$ as

$$[\mathbf{I} - \hat{\mathbf{R}}_{\uparrow\downarrow}^{(0)}(\omega)\hat{\mathbf{R}}_{\downarrow\uparrow}^{(1)}(\omega)]^{-1} = \sum_{m=-M}^{M}\sum_{n=-M}^{M} B_{mn}(\omega)\mathbf{q}_m(\omega)[\mathbf{q}_n(\omega)^*]^T. \quad (D.11)$$

To determine the expansion coefficients $B_{mn}(\omega)$, we shall use

$$[\mathbf{I} - \hat{\mathbf{R}}_{\uparrow\downarrow}^{(0)}(\omega)\hat{\mathbf{R}}_{\downarrow\uparrow}^{(1)}(\omega)]^{-1}[\mathbf{I} - \hat{\mathbf{R}}_{\uparrow\downarrow}^{(0)}(\omega)\hat{\mathbf{R}}_{\downarrow\uparrow}^{(1)}(\omega)] = \mathbf{I}, \quad (D.12)$$

and

$$H_{lk} = \{[\mathbf{q}_l(\omega)^*]^T \mathbf{q}_k(\omega)\} = \{[\mathbf{q}_l(\omega)^*]^T\}\mathbf{I}\{\mathbf{q}_k(\omega)\}. \quad (D.13)$$

Substituting Eqs. (D.11) and (D.1) into (D.12), and using Eq. (D.13), we obtain

HBHη**ZH = H**.

Using (D.10), the above equation can be reduced to

$$\mathbf{B} = \eta^{-1}\mathbf{Z}. \quad (D.14)$$

Substituting Eq. (D.14) into (D.11) and considering that η is a diagonal matrix, we finally obtain,

$$[\mathbf{I} - \hat{\mathbf{R}}_{\uparrow\downarrow}^{(0)}(\omega)\hat{\mathbf{R}}_{\downarrow\uparrow}^{(1)}(\omega)]^{-1} = \sum_{m=-M}^{M}\sum_{n=-M}^{M} \frac{Z_{mn}(\omega)}{\eta_m(\omega)}\mathbf{q}_m(\omega)[\mathbf{q}_n(\omega)^*]^T.$$

This is Eq. (42) used in the text. □

REFERENCES

Aki, K., Larner, K.L. (1970). Surface motion of a layered medium having an irregular interface due to incident plane SH waves. *J. Geophys. Res.* **70**, 933–954.

Aki, K., Richards, P.G. (1980). Quantitative Seismology: Theory and Methods, vol. 1. W.H. Freeman, San Francisco.

Arfken, G.B., Weber, H.J. (2005). Mathematical Methods for Physicists, sixth ed. Academic Press.

Bard, P.Y., Bouchon, M. (1980a). The seismic response of sediment-filled valleys: 1. The case of incident SH-waves. *Bull. Seismol. Soc. Am.* **70**, 1263–1286.

Bard, P.Y., Bouchon, M. (1980b). The seismic response of sediment-filled valleys: 2. The case of incident P-waves and SV waves. *Bull. Seismol. Soc. Am.* **70**, 1921–1941.

Boore, D.M. (1972). Finite Difference Methods for Seismic Wave Propagation in Heterogeneous Material. *Methods in Computational Physics*, vol. 11.

Bouchon, M. (1973). Effect of topography on surface motion. *Bull. Seismol. Soc. Am.* **63**, 615–632.
Bouchon, M. (1985). A simple complete numerical solution to the problem of diffraction of SH waves by an irregular interface. *J. Acoust. Soc. Am.* **77**, 1–5.
Bouchon, M., Aki, K. (1977). Discrete wavenumber representation of seismic source wave fields. *Bull. Seismol. Soc. Am.* **67**, 259–277.
Bouchon, M., Campillo, M., Gaffe, S. (1989). A boundary integral equation-discrete wave-number representation method to study wave propagation in multi-layered media with irregular interfaces. *Geophysics* **54**, 1134–1140.
Bouchon, M., Sanchez-Sesma, F. (2006). Boundary integral equations and boundary element methods in elastodynamics. In this Monograph.
Campillo, M., Bouchon, M. (1985). Synthetic SH seismograms in a laterally varying medium by the discrete wavenumber method. *Geophysics* **54**, 1134–1140.
Cao, J., Zhang, J., Ge, Z.X., Chen, X.F. (2004). A comparative study on seismic wave methods for multi-layered media with irregular interfaces: Irregular topography problem. *Chinese J. Geophys.* **47**, 495–503.
Cerveny, V., Molotkov, I.A., Psencik, I. (1977). Ray method in seismology. Karlova University, Praha.
Cerveny, V., Klimes, L., Psencik, I. (2006). Seismic ray method: Recent developments. In this Monograph.
Chaljub, E., Komatitsch, D., Vilotte, J., Capdeville, Y., Valette, B., Festa, G. (2006). Spectral element analysis in seismology. In this Monograph.
Chapman, C.H., Drummond, R. (1982). Body wave seismograms in inhomogeneous media using Maslov asymptotic theory. *Bull. Seismol. Soc. Am.* **72**, 5277–5317.
Chen, X.F. (1990). Seismograms synthesis for multi-layered media with irregular interfaces by global generalized reflection/transmission matrices method. Part I. Theory of 2-D SH case. *Bull. Seismol. Soc. Am.* **80**, 1696–1724.
Chen, X.F. (1991). Seismograms synthesis for multi-layered media with irregular interfaces by global generalized reflection/transmission matrices method. Ph.D. Dissertation, University of Southern California, Los Angeles.
Chen, X.F. (1993). A systematic and efficient method of computing normal mode for multi-layered half-space. *Geophys. J. Int.* **115**, 391–409.
Chen, X.F. (1995). Seismograms synthesis for multi-layered media with irregular interfaces by global generalized reflection/transmission matrices method. Part II. Applications of 2-D SH case. *Bull. Seismol. Soc. Am.* **85**, 1094–1106.
Chen, X.F. (1996). Seismograms synthesis for multi-layered media with irregular interfaces by global generalized reflection/transmission matrices method. Part III. Theory of 2-D P-SV case. *Bull. Seismol. Soc. Am.* **86**, 389–405.
Chen, X.F. (1999a). Love waves in multi-layered media with irregular interfaces: I. Modal solution and excitation formulation. *Bull. Seismol. Soc. Am.* **89**, 1519–1534.
Chen, X.F. (1999b). Seismograms synthesis in multi-layered half-space (I). Theoretical formulation. *Earthquake Res. China* **13**, 149–174.
Day, M.S. (1977). Finite element analysis of seismic scattering problems, Ph.D. Thesis. University of California, San Diego, California.
Dravinski, M. (1982). Influence of interface depth upon strong ground motion. *Bull. Seismol. Soc. Am.* **72**, 597–614.
Frazer, L.N., Sen, M.K. (1985). Kirchoff-Helmholtz reflection seismograms in a laterally inhomogeneous multi-layered elastic medium, part I. Theory. *Geophys. J. R. Astron. Soc.* **78**, 413–429.
Frazer, L.N. (1987). Synthetic seismograms using multifold path integral. Part I. Theory. *Geophys. J. R. Astron. Soc.* **88**, 621–646.
Fuchs, K., Muller, G. (1971). Computation of synthetic seismograms with the reflectivity method and comparison of observations. *Geophys. J. R. Astron. Soc.* **23**, 417–433.

Furumura, T., Kennett, B.L.N., Furumura, M. (1998). Seismic wave field calculation for laterally heterogeneous whole earth models using the pseudospectral method. *Geophys. J. Int.* **135**, 845–860.

Gjevik, B. (1973). A variational method for Love waves in non-horizontal layered structures. *Bull. Seismol. Soc. Am.* **63**, 1013–1023.

Harkrider, D.G., Anderson, D.L. (1966). Surface wave energy from point sources in plane layered Earth models. *J. Geophys. Res.* **71**, 2967–2980.

Haskell, N.A. (1953). The dispersion of surface waves on multi-layered media. *Bull. Seismol. Soc. Am.* **43**, 17–34.

Helmberger, D.V., Harkrider, D.A. (1978). Modeling earthquakes with generalized ray theory. In: Miklowitz, J., Achenbach, J.D. (Eds.), *Modern Problems in Elastic Wave Propagation*. New York.

Helmberger, D.V. (1983). Theory and application of synthetic seismograms. In: Kanamori, H., Boschi, E. (Eds.), *Earthquakes: Observation, Theory and Interpretation*. North-Holland, New York.

Hong, T.L., Helmberger, D.V. (1978). Glorified optics and wave propagation in nonplanar structure. *Bull. Seismol. Soc. Am.* **68**, 1313–1358.

Kawase, H. (1988). Time-domain response of a semi-circular canyon for incident SV, P and Rayleigh waves calculated by the discrete wave number boundary element method. *Bull. Seismol. Soc. Am.* **78**, 1415–1437.

Keilis-Borok, V.I., Yanoskaya, T.B. (1962). Dependence of the spectrum of surface waves on the depth of the focus within the Earth's crust. *Bull. Acad. Sci. USSR, Geophys. Ser.* **11**, 1532–1539. (English Translation).

Kennett, B.L.N. (1983). Seismic Wave Propagation in Stratified Media. Cambridge Univ. Press, New York.

Knopoff, L. (1964). A matrix method for elastic wave problems. *Bull. Seismol. Soc. Am.* **54**, 431–438.

Kohketsu, K. (1987). 2-D reflectivity method and synthetic seismograms for irregularly layered structures, part I. SH wave generation. *Geophys. J. R. Astron. Soc.* **89**, 821–838.

Kohketsu, K., Kennett, B.L., Takenaka, H. (1991). 2-D reflectivity method and synthetic seismograms for irregularly layered structures-II. Invariant embedding approach. *Geophys. J. Int.* **105**, 119–130.

Komatitsch, D., Vilotte, J.P. (1998). The spectral element method: An efficient tool to simulate the seismic response of 2-D and 3-D geological structures. *Bull. Seismol. Soc. Am.* **88**, 368–392.

Kosloff, D., Baysal, E. (1982). Forward modeling by a Fourier method. *Geophysics* **47**, 1402–1412.

Levander, A.R. (1988). 4th order finite difference P-SV seismograms. *Geophysics* **53**, 1425–1436.

Luco, J.E., Apsel, R.J. (1983). On the Green's function for a layered half-space, Part I. *Bull. Seismol. Soc. Am.* **73**, 909–929.

Maupin, V. (2006). Introduction to mode coupling method for surface waves. In this Monograph.

Moczo, P., Kristek, J., Galis, M. (2004). Simulation of the planar free surface with near-surface lateral discontinuities in the finite difference modeling of seismic motion. *Bull. Seismol. Soc. Am.* **94**, 760–768.

Moczo, P., Robertson, J., Eisner, L. (2006). The finite-difference time-domain method for modeling of seismic wave propagation. In this Monograph.

Morse, P.M., Feshbach, H. (1953). Methods of Theoretical Physics. McGraw–Hill, New York.

Rial, J.A., Ling, H. (1992). Theoretical estimation of the eigenfrequencies of 2-D resonant sedimentary basins: numerical computations and analytic approximations to the elastic problem. *Bull. Seismol. Soc. Am.* **82**, 2350–2367.

Ricker, N.H. (1977). Transient Waves in Visco-Elastic Media. Elsevier, Amsterdam, Holland.

Saito, M. (1967). Excitation of free oscillations and surface waves by a point source in a vertically heterogeneous Earth. *J. Geophys. Res.* **72**, 3689–3699.

Sanchez-Sesma, F.J., Rosenbluth, E. (1979). Ground motions canyons of arbitrary shapes under incident SH waves. *Earthquake Engrg. Struct. Dyn.* **7**, 441–450.

Smith, W.D. (1974). A non-reflecting plane boundary for wave propagation problems. *J. Comput. Phys.* **15**, 267–281.

Snieder, R. (1987). Surface Wave Holography. In: Nolet, G. (Ed.), Reidel, Boston, pp. 323–337.

Takenaka, H., Ohori, M., Koketsu, K., Kennett, B.L.N. (1996). An efficient approach to the seismogram synthesis for a basin structure using propagation invariants. *Bull. Seismol. Soc. Am.* **86**, 379–388.

Takeuchi, H., Saito, M. (1972). Seismic Surface Waves. In: Bolt, B.A. (Ed.), *Seismology: Surface Waves and Earth Oscillations*. In: *Methods in Computational Physics*, vol. 11. New York, Academic Press.

Trifunac, M.D. (1973). Scattering of plane SH waves by a semi-cylindrical canyon. *Earthquake Engrg. Struct. Dyn.* **1**, 267–281.

Wang, Y.B., Takenaka, H., Furumura, T. (2001). Modelling seismic wave propagation in a 2-D cylindrical whole-earth model using the pseudospectral method. *Geophys. J. Int.* **145**, 689–708.

Waterman, P.C. (1969). New formulation of acoustic scattering. *J. Acoustic Soc. Am.* **45**, 1417–1429.

Waterman, P.C. (1975). Scattering by periodic surfaces. *J. Acoustic Soc. Am.* **57**, 791–802.

Woodhouse, J.H. (1974). Surface waves in a laterally varying layered media structure. *Geophys. J. R. Astron. Soc.* **37**, 461–490.

Yao, Z., Harkrider, D.G. (1983). Generalized reflection-transmission matrix and discrete wavenumber method for synthetic seismograms. *Bull. Seismol. Soc. Am.* **73**, 1685–1699.

Yomogida, K. (1985). Gaussian Beams for surface waves in laterally slowly varying media. *Geophys. J. R. Astron. Soc.* **82**, 511–533.

Zhu, L.B., Zeng, R.S. (1993). Evaluation of seismic parameters by Maslov's surface wave method. In: Chen, Y.T., et al. (Eds.), *Advances in Solid Earth Geophysics in China*. Beijing.

ADVANCES IN GEOPHYSICS, VOL. 48, CHAPTER 5

ONE-WAY AND ONE-RETURN APPROXIMATIONS (DE WOLF APPROXIMATION) FOR FAST ELASTIC WAVE MODELING IN COMPLEX MEDIA

RU-SHAN WU, XIAO-BI XIE AND XIAN-YUN WU

Modeling and Imaging Laboratory, Institute of Geophysics and Planetary Physics, University of California, Santa Cruz, California, USA

ABSTRACT

The De Wolf approximation has been introduced to overcome the limitation of the Born and Rytov approximations in long range forward propagation and backscattering calculations. The De Wolf approximation is a multiple-forescattering-single-backscattering (MFSB) approximation, which can be implemented by using an iterative marching algorithm with a single backscattering calculation for each marching step (a thin-slab). Therefore, it is also called a one-return approximation. The marching algorithm not only updates the incident field step-by-step, in the forward direction, but also the Green's function when propagating the backscattered waves to the receivers. This distinguishes it from the first order approximation of the asymptotic multiple scattering series, such as the generalized Bremmer series, where the Green's function is approximated by an asymptotic solution. The De Wolf approximation neglects the reverberations (internal multiples) inside thin-slabs, but can model all the forward scattering phenomena, such as focusing/defocusing, diffraction, refraction, interference, as well as the primary reflections.

In this chapter, renormalized MFSB (multiple-forescattering–single-backscattering) equations and the dual-domain expressions for scalar, acoustic and elastic waves are derived by using a unified approach. Two versions of the one-return method (using MFSB approximation) are given: one is the wide-angle, dual-domain formulation (thin-slab approximation) (compared to the screen approximation, no small-angle approximation is made in the derivation); the other is the screen approximation. In the screen approximation, which involves a small-angle approximation for the wave-medium interaction, it can be clearly seen that the forward scattered, or transmitted waves are mainly controlled by velocity perturbations; while the backscattered or reflected waves, are mainly controlled by impedance perturbations. Later in this chapter the validity of the thin-slab and screen methods, and the wide-angle capability of the dual-domain implementation are demonstrated by numerical examples. Reflection coefficients of a plane interface, derived from numerical simulations by the wide-angle method, are shown to match the theoretical curves well up to critical angles. The methods are applied to the fast calculation of synthetic seismograms. The results are compared with finite difference (FD) calculations for the elastic French model. For weak heterogeneities ($\pm 15\%$ perturbation), good agreement between the two methods verifies the validity of the one-return approach. However, the one-return approach is about 2–3 orders of magnitude faster than the elastic FD algorithm. The other example of application is the modeling of amplitude variation with angle (AVA) responses for a complex reservoir with heterogeneous overburdens. In addition to its fast computation speed, the one return method (thin-slab and complex-screen propagators) has some special advantages when applied to the thin-bed and random layer responses.

Keywords: One-way wave equation, Generalized screen propagator, Seismic wave modeling, AVO

1. Introduction

One-way approximation for wave propagation has been introduced and widely used as propagators in forward and inverse problems of scalar, acoustic and elastic waves (e.g., Claerbout, 1970, 1976; Landers and Claerbout, 1972; Flatté and Tappert, 1975; Corones, 1975; Tappert, 1977; McCoy, 1977; Hudson, 1980; Ma, 1982; Wales and McCoy, 1983; Fishman and McCoy, 1984, 1985; Wales, 1986; McCoy and Frazer, 1986; Collins, 1989, 1993; Collins and Westwood, 1991; Stoffa *et al.*, 1990; Fisk and McCartor, 1991; Wu and Huang, 1992; Ristow and Ruhl, 1994; Wu, 1994, 1996, 2003; Wu and Xie, 1994; Huang and Wu, 1996; Wu and Jin, 1997; Grimbergen *et al.*, 1998; Van Stralen *et al.*, 1998; Wild and Hudson, 1998; Huang *et al.*, 1999a, 1999b; Thomson, 1999, 2005; De Hoop *et al.*, 2000; Lee *et al.*, 2000; Wild *et al.*, 2000; Wu *et al.*, 2000a, 2000b; Le Rousseau and De Hoop, 2001; Wu and Wu, 2001, 2006; Xie and Wu, 2001, 2005; Han and Wu, 2005). The great advantages of one-way propagation methods are the fast speed of computation, often by several orders of magnitudes faster than the full wave finite difference and finite element methods, and the huge saving in internal memory. The successful extension and applications of one-way elastic wave propagation methods has stimulated the research interest in developing similar theory and techniques for reflected or backscattered wave calculation. There are several approaches in extending the one-way propagation method to include backscattering and multiple scattering calculations. The key difference between these approaches is how to define a reference Green's function for constructing one-way propagators. The generalized Bremmer series (GBS) approach (Corones, 1975; De Hoop, 1996; Wapenaar, 1996, 1998; Van Stralen *et al.*, 1998; Thomson, 1999; Le Rousseau and De Hoop, 2001) adopts an asymptotic solution of the acoustic or elastic wave equation in the heterogeneous medium as the Green's function, i.e. the one-way propagator in the preferred direction. The multiple scattering series is based on the interaction of Green's field (incident field) with the medium heterogeneities. The other approach, i.e. the generalized screen propagator (GSP) approach (Wu, 1994, 1996, 2003; Wu and Xie, 1993, 1994; Wu *et al.*,

1995; Wild and Hudson, 1998; De Hoop *et al.*, 2000; Wild *et al.*, 2000; Xie *et al.*, 2000; Xie and Wu, 2001, 2005), on the other hand, does not use asymptotic solutions. Instead, the approach uses the multiple-forward-scattering (MFS) corrected one-way propagator as the Green's function. When the backscattered field, calculated at each thin-slab, is propagated to the backward direction, the same MFS corrected one-way propagator is used. In surface wave modeling, Friederich *et al.* (1993) and Friederich (1999) adopted a similar approach of MFS approximation (see Chapter 2 by Maupin in this book). However, the approach did not apply the MFS correction to the one-way propagator in the backward direction for the backscattered waves, and therefore did not take the full advantages of the De Wolf approximation. In Section 2, we will compare GBS and GSP approaches after introducing the De Wolf multiple scattering series and the related approximation. In the rest of this chapter we will concentrate on the formulation and applications of the generalized screen approach.

In the generalized screen approach, Wu and Huang (1995) introduced a wide-angle modeling method for backscattered acoustic waves using the multiple-forward-scattering approximation and a phase-screen propagator. Xie and Wu (1995; 2001) extended the complex screen method to include the calculation of backscattered elastic waves under the small-angle approximation. Wu (1996) derived a more general theory for acoustic and elastic waves using the De Wolf approximation, and the theory provided two versions of algorithms: the thin-slab method and the complex-screen method. Later, Wu and Wu (1999, 2003a) introduced a fast implementation of the thin-slab method and a second order improvement for the complex-screen method. In Section 3, the dual-domain thin-slab formulations for the case of scalar, acoustic and elastic media are derived. In Section 4, a fast implementation of the thin-slab propagator is presented with numerical examples. The excellent agreement between the thin-slab and the elastic FD method in the numerical examples demonstrates the validity and efficiency of the theory and method. In Section 5, the small-angle approximation is introduced to derive the screen approximation, which is less accurate for wide angle scattering but is more efficient than the thin-slab method. The validity and potential of the one return approach and the wide-angle capability for the dual-domain implementation are demonstrated by numerical examples for both thin-slab and screen methods applying to the calculation of synthetic seismic sections. In Section 6, the thin-slab method is applied to wave field and amplitude variation with offset (AVO) modeling in exploration seismology.

2. BORN, RYTOV, DE WOLF APPROXIMATIONS AND MULTIPLE SCATTERING SERIES

The perturbation approach is one of the well-known approaches for wave propagation, scattering and imaging (see Chapter 9 of Morse and Feshbach, 1953;

Chapter 13 of Aki and Richards, 1980; Wu, 1989). Traditionally, the perturbation method was used only for weakly inhomogeneous media and short propagation distance. However, recent progress in this direction has led to the development of iterative perturbation solutions in the form of a one-way marching algorithm for scattering and imaging problems in strongly heterogeneous media. For a historical review, see Section 3.1 of Wu (2003). In this section, we give a theoretical analysis of the perturbation approach, including the Born, Rytov and De Wolf approximations, as well as the multiple scattering series. The relatively strong and weak points of the Born and Rytov approximations are analyzed. Since the Born approximation is a weak scattering approximation, it is not suitable for large volume or long-range numerical simulations. The Rytov approximation is a smooth scattering approximation, which works well for long-range small-angle propagation problems, but is not applicable to large-angle scattering and backscattering. Then, the De Wolf approximation (multiple forescattering single backscattering, or "one-return approximation") is introduced to overcome the limitations of the Born and Rytov approximations in long range forward propagation and backscattering calculations, which can serve as the theoretical basis of the new dual-domain propagators.

2.1. Born Approximation and Rytov Approximation: Their Strong and Weak Points

For the sake of simplicity, we consider the scalar wave case as an example. The scalar wave equation in inhomogeneous media can be written as

$$\left(\nabla^2 + \frac{\omega^2}{c^2(\vec{r})}\right) u(\vec{r}) = 0, \tag{1}$$

where ω is the circular frequency, $\vec{r}$ is the position vector, and $c(\vec{r})$ is wave velocity at $\vec{r}$. Define c_0 as the background velocity of the medium, resulting in

$$\left(\nabla^2 + k^2\right) u(\vec{r}) = -k^2 \varepsilon(\vec{r}) u(\vec{r}), \tag{2}$$

where $k = \omega/c_0$ is the background wavenumber and

$$\varepsilon(\vec{r}) = \frac{c_0^2}{c^2(\vec{r})} - 1 \tag{3}$$

is the perturbation function (dimensionless force). Set

$$u(\vec{r}) = u^0(\vec{r}) + U(\vec{r}), \tag{4}$$

where $u^0(\vec{r})$ is the unperturbed wave field or "incident wave field" (field in the homogeneous background medium), and $U(\vec{r})$ is the scattered wave field. Substitute (4) into (2) and notice that $u^0(\vec{r})$ satisfies the homogeneous wave equation, resulting in

$$u(\vec{r}) = u^0(\vec{r}) + k^2 \int_V d^3\vec{r}'\, g(\vec{r}; \vec{r}')\varepsilon(\vec{r}')u(\vec{r}'), \tag{5}$$

where $g(\vec{r}; \vec{r}')$ is the Green's function in the reference (background) medium, and the integral is over the whole volume of medium. This is the Lippmann–Schwinger integral equation. Since the field $u(\vec{r})$ under the integral is the total field which is unknown, Eq. (5) is not an explicit solution but an integral equation.

2.1.1. Born Approximation

Approximating the total field under the integral with the incident field $u^0(\vec{r})$, we obtain the Born approximation

$$u(\vec{r}) = u^0(\vec{r}) + k^2 \int_V d^3\vec{r}'\, g(\vec{r}; \vec{r}')\varepsilon(\vec{r}')u^0(\vec{r}'). \tag{6}$$

In general, the Born approximation is a weak scattering approximation, which is only valid when the scattered field is much smaller than the incident field. This implies that the heterogeneities are weak and the propagation distance is short. However, the valid regions of the Born approximation are very different for forward scattering than for backscattering. Forward scattering divergence or catastrophe is the weakest point of Born approximation. For simplicity, we use "forescattering" to stand for "forward scattering". As can be seen from Eq. (6), the total scattering field is the sum of scattered fields from all parts of the scattering volume. Each contribution is independent from other contributions since the incident field is not updated by the scattering process. In the forward direction, the scattered fields from each part propagate with the same speed as the incident field, so they will be coherently superposed, leading to the linear increase of the total field. The Born approximation does not obey energy conservation. The energy increase will be the fastest in the forward direction, resulting in a catastrophic divergence for long distance propagation. On the contrary, backscattering behaves quite differently from forescattering. Since there is no incident wave in the backward direction, the total observed field is the sum of all the backscattered fields from all the scatterers. However, the size of coherent stacking for backscattered waves is about $\lambda/4$ because of the two-way travel time difference. Beyond this coherent region, all other contributions will be cancelled out. For this reason, backscattering does not have the catastrophic divergence even when the Born approximation is used. This can be further explained with the spectral responses of heterogeneities to scatterings with different scattering angles.

From the analysis of scattering characteristics, we know that the forescattering is controlled by the d.c. component of the medium spectrum $W(0)$, but the backscattering is determined by the spectral component at spatial frequency $2k$, i.e. $W(2k)$, where k is the background wavenumber (see Wu and Aki, 1985;

Wu, 1989). The d.c. component of the medium spectrum is linearly increasing along the propagation distance, while the contribution from $W(2k)$ is usually much smaller and increases much slower than $W(0)$. The validity condition for the Born approximation is the smallness for the scattered field compared with the incident field. Therefore, the region of validity for the Born approximation for backscattering is much larger than that for forescattering. The other difference between backscattering and forescattering is their responses to different types of heterogeneities. The backscattering is sensitive to the impedance type of heterogeneities, while forescattering mainly responds to velocity type of heterogeneities. Velocity perturbation will produce travel-time or phase change, which can accumulate to quite large values, causing the breakdown of the Born approximation. This kind of phase-change accumulation can be easily handled by the Rytov transformation. This is why the Rytov approximation performs better than the Born approximation for forescattering and has been widely used for long range propagation in the case of forescattering or small-angle scattering dominance.

2.1.2. Rytov Approximation

Let $u^0(\vec{r})$ be the solution in the absence of perturbations, i.e.,

$$(\nabla^2 + k^2)u^0 = 0, \tag{7}$$

and the perturbed wave field after interaction with the heterogeneity as $u(\vec{r})$. We normalize $u(\vec{r})$ by the unperturbed field $u^0(\vec{r})$ and express the perturbation of the field by a complex phase perturbation function $\psi(\vec{r})$, i.e.,

$$\frac{u(\vec{r})}{u^0(\vec{r})} = e^{\psi(\vec{r})}. \tag{8}$$

This is the Rytov Transformation (see Tatarskii, 1971; or Ishimaru, 1978, p. 349). $\psi(\vec{r})$ denotes the phase and log-amplitude deviations from the incident field:

$$\psi = \log u - \log u^0 = \log\left[\frac{A}{A^0}\right] + i(\phi - \phi^0), \tag{9}$$

where ϕ and ϕ^0 are phases of perturbed and unperturbed waves. Combining (2), (7) and (8) yields

$$2\nabla u^0 \cdot \nabla \psi + u^0 \nabla^2 \psi = -u^0(\nabla \psi \cdot \nabla \psi + k^2 \varepsilon). \tag{10}$$

The simple identity

$$\nabla^2(u^0 \psi) = \psi \nabla^2 u^0 + 2\nabla u^0 \cdot \nabla \psi + u^0 \nabla^2 \psi,$$

together with (7) results in

$$2\nabla u^0 \cdot \nabla \psi + u^0 \nabla^2 \psi = (\nabla^2 + k^2)(u^0 \psi). \tag{11}$$

From (10) and (11) we obtain

$$(\nabla^2 + k^2)(u^0 \psi) = -u^0(\nabla \psi \cdot \nabla \psi + k^2 \varepsilon). \tag{12}$$

The solution of (12) can be expressed as an integral equation:

$$u^0(\vec{r})\psi(\vec{r}) = \int_V d^3\vec{r}' \, g(\vec{r}; \vec{r}') u^0(\vec{r}')[\nabla \psi(\vec{r}') \cdot \nabla \psi(\vec{r}') + k^2 \varepsilon(\vec{r}')], \tag{13}$$

where $g(\vec{r}; \vec{r}')$ is the Green's function for the background medium, $u^0(\vec{r}')$ and $u^0(\vec{r})$ are the incident field at $\vec{r}'$ and $\vec{r}$, respectively.

Equation (13) is a nonlinear (Ricatti) equation. Assuming $|\nabla \psi \cdot \nabla \psi|$ is small with respect to $k^2|\varepsilon|$, we can neglect the term $\nabla \psi \cdot \nabla \psi$ and obtain a solution known as the *Rytov approximation*:

$$\psi(\vec{r}) = \frac{k^2}{u^0(\vec{r})} \int_V d^3\vec{r}' \, g(\vec{r}; \vec{r}')\varepsilon(\vec{r}')u^0(\vec{r}'). \tag{14}$$

Now, we discuss the relationship between the Rytov and Born approximations, and their strong and weak points, respectively. By expanding e^ψ into a power series, the scattered field can be written as

$$u - u^0 = u^0(e^\psi - 1) = u^0 \psi + \frac{1}{2} u^0 \psi^2 + \cdots. \tag{15}$$

When $\psi \ll 1$, i.e., the accumulated phase change is less than one radian (corresponding to about one sixth of the wave period), the terms of ψ^2 and higher terms can be neglected, and

$$u - u^0 = u^0 \psi = k^2 \int_V d^3\vec{r}' \, g(\vec{r}; \vec{r}')\varepsilon(\vec{r}')u^0(\vec{r}'), \tag{16}$$

which is the Born approximation. This indicates that when $\psi \ll 1$, the Rytov approximation reduces to the Born approximation. In the case of large phase-change accumulation, the Born approximation is no longer valid. The Rytov approximation still holds as long as the condition $|\nabla \psi \cdot \nabla \psi| \ll k^2|\varepsilon|$ is satisfied.

Let us look at the implication of the condition $|\nabla \psi \cdot \nabla \psi| \ll k^2|\varepsilon|$ for the Rytov approximation. Assume that the observed total field after wave interacted with the heterogeneities is nearly a plane wave:

$$u = A e^{i\vec{k}_1 \cdot \vec{r}},$$

which could be the refracted wave in the forward direction, or the backscattered field, where $\vec{k}_1 = k\hat{k}_1$ and $\hat{k}_1$ is a unit vector. Since the incident wave is

$$u^0 = A_0 e^{i\vec{k}_0 \cdot \vec{r}},$$

the complex phase field ψ can be written as

$$\psi = \log(A/A_0) + i(\vec{k}_1 - \vec{k}_0) \cdot \vec{r}, \tag{17}$$

and

$$\nabla \psi = \nabla \log(A/A_0) + i(\vec{k}_1 - \vec{k}_0), \tag{18}$$

$$\nabla \psi \cdot \nabla \psi = |\nabla \log(A/A_0)|^2 - |\vec{k}_1 - \vec{k}_0|^2 \\ + 2i(\vec{k}_1 - \vec{k}_0) \cdot \nabla \log(A/A_0). \tag{19}$$

Normally wave amplitudes vary much slower than the phases, so the major contribution to $\nabla \psi \cdot \nabla \psi$ in (19) is from the phase term $|\vec{k}_1 - \vec{k}_0|^2$. Therefore, the condition for Rytov approximation can be approximately stated as

$$|\vec{k}_1 - \vec{k}_0|^2 = 4k^2 \sin^2 \frac{\theta}{2} \ll k^2 |\varepsilon|, \tag{20}$$

where θ is the scattering angle. Therefore the Rytov approximation is only valid when the scattering angle (deflection angle) is small enough to satisfy

$$\sin \frac{\theta}{2} \ll \sqrt{\frac{1}{4}\varepsilon} = \frac{1}{2}\sqrt{\frac{c_0^2 - c^2(\vec{r})}{c^2(\vec{r})}}. \tag{21}$$

This is a point-to-point analysis of the contributions from different terms (for example, the terms in the differential equation (12)). For the integral equation (13), one needs estimate the integral effects of $\nabla \psi \cdot \nabla \psi$ and $k^2 \varepsilon$. The heterogeneities need to be smooth enough to guarantee the smallness of the integral of $\nabla \psi \cdot \nabla \psi$ which is related to scattering angles, in comparison with the total scattering contribution $k^2 \varepsilon$. In any case, the Rytov approximation is totally inappropriate for backscattering. In the exactly backward direction, $\theta = 180°$ and $\sin \frac{\theta}{2} = 1$, inequality (21) is hardly satisfied. Therefore, although not explicitly specified, the Rytov approximation is a kind of small angle approximation. Together with the parabolic approximation, they formed a set of analytical tools widely used for the forward propagation and scattering problems, such as the line-of-sight propagation problem (e.g., Flatté et al., 1979; Ishimaru, 1978; Tatarskii, 1971). The Rytov approximation is also used in modeling transmission fluctuation for seismic array data (Flatté and Wu, 1988; Wu and Flatté, 1990), diffraction tomography (Devaney, 1982, 1984; Wu and Toksöz, 1987). Tatarskii (1971, Chapter 3B) has some discussions on the relation between the Rytov approximation and the parabolic approximation.

2.2. De Wolf Approximation

We see the limitations of both the Born and Rytov approximations. Even in weakly inhomogeneous media, we need better tools for wave modeling and imaging for long distance propagation. Higher order terms of the Born series (defined later in this section) may help in some cases. However, for strong scattering media, Born series will either converge very slowly, or become divergent. That is because the Born series is a global interaction series, each term in the series is global in nature. The first term of the Born approximation is a global response, and the higher terms are just global corrections. If the first term has a big error, it will be hard to correct it with higher order terms. One solution to the divergence of the scattering series is the renormalization procedure. Renormalization methods try to split the operations so that the scattering series can be reordered into many sub-series. We hope that some sub-series can be summed up theoretically so that the divergent elements of the series can be removed. The De Wolf approximation splits the scattering potential into forescattering and backscattering parts and renormalizes the incident field and Green's function into the forward propagated field and forward propagated Green's function (forward propagator), respectively (De Wolf, 1971, 1985). The forward propagated field u_f is the sum of an infinite sub-series including all the multiple forescattered fields. The forward propagator G_f is the sum of a similar sub-series including multiple forescattering corrections to the Green's function. For backscattering, only the single backscattered field is calculated at each step, and then propagated in the backward direction using the renormalized forward propagator (Green's function) G_f. The *De Wolf approximation* is also called the *"one-return approximation"* (Wu, 1996, 2003; Wu and Huang, 1995; Wu et al., 2000a, 2000b), since it is a *multiple-forescattering–single-backscattering (MFSB) approximation*. It is also a kind of *local* Born approximation since the Born approximation applies only locally to the individual thin-slabs. From previous sections we know that the Born approximation works well for backscattering. With the renormalized incident field and Green's function, the local Born (MFSB) proved to work surprisingly well for many practical applications. The key is to have good forward propagators. Rino (1988) has obtained better approximation than MFSB in the wavenumber domain and pointed out the error of the De Wolf approximation in the calculation of backscattering enhancement. The error (overestimate) is again due to the violation of the energy conservation law by the Born approximation. Even with a forescattering correction, the backscattered energy is still not removed from the forward propagated waves for local Born approximation. However, for short propagation distances in exploration seismology and some other applications, the errors in reflection amplitudes may not become a serious problem.

In the following, we will adopt an intuitive approach of derivation to see the physical meaning of the approximation. The De Wolf approximation bears some similarity to the Twersky approximation in the case of discrete scatterers (Twersky, 1964; Ishimaru, 1978). The Twersky approximation includes all the multiple scattering, except the reverberations between pairs of scatterers that excludes the paths which connect the two neighboring scatterers more than once. The Twersky approximation has less restrictions and therefore a wider range of applications than the De Wolf approximation. The latter needs define the split of forward and back scatterings. We define the scattering to the forward hemisphere as forescattering and its complement as backscattering.

The Lippmann–Schwinger equation (5) can be written symbolically as

$$u = u^0 + G_0 \varepsilon u, \tag{22}$$

where ε is a diagonal operator in space domain, and G_0 is a nondiagonal integral operator. If the reference medium is homogeneous, G_0 will be the volume integral with the Green's function $g_0(\vec{r}; \vec{r}')$ as the kernel. Formally, (22) can be expanded into infinite scattering series (Born series)

$$u = u^0 + G_0 \varepsilon u^0 + G_0 \varepsilon G_0 \varepsilon u^0 + \cdots. \tag{23}$$

If we split the scattering potential into the forescattering and backscattering parts

$$\varepsilon = \varepsilon_f + \varepsilon_b \tag{24}$$

and substitute it into (23), we can have all combinations of multiple forescattering and backscattering. We neglect the multiple backscattering (reverberations), i.e., drop all the terms containing two or more backscattering potentials, resulting in a multiple scattering series which contains terms with only one ε_b.

The general term will look like

$$G_0 \varepsilon_f G_0 \varepsilon_f \cdots G_0 \varepsilon_b G_0 \varepsilon_f \cdots G_0 \varepsilon_f u^0. \tag{25}$$

The multiple forescattering on the left side of ε_b can be written as

$$G_f^m = [G_0 \varepsilon_f]^m G_0, \tag{26}$$

and on its right side,

$$u_f^n = [G_0 \varepsilon_f]^n u^0. \tag{27}$$

Collecting all the terms of G_f^m and u_f^n, respectively, we have

$$G_f^M = \sum_{m=0}^{M} [G_0 \varepsilon_f]^m G_0,$$

$$u_f^N = \sum_{n=0}^{N} [G_0 \varepsilon_f]^n u^0. \tag{28}$$

Let M and N go to infinite, then the renormalized G_f (forward propagator) and u_f (forescattering corrected incident field) are:

$$G_f = \sum_{m=0}^{\infty} [G_0 \varepsilon_f]^m G_0,$$

$$u_f = \sum_{n=0}^{\infty}[G_0\varepsilon_f]^n u^0, \tag{29}$$

and the De Wolf approximation (in operator form) becomes

$$u = u_f + G_f \varepsilon_b u_f. \tag{30}$$

The observed total field u in (30) is different for different observation geometries. For transmission problems, the backscattering potential does not have any effect under the De Wolf approximation,

$$u_{\text{transmission}} = u_f. \tag{31}$$

On the other hand, for reflection measurement, that is, when the observations are in the same level as or behind the source with respect to the propagation direction, there is no u_f in the total field,

$$u_{\text{reflection}} = G_f \varepsilon_b u_f. \tag{32}$$

Write it into integral form, (30) becomes

$$u(\vec{r}) = u_f(\vec{r}) + \int_V d^3\vec{r}' \, g_f(\vec{r},\vec{r}')\varepsilon_b(\vec{r}')u_f(\vec{r}'). \tag{33}$$

Note that both the incident field and the Green's function have been renormalized by the multiple forescattering process through the multiple interactions with the forward scattering potential ε_f.

2.3. The De Wolf Series (DWS) of Multiple Scattering

De Wolf approximation can be considered as the first term of a multiple scattering series: the De Wolf series. After substituting the decomposition (24) into the Born series (23), we rearrange and recompose the scattering series into a series according to the power of the backscattering potential ε_b. The first order in ε_b will be the De Wolf approximation (one-return approximation). The second order corresponds to the double backscattering (double reflection, double return) term. The higher terms represent the multiple backscattering series. The whole multiple scattering series (23) can be reorganized into

$$\begin{aligned} u &= u_f + G_f\varepsilon_b u_f + G_f\varepsilon_b G_f\varepsilon_b u_f + \cdots \\ &= \sum_{m=0}^{M}[G_f\varepsilon_b]^m u_f. \end{aligned} \tag{34}$$

We see that the zero order term is the forward scattering approximated direct wavefield. It is the direct transmitted wave in the real media. The first order term is the De Wolf approximation, which corresponds to the single backscattering signal, or the primary reflections, as called in exploration seismology. This single backscattering signal is different from the Born approximation where the incident field and the Green's function are both defined in the background medium (a homogeneous medium).

The De Wolf Series and Generalized Bremmer Series

Here we point out the differences between the De Wolf series (DWS) and the Generalized Bremmer Series (GBS). The original Bremmer series (Bremmer, 1951) is a geometric-optical series for stratified media, which can be considered as a higher order extension to the regular WKBJ solution (the first order term). Later it was generalized to 3-D inhomogeneous media, and was named the generalized Bremmer series (Corones, 1975; De Hoop, 1996; Wapenaar, 1996, 1998; Van Stralen *et al.*, 1998; Thomson, 1999; Le Rousseau and De Hoop, 2001). The zero order term (the leading term) of the GBS is a high-frequency asymptotic solution (a WKBJ-like solution or Rytov-like solution) (De Hoop, 1996), and used as the Green's function for deriving the higher order terms. The Green's function is not updated when calculating the higher order scattering. Therefore, it is similar to the Born series in a certain sense. Unlike the DWS, which is a series in terms of medium velocity variation, the GBS is in terms of the spatial derivatives of the medium properties (De Hoop, 1996). Because of the asymptotic nature of its Green's function, the media need to be "smooth" on the scale of the irradiating pulse (De Hoop, 1996; Van Stralen *et al.*, 1998; Thomson, 1999). Some authors used an equivalent medium averaging process to smooth the medium before the application of the method (Van Stralen *et al.*, 1998). Wapenaar's approach (1996, 1998) is to get an asymptotic solution, without averaging the medium, using the flux normalized decomposition of wavefield. Thomson (1999, 2005) included the second term of the asymptotic series to the asymptotic Green's function to improve the amplitude accuracy. This zero-order term solution for generally inhomogeneous media can be useful for seismic imaging and inversion (e.g. Berkhout, 1982; Berkhout and Wapenaar, 1989). With careful amplitude correction, this type of Green's function is an energy-flux conserved propagator in any heterogeneous media (including discontinuities), and is called (Wu *et al.*, 2004; Wu and Cao, 2005) the "transparent propagator". The first term in the GBS (De Hoop, 1996; Wapenaar, 1996) in fact is not the primary reflection from the real media, but a "primary reflection" on the basis of the asymptotic Green's function. Because the incident field and Green's function are not updated in deriving the first order term, the "primary reflection" thus obtained is similar to a "distorted primary reflection", following the term of "distorted Born approximation" in the physics literature. The real primary reflection, which is the first term in the De Wolf series, includes some multiple scattering terms in the GBS.

3. A DUAL-DOMAIN THIN-SLAB FORMULATION FOR ONE-RETURN (MFSB) SYNTHETICS

The physical meaning of the one-return approximation is shown in Fig. 1. First a preferred direction needs to be selected for the forward/backward scattering decomposition. In Fig. 1 we choose the z-direction as the preferred one. We see that all the solid wave paths have only one backscattering with respect to the z-direction, and therefore will be included in the simulation using the one-return approximation. On the other hand, the dashed line has three backscattering points and cannot be modeled by the one-return approximation. Figure 2 illustrates schematically the numerical implementation of one-return approximation using a thin-slab marching algorithm. The heterogeneous medium is sliced into numerous thin-slabs. Within each thin-slab, the local Born approximation can be utilized for the calculation of the forward and backward scatterings. The forescattered field is used to update the incident field (the forward propagated wavefield u_f) and the backscattered field u_b is stored for later use. This procedure is iteratively done slab by slab, until the end of the model is reached. At the bottom of the model, u_f will be the primary transmitted wave. To get the primary reflected waves, the marching will be done in the reversed direction, i.e. from the bottom to the top of the model. At each thin-slab, the stored backscattered wavefield will be picked up and propagated to the receiving point with the forescattering updated Green's function G_f. In the next few sections, the derivation and formulations of the thin-slab and screen approximations for scalar, acoustic and elastic waves will be given.

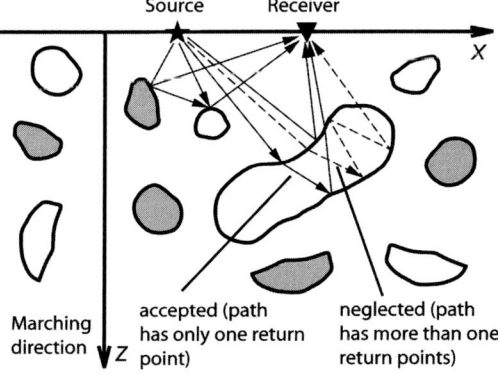

FIG. 1. Sketch showing the meaning of the De Wolf (one-return) approximation.

3.1. The Case of Scalar Media

Based on the De Wolf approximation (33) and the marching algorithm shown in Fig. 2, at each step we need calculate the fore- and backscattered wave fields caused by the thin-slab. We can choose the thickness of the thin-slab to be thin enough so that the local Born approximation can be applied to the calculation. Replacing u^0 in the Born approximation (6) with the updated local incident field u^f, the scalar pressure field at an observation point $\mathbf{x}^*$ can be expressed as

$$p(\mathbf{x}^*) = p^f(\mathbf{x}^*) + k^2 \int_V d^3\mathbf{x}\, g(\mathbf{x}^*; \mathbf{x}) F(\mathbf{x}) p^f(\mathbf{x}), \tag{35}$$

where p^f is the local incident pressure field and g, the Green's function in the thin-slab. $F(\mathbf{x})$ is the perturbation function (scattering potential),

$$F(\mathbf{x}) = \frac{c_0^2}{c^2(\mathbf{x})} - 1 = \frac{s^2(\mathbf{x}) - s_0^2}{s_0^2}, \tag{36}$$

with $s = 1/c$ as the slowness, where c is the velocity. The second term in the right-hand side of (35) is the scattered field and the volume integration is over the thin-

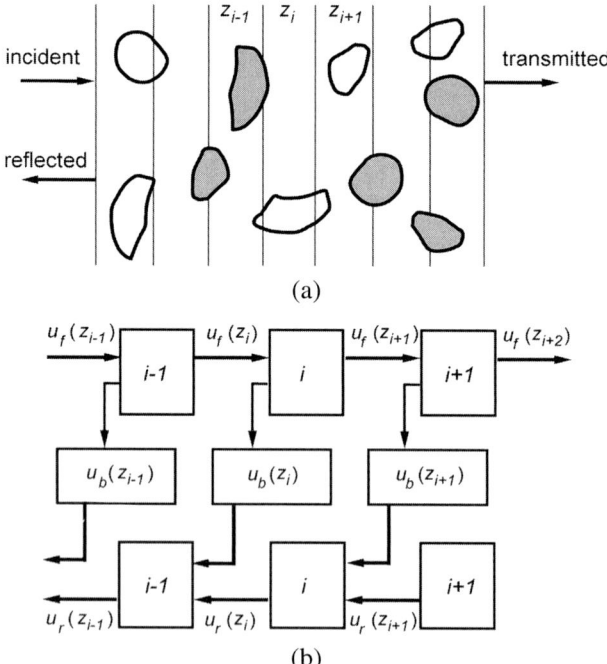

FIG. 2. Schematic illustration of the thin-slab method for implementing the one-return approximation: (a) original medium is sliced into thin-slabs, (b) iterative procedure of transmitted and reflected wave calculations by the one-return approximation.

slab. Choosing the z-direction as the main propagation direction, the scattered field at a receiving point $(\mathbf{x}_T^*, z^*)$ can be calculated as

$$P(\mathbf{x}_T^*, z^*) = k^2 \int_V d^3\mathbf{x}\, g(\mathbf{x}_T^*, z^*; \mathbf{x}) F(\mathbf{x}) p^f(\mathbf{x}), \tag{37}$$

where $\mathbf{x}_T^*$ is the horizontal position in the receiver plane at depth z^*. In the derivations of the next few sections, we use a thin-slab geometry as illustrated in Fig. 3. Within the slab the Green's function is assumed to be a constant medium Green's function g^0. Set z' and z_1 as the slab entrance (top) and exit (bottom), respectively, and Fourier-transform equation (37) with respect to $\mathbf{x}_T^*$, resulting in

$$P(\mathbf{K}_T, z^*) = k^2 \int_{z'}^{z_1} dz \iint d^2\mathbf{x}_T\, g^0(\mathbf{K}_T, z^*; \mathbf{x}) F(\mathbf{x}) p^f(\mathbf{x}), \tag{38}$$

where

$$g^0(\mathbf{K}_T, z^*; \mathbf{x}_T, z) = \frac{i}{2\gamma} e^{i\gamma|z^*-z|} e^{-i\mathbf{K}_T \cdot \mathbf{x}_T} \tag{39}$$

is the wavenumber-domain Green's function in constant media (see Berkhout, 1987; Wu, 1994), and

$$\gamma = \sqrt{k^2 - \mathbf{K}_T^2} \tag{40}$$

is the vertical wavenumber (or the propagating wavenumber). Substituting (39) into (38) yields

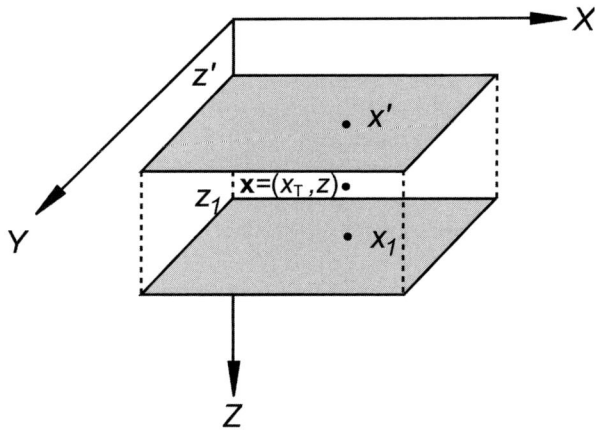

FIG. 3. Geometry for the derivation of the thin-slab method.

$$P(\mathbf{K}_T, z^*) = \frac{i}{2\gamma}k^2 \int_{z'}^{z_1} dz\, e^{i\gamma|z^*-z|} \iint d^2x_T\, e^{-i\mathbf{K}_T\cdot\mathbf{x}_T}$$
$$\times \left[F(\mathbf{x}_T, z)p^f(\mathbf{x}_T, z)\right]. \tag{41}$$

Note that the two dimensional inner integral is a 2-D Fourier transform. Therefore, the dual-domain technique can be used to implement (41). If $\mathbf{x}^* = \mathbf{x}_1$, (41) is used to calculate the forward scattered field and update the incident field (transmitted waves) (35); on the other hand, if $\mathbf{x}^* = \mathbf{x}'$, (41) is for the calculation of backscattered waves. In the case that only forward propagation is concerned, the iterative implementation of (35) and (41) composes a *one-way propagator*. The derivation of (35) and (41) is based on the local Born approximation, however, the implementation in dual domains is similar to the classical phase-screen approach of a one-way propagator. It is known that the Born approximation is basically a low-frequency approximation, and has severe phase errors for strong contrast and high-frequencies. In order to have better phase accuracy, which is important for imaging (migration), certain high-frequency asymptotic phase-matching has been applied to the local Born solution such that the travel time of the solution match exactly the geometric-optical (ray) travel time in the forward direction. Even a zero-order matching leads to a solution better than the classic phase-screen method, i.e. the spit-step Fourier method (e.g. Stoffa *et al.*, 1990). The method was originally called the "pseudo-screen" method (Wu and De Hoop, 1996; Huang and Wu, 1996; Jin *et al.*, 1998, 1999) to distinguish the new form of screen propagator from the classic phase-screen propagator. The phase-screen has operations only in the space domain so that the phase correction is accurate only for small-angle waves; while the pseudo-screen has operations in both the space and wavenumber domains to improve the accuracy for large-angle waves. The asymptotic phase-matching method used by Jin and Wu (1999) and Jin *et al.* (1998, 1999, 2002) in the hybrid pseudo-screen propagator applies a wavenumber filter in the form of continued fraction expansion and can improve the large-angle wave response significantly. In the method, the wide-angle correction is implemented with an implicit finite difference scheme and the expansion coefficients are optimized by phase-matching. The pseudo-screen propagator belongs to a more general category of generalized screen propagators (GSP) (Wu, 1994, 1996; Xie and Wu, 1998; De Hoop *et al.*, 2000; Le Rousseau and De Hoop, 2001). The hybrid GSP with finite difference wide-angle correction is similar to the Fourier-finite difference (FFD) method (Ristow and Ruhl, 1994).

3.2. The Case of Acoustic Media

For a linear isotropic acoustic medium, the wave equation in frequency domain is

$$\nabla\cdot\frac{1}{\rho}\nabla p + \frac{\omega^2}{\kappa}p = 0, \tag{42}$$

where p is the pressure field, ρ and κ are the density and bulk module of the medium, respectively. Assuming ρ_0 and κ_0 as the parameters of the background medium, equation (42) can be written as

$$\frac{1}{\rho_0}\nabla^2 p + \frac{\omega^2}{\kappa_0} p = -\left[\omega^2\left(\frac{1}{\kappa} - \frac{1}{\kappa_0}\right)p + \nabla\cdot\left(\frac{1}{\rho} - \frac{1}{\rho_0}\right)\nabla p\right], \tag{43}$$

or

$$(\nabla^2 + k^2)p(\mathbf{x}) = -k^2 F(\mathbf{x})p(\mathbf{x}), \tag{44}$$

which is the same as the case of scalar media except

$$F(\mathbf{x}) = \delta_\kappa(\mathbf{x}) + \frac{1}{k^2}\nabla\cdot\delta_\rho\nabla, \tag{45}$$

where $F(\mathbf{x})$ is an operator instead of a scalar function, with

$$\delta_\kappa(\mathbf{x}) = \frac{\kappa_0}{\kappa(\mathbf{x})} - 1, \tag{46}$$

and

$$\delta_\rho(\mathbf{x}) = \frac{\rho_0}{\rho(\mathbf{x})} - 1. \tag{47}$$

If ρ is kept constant ($\rho = \rho_0$), then $\delta_\kappa = c_0^2/c^2 - 1$, going back to the scalar medium case. Following derivation of Eq. (41), using the thin-slab geometry and the dual-domain expression for the acoustic media, the scattered pressure field at the receiving depth z^* can be written explicitly as

$$P(\mathbf{K}_T, z^*) = \frac{i}{2\gamma}k^2 \int_{z'}^{z_1} dz\, e^{i\gamma|z^*-z|}$$
$$\times \left\{\iint d^2\mathbf{x}_T\, e^{-i\mathbf{K}_T\cdot\mathbf{x}_T}\left[\delta_\kappa(\mathbf{x}_T, z)p^f(\mathbf{x}_T, z)\right]\right.$$
$$\left. + \frac{i}{k}\hat{\mathbf{k}}\cdot\iint d^2\mathbf{x}_T\, e^{-i\mathbf{K}_T\cdot\mathbf{x}_T}\left[\delta_\rho(\mathbf{x}_T, z)\nabla p^f(\mathbf{x}_T, z)\right]\right\}, \tag{48}$$

where

$$\hat{k} = \frac{1}{k}(\mathbf{K}_T, k_z), \tag{49}$$

and $k_z = \pm\gamma$ for forescattering and backscattering, respectively. The incident field $p^f(\mathbf{x}_T, z)$ and its gradient $\nabla p^f(\mathbf{x}_T, z)$ at depth z can be calculated from the field $p^0(\mathbf{x}'_T, z')$ at the slab entrance

$$p^f(\mathbf{x}) = p^f(\mathbf{x}_T, z) = \frac{1}{4\pi^2}\iint d^2\mathbf{K}'_T\, p^0(\mathbf{K}'_T)e^{i\gamma'(z-z')}e^{i\mathbf{K}'_T\cdot\mathbf{x}_T}, \tag{50}$$

and

$$\nabla p^f(\mathbf{x}) = \frac{ik}{4\pi^2} \iint d^2\mathbf{K}'_T \, \hat{k}' p^0(\mathbf{K}'_T) e^{i\gamma'(z-z')} e^{i\mathbf{K}'_T \cdot \mathbf{x}_T}, \quad (51)$$

where

$$\hat{k}' = \frac{1}{k}(\mathbf{K}'_T, \gamma'). \quad (52)$$

Numerical Tests for Reflected Waves in Acoustic Media

In the dual-domain thin-slab formulation, no small-angle approximation is made. The only approximation is the smallness of perturbations within each thin-slab so that the background Green's function can be applied and the incident waves can be treated as propagated in the background media. In the limiting case, the thickness of the thin-slab can be shrunk to a one grid step. In that case, the only approximation involved is the dual-domain implementation (or split step algorithm). Numerical tests on the wide-angle version (thin-slab approximation) of the acoustic one-return method showed that the reflection coefficients calculated from the synthetic acoustic records agree well with the theoretical predictions when the incidence angles are smaller than the critical angle in the case of high-velocity layer reflection, and to approximately 70° in the case of low-velocity layer reflection (Wu and Huang, 1992, 1995; Wu, 1996).

3.3. The Case of Elastic Media

The equation of motion in a linear, heterogeneous elastic medium can be written as (Aki and Richards, 1980)

$$-\omega^2 \rho(\mathbf{x})\mathbf{u}(\mathbf{x}) = \nabla \cdot \sigma(\mathbf{x}), \quad (53)$$

where $\mathbf{u}$ is the displacement vector, σ is the stress tensor (dyadic) and ρ is the density of the medium. Here we assume no body force exists in the medium. We know the stress-displacement relation

$$\sigma(\mathbf{x}) = \mathbf{c}(\mathbf{x}) : \varepsilon(\mathbf{x}) = \frac{1}{2}\mathbf{c} : (\nabla \mathbf{u} + \mathbf{u}\nabla), \quad (54)$$

where $\mathbf{c}$ is the elastic constant tensor of the medium, ε is the strain field, $\mathbf{u}\nabla$ stands for the transpose of $\nabla \mathbf{u}$, and ":" stands for double scalar product of tensors defined through $(\mathbf{ab}) : (\mathbf{cd}) = (\mathbf{b} \cdot \mathbf{c})(\mathbf{a} \cdot \mathbf{d})$. Equation (53) can be written as a wave equation of the displacement field:

$$-\omega^2 \rho(\mathbf{x})\mathbf{u}(\mathbf{x}) = \nabla \cdot \left[\frac{1}{2}\mathbf{c} : (\nabla \mathbf{u} + \mathbf{u}\nabla)\right]. \quad (55)$$

If the parameters of the elastic medium and the total wave field can be decomposed as

$$\rho(\mathbf{x}) = \rho_0 + \delta\rho(\mathbf{x}), \qquad \mathbf{c}(\mathbf{x}) = \mathbf{c}_0 + \delta\mathbf{c}(\mathbf{x}),$$
$$\mathbf{u}(\mathbf{x}) = \mathbf{u}^0(\mathbf{x}) + \mathbf{U}(\mathbf{x}), \tag{56}$$

where ρ_0 and $\mathbf{c}_0$ are the parameters of the background medium, $\delta\rho$ and $\delta\mathbf{c}$ are the corresponding perturbations, $\mathbf{u}^0$ is the incident field and $\mathbf{U}$ is the scattered field, then (55) can be rewritten as:

$$-\omega^2 \rho_0 \mathbf{U} - \nabla \cdot \left[\frac{1}{2} \mathbf{c}_0 : (\nabla \mathbf{U} + \mathbf{U}\nabla) \right] = \mathbf{F}, \tag{57}$$

where

$$\mathbf{F} = \omega^2 \delta\rho \mathbf{u} + \nabla \cdot [\delta\mathbf{c} : \varepsilon]$$

is the equivalent body force due to scattering.

Similar to (35), we can write the equation of the De Wolf approximation for elastic displacement field as:

$$\mathbf{u}(\mathbf{x}_T^*, z^*) = \mathbf{u}^f(\mathbf{x}_T^*, z^*) + \int_V d^3\mathbf{x} \left\{ \delta\rho\omega^2 \mathbf{u}^f(\mathbf{x}_T, z) \right.$$
$$\left. + \nabla \cdot [\delta\mathbf{c} : \varepsilon^f(\mathbf{x}_T, z)] \right\} \cdot \mathbf{G}^f(\mathbf{x}_T^*, z^*; \mathbf{x}_T, z). \tag{58}$$

Following the derivation of Eq. (38), we can express the scattered displacement field for a thin-slab in the horizontal wavenumber domain using local Born approximation as:

$$\mathbf{U}(\mathbf{K}_T, z^*) = \int_{z'}^{z_1} dz \iint d^2\mathbf{x}_T \left\{ \delta\rho\omega^2 \mathbf{u}^f(\mathbf{x}_T, z) \right.$$
$$\left. + \nabla \cdot [\delta\mathbf{c} : \varepsilon^f(\mathbf{x}_T, z)] \right\} \cdot \mathbf{G}^0(\mathbf{K}_T, z^*; \mathbf{x}_T, z), \tag{59}$$

where

$$\mathbf{G}^0(z^*, \mathbf{K}_T; z, \mathbf{x}_T)$$
$$= \frac{ik_\alpha^2}{2\rho_0\omega^2} \hat{\mathbf{k}}_\alpha \hat{\mathbf{k}}_\alpha \frac{1}{\gamma_\alpha} e^{i\mathbf{k}_\alpha \cdot \mathbf{r}} + \frac{ik_\beta^2}{2\rho_0\omega^2} (\mathbf{I} - \hat{\mathbf{k}}_\beta \hat{\mathbf{k}}_\beta) \frac{1}{\gamma_\beta} e^{i\mathbf{k}_\beta \cdot \mathbf{r}}, \tag{60}$$

where $\mathbf{I}$ is the unit dyadic, and

$$\gamma_\alpha = \sqrt{k_\alpha^2 - K_T^2},$$
$$\gamma_\beta = \sqrt{k_\beta^2 - K_T^2}, \tag{61}$$

where $k_\alpha = \omega/\alpha_0$ and $k_\beta = \omega/\beta_0$ are P and S wavenumbers with α_0 and β_0 as the P and S wave background velocities of the thin-slab, respectively. For isotropic media,

$$\delta \mathbf{c}(\mathbf{x}) : \varepsilon(\mathbf{x}) = \delta\lambda(\mathbf{x})|\varepsilon|\mathbf{I} + 2\delta\mu(\mathbf{x})\varepsilon(\mathbf{x}). \tag{62}$$

Substituting (60) into (59), we can derive the dual-domain expressions for scattered displacement fields in isotropic elastic media.

For P to P scattering:

$$\begin{aligned}
\mathbf{U}^{PP}&(\mathbf{K}_T, z^*) \\
&= \frac{ik_\alpha^2}{2\gamma_\alpha} \int_{z'}^{z_1} dz\, e^{ik_z^\alpha(z^*-z)} \left\{ \hat{k}_\alpha \hat{k}_\alpha \cdot \iint d^2\mathbf{x}_T\, e^{-i\mathbf{K}_T \cdot \mathbf{x}_T} \frac{\delta\rho(\mathbf{x}_T, z)}{\rho} \mathbf{u}_\alpha^f(\mathbf{x}_T, z) \right. \\
&\quad - \hat{k}_\alpha \iint d^2\mathbf{x}_T\, e^{-i\mathbf{K}_T \cdot \mathbf{x}_T} \frac{\delta\lambda(\mathbf{x}_T, z)}{\lambda + 2\mu} \frac{1}{ik_\alpha} \nabla \cdot \mathbf{u}_\alpha^f(\mathbf{x}_T, z) \\
&\quad \left. - \hat{k}_\alpha(\hat{k}_\alpha \hat{k}_\alpha) : \iint d^2\mathbf{x}_T\, e^{-i\mathbf{K}_T \cdot \mathbf{x}_T} \frac{\delta\mu(\mathbf{x}_T, z)}{\lambda + 2\mu} \frac{1}{ik_\alpha} \varepsilon_\alpha^f(\mathbf{x}_T, z) \right\}, \tag{63}
\end{aligned}$$

with $k_z^\alpha = +\gamma_\alpha$ for forescattering and $k_z^\alpha = -\gamma_\alpha$ for backscattering, and $\hat{k}_\alpha = (\mathbf{K}_T, k_z^\alpha)/k_\alpha$. Note that we replaced ρ_0, λ_0, μ_0 in denominators by $\rho = \rho_0 + \delta\rho, \lambda = \lambda_0 + \delta\lambda$ and $\mu = \mu_0 + \delta\mu$. This replacement is the result of asymptotic matching between the Born approximation for large-angle scattering and the $h - f$ asymptotic travel-time (phase) for forward propagation. It is proved (Wu and Wu, 2003a, 2003b) that with this replacement (asymptotic matching), the phase-shift in the exact forward direction is accurate and the phase error for small angles is reduced compared with the Born approximation. In the meanwhile, the phase error for large angle scattering is much smaller than that of the phase-screen approximation.

In (63) $\mathbf{u}_\alpha^f(\mathbf{x}_T, z)$, $\nabla \cdot \mathbf{u}_\alpha^f(\mathbf{x}_T, z)$ and $\varepsilon_\alpha^f(\mathbf{x}_T, z)$ can be calculated by:

$$\mathbf{u}_\alpha^f(\mathbf{x}_T, z) = \frac{1}{4\pi} \iint d^2\mathbf{K}_T'\, e^{i\mathbf{K}_T' \cdot \mathbf{x}_T} u_\alpha{}^0(\mathbf{K}_T') e^{i\gamma_\alpha'(z-z')},$$

$$\frac{1}{ik_\alpha} \nabla \cdot \mathbf{u}_\alpha^f(\mathbf{x}_T, z) = \frac{1}{4\pi} \iint d^2\mathbf{K}_T'\, e^{i\mathbf{K}_T' \cdot \mathbf{x}_T} \hat{k}_\alpha' \cdot \mathbf{u}_\alpha^0(\mathbf{K}_T') e^{i\gamma_\alpha'(z-z')},$$

$$\begin{aligned}
\frac{1}{ik_\alpha} &\varepsilon_\alpha^f(\mathbf{x}_T, z) \\
&= \frac{1}{4\pi} \iint d^2\mathbf{K}_T'\, e^{i\mathbf{K}_T' \cdot \mathbf{x}_T} \frac{1}{2}[\hat{k}_\alpha' \mathbf{u}_\alpha^0(\mathbf{K}_T') + \hat{k}_\alpha' \mathbf{u}_\alpha^0(\mathbf{K}_T')] e^{i\gamma_\alpha'(z-z')} \\
&= \frac{1}{4\pi} \iint d^2\mathbf{K}_T'\, e^{i\mathbf{K}_T' \cdot \mathbf{x}_T} \hat{k}_\alpha' \hat{k}_\alpha' u_\alpha^0(\mathbf{K}_T') e^{i\gamma_\alpha'(z-z')}, \tag{64}
\end{aligned}$$

where $u_\alpha^0(\mathbf{K}_T') = |\mathbf{u}_\alpha^0(\mathbf{K}_T')|$ and $\hat{k}_\alpha' = (\mathbf{K}_T', \gamma_\alpha')/k_\alpha$.

For P to S scattering:

$$\mathbf{U}^{PS}(\mathbf{K}_T, z^*)$$
$$= \frac{ik_\beta^2}{2\gamma_\beta} \int_{z'}^{z_1} dz\, e^{ik_z^\beta(z^*-z)}$$
$$\times \left\{ (\mathbf{I} - \hat{k}_\beta \hat{k}_\beta) \cdot \iint d^2\mathbf{x}_T\, e^{-i\mathbf{K}_T \cdot \mathbf{x}_T} \frac{\delta\rho(\mathbf{x}_T, z)}{\rho} \mathbf{u}_\alpha^f(\mathbf{x}_T, z) \right.$$
$$\left. - (\mathbf{I} - \hat{k}_\beta \hat{k}_\beta) \cdot \left[\hat{k}_\beta \cdot \iint d^2\mathbf{x}_T\, e^{-i\mathbf{K}_T \cdot \mathbf{x}_T} 2 \frac{\delta\mu(\mathbf{x}_T, z)}{\mu} \frac{1}{ik_\beta} \varepsilon_\alpha^f(\mathbf{x}_T, z) \right] \right\}, \quad (65)$$

where $\hat{k}_\beta = (\mathbf{K}_T, k_z^\beta)/k_\beta$.

For S to P scattering:

$$\mathbf{U}^{SP}(\mathbf{K}_T, z^*)$$
$$= \frac{ik_\alpha^2}{2\gamma_\alpha} \int_{z'}^{z_1} dz\, e^{ik_z^\alpha(z^*-z)} \left\{ \hat{k}_\alpha \hat{k}_\alpha \cdot \iint d^2\mathbf{x}_T\, e^{-i\mathbf{K}_T \cdot \mathbf{x}_T} \frac{\delta\rho(\mathbf{x}_T, z)}{\rho} \mathbf{u}_\beta^f(\mathbf{x}_T, z) \right.$$
$$\left. - \left(\frac{k_\alpha}{k_\beta} \right) \hat{k}_\alpha (\hat{k}_\alpha \hat{k}_\alpha) : \iint d^2\mathbf{x}_T\, e^{-i\mathbf{K}_T \cdot \mathbf{x}_T} 2 \frac{\delta\mu(\mathbf{x}_T, z)}{\mu} \frac{1}{ik_\beta} \varepsilon_\beta^f(\mathbf{x}_T, z) \right\}. \quad (66)$$

For S to S scattering:

$$\mathbf{U}^{SS}(\mathbf{K}_T, z^*)$$
$$= \frac{ik_\beta^2}{2\gamma_\beta} \int_{z'}^{z_1} dz\, e^{ik_z^\beta(z^*-z)}$$
$$\times \left\{ (\mathbf{I} - \hat{k}_\beta \hat{k}_\beta) \cdot \iint d^2\mathbf{x}_T\, e^{-i\mathbf{K}_T \cdot \mathbf{x}_T} \frac{\delta\rho(\mathbf{x}_T, z)}{\rho} \mathbf{u}_\beta^f(\mathbf{x}_T, z) \right.$$
$$\left. - (\mathbf{I} - \hat{k}_\beta \hat{k}_\beta) \cdot \left[\hat{k}_\beta \cdot \iint d^2\mathbf{x}_T\, e^{-i\mathbf{K}_T \cdot \mathbf{x}_T} 2 \frac{\delta\mu(\mathbf{x}_T, z)}{\mu} \frac{1}{ik_\beta} \varepsilon_\beta^f(\mathbf{x}_T, z) \right] \right\}. \quad (67)$$

In Eqs. (66) and (67), $\mathbf{u}_\beta^f(\mathbf{x}_T, z)$ and $\varepsilon_\beta^f(\mathbf{x}_T, z)$ can be calculated by

$$\mathbf{u}_\beta^f(\mathbf{x}_T, z) = \frac{1}{4\pi^2} \iint d^2\mathbf{K}_T'\, e^{i\mathbf{K}_T' \cdot \mathbf{x}_T} \mathbf{u}_\beta^0(\mathbf{K}_T') e^{i\gamma_\beta'(z-z')},$$

$$\frac{1}{ik_\beta} \varepsilon_\beta^f(\mathbf{x}_T, z)$$
$$= \frac{1}{4\pi^2} \iint d^2\mathbf{K}_T'\, e^{i\mathbf{K}_T' \cdot \mathbf{x}_T} \frac{1}{2} [\hat{k}_\beta' \mathbf{u}_\beta^0(\mathbf{K}_T') + \mathbf{u}_\beta^0(\mathbf{K}_T') \hat{k}_\beta'] e^{i\gamma_\beta'(z-z')}, \quad (68)$$

where $\hat{k}_\beta' = (\mathbf{K}_T', \gamma_\beta')/k_\beta$.

3.4. Implementation Procedure of the One-Return Simulation

Under the MFSB approximation we can update the total field with a marching algorithm in the forward direction. We can slice the whole medium into thin-slabs perpendicular to the propagation direction. Weak scattering condition holds for each thin-slab. For each step forward, the forward and backward scattered fields by a thin-slab between z' and z_1 are calculated. The forescattered field is added to the incident field so that the updated field becomes the incident field for the next thin-slab. The procedure for acoustic and elastic media can be summarized as follows (see Fig. 2). The simplification for the case of scalar media is straightforward.

1. Fourier transform (FT) the incident fields into wavenumber domain at the entrance of each thin-slab.
2. Free propagate in wavenumber domain and calculate the primary fields and its gradients (including strain fields for the case of elastic media) within the slab.
3. Inverse FT these primary fields and its gradients into space domain, and then interact with the medium perturbations: calculation of the distorted fields.
4. FT the distorted fields into wavenumber domain and perform the divergence (and curl, in the case of elastic media) operations to get the backscattered fields. Sum up the scattered fields by all perturbation parameters and multiply it with a weighting factor $i/2\gamma$, then propagate back to the entrance of the slab.
5. Calculate the forescattered field at the slab exit and add to the primary field to form the total field as the incident field at the entrance of the next thin-slab.
6. Continue the procedure iteratively until the bottom of the model is reached.
7. Propagate the backscattered waves from the bottom up to the surface and sum up the contributions of all the thin-slabs during the propagation.

Note that medium-wave interaction, for the case of acoustic waves, involves vector operations and needs 3 pairs of fast Fourier transforms (FFTs) for each step, while for the case of elastic waves, tensor (strain fields) operations are involved. Due to the symmetric properties of the strain tensors, there are only 6 independent components for each tensor. From (63)–(68) we see that many pairs of FFTs are required for each step and therefore the computation for elastic wave scattering is rather intensive.

4. Fast Algorithm of the Elastic Thin-Slab Propagator and Some Practical Issues

4.1. Fast Implementation in Dual Domains

From Eqs. (63) to (68), we see that the leading-order interactions between incident fields and heterogeneities are expressed in three-dimensional volume integrals. Also the scattered and incident wavenumbers are coupled with each other. So the computation of these equations is still intensive. In this section, the parts of the integration over z in the equations are analytically estimated. Assume that the slab for each marching step is thin enough that the parameters (velocity and density) can be approximately taken as invariant along z, the integration with respect to z in Eq. (63) can be calculated as

$$\int_{z'}^{z_1} dz\, e^{ik_z^\alpha(z^*-z)+i\gamma_\alpha'(z-z')}$$

$$= \begin{cases} \Delta z\, e^{i(\gamma_\alpha+\gamma_\alpha')\Delta z/2} \operatorname{sinc}\left[\frac{\gamma_\alpha-\gamma_\alpha'}{2}\Delta z\right] & \text{for forescattering } (z^*=z_1), \\ \Delta z\, e^{i(\gamma_\alpha+\gamma_\alpha')\Delta z/2} \operatorname{sinc}\left[\frac{\gamma_\alpha+\gamma_\alpha'}{2}\Delta z\right] & \text{for backscattering } (z^*=z'). \end{cases} \quad (69)$$

We see that the integration over z has been done analytically; however, γ_α and γ_α' are still coupled, which prevents the fast computation of the thin-slab method. To decouple γ_α and γ_α', we neglect the angular variation of amplitude factors but keep the phase information untouched by taking the approximation $\gamma_\alpha = \gamma_\alpha' = k_\alpha$ for the amplitude factors in Eq. (69). This assumption is valid for the case where the small-angle scattering is dominant, and therefore the direction of the scattered waves are not far from the incident direction. Under this approximation, Eq. (69) becomes

$$\int_{z'}^{z_1} dz\, e^{ik_z^\alpha(z^*-z)+i\gamma_\alpha'(z-z')}$$

$$\approx \begin{cases} \Delta z\, e^{i(\gamma_\alpha+\gamma_\alpha')\Delta z/2} & \text{for forescattering } (z^*=z_1), \\ \Delta z\, e^{i(\gamma_\alpha+\gamma_\alpha')\Delta z/2} \operatorname{sinc}(k_\alpha \Delta z) & \text{for backscattering } (z^*=z'). \end{cases} \quad (70)$$

For the scattered fields P–S, S–P and S–S, similar approximations can be obtained as follows. For P–S or S–P scattering,

$$\int_{z'}^{z_1} dz\, e^{ik_z^\beta(z^*-z)+i\gamma_\alpha'(z-z')}$$

$$\approx \begin{cases} \Delta z\, e^{i(\gamma_\alpha'+\gamma_\beta)\Delta z/2} \operatorname{sinc}\left[(k_\alpha - k_\beta)\Delta z/2\right] & \text{for forescattering } (z^*=z_1), \\ \Delta z\, e^{i(\gamma_\alpha'+\gamma_\beta)\Delta z/2} \operatorname{sinc}\left[(k_\alpha + k_\beta)\Delta z/2\right] & \text{for backscattering } (z^*=z'). \end{cases} \quad (71)$$

For S–S scattering,

$$\int_{z'}^{z_1} dz\, e^{ik_z^\beta(z^*-z)+i\gamma_\beta'(z-z')}$$

$$\approx \begin{cases} \Delta z\, e^{i(\gamma_\beta+\gamma_\beta')\Delta z/2} & \text{for forescattering } (z^* = z_1), \\ \Delta z\, e^{i(\gamma_\beta+\gamma_\beta')\Delta z/2} \operatorname{sinc}(k_\beta \Delta z) & \text{for backscattering } (z^* = z'). \end{cases} \quad (72)$$

After integration over z, the integration over transverse plane $\mathbf{x}_T$ in Eqs. (63)–(68) can be carried out by the FFT. In order to further expedite the computation, we can group the scattered field equations (63) to (68) into $\mathbf{U}^P(\mathbf{K}_T, z^*) = \mathbf{U}^{PP}(\mathbf{K}_T, z^*) + \mathbf{U}^{SP}(\mathbf{K}_T, z^*)$, and $\mathbf{U}^S(\mathbf{K}_T, z^*) = \mathbf{U}^{PS}(\mathbf{K}_T, z^*) + \mathbf{U}^{SS}(\mathbf{K}_T, z^*)$, i.e.,

$$\mathbf{U}^P(\mathbf{K}_T, z^*)$$

$$= \frac{ik_\alpha^2}{2\gamma_\alpha} e^{i\gamma_\alpha \Delta z/2} \Delta z \hat{k}_\alpha$$

$$\times \left\{ \hat{k}_\alpha \cdot \iint d^2\mathbf{x}_T\, e^{-i\mathbf{K}_T \cdot \mathbf{x}_T} \frac{\delta\rho(\mathbf{x}_T)}{\rho} \left[\eta^{PP} \mathbf{u}_\alpha^f(\mathbf{x}_T) + \eta^{SP} \mathbf{u}_\beta^f(\mathbf{x}_T) \right] \right.$$

$$- \iint d^2\mathbf{x}_T\, e^{-i\mathbf{K}_T \cdot \mathbf{x}_T} \frac{\delta\lambda(\mathbf{x}_T)}{\lambda+2\mu} \frac{1}{ik_\alpha} \nabla \cdot \left[\eta^{PP} \mathbf{u}_\alpha^f(\mathbf{x}_T) \right]$$

$$- (\hat{k}_\alpha \hat{k}_\alpha) : \iint d^2\mathbf{x}_T\, e^{-i\mathbf{K}_T \cdot \mathbf{x}_T} \frac{2\delta\mu(\mathbf{x}_T)}{\lambda+2\mu} \frac{1}{ik_\alpha}$$

$$\left. \times \left[\eta^{PP} \varepsilon_\alpha^f(\mathbf{x}_T) + \eta^{SP} \varepsilon_\beta^f(\mathbf{x}_T) \right] \right\}, \quad (73)$$

$$\mathbf{U}^S(\mathbf{K}_T, z^*)$$

$$= \frac{ik_\beta^2}{2\gamma_\beta} e^{i\gamma_\beta \Delta z/2} \Delta z (\mathbf{I} - \hat{k}_\beta \hat{k}_\beta)$$

$$\times \left\{ \iint d^2\mathbf{x}_T\, e^{-i\mathbf{K}_T \cdot \mathbf{x}_T} \frac{\delta\rho(\mathbf{x}_T)}{\rho} \left[\eta^{PS} \mathbf{u}_\alpha^f(\mathbf{x}_T) + \eta^{SS} \mathbf{u}_\beta^f(\mathbf{x}_T) \right] \right.$$

$$- \hat{k}_\beta \cdot \iint d^2\mathbf{x}_T\, e^{-i\mathbf{K}_T \cdot \mathbf{x}_T} \frac{2\delta\mu(\mathbf{x}_T)}{\mu} \frac{1}{ik_\beta}$$

$$\left. \times \left[\eta^{PS} \varepsilon_\alpha^f(\mathbf{x}_T) + \eta^{SS} \varepsilon_\beta^f(\mathbf{x}_T) \right] \right\}, \quad (74)$$

where z^* ($z^* = z'$ or $z^* = z_1$) indicates the position of the receiver plane. The modulation factors η^{PP}, $\eta^{SP} = \eta^{PS}$ and η^{SS} are

$$\eta^{PP} = \begin{cases} 1 & \text{for forescattering}, \\ \operatorname{sinc}(k_\alpha \Delta z) & \text{for backscattering}, \end{cases} \quad (75)$$

$$\eta^{PS} = \begin{cases} \text{sinc}\left[(k_\alpha - k_\beta)\Delta z/2\right] & \text{for forescattering,} \\ \text{sinc}\left[(k_\alpha + k_\beta)\Delta z/2\right] & \text{for backscattering,} \end{cases} \quad (76)$$

$$\eta^{SS} = \begin{cases} 1 & \text{for forescattering,} \\ \text{sinc}(k_\beta \Delta z) & \text{for backscattering.} \end{cases} \quad (77)$$

Note that the factors $e^{i\gamma_\alpha(z-z')}$ and $e^{i\gamma_\beta(z-z')}$ have been replaced by $e^{i\gamma_\alpha \Delta z/2}$ and $e^{i\gamma_\beta \Delta z/2}$ for calculating the background fields. The phase matching (asymptotic matching) has been applied in Eqs. (73) and (74).

Under the above approximations, the thin-slab formulas may be implemented by the procedures as used in the complex screen method (Wu, 1994; Xie and Wu, 2001). First, the whole medium is sliced into appropriate thin-slabs along the overall propagation direction. Weak scattering conditions hold for each thin-slab and the parameters can be considered invariant within each thin-slab in the preferred propagation direction. Suppose that all incident fields, at the entrance of each thin-slab, are given in the wavenumber domain. The implementation procedures may be summarized as follows:

1. Freely propagate in the wavenumber domain and calculate the primary fields, the divergence of incident P wave, and strains (Eqs. (64) and (68)).
2. Inverse-FFT the primary fields, the divergence and strains into the space domain, and then calculate the distorted fields (the space domain functions before the FT in Eqs. (73)–(74)).
3. Calculate the forescattered fields at the thin-slab exit, and add the forescattered fields to the primary fields to form the total fields as the incident fields for the next thin-slab. In the same time backscattered fields at the thin-slab entrance are also calculated.
4. Continue the procedures 1 to 3 iteratively until the last thin-slab, and the total transmitted fields are obtained at the exit of the last thin-slab.
5. Propagate back the backscattered fields from the last to the first thin-slabs using a similar iterative procedure as step 1 to 3, and sum up all backscattered fields generated by each thin-slab to get the total reflected fields at the surface.

Let us estimate the computation speed. Most calculations involve only fast Fourier transforms. The total computation time can be estimated from the time used in calling FFTs. Taking the complex screen method as a reference, for a 2-D case, the thin-slab method needs 11 inverse FFTs and 11 forward FFTs, while the complex screen method needs 5 inverse FFTs and 7 forward FFTs. For a 3-D case, the thin-slab method needs 19 inverse FFTs and 19 forward FFTs, while the complex screen method needs 7 inverse FFTs and 10 forward FFTs. The thin-slab method takes about twice as much time as the complex screen method does, but is still much faster than the full wave methods. We will see a comparison of computation time between the thin-slab method and finite difference method in Section 4.4.

4.2. Incorporation of Boundary Transmission/Reflection into the One-Return Method

For the one-return elastic wave propagators mentioned above, the boundary transmission and reflection for a thick-layer (much thicker than the wavelength) is formed by the superposition and interference of numerous thin-slab scatterings. This gives the flexibility of the thin-slab propagator to treat arbitrarily irregular interface. However, for a homogeneous thick-layer, the calculation of boundary reflection/transmission (R/T) using the reflectivity method (see Aki and Richards, Chapter 5, 1980) is very efficient and accurate. Therefore the reflectivity method has been incorporated into the thin-slab propagator (Wu and Wu, 2003a, 2003b). In addition to the increase of efficiency, this can also reduce the accumulated errors of the thin-slab propagator when propagating in a thick layer with strong contrast of parameters from the surrounding medium. Since the thin-slab method is based on a perturbation approach, the choice of the background medium is an important issue. To make the perturbation small, the background medium parameters are changed as soon as the waves enter a laterally homogeneous region. Take the model shown on Fig. 10 as an example. On top of the irregular structure (with grey color) is a homogeneous background. When the wave enters the laterally heterogeneous part of the medium, the thin-slab propagator is used for the propagation. We can put in an artificial boundary as shown in the model with bold lines. At that artificial boundary, the reference parameters will jump from the top medium to the grey structure, a -10% jump. The reflection/transmission at that boundary can be calculated using the analytical formulation, such as the Zoeppritz equation, and then the propagation can be done, with an elastic propagator in homogeneous media, to the bottom by one big step. Note that the artificial boundary is added only to facilitate the calculation. The reflected field generated by the reflectivity calculation will be cancelled by the accumulated thin-slab scattered field. Therefore the artificial boundary will not generate spurious arrivals.

The thin-slab propagator is a dual-domain (space-wavenumber domains) approach. In a wavenumber domain, the wavefield can be expressed by a superposition of plane waves. Therefore, incorporating a reflectivity calculation into the thin-slab propagator is straightforward. After plane wave decomposition, the spectra of incident fields can be decomposed into P- and SV- and SH-components, and the R/T formulation can be applied directly to these components. In Section 4.4 numerical examples will be given to show the validity and efficiency of this approach.

4.3. Treatment of Anelasticity: The Q-Factor

Spatially varying quality factors (Q_P and Q_S) can also be incorporated into the thin-slab propagator to study the effects of intrinsic attenuation in visco-elastic media. Spatially varying intrinsic attenuation can cause not only wavefield attenuation but also scattering and frequency-dependent reflections. The thin-slab propagator, with spatially varying Q-factor, is an efficient tool for such study.

For elastic, isotropic media, Lamé constants λ, μ, λ_0, and μ_0 are related with elastic parameters by

$$\lambda = \rho\alpha^2 - 2\rho\beta^2, \qquad \mu = \rho\beta^2, \tag{78}$$

$$\lambda_0 = \rho_0\alpha_0^2 - 2\rho_0\beta_0^2, \qquad \mu_0 = \rho_0\beta_0^2, \tag{79}$$

where α_0, and β_0 are compressional- and shear-wave velocities of background medium. We introduce complex velocities by performing the following transforms:

$$\alpha \to \alpha(1 - i/2Q_P), \qquad \beta \to \beta(1 - i/2Q_S), \tag{80}$$

$$\alpha_0 \to \alpha_0(1 - i/2Q_P^0), \qquad \beta_0 \to \beta_0(1 - i/2Q_S^0), \tag{81}$$

where Q_P^0 and Q_S^0 are compressional- and shear-wave quality factors of background medium. i is the imaginary unit. Once all parameters in Eqs. (80)–(81) are known, we can calculate the perturbations $\delta\lambda$ and $\delta\mu$ using Eqs. (78)–(79). Now λ_0, and μ_0 become complex. As a result, the reference wavenumbers of P- and S-waves also become complex. The heterogeneities of quality factors (Q_P and Q_S) have been included in the complex $\delta\lambda$ and $\delta\mu$. With the above extension, the dual-domain thin-slab propagators can be used to model visco-elastic seismic responses. In Section 6.3 some numerical examples will be given.

4.4. Numerical Tests for the Elastic Thin-Slab Method

4.4.1. Reflection Coefficient Calculations

The following examples show the angle dependence of amplitudes (reflection coefficients) calculated by the thin-slab method. The model used is defined on a 2048 × 200 rectangular grid. The grid spacing in the horizontal direction is 16 m and in the vertical direction is 4 m. A horizontal interface is introduced in the middle of the model. The upper layer has $\alpha = 3.6$ km/s, $\beta = 2.08$ km/s and $\rho = 2$ g/cm^3, which is taken as background medium. The lower layer has different P and S wave velocities relative to the upper layer. A 15 Hz plane P wave (or S wave) is incident on the interface at different angles. To enhance the stability in the calculation of reflection coefficients, we chose 500 samples (displacement amplitudes in the space domain) at the center of the model for both the incident and reflected fields and calculate their means, respectively. Reflection/conversion coefficients are defined by the ratio of the reflected amplitudes to the incident amplitudes at the same receiver plane. Figure 4 shows the reflection/conversion coefficients versus angle of incidence with a perturbation of 10% for both P and S wave velocities. The theoretical reflection/conversion coefficients (dashed lines)

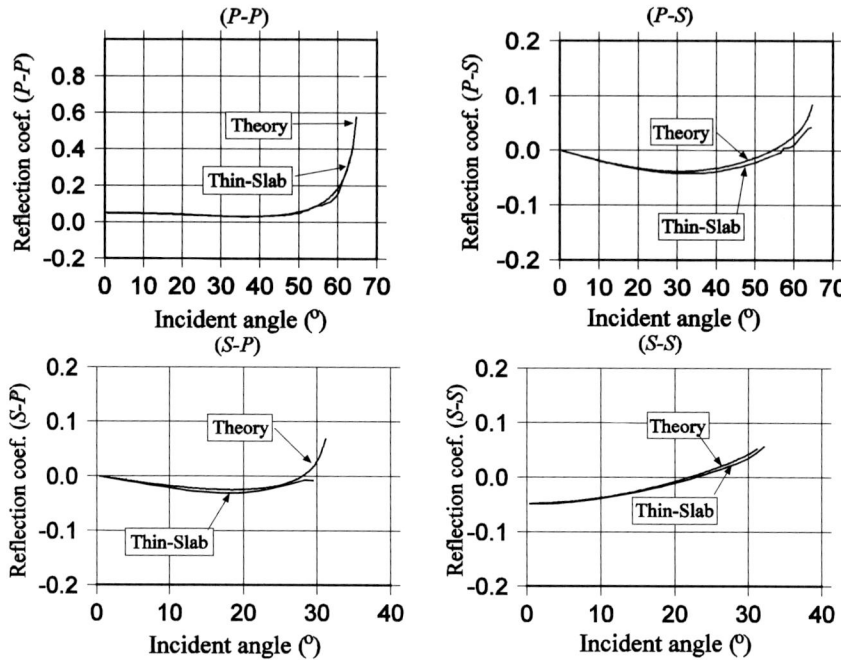

FIG. 4. Comparisons of reflection coefficients calculated by the thin-slab method and the exact solutions. The upper panel corresponds to P wave and the lower panel to S wave incidences respectively. The bottom layer of the model has 10% velocity perturbations for both P and S waves with respect to the top layer.

are also given as references. The upper panel corresponds to P wave incidence and the lower panel, S wave incidence. The angles of incidence of S wave are limited to the critical angle of S–P converted wave. Both results agree well with the theory for a wide range of incident angles (55° for P wave incidence, and near critical angle 32° for S wave incidence).

Figures 5 to 7 show similar results as shown in Fig. 4 but with different perturbations of −10%, 20% and −20%, respectively. Figure 5 corresponds to a negative velocity perturbation. For P wave incidence, no critical angle exists. However, errors occur for large angles of incidence (>65° for P wave incidence). This is limited by the ability of the one-way propagator to handle wide-angle waves. For S wave incidence, both results are in good agreement up to the critical angle. Figure 6 corresponds to a perturbation of 20% for both P and S wave velocities. For a small angle of incidence (40° for P wave incidence and 20° for S wave incidence), the thin-slab results match the theoretical values. Comparing Figs. 6 and 4, we see that the wide-angle capacity of the thin-slab method decreases as perturbations increase. Figure 7 corresponds to a perturbation of −20% for both P and S wave velocities.

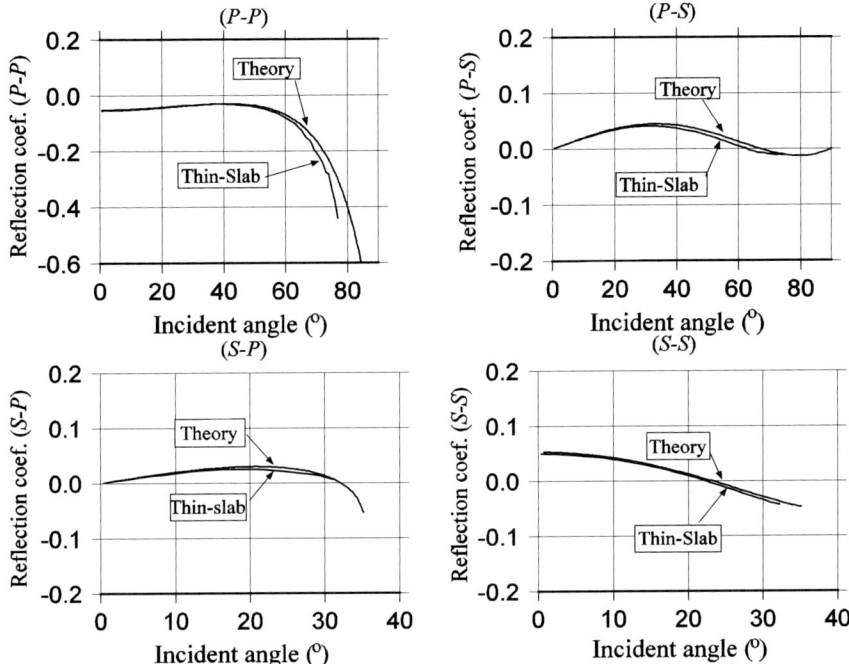

FIG. 5. Comparisons of reflection coefficients calculated by the thin-slab method and the exact solutions. The perturbation of the bottom layer is −10% with respect to the top layer for both P and S wave velocities.

4.4.2. 2-D Synthetic Seismograms

Synthetic seismograms are also generated to further demonstrate the capability of the thin-slab method. Figure 8 shows a 2-D model that is a vertical slice cut from the elastic French model (French, 1974). The model has a strong irregular interface that will generate large-angle reflections and scattering. The parameters of the background medium are taken as $\alpha_0 = 3.6$ km/s, $\beta_0 = 2.08$ km/s and $\rho_0 = 2.2$ g/cm^3. The layer in black color has a perturbation of −20% for both P and S wave velocities. Source and receiver geometry are also shown in the figure. A Ricker wavelet with a dominant frequency of 20 Hz is used. For the thin-slab method, the spacing interval is 8 m in both horizontal and vertical directions. For the finite difference method, the spacing interval is 4 m and time interval is 0.5 ms. The stability criterion is satisfied. The direct arrivals have been properly removed from the finite difference results. Figure 9 shows a comparison of the synthetic seismograms between the thin-slab method and finite difference method. Both amplitudes and arrival times are in good agreement up to large offsets (∼1.4 km) compared to the depth (∼1 km). For this

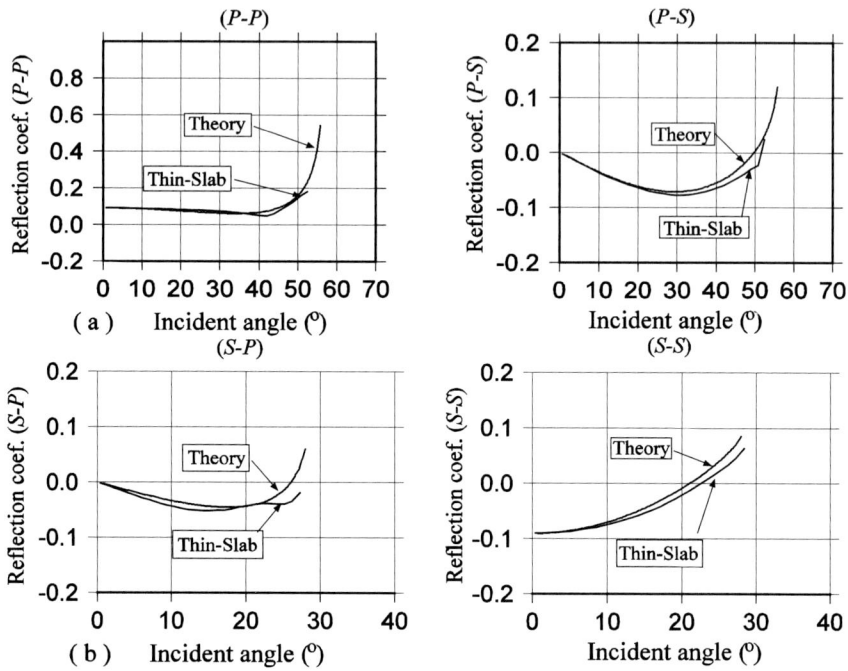

FIG. 6. Comparisons of reflection coefficients calculated by the thin-slab method and the exact solutions. The perturbation of the bottom layer is 20% with respect to the top layer for both P and S wave velocities.

example, the thin-slab method is about 57 times faster than the finite difference method. The factor will increase as the size of model increases, especially for 3-D cases.

4.4.3. 3-D Synthetic Seismograms

Figure 10 shows a 3-D French model. The parameters of the background medium are taken as $\alpha_0 = 3.6$ km/s, $\beta_0 = 2.08$ km/s and $\rho_0 = 2.2$ g/cm^3. The Grey structure has a perturbation of -10% for both P- and S-wave velocities. A Ricker wavelet with a dominant frequency of 10 Hz is used. For the 8th-order 3-D elastic finite-difference method (Yoon and McMechan, 1996), the spacing interval is 20 m. The actual grid size used is 250 × 250 × 250 including 25 grids of absorbing boundary for each face of the model. Time interval used is 0.001 s and 2500 time steps are calculated. It took about 28 hours. For thin-slab method, the spacing interval used is 20 m in transversal plane, which is the same as used for the finite-difference method. But a fine grid size of 5 m is used in propagation direction. We did the same calculation on the same machine using the

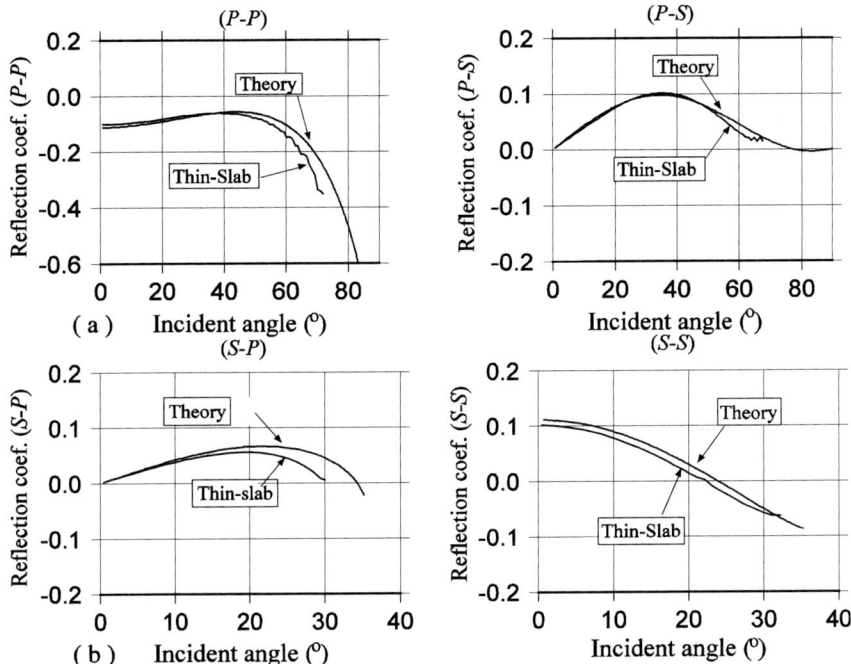

FIG. 7. Comparisons of reflection coefficients calculated by the thin-slab method and the exact solutions. The perturbation of the bottom layer is -20% with respect to the top layer for both P and S wave velocities.

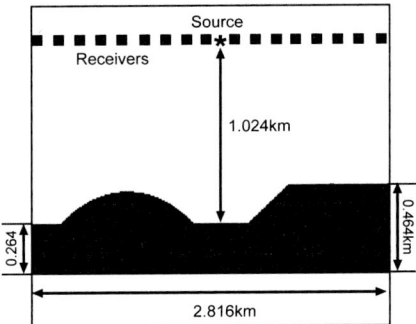

FIG. 8. A 2-D heterogeneous model (French, 1974) with irregular interface used to test the validity and accuracy of the thin-slab method. The background medium has $\alpha_0 = 3.6$ km/s, $\beta_0 = 2.08$ km/s and $\rho_0 = 2.2$ g/cm^3. The layer in black color has a perturbation of -20% for both P and S wave velocities.

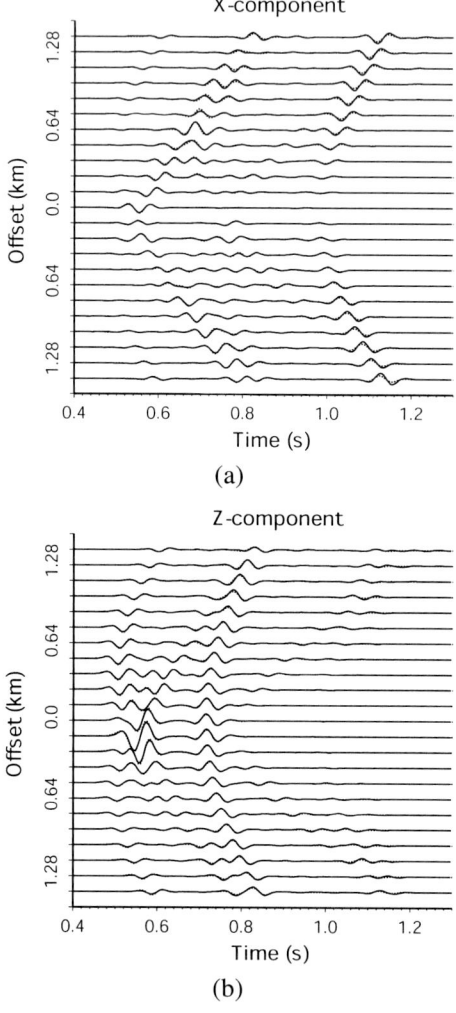

FIG. 9. Comparison of synthetic seismograms calculated by finite difference (solid) and by thin-slab methods (dashed). Curves in (a) and (b) are horizontal and vertical components of displacement, respectively.

thin-slab propagator. It took 2.7 hours. Thin-slab is about 10 times faster than the finite-difference method. Figure 11 gives the 3 components of the synthetic seismograms calculated using the finite-difference method (solid) and by the thin-slab method (dotted). The Y-component in Fig. 11 has been multiplied by a factor of 3. We see that the results of the two methods are in excellent agreements for small to mild angle scatterings. Especially for the reflected/converted waves generated

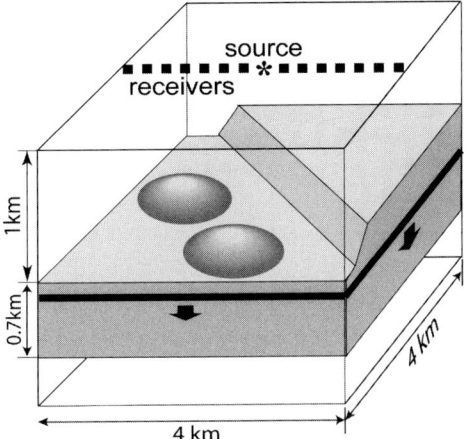

FIG. 10. The 3-D French model for the numerical tests. The fat solid lines delineated the horizontal surface at which the thin-slab method is connected to the reflectivity method. The background medium has the same parameters as those in Fig. 8. The structure in grey color has a perturbation of -10% for both P and S velocities.

at the lower interface of the model, the use of reflectivity method improved the accuracy of the arrival times of those events, and matched well with those by the finite-difference method.

5. THE SCREEN APPROXIMATION

For some special applications, such as slowly varying media, the synthetics only involve small-angle scattering. In this case the screen approximation can be applied to accelerate the computation.

Under small-angle scattering approximation, we can compress the thin-slab into an equivalent screen and therefore change the 3-D spectrum into a 2-D spectrum. Dual-domain implementation of the screen approximation will make the modeling of backscattering very efficient. In the following, the cases of acoustic and elastic media will be given respectively.

5.1. Screen Propagators for Acoustic Media

5.1.1. Thin-Slab Formulation in Wavenumber Domain

In order to further accelerate the computation, approximations to the interaction between the thin-slab and incident waves can be applied. First, we discuss

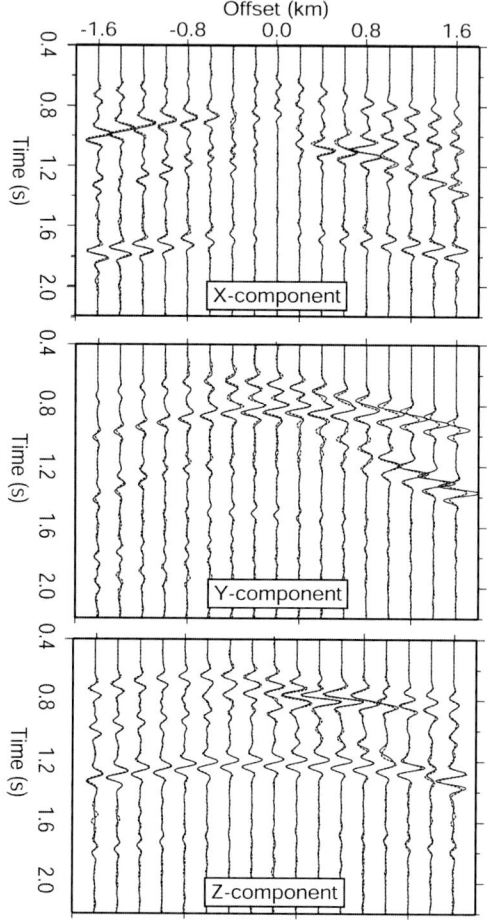

FIG. 11. Comparison of synthetic seismograms calculated by finite difference (solid lines) and by thin-slab methods (dashed lines). From top to the bottom are X-, Y- and Z-components of displacement, respectively.

the thin-slab formulation in wavenumber domain, and in the next section, the screen approximation will be made based on the wavenumber domain formulation.

To obtain the wavenumber domain formulation, we carry out analytically the integration along z-direction between the slab entrance z' and the exit z_1 (see Fig. 3). In the case of acoustic media, we substitute (50) and (51) into Eq. (48) and perform the moving frame coordinate transform $z \to z - z'$, resulting in

$$P(\mathbf{K}_T, z^*)$$
$$= \frac{i}{2\gamma} \frac{k^2}{4\pi^2} e^{i(\pm\gamma)(z^*-z')} \iint d\mathbf{K}'_T$$
$$\times \left\{ \int_0^{\Delta z} dz \iint d^2\mathbf{x}_T \, \varepsilon_\kappa(z, \mathbf{x}_T) p^0(\mathbf{K}'_T) e^{-i(\pm\gamma-\gamma')z} e^{-i(\mathbf{K}_T-\mathbf{K}'_T)\cdot\mathbf{x}_T} \right.$$
$$- (\hat{k} \cdot \hat{k}') \int_0^{\Delta z} dz \iint d^2\mathbf{x}_T \, \varepsilon_\rho(z, \mathbf{x}_T) p^0(\mathbf{K}'_T) e^{-i(\pm\gamma-\gamma')z}$$
$$\left. \times e^{-i(\mathbf{K}_T-\mathbf{K}'_T)\cdot\mathbf{x}_T} \right\}, \tag{82}$$

where
$$p^0(\mathbf{K}'_T) = p^f(z', \mathbf{K}'_T) \tag{83}$$

is the incident field at the slab entrance, $\Delta z = z_1 - z'$, and $\pm\gamma$ correspond to forescattering and backscattering respectively. Note that

$$\int_0^{\Delta z} dz \iint d^2\mathbf{x}_T \, F(\mathbf{x}) e^{-i(\mathbf{k}-\mathbf{k}')\cdot\mathbf{x}} = \tilde{F}(\mathbf{k} - \mathbf{k}'), \tag{84}$$

where $\tilde{F}(\mathbf{k})$ is the 3-D Fourier transform of the thin-slab, i.e., the Slab-Spectrum, $\mathbf{k}'$ is the incident wavenumber (52), and $\mathbf{k}$ is the outgoing wavenumber (scattering wavenumber) defined as

$$\mathbf{k} = \mathbf{k}^f = \mathbf{K}_T + \gamma \hat{e}_z \tag{85}$$

for forescattered field and

$$\mathbf{k} = \mathbf{k}^b = \mathbf{K}_T - \gamma \hat{e}_z \tag{86}$$

for backscattered field. Therefore the local Born scattering in wavenumber domain can be written as

$$P(\mathbf{K}_T, z^*) = i \frac{k^2}{8\pi^2\gamma} e^{ik_z(z^*-z')}$$
$$\times \iint d\mathbf{K}'_T \big[\varepsilon_\kappa(\mathbf{K}_T - \mathbf{K}'_T, k_z - \gamma')$$
$$- (\hat{k} \cdot \hat{k}') \varepsilon_\rho(\mathbf{K}_T - \mathbf{K}'_T, k_z - \gamma') \big] p^0(\mathbf{K}'_T), \tag{87}$$

with $k_z = \gamma$ for forescattering and $k_z = -\gamma$ for backscattering. When the receiving level is at the bottom of the thin-slab (forescattering), $z^* = z_1$; while $z^* = z'$ is for the backscattered field at the entrance of the thin-slab. The total transmitted field at the slab exit can be calculated as the sum of the forescattered field and the primary field, which can be approximated as

$$p^0(\mathbf{x}'_T, z^*) = \frac{1}{4\pi^2} \iint d\mathbf{K}_T \, p^0(\mathbf{K}_T) e^{i\gamma(z^*-z')} e^{i\mathbf{K}_T \cdot \mathbf{x}'_T}. \tag{88}$$

We see that the scattering characteristics depend on the spectral properties of heterogeneities. In the case of large-scale heterogeneities, where lateral sizes of heterogeneities are much larger than wavelength, the major forescattered energy is concentrated in a small cone towards the forward direction. For forescattering, the outgoing wavenumbers $\mathbf{k}$ are nearly in the same direction as the incoming wavenumbers $\mathbf{k}'$. Within the small cone, $\mathbf{k} - \mathbf{k}'$ stays small. Therefore, the scattered waves are controlled by the low lateral spatial-frequency components of heterogeneities. In the meanwhile, $\mathbf{k}$ and $\mathbf{k}'$ have opposite directions for backscattered waves and the backscattering is most sensitive to those vertical spectral components which are comparable to the half-wavelength (see Wu and Aki, 1985). Also, we know that forescattering is controlled by velocity heterogeneities, while backscattering responses mostly to impedance heterogeneities (ibid). This point will be seen much more clearly in the next section.

Note that the wave-slab interaction in wavenumber domain [Eq. (87)] is not a convolution and therefore the operation in space domain is not local. Therefore, the wave-slab interaction in wavenumber domain involves matrix multiplication and is computationally intensive.

5.1.2. Small-Angle Approximation and the Screen Propagators

Under this approximation, both incoming and outgoing wavenumbers have small transversal components K_T compared to the longitudinal component γ and therefore

$$\gamma = \sqrt{k^2 - K_T^2} \approx k\left(1 - \frac{K_T^2}{2k}\right). \tag{89}$$

Then

$$\begin{aligned}
\mathbf{k}^f - \mathbf{k}' &= (\mathbf{K}_T - \mathbf{K}'_T) + (\gamma - \gamma')\hat{e}_z \\
&\approx (\mathbf{K}_T - \mathbf{K}'_T) + \left(\frac{K_T'^2}{2k} - \frac{K_T^2}{2k}\right)\hat{e}_z \\
&\approx (\mathbf{K}_T - \mathbf{K}'_T) + 0\hat{e}_z
\end{aligned} \tag{90}$$

and

$$\mathbf{k}^b - \mathbf{k}' = (\mathbf{K}_T - \mathbf{K}'_T) - (\gamma + \gamma')\hat{e}_z \approx (\mathbf{K}_T - \mathbf{K}'_T) - 2k\hat{e}_z. \tag{91}$$

5.1.3. Screen Approximation in Acoustic Media

With the small-angle approximation, the 3-D thin-slab spectrum (84) can be approximated by

$$\tilde{F}(\mathbf{k}^f - \mathbf{k}) \approx \tilde{F}(\mathbf{K}_T - \mathbf{K}'_T, K_z = 0)$$

$$= \iint d^2\mathbf{x}_T \, e^{-i(\mathbf{K}_T - \mathbf{K}'_T)\cdot\mathbf{x}_T} \int_0^{\Delta z} dz \left[\varepsilon_\kappa(\mathbf{x}_T, z) - \varepsilon_\rho(\mathbf{x}_T, z)\right]$$

$$= 2 \iint d^2\mathbf{x}_T \, e^{-i(\mathbf{K}_T - \mathbf{K}'_T)\cdot\mathbf{x}_T} S_V(\mathbf{x}_T), \qquad (92)$$

where S_V is a 2-D screen of velocity perturbation, and

$$\tilde{F}(\mathbf{k}^b - \mathbf{k}) \approx \tilde{F}(\mathbf{K}_T - \mathbf{K}'_T, K_z = -2k)$$

$$= \iint d^2\mathbf{x}_T \, e^{-i(\mathbf{K}_T - \mathbf{K}'_T)\cdot\mathbf{x}_T} \int_0^{\Delta z} dz \, e^{i2kz}\left[\varepsilon_\kappa(\mathbf{x}_T, z) + \varepsilon_\rho(\mathbf{x}_T, z)\right]$$

$$= 2 \iint d^2\mathbf{x}_T \, e^{-i(\mathbf{K}_T - \mathbf{K}'_T)\cdot\mathbf{x}_T} S_I(\mathbf{x}_T), \qquad (93)$$

where S_I is a 2-D screen of impedance perturbation. We see that with the above approximation, 3-D thin-slab spectra have been replaced by 2-D screen spectra that are slices of the 3-D spectra. In the case of forescattering, the slice is from a velocity spectrum at $K_z = 0$, where K_z is the spatial frequency in the z-direction; while for backscattering, from an impedance spectrum, a slice at $K_z = -2k$. In the special case when $F(\mathbf{x}_T, z)$ varies very little along z within the thin-slab, the screen spectra can be further approximated as

$$\tilde{F}(\mathbf{k}^f - \mathbf{k}) \approx 2\tilde{S}_V(\mathbf{K}_T - \mathbf{K}'_T) = \left[\delta_\kappa(\mathbf{K}_T - \mathbf{K}'_T) - \delta_\rho(\mathbf{K}_T - \mathbf{K}'_T)\right]\Delta z,$$

$$\tilde{F}(\mathbf{k}^b - \mathbf{k}) \approx 2\tilde{S}_I(\mathbf{K}_T - \mathbf{K}'_T)$$

$$= \left[\delta_\kappa(\mathbf{K}_T - \mathbf{K}'_T) + \delta_\rho(\mathbf{K}_T - \mathbf{K}'_T)\right]\Delta z \, \text{sinc}(k\Delta z) e^{ik\Delta z}. \qquad (94)$$

The scattered fields (74) under the screen approximation become

$$P(\mathbf{K}_T, z^*) \approx i \frac{k^2}{4\pi^2 \gamma} e^{ik_z(z^* - z')} \iint d\mathbf{K}'_T \, \tilde{S}(\mathbf{K}_T - \mathbf{K}'_T) p^0(\mathbf{K}'_T). \qquad (95)$$

The above equation is a convolution integral in wavenumber domain and the corresponding operation in space domain is a local one. The dual-domain technique can be used to speed up the computation:

$$P(\mathbf{K}_T, z^*) \approx \frac{ik^2}{\gamma} e^{ik_z(z^*-z')} \iint d\mathbf{x}_T \, e^{-i\mathbf{K}_T \cdot \mathbf{x}_T} S(\mathbf{x}_T) p^0(\mathbf{x}_T), \qquad (96)$$

where

$$S(\mathbf{x}_T) = S_V(\mathbf{x}_T)$$
$$= \frac{1}{4} \int_0^{\Delta z} dz \, [\delta_\kappa(\mathbf{x}_T, z) - \delta_\rho(\mathbf{x}_T, z)] \quad \text{for forescattering}, \qquad (97)$$

$$S(\mathbf{x}_T) = S_I(\mathbf{x}_T)$$
$$= \frac{1}{4} \int_0^{\Delta z} dz \, e^{i2kz} [\delta_\kappa(\mathbf{x}_T, z) + \delta_\rho(\mathbf{x}_T, z)] \quad \text{for backscattering}. \qquad (98)$$

The total transmitted field at z_1 is

$$P^f(\mathbf{K}_T, z_1) = p^0(\mathbf{K}_T, z_1) + P^f(\mathbf{K}_T, z_1)$$
$$= e^{ik_z(z_1-z')} \iint d\mathbf{x}_T \, e^{-i\mathbf{K}_T \cdot \mathbf{x}_T} p^0(\mathbf{x}_T)[1 + ikS_V(\mathbf{x}_T)]$$
$$= e^{ik_Z(z_1-z')} \iint d\mathbf{x}_T \, e^{-i\mathbf{K}_T \cdot \mathbf{x}_T} p^0(\mathbf{x}_T) \exp[ikS_V(\mathbf{x}_T)], \qquad (99)$$

where $k/\gamma_\alpha \approx 1$ has been used for the scattered field based on the small-angle scattering approximation. The above equation is the dual-domain implementation of phase-screen propagation.

5.1.4. Procedure of MFSB Using the Screen Approximation

1. Fourier transform the incident field at the starting plane into wavenumber domain and propagate to the screen using a constant velocity propagator.
2. Inverse Fourier transform the incident field into space domain. Interact with the impedance screen (complex-screen) to get the backscattered field and interact with the velocity screen (phase-screen) to get the transmitted field.
3. Fourier transform the transmitted field into wavenumber domain and propagate to the next screen using a constant velocity propagator.
4. Repeat the propagation and interaction screen-by-screen to the bottom of the model space.
5. Backpropagate and stack the stored backscattered field screen by screen from the bottom to the top to get the total backscattered field on the surface.

5.2. Screen Approximation for Elastic Media: Elastic complex Screen Method

Similar to the derivation for acoustic media, the wavenumber domain formulation can be obtained by substituting Eqs. (64), (68) into Eqs. (63) and (65)–(67).

$$\mathbf{U}^{PP}(\mathbf{K}_T, \mathbf{K}'_T) = \frac{i}{2\gamma_\alpha} k_\alpha^2 u_0^P \hat{k}_\alpha$$
$$\times \left[(\hat{k}_\alpha \cdot \hat{k}'_\alpha) \frac{\delta\rho(\tilde{\mathbf{k}})}{\rho} - \frac{\delta\lambda(\tilde{\mathbf{k}})}{\lambda + 2\mu} - (\hat{k}_\alpha \cdot \hat{k}'_\alpha)^2 \frac{2\delta\mu(\tilde{\mathbf{k}})}{\lambda + 2\mu} \right],$$

$$\mathbf{U}^{PS}(\mathbf{K}_T, \mathbf{K}'_T) = \frac{i}{2\gamma_\beta} k_\beta^2 (\mathbf{I} - \hat{k}_\beta \hat{k}_\beta) \cdot \mathbf{u}_0^P \left[\frac{\delta\rho(\tilde{\mathbf{k}})}{\rho} - 2\frac{\beta_0}{\alpha_0} (\hat{k}_\alpha \cdot \hat{k}'_\beta) \frac{\delta\mu(\tilde{\mathbf{k}})}{\mu} \right],$$

$$\mathbf{U}^{SP}(\mathbf{K}_T, \mathbf{K}'_T) = \frac{i}{2\gamma_\alpha} k_\alpha^2 (\mathbf{u}_0^S \cdot \hat{k}_\alpha) \hat{k}_\alpha \left[\frac{\delta\rho(\tilde{\mathbf{k}})}{\rho} - 2\frac{\beta_0}{\alpha_0} (\hat{k}_\beta \cdot \hat{k}'_\alpha) \frac{\delta\mu(\tilde{\mathbf{k}})}{\mu} \right],$$

$$\mathbf{U}^{SS}(\mathbf{K}_T, \mathbf{K}'_T) = \frac{i}{2\gamma_\beta} k_\beta^2 (\mathbf{I} - \hat{k}_\beta \hat{k}_\beta) \cdot \left\{ \mathbf{u}_0^S \left[\frac{\delta\rho(\tilde{\mathbf{k}})}{\rho} - (\hat{k}_\beta \cdot \hat{k}'_\beta) \frac{\delta\mu(\tilde{\mathbf{k}})}{\mu} \right] \right.$$
$$\left. - \hat{k}'_\beta (\mathbf{u}_0^S \cdot \hat{k}_\beta) \frac{\delta\mu(\tilde{\mathbf{k}})}{\mu} \right\}, \qquad (100)$$

where $\mathbf{u}_0^P$ is the spectral field of the incident P wave, and $\delta\rho(\tilde{\mathbf{k}})$, $\delta\lambda(\tilde{\mathbf{k}})$ and $\delta\mu(\tilde{\mathbf{k}})$ are the three-dimensional Fourier transforms of medium perturbations, and $\tilde{\mathbf{k}} = \hat{k} - \hat{k}'$ is the exchange wavenumber with $\hat{k}'$ and $\hat{k}$ as incident and scattering wavenumber vectors, respectively.

From the thin-slab formulation, under the small-angle approximation, both incident and scattered wavenumbers have small lateral components K_T and K'_T compared to vertical components and therefore

$$\gamma_\alpha \approx k_\alpha (1 - K_T^2/2k_\alpha^2), \qquad \gamma_\beta \approx k_\beta (1 - K_T^2/2k_\beta^2),$$
$$\gamma'_\alpha \approx k_\alpha (1 - K_T'^2/2k_\alpha^2), \qquad \gamma'_\beta \approx k_\beta (1 - K_T'^2/2k_\beta^2). \qquad (101)$$

Using these approximations and neglecting higher-order (greater than second order) small quantities, the scattering patterns can be obtained as

$$(\hat{k}_\alpha \cdot \hat{k}'_\alpha) \approx \begin{cases} +1 + (|\mathbf{K}_T - \mathbf{K}'_T|)^2 / 2k_\alpha^2 & \text{for forescattering,} \\ -1 + (|\mathbf{K}_T + \mathbf{K}'_T|)^2 / 2k_\alpha^2 & \text{for backscattering,} \end{cases}$$

$$(\hat{k}_\alpha \cdot \hat{k}'_\beta) \approx \begin{cases} +1 - (|k_\beta \mathbf{K}_T - k_\alpha \mathbf{K}'_T|)^2 / (2k_\alpha^2 k_\beta^2) & \text{for forescattering,} \\ -1 + (|k_\beta \mathbf{K}_T + k_\alpha \mathbf{K}'_T|)^2 / (2k_\alpha^2 k_\beta^2) & \text{for backscattering,} \end{cases}$$

$$(\hat{k}_\beta \cdot \hat{k}'_\alpha) \approx \begin{cases} +1 - (|k_\alpha \mathbf{K}_T - k_\beta \mathbf{K}'_T|)^2 / (2k_\alpha^2 k_\beta^2) & \text{for forescattering,} \\ -1 + (|k_\alpha \mathbf{K}_T + k_\beta \mathbf{K}'_T|)^2 / (2k_\alpha^2 k_\beta^2) & \text{for backscattering,} \end{cases}$$

$$(\hat{k}_\beta \cdot \hat{k}'_\beta) \approx \begin{cases} +1 + (|\mathbf{K}_T - \mathbf{K}'_T|)^2 / 2k_\beta^2 & \text{for forescattering,} \\ -1 + (|\mathbf{K}_T + \mathbf{K}'_T|)^2 / 2k_\beta^2 & \text{for backscattering.} \end{cases} \qquad (102)$$

To decouple the incident wavenumber $\mathbf{K}_T$ and the scattered wavenumber $\mathbf{K}'_T$ in Eqs. (102), suppose that the medium heterogeneities are smooth enough that the scattering wavenumbers $\hat{k}_\alpha$ or $\hat{k}_\beta$ deviates not too far from the incident wavenumbers $\hat{k}'_\alpha$ or $\hat{k}'_\beta$. We can take $\mathbf{K}'_T$ to be approximately equal to $\mathbf{K}_T$, the wavenumber couplings may be simplified further by

$$(\hat{k}_\alpha \cdot \hat{k}'_\alpha) \approx \begin{cases} +1 & \text{for forescattering,} \\ -1 + 2K_T^2/k_\alpha^2 & \text{for backscattering,} \end{cases}$$

$$(\hat{k}_\alpha \cdot \hat{k}'_\beta) \approx \begin{cases} +1 - (K_T/k_\beta - K_T/k_\alpha)^2/2 & \text{for forescattering,} \\ -1 + (K_T/k_\beta + K_T/k_\alpha)^2/2 & \text{for backscattering,} \end{cases}$$

$$(\hat{k}_\beta \cdot \hat{k}'_\alpha) \approx \begin{cases} +1 - (K_T/k_\alpha - K_T/k_\beta)^2/2 & \text{for forescattering,} \\ -1 + (K_T/k_\alpha + K_T/k_\beta)^2/2 & \text{for backscattering,} \end{cases}$$

$$(\hat{k}_\beta \cdot \hat{k}'_\beta) \approx \begin{cases} +1 & \text{for forescattering,} \\ -1 + 2K_T^2/k_\beta^2 & \text{for backscattering.} \end{cases} \quad (103)$$

It can be seen from Eqs. (102)–(103) that the first-order corrections have relatively stronger effects on the backscattering than on the forescattering. The scattering pattern $(\mathbf{I} - \hat{k}_\beta \hat{k}_\beta) \cdot \hat{k}'_\beta$ appearing in Eqs. (103) is zero for forward scattering, while for backward scattering it is

$$\begin{aligned}\left[(\mathbf{I} - \hat{k}_\beta \hat{k}_\beta) \cdot \hat{k}'_\beta\right]_{\text{backward}} &= \left[\hat{k}'_\beta - (\hat{k}_\beta \cdot \hat{k}'_\beta)\hat{k}_\beta\right]_{\text{backward}} \\ &\approx \left[\hat{k}'_\beta - (-1 + 2K_T^2/k_\beta^2)\hat{k}_\beta\right]_{\text{backward}} \\ &\approx \left(2K_T/k_\beta, 2K_T^2/k_\beta^2\right), \end{aligned} \quad (104)$$

up to the first-order correction.

5.3. Complex Screen Method with First-Order Corrections

Numerical tests show that the effect of the first-order corrections of wave couplings for P–S and S–P forescattering can be neglected. So the first-order corrections are introduced only for backward propagators. Substituting Eqs. (103)–(104) into Eqs. (100) and integrating over incident wavenumber $\hat{k}'$, we can obtain the following formulas

$$\mathbf{U}_f^{PP}(\mathbf{K}'_T, \mathbf{K}_T) = -ik_\alpha \Delta z \hat{k}'_\alpha u_0^P(\mathbf{K}_T) \frac{\delta\alpha(\tilde{\mathbf{K}}_T)}{\alpha_0} \eta_f^{PP}, \quad (105)$$

$$\mathbf{U}_f^{PS}(\mathbf{K}'_T, \mathbf{K}_T) = -ik_\beta \Delta z u_0^P(\mathbf{K}_T) [\hat{k}_\alpha - \hat{k}'_\beta(\hat{k}_\alpha \cdot \hat{k}'_\beta)]$$
$$\times \left[\frac{\delta\beta(\tilde{\mathbf{K}}_T)}{\beta_0} + \left(\frac{\beta_0}{\alpha_0} - \frac{1}{2}\right) \frac{\delta\mu(\tilde{\mathbf{K}}_T)}{\mu_0}\right] \eta_f^{PS}, \quad (106)$$

$$\mathbf{U}_f^{SP}(\mathbf{K}'_T, \mathbf{K}_T) = -ik_\alpha \Delta z (\mathbf{u}_0^S(\mathbf{K}_T) \cdot \hat{k}_\alpha) \hat{k}_\alpha$$
$$\times \left[\frac{\delta\beta(\tilde{\mathbf{K}}_T)}{\beta_0} + \left(\frac{\beta_0}{\alpha_0} - \frac{1}{2}\right) \frac{\delta\mu(\tilde{\mathbf{K}}_T)}{\mu_0}\right] \eta_f^{SP}, \quad (107)$$

$$\mathbf{U}_f^{SS}(\mathbf{K}_T', \mathbf{K}_T) = -ik_\beta \Delta z [\mathbf{u}_0^S(\mathbf{K}_T) - \hat{k}_\beta'(\mathbf{u}_0^S(\mathbf{K}_T) \cdot \hat{k}_\beta')] \frac{\delta\beta(\tilde{\mathbf{K}}_T)}{\beta_0} \eta_f^{SS}, \tag{108}$$

$$\mathbf{U}_b^{PP}(\mathbf{K}_T, z_1) = \frac{i}{2} k_\alpha \hat{k}_a \Delta z e^{ik_\alpha \Delta z} \eta_b^{PP}$$
$$\times \left[\left(-1 + \frac{2K_T^2}{k_\alpha^2}\right) V_\rho - V_\lambda - \left(1 - \frac{2K_T^2}{k_\alpha^2}\right)^2 V_\mu \right], \tag{109}$$

$$\mathbf{U}_b^{PS}(\mathbf{K}_T, z_1) = \frac{i}{2} k_\beta \Delta z e^{i(k_\alpha + k_\beta)\Delta z/2} \eta_b^{PS} (\mathbf{I} - \hat{k}_\beta \hat{k}_\beta)$$
$$\times \left\{ \mathbf{V}_\rho^P - 2\frac{\beta_0}{\alpha_0} \left[-1 + \frac{1}{2}\left(\frac{K_T}{k_\alpha} + \frac{K_T}{k_\beta}\right)^2 \right] \mathbf{V}_\mu^P \right\}, \tag{110}$$

$$\mathbf{U}_b^{SP}(\mathbf{K}_T, z_1) = \frac{i}{2} k_\alpha \Delta z e^{i(k_\alpha + k_\beta)\Delta z/2} \eta_b^{SP} \hat{k}_\alpha \hat{k}_\alpha$$
$$\times \left\{ \mathbf{V}_\rho^S - 2\frac{\beta_0}{\alpha_0} \left[-1 + \frac{1}{2}\left(\frac{K_T}{k_\alpha} + \frac{K_T}{k_\beta}\right)^2 \right] \mathbf{V}_\mu^S \right\}, \tag{111}$$

$$\mathbf{U}_b^{SS}(\mathbf{K}_T, z_1) = \frac{i}{2} k_\beta \Delta z e^{ik_\beta \Delta z} \eta_b^{SS}$$
$$\times \left\{ (\mathbf{I} - \hat{k}_\beta \hat{k}_\beta) \cdot \left[\mathbf{V}_\rho^S + \left(1 - \frac{2K_T^2}{k_\beta^2}\right) \mathbf{V}_\mu^S \right] - \tilde{K}_\beta(\hat{k}_\beta \cdot \mathbf{V}_\mu^S) \right\}, \tag{112}$$

where V_ρ is the distorted P wave amplitude by density perturbation, $\mathbf{V}_\rho^P$ is the distorted P wave vector field (displacement) by density perturbation, etc., defined as

$$V_\rho(\mathbf{K}_T) = \iint d^2\mathbf{x}_T\, e^{-i\mathbf{K}_T \cdot \mathbf{x}_T} u_0^P(\mathbf{x}_T, z') \frac{\delta\rho(\mathbf{x}_T)}{\rho},$$

$$V_\lambda(\mathbf{K}_T) = \iint d^2\mathbf{x}_T\, e^{-i\mathbf{K}_T \cdot \mathbf{x}_T} u_0^P(\mathbf{x}_T, z') \frac{\delta\lambda(\mathbf{x}_T)}{\lambda + 2\mu},$$

$$V_\mu(\mathbf{K}_T) = \iint d^2\mathbf{x}_T\, e^{-i\mathbf{K}_T \cdot \mathbf{x}_T} u_0^P(\mathbf{x}_T, z') \frac{2\delta\mu(\mathbf{x}_T)}{\lambda + 2\mu},$$

$$\mathbf{V}_\rho^P(\mathbf{K}_T) = \iint d^2\mathbf{x}_T\, e^{-i\mathbf{K}_T \cdot \mathbf{x}_T} \mathbf{u}_0^P(\mathbf{x}_T, z') \frac{\delta\rho(\mathbf{x}_T)}{\rho},$$

$$\mathbf{V}_\mu^P(\mathbf{K}_T) = \iint d^2\mathbf{x}_T\, e^{-i\mathbf{K}_T \cdot \mathbf{x}_T} \mathbf{u}_0^P(\mathbf{x}_T, z') \frac{\delta\mu(\mathbf{x}_T)}{\mu},$$

$$\mathbf{V}_\rho^S(\mathbf{K}_T) = \iint d^2\mathbf{x}_T\, e^{-i\mathbf{K}_T \cdot \mathbf{x}_T} \mathbf{u}_0^S(\mathbf{x}_T, z') \frac{\delta\rho(\mathbf{x}_T)}{\rho},$$

$$\mathbf{V}_\mu^S(\mathbf{K}_T) = \iint d^2\mathbf{x}_T\, e^{-i\mathbf{K}_T \cdot \mathbf{x}_T} \mathbf{u}_0^S(\mathbf{x}_T, z') \frac{\delta\mu(\mathbf{x}_T)}{\mu},$$

$$\tilde{K}_\beta(K_T) = \left(2\mathbf{K}_T/k_\beta,\, 2K_T^2/k_\beta^2\right),$$

and the factors η_f^{PP}, η_b^{PP}, η_f^{PS}, η_b^{PS}, η_f^{SP}, η_b^{SP} and η_f^{SS}, η_b^{SS} are given by Eqs. (75)–(77). In the above derivation, we assume that the thin-slab is thin enough so that the parameters can be approximated as unchanging along the z direction, and we also approximate γ_α and γ_β in the denominators of Eqs. (100) by k_α and k_β, respectively. Although Eqs. (109)–(112) look a little more complicated in form than those in zero-order method, they can all be implemented using the FFT and the efficiency is not compromised too much. It has been shown that the zero-order complex screen results agree with the thin-slab results only for small incidence angles. As incidence angle increases, the amplitude of the reflected waves deviates gradually from that by the thin-slab method. However, the first order complex screen results are in good agreement with the thin-slab results for fairly wide angles (40° for P wave incidence and 20° for S wave incidence for the French model).

6. REFLECTED WAVE FIELD MODELING USING THIN-SLAB METHOD

AVO analysis plays an important role in modern seismic data interpretation in exploration seismology. AVO measures the angle-dependent reflection response of an interface and relates the response to in-situ elastic parameters and fluid contents of the target intervals. For flat interfaces in homogeneous elastic media, AVO curves can be easily calculated theoretically. However, the observed reflection responses of a seismic target are significantly affected by many other factors, such as data collection, data processing and wave propagation effects in heterogeneous media. Before analyzing the AVO responses, these effects should be studied and compensated.

Forward modeling can be useful in understanding wave propagation effects on AVO. In addition, forward modeling can also be useful in interpreting complicated AVO measurements, providing appropriate model parameters for data processing, and developing algorithms of frequency-dependent AVO inversion (Dey-Sarkar and Svatek, 1993). There are several modeling algorithms available for AVO calculation. The reflectivity method is one of the most common methods used for modeling AVO responses in layered media (Fuchs and Müller, 1971; Simmons and Backus, 1994; Wapenaar et al., 1999). It can generate an exact AVO response in arbitrarily layered medium, but cannot be used in structures with lateral velocity variations. Ray methods based on various approximations of the Zoeppritz equations are also very common for AVO analysis (Widess, 1973; Simmons and Backus, 1994; Bakke and Ursin, 1998). However, in the presence of thin layers, primaries-only Zoeppritz modeling may produce incorrect results. Another intrinsic limit of the ray methods is its inability in dealing with frequency-dependent scattering associated with heterogeneities. A number of authors applied pseudospectral and finite-difference methods to more complicated geologic models including anelasticity, overburden structure, scattering attenuation, and anisotropy (Chang and McMechan, 1996; Adriansyah and McMechan, 1998; Yoon and McMechan, 1996). In principle, these methods can deal with arbitrarily complicated geologic models. However, they are very time-consuming and memory demanding especially for 3-D models. This is also true for handling structures with thin layers where fine grids must be used.

In this section we apply the elastic thin-slab method to AVO modeling in sedimentary rocks. Several numerical experiments, including reflection coefficient calculations, reflection synthetics with lateral parameter variations, and thin-bed AVO, have been conducted and compared with reflectivity and finite-difference methods. The accuracy and wide-angle capacity of the thin-slab method are demonstrated. Some examples showing the effects of lateral structure variations and random heterogeneities on AVO in sedimentary rocks are presented and analyzed.

6.1. Reflection Coefficients of Sedimentary Interfaces

Reflection coefficients vary as a function of offset from the source (or reflection angle) across an interface. This information is the core of AVO analysis. For either forward modeling or inversion, accurate calculation of reflection coefficients is crucial. Here, we use the thin-slab method to calculate the reflection coefficients at the shale/gas, shale/oil and shale/brine interfaces. A rectangular grid of 1024×200 is used in the calculation with its parameters given in Table 1. The grid spacings used are 16 m in horizontal direction and 4 m in vertical direction. The interface is located at the depth of 200 m. A taper function is applied to the bottom of the model for eliminating the reflection from the artificial truncation at bottom. Only P–P reflection coefficients are displayed. For reflection coefficient calculation, we take 200 samples (displacement amplitudes in space domain) in the middle of the model for both incident and reflected waves, and define the ratio of their average displacement amplitudes as the reflection coefficient. Figure 12 shows the P–P reflection coefficients for all sets of parameters given in Table 1. The dotted curves represent the corresponding theoretical results. ϕ is the percentage porosity, and σ_1 and σ_2 are the Poisson's ratios for the shale and sand,

TABLE 1. Reservoir model

	$\phi = 20\%$				$\phi = 23\%$				$\phi = 25\%$			
	α	β	ρ	σ	α	β	ρ	σ	α	β	ρ	σ
Shale	3170	1608	2.36	0.31	3170	1668	2.36	0.31	3170	1668	2.36	0.31
Gas	3500	2374	2.10	0.10	3350	2231	2.12	0.10	3188	2124	1.96	0.10
Oil	3734	2280	2.27	0.20	3527	2131	2.22	0.21	3362	2015	2.18	0.22
Brine	3749	2262	2.31	0.21	3551	2109	2.27	0.23	3399	1993	2.23	0.24

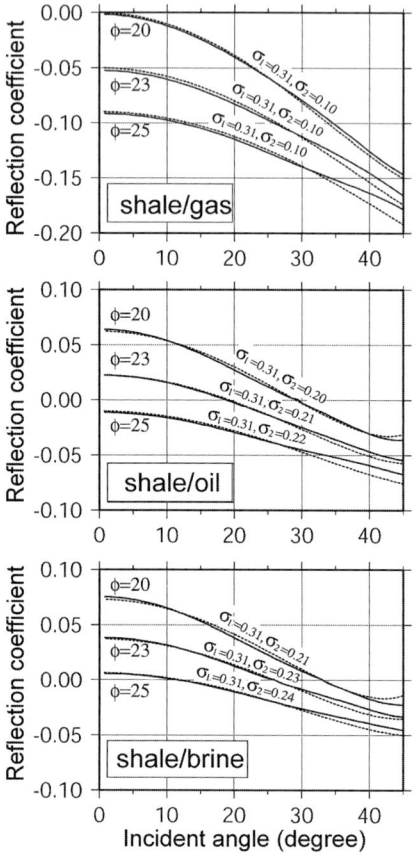

FIG. 12. P–P reflection coefficients at different type of interfaces: shale/gas (top), shale/oil (middle), and shale/brine (bottom). All formation parameters are listed in Table 1. The solid curves are calculated by the thin-slab method and the dotted lines, by the Zoeppritz equations.

respectively. The top panel corresponds to shale/gas interface, the middle panel, shale/oil interface, and the bottom panel, shale/brine interface. We see that the reflection coefficients calculated by the thin-slab method are in excellent agreement with theoretical values for small and medium incident angles ($<30°$) for all cases, and up to a wide angle of incidence ($<40°$) for the 20-percent-porosity gas, oil, and brine sands.

Although the 20-percent-porosity sand has relatively large velocity perturbations, the thin-slab results show better matches to theoretical reflection coefficients for wide angles of incidence than those for 23-percent-porosity and 25-percent-porosity sands at wide angles. It implies that the accuracy and wide-angle

TABLE 2. Perturbations of reservoir parameter

	$\phi = 20\%$				$\phi = 23\%$				$\phi = 25\%$			
	$\frac{\delta\alpha}{\alpha}\%$	$\frac{\delta\beta}{\beta}\%$	$\frac{\delta\rho}{\rho}\%$	σ	$\frac{\delta\alpha}{\alpha}\%$	$\frac{\delta\beta}{\beta}\%$	$\frac{\delta\rho}{\rho}\%$	σ	$\frac{\delta\alpha}{\alpha}\%$	$\frac{\delta\beta}{\beta}\%$	$\frac{\delta\rho}{\rho}\%$	σ
Gas	12.3	42.3	−11.0	0.10	5.7	33.7	−14.4	0.10	0.6	27.0	−17.0	0.10
Oil	17.8	36.7	−3.80	0.20	11.3	27.8	−5.9	0.21	6.1	20.8	−7.6	0.22
Brine	18.3	35.6	−2.10	0.21	12.0	26.4	−3.8	0.23	7.2	19.5	−5.5	0.24

capacity of the thin-slab method are related not only to velocity and density perturbations but also to their combinations (the impedance).

The computational efficiency, accuracy, and wide-angle capacity of the thin-slab method are closely related to the amount of perturbations. Generally, the accuracy and wide-angle capacity decrease as perturbations (velocity and/or density) increase. Its computational efficiency also decreases because finer forward steps must be utilized to guarantee the convergence. To see the perturbations of reservoir parameters shown in Table 1, we calculate velocity and density fluctuations relative to the shale (shown in Table 2). The perturbations of the physical parameters used in the models are typical of most sedimentary rocks.

6.2. Reflections from a dipping sandstone reservoir

Figure 13 shows a model of a dipping sandstone reservoir bearing gas, oil, and brine. The dipping angle of the reservoir is 10° to horizontal plane. The reservoir is thick enough so that the reflections from the top and base are separated in seismograms. Simmons and Backus (1994) used these models to investigate AVO responses associated with angles of incidence, different type interfaces, and porosities of interest. In this section we will use these models to show the accuracy of the thin-slab method for modeling reflections including primary reflections, conversions and diffractions.

Figure 14 displays the synthetic seismograms with incident plane waves for three different porosities (labeled in each panel). The plane P-wave is vertically downward incident to the model. The left column corresponds to the finite-difference results and the right panel, to the thin-slab results. The last two arrivals are the converted shear waves produced at the top and base of the reservoir and propagated in the shale. Single-leg (one converted wave path in the layer) and double-leg (two converted wave paths in the layer) converted shear waves are weak at a small angle of incidence and overlap with the primary reflections from the base. Note that in the thin-slab results (the right column in Fig. 14) the multiples within the sandstone are neglected. Comparing the two columns, we see that the thin-slab results are in good agreement with the finite-difference results, implying that the multiples are not significant in sedimentary rocks.

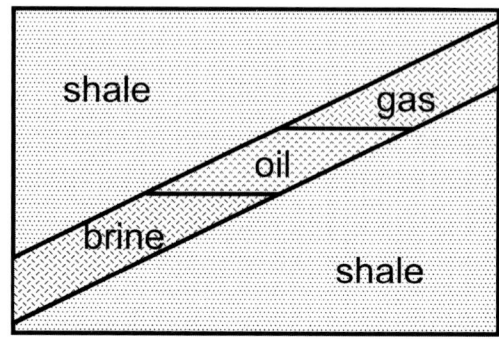

FIG. 13. Model of a dipping sandstone reservoir filled with gas, oil, and brine. The reservoir is thick enough so that the reflections from the top and base can be separated.

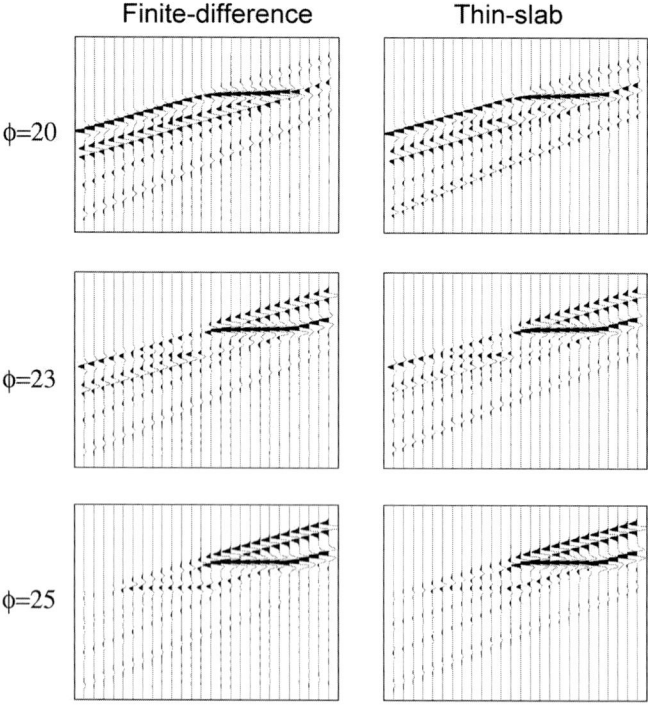

FIG. 14. Comparisons of plane wave seismograms calculated by finite-difference (left column) and thin-slab (right column) methods for a dipping reservoir model shown in Fig. 13.

6.3. Reservoir Reflections with Scattering and Attenuation from a Heterogeneous Overburden

We examine the effect of scattering and attenuation associated with heterogeneities in sedimentary rocks on AVO using the thin-slab modeling method. First we study the effects of random scattering. The reservoir is a flat model generated by rotating the model in Fig. 13. A 2-D random field with exponential correlation function is used to perturb the velocity and density parameters of the sedimentary rocks (Fig. 15). The correlation lengths of the random field are 100 m in horizontal direction and 40 m in depth. The rms values used are 1%, 2% and 3%. Note that both P- and S-wave velocities have the same distributions and rms perturbations; while the density has the same distribution but only one half of the

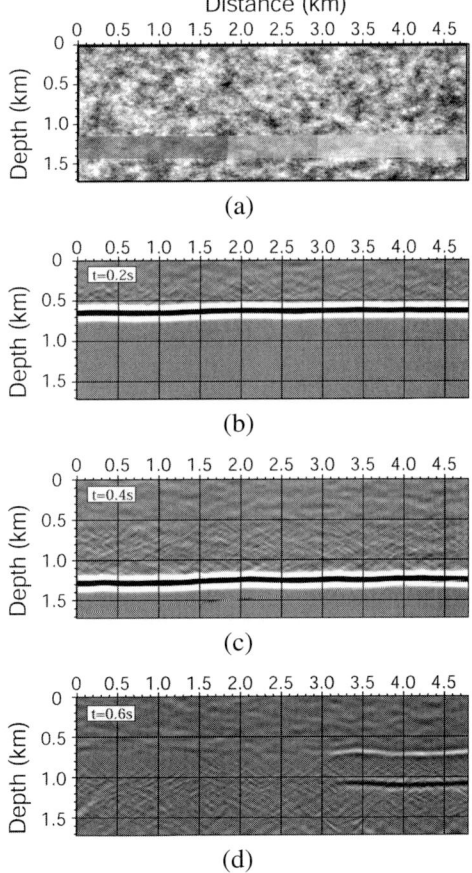

FIG. 15. (a) Sandstone model with heterogeneities, (b) to (d) are Snapshots at $t = 0.2$ s, 0.4 s, and 0.6 s, respectively. A 30 Hz plane P-wave source is vertically incident from the top of the model.

rms value of velocity perturbation. Only velocity rms values are indicated in the text and figures. For simplicity, we consider a plane P-wave vertically incident on the reservoir. Figure 15 shows the model and the snapshots at $t = 0.2$ s, 0.4 s and 0.6 s, respectively. The porosity of the sand is 25% and rms velocity perturbation is 2%. In Fig. 15 we see that abundant coda waves are produced. The wavefronts of both forward and reflected waves are distorted. For the reflected waves from shale/oil and shale/brine interfaces, serious distortions can be seen. Figure 16 shows the maximum magnitudes of the responses from the top interface of the sand. The vertical axis is the logarithmic amplitude, and (a), (b) and (c) correspond to porosities of 20%, 23%, and 25%, respectively. Solid lines correspond to calculations in models without overburden heterogeneities, in which the fluctuations in amplitudes are caused by the interference of boundary diffractions. The dotted, dashed and dotted-dashed lines correspond to overburden models having rms random velocity fluctuations of 1%, 2% and 3%, respectively. The vertical interfaces separating gas, oil and brine are indicated by arrows. Note that for an overlying shale with heterogeneities as small as rms $= 1\%$, the piece-wise uniform reflection amplitudes become fluctuating due to the focusing and defocusing of waves. As the velocity fluctuation increases, stronger amplitude fluctuations are generated. The amplitude fluctuations are closely related to the statistical properties of heterogeneities. The existence of a laterally varying overburden layer has serious effects on reflecting waves from reservoirs. It generates scattered waves and affects the reflection characteristics of local interfaces. For weak reflection sands, the scattering effect from heterogeneous overburden could be important and must be taken into account for AVO analysis. The top panel shows reservoir model bearing gas, oil and brine, respectively. The formation is anelastic and heterogeneous. The lower three panels show AVOs for various kinds of interfaces: (a) shale/gas, (b) shale/oil, and (c) shale/brine. For each kind of interface, three different constant Q's ($Q =$ infinity, 150, 50) are given to shale. The sand has constant $Qp = Qs = 10$. The correlation lengths of the random field for perturbing Q and elastic parameters are the same. The rms values are 4% for elastic parameters and 25% for Q.

Next, we show combined effects of random scattering and intrinsic attenuation. In practice, the geologic models may contain arbitrary spatial variations in compressional- and shear-wave quality factors, as well as density and velocities (see the top panel in Fig. 17). For each kind of interface, three different averaged Q's (i.e., $Q = 50, 150,$ infinity) are given to shale, and the sands have quality factors of $Q_P = Q_S = 10$. The correlation lengths of the random field for perturbing Q and elastic parameters are the same. The rms values are 4% for elastic parameters and 25% for Q the source and receiver array are located in shale and 1200 m away from the interface. The dotted lines in Fig. 17 correspond to the homogeneous cases with constant Q's. We see that intrinsic attenuation mainly affect the absolute reflected amplitudes and heterogeneities in both Q and elastic parameters affect local amplitude fluctuation with offset. In summary, the AVO

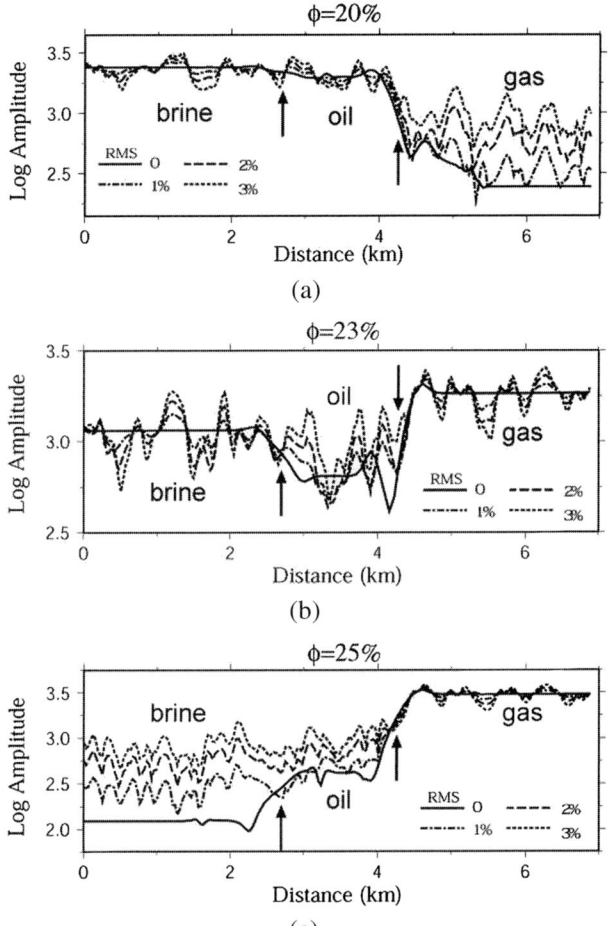

FIG. 16. Effect of scattering by heterogeneities on reflected amplitudes of reservoirs shown in Fig. 15.

responses of the target subsurface have been significantly deformed due to both the heterogeneities and intrinsic attenuation.

6.4. Thin Layer AVO Response

The amplitude response of a thin-bed has drawn increasing interest in hydrocarbon interpretation because large quantities of gas reserves are found to be trapped with thin sands. The AVO response of a thin-bed is different from that of a thick bed because of the effects of wave interference, conversion and tunneling. For a thin-bed with thickness much less than the predominant wavelength, the grid methods such as finite-difference and finite-element methods are too costly for

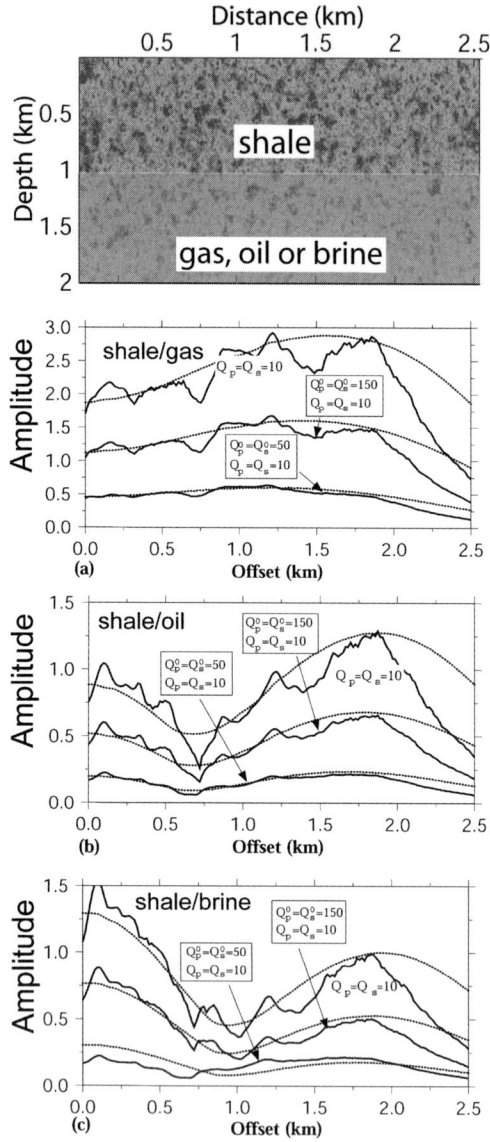

FIG. 17. The top panel shows reservoir model bearing gas, oil and brine, respectively. The formation is anelastic and heterogeneous. The lower three panels show AVOs for various kinds of interfaces: (a) shale/gas, (b) shale/oil, and (c) shale/brine. For each kind of interface, three different constant Q's (Q = infinity, 150, 50) are given to shale. The sand has constant $Q_P = Q_S = 10$. The correlation lengths of the random field for perturbing Q and elastic parameters are the same. The rms values are 4% for elastic parameters and 25% for Q.

modeling reflections in AVO analysis. We investigate the ability of the thin-slab method for handling elastic thin-bed reflections. The thin-bed used is 5 m thick, located at a depth of 1500 m and is filled with an oil sand of porosity 20%. The predominant wavelength is 125 m, being 25 times greater than the thickness of the thin-bed. The surrounding medium is shale. Source and receivers are on the top of the model. For the calculation using the thin-slab method, the grid spacing in the horizontal direction is 10 m and in the vertical direction is 10 m for the background area and 0.5 m for the thin-bed. Figure 18 displays reflection seismograms calculated by the reflectivity method (a) and thin-slab method (b). NMO corrections are applied to the top of the sand and reduced time is used in showing seismograms. In Fig. 18, A represents reflections from the top of the model, B gives reflections from both the top and base but no converted waves are involved, C includes all waves (exact solution) and D gives reflections from both the top and base including both single-leg and double-leg converted shear waves. Comparing A and B, we see the big difference between AVOs of a single impedance interface and two closely located impedance interfaces. B is equivalent to primary-only reflections. Comparing B with C and D, we see the important effect of locally converted shear waves that can alter amplitude variation with offset. Comparing C and D, only small differences exist at large offsets. In this case, the effect of multiples is negligible.

For a homogeneous thin-slab under small-angle approximation, equation for one-return reflection can be simplified into (Wu, 1996; Xie and Wu, 2001)

$$\mathbf{U}^P(\mathbf{K}_T) = -ik_\alpha \Delta z \frac{\delta Z_\alpha}{Z_\alpha} \mathbf{u}_\alpha^0(\mathbf{K}_T),$$

where Z_α is the P-wave impedance. The above equation presents not only the amplitude of the thin-bed reflection but also the change in wavelet. Therefore,

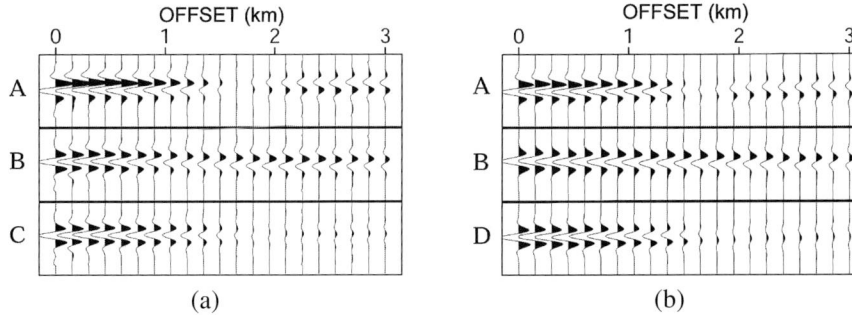

FIG. 18. Reflection seismograms for an oil sand model. The oil sand is 5 m thick and at depth of 1500 m. (a) By reflectivity method, (b) by thin-slab method. In the figure, A represents reflections from the top of the model (thick layer), B, reflections from both top and base but no converted waves included, C, all waves included (exact solution), and D, reflections from both top and base including single-leg and double-leg converted shear waves.

the thin-slab propagators represent the response of an elastic heterogeneous thin-layer. Since the choice of Δz or step interval in wavefield extrapolation is flexible and can vary according to local heterogeneities, the thin-slab method can naturally handle arbitrarily thin layers.

6.5. AVO Response in Laterally Varying Media

In this section we use a simple model containing both a truncated salt layer and a thin gas sand, which is located 200 m below the salt layer (Fig. 19a), to examine the AVO response using the thin-slab method.

The parameters for the gas sand are given in Table 1 and the corresponding porosity is 20%. The parameters for salt are $V_p = 4.48$ km/s, $V_s = 2.594$ km/s, $\rho = 2.1$ g/cm^3. The thicknesses of the salt layers are 20 m, 40 m and 80 m, respectively. The grid spacing is the same as that used in Fig. 18 except for inside the salt body, where a 0.1 m grid space is used. In Fig. 19, the maximum negative amplitudes of the thin-bed responses are picked and plotted versus offset. The solid curve represents the thin-bed (without salt) AVO. The AVO trend of a thin-bed is different from that shown in Fig. 12 (shale/gas, $\phi = 20$) where the

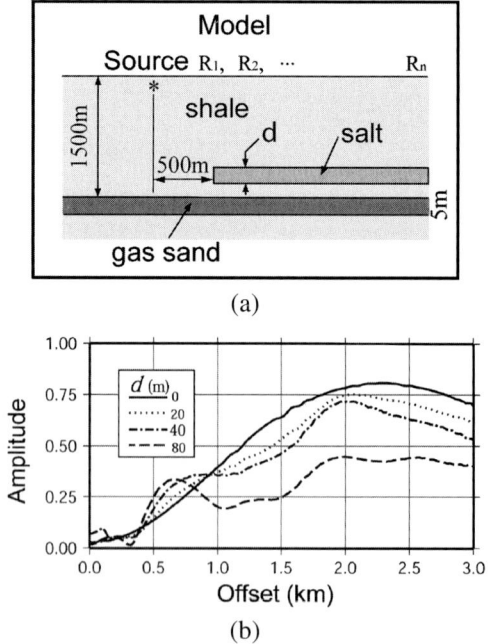

FIG. 19. (a) Model containing a truncated salt layer with thickness d and a thin gas sand layer below the salt, (b) angle-dependent reflections (AVO) of a thin gas layer with reflections from the salt layer of different thickness.

reflection coefficient increases with incident angles up to 45° or an offset of 3 km. The dotted, dashed and dotted-dashed curves correspond to salt layer thicknesses of 20 m, 40 m, and 80 m, respectively. We see that AVOs are altered due to the presence of the salt layer. The salt layer can cause diffraction, defocusing, conversion and transmission loss. The diffraction affects AVO by causing interference between diffracted waves from the left side of the salt layer and reflected waves from the thin-bed at receiver locations. This interference causes local variations in AVO measurements. The transmission loss affects all the reflections passing through the salt body and causes a systematic decrease in AVO.

Figure 20 shows the combined effects of a truncated salt layer and random heterogeneities on AVO. Heterogeneities are introduced into the model shown in Fig. 19a. The correlation lengths of the 2-D random field are 100 m in horizontal direction and 40 m in vertical direction. The rms fluctuations used are 1%, 2%, and 3%, respectively. The four panels in Fig. 20 correspond to the cases of zero-salt, 20 m, 40 m and 80 m-thick salt (all labeled in the figure). The case of zero-salt shows the effect from heterogeneities alone. The focusing and defocusing of the spatially correlated heterogeneities produce the local variation in reflected amplitudes versus offset which becomes significant for sedimentary models with weak reflections. Interpretation of AVO observations based on homogeneous elastic models will therefore bias from actual properties of the target. The frequency and amplitude variations of reflections are closely related to the source spectrum and the statistical properties of velocity perturbations, although the overall AVO trends are still controlled by the target properties and overburden structures. To improve AVO analysis in sedimentary rocks, it is necessary to have

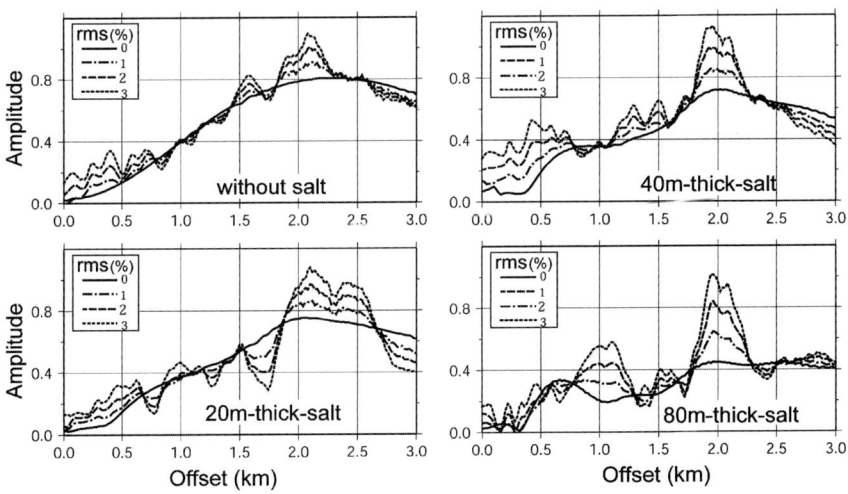

FIG. 20. AVOs of the thin gas layer with the combined effects of lateral structure variation and random heterogeneities.

sufficient acquisition apertures and take into account of the overburden structures including random heterogeneities and thin-bed effects.

7. Conclusions

In this chapter renormalized MFSB (multiple-forescattering–single-backscattering) equations and the dual-domain expressions for scalar, acoustic and elastic waves are treated in a unified approach. The De Wolf approximation neglects the reverberations (internal multiples) inside, but can model all the forward scattering phenomena, such as focusing/defocusing, diffraction, refraction, interference, etc., and the primary reflections. The De Wolf approximation can be considered as the first order term in a De Wolf multiple scattering series. This single backscattering signal defined this way is the *primary reflections* from the *real medium*, and is different from the Born approximation where the incident field and the Green's function are both defined in the reference medium (a homogeneous medium). It is also different from the first term of the generalized Bremmer series, which uses a high-frequency asymptotic solution (a WKBJ-like solution) as the Green's function (in an equivalent reference medium).

Two versions of the one-return method (using MFSB approximation) are given: one is the wide-angle dual-domain thin-slab approximation; the other is the screen approximation. The latter involves a small-angle approximation for the wave-medium interaction. Q-factor from intrinsic attenuations has been incorporated into the algorithm by the introduction of complex velocities. The reflectivity method has also been incorporated to the thin-slab modeling to increase the efficiency.

The theory and methods are applied to the fast calculation of synthetic seismograms. For weak heterogeneities ($\pm 15\%$ perturbation), good agreement between the one-return method and finite difference simulations verifies the validity of the one-return approach. However, the one-return approach is about 2–3 orders of magnitude faster than the elastic FD algorithm. The method can be applied to the fast modeling of AVA responses for a complex reservoir with heterogeneous overburdens. The method can be applied to most of the real cases where the perturbations of P- and S-wave velocities are around or smaller than 30%. The influences of heterogeneous or random overburdens, thin-bedding, Q-variation, irregular salt layer, etc., can be studied using this modeling technique.

Acknowledgements

Helpful discussions with and comments from M. Fehler, L.J. Huang, R. Hobbs, S. Jin, C. Mosher, and K. Wapenaar are greatly appreciated. This work was supported by the Department of Energy through various contracts and by the WTOPI (Wavelet Transform on Propagation and Imaging for seismic exploration) Research Consortium at the University of California, Santa Cruz. Facility support from the W.M. Keck Foundation is also acknowledged. Contribution number 488 of CSIDE, IGPP, University of California, Santa Cruz.

REFERENCES

Aki, K., Richards, P. (1980). Quantitative Seismology: Theory and Methods, vols. 1 and 2. Freeman, New York.
Adriansyah, O., McMechan, G.A. (1998). Effects of attenuation and scattering on AVO measurements. *Geophysics* **63**, 2025–2034.
Bakke, N.E., Ursin, B. (1998). Thin-bed AVO effects. *Geophys. Prosp.* **46**, 571–587.
Berkhout, A.J. (1982). Seismic Migration. Elsevier, Amsterdam.
Berkhout, A.J. (1987). Applied Seismic Wave Theory. Elsevier, Amsterdam.
Berkhout, A.J., Wapenaar, C.P.A. (1989). One-way versions of the Kirchhoff integral. *Geophysics* **54**, 460–467.
Bremmer, H. (1951). The W.K.B. approximation as a first order term of a geometric-optical series. *Commun. Pure Appl. Math.* **4**, 105.
Chang, H., McMechan, G.A. (1996). Numerical simulation of multi-parameter scattering. *Bull. Seismol. Soc. Am.* **86**, 1820–1829.
Claerbout, J.F. (1970). Coarse grid calculations of waves in inhomogeneous media with applications to delineation of complicated seismic structure. *Geophysics* **35**, 407–418.
Claerbout, J.F. (1976). Fundamentals of Geophysical Data Processing. McGraw–Hill, New York.
Collins, M.D. (1989). Applications and time-domain solution of higher-order parabolic equation in underwater acoustics. *J. Acoust. Soc. Am.* **86**, 1097–1102.
Collins, M.D. (1993). A two-way parabolic equation for elastic media. *J. Acoust. Soc. Am.* **93**, 1815–1825.
Collins, M.D., Westwood, E.K. (1991). A higher-order energy-conserving parabolic equation for range-dependent ocean depth, sound speed, and density. *J. Acoust. Soc. Am.* **89**, 1068–1175.
Corones, J. (1975). Bremmer series that correct parabolic approximations. *J. Math. Anal. Appl.* **50**, 361–372.
De Hoop, M.V. (1996). Generalization of the Bremmer coupling series. *J. Math. Phys.* **37**, 3246–3282.
De Hoop, M., Rousseau, J., Wu, R.S. (2000). Generalization of the phase-screen approximation for the scattering of acoustic waves. *Wave Motion* **31**, 43–70.
Devaney, A.J. (1982). A filtered back propagation algorithm for diffraction tomography. *Ultrasonic Img.* **4**, 336–350.
Devaney, A.J. (1984). Geophysical diffraction tomography. *Trans. IEEE* **GE-22**, 3–13.
De Wolf, D.A. (1971). Electromagnetic reflection from an extended turbulent medium: Cumulative forward-scatter single-backscatter approximation. *IEEE Trans. Antennas and Propagations* **AP-19**, 254–262.
De Wolf, D.A. (1985). Renormalization of EM fields in application to large-angle scattering from randomly continuous media and sparse particle distributions. *IEEE Trans. Antennas and Propagations* **AP-33**, 608–615.
Dey-Sarkar, S.K., Svatek, S.V. (1993). Prestack analysis-An integrated approach for seismic interpretation in elastic basin. In: Castagna, J.P., Backus, M.M. (Eds.), *Offset Dependent Reflectivity: Theory and Practice of AVO Analysis*. In: *Soc. Exp. Geophys. Invest. Geophys.*, vol. 8, pp. 57–77.
Fishman, L., McCoy, J.J. (1984). Derivation and application of extended parabolic wave theories II. Path integral representations. *J. Math. Phys.* **25**, 297–308.

Fishman, L., McCoy, J.J. (1985). A new class of propagation models based on a factorization of the Helmholtz equation. *Geophys. J. Astron. Soc.* **80**, 439–461.

Fisk, M.D., McCartor, G.D. (1991). The phase screen method for vector elastic waves. *J. Geophys. Res.* **96**, 5985–6010.

Flatté, S.M., Dashen, R., Munk, W., Watson, K., Zachariasen, F. (1979). Sound Transmission through a Fluctuating Ocean. Cambridge Univ. Press, Cambridge.

Flatté, S.M., Tappert, E.D. (1975). Calculation of the effect of internal waves on oceanic sound transmission. *J. Acoust. Soc. Am.* **58**, 1151–1159.

Flatté, S.M., Wu, R.S. (1988). Small-scale structure in the lithosphere and asthenosphere deduced from arrival-time and amplitude fluctuations at NORSAR. *J. Geophys. Res.* **93**, 6601–6614.

French, W.S. (1974). Two-dimensional and three-dimensional migration of model-experiment reflection profiles. *Geophysics* **39**, 265–277.

Friederich, W. (1999). Propagation of shear wave and surface waves in a laterally heterogeneous mantle by multiple forward scattering. *Geophys. J. Int.* **136**, 180–204.

Friederich, W., Wielandt, E., Stange, S. (1993). Multiple forward scattering of surface waves: Comparison with an exact solution and Born single-scattering methods. *Geophys. J. Int.* **112**, 264–275.

Fuchs, K., Müller, G. (1971). Computation of synthetic seismograms with the reflectivity method and comparison with observations. *Geophys. J. R. Astron. Soc.* **23**, 417–433.

Grimbergen, J.L.T., Dessing, F.J., Wapenaar, K. (1998). Modal expansion of one-way operators in laterally varying media. *Geophysics* **63**, 995–1005.

Han, Q., Wu, R.S. (2005). One-way dual-domain propagators for scalar qP-wave in VTI media. *Geophysics* **70**, D9–D17.

Huang, L.J., Fehler, M.C., Wu, R.S. (1999a). Extended local Born Fourier migration method. *Geophysics* **64**, 1524–1534.

Huang, L.J., Fehler, M.C., Roberts, P.M., Burch, C.C. (1999b). Extended local Rytov Fourier migration method. *Geophysics* **64**, 1535–1545.

Huang, L.J., Wu, R.S. (1996). Prestack depth migration with acoustic pseudo-screen propagators. In: *Mathematical Methods in Geophysical Imaging IV*. In: *Proceedings of SPIE*, vol. 2822, pp. 40–51.

Hudson, A. (1980). A parabolic approximation for elastic waves. *Wave Motion* **2**, 207–214.

Ishimaru, A. (1978). Wave Propagation and Scattering in Random Media, vol. II. Academic Press, New York.

Jin, S., Wu, R.S. (1999). Common-offset pseudo-screen depth migration. In: *SEG 69th Annual Meeting*, pp. 1516–1519. Expanded Abstracts.

Jin, S., Mosher, C.C., Wu, R.S. (2002). Offset-domain pseudoscreen prestack depth migration. *Geophysics* **67**, 1895–1902.

Jin, S., Wu, R.S., Peng, C. (1998). Prestack depth migration using a hybrid pseudo-screen propagator. In: *SEG 68th Annual Meeting*, pp. 1819–1822. Expanded Abstracts.

Jin, S., Wu, R.S., Peng, C. (1999). Seismic depth migration with screen propagators. *Comput. Geosci.* **3**, 321–335.

Landers, T., Claerbout, J.F. (1972). Numerical calculation of elastic waves in laterally inhomogeneous media. *J. Geophys. Res.* **77**, 1476–1482.

Le Rousseau, J.H., De Hoop, M.V. (2001). Scalar generalized-screen algorithms in transversely isotropic media with a vertical symmetry axis. *Geophysics* **66**, 1538–1550.

Lee, D., Shang, E.-C., Buckingham, M.J. (2000). Parabolic equation development in the twentieth century. *J. Comput. Acoust.* **8**, 527–637.

Ma, Z. (1982). Finite-difference migration with higher order approximation. *Oil Geophys. Prosp. China* **1**, 6–15.

McCoy, J.J. (1977). A parabolic theory of stress wave propagation through inhomogeneous linearly elastic solids. *J. Appl. Mech.* **44**, 462–468.

McCoy, J.J., Frazer, L.N. (1986). Propagation modeling based on wavefield factorization and invariant embedding. *Geophys. J. R. Astron. Soc.* **86**, 703–717.

Morse, P.M., Feshbach, H. (1953). Methods of Theoretical Physics. McGraw–Hill, New York.
Rino, C.L. (1988). A spectral-domain method for multiple scattering in continuous randomly irregular media. *IEEE Trans. Antennas and Propagations* **36**, 1114–1128.
Ristow, D., Ruhl, T. (1994). Fourier finite-difference migration. *Geophysics* **59**, 1882–1893.
Simmons, J.L., Backus, M.M. (1994). AVO modeling and the locally converted shear wave. *Geophysics* **59**, 1237–1248.
Stoffa, P.L., Fokkema, J.T., Freire, R.M.D., Kessinger, W.P. (1990). Split-step Fourier migration. *Geophysics* **55**, 410–421.
Tappert, F.D. (1977). The parabolic equation method. In: Keller, J.B., Papadakis, J.S. (Eds.), *Wave Propagation and Underwater Acoustics*. Springer, New York.
Tatarskii, V.L. (1971). The Effects of the Turbulent Atmosphere on Wave Propagation (translated from Russian). National Technical Information Service.
Thomson, C.J. (1999). The 'gap' between seismic ray theory and 'full' wavefield extrapolation. *Geophys. J. Int.* **137**, 364–380.
Thomson, C.J. (2005). Accuracy and efficiency considerations for wide-angle wavefield extrapolators and scattering operators. *Geophys. J. Int.* **163**, 308–323.
Twersky, V. (1964). On propagation in random media of discrete scatterers. *Proc. Am. Math Soc. Symp. Stochas. Proc. Math. Phys. Engrg.* **16**, 84–116.
Van Stralen, M.J.N., De Hoop, M., Blok, H. (1998). Generalized Bremmer series with rational approximation for the scattering of waves in inhomogeneous media. *J. Acoust. Soc. Am.* **104**, 1943–1963.
Wales, S.C. (1986). A vector parabolic equation model for elastic propagation. In: Akal, T., Berkson, M.J. (Eds.), *Ocean Seismo-Acoustics*. Plenum, New York.
Wales, S.C., McCoy, J.J. (1983). A comparison of parabolic wave theories for linearly elastic solids. *Wave Motion* **5**, 99–113.
Wapenaar, C.P.A. (1996). One-way representation of seismic data. *Geophys. J. Int.* **127**, 178–188.
Wapenaar, C.P.A. (1998). Reciprocity properties of one-way propagators. *Geophysics* **63**, 1795–1798.
Wapenaar, K., van Wijngaarden, A.J., van Geloven, W., van der Leij, E. (1999). Apparent AVA effects of fine layering. *Geophysics* **64**, 1939–1948.
Widess, M.B. (1973). How thin is a thin-bed. *Geophysics* **38**, 1176–1180.
Wild, A.J., Hobbs, R.W., Frenje, L. (2000). Modeling complex media: An introduction to the phase-screen method. *Phys. Earth Planet. Int.* **120**, 219–226.
Wild, A.J., Hudson, J.A. (1998). A geometrical approach to the elastic complex screen. *J. Geophys. Res.* **103**, 707–726.
Wu, R.S. (1989). The perturbation method for elastic wave scattering. In: Wu, R.S., Aki, K. (Eds.), *Seismic Wave Scattering and Attenuation*. In: *Pure Appl. Geophys.*, vol. 131, pp. 605–637.
Wu, R.S. (1994). Wide-angle elastic wave one-way propagator in heterogeneous media and an elastic wave complex-screen method. *J. Geophys. Res.* **99**, 751–766.
Wu, R.S. (1996). Synthetic seismograms in heterogeneous media by one-return approximation. *Pure Appl. Geophys.* **148**, 155–173.
Wu, R.S. (2003). Wave propagation, scattering and imaging using dual-domain one-way and one-return propagators. *Pure Appl. Geophys.* **160**, 509–539.
Wu, R.S., Aki, K. (1985). Scattering characteristics of elastic waves an elastic heterogeneity. *Geophysics* **50**, 582–595.
Wu, R.S., Cao, J. (2005). WKBJ solution and transparent propagators. In: *67th Annual International Meeting*. EAGE. Expanded Abstracts.
Wu, R.S., De Hoop, M.V. (1996). Accuracy analysis and numerical tests of screen propagators for wave extrapolation. In: *Mathematical Methods in Geophysical Imaging IV*. In: *Proceedings of SPIE*, vol. 2822, pp. 196–209.
Wu, R.S., Flatté, S.M. (1990). Transmission fluctuations across an array and heterogeneities in the crust and upper mantle. *Pure Appl. Geophys.* **132**, 175–196.
Wu, R.S., Huang, L.J. (1992). Scattered field calculation in heterogeneous media using phase-screen propagator. In: *SEG 62nd Annual Meeting*, pp. 1289–1292. Expanded Abstracts.

Wu, R.S., Huang, L.J. (1995). Reflected wave modeling in heterogeneous acoustic media using De Wolf approximation. In: *Mathematical Methods in Geophysical Imaging III*. In: *Proceedings of SPIE*, vol. 2571, pp. 176–193.

Wu, R.S., Huang, L.J., Xie, X.B. (1995). Backscattered wave calculation using the De Wolf approximation and a phase-screen propagator. In: *SEG 65th Annual Meeting*, pp. 1293–1296. Expanded Abstracts.

Wu, R.S., Jin, S. (1997). Windowed GSP (generalized screen propagators) migration applied to SEG-EAEG salt model data. In: *SEG 67th Annual Meeting*, pp. 1746–1749. Expanded Abstracts.

Wu, R.-S., Toksöz, M.N. (1987). Diffraction tomography and multisource holography applied to seismic imaging. *Geophysics* **52**, 11–25.

Wu, R.S., Jin, S., Xie, X.B. (2000a). Energy partition and attenuation Lg waves by numerical simulations using screen propagators. *Phys. Earth Planet. Int.* **120**, 227–244.

Wu, R.S., Jin, S., Xie, X.B. (2000b). Seismic wave propagation and scattering in heterogeneous crustal waveguides using screen propagators: I SH waves. *Bull. Seismol. Soc. Am.* **90**, 401–413.

Wu, R.S., Luo, M., Chen, S., Xie, X.B. (2004). Acquisition aperture correction in angle-domain and true-amplitude imaging for wave equation migration. In: *SEG 74th Annual Meeting*, pp. 937–940. Expanded Abstracts.

Wu, X.Y., Wu, R.S. (1999). Wide-angle thin-slab propagator with phase matching for elastic modeling. In: *SEG 69th Annual Meeting*, pp. 1867–1870. Expanded Abstracts.

Wu, R.S., Xie, X.B. (1993). A complex-screen method for elastic wave one-way propagation in heterogeneous media. In: *3rd International Congress of the Brazilian Geophysical Society*. Expanded Abstracts.

Wu, R.S., Xie, X.B. (1994). Multi-screen backpropagator for fast 3-D elastic prestack migration. In: *Mathematical Methods in Geophysical Imaging II*. In: *Proceedings of SPIE*, vol. 2301, pp. 181–193.

Wu, X.Y., Wu, R.S. (2001). Lg-wave simulation in heterogeneous crusts with surface topography using screen propagators. *Geophys. J. Int.* **146**, 670–678.

Wu, X.Y., Wu, R.S. (2003a). Fast modeling of 2-D/3-D elastic reflections using thin-slab method. In: *SEG 73rd Annual Meeting*, pp. 1865–1868. Expanded Abstracts.

Wu, X.Y., Wu, R.S. (2003b). Synthesizing AVO responses in visco-elastic media using fast one-way elastic propagators. In: *SEG 73rd Annual Meeting*, pp. 208–210. Expanded Abstracts.

Wu, X.Y., Wu, R.S. (2006). AVO modeling using elastic thin-slab method. *Geophysics*, in press.

Xie, X.B., Wu, R.S. (1995). A complex-screen method for modeling elastic wave reflections. In: *SEG 65th Annual Meeting*, pp. 1269–1272. Expanded Abstracts.

Xie, X.B., Wu, R.S. (1998). Improve the wide angle accuracy of screen method under large contrast. In: *SEG 68th Annual Meeting*, pp. 1811–1814. Expanded Abstracts.

Xie, X.B., Mosher, C.C., Wu, R.S. (2000). The application of wide angle screen propagator to 2-D and 3-D depth migrations. In: *SEG 70th Annual Meeting*, pp. 878–881. Expanded Abstracts.

Xie, X.B., Wu, R.S. (2001). Modeling elastic wave forward propagation and reflection using the complex-screen method. *J. Acoust. Soc. Am.* **109**, 2629–2635.

Xie, X.B., Wu, R.S. (2005). Multicomponent prestack depth migration using elastic screen method. *Geophysics* **70**, S30–S37.

Yoon, K.H., McMechan, G.A. (1996). 3-D eighth-order elastic finite-difference modeling of refraction and strong-motion data from the Coyote Lake region, California. *Bull. Seismol. Soc. Am.* **86**, 616–626.

SIMULATION OF HIGH-FREQUENCY WAVE PROPAGATION IN COMPLEX CRUSTAL WAVEGUIDES USING GENERALIZED SCREEN PROPAGATORS

RU-SHAN WU, XIAN-YUN WU AND XIAO-BI XIE

Modeling and Imaging Laboratory, Institute of Geophysics and Planetary Physics, University of California, Santa Cruz, California, USA

ABSTRACT

In the crustal waveguide environment, the major part of wave energy is carried by forward propagating waves, including forward scattered waves. Therefore, neglecting backscattered waves in numerical modeling will not modify the main features of regional waves in most cases. By neglecting backscattering in the theory, the wave modeling becomes a forward marching problem in which the next step of propagation depends only on the present values of the wavefield in a transverse cross-section and the heterogeneities between the present cross-section and the next one (wavefield extrapolation interval). The saving of computation time and computer memory is enormous. A half-space screen propagator (generalized screen propagator) has been developed to accommodate the free-surface boundary condition for modeling SH wave propagation in complex crustal waveguides. The SH screen propagator has also been extended to handle irregular surface topography using conformal or nonconformal topographic transforms. The screen propagator for modeling regional SH waves has been calibrated extensively against some full-wave methods, such as the wavenumber integration, finite-difference and boundary element methods, for different crustal models. Excellent agreement with these full-wave methods demonstrated the validity and accuracy of the new one-way propagator method. For medium size problems, the screen-propagator method is 2–3 orders of magnitude faster than finite-difference methods. It has been used for the simulation of Lg propagation in crustal models with random heterogeneities and the related energy partition, attenuation and blockage. It is found that the leakage attenuation of Lg waves caused by large-angle forward scattering by random heterogeneities, which scatters the guided waves out of the trapped modes and leaking into the mantle, may contribute significantly to Lg attenuation and blockage in some regions. In the case of P-SV elastic screen propagators, plane wave reflection calculations have been incorporated into the elastic screen method to handle the free surface. Body waves, including the reflected and converted waves, can be calculated by real wavenumber integration; while surface waves (Rayleigh waves) can be obtained with imaginary wavenumber integration. Numerical tests proved the validity of the theory and methods.

Keywords: Lg-wave, Crustal wave guide, One-way propagator, Seismic wave scattering

1. INTRODUCTION

High-frequency regional wave propagation in complex crustal waveguides is one of the most challenging problems in theoretical and computational seismology. A good understanding of propagation, scattering, attenuation and wave-type conversion of regional waves and the availability of analytical/numerical tools to simulate and analyze these phenomena for complex crustal structures, including rough surface, Moho topography and small-scale heterogeneities, are crucial to the applications of regional waves to various geophysical problems. Regional wave tomography for crustal structures, path correction for discrimination and yield estimation of low-yield nuclear tests, location determination of earthquakes, or underground explosions using regional phases are examples among the possible applications. Nuclear explosion monitoring at regional distances is even more demanding for the simulation and analyzing tools. For this purpose, simulation algorithms are desirable for generating synthetic waveforms for high frequencies up to 25 Hz at distances greater than 1000 km.

Substantial efforts have been made in modeling regional wave propagation. Methods based on layered earth models, such as the reflectivity and mode summation methods (e.g., Bouchon *et al.*, 1985; Kennett, 1989, 1990; Maupin, 1989; Baumgardt, 1990; Campillo, 1990; Campillo and Paul, 1992; Campillo *et al.*, 1993; Gibson and Campillo, 1994) have very high efficiency and can be applied to relatively high frequencies, but they can be used only for very simplified cases with layered or smoothly varying layered models, or be applied to part of the wavefield. Modeling techniques that can treat realistic 3-D heterogeneous media, rather than smoothly varying layered media, are needed to test and study many observations and hypotheses. New modeling methods are needed to handle sudden changes of crustal thickness, strong lateral variations and irregular 3-D heterogeneities. As pointed out by Campillo *et al.* (1993), actual Lg amplitudes are reduced more than 10 times for paths passing through an anomalous zone on the east side of the Alpine range, while the modeling results using existing methods (including the effect of known large-scale lateral structural variation) only account for 20–30% of the amplitude reduction. Other attenuation mechanisms such as the scattering and attenuation by small-scale heterogeneities must be taken into account.

Kennett (1984, 1998) and Maupin and Kennett (1987) developed a coupled mode method for calculating guided seismic waves in horizontally varying structures. The method works well for relatively low frequency waves in moderately heterogeneous models (for a summary of the coupling mode method, see Chapter 2 of this book by Maupin). However, the implementation of the method for high frequency 3-D models still requires formidable computational efforts.

Chen (1990, 1995) developed a global R/T (Reflection/Transmission) matrices method, for simulating the seismic wave excitation and propagation in an arbitrarily multi-layered medium with irregular interfaces, which can be regarded as an extension of the generalized R/T coefficients method (reflectivity method) for the horizontally layered case by incorporating the T-matrix approach (for a summary, see Chapter 4 of this book by Chen). Again the application of the method is limited to low frequencies and short propagation distances.

Cormier and Anderson (1996, 1997) applied elastic Born scattering (in the regime of Rayleigh scattering) to the locked-mode solution for plane layered media to calculate the effects of small-scale heterogeneities. However, the approximation is limited to single scattering and is only good for heterogeneities with scales much smaller than the wavelength. Ray method has very limited success in modeling regional waves due to the chaotic behavior of rays caused by the multiple reflections from the free-surface and Moho. Keers *et al.* (1996a, 1996b) applied the Maslov integral method to avoid the caustics and pseudo-caustics (caustics of plane waves) by working in the phase-space. However, when chaos develops in the ray system, more complicated caustics arise for which the Maslov method does not work. In addition, ray-tracing computation is very time-consuming in this case. An alternative and flexible approach using ray-tracing has been developed by Kennett (1986), Bostock and Kennett (1990) and Kennett *et al.* (1990), in which ray diagrams are used to study Lg waves crossing structural boundaries. The method agrees well with modal calculations and can be applied to surface topography, 3-D crustal structures and other cases. However, the method cannot provide information on wave phenomena for complicated waveguides.

Finite-difference methods (e.g., Xie and Lay, 1994; App *et al.*, 1996; Goldstein *et al.*, 1996, 1997, 1999; Husebye and Ruud, 1996; Jih, 1996; Nolte *et al.*, 1996; Jones *et al.*, 1997; McLaughlin and Wilkins, 1997; Bradley and Jones, 1998, 1999; Xie *et al.*, 2005) and pseudo-spectral methods (e.g., Kosloff *et al.*, 1990; Archambeau *et al.*, 1996; Schatzman, 1996; Furumura and Kennett, 1997) are commonly-used numerical methods that have been extensively applied to regional wave propagation. Theoretically, these methods can deal with arbitrarily heterogeneous media. However, it is necessary to use very dense spatial sampling to avoid grid dispersion for long distance regional wave propagation (for grid dispersion problem, see Fornberg, 1987). The capability of the present-day computers usually restricts them to short propagation ranges and relatively low frequencies, which prevents them from being applied to more realistic cases.

The state-of-the-art of the traditional simulation techniques for regional waves has its application to relatively low frequencies and short propagation distances. Correspondingly, the volume heterogeneities and surface irregularity in the crustal models are limited to rather large scales. However, high-frequency regional waves up to 20 Hz or higher have been observed over different distances, ranging from a few hundred kilometers to more than one thousand kilometers (e.g., Ni *et al.*, 1996; Herrmann *et al.*, 1997; Lay *et al.*, 1999). Since high-frequency waves can be used for event locations with high accuracy, simulation and modeling of high-frequency regional wave propagation are very desirable for many applications. For high-frequency wave propagation, scattering and attenuation, the role of small-scale heterogeneities and surface roughness are all important.

The existence of small-scale heterogeneities in the crust and the associated seismic wave scattering have been known among seismologists (e.g., Aki and Richards, 1980; Wu and Aki, 1988, 1989, 1990; Sato and Fehler, 1998). However, the effects of these heterogeneities on guided wave (Lg) propagation in the crust have not been explored extensively. The reasons may be the following. First, the spectra, strength and distribution of the small-scale heterogeneities in different regions are not well-known. Very few data sets can be used to characterize the paths concerned. Second, there lack analytical and numerical tools to model or analyze their influence on the guided wave propagation. The theory of wave propagation in unbounded random media has been well developed. However, for waves in complex crustal waveguides with random heterogeneities, the theoretical difficulties are overwhelming, and no analytical tools are available for performing realistic calculations. Therefore, numerical methods for simulating regional wave propagation in complex waveguides with small-scale heterogeneities are highly desirable. It has become clear that small-scale heterogeneities are widely distributed in tectonically active regions. Strong topographic variation is the manifestation of tectonically active regions and often the indication of small-scale heterogeneities. Figure 1 gives a topographic profile (top panel) and its power spectrum (bottom panel) for a path crossing the Tibet region. The slope of the spectral roll-off is close to $1/k$, a flicker noise spectrum, very rich in small-scale variations. This spectrum is similar to the observations of the sonic well-log in the KTB super-deep continental drilling well (Wu et al., 1994; Jones and Holliger, 1997; Goff and Holliger, 1999), where the spectrum also has a $1/k$ slope. Recently, Goff and Holliger (1999) explained the $1/k$ spectra as a combination of hierarchical, multi-scale heterogeneities. Overall, the $1/k$ spectra demonstrate the richness of small-scale heterogeneities.

Recently, the generalized screen method has been introduced into seismic wave simulations and applied to the problems of both exploration and theoretical seismology. The generalized screen method is based on the one-way wave equation and the one-return approximation. The one-way generalized screen propagator (GSP) neglects backscattered waves, but correctly handles all the forward multiple-scattering effects, e.g., focusing/defocusing, diffraction, interference, and conversion between different wave types. The one-return approximation is also called the De Wolf approximation (De Wolf, 1971, 1985), which neglects the reverberation between screens and can simulate multiple-forescattering–single-backscattering (MFSB). Significant progress has been made on the development of an elastic complex screen (ECS) method for modeling elastic wave propagation and scattering in arbitrarily complicated structures (Wu, 1994, 1996; Xie and Wu, 1995, 2001; Wild and Hudson, 1998; Wu and Wu, 1999). The method is two

SIMULATION OF HIGH-FREQUENCY REGIONAL WAVE

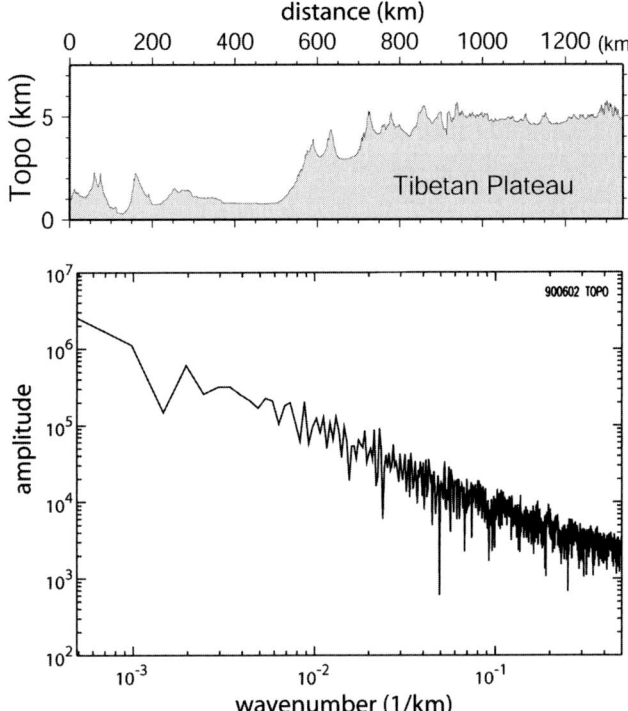

FIG. 1. Topographic profile (top) and its power spectrum (bottom) for a path crossing the Tibet region.

to three orders of magnitude faster than the elastic finite-difference method for a medium sized 3-D problem. For detailed derivation and the physical meaning of the MFSB approximation and the generalized screen method, see Chapter 5 of this volume. The screen method has been successfully used in forward modeling (Wu, 1994; Wu and Huang, 1995; Wu et al., 1995; Xie and Wu, 1995, 1996, 1999, 2001; Wu and Wu, 2001, 2005) and as backpropagators for seismic wave imaging/migration in both acoustic and elastic media (e.g., Wu and Xie, 1994; Huang and Wu, 1996; Huang et al., 1999a, 1999b; Jin and Wu, 1999; Jin et al., 1999; Xie and Wu, 1998, 2005; Xie et al., 2000).

The rest of this chapter is divided into three sections. We first briefly describe the basic concept of the one-way propagator for simulating elastic wave propagation in crustal wave guide. In the second part, we give a systematic review of the screen propagator for the SH wave propagation in complex crustal waveguides including irregular topography. In the last part, we present the P-SV Lg screen propagator.

2. A Brief Description of the Generalized Screen Propagator for Guided Waves

In the crustal wave guide environment, major part of wave energy is carried by forward propagating waves, including forward scattered waves. The Lg energy, which is in the form of guided waves, is carried by forward propagating waves bouncing up and down between the free surface and major geophysical discontinuities such as the Moho and Conrad discontinuities. Beyond the critical reflection angle, these waves are systematically dominated by small-angle waves (relative to the main propagation direction) trapped in the crustal waveguides. Therefore, neglecting backscattered waves in the propagation during numerical modeling will not modify the main features of regional waves in most cases. With this approximation, the modeling method becomes a forward marching algorithm in which the next step of propagation depends only on the present values of the wavefield in a transverse cross-section and the heterogeneities between the present cross-section and the next one (within an extrapolation interval). To formulate the problem, we divide the crustal wave guide into a sequence of vertical slabs. The horizontal direction is chosen as the main propagation direction. The geometry of the model is shown in Fig. 2a. Choosing one slab as the example, Fig. 2b shows the interaction between the incident waves and the slab. By introducing the local Born approximation, both wavefields and the elastic parameters can be separated into two parts, the background values and the perturbations. The "thin-slab" must be thin enough to satisfy the local Born approximation: the scattered field due to the heterogeneities in the slab be much smaller than the incident field. The incident P- and S-waves $\mathbf{u}_0^P$ and $\mathbf{u}_0^S$ enter the slab from the vertical plane at x_0. After the incident waves pass through the thin-slab between x_0 and x_1, and interacting with the heterogeneities within it, there will be both incident

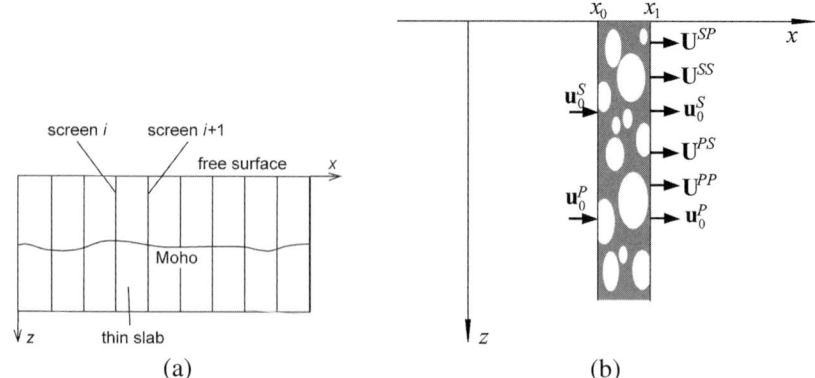

FIG. 2. (a) Geometry using screen method to simulate Lg wave; (b) Sketch showing the interaction between the incident waves and a thin slab.

waves and different types of forward scattered waves at the exit plane at x_1. The new P-wave $\mathbf{u}^P = \mathbf{u}_0^P + \mathbf{U}^{PP} + \mathbf{U}^{SP}$ is composed of incident P-wave and scattered P-waves $\mathbf{U}^{PP}$ and $\mathbf{U}^{SP}$, respectively from incident P- and S-waves, and the new S-wave $\mathbf{u}^S = \mathbf{u}_0^S + \mathbf{U}^{PS} + \mathbf{U}^{SS}$ is composed of incident S-wave and scattered S-waves $\mathbf{U}^{PS}$ and $\mathbf{U}^{SS}$, respectively. The propagation and scattering in the thin-slab can be formulated using the perturbation theory and calculated by two separated steps: (1) The interactions between the incoming waves and the heterogeneities are conducted in the spatial domain, accounting for the scattering and the coupling between different wave types. (2) Plane wave propagation through the background medium is conducted in the wavenumber domain by simple phase-shift. In both domains, the calculations are local and highly efficient. There is no time-consuming spatial or wavenumber domain convolution involved. Forward and inverse fast Fourier transforms (FFT) shuttle the wavefield between the two domains. By iteratively using this process and making the output from one slab as the input of the next slab, the wavefield can be propagated through the entire model.

The conventional wavenumber integral method is for the horizontally layered model and the integral is along horizontal wavenumber k_x. By contrast, the elastic screen method for propagating guided waves in crustal environment uses vertical screens and the wavenumber integration is carried out along the vertical k_z axis. Under this geometry, the postcritical reflections become small-angle events with respect to the main propagation direction (the x-direction), therefore, the generalized screen propagator (GSP) methods based on small angle approximation is suitable for modeling Lg waves. Our discretized model is composed of vertical thin-slabs and therefore the wavenumber integral is along vertical wavenumber k_z, resulting in different features compared with the traditional wavenumber integration. Figure 3 is a sketch showing the difference between two integral axes. For the reflected P- and S-waves coupled at a horizontal free surface, both waves have the same horizontal wavenumber (Fig. 3a). Choosing the horizontal wavenumber as the integral variable, P- and S-waves have the same sampling points in the k_x axis. The Rayleigh pole, which contributes to the generation of Rayleigh wave, has a unique location on the k_x axis (Figs. 3a and b). However, choosing the vertical wavenumber k_z as the integral variable, the P- and S-waves have different sampling points. The Rayleigh pole in the real k_x axis splits into two points in the imaginary k_z axis (Fig. 3c). This makes resampling necessary when calculating converted waves. Fast Fourier transform with regular sampling interval cannot be used in the case of vertical wavenumber integration for converted waves. All these introduce additional complexity in wavenumber integration and special treatment is required. However, the general principle of plane wave (including the inhomogeneous plane waves) superposition for representing point sources still holds.

GSP is accurate for small-angle propagation and scattering (near horizontal for the crustal waveguide environment). A half-space screen propagator has

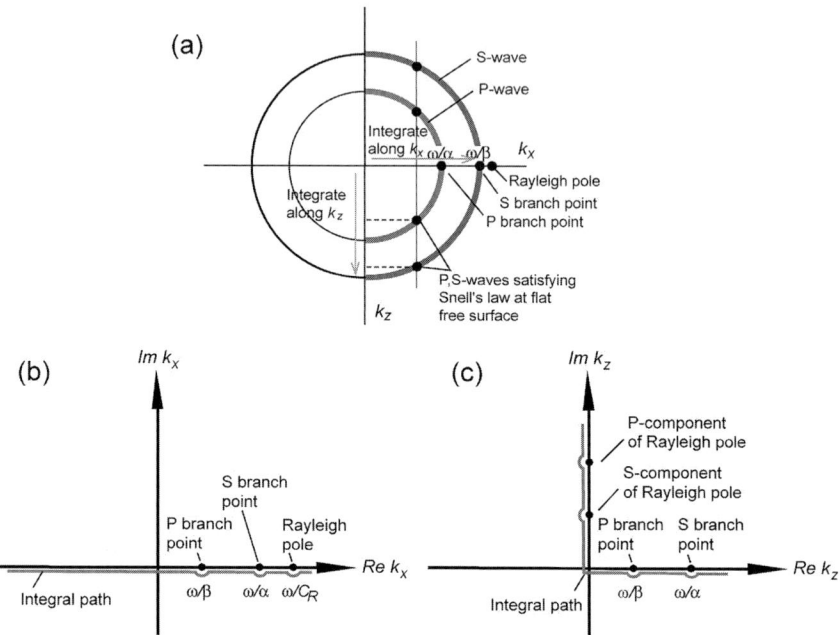

FIG. 3. Sketch showing different wavenumber domain integrals: (a) Dispersion relation in the k_x–k_z plane; (b) Traditional wavenumber integration in the complex k_x plane; (c) Wavenumber integration in the complex k_z plane for the generalized screen method using vertical screens in a half-space.

been introduced by Wu *et al.* (1996, 1997, 1998, 2000a, 2000b) to accommodate the free-surface boundary condition and treat the SH wave propagation in complex crustal waveguides. The new one-way method for modeling regional SH waves has been calibrated extensively with various full-wave methods for different crustal models, such as the wavenumber integration method for flat structures and full-wave finite-difference method for heterogeneous crustal waveguides. Excellent agreement with these methods demonstrated the validity and accuracy of the new one-way method. For a model with propagation distance of 250 km, dominant frequency at 0.5 Hz and with similar accuracy, the GSP method is about 300 times faster than the finite-difference method. The GSP method has been applied to the simulation of Lg propagation in random media for calculating the related energy partition and attenuation (Wu *et al.*, 2000b). It is found that the leakage attenuation of Lg waves caused by forward large-angle scattering from random heterogeneities, which scatters the guided waves out of the trapped modes resulting in energy leaking into the mantle, may contribute significantly to Lg attenuation and blockage in some regions. The apparent Q for leakage attenuation, as a function of normalized scale length

ka of the random heterogeneities, agrees well with the scattering theory. Later, the SH screen propagator is extended to the case of irregular surface topography by conformal or non-conformal topographic transforms (Wu et al., 1999; Wu and Wu, 2001). In the conformal transform method, the coordinate system is rotated according to the local topographic slope, and the mirror image method can be applied to the local plane surface; the non-conformal method is a surface flattening transform which turns the free surface topography into modified volume perturbations of elastic parameters. The former method is suitable to deal with smoothly varying topography, while the latter method can treat rough but moderate topography.

In the P-SV wave case, the derivation and application of one-way GSP screen propagators are much more complicated. Unlike for SH waves, the mirror image method of generating the half-space Green's function cannot be used to account for the effect of the free surface. Plane wave reflection calculations are incorporated into the elastic screen method (Wu et al., 2000c). Body waves, including the reflected and converted waves, can be calculated by real wavenumber integration; while surface waves (Rayleigh waves) can be obtained with imaginary wavenumber integration. Numerical tests show good agreement with the theory. More work has to be conducted for the coupling among the body waves, guided wave and surface wave caused by lateral heterogeneities and irregular topography.

3. SH Wave Case

3.1. Half-Space Screen Propagator

For a 2-D SH problem, only the y-component of the displacement field, noted as u, exists. With the perturbation method, the medium and the wave field are decomposed into

$$\rho = \rho_0 + \delta\rho,$$
$$\mu = \mu_0 + \delta\mu,$$
$$u = u^0 + U,$$

where ρ_0 and μ_0 are the density and shear rigidity of the background medium, $\delta\rho$ and $\delta\mu$ are their corresponding perturbations, u^0 is the primary field and U is the scattered field. The SH wave equation in the frequency domain can be written as

$$\mu_0 \nabla^2 u + \omega^2 \rho_0 u = -\lfloor \omega^2 \delta\rho u + \nabla \cdot \delta\mu \nabla u \rfloor, \tag{1}$$

where ∇ is the 2-D gradient operator and "·" stands for inner product. Equation (1) is a scalar Helmholtz equation. With a half-space scalar Green's function g^h, the scattered field U can be written as

$$U(\mathbf{r}_1) = k^2 \int_v d^2r \left\{ g^h(\mathbf{r}_1; \mathbf{r})\varepsilon_\rho(\mathbf{r})u(\mathbf{r}) - \frac{1}{k^2}\nabla g^h(\mathbf{r}_1; \mathbf{r}) \cdot \varepsilon_\mu(\mathbf{r})\nabla u(\mathbf{r}) \right\}. \tag{2}$$

Under the forward-scattering approximation, or more generally the multiple-forescattering–single-backscattering (MFSB) (De Wolf, 1971, 1985; Wu and Huang, 1995; Wu, 1996), the total field and Green's function under the integration in the above equation can be replaced by their forward-scattering approximated counterparts, and the field can be calculated by a one-way marching algorithm along the x-direction using a dual domain technique (see Chapter 5 of this book). Note that the half-space Green's function must be used here to account for the free surface effect.

For each step of the marching algorithm under the forward-scattering approximation, the total field at x_1 is calculated as the sum of the primary field which is the field propagating in the half-space from x' to x_1, and the scattered field caused by the heterogeneities in the thin-slab between x' and x_1. The thickness of the slab should be made thin enough to ensure the validity of the local Born approximation. The Green's function in the homogeneous half-space can be obtained by the mirror image method. The stress should vanish at the free surface $z = 0$. Therefore we have

$$g_0^h(\mathbf{r}_1; \mathbf{r}) = g_0(\mathbf{r}_1; \mathbf{r}) + g_0(\mathbf{r}_1; \mathbf{r}^*), \tag{3}$$

where g_0 is the infinite homogeneous Green's function and $\mathbf{r}^*$ is the mirror image point of $\mathbf{r}$ with respect to the free surface.

The free space Green's function in wavenumber domain is (Wu, 1996)

$$g_0(x_1, K_z; x, z) = \frac{1}{2\gamma} e^{i\gamma|x_1-x|} e^{-iK_z z} \tag{4}$$

with

$$\gamma = \sqrt{k^2 - K_z^2}. \tag{5}$$

Therefore,

$$g_0^h(x_1, K_z; x, z) = \frac{1}{2\gamma} e^{i\gamma(x_1-x)} 2\cos(K_z z). \tag{6}$$

In a similar way we can obtain

$$\nabla g_0^h = \frac{1}{2} e^{i\gamma(x_1-x)} \left\{ \hat{e}_x 2\cos(K_z z) - \hat{e}_z (K_z/\gamma) 2i\sin(K_z z) \right\}, \tag{7}$$

where $\hat{e}_x$ and $\hat{e}_z$ are the unit vectors in the x- and z-directions, respectively.

Taking Fourier transform of Eq. (2) along z_1 for the case of a thin-slab perpendicular to the x-direction, and substitute the half-space Green's functions into it, the scattered field by the thin-slab can be calculated by (for details see Wu et al.,

2000a)

$$U(x_1, K_z) = U_\rho(x_1, K_z) + U_\mu(x_1, K_z),$$

$$U_\rho(x_1, K_z) = ik \int_{x'}^{x_1} dx\, e^{i\gamma(x_1-x)} C\left[\frac{k}{\gamma}\varepsilon_\rho(z)u_0(z)\right],$$

$$U_\mu(x_1, K_z) = ik \int_{x'}^{x_1} dx\, e^{i\gamma(x_1-x)}$$
$$\times \left\{ C[\varepsilon_\mu(z)\bar{\partial}_x u_0(z)] - iS\left[\frac{K_z}{\gamma}\varepsilon_\mu(z)\bar{\partial}_z u_0(z)\right]\right\}, \quad (8)$$

where

$$\varepsilon_\rho(\mathbf{r}) = \frac{\delta\rho(\mathbf{r})}{\rho_0}, \qquad \varepsilon_\mu(\mathbf{r}) = \frac{\delta\mu(\mathbf{r})}{\mu_0},$$

and $\gamma = \sqrt{k^2 - K_z^2}$ is the propagating wavenumber in the x-direction, K_z is transverse wavenumber along the z-axis, and

$$\bar{\partial}_x = \frac{1}{ik}\frac{\partial}{\partial x}$$

are dimensionless partial derivatives. In the above equations, $C[f(z)]$ and $S[f(z)]$ are the cosine and sine transforms:

$$C[f(z)] = \int_0^\infty dz\, 2\cos(K_z z) f(z),$$

$$S[f(z)] = \int_0^\infty dz\, 2\sin(K_z z) f(z), \quad (9)$$

and u_0, $\bar{\partial}_x u_0$ and $\bar{\partial}_z u_0$ can be calculated by

$$u_0(x,z) = \frac{1}{2\pi}\int_{-\infty}^\infty dK_z'\, e^{iK_z' z} e^{i\gamma'(x-x')} u_{0(x',K_z')}$$
$$= C^{-1}\left[e^{i\gamma'(x-x')} u_0(x', K_z')\right], \quad (10)$$

$$\bar{\partial}_x u_0(x,z) = C^{-1}\left[e^{i\gamma'(x-x')}\frac{\gamma'}{k} u_0(x', K_z')\right],$$

$$\bar{\partial}_z u_0(x,z) = iS^{-1}\left[e^{i\gamma'(x-x')}\frac{K_z'}{k} u_0(x', K_z')\right]. \quad (11)$$

The above equations are the general wide-angle formulation. When the energy of crustal guided waves is carried mainly by small-angle waves (with respect to the horizontal direction), the phase-screen approximation can be invoked to simplify the theory and calculations. Summing up the primary and scattered fields and invoking the Rytov transform results in the dual-domain expression of phase-screen propagator

$$u(x_1, K_z) \approx e^{i\gamma(x_1-x')} C\left[e^{ikS_s(z)} u_{0(x',z)}\right], \tag{12}$$

where $e^{ikS_s(z)}$ is the phase delay operator with

$$S_s(z) = \frac{1}{2} \int_{x'}^{x_1} dx \left[\varepsilon_\rho(x, z) - \varepsilon_\mu(x, z)\right] \approx \Delta x \bar{\varepsilon}_s(z), \tag{13}$$

where $\bar{\varepsilon}_s(z)$ is the average S-wave slowness perturbation over the thin-slab at depth z,

$$\bar{\varepsilon}_s(z) = \frac{1}{x_1 - x'} \int_{x'}^{x_1} dx \frac{s(x, z) - s_0}{s_0}, \tag{14}$$

with $s(x, z) = 1/v(x, z)$ and $\Delta x = (x_1 - x')$ is the thin-slab thickness. Equation (12) is the SH phase-screen propagator for the half space. It has a similar form as the whole space propagator with the Fourier transform replaced by a cosine transform.

The phase-screen propagator has long been used in ocean acoustics to simulate long range acoustic wave propagation in the heterogeneous ocean due to internal waves. Most work in the literature deals with the stochastic treatment of waves in random media. For an introduction and brief summary of the work in that field, the reader is referred to Flatté et al. (1979). However, in this work we will use the half-space screen propagator for deterministic modeling in heterogeneous crustal waveguides.

3.2. Treatment of the Moho Discontinuity

The Moho discontinuity can be treated in two ways. One way is to put the impedance boundary conditions in the formulation, the other is to treat the parameter changes as perturbations and therefore incorporate the discontinuity into the screen interaction. In this paper, we adopt the latter approach because of its flexibility in treating irregular Moho discontinuity. The validity of the perturbation approach for the Moho discontinuity is verified by the comparison with wavenumber integration and full-wave finite-difference algorithms. Since for guided waves, or crustal waves with critical or post-critical reflections, the related mantle waves are nearly horizontal, the screen approximation is quite accurate in this case. The excellent agreements of the method with the wavenumber integration for flat Moho, and with the finite-difference method for irregular Moho demonstrate the validity of this approach (Wu et al., 2000a).

Figure 4 compares the reflection coefficients of the Moho discontinuity calculated using the theoretical equation (dotted line) and using phase-screen method (solid line). A constant velocity crust model ($v_c = 3.5$ km/s, $\rho_c = 2.8$ g/cm^3,

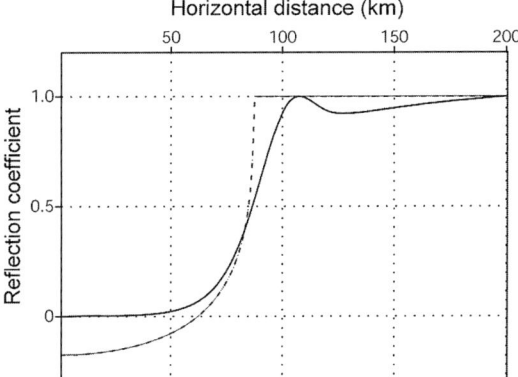

FIG. 4. Comparison of reflection coefficients at the Moho discontinuity. Dotted line denotes result from theoretical equation and solid line denotes result calculated using screen method. A constant velocity crust model is used in the calculation and the source is located 30 km above the Moho discontinuity.

$v_m = 4.5$ km/s, $\rho_m = 3.1$ g/cm^3) is used in the calculation and the source is located 30 km above the Moho. The curve from the screen method is obtained by using the root-mean-square (RMS) of the reflected waveforms. We see that the process of critical reflection is well matched, except that the transition from pre-critical to critical calculated by the screen method is not as sharp as the theoretical curve. This may be caused by the small phase error in the phase-screen approximation. In addition, the reflection for wide-angle incidence, especially for nearly vertical incidence, is not well modeled by the phase-screen method. This error results from the small angle approximation used in the screen formulation. However, this limitation occurs only at short range, well before the critical distance (around 80 km in this case). As can be verified using numerical simulations, the screen method behaves exceptionally well beyond the critical distance, making it a good candidate for guided wave simulation.

3.3. Numerical Verifications and Simulation Examples

In this section we show some examples demonstrating the validity of the method and its potential applications to various problems of regional wave propagation. First, we show a comparison between the screen method and a full-wave finite-difference method for a heterogeneous crustal model. Shown in Fig. 5a is a wave guide model with a crust necking. Figure 5b shows the synthetic seismograms along a vertical profile at an epicentral distance of 250 km. The thin lines are from the finite-difference method and the thick lines are from the screen method. The source is located at a depth of 2 km and the source time function has a dominant frequency of 0.5 Hz. Figure 5 demonstrates excellent agreement

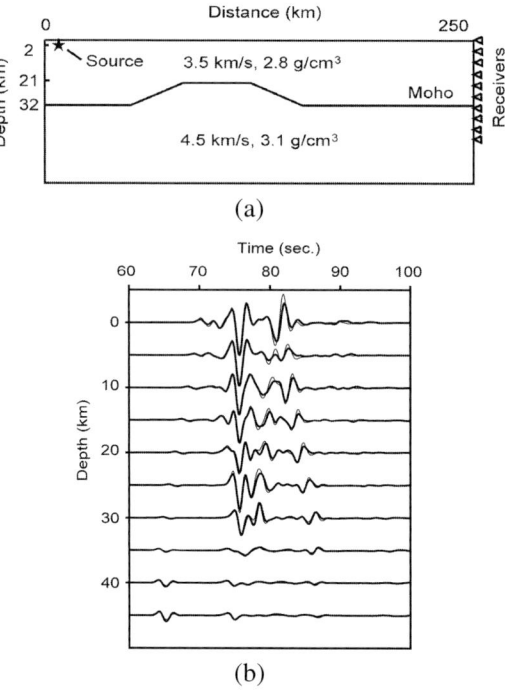

FIG. 5. Comparison of synthetic seismograms along a vertical profile at a distance of 250 km. Shown in (a) is the velocity model with a laterally varying crustal wave guide. Shown in (b) are synthetic seismograms calculated using the screen method (thick lines) and a finite-difference method (thin lines). The source depth is 2 km and the source time function is a Gaussian derivative with a dominant frequency of 0.5 Hz.

between the two methods. For this example, the GSP method is about 300 times faster than the finite-difference method. Note that the grid spacing used in the FD simulation was 3–4 times smaller than the stability requirement in order to reduce the numerical dispersion. Other comparisons with wavenumber integration and finite-difference methods can be found in Wu *et al.* (2000a, 2000b).

The importance of small-scale random heterogeneities to seismic wave propagation is well known. There are extensive publications on this subject in seismology. However, due to the complexity of the problem, the role of random heterogeneities in Lg excitation, propagation, attenuation and blockage is still unclear. For waves in complex crustal waveguides with random heterogeneities, there are still no analytical tools available for performing realistic calculations. Numerical simulation is a useful alternative to the theory. Some finite-difference simulations have been conducted (e.g., Frankel and Clayton, 1986; Frankel, 1989; Xie and Lay, 1994; Jih, 1996). Due to the limit of the computing power, the wave-propagation distances in the finite-difference simulations are relatively short.

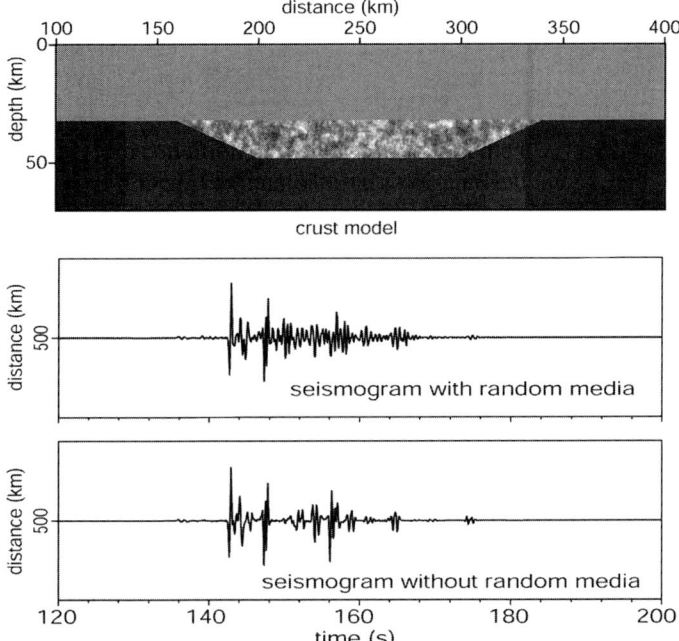

FIG. 6. A heterogeneous crustal model representing a mountain root with small-scale random heterogeneities (top panel). The comparisons between synthetic seismograms with and without random heterogeneities are shown on the middle and bottom panels, respectively.

Liu and Wu (1994) have done some numerical simulations using the phase-screen method, but the models simulated are limited to unbounded media. The development of the half-space GSP method enables us to simulate long distances, high-frequency wave propagation in complex crustal waveguides. We present here two numerical examples to demonstrate the capability of the GSP method.

Figure 6 shows a heterogeneous crustal model representing a "mountain root" with small-scale random heterogeneities. The top panel is the velocity model, and the comparisons between synthetic seismograms with and without random heterogeneities are shown on the middle and bottom panels, respectively. The heterogeneities have an exponential correlation function, with the scale length $a_x = a_z = 1.6$ km (in horizontal and vertical directions, respectively). The RMS velocity perturbation is 5%. The dominant frequency of the source time function is 2 Hz. Figures 7a and b show a comparison of wavefield snapshots between models with and without random heterogeneities. We see that random heterogeneities drastically increase the complexity of the wavefield and the energy leakage to the upper mantle.

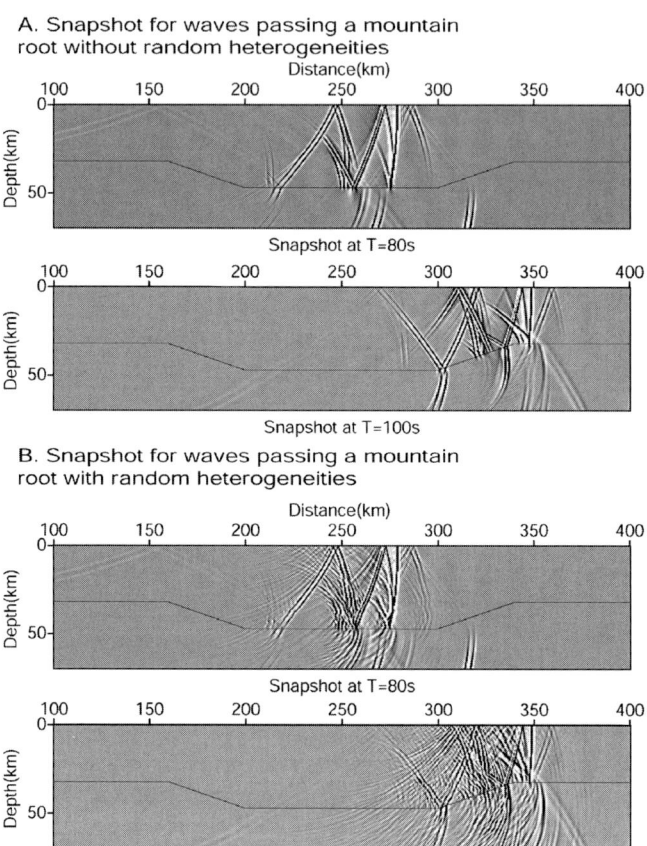

FIG. 7. Comparison between snapshots for waves passing through a "mountain root" with or without random heterogeneities, shown on A and B, respectively.

3.4. Application to Energy Partition and Attenuation in Crustal Waveguide with Random Heterogeneities

In heterogeneous crustal wave guides, the upper boundary is the free surface, which is a perfect reflector. The lower boundary of the wave guide is the Moho discontinuity. For waves incident on the Moho discontinuity, a part of the energy will leak into the upper mantle. However, for waves incident on the Moho with post-critical angles, total reflections occur and all the energy is reflected and trapped in the waveguide. Generally speaking, the guided wave energy can be expressed as

$$E_g = C \int_{K_z < K_c} |u(K_z)|^2 \, dK_z, \tag{15}$$

where C is a constant, K_z is the wavenumber in the z-direction, namely the transverse wavenumber, and K_c is the critical wavenumber. Scattering processes can redistribute the energy in wavenumber domain, causing the leak of trapped energy into the upper mantle. In addition to the leakage loss, the guided waves suffer also the anelastic loss and backscattering loss. Assuming a homogeneous mantle and neglecting reverberation in the x-direction, the energy balance after propagating a short distance dx in the x-direction is

$$E_g(x + dx) = E_g(x) - E_a(x) - E_b(x) - E_l(x), \qquad (16)$$

where E_g is the energy of guided crustal waves; E_a, energy lost due to absorption (anelastic loss); E_b, energy lost due to backscattering by random heterogeneities; E_l, energy lost due to leakage to the mantle caused by heterogeneities. In terms of different attenuation coefficients, it can be written as

$$\frac{dE_g}{dx} = -[\eta_a + \eta_b + \eta_l]E_g(x) = -\eta_g E_g(x), \qquad (17)$$

where $\eta_a = (E_a/E_g)/dx$, $\eta_b = (E_b/E_g)/dx$, and $\eta_l = (E_l/E_g)/dx$ are the apparent attenuation coefficients for guided crustal waves. Equivalently,

$$\eta_g = kQ_g^{-1} = k(Q_a^{-1} + Q_b^{-1} + Q_l^{-1}), \qquad (18)$$

where Q's are the corresponding apparent quality factors.

The leakage loss is the scattering loss due to the redistribution of Lg angular spectra. It is caused dominantly by large-angle forward scattering and therefore, it is several orders of magnitude larger than the backscattering loss, i.e. $\eta_l \gg \eta_b$. In the following, we will concentrate on the analysis of leakage loss of guided waves. For the leakage analysis, the angular spectrum representation or the energy distribution versus propagation angle (or vertical slowness) will be very useful and can show clearly which part of the energy would be trapped in the wave guide and which part of the energy would leak into the mantle.

In first-order approximation, the anelastic (intrinsic) attenuation is additive to the leakage loss, so that we can calculate and analyze these two attenuation mechanisms separately. For the Lg RMS amplitude attenuation, one more attenuation mechanism is involved:

$$b_g = kQ_g^{-1} = k(Q_a^{-1} + Q_b^{-1} + Q_l^{-1} + Q_d^{-1}) \qquad (19)$$

where Q_d^{-1} is the equivalent Q of diffusion loss, which represents the amplitude decrease of Lg due to the transfer of coherent energy into incoherent energy (Lg coda) by random heterogeneities.

Shown in Fig. 8 is a comparison between angular spectra from a waveguide model with 5% RMS random velocity perturbation in the crust and a reference flat crust without velocity perturbation. The perturbation has an exponential correlation function with horizontal and vertical characteristic scales (correlation lengths) of 5.0 and 3.0 km, respectively. From the top panel to the bottom panel

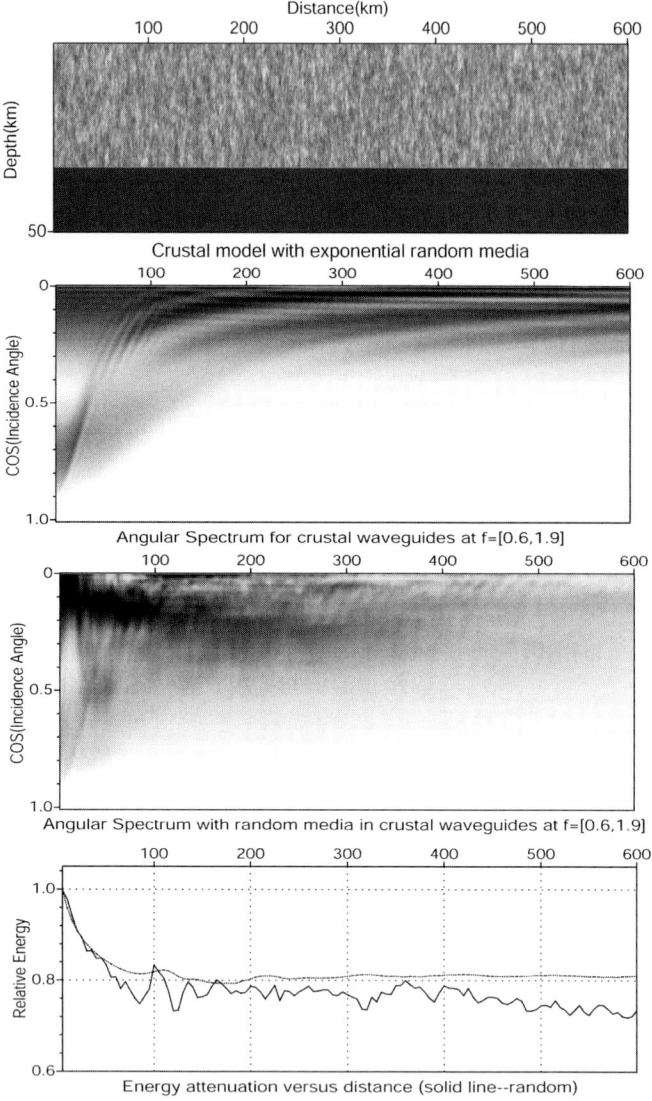

FIG. 8. Energy distribution for different crustal models. From top to bottom are: waveguide model with 5% RMS velocity perturbations in the crust; energy angular spectra versus distance for a flat crust; energy angular spectra versus distance for random crust; and relative energy attenuations versus distance. The dotted line is for the flat crust model and the solid line is for the random crust model.

in Fig. 8, they are a random velocity model, energy distribution for homogeneous crust, energy distribution for random crust; and energy attenuation curves versus distances, respectively. For energy distributions, the vertical coordinate is the normalized vertical slowness K_z/k, corresponding to the cosine of incident angles (or sine of the grazing angles). Note that zero vertical slowness means horizontal propagation. The frequency range is between 0.6 and 1.9 Hz. For the flat crust model, there is a considerable portion of energy with large vertical slowness (or steep angles) at the initial stage. After multiple reflections, energy with larger vertical slowness is depleted due to the leakage to the mantle, leaving the energy with small vertical slowness, i.e., the guided waves, propagating in the waveguide. For the model with random velocity perturbations, the distinct feature is the continuous energy repartition, moving from small (grazing) angle waves to large-angle waves due to scattering by small scale heterogeneities. The energy propagating with large angles tends to leak into the mantle and causes Lg-wave energy attenuation. The bottom panel of Fig. 8 is the wave energy attenuation versus the distance. The energy is calculated from synthetic seismograms on the free surface. The dotted line is for the reference (homogeneous) crust model. It can be seen that for this case, after passing 100 km or more, the energy is basically kept constant, which means that the trapped mode has been formed. The solid line is for the random waveguide. Due to the scattering, the energy is decreasing with distance.

Figure 9 gives the attenuation curves for different characteristic scales. The upper panel is the attenuation curve of total energy, which is the energy contained in the entire seismogram recorded on the surface. The thin solid line is for $ka = 1$, the thick solid line is for $ka = 10$, and the dashed line is for the reference (homogeneous) model. We see that for the reference model, the total energy remains constant beyond critical distance, which serves as a checking point for the numerical simulations. The middle panel gives the coherent Lg energy, which is calculated using waves within the Lg window (group velocity between 3.7 km/s and 3.2 km/s) versus distance. Again, the thin, thick and dashed lines are for $ka = 1$, $ka = 10$ and the reference model, respectively. In both measurements, the cases with $ka = 1$ are always associated with stronger attenuation than $ka = 10$ cases. We also see that the coherent Lg energy corresponding to the peak amplitude suffers more attenuations than the total energy. This is due to the extra attenuation, i.e. the diffusion loss which scatters the waves out of the Lg window and transfers them into incoherent waves (Lg coda). However, in these numerical simulations, there is no intrinsic attenuation, and leakage attenuation dominates. The difference between the coherent energy attenuation and the total energy attenuation is relatively small. In the bottom panel of Fig. 9, we plot the curve of apparent inverse quality factor for leakage attenuation Q_l^{-1} versus the normalized scale length (ka) of random heterogeneities, where $k = 2\pi/\lambda$ with λ being the wavelength of the dominant frequency, and a the correlation length. Since no intrinsic (anelastic) attenuation exists in the model and no backscattering is involved, the attenuation is solely caused by the leakage loss due to scattering. From the curve we see that Q_l^{-1} reaches its peak at $ka \approx 1.5$–2.0 and keeps flat until $ka \approx 8.0$. This is a feature of large-angle

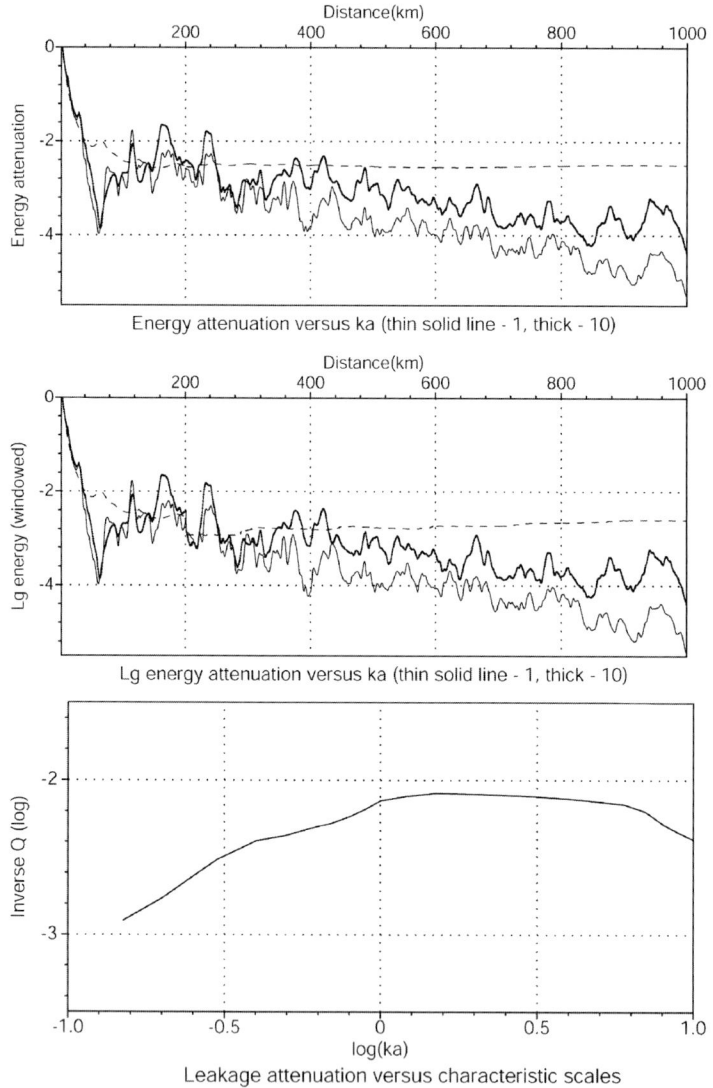

FIG. 9. Total energy attenuation (top panel), and windowed Lg energy attenuation between group velocities 3.1 km/s and 3.7 km/s (middle panel) versus distance for $ka = 1$ (thin lines) and $ka = 10$ (thick lines). The bottom panel shows the equivalent Q^{-1} for leakage attenuation versus the normalized scale length ka. The dashed line is for the reference model of a homogeneous crust.

forward-scattering dominance. For backscattering, the maximum scattering Q^{-1} is around $ka \approx 1.0$ and decreases rapidly at $ka > 1.0$ for exponential correlation functions; while for large-angle forescattering, the plateau is quite wide after

$ka = 1.0$ (Wu, 1982; Frankel and Clayton, 1986). The numerical simulations agree well with the scattering theory. The values of the equivalent Q (300–900 for $f_0 = 1$ Hz) are comparable with some observations (Xie and Mitchell, 1991; Xie, 1993). This suggests that the leakage attenuation caused by small-scale random heterogeneities may be responsible and even the dominant mechanism for some observed Lg attenuations and blockages.

3.5. SH-Waves in Crustal Waveguides with Irregular Surface Topography

Theoretical studies and observations show that surface topography is one of the important factors affecting Lg wave propagation. For example, irregular surface can cause anomalous variation of Lg amplitude along the propagation path (Sills, 1978; Geli et al., 1988; Bouchon and Barker, 1996). Methodologically, range-independent boundary conditions for flat surface must be replaced by range-dependent boundary condition for an irregular surface. In the case of surface topography, the global mirror symmetry no longer exists. To use the GSP method for solving range-dependent boundary condition problems, both conformal and non-conformal coordinate transforms were incorporated into the GSP method and their relative merits and accuracies were analyzed (Wu et al., 1999; Wu and Wu, 2001). The following is a summary of these two approaches.

3.5.1. Conformal Coordinate Transform Method for Smoothly Varying Topography

For a flat free surface, Wu et al. (2000a) derived a half-space GSP solution for Lg wave propagation. In the case of irregular topography, the global mirror symmetry for the problem no longer exists. However, taking a local plane-surface approximation for the topography, we can modify the mirror wavefield method to a local mirror wavefield method and apply the corresponding coordinate transform to obtain a GSP solution for the irregular topography.

Figure 10 shows the geometry of the derivation. Assume that $u_0^+(x', z)$ is the wavefield on the half-screen S^+ in the lower half-space. To calculate the wavefield in the next screen with the existence of a locally dipping surface, we first obtain the mirror wavefield $u_0^-(\tilde{x}', \tilde{z})$ on the half-screen S^- in the upper half-space. The total wavefield in the next screen is composed of contributions from incident waves $u_0^+(x', z)$ and $u_0^-(\tilde{x}', \tilde{z})$ plus the scattered field which is generated by the local heterogeneities in the thin-slab. The effects of the heterogeneities and the topography can be calculated separately for each step in the GSP method. The effect of the slant free-surface can be incorporated into the propagation integral. Assume $u_t(x, z)$ is the total field including the scattering effect of the volume heterogeneities. The wavefield $u(x_1, z_1)$ can be calculated by the Kirchhoff inte-

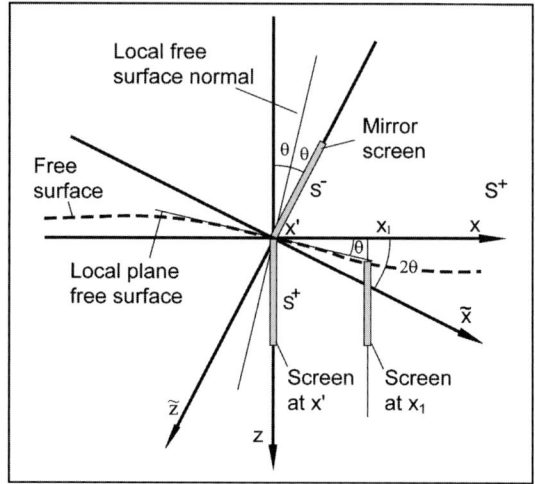

FIG. 10. Geometry of the conformal coordinate transform.

gral

$$u(x_1, z_1) = \int_S ds \left\{ g(x, z; x_1, z_1) \frac{\partial u_t(x, z)}{\partial n} - \frac{\partial g(x, z; x_1, z_1)}{\partial n} u_t(x, z) \right\}$$
$$= \int_{S^-} ds \{\cdots\} + \int_{S^+} ds \{\cdots\}, \quad (20)$$

where $g(\cdot)$ is the Green's function for the full space with the background velocity, $S = S^+ + S^-$ is the integration surface composed of lower and upper half-surfaces S^+ and S^-, respectively. The Rayleigh integral can be used to replace the Kirchhoff integral for each half surface integral. For the lower half-surface the contribution of S^+ is

$$u_t^+(x_1, z_1) = -2 \int_0^\infty dz\, u_t^+(x, z) \frac{\partial g(x, z; x_1, z_1)}{\partial n}$$
$$= \frac{1}{2\pi} \int dK_T\, e^{iK_T z_1} u_t^+(x_1, K_T), \quad (21)$$

where

$$u_t^+(x_1, K_T) = e^{i\gamma(x_1 - x)} \int_0^\infty dz_1\, u_t^+(x, z_1) e^{-iK_T z_1}. \quad (22)$$

Here $u_t^+(x, z)$ is the total wavefield composed of incident field $u_0^+(x, z)$ and the scattered field $U^+(x, z)$ caused by the heterogeneities within the slab (see Wu, 1994; Wu et al., 2000a). If we put the slab entrance at $x = x'$ and the wavefield on the entrance surface S^+ as $u_t^+(x', z')$, then

$$u_t^+(x', z') = u_0^+(x', z') + U^+(x', z'), \quad (23)$$

where

$$U^+(x', z') = k^2 \int_{x'}^{x_1} dx\, e^{-i\gamma(x_1-x')} \int_0^\infty dz \Big\{ g(x, z; x_1, z_1)\varepsilon_\rho(x, z)u_0(x, z)$$
$$- \frac{1}{k^2}\nabla g(x, z; x_1, z_1) \cdot \varepsilon_\mu(x, z)\nabla u_0(x, z) \Big\}. \tag{24}$$

For the bent upper half surface, we perform a coordinate transform by clockwise rotation of 2θ to a new coordinate system $(\tilde{x}, \tilde{z})$. Taking the downward direction as positive z-direction and the rotation angle from x to z as positive, the relation connecting the two systems is

$$\tilde{x} = x\cos 2\theta + z\sin 2\theta,$$
$$\tilde{z} = -x\sin 2\theta + z\cos 2\theta. \tag{25}$$

In the new system, the surface S^- is parallel to the $\tilde{z}$-axis, so that

$$u_t^-(\tilde{x}_1, \tilde{K}_T) = e^{i\tilde{\gamma}(\tilde{x}_1-\tilde{x})} \int_{-\infty}^0 d\tilde{z}'\, u_t^-(\tilde{x}', \tilde{z}')e^{-i\tilde{K}_T\tilde{z}'}, \tag{26}$$

where $u_t^-(\tilde{x}', -\tilde{z}') = u_t^+(x', z')$. The field in the space domain can be obtained by synthesizing the contributions from all plane waves

$$u_t^-(\tilde{x}_1, \tilde{z}) = \int d\tilde{K}_T\, e^{i\tilde{\gamma}(\tilde{x}_1-\tilde{x}')}e^{i\tilde{K}_T\tilde{z}'}u_t^-(\tilde{x}', \tilde{K}_T), \tag{27}$$

where

$$u_t^-(\tilde{x}', \tilde{K}_T) = \int_{-\infty}^0 d\tilde{z}'\, u_t^-(\tilde{x}', \tilde{z}')e^{-i\tilde{K}_T\tilde{z}'}. \tag{28}$$

Transform back to the original coordinate system, resulting in

$$u_t^-(x_1, z_1) = \int d\tilde{K}_T \exp\{i[(\tilde{\gamma}\cos 2\theta - \tilde{K}_T\sin 2\theta)(x_1 - x')$$
$$+ (\tilde{\gamma}\sin 2\theta + \tilde{K}_T\cos 2\theta)z_1]\}u_t^-(\tilde{x}', \tilde{K}_T). \tag{29}$$

We see that in the original coordinate system, the effective transversal and propagating wavenumbers are

$$K_T = \tilde{\gamma}\sin 2\theta + \tilde{K}_T\cos 2\theta,$$
$$\gamma = \tilde{\gamma}\cos 2\theta - \tilde{K}_T\sin 2\theta. \tag{30}$$

If we transform the $(\tilde{K}_T, \tilde{\gamma})$ system into (K_T, γ),

$$u_t^-(x_1, z_1) = \int dK_T\, u_t^-(\tilde{x}', K_T\cos 2\theta - \gamma\sin 2\theta)e^{i\gamma(x_1-x')}e^{iK_T z_1}. \tag{31}$$

The total field is a summation of the contributions from both $u_t^+(x_1, z_1)$ and $u_t^-(x_1, z_1)$

$$u(x_1, z_1) = \int dK_T \, e^{i\gamma(x_1-x')} e^{iK_T z_1}$$
$$\times \left[u_t^+(x', K_T) + u_t^-(\tilde{x}', K_T \cos 2\theta - \gamma \sin 2\theta) \right]. \quad (32)$$

When small-angle waves prevail such as in the case of Lg propagation, the spectral interpolation in Eq. (32) can be avoided and replaced by operations in the space domain using a narrow-angle approximation. From (23), it can be seen that to calculate the reflection response we need to find the spectral component $u_t^+(-\tilde{K}_T)$. We try to obtain the approximate space-domain operations corresponding to the wavenumber-domain interpolation. We know that

$$u_t^+(-\tilde{K}_T) = \int_0^\infty dz \, e^{i(-K_T \cos 2\theta + \gamma \sin 2\theta)z} u_t^+(z). \quad (33)$$

With narrow-angle approximation, $\gamma \approx k$, therefore,

$$u_t^+(-\tilde{K}_T) = \int_0^\infty dz' \, e^{iK_T z'} \left[\frac{1}{\cos 2\theta} e^{ik(\tan 2\theta)z'} u_t^+\left(\frac{z'}{\cos 2\theta}\right) \right], \quad (34)$$

where θ is the dipping angle of the free surface at $x = x'$. We see that the wavenumber-domain interpolation is transformed into a space-domain operation which is a modulation plus a coordinate stretching. For a flat surface, Eq. (32) reduces to the original half-space GSP method (Wu et al., 2000a).

Shown in Fig. 11 are the synthetic seismograms obtained using the conformal screen method for a Gaussian hill model (Fig. 11a). The Gaussian hill is represented by $h(x) = -h_0 \exp[-(x-x_0)^2/2\sigma^2]$ with $x_0 = 62.25$ km, $h_0 = 4$ km, and $\sigma = 9.129$ km. Synthetic seismograms calculated with a more accurate boundary element method (Fu and Wu, 2001) are also given as a reference. The solid lines are from the screen method and the dashed lines are from boundary element method. The comparison indicates that the screen method gives a satisfactory result. It correctly modeled waveforms between distance 60 and 70 km, where two reflections from the convex free surface interfere with each other and generate complex waveforms. Note that the coordinate stretch $z/\cos 2\theta$ increases very fast with large angle θ, the conformal screen method works only for smoothly varying topography.

3.5.2. Non-Conformal Coordinate Transform Method for Rough Topography

Another alternative approach is to incorporate surface flattening transform into the GSP method. The transform converts surface height perturbations into modified volume perturbations. In this way the range-dependent boundary condition becomes a stress release boundary condition on a flat surface in the new coordinate system where the half-space GSP method is applicable. The transform is

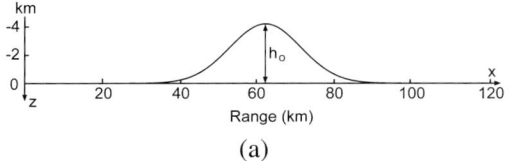

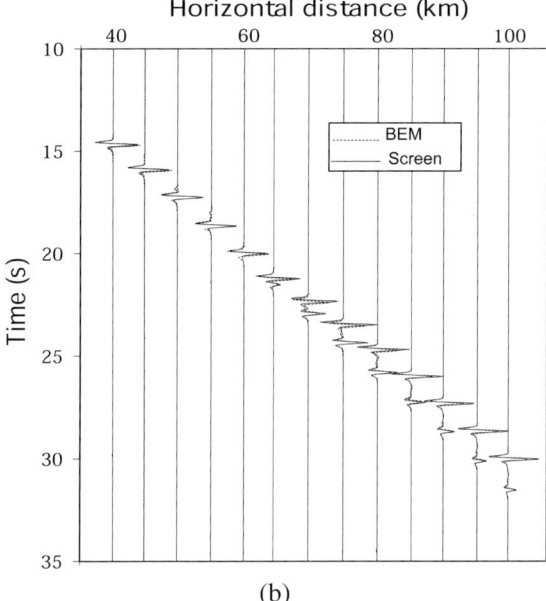

FIG. 11. (a) Velocity model with a Gaussian hill topography and (b) synthetic seismograms calculated from this model. For the calculation, $dx = dz = 0.25$ km and $dt = 0.05$ s. The source is located at a depth of 32 km and the dominant frequency of source time function is 3 Hz. Receivers are on the free surface. The solid lines are synthetic seismograms calculated using the screen method with a conformal transform, and the dashed lines are synthetic seismograms calculated with boundary element method (Fu and Wu, 2001).

defined as (Beillis and Tappert, 1979)

$$\begin{cases} \chi = x, \\ \zeta = z - h(x), \end{cases} \tag{35}$$

where $h(x)$ is the height function of free surface. Equation (35) shows that the transform gives only a shift to depth variable z, i.e., depth measurement starts from the free surface. Thus, it is a non-conformal transform. Using the above transform and perturbation theory, the original half-space screen propagator becomes (Wu and Wu, 2001)

$$\hat{u}(\chi_1, k_\varsigma) = e^{i\gamma \Delta \chi} C \left\{ e^{ik_0 S_s(\zeta)} C^{-1} \left[\hat{u}_0(\chi', k'_\zeta) \right] \right.$$
$$\left. - Z(\chi_1) \frac{\hat{\mu}(\zeta)}{\mu_0} S^{-1} \left[k'_\zeta \hat{u}_0(\chi', k'_\zeta) \right] \right\}, \tag{36}$$

where C and C^{-1} are the forward and inverse cosine transforms, and S^{-1} is the inverse sine transform defined by Eq. (9), μ_0 is shear modulus of background medium, S_S is the relative slowness perturbation of the thin-slab and given by Eq. (13), and

$$Z(\chi_1) = h(x_1) - h(x'). \tag{37}$$

Equation (36) is expressed in the new coordinates (χ, ζ). It is clear that the second term in the bracket in Eq. (36) comes from the roughness of topography, which is proportional to the height difference of the adjacent two screens for each forward step. For the upgoing slope, $Z(\chi) < 0$, the field scattered by topography is in-phase with the background field and strengthens the background field, while for downgoing slope $Z(\chi) > 0$, the field scattered by topography is out-phase with the background field and weakens the background field. Equation (36) is computationally efficient, in which all calculations can be done by FFT.

Shown in Fig. 12 are synthetic seismograms calculated using the non-conformal screen method for the Gaussian hill model shown in Fig. 11. The solid lines are from the screen method and the dashed lines are from the boundary element method. The excellent agreement between the two methods is clearly seen except at the vicinity of the hill top where a small discrepancy exists both in wave shapes and amplitudes. The error can be reduced by using a smaller step

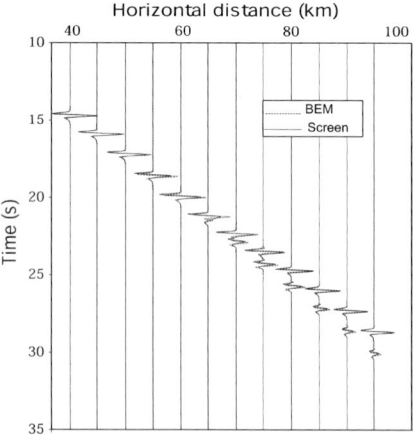

FIG. 12. Synthetic seismograms for a Gaussian hill model (Fig. 11a). The solid lines are calculated using the screen method with a non-conformal transform and the dashed lines are calculated using the boundary element method. The parameters for the calculation are the same as in Fig. 11.

SIMULATION OF HIGH-FREQUENCY REGIONAL WAVE

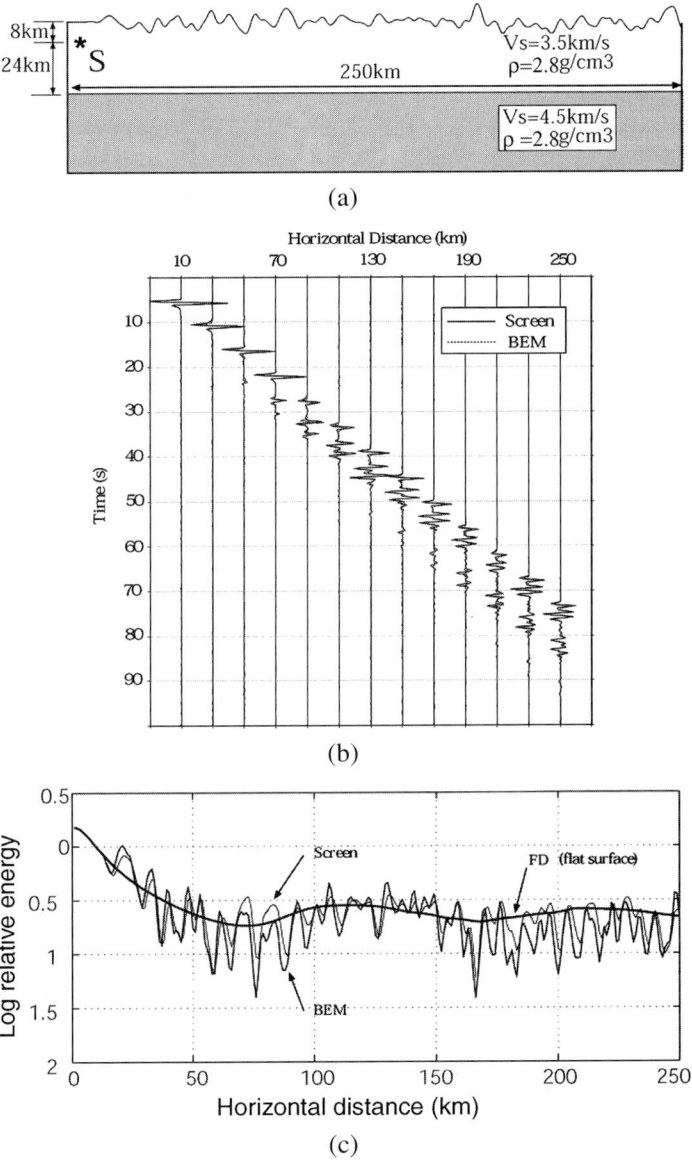

FIG. 13. (a) A crustal model with a rough random surface. The correlation length is 2.5 km, RMS perturbation is 0.6 km. (b) Synthetic seismograms, and (c) energy distribution versus horizontal distance. (b) and (c) show a comparison between the non-conformal screen method and BE method for a crustal waveguide with a rough random surface. The thick smoothly varying curve in (c) is calculated with finite difference method for a uniform waveguide. The source is located at the depth of 8 km, the dominant frequency of the source time function is 1 Hz.

length Δx. For forward marching algorithms, the step length Δx can be adjusted according to the roughness of topography. The more severe the topography is, the finer the step length Δx should be. Therefore, the non-conformal screen method can handle more severe topography than the conformal screen method. Figure 13a shows a crustal model with a rough random surface used for testing feasibility and accuracy of the non-conformal screen method. The correlation length is 2.5 km, RMS height fluctuation is 0.6 km. Figures 13b and c show a comparison of synthetic seismograms calculated by the non-conformal screen method and BE method, and the corresponding energy attenuation curves, respectively. The thick smoothly varying curve in Fig. 13c is the energy distribution for a similar waveguide but with a flat free surface. We see that the presence of a rough random surface makes the waveforms and attenuation curves more complicated. Except for large-angle Moho reflections, the results of the screen method agree well with those of the BE method. However, for this example, the screen method took about 35 minutes, while the BE method took about 72 hours.

Figure 14 shows an investigation of the combining effect of rough topography and volume heterogeneity on Lg wave propagation using the non-conformal screen method. The rough topography is the same as shown in Fig. 13. The heterogeneities are velocity variations only. The correlation lengths are 6 km in range and 4 km in depth, RMS velocity fluctuations are 5% and 10%, respectively. The thickly dashed line calculated by finite-difference method for a uniform crustal waveguide is used as a reference. We see from Fig. 14 that random heterogeneities combined with rough topography drastically increase the attenuation of high frequency Lg waves. This example shows that the non-conformal screen method can handle the effects of both volume heterogeneities and moderately rough topography on Lg wave propagation at long distances and high frequencies.

4. P-SV CASE

To introduce the P-SV elastic screen propagator for a flat free surface, the basic idea is to incorporate plane wave reflection calculation into elastic screen method (Wu, 1994; Wu *et al.*, 2000c). The half space is extended upward in vertical direction from free surface. The extended part has the parameters of background medium and will be used to keep records of upgoing waves which can be used for the calculation of reflected/converted waves by the free surface. The incident P and S waves at vertical profile $x = x'$ can be decomposed into a superposition of plane waves $\mathbf{u}_0^P(K_z, x')$ and $\mathbf{u}_0^S(K_z, x')$. The propagating waves are represented by the real vertical wavenumbers, and the imaginary vertical wavenumbers correspond to the surface waves (inhomogeneous waves). Reflection at the free surface can be calculated at each step, and the total field including the reflected waves will interact with the heterogeneities. We will first treat the propagating waves (homogeneous waves) and then discuss the calculation of the fundamental mode Rayleigh wave as an example of the surface wave modeling.

SIMULATION OF HIGH-FREQUENCY REGIONAL WAVE

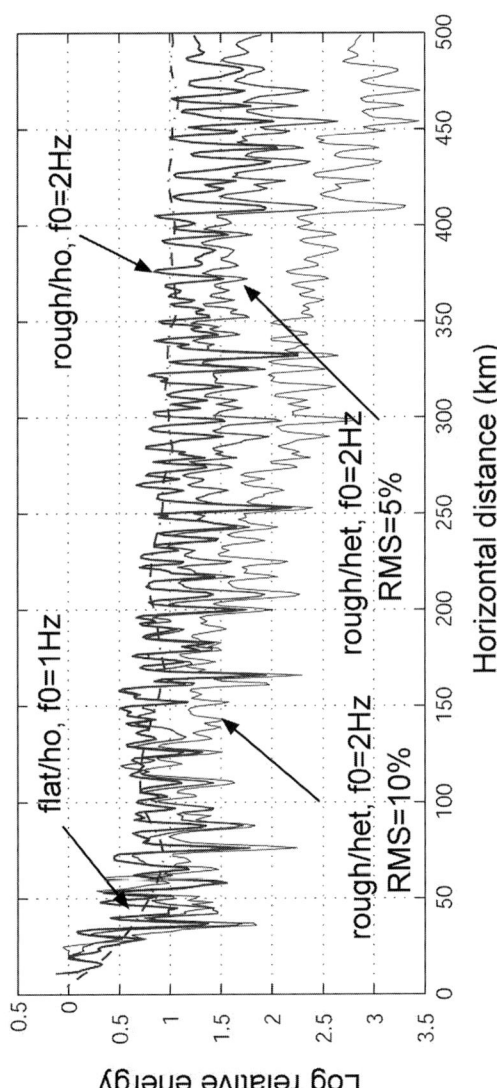

FIG. 14. Lg wave attenuation versus horizontal distances. A random medium whose correlation lengths are 6 km in range and 4 km in depth, and RMS velocity fluctuations are 5% and 10%, respectively. The source is located at a depth of 8 km, the dominant frequency (f_0) of source time function is 2 Hz. In the figure, "rough" means the crust with rough topography, "ho" and "het" denote homogeneous and heterogeneous crustal models, respectively.

Applying the reflection coefficients, the free surface reflected P and S waves due to incident P wave can be expressed by

$$\mathbf{u}^{PP}(x, z) = e^{i\gamma_\alpha(x-x')} \int dK_z \, |\mathbf{u}_0^P(K_z, x')| PP \hat{\mathbf{a}}_1 e^{-iK_z z}, \tag{38}$$

$$\mathbf{u}^{PS}(x, z) = e^{i\gamma_\alpha(x-x')} \int dK_z \, |\mathbf{u}_0^P(K_z, x')| PS \hat{\mathbf{a}}_2 e^{-iK_z^* z}, \tag{39}$$

where $\gamma_\alpha = \sqrt{k_\alpha^2 - K_z^2}$ is the propagating wavenumber for P waves (here in the x-direction) and $K_z^* = \sqrt{k_\beta^2 - k_\alpha^2 + K_z^2}$ is the transverse wavenumber of converted S waves determined by Snell's law. Unit vectors $\hat{\mathbf{a}}_1 = (\gamma_\alpha, -K_z)/k_\alpha$ and $\hat{\mathbf{a}}_2 = (K_z^*, \gamma_\alpha)/k_\beta \cdot |\mathbf{u}_0^P(K_z, x')|$ is the scalar spectrum of the incident P wave with a transverse wavenumber K_z (here in z-direction). The reflected P and S waves due to the incident plane S wave can be obtained by

$$\mathbf{u}^{SP}(x, z) = e^{i\gamma_\beta(x-x')} \int dK_z \, |\mathbf{u}_0^S(K_z, x')| SP \hat{\mathbf{a}}_3 e^{-iK_z'^* z}, \tag{40}$$

$$\mathbf{u}^{SS}(x, z) = e^{i\gamma_\beta(x-x')} \int dK_z \, |\mathbf{u}_0^S(K_z, x')| SS \hat{\mathbf{a}}_4 e^{-iK_z z}, \tag{41}$$

where $\gamma_\beta = \sqrt{k_\beta^2 - K_z^2}$ is the propagating wavenumber for S waves (here in the x-direction) and $K_z'^* = \sqrt{k_\alpha^2 - k_\beta^2 + K_z^2}$ is the transverse wavenumber of the reflected P wave. Unit vectors $\hat{\mathbf{a}}_3 = (\gamma_\beta, -K_z'^*)/k_\alpha$ and $\hat{\mathbf{a}}_4 = (K_z, \gamma_\beta)/k_\beta$. $|\mathbf{u}_0^S(K_z, x_1)|$ is the scalar spectrum of the incident S wave with a transverse wavenumber K_z. PP, PS, SP and SS in Eqs. (38)–(41) are reflection coefficients of different wave types at the free surface (Aki and Richards, 1980). Figure 15 is an example of those reflection coefficients versus horizontal slowness (ray parameter p). In Fig. 15, p_A corresponds to P slowness (inverse velocity) and p_B to S slowness. For $p < p_A$, P and S waves are both homogeneous waves, their transverse wavenumbers are real. For $p > p_B$, P and S waves are both inhomogeneous waves, their transverse wavenumbers are imaginary. For $p_A < p < p_B$, P wave is inhomogeneous while S wave is homogeneous. A Rayleigh pole is located in the region of $p > p_B$. In general, we can calculate all reflected waves using Eqs. (38)–(41), once the incident fields $|\mathbf{u}_0^P|$ and $|\mathbf{u}_0^S|$ are known. However, numerically, it is more convenient to separate the calculation of Eqs. (38)–(41) into homogeneous and inhomogeneous waves, respectively.

For homogeneous waves, Eqs. (38) and (41) (common-type) can be implemented by FFT. However, the reflected waves of converted-type cannot be directly implemented by FFT because the nonlinear relationship exists between K_z and K_z^* for P–S conversion (or K_z and $K_z'^*$ for S–P conversion). Although we can obtain uniform samples with respect to K_z and K_z^* (or K_z and $K_z'^*$) by complex variable interpolation to make FFT applicable, numerical tests have shown that

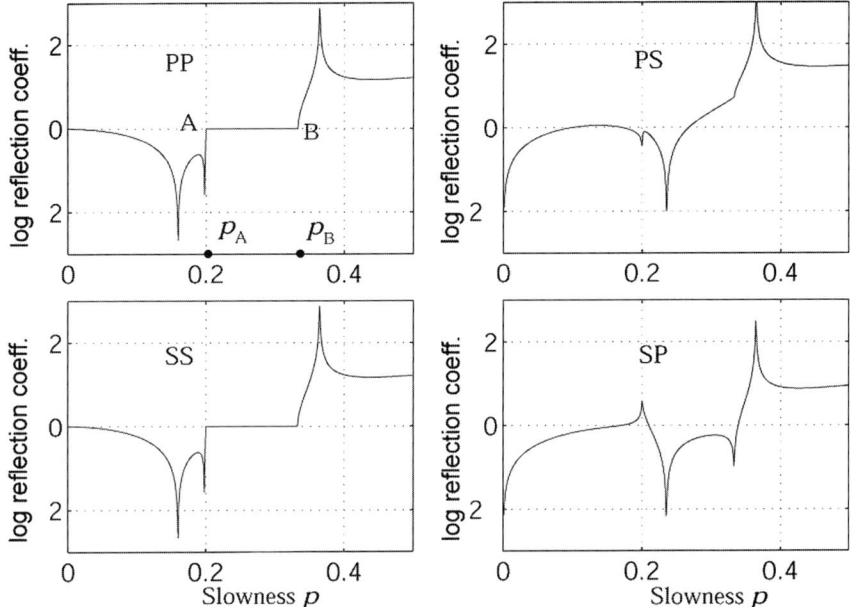

FIG. 15. The free surface reflection coefficients (in logarithmic scale) versus horizontal slowness. The P and S wave velocities for the elastic half-space are $\alpha = 5$ km/s and $\beta = 3.5$ km/s. The p_A and p_S denote P and S slownesses.

the noise due to the interpolation is so strong that the accumulated errors increase very fast for multiple step propagation. In our study, the direct summations over the incident waves ($p < p_A$ for P incidence or $p < p_B$ for S incidence) are performed for calculating the converted reflections. Figure 16 shows synthetic seismograms calculated with Eqs. (38)–(41) for an elastic half-space with only homogeneous waves. The results calculated with wavenumber integration (WI) method (dashed lines) are also shown as references. Since the source is deep compared with the propagation distance, Rayleigh wave is very weak in the exact solution. Figure 16a shows the vertical component of the displacement and Fig. 16b shows the horizontal component. From Fig. 16 we see that the calculations of the reflection and conversion by the free surface are in excellent agreement with the theory. Figure 17 shows synthetic seismograms for Flora–Asnes crustal model (see, Fig. 18) using elastic screen method. A double-couple source is located at a depth of 16 km and has a dominant frequency of 2 Hz. We see that both P and S waves are well excited. Figure 18 is the corresponding snapshots. From Figs. 17 and 18, the short-period phases Pn, Sn, Lg, etc., can be identified. For the elastic screen method at its current stage, only real transverse wavenumbers are used in FFT, which can only handle propagating waves (homogeneous waves).

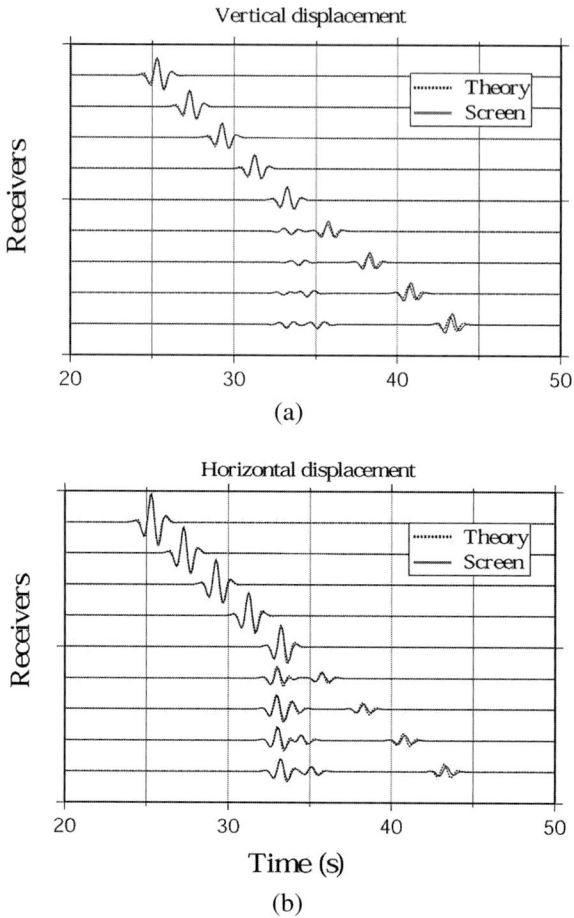

FIG. 16. Synthetic seismograms calculated by the elastic screen method (solid lines) and wavenumber integration method (dashed lines) for an elastic halfspace. Only homogeneous waves are included in the results of elastic screen method. (a) Shows the vertical components of displacement; (b) Shows the horizontal components. A point explosion source is located at the depth of 16 km and the dominant frequency of source time function is 1 Hz. The first 4 receivers are placed along the free surface separated from the source by 100–124 km, and the last 5 receivers are placed in a vertical profile at an epicenter distance of 132 km and with depths ranging from 0–32 km.

For inhomogeneous waves, their transverse wavenumbers are imaginary so that Eqs. (38)–(41) cannot be calculated by FFT. However, the imaginary transverse wavenumber makes the propagation of inhomogeneous waves simple. The phase advance takes place only along the horizontal direction. It can be easily incorporated into the screen method, once the spectra of inhomogeneous waves

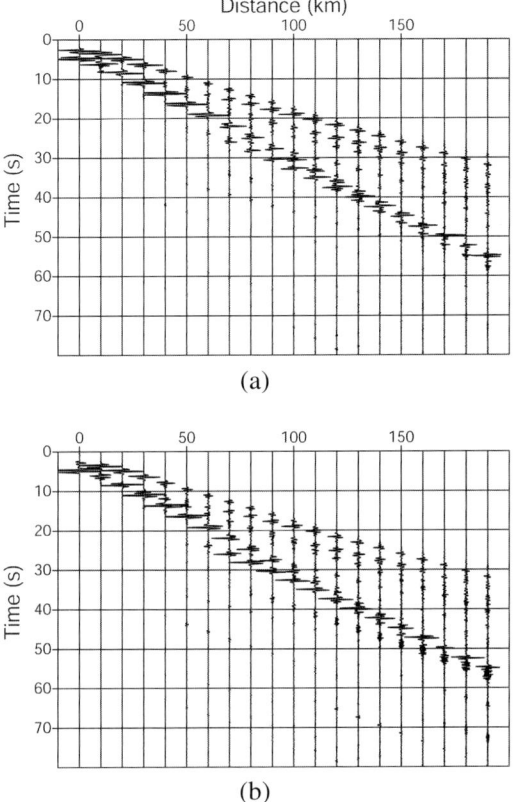

FIG. 17. Synthetic seismograms for Flora–Asnes crustal model (see Fig. 18) using P-SV elastic screen method. Only homogeneous wave are involved. (a) Shows the vertical components of displacement and (b) shows the horizontal components. A double-couple source is located at the depth of 16 km and has a dominant frequency of 2 Hz. Receivers are on the surface.

are known. Another important feature of inhomogeneous waves is the exponential decay only in the direction perpendicular to propagation direction. Then the spectra of inhomogeneous waves can be calculated with Laplace transform. Figure 19 shows an example of such a treatment for Rayleigh wave propagating in homogeneous elastic half-space. The source is located at a depth of 2 km and has a dominant frequency of 0.5 Hz. The vertical receiver array is located at a distance of 100 km. Figure 19a shows the horizontal component of Rayleigh wave synthetic seismograms and Fig. 19b shows the vertical component of synthetic seismograms. The solid lines are exact solution. The agreement between the screen calculation and the exact solution is excellent. The interaction between inhomogeneous waves and heterogeneities and the conversion between body waves and surface wave are still on-going research.

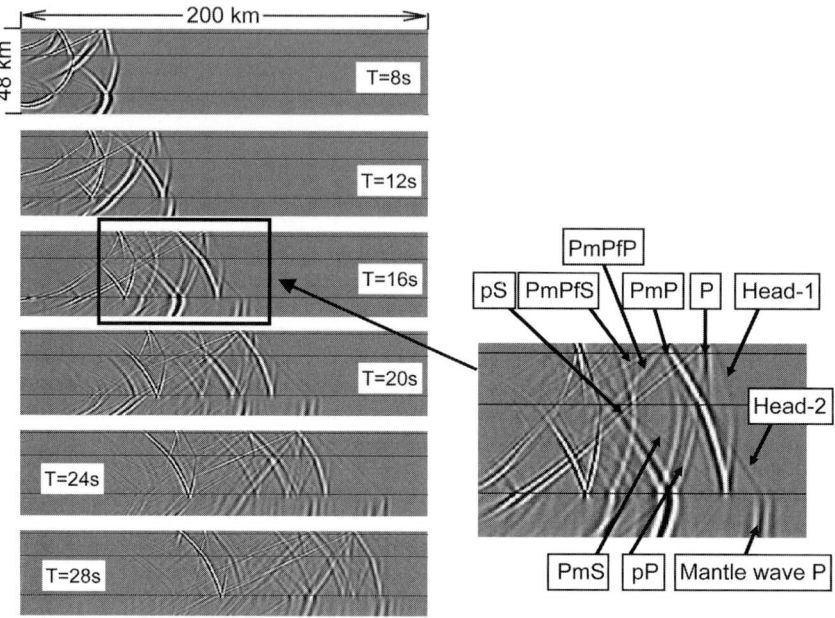

FIG. 18. Snapshots (horizontal component of displacement) for Flora–Asnes crustal model using P-SV elastic screen method. A double-couple source is located at the depth of 16 km and has a dominant frequency of 2 Hz. The thicknesses of layers (from top to bottom) are 1 km, 14 km, 22 km and infinity, respectively. Their velocity and density parameters are $\alpha_1 = 5.2$ km/s, $\beta_1 = 3$ km/s, $\rho_1 = 2.6$ g/cm^3; $\alpha_2 = 6.0$ km/s, $\beta_2 = 3.46$ km/s, $\rho_2 = 2.8$ g/cm^3; $\alpha_3 = 6.51$ km/s, $\beta_3 = 3.76$ km/s, $\rho_3 = 3$ g/cm^3; $\alpha_4 = 8.05$ km/s, $\beta_4 = 4.65$ km/s, $\rho_4 = 3.3$ g/cm^3. The major phases are labeled in the figure.

5. Conclusion

In the crustal waveguide environment, the major part of wave energy is carried by forward propagating waves, including forward scattered waves. Therefore, the neglect of backscattered waves in the modeling can still simulate the main features of regional waves in most cases. By neglecting backscattering in the theory, the method becomes a forward marching algorithm. A half-space screen propagator (generalized screen propagator) has been developed to accommodate the free-surface boundary condition and treat the SH wave propagation in complex crustal waveguides. The SH screen propagator has also been extended to the case of irregular surface topography by conformal or non-conformal topographic transforms. For medium sized problems, the screen-propagator method is two to three orders of magnitude faster than the finite-difference methods.

SIMULATION OF HIGH-FREQUENCY REGIONAL WAVE 357

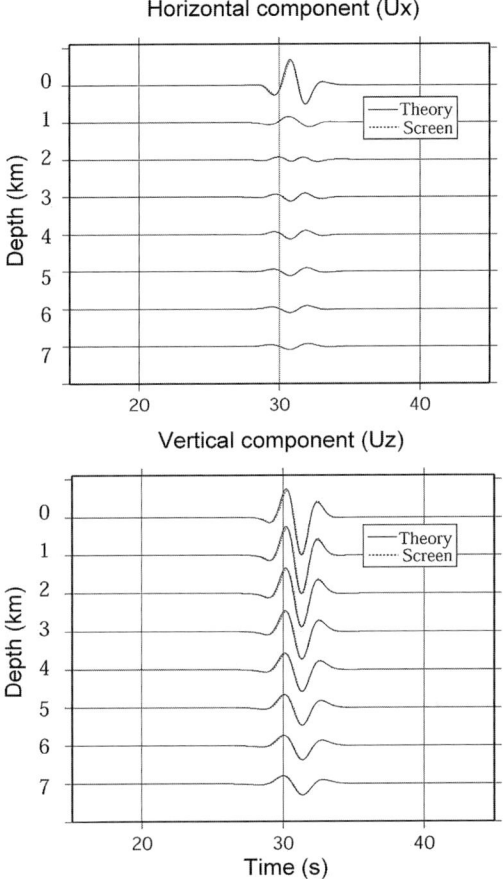

FIG. 19. Comparison of synthetic Rayleigh wave calculated using the screen method (dotted lines) with those calculated using the exact solution (solid lines). The source is located at a depth of 2 km and has a dominant frequency of 0.5 Hz. (a) Shows the horizontal components of displacement of Rayleigh wave and (b) shows the vertical components. The half-space parameters are $\alpha = 6$ km/s and $\beta = 3.5$ km/s.

In the case of P-SV elastic screen propagators, plane wave reflection calculations have been incorporated into the elastic screen method. Body waves including the reflected and converted waves can be calculated by real wavenumber integration, while surface waves (Rayleigh waves) can be obtained with imaginary wavenumber integration. Numerical tests show good agreement with the theory.

From the theoretical developments and numerical tests of both SH and P-SV screen-propagators, we see that the one-way screen propagator approach for regional wave simulation is a viable approach and the savings in computation time and computer storage are enormous, especially for large 3-D elastic, long-range wave propagation problems. For the SH screen propagators, the theory and method are rather complete and can treat various crustal models including large-scale structures, random heterogeneities, and irregular or rough surface topography. The method has been applied to the simulation of Lg propagation in random media and the related energy partition and attenuation. It is found that the leakage attenuation of Lg waves caused by large-angle forward scattering from random heterogeneities may contribute significantly to Lg attenuation and blockage in some regions. The apparent Q for leakage attenuation as a function of normalized scale length (ka) of the random heterogeneities agrees well with the scattering theory. More work can be done along this direction and comparison with observations may reveal the real mechanisms for Lg blockage in different regions.

More work has to be done for the P-SV problems, e.g., the conversion between body wave, guided wave and surface wave caused by crustal heterogeneities and irregular topography. Then the full 3-D elastic screen propagator will provide the full power of simulating regional wave propagation in complex crustal models.

Acknowledgements

The helpful discussions with T. Lay, S. Jin and G. Fan are greatly appreciated. This work was supported by the Department of Energy, the Defense Threat Reduction Agency and the Air Force Research Laboratory through various contracts. Facility support from the W.M. Keck Foundation is also acknowledged. Contribution number 432 of CSIDE, IGPP, University of California, Santa Cruz.

References

Aki, K., Richards, P.G. (1980). Quantitative Seismology: Theory and Methods, vols. 1 and 2. W.H. Freeman, New York.

App, F.N., Bos, R.J., Kamm, J.R. (1996). Synthetic seismograms at regional distances for May 1995 Earthquake and Explosion source in Western China. In: *Proceedings of the 18th Annual Seismic Research Symposium on Monitoring a Comprehensive Nuclear-Test-Ban Treaty.* 4–6 September, pp. 119–128.

Archambeau, C., Orrey, J., Kohl, B. (1996). 3-D seismic wave synthesis and full wavefield inversion. In: *Proceedings of the 18th Annual Seismic Research Symposium on Monitoring a Comprehensive Nuclear-Test-Ban Treaty.* 4–6 September, pp. 129–138.

Baumgardt, D.R. (1990). Investigation of teleseismic Lg blockage and scattering using regional arrays. *Bull. Seismol. Soc. Am.* **80**, 2261–2281.

Beillis, A., Tappert, F.D. (1979). Coupled mode analysis of multiple rough surface scattering. *J. Acoust. Soc. Am.* **66**, 811–826.

Bostock, M.G., Kennett, B.L.N. (1990). The effect of 3-D structure on Lg propagation patterns. *Geophys. J. Int.* **101**, 355–365.

Bouchon, M., et al. (1985). Theoretical modeling of Lg wave attenuation. In: *The VELA Program: A Twenty-five Year Review of Basic Research*.
Bouchon, M., Barker, J. (1996). Seismic response of a hill: The example of Tarzana, California. *Bull. Seismol. Soc. Am.* **86**, 66–72.
Bradley, C.R., Jones, E.M. (1998). Modeling propagation effects from explosion in Western China and India. In: *Proceedings of the 20th Annual Seismic Research Symposium on Monitoring a Comprehensive Nuclear-Test-Ban Treaty*. 21–23 September, pp. 173–181.
Bradley, C.R., Jones, L.E. (1999). Full waveform modeling of the effects of Q and structure over subregional paths in Western China. In: *Proceedings of the 21st Annual Seismic Research Symposium on Monitoring a Comprehensive Nuclear-Test-Ban Treaty*. 21–24 September, pp. 28–38.
Campillo, M. (1990). Propagation and attenuation characteristics of the crustal phase Lg. *Pure Appl. Geophys.* **132**, 1–19.
Campillo, M., Paul, A. (1992). Influence of lower crustal structure on the early coda of regional seismograms. *J. Geophys. Res.* **97**, 3405–3416.
Campillo, M., Feignier, B., Bouchon, M., Bethoux, N. (1993). Attenuation of crustal waves across the Alpine range. *J. Geophys. Res.* **98**, 1987–1996.
Chen, X. (1990). Seismograms synthesis for multi-layered media with irregular interfaces by global generalized reflection/transmission matrices method, Part I. Theory of 2-D SH case. *Bull. Seismol. Soc. Am.* **80**, 1696–1724.
Chen, X. (1995). Seismograms synthesis for multi-layered media with irregular interfaces by global generalized reflection/transmission matrices method, Part II. Applications of 2-D SH case. *Bull. Seismol. Soc. Am.* **85**, 1094–1106.
Cormier, V.F., Anderson, T. (1996). Lg blockage and scattering at CNET and KEET. In: *Proceedings of the 18th Annual Seismic Research Symposium on Monitoring a Comprehensive Nuclear-Test-Ban Treaty*. 4–6 September, pp. 159–168.
Cormier, V.F., Anderson, T. (1997). Lg blockage and scattering at Central Eurasian Arrays CNET and ILPA. In: *Proceedings of the 19th Annual Seismic Research Symposium on Monitoring a Comprehensive Nuclear-Test-Ban Treaty*. 23–25 September, pp. 479–485.
De Wolf, D.A. (1971). Electromagnetic reflection from an extended turbulent medium: Cumulative forward-scatter single-backscatter approximation. *IEEE Trans. Antennas and Propagation* **AP-19**, 254–262.
De Wolf, D.A. (1985). Renormalization of EM fields in application to large-angle scattering from randomly continuous media and sparse particle distributions. *IEEE Trans. Antennas and Propagation* **AP-33**, 608–615.
Flatté, S.M., Dashen, R., Munk, W., Watson, K., Zachariasen, F. (1979). *Sound Transmission Through a Fluctuating Ocean*. Cambridge Univ. Press, Cambridge.
Fornberg, B. (1987). The pseudospectral method: Comparisons with finite differences for the elastic wave equation. *Geophysics* **52**, 482–501.
Frankel, A. (1989). A review of numerical experiments on seismic wave scattering. In: Wu, R.S., Aki, K. (Eds.), In: *Scattering and Attenuation of Seismic Waves*, vol. 2. Birkhauser, Berlin, pp. 639–686.
Frankel, A., Clayton, R.W. (1986). Finite difference simulations of seismic scattering: Implications for propagation of short-period seismic waves in the crust and models of crustal heterogeneity. *J. Geophys. Res.* **91**, 6465–6489.
Fu, L.Y., Wu, R.S. (2001). A hybrid BE-GS method for modeling regional wave propagation. *Pure Appl. Geophys.* **158**, 1251–1277.
Furumura, T., Kennett, B.L.N. (1997). On the nature of regional seismic phases-II. On the influence of structural barriers. *Geophys. J. Int.* **129**, 221–234.
Geli, L., Bard, P.Y., Jullien, B. (1988). The effects of topography on earthquake ground motion: A review and new results. *Bull. Seismol. Soc. Am.* **78**, 42–63.
Gibson, R.L., Campillo, M. (1994). Numerical simulation of high- and low-frequency Lg-wave propagation. *Geophys. J. Int.* **118**, 47–56.

Goff, J., Holliger, K. (1999). Nature and origin of upper crustal seismic velocity fluctuations and associated scaling properties: Combined stochastic analyses of KTB velocity and lithology logs. *J. Geophys. Res.* **104**, 0148–0227.

Goldstein, P., Bhattacharyya, J., Ichinose, G., Leach, R. (1999). On the sensitivity of broad band regional seismic phases to multi-dimensional earth structure: Implications for phase identification. In: *Proceedings of the 21st Annual Seismic Research Symposium on Monitoring a Comprehensive Nuclear-Test-Ban Treaty*. 21–24 September, pp. 58–63.

Goldstein, P., Schultz, C., Larsen, S., Minner, L. (1996). Modeling of regional wave propagation phenomena in the middle east and north Africa and new analysis capabilities in SAC2000. In: *Proceedings of the 18th Annual Seismic Research Symposium on Monitoring a Comprehensive Nuclear-Test-Ban Treaty*. 4–6 September, pp. 165–171.

Goldstein, P., Schultz, C., Larsen, S. (1997). The influence of deep sedimentary basins, crustal thinning, attenuation, and topography on regional phases: Selected examples from the Eastern Mediterranean and the Caspian sea regions. In: *Proceedings of the 19th Annual Seismic Research Symposium on Monitoring a Comprehensive Nuclear-Test-Ban Treaty*. 23–25 September, pp. 486–494.

Herrmann, R.B., Mokhtar, T.A., Raoof, M., Ammon, C. (1997). Wave propagation-16 Hz to 60 s. In: *Proceedings of the 19th Annual Seismic Research Symposium on Monitoring a Comprehensive Nuclear-Test-Ban Treaty*. 23–25 September, pp. 495–503.

Huang, L.J., Fehler, M.C., Wu, R.S. (1999a). Extended local Born Fourier migration method. *Geophysics* **64**, 1524–1534.

Huang, L.J., Fehler, M.C., Roberts, P.M., Burch, C.C. (1999b). Extended local Rytov Fourier migration method. *Geophysics* **64**, 1535–1545.

Huang, L.J., Wu, R.S. (1996). 3-D prestack depth migration with acoustic pseudo-screen propagators. In: *Mathematical Methods in Geophysical Imaging IV*. In: *Proceedings of SPIE*, vol. 2822, pp. 40–51.

Husebye, E.S., Ruud, B.O. (1996). Wave propagation in complex crust-CTBT implications. In: *Proceedings of the 18th Annual Seismic Research Symposium on Monitoring a Comprehensive Nuclear-Test-Ban Treaty*. 4–6 September, pp. 172–181.

Jih, R.S. (1996). Waveguide Effects of large-scale structural variation, anelastic attenuation, and random heterogeneity on SV Lg propagation: A finite-difference modeling study. In: *Proceedings of the 18th Annual Seismic Research Symposium on Monitoring a Comprehensive Nuclear-Test-Ban Treaty*. 4–6 September, pp. 182–194.

Jin, S., Wu, R.S. (1999). Depth migration with a windowed screen propagator. *J. Seismic Exp.* **8**, 27–38.

Jin, S., Wu, R.S., Peng, C. (1999). Seismic depth migration with screen propagators. *Comput. Geosci.* **3**, 321–335.

Jones, E.M., App, F.N., Bos, R.J. (1997). The effects of major structural features in Western China on explosion seismograms. In: *Proceedings of the 19th Annual Seismic Research Symposium on Monitoring a Comprehensive Nuclear-Test-Ban Treaty*. 23–25 September, pp. 504–513.

Jones, E.M., Holliger, K. (1997). Spectral analyses of the KTB sonic and density logs using robust non-parametric methods. *J. Geophys. Res.* **102**, 18391–18403.

Keers, H., Nolet, G., Dahlen, F.A. (1996a). Ray theoretical analysis of Lg. *Bull. Seismol. Soc. Am.* **86**, 726–736.

Keers, H., Vogfjord, G., Nolet, G., Dahlen, F.A. (1996b). High frequency propagation of crustal waves. In: *Proceedings of the 18th Annual Seismic Research Symposium o Monitoring a Comprehensive Nuclear-Test-Ban Treaty*. 4–6 September, pp. 199–205.

Kennett, B.L.N. (1984). Guided wave propagation in laterally varying media-I. Theoretical development. *Geophys. J. R. Astron. Soc.* **79**, 235–255.

Kennett, B.L.N. (1986). Lg waves and structural boundaries. *Bull. Seismol. Soc. Am.* **76**, 1133–1141.

Kennett, B.L.N. (1989). Lg-wave propagation in heterogeneous media. *Bull. Seismol. Soc. Am.* **79**, 860–872.

Kennett, B.L.N. (1990). Guided wave attenuation in laterally varying media. *Geophys. J. Int.* **100**, 415–422.
Kennett, B.L.N. (1998). Guided waves in three-dimensional structures. *Geophys. J. Int.* **133**, 159–174.
Kennett, B.L.N., Bostock, M.G., Xie, J.K. (1990). Guided-wave tracking in 3-D: A tool for interpreting complex regional seismograms. *Bull. Seismol. Soc. Am.* **80**, 633–642.
Kosloff, D., Kessler, D., Quieroz, A., Tessmer, E. (1990). Solution of the equations of Dynamic elasticity by a Chebychev spectral method. *Geophysics* **55**, 734–748.
Lay, T., Fan, G., Wu, R.S., Xie, X.B. (1999). Path corrections for regional phase discriminants. In: *Proceedings of the 21st Annual Seismic Research Symposium on Monitoring a Comprehensive Nuclear-Test-Ban Treaty*. 21–24 September, pp. 510–519.
Liu, Y.B., Wu, R.S. (1994). A comparison between phase-screen, finite difference and eigenfunction expansion calculations for scalar waves in inhomogeneous media. *Bull. Seismol. Soc. Am.* **84**, 1154–1168.
Maupin, V. (1989). Numerical modeling of Lg wave propagation across the North Sea central graben. *Geophys. J. Int.* **99**, 273–283.
Maupin, V., Kennett, B.L.N. (1987). On the use of truncated model expansion in laterally varying media. *Geophys. J. R. Astron. Soc.* **91**, 837–851.
McLaughlin, K.L., Wilkins, D. (1997). Progress in numerical methods for wave propagation. In: *Proceedings of the 19th Annual Seismic Research Symposium on Monitoring a Comprehensive Nuclear-Test-Ban Treaty*. 23–25 September, pp. 514–523.
Ni, J., Reese, C., Wu, J., Zhao, L.S. (1996). Crustal structure and attenuation in Southern Tibet. In: *Proceedings of the 18th Annual Seismic Research Symposium on Monitoring a Comprehensive Nuclear-Test-Ban Treaty*. 4–6 September, pp. 390–399.
Nolte, B., Gibson, R.L., Toksöz, M.N. (1996). Irregular-grid modeling of regional wave propagation. In: *Proceedings of the 18th Annual Seismic Research Symposium on Monitoring a Comprehensive Nuclear-Test-Ban Treaty*. 4–6 September, pp. 231–240.
Sato, H., Fehler, M.C. (1998). Seismic Wave Propagation and Scattering in the Heterogeneous Earth. Springer-Verlag, New York.
Schatzman, J.C. (1996). A pseudo-spectral scheme for viscoelastic seismic modeling. In: *Proceedings of the 18th Annual Seismic Research Symposium on Monitoring a Comprehensive Nuclear-Test-Ban Treaty*. 4–6 September, pp. 261–270.
Sills, L.B. (1978). Scattering of horizontally-polarized shear waves by surface irregularities. *Geophys. J. R. Astron. Soc.* **54**, 319–348.
Wild, A.J., Hudson, J.A. (1998). A geometrical approach to the elastic complex screen. *J. Geophys. Res.* **103**, 707–725.
Wu, R.S. (1982). Attenuation of short period seismic wave due to scattering. *Geophys. Res. Lett.* **9**, 9–12.
Wu, R.S. (1994). Wide-angle elastic wave one-way propagation in heterogeneous media and an elastic wave complex-screen method. *J. Geophys. Res.* **99**, 751–766.
Wu, R.S. (1996). Synthetic seismograms in heterogeneous media by one-return approximation. *Pure Appl. Geophys.* **148**, 155–173.
Wu, R.S., Aki, K. (Eds.) (1988). *Scattering and Attenuation of Seismic Waves*, vol. I. Birkhauser Verlag, Boston.
Wu, R.S., Aki, K. (Eds.) (1989). *Scattering and Attenuation of Seismic Waves*, vol. II. Birkhauser Verlag, Boston.
Wu, R.S., Aki, K. (Eds.) (1990). *Scattering and Attenuation of Seismic Waves*, vol. III. Birkhauser Verlag, Boston.
Wu, R.S., Huang, L.J. (1995). Reflected wave modeling in heterogeneous acoustic media using the de Wolf approximation. In: *Mathematical Methods in Geophysical Imaging-III*. In: *Proceedings of SPIE*, vol. 2571, pp. 176–186.

Wu, R.S., Huang, L.J., Xie, X.B. (1995). Backscattered wave calculation using the De Wolf approximation and a phase-screen propagator. In: *SEG 65th Annual Meeting*, pp. 1293–1296. Expanded Abstracts.

Wu, R.S., Jin, S., Xie, X.B. (1996). Synthetic seismograms in heterogeneous crustal waveguides using screen propagators. In: *Proceedings of the 18th Annual Seismic Research Symposium on Monitoring a Comprehensive Nuclear-Test-Ban Treaty*. 4–6 September, pp. 291–300.

Wu, R.S., Jin, S., Xie, X.B. (2000a). Seismic wave propagation and scattering in heterogeneous crustal waveguides using screen propagators: I SH waves. *Bull. Seismol. Soc. Am.* **90**, 401–413.

Wu, R.S., Jin, S., Xie, X.B. (2000b). Energy partition and attenuation of Lg waves by numerical simulations using screen propagators. *Phys. Earth Planet. Int.* **120**, 227–243.

Wu, R.S., Jin, S., Xie, X.B., Lay, T. (1997). Verification and applications of GSP (Generalized screen propagators) method for regional wave propagation. In: *Proceeding of 19th Annual Seismic Research Symposium on Monitoring a Comprehensive Nuclear-Test-Ban Treaty*, pp. 552–561.

Wu, R.S., Xie, X.B. (1994). Multi-screen backpropagator for fast 3-D elastic prestack migration. In: *Mathematical Methods in Geophysical Imaging-II*. In: *Proceedings of SPIE*, vol. 2301, pp. 181–193.

Wu, R.S., Xie, X.B., Jin, S., Fu, L., Lay, T. (1998). Seismic wave propagation and scattering in heterogeneous crustal waveguides using screen propagators. In: *Proceedings of the 20th Annual Seismic Research Symposium on Monitoring a Comprehensive Nuclear-Test-Ban Treaty*. 21–23 September, pp. 201–210.

Wu, R.S., Xie, X.B., Wu, X.Y. (1999). Lg wave simulations in heterogeneous crusts with irregular topography suing half-space screen propagators. In: *Proceedings of the 21st Annual Seismic Research Symposium on Monitoring a Comprehensive Nuclear-Test-Ban Treaty*. 21–24 September, pp. 683–693.

Wu, R.S., Xie, X.B., Wu, X.Y. (2000c). Lg wave propagation using SH and P-SV screen propagators in heterogeneous crusts with irregular topography. In: *Proceedings of the 22nd Annual Seismic Research Symposium on Monitoring a Comprehensive Nuclear-Test-Ban Treaty*.

Wu, R.S., Xu, Z., Li, X.P. (1994). Heterogeneity spectrum and scale-anisotropy in the upper crust revealed by the German continental deep-drilling (KTB) holes. *Geophys. Res. Lett.* **21**, 911–914.

Wu, X.Y., Wu, R.S. (1999). Wide-angle thin-slab propagator with phase matching for elastic wave modeling. In: *SEG 69th Annual Meeting*, pp. 1867–1870. Expanded Abstracts.

Wu, X.Y., Wu, R.S. (2001). Lg wave simulation in heterogeneous crusts with surface topography using screen propagators. *Geophys. J. Int.* **146**, 670–678.

Wu, X.Y., Wu, R.S. (2005). AVO modeling using elastic thin-slab method. *Geophysics*, in press.

Xie, J. (1993). Simultaneous inversion for source spectrum and path Q using Lg with application to three Semipalatinsk explosions. *Bull. Seismol. Soc. Am.* **83**, 1547–1562.

Xie, J., Mitchell, B.J. (1991). Lg coda and Q across Eurasia. In: Mitchell, B.J. (Ed.), *Yield and Discrimination Studies in Stable Continental Regions*. Phillips Laboratory, Hanscom Air Force Base, MA, pp. 77–91. Report PL-TR-91-2286.

Xie, X.B., Ge, Z., Lay, T. (2005). Investigating explosion source energy partitioning and Lg-wave excitation using a finite-difference plus slowness analysis method. *Bull. Seismol. Soc. Am.* **95**, 2412–2427.

Xie, X.B., Mosher, C.C., Wu, R.S. (2000). The application of wide-angle screen propagator to 2-D and 3-D depth migrations. In: *70th Annual International Meeting, SEG*, pp. 878–881. Expanded Abstracts.

Xie, X.B., Lay, T. (1994). The excitation of explosion Lg, a finite-difference investigation. *Bull. Seismol. Soc. Am.* **84**, 324–342. Expanded Abstracts.

Xie, X.B., Wu, R.S. (1995). A complex-screen method for modeling elastic wave reflections. In: *SEG 65th Annual Meeting*, pp. 1269–1272. Expanded Abstracts.

Xie, X.B., Wu, R.S. (1996). 3-D elastic wave modeling using the complex screen method. In: *SEG 66th Annual Meeting*, pp. 1247–1250. Expanded Abstracts.

Xie, X.B., Wu, R.S. (1998). Improving the wide angle accuracy of the screen method under large contrast. In: *SEG 68th Annual Meeting*, pp. 1811–1814. Expanded Abstracts.

Xie, X.B., Wu, R.S. (1999). Improving the wide angle accuracy of the screen propagator for elastic wave propagation. In: *SEG 69th Annual Meeting*, pp. 1863–1866. Expanded Abstracts.

Xie, X.B., Wu, R.S. (2001). Modeling elastic wave forward propagation and reflection using the complex-screen method. *J. Acoust. Soc. Am.* **109**, 2629–2635.

Xie, X.B., Wu, R.S. (2005). Multicomponent prestack depth migration using elastic screen method. *Geophysics* **70**, S30–S37.

SPECTRAL-ELEMENT ANALYSIS IN SEISMOLOGY

EMMANUEL CHALJUB[1], DIMITRI KOMATITSCH[2,4], JEAN-PIERRE VILOTTE[3],
YANN CAPDEVILLE[3], BERNARD VALETTE[5] AND GAETANO FESTA[3]

[1] Laboratoire de Géophysique Interne et de Tectonophysique,
BP 53, 38041 Grenoble Cedex 9, France
[2] Seismological Laboratory, California Institute of Technology, 1200 East California Boulevard,
Pasadena, California 91125, USA
[3] Institut de Physique du Globe, 4 place Jussieu, 75252 Paris Cedex 05, France
[4] Now at: Laboratoire de Modélisation et d'Imagerie en Géosciences, CNRS UMR 5212 and Magique
3D INRIA Futurs, Université de Pau et des Pays de l'Adour, BP 1155, 64013 Pau Cedex, France
[5] Laboratoire de Géophysique Interne et de Tectonophysique, IRD, Université de Savoie,
73376 Le Bourget-du-Lac Cedex, France

ABSTRACT

We present a review of the application of the spectral-element method to regional and global seismology. This technique is a high-order variational method that allows one to compute accurate synthetic seismograms in three-dimensional heterogeneous Earth models with deformed geometry. We first recall the strong and weak forms of the seismic wave equation with a particular emphasis set on fluid regions. We then discuss in detail how the conditions that hold on the boundaries, including coupling boundaries, are honored. We briefly outline the spectral-element discretization procedure and present the time-marching algorithm that makes use of the diagonal structure of the mass matrix. We show examples that illustrate the capabilities of the method and its interest in the context of the computation of three-dimensional synthetic seismograms.

Keywords: DtN operator, Elastodynamics, Global seismology, Regional seismology, Numerical modeling, Perfectly Matched Layers, Potential formulation, Self-gravitation, Spectral-element method, Surface waves, Synthetic seismograms, Topography

1. INTRODUCTION

Coming along with the tremendous increase of computational power, the development of numerical methods for the accurate calculation of synthetic seismograms in three-dimensional (3-D) models of the Earth has been the subject of a continuous effort in the past thirty years.

Among the different numerical approaches used to model the propagation of seismic waves, the most popular is probably the finite-difference (FD) method, in which partial derivatives are approximated by discrete operators involving differences between adjacent grid points. The literature about FD methods is huge, from the historical articles of Alterman and Karal (1968), Alford *et al.* (1974), Kelly *et al.* (1976), Madariaga (1976) and Virieux (1986) to realistic applications to 3-D strong ground motion (e.g., Frankel and Vidale, 1992; Frankel, 1993; Olsen and Archuleta, 1996; Pitarka *et al.*, 1998; Olsen, 2000). However, in spite of recent improvements such as optimal or compact operators (e.g., Zingg *et al.*, 1996; Zingg, 2000), FD methods still suffer from severe limitations regarding (i) the calculation of accurate surface waves; (ii) the ability to account for deformed geometries, in particular to deal with topography at the free surface; and (iii) the possibility to adapt the size of the grid to the seismic wavelengths under consideration in order to save computing time and memory. As an illustration of such limitations, it is worth noting that FD methods have not been able to tackle the challenge posed by global seismology, except in simplified geometries (e.g., Igel and Weber, 1995, 1996; Chaljub and Tarantola, 1997). The reader is referred to Chapter 8 of this book for a detailed presentation of the FD method.

Boundary-element methods based upon integral representation theorems combined with discrete wavenumber expansions of Green's functions have been proposed to incorporate realistic surface and interface topography (see e.g., Bouchon (1996) and references therein, Bouchon and Sánchez-Sesma (2006) (Chapter 3 of this book), Chen (2006) (Chapter 4 of this book)). However, these methods are restricted to media consisting of a finite number of homogeneous layers, and lead to large ill-conditioned linear matrix systems in 3-D applications.

Spectral and pseudo-spectral methods have also been applied to regional (e.g., Tessmer and Kosloff, 1994; Carcione, 1994) and (axisymmetric) global (e.g., Furumura *et al.*, 1998, 1999) elastodynamics, outperforming FD methods regarding the very small amount of numerical dispersion obtained with only a small number of grid points per wavelength that can be chosen very close to the theoretical Nyquist sampling limit. However, because of the global nature of the polynomial basis chosen, these methods are restricted to smooth geological media and cannot be applied to complicated geometries or sharp heterogeneities. Like FD methods, they are also unable to model surface waves with the same accuracy as body waves because of the one-way treatment that needs to be performed in order to implement the free-surface condition (e.g., Carcione, 1994).

Although well suited to handle complex geometries and interface conditions, classical finite-element methods (FEM) have not very often been applied to elastodynamics (e.g., Lysmer and Drake, 1972; Toshinawa and Ohmachi, 1992; Bao *et al.*, 1998) because of the large amount of numerical dispersion related to the low-order polynomial bases used (Marfurt, 1984), and also because of the computational effort needed to solve the resulting large linear systems on parallel computers with distributed memory.

The scope of this article is to review recent progress in the application of the spectral-element method (SEM) to regional and global seismology. The SEM was introduced twenty years ago in computational fluid mechanics by Patera (1984) and Maday and Patera (1989). It is a high-order variational method that retains the ability of FEMs to handle complicated geometries while keeping the exponential convergence rate of spectral methods. This last property explains the term 'spectral element' that was chosen by Patera (1984), which can be confusing because the method works in the time domain, not in the frequency domain. Applications of the SEM to two-dimensional (2-D) (e.g., Cohen *et al.*, 1993; Priolo *et al.*, 1994) and 3-D elastodynamics (e.g., Komatitsch, 1997; Faccioli *et al.*, 1997; Komatitsch and Vilotte, 1998; Seriani, 1998; Komatitsch and Tromp, 1999; Komatitsch *et al.*, 2004; Liu *et al.*, 2004) have shown that very high accuracy and low numerical dispersion are obtained and that an efficient parallel implementation is possible. The SEM has further been applied successfully to global seismology in fully 3-D Earth models (Chaljub, 2000; Komatitsch and Tromp, 2002a, 2002b; Komatitsch *et al.*, 2002; Chaljub *et al.*, 2003; Komatitsch *et al.*, 2003; Chaljub and Valette, 2004) by taking advantage of its great geometrical flexibility. It is also worth mentioning other geophysical applications of the method to spherical geometry, such as the resolution of the shallow water equation (Taylor *et al.*, 1997).

In this article we first recall the wave equation, with particular emphasis on the potential formulation that is used in the fluid parts of the model. We present the method used to couple the SEM with other methods for static or dynamic problems, and to design adapted interface operators for specific problems such as the computation of the exterior gravity potential and the coupling with a modal solution in the frequency domain. We further recall the weak form of the equations and detail how boundary conditions are honored, in particular at fluid–solid interfaces. We then briefly outline the spectral-element discretization and review technical aspects relevant to mesh generation and to the choice of parameters for accurate calculations. Details are also given on how to adapt the time-marching algorithm in order to include Perfectly Matched Layers, attenuation, rotation and self-gravitation. Examples of application to regional and global seismology are finally presented, which illustrate the potentiality of the method in the context of 3-D geophysical problems.

2. The Wave Equation in Regional and Global Seismology

In this section we recall the seismic wave equation in the general framework of a self-gravitating, rotating Earth as well as the simpler formulation relevant at a regional scale. We provide a potential decomposition of the displacement field in the fluid regions. We also recall the continuity conditions that hold on the

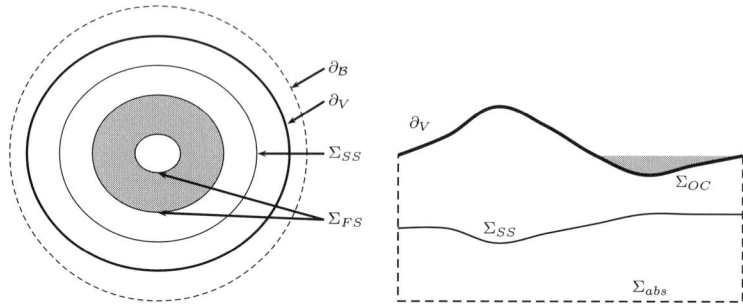

FIG. 1. Sketch of a typical problem geometry for global (left) and regional (right) simulations of wave propagation. Shaded areas denote fluid regions, i.e., the outer core and the oceans. Dashed lines are specific to each problem: the spherical outer boundary ∂_B is relevant for the treatment of self-gravitation at the global scale, while the absorbing boundary Σ_{abs} only matters for regional applications.

internal interfaces and we define a set of boundaries through which the spectral-element solution can be coupled to solutions obtained based upon other methods. Finally, we derive the weak form of the equations in the fluid and solid regions. The reader is referred to e.g. Aki and Richards (1980) for a complete treatment of seismic wave propagation problems and to e.g. Dahlen and Tromp (1998) for a general discussion of global seismology.

2.1. General Framework

Let us start with a description of the geometry of the problem and define some notations. Let V denote the spatial domain in which we are interested in modeling seismic wave propagation. V consists either of the whole Earth or of a portion of it, as depicted in Fig. 1. The solid and fluid regions of V are denoted by V_S and V_F, respectively, and Σ_{SS} (respectively Σ_{FS}) stands for the set of solid–solid (respectively fluid–solid) interfaces. We consider the ocean–crust boundary as a particular interface, which we denote by Σ_{OC}.

Finally, two interfaces are introduced depending on the size of the problem: the outer spherical boundary ∂_B that will be used when accounting for self-gravitation at the global scale, and the absorbing boundary Σ_{abs} that will be needed in regional applications.

2.2. Reference Configuration

A seismic wave is assumed to be a small perturbation about a rotating reference configuration in which the elastic medium is stressed due to self-gravitation. The effect of deviatoric prestress on wave propagation being small (e.g., Valette, 1986), the initial stress is usually assumed to be hydrostatic. Under this assumption, the static equilibrium in the reference configuration is

$$\nabla \cdot \mathbf{T}^0 = -\nabla p^0 = \rho \nabla \phi, \tag{1}$$

where $\mathbf{T}^0$ is the initial Cauchy stress tensor, p^0 is the initial hydrostatic pressure, ρ is density and ϕ is the geopotential, which is related to the non-rotating gravity potential ϕ' by

$$\phi(\mathbf{r}) = \phi'(\mathbf{r}) - \frac{1}{2}(\Omega^2 r^2 - (\boldsymbol{\Omega} \cdot \mathbf{r})^2), \tag{2}$$

where $\mathbf{r}$ denotes position and $\boldsymbol{\Omega}$ is the Earth's angular velocity vector. Gravity is defined from the geopotential by $\mathbf{g} = -\nabla \phi$, and the non-rotating gravity potential is obtained by solving Poisson's equation:

$$\nabla^2 \phi' = \begin{cases} 4\pi G \rho & \text{in } V, \\ 0 & \text{outside } V, \end{cases} \tag{3}$$

together with the additional requirements that both ϕ' and $\nabla \phi' \cdot \hat{\mathbf{n}}$ be continuous and that the regularity condition

$$\lim_{\|\mathbf{r}\| \to \infty} \phi'(\mathbf{r}) = 0, \tag{4}$$

hold.

The equilibrium equation (1) implies that the level surfaces of density, geopotential and pressure coincide, which imposes strong constraints on admissible distributions of density within the medium. In what follows, we use the so-called quasi-hydrostatic approximation, in which lateral variations of density are allowed but where the effects of deviatoric prestress are ignored (see Dahlen and Tromp, 1998, p. 101).

2.3. The Gravito-Elastodynamic Equations

The wave equation is obtained through a first-order Lagrangian perturbation of the equilibrium equation (1). We write it under the general form:

$$\ddot{\mathbf{u}} + 2\boldsymbol{\Omega} \times \dot{\mathbf{u}} + \mathcal{A}(\dot{\mathbf{u}}, \mathbf{u}) = \frac{1}{\rho}\mathbf{f}, \tag{5}$$

where a dot over a symbol indicates time differentiation. In Eq. (5), $\mathbf{u}$ is the Lagrangian perturbation of displacement and $\mathcal{A}$ denotes the elastic-gravitational operator. The dependence of $\mathcal{A}$ on velocity accounts for possible attenuation, the treatment of which is postponed to Appendix A.1.

In the case of global seismology, the seismic source $\mathbf{f}$ is usually represented by a point moment tensor:

$$\mathbf{f}(\mathbf{r}, t) = -\nabla \cdot \big(\mathbf{m}_0 \delta(\mathbf{r} - \mathbf{r}_s)\big) g(t), \tag{6}$$

where $\mathbf{r}_s$ is the position of the source, $\mathbf{m}_0$ is a given seismic moment tensor, and $g(t)$ is the source time function. The spatial extent and the kinematics of the source can be accounted for by considering a superimposition of punctual sources with different onset time, moment tensor and time history (e.g., Tsuboi et al., 2003).

The most general form of the elastic-gravitational operator is

$$\rho\mathcal{A}(\mathbf{u}) = -\nabla \cdot \mathbf{T}(\mathbf{u}) - \nabla(\rho\mathbf{u}\cdot\mathbf{g}) + \{\nabla\cdot(\rho\mathbf{u})\}\mathbf{g} + \rho\nabla\psi(\mathbf{u}), \quad (7)$$

where $\mathbf{T}$ is the Lagrangian incremental stress tensor and $\psi(\mathbf{u})$ is the Eulerian perturbation of the gravitational potential ϕ', also referred to as the mass redistribution potential. The first term on the right-hand side of Eq. (7) represents the internal elastic forces, while the second term accounts for hydrostatic prestress. The remaining terms reflect the perturbation of gravity, which is twofold: a local term in which only density is perturbed and a global term that accounts for the perturbation of the gravitational field.

The Lagrangian incremental stress tensor is related to the infinitesimal strain tensor $\boldsymbol{\varepsilon}(\mathbf{u}) = \frac{1}{2}(\nabla\mathbf{u} + \nabla^T\mathbf{u})$ by Hooke's law:

$$\mathbf{T}_{ij}(\mathbf{u}) = d_{ijkl}\varepsilon_{kl}(\mathbf{u}), \quad (8)$$

in which the fourth-order tensor d is obtained from the classical elastic tensor c (i.e., that relating the Eulerian Cauchy stress tensor to the infinitesimal strain tensor) by

$$d_{ijkl} = c_{ijkl} - \mathbf{T}^0_{ij}\delta_{kl} + \mathbf{T}^0_{ik}\delta_{jl} + \mathbf{T}^0_{jk}\delta_{il}, \quad (9)$$

where δ is the Kronecker delta.

Whenever intrinsic attenuation is included, or when Perfectly Matched Layers are used in order to absorb energy in the context of regional applications, the stress–strain relation (8) has to be modified. The reader is referred to Appendix A for details on how to derive the modified Hooke's law that needs to be used in such a case.

Finally, the mass redistribution potential is obtained by solving a Poisson–Laplace equation over the entire space:

$$\nabla^2\psi(\mathbf{u}) = \begin{cases} -4\pi G \nabla\cdot(\rho\mathbf{u}) & \text{in } V, \\ 0 & \text{outside } V. \end{cases} \quad (10)$$

2.3.1. Cowling Approximation

Ignoring the perturbation of the gravitational potential in Eq. (7) yields the so-called Cowling approximation (Cowling, 1941), in which:

$$\rho\mathcal{A}(\mathbf{u}) = -\nabla\cdot\mathbf{T}(\mathbf{u}) - \nabla(\rho\mathbf{u}\cdot\mathbf{g}) + \{\nabla\cdot(\rho\mathbf{u})\}\mathbf{g}. \quad (11)$$

The latter form has been used by Chaljub (2000); Komatitsch and Tromp (2002b); Capdeville *et al.* (2003a, 2003b) and Chaljub *et al.* (2003) because in that context it is not necessary to solve the unbounded Poisson–Laplace equation (10).

Note that even in the Cowling approximation, one needs to solve Eq. (3) in the entire space for the 3-D gravitational potential. This step was neglected by Komatitsch and Tromp (2002b) and Capdeville *et al.* (2003b), who used the alternative approximation:

$$\rho \mathcal{A}(\mathbf{u}) = -\nabla \cdot \mathbf{T}(\mathbf{u}) - \nabla(\rho \mathbf{u} \cdot \mathbf{g}') + \{\nabla \cdot (\rho \mathbf{u})\}\mathbf{g}', \qquad (12)$$

where $\rho = \rho' + \delta\rho$ consists of lateral variations $\delta\rho$ superimposed to a one-dimensional (1-D) spherically-symmetric profile ρ', and where $\mathbf{g}'$ is the (analytical) gravity vector computed from ρ'.

A consistent implementation of the Cowling approximation, i.e., that includes lateral variations in gravity, is proposed in Appendix B.2 based upon the work of Chaljub and Valette (2004).

2.3.2. Regional Seismology

Self-gravitation and rotation can be safely neglected at the periods considered in regional applications (see e.g., Dahlen and Tromp, 1998, p. 142). The wave equation can then be written directly in Eulerian form using Eq. (1), without the Coriolis term on the left-hand side and with

$$\rho \mathcal{A}(\mathbf{u}) = -\nabla \cdot \mathbf{T}(\mathbf{u}). \qquad (13)$$

In Eq. (13), $\mathbf{T}$ is the Cauchy stress tensor, which satisfies Hooke's law (8) in which d is replaced with the usual elastic tensor c by virtue of (9).

2.4. Potential Formulation in Fluid Regions

Accounting for the propagation of seismic waves in media with both fluid and solid regions is known to be a difficult task in the finite-element method: one cannot simply set the shear modulus to zero in the fluid because spurious (shear) oscillations appear and evolve without any numerical control (e.g., Hamdi *et al.*, 1978) due to the fact the wave equation, when formulated in displacement, has too many degrees of freedom in the fluid. One must therefore impose an additional constraint to prevent spurious modes from developing. From a numerical point of view, the critical aspect is to correctly discretize the null space $\mathcal{N}(\mathcal{A})$ of the elastic-gravitational operator (Bermúdez and Rodríguez, 1994). An alternative to the discretization of $\mathcal{N}(\mathcal{A})$ is to resort to a potential formulation in order to eliminate the shear part of displacement.

In what follows, we recall the wave equation in the fluid and present a general approach to derive the potential formulation to be used in global or regional seismology, which generalizes those of Komatitsch *et al.* (2000), Komatitsch and Tromp (2002b), Chaljub *et al.* (2003) and Chaljub and Valette (2004) to the case of a self-gravitating, rotating, arbitrarily-stratified fluid.

2.4.1. Fluid Regions

In the fluid regions, the stress tensor $\mathbf{T}(\mathbf{u})$ takes the form:

$$\mathbf{T}(\mathbf{u}) = -\delta p \mathbf{I} = \rho c^2 \nabla \cdot \mathbf{u} \mathbf{I}, \tag{14}$$

where δp is the Lagrangian pressure perturbation, c is the P-wave velocity and $\mathbf{I}$ denotes the second-order identity tensor. The gravito-elastic operator can be rewritten in the general form

$$-\mathcal{A}(\mathbf{u}) = \nabla \left[c^2 \nabla \cdot \mathbf{u} + \mathbf{u} \cdot \mathbf{g} - \psi(\mathbf{u}) \right] + c^2 \nabla \cdot \mathbf{u} \mathbf{s}, \tag{15}$$

where $\mathbf{s}$ is defined by

$$\mathbf{s} = \frac{\nabla \rho}{\rho} - \frac{\mathbf{g}}{c^2}, \tag{16}$$

and is related to the so-called Brunt–Väisälä squared frequency N^2 by

$$N^2 = \mathbf{s} \cdot \mathbf{g} = \frac{1}{\rho} \left(\nabla \rho - \frac{\rho}{c^2} \mathbf{g} \right) \cdot \mathbf{g}. \tag{17}$$

A fluid is said to be stably stratified if N^2 is positive. In this case, the motion of a small fluid region moved away from its equilibrium position is oscillatory, with a characteristic frequency equal to N. The restoring force in this process being buoyancy, N is often referred to as the buoyancy frequency. The corresponding oscillations are referred to as internal gravity waves in the ocean or in the atmosphere and as the core undertones in the Earth's outer core. Wherever $N^2 < 0$, the stratification is said to be unstable because the motion of a small fluid region can diverge locally. This situation corresponds to the onset of convection. Finally, a fluid that has $N^2 = 0$ is said to be neutrally stratified. Note that the spectrum of the elastic-gravitational operator acquires a continuous non-seismic low-frequency part when $N^2 \neq 0$ (Valette, 1989), which makes a numerical solution difficult to design. In what follows, we do not make any assumption about the stratification within the fluid, but let us mention that we do not intent to compute the non-seismic part of the spectrum.

Upon substituting Eq. (15) into Eq. (1), we obtain the wave equation in the fluid regions:

$$\ddot{\mathbf{u}} = \nabla \left[c^2 \nabla \cdot \mathbf{u} + \mathbf{u} \cdot \mathbf{g} - \psi(\mathbf{u}) \right] + c^2 \nabla \cdot \mathbf{u} \mathbf{s} - 2 \mathbf{\Omega} \times \dot{\mathbf{u}}. \tag{18}$$

2.4.2. Potential Formulation

Considering the wave equation (18), we assume that the displacement field in the fluid takes the form:

$$\mathbf{u} = \nabla\chi + \zeta\mathbf{s} + \mathbf{w}, \tag{19}$$

where χ and ζ denote two arbitrary potentials and $\mathbf{w}$ is an arbitrary vector field. Differentiating twice with respect to time and identifying each term with the right-hand side of Eq. (18), we obtain a system of two scalar and one vector wave equations:

$$\ddot{\zeta} = c^2 \nabla \cdot (\nabla\chi + \zeta\mathbf{s} + \mathbf{w}), \tag{20}$$

$$\ddot{\chi} = \ddot{\zeta} + (\nabla\chi + \mathbf{w}) \cdot \mathbf{g} + N^2\zeta - \psi(\mathbf{u}), \tag{21}$$

$$\ddot{\mathbf{w}} = -2\mathbf{\Omega} \times (\nabla\dot{\chi} + \dot{\zeta}\mathbf{s} - 2\mathbf{\Omega} \times \dot{\mathbf{w}}). \tag{22}$$

Note that the displacement field defined by Eq. (19) with χ, ζ and $\mathbf{w}$ satisfying Eqs. (20)–(22) is the unique solution to the wave equation (18), even though the choice of the potential decomposition is arbitrary. With the choice made in Eq. (19), the two potentials ζ and χ are respectively related to the Lagrangian and Eulerian perturbations of pressure through

$$\delta p = -\rho\ddot{\zeta} \tag{23}$$

and

$$p' = \delta p - \rho\mathbf{u} \cdot \mathbf{g} = -\rho(\ddot{\chi} + \psi(\mathbf{u})). \tag{24}$$

Note that for a neutrally stratified fluid ($\mathbf{s} = \mathbf{0}$), only one scalar potential is required. Eqs. (20) and (21) can indeed be combined into

$$\ddot{\chi} = c^2 \nabla \cdot (\nabla\chi + \mathbf{w}) + (\nabla\chi + \mathbf{w}) \cdot \mathbf{g} - \psi(\mathbf{u}). \tag{25}$$

Setting $\mathbf{w} = \mathbf{0}$ and $\psi(\mathbf{u}) = 0$ in (25), and noting that $\mathbf{g} = (c^2/\rho)\nabla\rho$ for a neutrally stratified fluid, one obtains the same equation as in Komatitsch and Tromp (2002b) and Chaljub et al. (2003) for the velocity potential. For the sake of generality, we introduce a second potential such that $\ddot{\zeta} = \ddot{\chi} - (\nabla\chi + \mathbf{w}) \cdot \mathbf{g} + \psi(\mathbf{u})$, so that Eq. (23) defining the Lagrangian perturbation of pressure still holds.

Finally, note that the expression of displacement (19) also has to be used in Eq. (10) when solving for the mass redistribution potential in the fluid regions V_F.

2.4.3. Regional Seismology

When self-gravitation is neglected, it is not convenient to deal with the general expression (15). Instead, we consider the elastic operator obtained by inserting

(14) into (13):

$$-\rho \mathcal{A}(\mathbf{u}) = \nabla[\rho c^2 \nabla \cdot \mathbf{u}]. \tag{26}$$

The process used to derive the potential formulation is the same as above: we write the wave equation

$$\ddot{\mathbf{u}} = \frac{1}{\rho}\nabla[\rho c^2 \nabla \cdot \mathbf{u}], \tag{27}$$

from which we assume that the displacement field should have the form

$$\mathbf{u} = \frac{1}{\rho}\nabla(\rho \zeta), \tag{28}$$

where ζ is a scalar potential. The equation governing the evolution of ζ is obtained by differentiating Eq. (28) twice with respect to time and identifying it with the right-hand side of Eq. (27):

$$\ddot{\zeta} = c^2 \nabla \cdot \left(\frac{1}{\rho}\nabla(\rho \zeta)\right). \tag{29}$$

Note that we have chosen the potential in Eq. (28) so that pressure is $p = -\rho \ddot{\zeta}$, as in the self-gravitating case, and therefore our potential differs from that used by Komatitsch and Tromp (1999). As mentioned previously, this is not a problem because the decomposition is not unique, only the solution to the wave equation is.

2.5. Initial and Boundary Conditions

In order to define a unique solution to the wave equation (5) and (10), we must provide both initial and boundary conditions.

2.5.1. Initial Conditions

According to the assumption that the medium is initially at rest, the initial conditions require that displacement and velocity vanish in the reference configuration. We thus set $\mathbf{u} = \dot{\mathbf{u}} = \mathbf{0}$ in the solid and $\chi = \dot{\chi} = \zeta = \dot{\zeta} = 0$ and $\mathbf{w} = \dot{\mathbf{w}} = \mathbf{0}$ in the fluid at $t = 0$. There is no perturbation of the gravitational potential in the initial configuration, therefore $\psi(\mathbf{u})$ is also set to zero.

Taking the initial conditions into account, the precession equation (22) governing the evolution of $\mathbf{w}$ can be integrated in time in order to get a form closer to that given by Komatitsch and Tromp (2002b):

$$\dot{\mathbf{w}} + 2\mathbf{\Omega} \times \mathbf{w} = -2\mathbf{\Omega} \times (\nabla \chi + \zeta \mathbf{s}). \tag{30}$$

2.5.2. Internal Boundary Conditions

First, for the solid regions to remain in welded contact, the displacement must be continuous across the solid–solid interfaces that are present at the contact between layers in the geological medium. This can be written as

$$[\mathbf{u}]_{\Sigma_{SS}} = \mathbf{0}, \tag{31}$$

where $[\ ]_{\Sigma_{SS}}$ stands for the jump operator across Σ_{SS}, defined in accordance with the unit normal vector $\hat{\mathbf{n}}$: $[\mathbf{u}]_{\Sigma_{SS}} = \mathbf{u}^+ - \mathbf{u}^-$, with $\hat{\mathbf{n}}$ pointing from the '$-$' to the '$+$' side.

Then, because we assume that we deal with inviscid fluids only, any tangential slip is allowed at the fluid–solid interfaces and therefore the kinematic condition to be used is the continuity of normal displacement:

$$[\mathbf{u} \cdot \hat{\mathbf{n}}]_{\Sigma_{FS}} = 0, \tag{32}$$

which, using the decomposition (19), gives

$$\mathbf{u} \cdot \hat{\mathbf{n}} = (\nabla \chi + \zeta \mathbf{s} + \mathbf{w}) \cdot \hat{\mathbf{n}} \quad \text{on } \Sigma_{FS}. \tag{33}$$

Traction is continuous across every interface of the medium in virtue of the action-reaction principle. Let Σ stand for a particular internal boundary, we have

$$\left[\mathbf{T}(\mathbf{u}) \cdot \hat{\mathbf{n}}\right]_{\Sigma} = \mathbf{0}. \tag{34}$$

According to (23), this can be rewritten on the fluid–solid interfaces Σ_{FS} as

$$\mathbf{T}(\mathbf{u}) \cdot \hat{\mathbf{n}} = \rho \ddot{\zeta} \hat{\mathbf{n}}. \tag{35}$$

Finally, the mass redistribution potential is continuous across every interface Σ:

$$\left[\psi(\mathbf{u})\right]_{\Sigma} = 0, \tag{36}$$

but its normal derivative can be discontinuous if a density contrast is present:

$$\left[\nabla \psi(\mathbf{u}) \cdot \hat{\mathbf{n}}\right]_{\Sigma} = -4\pi G [\rho]_{\Sigma} \mathbf{u} \cdot \hat{\mathbf{n}}. \tag{37}$$

2.6. Coupling Boundaries

In some situations, we may be interested in solving the wave equation outside the computational domain V based on another method than the SEM. This is typically the case for applications in unbounded domains, for example, when computing the perturbation of the gravitational potential at the global scale or when imposing absorbing boundary conditions at the regional scale. This situation also arises when a direct method, either exact or approximate, is available to solve the wave equation outside V. In some applications in global seismology for example, one may assume that the lower mantle and the core are radially stratified and that lateral heterogeneities are only present in the upper mantle. In such a case, it is interesting to use a simpler method to compute the solution in the 1-D regions of the model and to restrict the use of the SEM to the mantle and the crust, where 3-D heterogeneities are present. Another example arises when accounting for the presence of the oceans in global seismology applications: in many (but not all) seismological studies performed at the global scale, the thickness of the oceans is usually one order of magnitude smaller than the seismic wavelengths of interest, allowing one to use an approximation of the wave equation in the water layer. In all cases, the exterior domain, say W, can be accounted for by means of a boundary condition defined on the interface between V and W. In what follows, we explain how to derive such a condition by introducing a so-called Dirichlet-to-Neumann operator on the coupling boundary. Detailed expressions of the coupling operators needed in practical situations are given in Appendix B.

2.6.1. Coupling Principle

Suppose that we know how to solve the wave equation (5) and (10) in a region W, and let Γ stand for the interface between W and the computational domain V. We can assume without loss of generality that Γ is a solid–solid boundary. Then, according to Eqs. (31) and (34), both displacement and traction must match on Γ at all times. This means that $\forall t \in [0, T]$:

$$\mathbf{u}_t^V = \mathbf{u}_t^W, \tag{38}$$

$$\mathbf{T}(\mathbf{u}_t^V) \cdot \hat{\mathbf{n}} = \mathbf{T}(\mathbf{u}_t^W) \cdot \hat{\mathbf{n}}, \tag{39}$$

where the superscript denotes the region where the solution is defined. Let us now recall that for the wave equation to be well-posed in region V or W, it is sufficient to know either traction or displacement on Γ at all times. The principle of the coupling is to mix these conditions: for example to solve in V with the prescribed traction $\mathbf{T}(\mathbf{u}_t^W) \cdot \hat{\mathbf{n}}$ on Γ and to solve in W with the prescribed displacement $\mathbf{u}_t^V$ on Γ. The key issue to apply this strategy is to compute the surface traction from a given displacement on Γ. This process is referred to as a Dirichlet-to-Neumann (DtN) operator, in reference to the nomenclature of classical boundary conditions in elasticity (e.g., Givoli, 1992).

Let then $\mathcal{D}_t^W$ denote the DtN operator that computes the surface traction on Γ at time t. The superscript W indicates that the operator is based upon the knowledge of the solution in region W. In order to be properly defined, $\mathcal{D}_t^W$ must be applied to the entire history of surface displacements:

$$\mathcal{D}_t^W(\mathbf{u}_{t' \leq t}^W) = \mathbf{T}(\mathbf{u}_t^W) \cdot \hat{\mathbf{n}}. \tag{40}$$

The boundary condition to be used in order to solve the wave equation in V can then be expressed formally as

$$\mathbf{T}(\mathbf{u}_t^V) \cdot \hat{\mathbf{n}} = \mathcal{D}_t^W(\mathbf{u}_{t' \leq t}^V). \tag{41}$$

The reader is referred to Appendix B for detailed expressions of DtN operators needed in practical applications to (i) couple the SEM with a normal-mode solution in the frequency domain (see Appendix B.1), (ii) take into account gravity perturbations induced by low-frequency seismic waves (see Appendix B.2) or (iii) approximate the elastic response of the ocean in global seismology applications (see Appendix B.3).

2.7. Weak Formulation

The SEM is a particular case of the finite-element method and is therefore based on the variational, or weak form of the wave equation. Let us recall the formulation in the solid and fluid regions with a particular emphasis on how to honor the boundary conditions.

2.7.1. Solid Regions

To obtain the weak form of the wave equation in the solid we first take the $L^2(V_S)$ scalar product of Eq. (5) with an admissible displacement field $\tilde{\mathbf{u}}$:

$$\int_{V_S} \rho(\ddot{\mathbf{u}} + 2\mathbf{\Omega} \times \dot{\mathbf{u}}) \cdot \tilde{\mathbf{u}} \, dV + \int_{V_S} \rho \mathcal{A}(\mathbf{u}) \cdot \tilde{\mathbf{u}} \, dV = \int_{V_S} \mathbf{f} \cdot \tilde{\mathbf{u}} \, dV. \tag{42}$$

Then, in the last term of the left-hand side we integrate by parts the divergence of the stress tensor to remove the second-order spatial derivatives of the displacement. We obtain

$$-\int_{V_S} \nabla \cdot \mathbf{T}(\mathbf{u}) \cdot \tilde{\mathbf{u}} \, dV = \int_{V_S} \mathbf{T}(\mathbf{u}) \cdot \nabla \tilde{\mathbf{u}} \, dV - \int_{\Sigma} \mathbf{T}(\mathbf{u}) \cdot \hat{\mathbf{n}} \cdot \tilde{\mathbf{u}} \, dS, \tag{43}$$

where Σ comprises the free surface ∂_V, the fluid–solid interfaces Σ_{FS}, the ocean–crust boundary Σ_{OC}, the coupling boundaries Γ, and the absorbing boundaries Σ_{abs}. We then impose the free surface condition by setting the traction to zero on $\partial_V \cap \partial_{V_S}$, which gives:

$$\int_{\Sigma} \mathbf{T}(\mathbf{u}) \cdot \hat{\mathbf{n}} \cdot \tilde{\mathbf{u}} \, dS = \int_{\Sigma_{FS}} \mathbf{T}(\mathbf{u}) \cdot \hat{\mathbf{n}} \cdot \tilde{\mathbf{u}} \, dS + \int_{\Sigma_{OC}} \mathbf{T}(\mathbf{u}) \cdot \hat{\mathbf{n}} \cdot \tilde{\mathbf{u}} \, dS$$
$$+ \int_{\Gamma} \mathbf{T}(\mathbf{u}) \cdot \hat{\mathbf{n}} \cdot \tilde{\mathbf{u}} \, dS. \tag{44}$$

Upon substituting the expressions of the different tractions given by Eqs. (35), (B.11) and (41) we obtain:

$$\int_{\Sigma} \mathbf{T}(\mathbf{u}) \cdot \hat{\mathbf{n}} \cdot \tilde{\mathbf{u}} \, dS = \int_{\Sigma_{FS}} \rho \ddot{\zeta} \tilde{\mathbf{u}} \cdot \hat{\mathbf{n}} \, dS - \int_{\Sigma_{OC}} \rho_w h_w (\ddot{\mathbf{u}} \cdot \hat{\mathbf{n}})(\tilde{\mathbf{u}} \cdot \hat{\mathbf{n}}) \, dS$$
$$+ \int_{\Gamma} \mathcal{D}_t^W (\mathbf{u}_{t' \leq t}^V) \cdot \tilde{\mathbf{u}} \, dS. \tag{45}$$

The weak form (42) must be satisfied by every admissible displacement $\tilde{\mathbf{u}} \in \mathbf{H}^1(V_S)$, where $\mathbf{H}^1(V_S)$ denotes the space of vector fields that, together with their first derivatives, are square-integrable on V_S. It can be shown that the vector fields in $\mathbf{H}^1(V_S)$ are continuous in a weak sense in V_S, which makes $\mathbf{H}^1(V_S)$ a suitable space for describing admissible displacements in the solid. On the contrary, the continuity of traction is a natural property of weak formulations in the sense that it is satisfied implicitly, just as the free surface condition.

2.7.2. Fluid Regions

The weak form of the wave equation in the fluid regions is obtained by taking the $L^2(V_F)$ scalar product of each side of Eqs. (21), (20) and (30) with admissible potentials $\tilde{\chi} \in H^1(V_F)$, $\tilde{\zeta} \in H^1(V_F)$ and an admissible vector $\tilde{\mathbf{w}} \in \mathbf{L}^2(V_F)$, respectively, and integrating by parts over V_F to remove the second-order space derivatives, which gives:

$$\int_{V_F} \frac{1}{c^2} \ddot{\zeta} \tilde{\zeta} \, dV = -\int_{V_F} (\nabla\chi + \zeta\mathbf{s} + \mathbf{w}) \cdot \nabla\tilde{\zeta} \, dV$$
$$+ \int_{\Sigma_{FS}} (\nabla\chi + \zeta\mathbf{s} + \mathbf{w}) \cdot \hat{\mathbf{n}} \tilde{\zeta} \, dS, \quad (46)$$

$$\int_{V_F} \frac{1}{c^2} \ddot{\chi} \tilde{\chi} \, dV = \int_{V_F} \frac{1}{c^2} \left(\ddot{\zeta} + (\nabla\chi + \mathbf{w}) \cdot \mathbf{g} + N^2 \zeta - \psi(\mathbf{u}) \right) \tilde{\chi} \, dV, \quad (47)$$

and

$$\int_{V_F} (\dot{\mathbf{w}} + 2\boldsymbol{\Omega} \times \mathbf{w}) \cdot \tilde{\mathbf{w}} \, dV = -\int_{V_F} 2\boldsymbol{\Omega} \times (\nabla\chi + \zeta\mathbf{s}) \cdot \tilde{\mathbf{w}} \, dV. \quad (48)$$

Finally, we enforce the continuity of normal displacement across the fluid–solid interfaces, so that Eq. (46) becomes:

$$\int_{V_F} \frac{1}{c^2} \ddot{\zeta} \tilde{\zeta} \, dV = -\int_{V_F} (\nabla\chi + \zeta\mathbf{s} + \mathbf{w}) \cdot \nabla\tilde{\zeta} \, dV + \int_{\Sigma_{FS}} \mathbf{u} \cdot \hat{\mathbf{n}} \tilde{\zeta} \, dS. \quad (49)$$

Note that the way we couple the fluid and the solid regions is very similar to the DtN coupling strategy detailed in the previous section. The key issue is again to solve the equation in the solid with a boundary condition in traction and to solve the equation in the fluid with normal displacement imposed on Σ_{FS}. The only difference is semantic: the normal displacement is now a Neumann boundary condition from the fluid side because of the potential decomposition (19). The fluid–solid coupling strategy therefore consists in building a Neumann-to-Dirichlet operator on Σ_{FS}.

In many cases of practical interest, for example in applications involving offshore seismic acquisition in the petroleum industry, we may be interested in a medium in which the fluid layer comprises a free surface (typically the surface of the ocean). In this case, it is worth mentioning that the free-surface condition for the potential formulation we use in the fluid is a Dirichlet condition, not a Neumann condition as in the case of a solid, because the potentials should vanish at the surface to implement zero pressure, as can be deduced from (23) and (24).

2.7.3. Self-Gravitation

In order to establish the weak form of Eq. (10), we consider Poisson's equation within the finite volume $\mathcal{B}$. Multiplying with an admissible potential $\tilde{\psi} \in H^1(\mathcal{B})$ and integrating the Laplacian and the divergence terms by parts gives

$$\int_{\mathcal{B}} \nabla \psi(\mathbf{u}) \cdot \nabla \tilde{\psi} \, dV - \int_{\partial \mathcal{B}} \nabla \psi(\mathbf{u}) \cdot \hat{\mathbf{n}} \tilde{\psi} \, dS$$
$$= -4\pi G \left\{ \int_{V_S} \rho \mathbf{u} \cdot \nabla \tilde{\psi} \, dV - \int_{\partial \mathcal{B}} \rho \mathbf{u} \cdot \hat{\mathbf{n}} \tilde{\psi} \, dS \right.$$
$$\left. + \int_{V_F} \rho (\nabla \chi + \zeta \mathbf{s} + \mathbf{w}) \cdot \nabla \tilde{\psi} \, dV \right\}, \tag{50}$$

where the jump condition (37) across fluid–solid interfaces is taken into account naturally. Then, upon applying the exterior boundary condition (B.8), we obtain

$$\int_{\mathcal{B}} \nabla \psi(\mathbf{u}) \cdot \nabla \tilde{\psi} \, dV - \int_{\partial \mathcal{B}} \mathcal{D}^W (\psi(\mathbf{u})) \tilde{\psi} \, dS$$
$$= -4\pi G \left\{ \int_{V_S} \rho \mathbf{u} \cdot \nabla \tilde{\psi} \, dV + \int_{V_F} \rho (\nabla \chi + \zeta \mathbf{s} + \mathbf{w}) \cdot \nabla \tilde{\psi} \, dV \right\}, \tag{51}$$

where the DtN operator $\mathcal{D}^W$ is given by (B.6).

3. Spectral-Element Approximation

3.1. Spatial Discretization

The SEM is based upon a generalized Galerkin approximation (see e.g., Deville et al., 2002, p. 51) of the weak forms (42), (47), (49), (48) and (51), which we outline in this section.

Let h stand for a generic parameter related to the geometrical discretization process, for example the typical size of an element or the distance between two adjacent grid points. The discrete problem can be written as: find $\mathbf{u}_h \in \mathbb{D}_h$, $\zeta_h \in \mathbb{P}_h$, $\chi_h \in \mathbb{P}_h$, $\mathbf{w}_h \in \mathbb{W}_h$ and $\psi_h \in \mathbb{G}_h$ such that $\forall \tilde{\mathbf{u}} \in \mathbb{D}_h$, $\forall \tilde{\zeta} \in \mathbb{P}_h$, $\forall \tilde{\chi} \in \mathbb{P}_h$, $\forall \tilde{\mathbf{w}} \in \mathbb{W}_h$, $\forall \tilde{\psi} \in \mathbb{G}_h$:

$$(\rho \ddot{\mathbf{u}}_h; \tilde{\mathbf{u}})_h^S + (2\rho \mathbf{\Omega} \times \dot{\mathbf{u}}_h; \tilde{\mathbf{u}})_h^S + \mathcal{A}_h^S(\mathbf{u}_h, \psi_h; \tilde{\mathbf{u}})$$
$$= \mathcal{C}_h^{SF}(\ddot{\zeta}_h; \tilde{\mathbf{u}}) + \mathcal{C}_h^{OC}(\ddot{\mathbf{u}}_h; \tilde{\mathbf{u}}) + \mathcal{C}_h^{\Gamma}(\mathbf{u}_h; \tilde{\mathbf{u}}) + (\mathbf{f}_h; \tilde{\mathbf{u}})_h^S, \tag{52}$$

$$\left(\frac{1}{c^2}\ddot{\zeta}_h; \tilde{\zeta}\right)_h^F + \mathcal{A}_h^{F,\zeta}(\zeta_h, \chi_h, \mathbf{w}_h; \tilde{\zeta}) = \mathcal{C}_h^{\text{FS}}(\mathbf{u}_h; \tilde{\zeta}), \tag{53}$$

$$\left(\frac{1}{c^2}\ddot{\chi}_h; \tilde{\chi}\right)_h^F + \mathcal{A}_h^{F,\chi}(\ddot{\zeta}_h, \zeta_h, \chi_h, \mathbf{w}_h; \tilde{\chi}) = 0, \tag{54}$$

$$(\dot{\mathbf{w}}_h; \tilde{\mathbf{w}})_h^F + \mathcal{R}_h(\zeta_h, \chi_h, \mathbf{w}_h; \tilde{\mathbf{w}}) = \mathbf{0}, \tag{55}$$

$$\mathcal{P}_h(\psi_h; \tilde{\psi}) = \mathcal{D}_h(\mathbf{u}_h, \zeta_h, \chi_h, \mathbf{w}_h; \tilde{\psi}), \tag{56}$$

where $\mathbb{D}_h, \mathbb{P}_h, \mathbb{W}_h$ and $\mathbb{G}_h$ denote some finite-dimensional subspaces of $\mathbf{H}^1(V_S)$, $H^1(V_F), \mathbf{L}^2(V_F)$ and $H^1(V)$, respectively, and $()_h^S$ (respectively $()_h^F$) stands for an approximate scalar product on $\mathbf{L}^2(V_S)$ (respectively, $L^2(V_F)$). $\mathcal{A}_h^S$ results from the approximation of Eq. (7), which involves the elastic-gravitational operator in the solid regions. By analogy, we have denoted $\mathcal{A}_h^{F,\zeta}$ and $\mathcal{A}_h^{F,\chi}$ the result of the approximation of the right-hand sides of Eqs. (47) and (49). Approximate coupling operators defined on the interfaces of the medium are labeled $\mathcal{C}_h$, with a superscript referring to the different boundary terms in Eqs. (45) and (49). $\mathcal{R}_h$ arises from the discretization of the precession equation (48) in the rotating fluid regions. Finally, $\mathcal{P}_h$ refers to the approximation of the left-hand side of the Poisson–Laplace equation (51), while $\mathcal{D}_h$ arises from the discretization of the right-hand side (i.e., the divergence term).

3.1.1. Geometrical Approximation and Mesh Design

The first step of the spatial discretization process is to divide the computational domain into a mesh of E non-overlapping elements:

$$\overline{V} = \bigcup_{e=1}^{E} \overline{V^e}, \tag{57}$$

each of them being obtained from a reference square (in 2-D) or cube (in 3-D) based on an invertible mapping:

$$V^e = \mathcal{F}^e(\Lambda^d), \quad \Lambda = [-1, 1], \tag{58}$$

where $d = 2$ or 3 is the spatial dimension.

Note that the choice of elements with tensorized geometries such as quadrangles in 2-D and hexahedra in 3-D is a characteristic of the SEM that stems from the choice of tensorized polynomial bases, as will be illustrated in Section 3.1.2. Let us mention that one can generalize the SEM in 2-D to triangles (Cohen *et al.*, 2001; Giraldoa and Warburton, 2005; Bittencourt, 2005; Mercerat *et al.*, 2005) based upon the so-called Fekete points (Taylor *et al.*, 2000), and combine triangles and quadrangles within the same 2-D mesh (Komatitsch *et al.*, 2001; Pasquetti and Rapetti, 2004). In 3-D one could use this approach to couple hexahedra with tetrahedra based upon pyramidal transition elements to ensure the geometrical matching between quadrangular and triangular faces. However, it is worth noting that triangular spectral elements are both more expensive and less accurate than quadrangular spectral elements (Komatitsch *et al.*, 2001), therefore their main interest is to allow for more geometrical flexibility during the mesh generation process.

For the sake of simplicity, we will hereafter assume that the mesh is geometrically conforming, i.e., that two neighboring elements share either a single point or an entire edge or face, but not a partial edge.

Except in a few benchmark cases for which analytical transformations exist, each mapping $\mathcal{F}^e$ needs to be approximated, for example based on a polynomial function defined by a set of n_a so-called 'anchors' or 'control nodes'. If $\mathbf{x}^e$ and $\boldsymbol{\zeta}$ denote the respective generic vectors on V^e and Λ^d, this can be written as:

$$\mathbf{x}^e = \mathcal{F}^e(\boldsymbol{\zeta}) \simeq \sum_{a=1}^{n_a} N_a(\boldsymbol{\zeta})\mathbf{x}^e_a, \tag{59}$$

where N_a are the shape functions of the anchor nodes $\boldsymbol{\zeta}_a$ (with by definition $N_a(\boldsymbol{\zeta}_b) = \delta_{ab}$) and $\mathbf{x}^e_a = \mathcal{F}^e(\boldsymbol{\zeta}_a)$. In practice, the shape functions are chosen to be linear or quadratic, so that the number of anchors is at most 9 in 2-D (see Fig. 2) and 27 in 3-D. The reader is referred to Dhatt and Touzot (1984) for detailed formulas defining the shape functions of 2-D and 3-D elements. Note that because the polynomial order used to approximate the solution of the wave equation in the SEM (see next section) is higher than that defining the shape functions, the mapping is said to be sub-parametric.

The design of a mesh basically consists in choosing the values $\mathbf{x}^e_a$ in Eq. (59) in order to define each spectral element. During this process, it is important to honor the main geological layers and the surfaces across which the material properties of the geological medium are discontinuous. Indeed, as will be seen in Section 3.1.2, density and seismic velocities are approximated by smooth functions inside each element, which means that sharp local variations in the geological medium such as an interface between two layers can only be accounted for at element boundaries. Note also that because the interfaces are approximated based upon element-wise

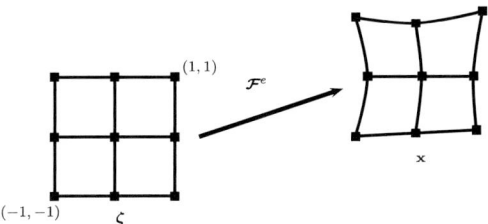

FIG. 2. Example of a 2-D mapping on the reference square based upon 9 control or 'anchor' points.

polynomials following Eq. (59), the number of elements must be chosen large enough to correctly describe the variations of topography and of the shape of the interfaces.

At the scale of the full Earth, topography is defined as a small and smooth variation with respect to a spherical or ellipsoidal reference. One must therefore first define a mesh of hexahedra that is suitable for a sphere or an ellipsoid. A classical way of proceeding is to tile the spherical (or ellipsoidal) interfaces based upon the so-called 'cubed sphere' analytical transformation (e.g., Sadourny, 1972; Ronchi *et al.*, 1996). This mapping, which consists in a central projection of the six faces of a cube onto its circumscribed sphere, provides a mesh with quasi-uniform grid sampling, as illustrated in Fig. 3. Once the different interfaces have been meshed with a set of quadrangles, they are connected in the radial direction to define a mesh of hexahedra. In order to circumvent the singularity at the center of the sphere, a cube is used instead inside the inner core, as shown in Fig. 4. The reader is referred to Chaljub *et al.* (2003) for the analytical expression defining the transition from the sphere to this central cube.

A proper mesh should be relatively regular, i.e., the seismic wavelengths should be sampled relatively uniformly throughout the medium. The motivation is twofold: first, oversampling the wavelengths would result in unnecessary calculations and memory storage from a computational point of view, and second, small grid elements would penalize the choice of the time step used in the explicit time-marching scheme because of the Courant–Friedrichs–Lewy (CFL) stability criterion (see Section 4).

The mesh shown in Fig. 4 does not satisfy this regularity property because the process of embedding spherical interfaces makes the lateral grid size decrease linearly with depth, while seismic velocities usually increase with depth. For geo-

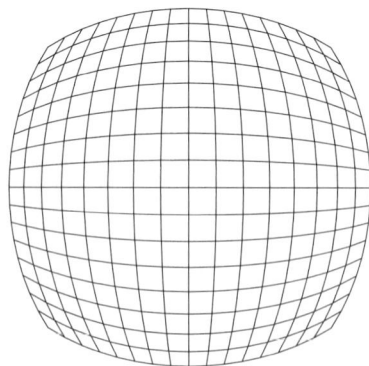 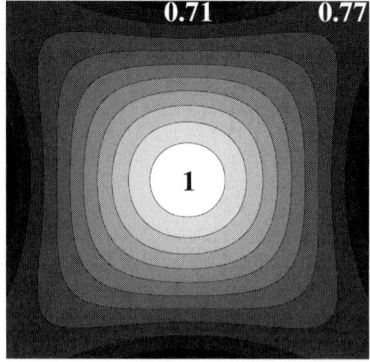

FIG. 3. Left: One-sixth of the surface of a sphere is tiled with a 16 × 16 quadrangular grid obtained by applying the 'cubed sphere' transformation. Right: Contour lines of area variation within the grid show that the area of each element is at most 29% smaller than a reference element that would be located at the center of the face.

metrical reasons, the resulting number of grid points per P-wavelength in such a grid would be about eight times greater in the inner core than in the crust. The same problem arises when studying the response of sedimentary basins, because there is typically at least a factor of 5 between the values of seismic velocities in the bedrock and in the sediments. To overcome this issue, the mesh must be coarsened with depth, either directly at the contact between two layers of elements, or over a few layers of elements in the radial direction. In the first case, geometrical conformity is lost, and one faces the difficulty of matching solutions defined on non-matching sub-grids. The second case is easier to handle in the sense that the implementation of the continuity conditions between elements is unchanged, the only issue being that the mesh becomes geometrically non-structured (which means that the number of elements that share a given grid point, i.e., the valence of that grid point, can in principle become arbitrarily large). For more details, the reader can refer to Chaljub et al. (2003) for an example, of a non-conforming implementation of the SEM and to Komatitsch and Tromp (1999, 2002a) for examples, of conforming implementations on non-structured grids.

Once a spectral-element code has been written (following the principles detailed in the next sections), in many cases of practical interest mesh design is going to be the most critical and time-consuming task before being able to run an accurate simulation. This task is difficult because the use of hexahedra as elementary mesh elements is not always convenient to capture geological features (in classical finite-element methods, one usually has the flexibility of using tetrahedra, pyramids and prisms in addition to hexahedra). For example, when discretizing basin edges in site effect studies, on the one hand one wants to have elements lying on the interface between the bedrock and the sediments in order to properly account for the mechanical contrast, but on the other hand, the

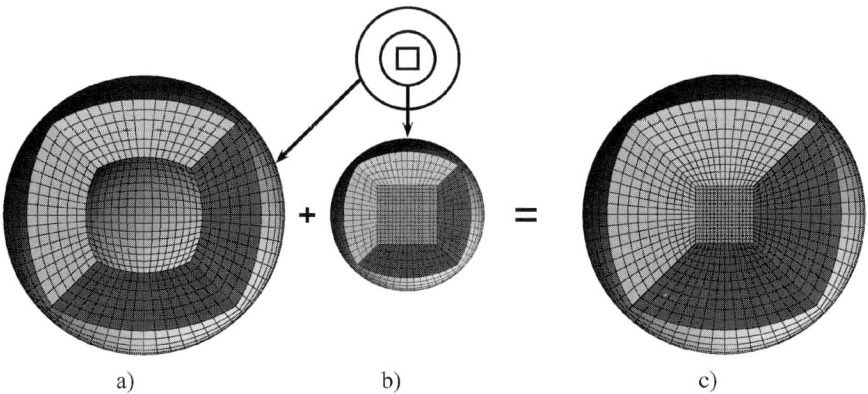

FIG. 4. Process used to build a mesh of hexahedra for a sphere: (a) the spherical interfaces are tiled based upon the cubed sphere transformation, and then connected in the radial direction; (b) a cube is inserted in the center; (c) the final 3-D mesh of the sphere is obtained by connecting (a) and (b).

deformation within the elements must be kept reasonably small for accuracy purposes. A trade-off can be found by defining an ad-hoc depth below which the interface is properly accounted for, while it is only approximated at shallower depths (e.g., see Komatitsch *et al.* (2004) and Section 6.1). What this example suggests is that in the SEM mesh design is always a difficult step that requires expertise and trial-and-error, and that usually cannot be fully automated. Let us mention however that the field of hexahedral mesh generation has significantly evolved in the last few years. Powerful mesh generation packages (also known as 'meshers') are now available, for instance CUBIT by SANDIA National Laboratories, which is based upon powerful mesh creation algorithms (Knupp, 1999; Tautges, 2004). The use of such packages greatly simplifies mesh design for complex 3-D models.

3.1.2. Functional Approximation

Based upon the set of spectral elements defined above, we approximate admissible displacements and potentials by continuous piecewise polynomials defined on the reference domain Λ^d and transported through the local mappings (58). This can be written as:

$$\mathbb{D}_h = \left\{ \tilde{\mathbf{u}} \in \left(\mathcal{C}^0(V_S)\right)^d, \ \tilde{\mathbf{u}}^e \in \left(\mathbb{Q}_N(\Lambda^d)\right)^d \ \forall e = 1, \ldots, E_S \right\}, \tag{60}$$

$$\mathbb{P}_h = \left\{ \tilde{\phi} \in \mathcal{C}^0(V_F), \ \tilde{\phi}^e \in \mathbb{Q}_N(\Lambda^d) \ \forall e = 1, \ldots, E_F \right\}, \tag{61}$$

$$\mathbb{G}_h = \left\{ \tilde{\psi} \in \mathcal{C}^0(V), \ \tilde{\psi}^e \in \mathbb{Q}_N(\Lambda^d) \ \forall e = 1, \ldots, E \right\}, \tag{62}$$

where E_S and E_F stand for the number of elements in the solid and fluid regions, respectively. The superscript indicates a composition with the local mapping: for example, $\tilde{\mathbf{u}}^e$ is defined on the reference domain by $\tilde{\mathbf{u}}^e = \tilde{\mathbf{u}}_{|V^e} \circ \mathcal{F}^e$. The space of continuous scalar fields on W (which stands for either V, V_S or V_F) is denoted $\mathcal{C}^0(W)$, and $\mathbb{Q}_N(\Lambda^d)$ is the space of polynomials of order at most N in each direction of space. In applications of the SEM to wave propagation problems, N is typically selected between 4 and 8 (see Section 5.1 for more details). Here we have assumed that the same order was used for all directions, but this is not a constraint imposed by the method. However, in practice the mesh is often non-structured, i.e., the elements are not regularly distributed along X, Y and Z, and the three directions are therefore coupled in the mesh, which implies that the same polynomial order must indeed be used for all directions.

The space of admissible precession vectors is discretized in the same way, except that the continuity requirement does not hold anymore:

$$\mathbb{W}_h = \left\{ \tilde{\mathbf{w}} \in \mathbf{L}^2(V_F), \ \tilde{\mathbf{w}}^e \in \left(\mathbb{Q}_N(\Lambda^d)\right)^d \ \forall e = 1, \ldots, E_F \right\}. \tag{63}$$

To make these definitions complete, a basis of $\mathbb{Q}_N(\Lambda^d)$ must be provided, and this is where tensorization arises. We start by introducing the set of Gauss–Lobatto–Legendre (GLL) points (e.g., Canuto *et al.*, 1988), which are by convention defined on the reference interval $\Lambda = [-1, 1]$ as the $(N + 1)$ zeroes of the polynomial of order N $Q(\xi) = (1 - \xi^2)L'_N(\xi)$, where L'_N is the derivative of the Legendre polynomial of order N on Λ:

$$Q(\xi_i^N) = 0, \quad \forall i = 0, \ldots, N. \tag{64}$$

Note that we choose the GLL points because they include the extremities of the reference interval, which will ease the implementation of continuity conditions between elements. Next, we define a basis of $\mathbb{Q}_N(\Lambda)$ by considering the Lagrange interpolants at the GLL points (Fig. 5), i.e., the $N + 1$ polynomials of order N that satisfy

$$h_i^N(\xi_j^N) = \delta_{ij}, \quad \forall (i, j) = \{0, \ldots, N\}^2, \tag{65}$$

where δ_{ij} is the Kronecker delta symbol.

As mentioned previously, the polynomial basis on the reference domain Λ^d is obtained by tensorization of these 1-D bases. Let Ξ^d be the set of points obtained by tensorization of the GLL points (Fig. 6). Assuming that we are in 3-D (i.e., $d = 3$), the points of Ξ^d are defined as $\zeta_{ijk}^N = (\xi_i^N, \xi_j^N, \xi_k^N)$ and the associated Lagrange interpolants are $h_{ijk}^N(\zeta) = h_i^N(\xi)h_j^N(\eta)h_k^N(\gamma)$, where $\zeta = (\xi, \eta, \gamma)$ is the generic coordinate vector on the reference cube.

Let for example $\tilde{\mathbf{u}} \in \mathbb{D}_h$ be a discrete displacement vector and let $\mathbf{x} = \mathcal{F}^e(\zeta) \in V_S^e$. For each component $\alpha = 1, \ldots, d$ we can write

$$\tilde{u}_\alpha(\mathbf{x}) = \tilde{u}_\alpha^e(\zeta) = \sum_{i=0}^{N} \sum_{j=0}^{N} \sum_{k=0}^{N} \tilde{u}_\alpha^e(\zeta_{ijk}^N) h_{ijk}^N(\zeta)$$

$$= \sum_{i=0}^{N} \sum_{j=0}^{N} \sum_{k=0}^{N} \tilde{u}_\alpha(\mathbf{x}_{ijk}^{e,N}) h_{ijk}^N(\zeta), \tag{66}$$

where we have introduced the collocation points $\mathbf{x}_{ijk}^{e,N}$ defined by $\mathbf{x}_{ijk}^{e,N} = \mathcal{F}^e(\zeta_{ijk}^N)$. Note that the interpolation formula (66) can handle local continuous variations of the elastic parameters inside the elements.

In order to build continuous bases for the displacements and potentials, it is customary to set up a global numbering of the collocation points over the whole domain. We first introduce the number of local collocation points $\mathcal{N}_L = E \times (N + 1)^d$, where the subscript L refers to the fact that only a numbering local to the elements is needed. We then obtain the number of global grid points in the entire mesh, $\mathcal{N}_G$, by counting only once the points that belong to several elements. Efficient mesh numbering algorithms are available for this purpose from the finite-element community. Let us consider for example a grid point with local coordinates (i, j, k) on the eth element, its global index $n \in \{1, \ldots, \mathcal{N}_G\}$ is defined formally as the surjective function $n = n(e, i, j, k)$. We then build a continuous basis of polynomials by considering shape functions $\phi_n(\mathbf{x})$ that we define

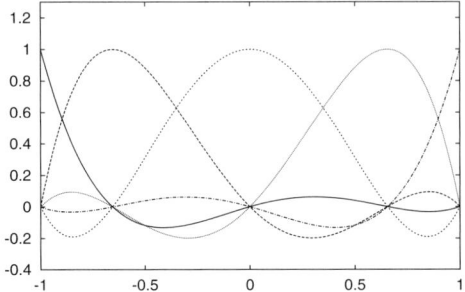

FIG. 5. Lagrange interpolants of degree $N = 4$ on the reference interval $\Lambda = [-1, 1]$. The corresponding $N + 1 = 5$ Gauss–Lobatto–Legendre points can be seen along the horizontal axis. All Lagrange polynomials are, by definition, equal to 1 or 0 at each of these points. Note that the first and last points are exactly -1 and 1.

FIG. 6. When a polynomial degree N is used to discretize the wave field, each 3-D spectral element contains a grid of $(N+1)^3$ Gauss–Lobatto–Legendre points, and each 2-D face of an element contains a grid of $(N + 1)^2$ Gauss–Lobatto–Legendre points, as illustrated here in the case of $N = 4$. By construction, these points are non-evenly spaced.

as follows: if the grid point is interior to an element then $\phi_n(\mathbf{x})$ is defined on V^e by $\phi_n(\mathbf{x}) = h_{ijk}^N(\boldsymbol{\zeta})$ and ϕ_n vanishes outside V^e. If the grid point is shared with the face, edge or corner of another element, say $n = n(e, i, j, k) = n(e', i', j', k')$, then $\phi_n(\mathbf{x})$ is extrapolated to $V^{e'}$ by $\phi_n(\mathbf{x}) = h_{i'j'k'}^N(\boldsymbol{\zeta})$. We can then assign to any discrete displacement field $\tilde{\mathbf{u}} \in \mathbb{D}_h$ a local vector $\tilde{\mathbf{U}}_L$ of dimension $d \times \mathcal{N}_L$ and a global vector $\tilde{\mathbf{U}}_G$ of dimension $d \times \mathcal{N}_G$ (see Deville et al. (2002) for details). Both vectors store the values of the components of $\tilde{\mathbf{u}}$ at the collocation points. The scatter operation that allows one to form the local vector from the global vector involves a rectangular Boolean connectivity matrix, $\mathbf{Q}$, and can be written $\tilde{\mathbf{U}}_L = \mathbf{Q}\tilde{\mathbf{U}}_G$. The transpose operation, i.e., the action of $\mathbf{Q}^T$, is a gather process that consists in summing the entries of the local vector into the global vector. The successive action of $\mathbf{Q}^T$ and $\mathbf{Q}$ is referred to as the assembly process, or direct stiffness summation, in the finite-element literature.

3.1.3. Numerical Integration

After discretization, the integrals that appear in the weak form of the equation of motion must be evaluated numerically. The choice made in the SEM is to use the GLL collocation points to perform this task. Let q be a function defined on the reference interval Λ, then the integral of q is approximated by

$$\int_\Lambda q(\xi)\, d\xi \simeq \sum_{i=0}^{N} \omega_i^N q(\xi_i^N), \tag{67}$$

where the weights ω_i^N are the integrals of the Lagrange interpolants at the GLL points: $\omega_i^N = \int_\Lambda h_i^N\, d\xi$. This quadrature formula is exact for polynomials of order up to $2N - 1$ (see e.g., Funaro, 1992).

The quadrature in a space of higher dimension (also known as cubature) is obtained in the SEM by a tensorization process, which is the main reason why we chose to restrict ourselves to quadrangles rather than triangles in 2-D, and to hexahedra rather than tetrahedra in 3-D. If q is a function defined on the reference domain Λ^d, its integral is approximated by:

$$\int_{\Lambda^d} q(\boldsymbol{\zeta})\, d\boldsymbol{\zeta} \simeq \sum_{i=0}^{N}\sum_{j=0}^{N}\sum_{k=0}^{N} \omega_i^N \omega_j^N \omega_k^N q(\xi_{ijk}^N). \tag{68}$$

An important consequence of choosing the integration points to be the collocation points is that the mass matrix, i.e., the matrix representation of the L^2 scalar product, is exactly diagonal. Because this is a key property of the SEM for wave propagation problems, let us briefly recall the proof. Let $n = n(e, i, j, k)$ and $n' = n(e', i', j', k')$ denote two global indices, we define the components of the scalar mass matrix M as

$$M_{nn'} = \int_V \phi_n(\mathbf{x}) \phi_{n'}(\mathbf{x})\, d\mathbf{x}, \tag{69}$$

where ϕ_n and $\phi_{n'}$ denote the shape functions introduced in the previous section. The integral is non-zero only if the two grid points belong to at least one common element. Let e denote such an element, and suppose that $n = n(e, i, j, k)$ and $n' = n(e, i', j', k')$. Expressing the integral back on the reference domain gives:

$$M_{nn'} = \int_{V^e} \phi_n(\mathbf{x}) \phi_{n'}(\mathbf{x})\, d\mathbf{x} = \int_{\Lambda^d} h_{ijk}^N(\boldsymbol{\zeta}) h_{i'j'k'}^N(\boldsymbol{\zeta}) \mathcal{J}^e(\boldsymbol{\zeta})\, d\boldsymbol{\zeta}, \tag{70}$$

where $\mathcal{J}^e$ is the Jacobian (i.e., the determinant of the Jacobian matrix) of the local mapping (58). Applying the numerical quadrature gives

$$M_{nn'} = \delta_{ii'}\delta_{jj'}\delta_{kk'} \omega_i^N \omega_j^N \omega_k^N \mathcal{J}^e(\xi_{ijk}^N), \tag{71}$$

which is the expected result since the right-hand side is non-zero only for $n' = n$.

3.1.4. System of Ordinary Differential Equations

To finish the discretization process, we expand the generalized Galerkin approximation (52)–(55) on the polynomial basis using (66) and compute the integrals with the numerical quadrature (68). This gives a system of ordinary differential equations in time that can be written as

$$\mathbf{M}_S\ddot{\mathbf{U}}(t) + \mathbf{Co}\dot{\mathbf{U}}(t)$$
$$= \mathbf{K}_S\mathbf{U}(t) + \mathbf{G}\boldsymbol{\psi}(t) + \mathbf{C}^{OC}\ddot{\mathbf{U}}(t) + \mathbf{C}^{SF}\ddot{\boldsymbol{\zeta}}(t) + \mathbf{C}^{\Gamma}\mathbf{U}(t) + \mathbf{F}^{\text{ext}}(t), \quad (72)$$
$$\tilde{\mathbf{M}}_F\ddot{\boldsymbol{\zeta}}(t) = \mathbf{K}_{F,\zeta}(\boldsymbol{\zeta}, \boldsymbol{\chi}, \mathbf{W})(t) + \mathbf{C}^{FS}\mathbf{U}(t), \quad (73)$$
$$\tilde{\mathbf{M}}_F\ddot{\boldsymbol{\chi}}(t) = \mathbf{K}_{F,\chi}(\ddot{\boldsymbol{\zeta}}, \boldsymbol{\zeta}, \boldsymbol{\chi}, \mathbf{W})(t), \quad (74)$$
$$\mathbf{M}_F\dot{\mathbf{W}}(t) = \mathbf{R}(\boldsymbol{\zeta}, \boldsymbol{\chi}, \mathbf{W})(t), \quad (75)$$
$$\mathbf{P}\boldsymbol{\psi}(t) = \mathbf{D}(\mathbf{U}, \boldsymbol{\zeta}, \boldsymbol{\chi}, \mathbf{W})(t), \quad (76)$$

where $\mathbf{U}, \boldsymbol{\zeta}, \boldsymbol{\chi}, \mathbf{W}$ and $\boldsymbol{\psi}$ denote the vectors that store the values of the unknowns at the $\mathcal{N}_G$ global collocation points; $\mathbf{M}_S$ is the mass matrix in the solid, modified to include density; $\mathbf{M}_F$ is the mass matrix in the fluid and $\tilde{\mathbf{M}}_F$ is modified to include the inverse of c^2; $\mathbf{Co}$ is the Coriolis matrix; $\mathbf{K}_S$ and $\mathbf{K}_{F,\zeta}$ are the stiffness matrices in the solid and fluid, respectively; $\mathbf{G}$ is the gradient matrix modified to include density; $\mathbf{C}^{OC}$ is the ocean–crust coupling matrix, $\mathbf{C}^{SF}$ and $\mathbf{C}^{FS}$ are the fluid–solid coupling matrices, and $\mathbf{C}^{\Gamma}$ is the DtN coupling matrix; $\mathbf{R}$ is the precession matrix, $\mathbf{P}$ and $\mathbf{D}$ are the matrix approximations of the Poisson–Laplace and divergence operators and $\mathbf{F}^{\text{ext}}$ is the vector of external forces.

It should be noted that none of the above global matrices needs to be explicitly built, because only their action on a vector is needed in order to perform computations. Furthermore, these matrix-vector products can be defined at the elemental level based upon the connectivity matrix $\mathbf{Q}$ introduced in Section 3.1.2. For example, the action of the mass matrix $\mathbf{M}_S$ can be decomposed into $\mathbf{M}_S\mathbf{U} = \mathbf{Q}^T\mathbf{M}_{S,L}\mathbf{U}_L$, where $\mathbf{M}_{S,L}$ is a set of elemental matrices that act on the local vector $\mathbf{U}_L = \mathbf{Q}\mathbf{U}$. Note that the coefficients of $\mathbf{M}_{S,L}$ are obtained after multiplying the right-hand side of Eq. (71) by the pointwise values of density.

The reader is referred to Komatitsch and Tromp (1999) for a detailed derivation of the source term $\mathbf{F}_L$ and of the stiffness matrix $\mathbf{K}_{S,L}$ in the absence of gravity. Regarding the effect of the ocean load, note that (45) involves a simple surface integral along the ocean–crust boundary, which shows that in the context of the approximation of Appendix B.3 we can efficiently take into account the effects of the oceans represented by the $\mathbf{C}^{OC}$ ocean–crust coupling matrix above by a simple modification of the mass matrix for the degrees of freedom located exactly at the ocean floor.

Finally, let us note that because of the use of a tensorized polynomial basis, the operation count of a product with the elemental stiffness matrix is only $\mathcal{O}(N^{d+1})$ in d dimensions instead of $\mathcal{O}(N^{2d})$ for a full matrix, which is another argument that makes the SEM efficient for 3-D calculations.

4. TIME DISCRETIZATION

The system of ordinary differential equations (72)–(76) is marched in time using a finite-difference scheme. Let us postpone to the end of this section the implementation of rotation, self-gravitation, attenuation and absorbing layers. For the sake of clarity, we write the previous system of differential equations in the compact form:

$$\mathbf{M}\ddot{\mathbf{q}}(t) = \mathbf{F}^{\text{ext}}(t) - \mathbf{F}^{\text{int}}(\mathbf{q}) - \mathbf{F}^{\Sigma}(\mathbf{q}) \tag{77}$$

where $\mathbf{q}$ is a global vector defined as $\mathbf{q} = (\mathbf{U}, \zeta, \chi)^T$. The vector of all internal forces is denoted by $\mathbf{F}^{\text{int}}$, and $\mathbf{F}^{\Sigma}$ stands for the surficial forces that result from the action of the coupling operators $\mathbf{C}^{\text{SF}}$, $\mathbf{C}^{\text{FS}}$ and $\mathbf{C}^{\Gamma}$. As mentioned in the previous section, the ocean–crust coupling term $\mathbf{C}^{\text{OC}}$ has been included in the mass matrix $\mathbf{M}$.

4.1. Newmark Time Stepping Method

The Newmark family of integrators (Newmark, 1959; Hughes, 1987) is written for a given time $t_n = n\Delta t$ as a map $(\mathbf{q}_n, \dot{\mathbf{q}}_n) \to (\mathbf{q}_{n+1}, \dot{\mathbf{q}}_{n+1})$ that is defined by enforcing the semi-discrete momentum equations at time t_{n+1}:

$$\ddot{\mathbf{q}}_{n+1} = \mathbf{M}^{-1}\left(\mathbf{F}^{\text{ext}}_{n+1} - \mathbf{F}^{\text{int}}(\mathbf{q}_{n+1}) - \mathbf{F}^{\Sigma}(\mathbf{q}_{n+1})\right), \tag{78}$$

$$\mathbf{q}_{n+1} = \mathbf{q}_n + \Delta t \dot{\mathbf{q}}_n + \Delta t^2 \left(\left(\frac{1}{2} - \beta\right)\ddot{\mathbf{q}}_n + \beta \ddot{\mathbf{q}}_{n+1}\right), \tag{79}$$

$$\dot{\mathbf{q}}_{n+1} = \dot{\mathbf{q}}_n + \Delta t \left((1-\gamma)\ddot{\mathbf{q}}_n + \gamma \ddot{\mathbf{q}}_{n+1}\right), \tag{80}$$

where one needs to select the two parameters $\gamma \in [0, 1]$ and $\beta \in [0, 1/2]$.

Second-order accuracy (instead of first-order) is obtained if and only if the velocity is updated based upon a centered scheme, i.e., if $\gamma = 1/2$. The particular choice of setting $\beta = 0$ and $\gamma = 1/2$ leads to the explicit central difference method, which is the most widely used in the context of spectral-element simulations of seismic wave propagation. It has the important property of conserving linear and angular momentum and having bounded energy errors (see Hughes (1987) and Kane et al. (2003) for theoretical considerations and Komatitsch and Vilotte (1998) for numerical illustrations of energy conservation). It is conditionally stable, the maximal time step being imposed by the Courant–Friedrichs–Lewy (CFL)

(Courant *et al.*, 1928) criterion that states that the speed at which information travels cannot exceed one grid cell per time step. This means that

$$\Delta t \leq C \left(\frac{\Delta x}{\alpha} \right)_{\min}, \qquad (81)$$

where Δx is the distance between adjacent grid points, α is the P-wave speed and C is a constant (typically ranging between 0.3 and 0.5) that depends on the geometry of the mesh and on the spatial dimension of the problem. The order in which operations are performed in the explicit scheme is the following: first, $\mathbf{q}$ (the displacement) is updated from $\dot{\mathbf{q}}$ (the velocity) and $\ddot{\mathbf{q}}$ (the acceleration) at time t_n using (79); second, acceleration at time t_{n+1} is computed by solving (78); finally, velocity is updated based upon the centered formula (80). The only linear system to invert during this process involves the mass matrix, which is exactly diagonal by construction. This is an important argument that favors the Legendre SEM when compared to other spectral methods such as the Chebyshev SEM (see Section 5.1).

An alternative formulation of the explicit central-difference method, more adapted to mixed velocity–stress formulations, is to write it in terms of displacement and velocity, provided that the two quantities be staggered in time (Festa and Vilotte, 2005). The scheme consists in defining the map $(\mathbf{q}_{n-\frac{1}{2}}, \dot{\mathbf{q}}_n) \rightarrow (\mathbf{q}_{n+\frac{1}{2}}, \dot{\mathbf{q}}_{n+1})$ based upon the mid-point rule

$$\mathbf{q}_{n+\frac{1}{2}} = \mathbf{q}_{n-\frac{1}{2}} + \Delta t \dot{\mathbf{q}}_n, \qquad (82)$$

$$\dot{\mathbf{q}}_{n+1} = \dot{\mathbf{q}}_n + \Delta t \mathbf{M}^{-1} \left(\mathbf{F}^{\text{ext}}_{n+\frac{1}{2}} - \mathbf{F}^{\text{int}}(\mathbf{q}_{n+\frac{1}{2}}) - \mathbf{F}^{\Sigma}(\mathbf{q}_{n+\frac{1}{2}}) \right). \qquad (83)$$

The slight difference between the two methods is that the mid-point rule evaluates forces at averaged times, whereas the Newmark scheme performs a time-average of the forces.

Finally, let us recall that coupling fluid and solid regions requires exchanging the normal displacement $\mathbf{u} \cdot \hat{\mathbf{n}}$ and the Lagrangian pressure perturbation $(-\rho \ddot{\zeta})$ from one side of the fluid–solid interface to the other. This can be done without iterating on the coupling condition because both quantities are known at the same time (through $\mathbf{q}$ and $\ddot{\mathbf{q}}$). As noticed by Chaljub and Valette (2004), this is a consequence of applying the potential decomposition (19) to the displacement instead of velocity. This has important consequences in practice because it greatly simplifies the implementation of the SEM algorithm and reduces the computational cost by completely suppressing iterations.

4.2. Rotation

Taking into account the effects of rotation requires adding the Coriolis force in the solid regions and solving (75) for the precession vector $\mathbf{W}$ in the fluid regions.

Note that because the Coriolis force involves velocity, in principle this term should be treated based upon an implicit time-marching scheme. However, because the Coriolis effect matters mostly at very long period (e.g., Dahlen and Tromp, 1998), in practice it is acceptable to purposely lose the second-order accuracy of the time-stepping method for the Coriolis term by handling it based upon an explicit scheme (Komatitsch and Tromp, 2002b) because the time step used is always very small compared to the periods involved. Likewise, the precession equation in the fluid regions can be solved based upon a simple first-order Euler scheme.

In Fig. 7, we show the effect of the Coriolis forcing on the frequencies of two eigenmodes of the Earth: the so-called football mode $_0S_2$ and the recently detected $_2S_1$ mode that involves the motion of the entire core (Rosat *et al.*, 2003). The spectra have been obtained by post-processing a spectral-element time series recorded at an epicentral distance of 90° of a point double-couple source in the PREM reference Earth model (Dziewonski and Anderson, 1981) without the ocean layer. Note that only the Coriolis forcing term has been included in the spectral-element calculations. Other effects of the rotation of the Earth (centrifugal potential, ellipticity) have been ignored. Because we needed a long time series to compute the spectra (about 100 hours), we also neglected the redistribution of mass to get results in a reasonable amount of compute time.

The effect of rotation is to split the degenerate eigenfrequencies $_n\omega_l$ of the spherically-symmetric model into $2l + 1$ singlets, which are visible on the plots. Also shown on Fig. 7 are the predictions of a first-order splitting theory (see e.g., Dahlen and Tromp, 1998, p. 605). The difference with the apparent eigenfrequencies derived from the spectral-element synthetic seismograms illustrates the magnitude of higher-order effects in the Coriolis splitting.

4.3. Self-Gravitation

Taking into account the perturbation of the gravitational potential does not affect the accuracy of the time-stepping method but reduces its efficiency because the algebraic system (76) must be solved at each time step. Because the Poisson–Laplace matrix **P** is symmetric, the system can be solved for instance based on a conjugate gradient (CG) method, the efficiency of which is governed by the condition number κ of the matrix (i.e., the ratio between the largest and the smallest of its eigenvalues). The number of CG iterations needed to reach a given level of accuracy scales as the square root of κ. Unfortunately, the Poisson–Laplace matrix arising from the spectral-element discretization is known for being poorly conditioned (see e.g., Deville *et al.*, 2002). A diagonal preconditioner is used by Chaljub and Valette (2004) but results in poor numerical efficiency (a number of iterations ranging between 50 and 100 is needed to obtain an acceptable level of accuracy). Additional work is therefore needed to build an efficient preconditioner and to be able to include the redistribution of mass in routine calculations.

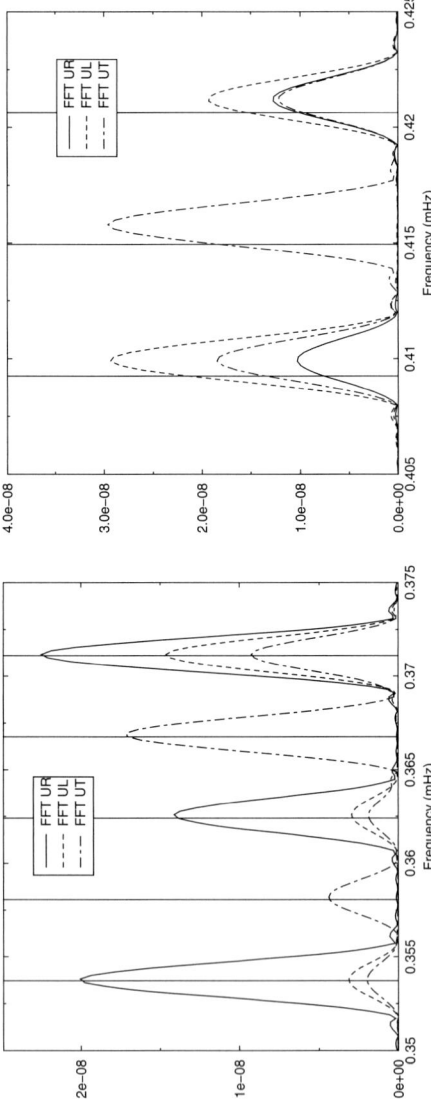

FIG. 7. Amplitude spectrum of spectral-element synthetic seismograms computed at a station located 90° away from the source. The left and right frequency windows display the $_0S_2$ and $_2S_1$ modes, respectively. The different singlets appear on the radial (solid line), longitudinal (dashed line) and transverse (dot-dashed line) components for this geometry. The vertical lines show the first-order theoretical prediction of the Coriolis splitting. The difference with the synthetic values illustrates the magnitude of higher-order effects. Note that the eigenvalues given here are slightly higher than real values because self-gravitation has not been accounted for in our calculations.

4.4. Attenuation

Including the effects of attenuation requires marching the memory variables in time according to Eq. (A.7). This can be achieved based upon a fourth-order Runge–Kutta method (see e.g., Komatitsch and Tromp, 1999, and references therein). Once the memory variables are updated, the stress tensor is given by Hooke's law (A.6), and the internal forces can be computed.

In global seismology, bulk attenuation is often ignored because Q_κ is typically one order of magnitude greater than Q_μ. Note that a complete treatment of attenuation should in principle include fluid regions, which implies that the potential formulation (19) should be modified to include appropriate memory variables. However, in global seismology, attenuation in the fluid is almost always completely negligible (e.g., Dahlen and Tromp, 1998).

4.5. Implementation of a Perfectly Matched Layer

As mentioned in Appendix A.2, a Perfectly Matched Layer (PML) absorbing region for the seismic wave equation is naturally formulated in terms of velocity and stress, i.e., for a system of first-order equations in time (e.g., Collino and Tsogka, 2001). In recent years, efforts have been made to derive PML formulations suitable for the seismic wave equation written as a second-order system in displacement (e.g., Komatitsch and Tromp, 2003; Basu and Chopra, 2003, 2004; Festa and Vilotte, 2005). To our knowledge, so far two of these approaches have been employed to adapt the PML to the SEM, which is based on the second-order system in displacement: Komatitsch and Tromp (2003) derived a formulation of the PML that is directly adapted to second-order equations, and Festa and Vilotte (2005) showed that the classical first-order PML formulation could be used as is, based upon a discrete equivalence between the Newmark time-stepping method and the mid-point rule applied to a staggered velocity–stress approximation of Eqs. (A.13) and (A.14). An alternative is to adapt the SEM to the PML by constructing a SEM formulation based on the mixed velocity–stress system, which is well adapted to the introduction of PML (Cohen and Fauqueux, 2005).

Both formulations are equivalent in the sense that they are equally efficient for absorbing body and surface waves and that they suffer from the same drawbacks as the original method: i.e., difficulties to absorb waves at grazing incidence (Collino and Monk, 1998; Winton and Rappaport, 2000) and intrinsic instabilities in some anisotropic media (Bécache et al., 2003). However, while Komatitsch and Tromp (2003) obtain a system of third-order equations in time (which is further decomposed into a second-order system plus a first-order equation solved by an additional Runge–Kutta scheme), the formulation of Festa and Vilotte (2005) fits naturally in the context of the Newmark scheme and can therefore be more easily added to existing spectral-element algorithms without major modifications of the time-stepping scheme.

5. Implementation

5.1. Choice of Parameters for Accurate Calculations

In a standard FEM, low-order polynomials (usually of order 1 or 2) are used to discretize functions, and therefore the accuracy of the method is mainly governed by the typical size of an element in the mesh, h. In a SEM, however, high-order Lagrange interpolants are used to express functions. Therefore, one can use the polynomial order N chosen to represent functions on an element as an additional parameter to control the accuracy of the method.

The GLL numerical integration rule is exact only for polynomials of order $2N - 1$ (e.g., Canuto *et al.*, 1988). This implies that because it involves the product of two polynomials of order N, the displacement and the test function, integration on the reference element is never exact, even in the simplest case of a reference cube with homogeneous material properties. For deformed and/or heterogeneous elements, there are additional errors related to curvature (Maday and Rønquist, 1990). Thus in the Legendre SEM we obtain a diagonal mass matrix by purposely selecting a numerical integration rule that is not exact (but of course still very accurate) for the polynomial basis chosen (in the finite-element literature, such a choice is known as sub-integration). A different choice is made in the Chebyshev SEM used by some authors (e.g., Priolo *et al.*, 1994; Seriani, 1998), in which an integration rule that is exact for the polynomial basis chosen is used, with the consequence that the mass matrix is no longer diagonal.

To our knowledge, no theoretical analysis of the accuracy of the SEM for elastic media is available in the literature, and only Tordjman (1995) and Cohen and Fauqueux (2000) have attempted such a theoretical accuracy study for the SEM in the acoustic case, and Seriani and Priolo (1994) have addressed the issue based on a numerical study, also in the acoustic case. Because of the lack of such theoretical criteria, we use heuristic rules in order to select the polynomial order to use in practice for an elastic SEM with a non-regular deformed mesh and a heterogeneous medium. These rules of thumb have been developed over the years by trial and error based upon a large number of numerical tests for which a quasi-analytical reference solution was known. In particular, we have used the main conclusions of Seriani and Priolo (1994) in the acoustic case and checked numerically for simple cases that these conclusions extend reasonably well to the elastic case. Based on these numerical experiments, we observe that if one uses a polynomial order lower than typically 4, one looses the advantages of using a SEM, because the calculations lead to a large amount of numerical dispersion, i.e., to similar inaccuracies as with standard FEMs (Marfurt, 1984). On the contrary, if one selects a very large polynomial order, e.g., greater than 10, the SEM is spatially very accurate, but the computational requirements become prohibitive because of the size of the calculations related to matrix multiplications involving the full stiffness matrix, a process with a cost of $O(N^4)$ in 3-D, i.e., the numerical cost of the technique becomes very high. Let us also recall that the GLL numerical integration points are non-evenly spaced and that in the case of a high order they become clustered toward the edges of each spectral element (e.g., Canuto *et al.*, 1988). More precisely, one can show that the spacing between the first two GLL points of the reference interval varies approximately as $O(N^{-2})$, and as a result of the small distance between these first two points, very small time steps have to be used to keep the explicit time-marching scheme stable (see Section 4), which drastically increases the cost of the Legendre SEM. Therefore, the empirical conclusion is that for most seismic wave propagation problems, polynomial orders between approximately 4 and 10 should be used in practice. In our numerical simulations, we often use a polynomial order $N = 4$. In order to obtain accurate results, we use another heuristic rule of thumb that says that for this polynomial order the average grid spacing h should be chosen such that the average number of points per minimum wavelength $\lambda_{\min}$ in an element, $(N+1)\lambda_{\min}/h$, be roughly equal to 5. Note that this value is of the same order as that obtained in the context of high-order finite-difference methods applied to wave propagation problems.

5.2. Parallel Implementation

Section 3.1.1 explained the basic concepts of mesh design in the SEM. However, real meshes designed to study large 3-D models are usually too large to fit on a single computer in terms of memory. One then needs to turn to parallel machines, for instance clusters of personal computers or grids of computers, which all have a distributed memory architecture. In order to program such parallel machines, one needs to use a message-passing methodology, usually based upon a library called MPI (e.g., Gropp *et al.*, 1994), an acronym for 'Message Passing Interface', which has the advantage of being freely available on most computer systems and therefore very portable.

Let us mention that the SEM can efficiently run on high-latency networks such as clusters or grids of computers because it is relatively insensitive to the speed of the network connecting the different processors. The main reason for that is that we can use a fully explicit time-marching scheme (see Section 4). In such schemes, one mostly needs to perform small local matrix-vector products in each spectral element, and as a consequence the processors spend most of their time doing actual calculations, and only a small amount of time exchanging information, which means that the algorithm is not very sensitive to the communication speed.

Once a mesh has been created, it needs to be split into slices if one wants to run the SEM on a parallel computer. Powerful algorithms called 'mesh partitioners' are available for this purpose, such as METIS, which is a set of programs for partitioning graphs and finite-element meshes. The algorithms implemented in METIS are based on sophisticated multilevel graph-partitioning schemes Karypis and Kumar (1998a, 1998b). The mesh should be decomposed into as many slices as the number of processors one plans to use on the machine. Calculations can then be performed locally by each processor on the spectral elements that constitute the mesh slice it carries after the mesh partitioning step. In the time-marching algorithm, one communication phase is then required at each time step in order to sum the internal forces computed at the common faces, edges, and corners shared by mesh slices carried by different processors. MPI communications are performed based upon a (constant) communication topology produced once and for all by the mesh partitioner. This topology represents the sequence of messages that needs to be exchanged between the slices at each time step. It is constant because there is no dynamic (i.e., adaptive) remeshing in the SEM applied to seismic wave propagation studies, which are small perturbation problems for which a constant mesh is sufficient.

6. Applications

6.1. 3-D Ground Motion in Alpine Valleys

As a first example of the application of the SEM to realistic wave propagation problems, we consider the simulation of ground motion in the alpine valley of Grenoble, French Alps, France. Like sedimentary basins, alpine valleys are known for producing strong site effects because of (i) stiff bedrock conditions leading to large impedance contrasts between bedrock and sediments and (ii) thick postglacial deposits of lacustrine and fluviatile sediments. The computation of site effects in alpine valleys has received considerable attention since the pioneering work of Bard and Bouchon (1980a, 1980b). It is out of the scope of this article to review all the studies and methods devoted to this topic. Let us mention however that most of these studies are restricted to two-dimensional geometries. The computation of 3-D seismic wave propagation in alpine valleys is challenging for two reasons: first, unlike in large sedimentary basins, seismic energy diffracted off the valley edges and transported by surface waves dominates the seismic signal (e.g., Cornou *et al.*, 2003) and is hardly separable from body wave reflections. Second, alpine valleys are often deeply embanked and exhibit strong topographic variations of the free surface and of the interface between sediments and bedrock. With its ability to handle 3-D geometries and its accuracy regarding the computation of surface waves, the SEM is therefore well suited to simulate wave propagation in alpine valleys.

We consider the city of Grenoble in the French Alps. Figure 8 shows the 'Y' shape of the valley, surrounded by three topographic units with a typical elevation of 1000 m to 1500 m on the western side (Vercors and Chartreuse massifs) and 2500 m to 3000 m on the eastern side (Belledonne chain). Figure 8 also displays the surface of the spectral-element mesh used to propagate seismic waves at frequencies up to 1 Hz, assuming a minimal shear wave velocity of 300 m s^{-1} in the sediments. The 3-D mesh shown in Fig. 9 honors the sediment/bedrock interface, which is well constrained by the inversion of gravimetric anomalies performed by Vallon (1999). As explained in Section 3.1.1, it is difficult to honor the interface near the edges of the valley because (i) this requires the use of a fully 3-D mesh generator and (ii) even with such a tool, the elements located near the edges are expected to be strongly distorted, thus leading to a loss of accuracy. Here we avoid this difficulty by defining a threshold depth above which the interface is not honored but rather interpolated within the spectral elements (see Fig. 9), as suggested by Komatitsch et al. (2004). This solution is not fully satisfactory because at shallow depths (in particular near valley edges) the contrast between bedrock and sediments is smeared out due to the use of high-order ($N = 4$) polynomial interpolation within each spectral element.

We choose to simulate the April 26, 2003 $M_L = 2.9$ Lancey event, which occurred on the Belledonne border fault less than 10 km from the valley edges (Thouvenot et al., 2003). Based upon measurements performed in a 550 m-deep borehole (Nicoud et al., 2002), the velocity model used in the sediments is defined as:

$$\begin{cases} \beta = 300.0 + 19.0\sqrt{d}, \\ \alpha = 1450.0 + 1.2d, \end{cases} \tag{84}$$

in which d is depth in m, α and β are the P and S velocities in m s^{-1}, and mass density in kg m^{-3} is defined by $\rho = 2140.0 + 0.125d$. Attenuation is taken into account by defining a constant shear quality factor $Q_S = 50$ in the sediments. The bedrock model is adapted from that used by Thouvenot et al. (2003), with P and S velocities in the uppermost layer equal to $\alpha = 5600$ m s^{-1} and $\beta = 3200$ m s^{-1}, respectively. We use a polynomial order $N = 4$ (i.e., $(N + 1)^3 = 125$ GLL points per spectral element) in the SEM, which yields a total of approximately 3.3 million grid points and about 10 millions of degrees of freedom.

The time evolution of synthetic vertical ground velocity is represented in Fig. 10. The calculation is accurate up to approximately 1 Hz, and the source we use does not contain energy above this cutoff limit. The source mechanism, a strike-slip event with almost vertical dip, is clearly seen in the first second of the recorded signals. In the next four seconds, one sees the distortion of P and S wavefronts by the low velocities in the sediments. Surface waves generated at the edges of the valley can be observed relatively early (from 3.5 s on) and tend to form planar features aligned with the borders (e.g., between 7.5 s and 8.5 s). The rest of the simulation shows the trapping of energy within the sediments, causing the progressive illumination of the whole basin structure. Note how the wave field becomes complex after only a dozen of seconds of propagation, due to multiple reflections of surface waves diffracted off the valley edges. To give an idea of the typical numerical cost of such a simulation, the calculations required about one gigabyte of memory, and took (in 2004) about 16 hours of CPU time on four 3 GHz Intel Xeon processors to run 60000 time steps of 1 ms each.

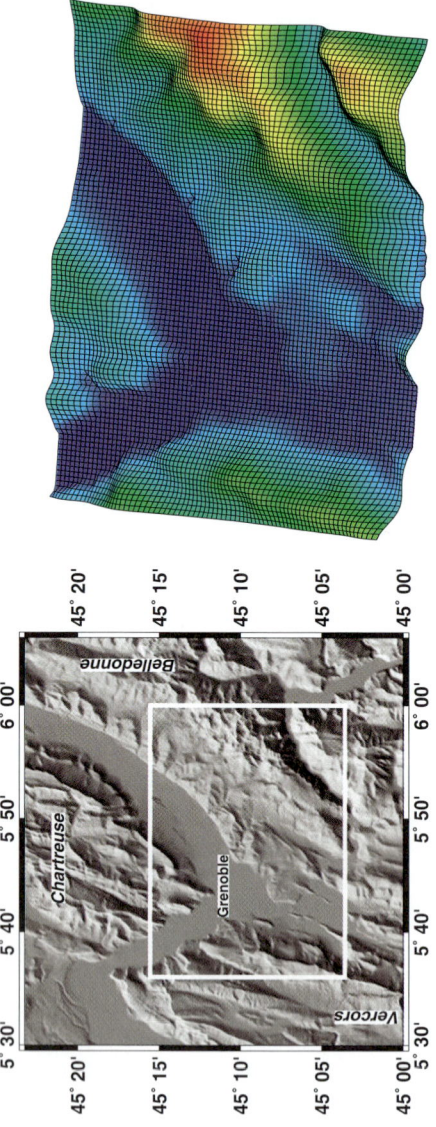

FIG. 8. Left: Topographic map of the Alps showing the city of Grenoble surrounded by the Vercors, Chartreuse and Belledonne massifs. The white square denotes the edges of the computing grid used to simulate ground motion. Right: Surface view of the spectral-element mesh used to compute synthetic waveforms accurate up to approximately 1 Hz. The color scale indicates surface elevation above sea level, which ranges from 200 m (blue) to approximately 2500 m (red). The mesh contains 96 × 80 surface elements. The vertical exaggeration factor is 5 and a spatial Gaussian filter has been applied to smooth sharp topographic variations.

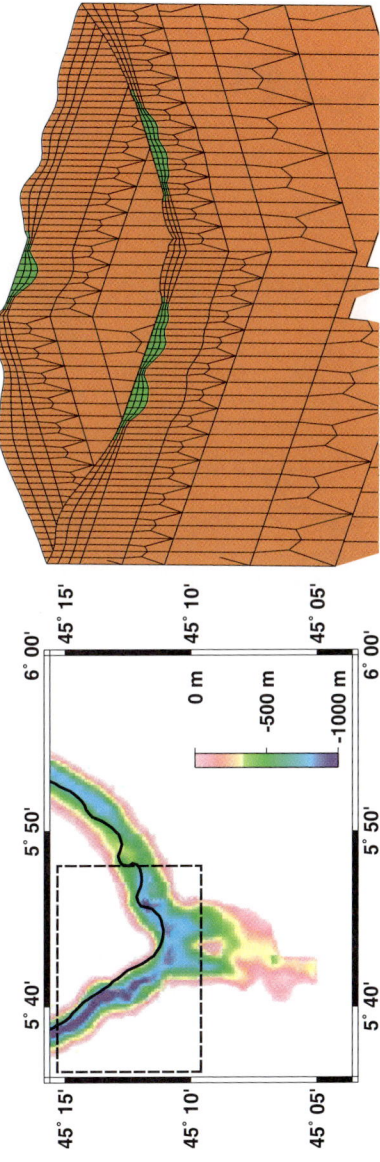

FIG. 9. Left: Map of the shape of the bedrock obtained by inverting gravimetric measurements over the Grenoble valley. The maximum sediment thickness, which is located in the western part of the 'Y' shaped valley, exceeds 1000 m. The black line represents the Isère river. Right: view from the south–east with no vertical exaggeration of a cross section through the spectral-element mesh (in the area indicated by the dashed lines in the left picture). Brown indicates elements that intersect the bedrock, whereas green represents elements located entirely within the sediments. The mesh is coarsened twice in depth to adapt it to the variations of seismic wavelengths within the model. The sediment/bedrock interface is honored exactly (i.e., the elastic parameters are discontinuous across the interface in the mesh) for depths greater than 220 m. At shallower depths (in particular near valley edges) the contrast between bedrock and sediments is smeared out due to the use of high-order ($N = 4$) polynomial interpolation within each spectral element.

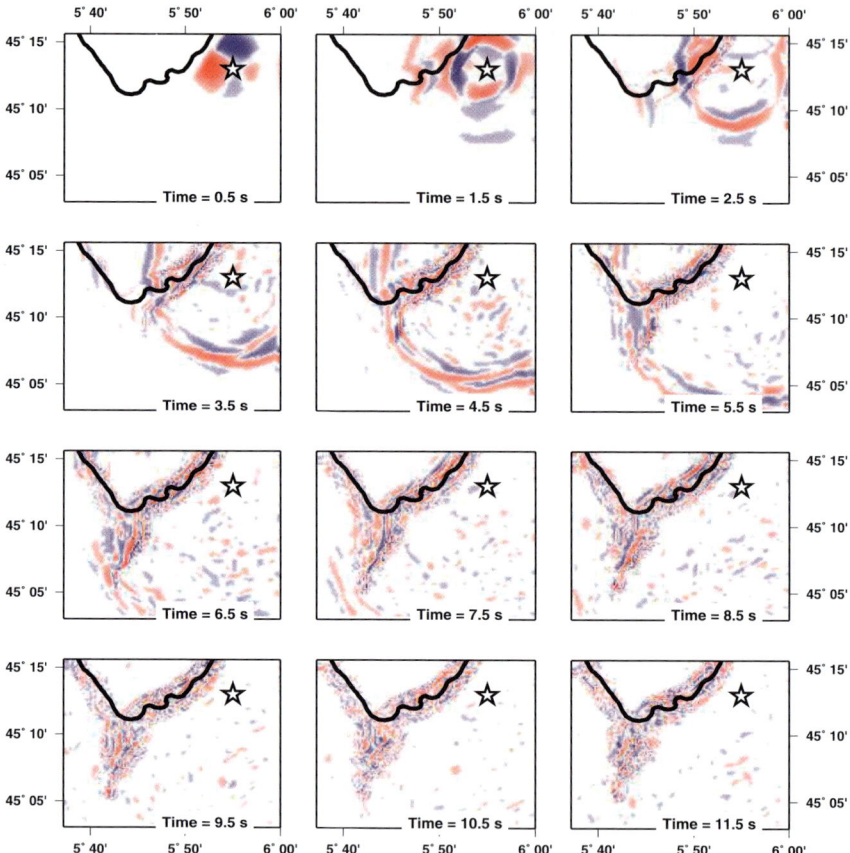

FIG. 10. Time evolution of the vertical component of synthetic ground velocity simulated for the April 26, 2003, $M_L = 2.9$, Lancey event, whose epicenter is indicated by the black and white star. Red and blue colors denote upward and downward motion, respectively. The radiation pattern of this strike-slip event is clearly seen at time 0.5 s. At times between 1.5 s and 3.5 s, the P and S wavefronts are deflected by the slow wave speeds in the basin. Energy diffracted off the edges of the basin and traveling as surface waves can be observed before 8.5 s. Note the complex shape of the wave field, with almost no coherent pattern after 10 s of propagation.

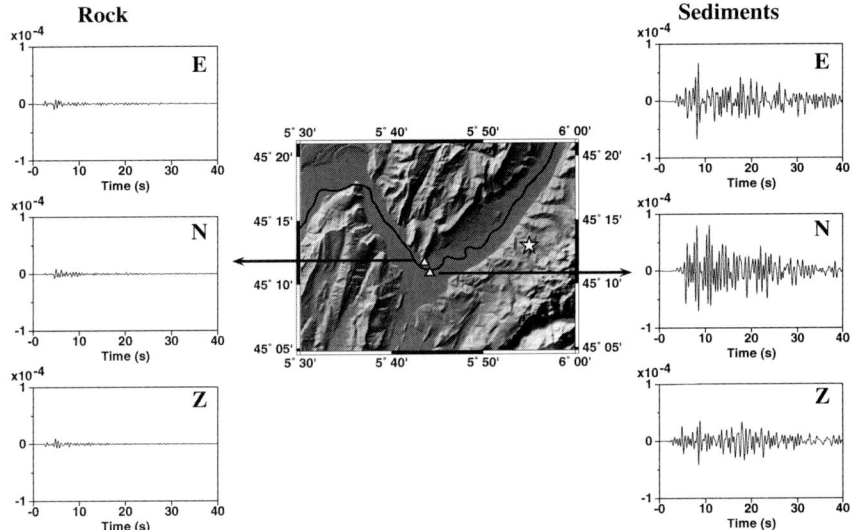

FIG. 11. Time series of ground velocity (in m s^{-1}) computed for the April 26, 2003, $M_L = 2.9$ Lancey event, whose epicenter is indicated by the black and white star, at two stations of the French accelerometric network (black and white triangles). The left records correspond to a rock site, and the right records to a site located in the center of the basin, where the sediment thickness exceeds 800 m. Both synthetic seismograms have been bandpass-filtered between 0.1 Hz and 2 Hz using a six-pole two-pass Butterworth filter. Note the strong amplification as well as the increased duration of the signal recorded in the basin even though the two sites are only separated by a distance of about 1 km.

Time series of ground velocity recorded at two sites with different soil conditions are displayed in Fig. 11. The source time function used is a smoothed version of a Heaviside step (in practice an error function with a short rise time $\tau = 0.01$ s). The synthetic seismograms are further bandpass-filtered between 0.1 Hz and 2 Hz. The reason why we can analyze frequencies higher than the limit used to design the mesh (1 Hz) is that the signal is dominated by surface waves that at low frequency are sensitive to the velocity model at depth, were velocity increases, and therefore travel at a velocity higher (typically about 600 m s^{-1} at 2 Hz) than the minimal S wave velocity in the model, which is found at the surface. The comparison shows a strong amplification (by about a factor of 10) and an increase in duration between the signal recorded in the sediments and on the rock site.

Comparison to data (Chaljub et al., 2005) shows a reasonable agreement for frequencies smaller than 1 Hz, confirming that the long-wavelength structure of the basin is well constrained. The perspective is now to better understand the variability of ground motion observed in high-frequency data (i.e., from 1 Hz to 10 Hz). The small size of the model, allowing computations up to 4 Hz to be performed with accessible computing facilities will help us to address this problem with the SEM.

6.2. 3-D Spectral-Element Reference Solution

One of the goals of seismologists is to constrain the physical properties of the Earth's interior by solving the inverse problem of finding Earth models for which synthetic seismograms match the observations. Most of the methods developed to solve the inverse problem involve the computation of a large number of synthetic seismograms, and different approximations have been derived to produce them in a reasonable amount of CPU time. These methods provide accurate results as long as they are used within their domain of validity, a condition unfortunately not always met in practice. A precise series of tests should therefore be performed to assess the effects of such approximations, but so far these have rarely been conducted due to the lack of an independent solution to compare with (Clévédé et al., 2000).

The application of the SEM to global seismology offers the possibility of overcoming this difficulty by providing a three-dimensional reference solution. As an example, let us show a comparison between the SEM and two methods widely used in global tomography: the Path Average Approximation (PAVA) (Woodhouse and Dziewonski, 1984) and Non-linear Asymptotic Coupling Theory (NACT) (Li and Tanimoto, 1993; Li and Romanowicz, 1995). Both methods are based on perturbation theory around a spherical reference Earth model. The reference solution, expanded on the basis of normal modes, is perturbed to the first order based on the Born approximation to obtain an approximate solution in the 3-D model. The solution is thus only valid for 'weak' and smooth-enough 3-D heterogeneities. In addition to the Born approximation, PAVA and NACT use the great-circle approximation, in which only 3-D heterogeneities lying under the great circle between the source and the receiver are taken into account. This approximation is valid when the wavelength of the model is much larger than the wavelength of the displacement field, i.e., for 'smooth-enough' models. Compared to NACT, PAVA makes the additional approximation that the coupling between dispersion branches is not accounted for in the perturbation. This results in a 1-D sensitivity kernel: the effect of an heterogeneity at a given depth does not depend on the position along the great circle. While this is acceptable for Love and Rayleigh waves, such an approximation cannot account for the ray-like sensitivity of body waves. We therefore expect NACT to give a more precise result than PAVA in a 3-D model, especially for higher modes.

Figure 12 shows an example of long-period (100 s and above) synthetic seismograms computed with PAVA, NACT and SEM in the tomographic model SAW24B16 (Mégnin and Romanowicz, 2000). The SEM synthetic seismograms are computed on a grid of 1782 spectral elements with 54936 collocation points, and the time step used is 1 s. The comparison of differential seismograms (defined as the difference between 3-D and 1-D synthetic seismograms) shows that both PAVA and NACT predictions are accurate for the Rayleigh wave R1, but only NACT retrieves the waveforms of higher modes, such as the X wave (Jobert et al.,

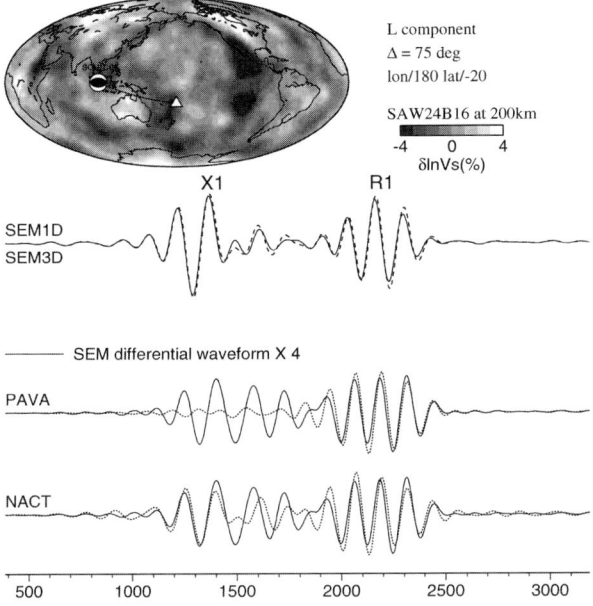

FIG. 12. Comparison of the two asymptotic methods PAVA and NACT on a 3-D synthetic example in tomographic model SAW24B16. The map shows the path selected. In this long-period test (100 s and above) only the longitudinal component is displayed. The top panel shows the SEM synthetic seismograms computed in the reference model (solid line) and in the 3-D model (dotted line). The last two panels show differential seismograms (3-D results minus 1-D results) obtained based upon the PAVA (middle) and NACT (bottom) methods and amplified by a factor of four. In each case, the solid line represents the differential synthetic seismograms obtained based upon the SEM. As expected, NACT performs better than PAVA in fitting the high-order harmonics (here the X wave). In this example, the spectral-element solution is used as a 3-D reference to estimate the fit obtained.

1977). Note that none of the results perfectly match the spectral-element solution, which is consistent with the approximations used in each method.

Another example in which a spectral-element calculation proved useful can be found in (Favier *et al.*, 2004), where a new sensitivity kernel to invert splitting measurements in anisotropic media is proposed, which includes the middle- and near-field terms of the Green's tensor generally neglected in classical theories. Using the spectral-element solution as a reference, the authors assessed the domain of validity of their new kernel and the error made by using classical far-field kernels.

6.3. 3-D Modeling in the D'' region

Another application of the SEM is to assess the quality of current tomographic models by providing an independent measure of their ability to fit the data. For example, Komatitsch *et al.* (2002) compute synthetic seismograms using tomographic mantle model S20RTS (Ritsema et al., 1999) and crustal model CRUST2.0 (Bassin et al., 2000). They show that this 3-D model explains body-wave travel times for periods up to approximately 20 s but that surface wave dispersion is not accurately accounted for at periods shorter than 40 s.

Here we show a similar comparison between data and synthetic seismograms obtained by coupling the SEM to a normal-mode solution, and we focus our attention on the D'' region. This region at the base of the Earth's mantle is known for being very heterogeneous and for significantly affecting the travel times and amplitudes of seismic waves, in particular those diffracted by the core-mantle boundary (CMB). We simulate the September 4, 1997, 621-km deep Fiji event ($M_w = 6.8$) using the PREM model everywhere except in a 370-km thick layer above the CMB (D'' region) in which we replace it by the 3-D tomographic S-velocity model SAW24B16 (Mégnin and Romanowicz, 2000). Perturbations in P-wave velocity and density are obtained from S-wave values by the linear scaling rules $\delta\alpha = 0.40\delta\beta$ and $\delta\rho = 0.25\delta\beta$ (Kumazawa and Anderson, 1969). The spectral-element solution is computed in D'' and coupled to two normal-mode solutions in the mantle and core, respectively. The spectral-element mesh used contains about 12000 elements, which corresponds to approximately 4 million grid points. This allows us to compute accurate synthetic seismograms for periods greater than about 12 s. Note that at these periods, using the SEM in the whole model would require us to run on a large parallel computer (Komatitsch et al., 2003), but because of the coupling with the normal-mode solution, the simulation only requires 13 gigabytes of computer memory and runs in less than a day on a cluster of 64 processors (Capdeville et al., 2003b).

Figure 13 displays the ten stations at which we compare synthetic seismograms to data. These stations are taken from the following networks: CNSN (the code used on the plot is CN), USAF/USGS (GT), GSN-IRIS/IDA (II), GSN-IRIS/USGS (IU), USNSN (US) and LODORE (XT). The stations are located at epicentral distances ranging between 95° and 127°, where the core-reflected waves ScS, $sScS$ and core-diffracted waves S_{diff}, sS_{diff} can easily be identified.

In Fig. 14 we compare data to synthetic seismograms computed in the 3-D hybrid model. We also display the 1-D synthetic seismograms obtained in PREM as a reference. For most of the stations, it is clear that the 3-D model performs better than PREM in explaining both time delays and amplitudes. This is particularly true for stations LMN and BOSA at large epicentral distance, where the effect on the amplitude is the largest. Note that slow regions are systematically associated with amplitudes higher than in PREM (e.g., YKW3, LMN) and fast regions with smaller amplitudes than in PREM (e.g., BOSA). Notice however that only first-order characteristics are reproduced by the 3-D synthetic seismograms. More detailed analysis reveals that time shifts are sometimes too large (e.g., WMQ) or that both amplitude and phase are poorly fit (e.g., BW06 and TLY). The second phase, sS_{diff}, is generally poorly modeled. One of the reasons for this discrepancy is that sS_{diff} spends significant time in the strongly heterogeneous upper mantle, which is not accounted for in our 3-D hybrid model.

SPECTRAL-ELEMENT ANALYSIS IN SEISMOLOGY 405

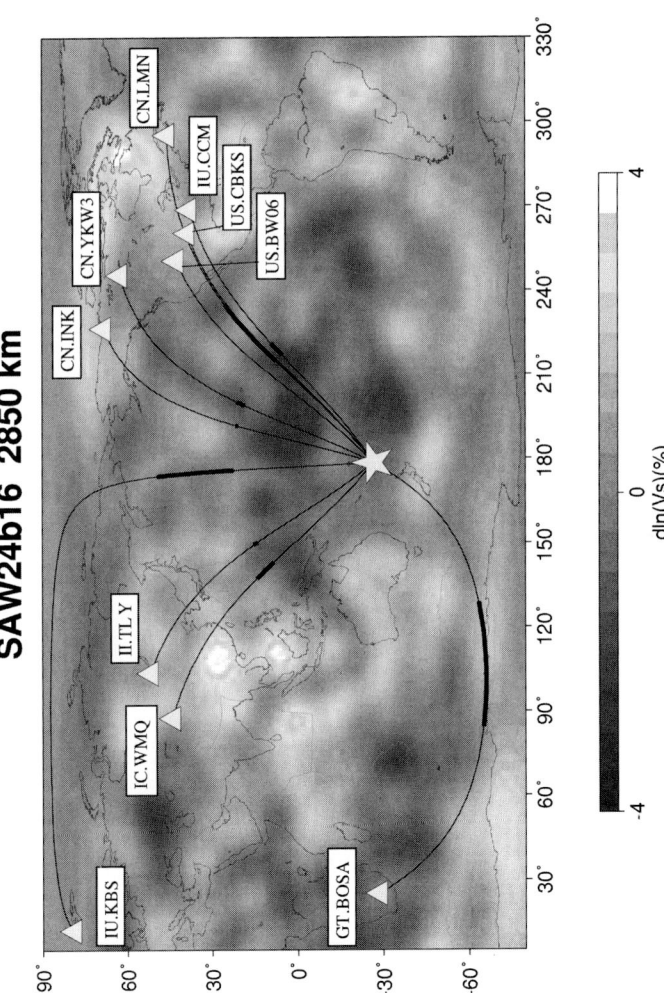

FIG. 13. Configuration of source and stations used to model wave propagation in the D″ layer. The background model is tomographic model SAW24B16 above the CMB. The paths between the source and the stations are plotted, and the part of the path that samples the D″ layer in PREM is represented in bold.

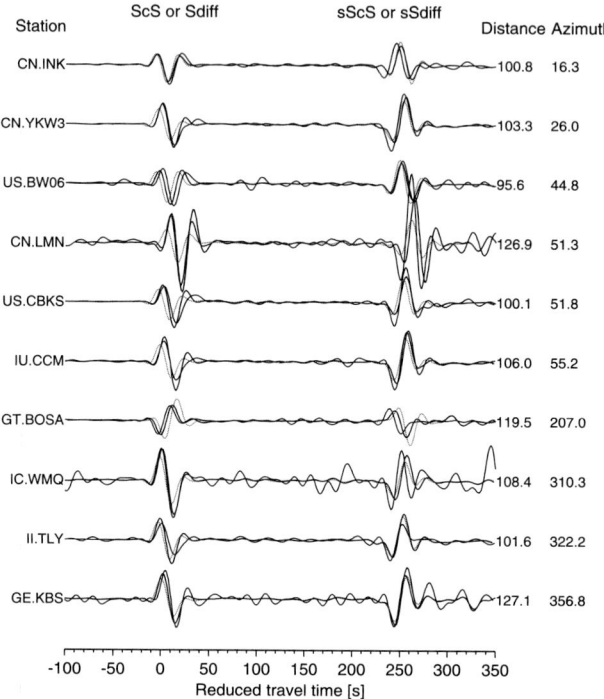

FIG. 14. Comparison between data (thin solid line), synthetic seismograms computed in PREM (dotted line) and synthetic seismograms computed in a velocity model that is heterogeneous in the 370 km above the CMB (bold solid line) for the transverse component of displacement. In the D″ layer, perturbations relative to PREM are taken from tomographic model SAW24B16. The traces are aligned with respect to the S_{diff} arrival time and represented with increasing azimuth. On each trace, the first energy arrival is a combination of the core reflected S wave (ScS) and the S wave diffracted at the CMB (S_{diff}). The second energy arrival corresponds to a similar interference between phases ($sScS$ and sS_{diff}) that have first been reflected off the Earth's surface. Except in a few places (e.g., at station TLY), the 3-D model generally improves the fit to data for the first energy arrival, both in amplitude and phase. The prediction of the 3-D model is not as good for the second energy arrival, because the effect of the crust and mantle (which is not accounted for in our calculation) represents an important contribution to the misfit between synthetic seismograms and data.

7. Conclusions and Perspectives

We have given an overview of the application of the spectral-element method (SEM) to realistic problems in regional and global seismology. We have shown examples of spectral-element calculations and have illustrated the ability of the method to provide a reference solution in 3-D heterogeneous models of the Earth. The method has been extended for example to include the dynamics of the rupture in seismic source studies (Ampuero, 2002; Vilotte and Festa, 2004). With the increasing computational power available, SEM-based inversions have started to appear, with the first SEM-based moment tensor inversions in a 3-D model (Liu *et al.*, 2004) as well as SEM adjoint formulations for seismic tomography (Tromp *et al.*, 2005) and SEM tomographic methods based on source stacking (Capdeville *et al.*, 2005).

The source code, called SPECFEM3-D, of our implementation of the SEM for parallel computers is freely available for academic, non-commercial research from http://www.geodynamics.org.

Acknowledgements

The authors would like to thank Barbara Romanowicz and Yuancheng Gung for providing the NACT and PAVA synthetic seismograms of Fig. 12, Akiko Toh for providing Fig. 13, and Christine Bernardi and Yvon Maday for fruitful discussions. Comments by two anonymous reviewers and by editor Valérie Maupin as well as careful reading by Alexandre Fournier improved the manuscript. The material presented in Section 6.1 is part of a working program supported by the European InterReg SISMOVALP project.

Appendix A. Modified Hooke's Law

A.1. Intrinsic Attenuation

In this appendix we briefly recall how to incorporate intrinsic attenuation into Spectral Element calculations. Because we handle the time evolution with a classical Finite Difference scheme, we refer to Moczo and Kristek (2005) and to Chapter 8 of this book for a detailed and accurate treatment of this matter.

In an anelastic medium, the stress tensor depends on the entire strain history and can be written as

$$\sigma(t) = \bigl([MH] \star \dot{\varepsilon}\bigr)(t) = \bigl([M\delta + \dot{M}H] \star \varepsilon\bigr)(t), \tag{A.1}$$

where σ stands for either hydrostatic or deviatoric stress, ε for the corresponding strain and M for the bulk or shear modulus. The star $\star$ denotes time convolution, δ is the Dirac delta distribution and H is the Heaviside distribution that ensures the causality of the stress–strain relation.

Following Liu *et al.* (1976), we simulate a constant Q attenuation mechanism by choosing a stress relaxation function $M(t)$ in (A.1) that is the superimposition of L standard linear solids. This can be formally written as

$$M(t) = M_R + \sum_{\ell=1}^{L} a_\ell e^{-t/\tau_\ell}, \tag{A.2}$$

or

$$M(t) = M_U - \sum_{\ell=1}^{L} a_\ell \left(1 - e^{-t/\tau_\ell}\right), \tag{A.3}$$

where M_R (respectively M_U) stands for the relaxed (respectively unrelaxed) elastic modulus and τ_ℓ denotes the relaxation time of the lth mechanism. The a_ℓ are coefficients that can be chosen in order to fit a given Q law. Their sum is the modulus defect: $\sum_{\ell=1}^{L} a_\ell = \delta M = M_U - M_R$.

A constant Q law is well approximated if the inverse relaxation times (i.e., the central frequencies of the L mechanisms) are chosen to be evenly distributed in logarithmic scale (e.g., Emmerich and Korn, 1987).

Next, we set $\Phi(t) = \dot{M} H(t) = \sum_{\ell=1}^{L} \Phi_\ell H(t)$ where

$$\Phi_\ell(t) = -\frac{a_\ell}{\tau_\ell} e^{-t/\tau_\ell}, \tag{A.4}$$

and we define a set of L memory variables (e.g., Carcione et al., 1988) by

$$R_l(t) = \bigl([\Phi_l H] \star \varepsilon\bigr)(t). \tag{A.5}$$

The stress–strain relation (A.1) then becomes

$$\sigma(t) = M_U \varepsilon(t) + \sum_{\ell=1}^{L} R_\ell(t). \tag{A.6}$$

The advantage of the modified stress–strain relation (A.6) is that it can be implemented in the time domain because each memory variable R_ℓ obeys the first-order differential equation:

$$\dot{R}_\ell(t) + \frac{1}{\tau_\ell} R_\ell(t) = -\frac{a_\ell}{\tau_\ell} \varepsilon(t). \tag{A.7}$$

In practice, three or four mechanisms are used to obtain a quality factor that is almost constant in the frequency range of interest (e.g., Emmerich and Korn, 1987; Komatitsch and Tromp, 2002a). In terms of implementation, the implication is that the memory variables need to be stored on the grid, and therefore the memory requirements increase substantially compared to purely elastic simulations. It is worth mentioning that to alleviate this burden, memory variables can be spread across an element, such that one carries only one memory variable at each grid point, obtaining the expected anelastic behavior in average (Zeng, 1996; Day, 1998; Day and Bradley, 2001; Graves and Day, 2003; Kristek and Moczo, 2003). In our implementation however, we do not spread the memory variables across the grid.

A.2. Fictitious Attenuation in Absorbing Layers

In regional applications, non-reflecting conditions must be defined at the boundaries of the computational domain to mimic an unbounded medium. Among the different methods that can be used, we focus here on the Perfectly Matched Layer (PML) method, originally developed by Bérenger (1994, 1996) in the context of electromagnetics, because it is the most efficient.

The PML can be viewed as an analytical continuation of the real coordinates in the complex space, where the stretching s is given by (e.g., Teixeira and Chew, 1999):

$$\tilde{x}_j = \int_0^{x_j} 1 + i \frac{s_j(\xi)}{\omega} \, d\xi. \tag{A.8}$$

Here ξ stands for the distance away from the interface between the PML and the regular domain in the j direction, and ω denotes angular frequency. With this definition, the amplitude of a plane wave defined by $\exp(i(\omega t - \mathbf{k} \cdot \mathbf{x}))$ decreases along the j direction by the frequency-independent factor $\exp(-\frac{k_j}{\omega} \int s_j(\xi) \, d\xi)$.

The mapping (A.8) induces a metric change $\tilde{\mathbf{G}} = \mathbf{\Lambda G \Lambda}$, where $\mathbf{G}$ is the Euclidean metric and

$$\mathbf{\Lambda} = \begin{pmatrix} s_1 & 0 & 0 \\ 0 & s_2 & 0 \\ 0 & 0 & s_3 \end{pmatrix}. \tag{A.9}$$

The conservation of momentum gives (see e.g., Festa and Vilotte, 2005):

$$i\omega\rho \det(\mathbf{\Lambda})\dot{\mathbf{u}} = \nabla \cdot (\tilde{\mathbf{T}}(\mathbf{u})) \tag{A.10}$$

where $\tilde{\mathbf{T}} = \det(\mathbf{\Lambda})\mathbf{\Lambda}^{-1}\mathbf{T}$ is the Piola transform of $\mathbf{T}$. The elastic properties of the medium are given by the symmetric second Piola tensor $\mathbf{T}' = \tilde{\mathbf{T}}\mathbf{\Lambda}^{-1}$, which satisfies Hooke's law:

$$\mathbf{T}' = \tilde{\mathbf{c}} : \nabla \mathbf{u} \tag{A.11}$$

where the fourth-order elastic tensor $\mathbf{c}$ is defined by

$$\tilde{c}_{ijkl} = \det(\mathbf{\Lambda}) c_{mnsl} \Lambda_{mi}^{-1} \Lambda_{nj}^{-1} \Lambda_{sk}^{-1}. \tag{A.12}$$

When going back to the time domain, the convolution can be avoided by splitting kinematic and dynamic fields into non-physical components following the directions of the derivatives normal and parallel to the interface between the PML and the regular domain. The original equations can then be written as (e.g., Festa and Nielsen, 2003):

$$\rho\left(\ddot{\mathbf{u}}_i^{(m)} + d_m \dot{\mathbf{u}}_i^{(m)}\right) = \frac{\partial \mathbf{T}_{ij}}{\partial x_j} \delta_{jm}, \tag{A.13}$$

$$\dot{\mathbf{T}}_{ij}^{(m)} + d_m \mathbf{T}_{ij}^{(m)} = c_{ijkl} \frac{\partial \dot{\mathbf{u}}_l}{\partial x_k} \delta_{km}, \tag{A.14}$$

where the superscripts refer to the only derivative that is kept on the right-hand side. The total fields are finally obtained by summing the split components:

$$\dot{\mathbf{u}}_i = \sum_{m=1}^{3} \dot{\mathbf{u}}_i^{(m)}, \qquad \mathbf{T}_{ij} = \sum_{m=1}^{3} \mathbf{T}_{ij}^{(m)}. \tag{A.15}$$

It is worth noting that unsplit PML formulations are currently under development for the seismic wave equation. Such formulations would have the advantage of reducing the memory cost of the technique because one would not need to define additional arrays to store the split components of the fields. They would also be simpler to implement in existing codes without PML because these are based on the total field.

APPENDIX B. DtN OPERATORS

B.1. Dynamic Coupling

Consider the coupling of a spectral-element solution with a normal-mode solution in global seismology. Such coupling is said to be dynamic because it depends on time (an example of static coupling is given in the next section). Such a coupling strategy is useful when some regions of the Earth can be assumed to be spherically symmetric. In those regions, the solution can therefore be sought in the frequency domain and expanded onto the basis of generalized spherical harmonics $\mathbf{Y}_{\ell m}$ (e.g., Phinney and Burridge, 1973), where ℓ and m stand for the angular and azimuthal degree, respectively. We thus face the problem of coupling a solution in the space–time domain with a solution in the spectral-frequency domain. Several options are available for that purpose. To derive the traction boundary condition on Γ, one can either convert the incoming displacement $\mathbf{u}^V$ in the spectral-frequency domain, apply the DtN operator to obtain the traction in the spectral-frequency domain, and then convert it back to the space–time domain. The DtN operator can also be converted to the space–time domain and applied to the incoming displacement $\mathbf{u}^V$, which implies a series of convolution in time and space. Even if not strictly equivalent numerically, both solutions require an infinite sum over the spherical harmonic degree ℓ and an unbounded integration over frequency, which both need to be truncated in order to be performed numerically. In the remainder of this section, we will assume that the DtN operator is applied in the space–time domain.

We start with the expression of the operator in the spectral-frequency domain. Denoting $\mathbf{u}_\omega(\mathbf{x}) = \mathbf{u}(\mathbf{x}, \omega)$, the counterpart of Eq. (40) in the frequency domain can be written as

$$\mathcal{D}_\omega^W\left(\mathbf{u}_\omega^W\right) = \mathbf{T}\left(\mathbf{u}_\omega^W\right) \cdot \hat{\mathbf{n}}, \tag{B.1}$$

which we expand on the basis of generalized spherical harmonics as

$$\mathcal{D}_\omega^W(\mathbf{u}_\omega^W) = \sum_{\ell=0}^{\infty} \sum_{m=-\ell}^{+\ell} {}_\ell\mathcal{D}_\omega^W(\mathbf{u}_\omega^W) : \mathbf{Y}_{\ell m}, \qquad (B.2)$$

where the colon denotes the dot product with the canonical basis (see e.g., Dahlen and Tromp, 1998, p. 901 for details). Note that the operator is local in the spectral-frequency domain, in the sense that the ℓm coefficients of traction at a given frequency ω only depend on the same coefficients of displacement at the same frequency. Note also that ${}_\ell\mathcal{D}_\omega^W$ has no dependence on the azimuthal degree m because of the spherical symmetry of W.

The DtN in the space–time domain is obtained by an inverse Fourier transform:

$$\mathcal{D}_t^W(\mathbf{u}_{t'\leq t}^V) = \sum_{\ell=0}^{\infty} \sum_{m=-\ell}^{+\ell} \int_{-\infty}^{+\infty} {}_\ell\mathcal{D}_\omega^W(\mathbf{u}_\omega^W) : \mathbf{Y}_{\ell m} e^{i\omega t} \, d\omega. \qquad (B.3)$$

In order to be performed numerically, the sum over ℓ needs to be truncated to a cutoff angular degree $\ell_{\max}$ large enough to provide an accurate approximation of the DtN operator in the space domain. In a spherically-symmetric model and far enough from the source, the relation between the corner frequency of the source and the maximum angular degree of the wave field is known from the dispersion relation. In practice, when dealing with a 3-D Earth model, we apply this 1-D criterion to the spherical average of the 3-D model in V.

Computing the inverse Fourier transform (B.3) requires one to circumvent two issues. First, because W is bounded, ${}_\ell\mathcal{D}_\omega^W$ is undefined for the set of frequencies corresponding to the eigenmodes of W with a homogeneous Dirichlet boundary condition on Γ. To perform the frequency integration, the DtN operator is split into a discrete spectrum for which the Cauchy theorem is used, and a continuous spectrum for which a fast Fourier transform is applied. Second, the infinite frequency range in Eq. (B.3) cannot be truncated without affecting the causality of the DtN operator. Let ω_c stand for a frequency greater than the corner frequency of the source, then the truncated DtN operator obtained by replacing the infinite frequency interval in (B.3) by $[-\omega_c, \omega_c]$ is not causal. To solve this problem, Capdeville et al. (2003a) introduced a regularized, causal operator based upon an asymptotic approximation at high frequency. This regularized operator ${}_r\mathcal{D}_t$ is defined by

$${}_r\mathcal{D}_\omega^W = \mathcal{D}_\omega^W - {}_a\mathcal{D}_\omega^W, \qquad (B.4)$$

where ${}_a\mathcal{D}_w^W$ stands for an asymptotic approximation of the operator based on an expansion in powers of $i\omega$. By definition, the regularized operator converges toward zero at high frequency. The final operator $\mathcal{D}_t^W$ is obtained by summing the inverse Fourier transforms:

$$\mathcal{D}_t^W = {}_r\mathcal{D}_t^W + {}_a\mathcal{D}_t^W. \qquad (B.5)$$

The reader is referred to Capdeville *et al.* (2003a) for details about the regularization procedure applied to the DtN operator, and to Capdeville *et al.* (2003b) for the extension to the case where the source is located within W.

B.2. Static Coupling

Another example of DtN coupling arises when accounting for the perturbation of the gravitational potential in the wave equation, or when computing the gravity field induced by a 3-D distribution of density. Both problems involve a scalar potential that is defined everywhere in space, whereas the computational domain is bounded. A coupling strategy with the analytical solution of the exterior problem allows one to circumvent this issue.

Let W denote the region exterior to $\mathcal{B}$: $W = \mathbb{R}^3 \setminus \mathcal{B}$, so that the coupling interface $\Gamma = \partial \mathcal{B}$ is the sphere of radius b (see Fig. 1). The coupling strategy is to solve Poisson's equation in $\mathcal{B}$ with a condition on Γ involving the normal derivative of the potential (i.e., the Neumann variable for this problem) and to solve Laplace's equation in W with a prescribed potential (the Dirichlet variable) on the boundary.

The DtN operator for the Poisson–Laplace equation is thus $\mathcal{D}^W(\chi^W) = \nabla \chi^W \cdot \hat{\mathbf{n}}$, where χ stands for either the mass redistribution potential $\psi(\mathbf{u})$ or the non-rotating gravity potential ϕ'. Because the coupling interface is spherical, the expression of the DtN operator is readily obtained in the spherical harmonics basis (see Chaljub and Valette, 2004):

$$\mathcal{D}^W(\chi^W) = -\frac{1}{b}\sum_{\ell=0}^{\infty}(\ell+1)\sum_{m=-\ell}^{\ell}\chi^W_{lm}(b)\mathcal{Y}_{\ell m}(\theta,\varphi), \tag{B.6}$$

where $\mathcal{Y}_{\ell m}$ denotes the real spherical harmonic of degree ℓ and order m, (θ, φ) are the spherical coordinates and $\chi^W_{\ell m}(b) = \int_\Gamma \chi^W \mathcal{Y}_{\ell m}\, dS$.

Finally, the boundary condition to be used to solve for the non-rotating gravity potential in $\mathcal{B}$ is

$$\nabla \phi'^{\mathcal{B}} \cdot \hat{\mathbf{n}} = \mathcal{D}^W(\phi'^{\mathcal{B}}), \tag{B.7}$$

and that for the mass redistribution potential is

$$\nabla \psi(\mathbf{u})^{\mathcal{B}} \cdot \hat{\mathbf{n}} = \mathcal{D}^W(\psi(\mathbf{u})^{\mathcal{B}}) + 4\pi G \rho^{\mathcal{B}} \mathbf{u}^{\mathcal{B}} \cdot \hat{\mathbf{n}}, \tag{B.8}$$

according to the jump condition (37) and assuming that the exterior domain W is a vacuum, i.e., $\rho^W = 0$. Note that the second term on the right-hand side of Eq. (B.8) only matters when topography or ellipticity is considered at the surface of the Earth.

B.3. Ocean Load Approximation

More than 71% of the surface of the Earth is covered by oceans. At short period, one may want to take into account the effect of these oceans on global seismic wave propagation (e.g., their effect on the reflection coefficient of phases reflected at the ocean bottom, such as PP, SS and SP, as well as on the dispersion of Rayleigh waves).

Following Komatitsch and Tromp (2002b), we assume that the oceans are incompressible, i.e., that the entire water column moves as a whole as a result of the normal displacement $\mathbf{u} \cdot \hat{\mathbf{n}}$ of the sea floor. This approximation, which is accurate for seismic wavelengths larger than the thickness of the oceans, is valid down to periods of approximately 20 s, but not below. With such an assumption, we can only reproduce the effect of the load at the ocean floor, but we cannot model phases such as the tsunami that propagate in the oceans themselves.

Assuming a homogeneous fluid layer (therefore neutrally stratified, i.e., its Brunt–Väisälä frequency N is zero in Eq. (17)) and neglecting the redistribution of mass, the wave equation (5) in the oceans reduces to

$$\ddot{\mathbf{u}} + 2\mathbf{\Omega} \times \dot{\mathbf{u}} = -\nabla\left(\frac{\delta p}{\rho} - \mathbf{u} \cdot \mathbf{g}\right), \tag{B.9}$$

where δp is the Lagrangian perturbation of pressure defined by (14).

Let h_w denote the local thickness of the oceans (i.e., of the equivalent local water column), which can be taken from a bathymetry map. If one considers the normal component of Eq. (B.9), integrates over the column of water and takes into account the variations of gravity with radius, one can show that locally

$$\delta p = \rho_w h_w \ddot{\mathbf{u}} \cdot \hat{\mathbf{n}} + 2\rho_w h_w (\mathbf{\Omega} \times \dot{\mathbf{u}}) \cdot \hat{\mathbf{n}} + 4\pi G \rho_w^2 h_w \mathbf{u} \cdot \hat{\mathbf{n}}, \tag{B.10}$$

where ρ_w denotes the density of sea water ($\rho_w \simeq 1020 \text{ kg m}^{-3}$). In practice, at periods above typically 20 s the ocean load is a small effect, and the gravity and rotation terms in (B.10) are even smaller corrections that can be safely ignored. We therefore get the following equation at the ocean floor

$$\mathbf{T}(\mathbf{u}) \cdot \hat{\mathbf{n}} = -\delta p \hat{\mathbf{n}} \simeq -\rho_w h_w (\ddot{\mathbf{u}} \cdot \hat{\mathbf{n}}) \hat{\mathbf{n}}. \tag{B.11}$$

References

Aki, K., Richards, P.G. (1980). Quantitative Seismology, Theory and Methods. Freeman, San Francisco.

Alford, R., Kelly, K., Boore, D. (1974). Accuracy of finite difference modeling of the acoustic wave equation. *Geophysics* **39**, 834–842.

Alterman, Z., Karal, F.C. (1968). Propagation of elastic waves in layered media by finite difference methods. *Bull. Seismol. Soc. Am.* **58**, 367–398.

Ampuero, J.-P. (2002). Étude physique et numérique de la nucléation des séismes (A physical and numerical study of earthquake nucleation), Ph.D. Thesis, Institut de Physique du Globe, Paris, France.

Bao, H., Bielak, J., Ghattas, O., Kallivokas, L.F., O'Hallaron, D.R., Shewchuk, J.R., Xu, J. (1998). Large-scale simulation of elastic wave propagation in heterogeneous media on parallel computers. *Comput. Methods Appl. Mech. Engrg.* **152**, 85–102.

Bard, P.-Y., Bouchon, M. (1980a). The seismic response of sediment-filled valleys, Part 1: The case of incident SH waves. *Bull. Seismol. Soc. Am.* **70**, 1263–1286.

Bard, P.-Y., Bouchon, M. (1980b). The seismic response of sediment-filled valleys, Part 2: The case of incident P and SV waves. *Bull. Seismol. Soc. Am.* **70**, 1921–1941.

Bassin, C., Laske, G., Masters, G. (2000). The current limits of resolution for surface wave tomography in North America. *EOS* **81**, F897.

Basu, U., Chopra, A.K. (2003). Perfectly matched layers for time-harmonic elastodynamics of unbounded domains: Theory and finite-element implementation. *Comput. Methods Appl. Mech. Engrg.* **192**, 1337–1375.

Basu, U., Chopra, A.K. (2004). Perfectly matched layers for transient elastodynamics of unbounded domains. *Int. J. Numer. Methods Engrg.* **59**, 1039–1074.

Bécache, E., Fauqueux, S., Joly, P. (2003). Stability of Perfectly Matched Layers, group velocities and anisotropic waves. *J. Comput. Phys.* **188** (2), 399–433.

Bérenger, J.P. (1994). A Perfectly Matched Layer for the absorption of electromagnetic waves. *J. Comput. Phys.* **114**, 185–200.

Bérenger, J.P. (1996). Three-dimensional Perfectly Matched Layer for the absorption of electromagnetic waves. *J. Comput. Phys.* **127**, 363–379.

Bermúdez, A., Rodríguez, R. (1994). Finite element computation of the vibration modes of a fluid–solid system. *Comput. Methods Appl. Mech. Engrg.* **119**, 355–370.

Bittencourt, M.L. (2005). Fully tensorial nodal and modal shape functions for triangles and tetrahedra. *Int. J. Numer. Methods Engrg.* **63** (11), 1530–1558.

Bouchon, M. (1996). The discrete wave number formulation of boundary integral equations and boundary element methods: A review with applications to the simulation of seismic wave propagation in complex geological structures. *Pure Appl. Geophys.* **148** (1), 3–20.

Bouchon, M., Sánchez-Sesma, F.J. (2006). Boundary-integral equations and boundary-element methods in elastodynamics. In: Wu, R.-S., Maupin, V. (Eds.), *Advances in Wave Propagation in Heterogeneous Media*. In: *Advances in Geophysics*, vol. 48. Elsevier, pp. 157–189 (this book).

Canuto, C., Hussaini, M.Y., Quarteroni, A., Zang, T.A. (1988). *Spectral Methods in Fluid Dynamics*. Springer-Verlag, New York.

Capdeville, Y., Chaljub, E., Vilotte, J.P., Montagner, J.P. (2003a). Coupling the spectral element method with a modal solution for elastic wave propagation in global Earth models. *Geophys. J. Int.* **152**, 34–67.

Capdeville, Y., To, A., Romanowicz, B. (2003b). Coupling spectral elements and modes in a spherical Earth: An extension to the 'sandwich' case. *Geophys. J. Int.* **154**, 44–57.

Capdeville, Y., Gung, Y., Romanowicz, B. (2005). Towards global Earth tomography using the spectral element method: A technique based on source stacking. *Geophys. J. Int.* **162**, 541–554.

Carcione, J.M. (1994). The wave equation in generalized coordinates. *Geophysics* **59**, 1911–1919.

Carcione, J.M., Kosloff, D., Kosloff, R. (1988). Wave propagation simulation in an elastic anisotropic (transversely isotropic) solid. *Quantum J. Mech. Appl. Math.* **41** (3), 319–345.

Chaljub, E. (2000). Modélisation numérique de la propagation d'ondes sismiques en géométrie sphérique: Application à la sismologie globale (Numerical modeling of the propagation of seismic waves in spherical geometry: Application to global seismology), Ph.D. Thesis, Université Paris VII Denis Diderot, Paris, France, in French.

Chaljub, E., Tarantola, A. (1997). Sensitivity of SS precursors to topography on the upper-mantle 660-km discontinuity. *Geophys. Res. Lett.* **24** (21), 2613–2616.

Chaljub, E., Valette, B. (2004). Spectral element modelling of three-dimensional wave propagation in a self-gravitating Earth with an arbitrarily stratified outer core. *Geophys. J. Int.* **158**, 131–141.

Chaljub, E., Capdeville, Y., Vilotte, J.P. (2003). Solving elastodynamics in a fluid–solid heterogeneous sphere: A parallel spectral element approximation on non-conforming grids. *J. Comput. Phys.* **187** (2), 457–491.

Chaljub, E., Cornou, C., Guéguen, P., Causse, M., Komatitsch, D. (2005). Spectral-element modeling of 3-D wave propagation in the alpine valley of Grenoble, France. In: *Geophysical Research Abstracts*, vol. 7, p. 05225. EGU 2nd general assembly, Wien, Austria.

Chen, X. (2006). Generation and propagation of seismic SH waves in multilayered media with irregular interfaces. In: Wu, R.-S., Maupin, V. (Eds.), *Advances in Wave Propagation in Heterogeneous Media*. In: *Advances in Geophysics*, vol. 48. Elsevier, pp. 191–264 (this book).

Clévédé, E., Mégnin, C., Romanowicz, B., Lognonné, P. (2000). Seismic waveform modeling and surface wave tomography in a three-dimensional Earth: Asymptotic and non-asymptotic approaches. *Phys. Earth Planet. Int.* **119** (1), 37–56.

Cohen, G., Fauqueux, S. (2000). Mixed finite elements with mass lumping for the transient wave equation. *J. Comput. Acoust.* **8** (1), 171–188.

Cohen, G., Fauqueux, S. (2005). Mixed spectral finite elements for the linear elasticity system in unbounded domains. *SIAM J. Sci. Comput.* **26** (3), 864–884.

Cohen, G., Joly, P., Tordjman, N. (1993). Construction and analysis of higher-order finite elements with mass lumping for the wave equation. In: Kleinman, R. (Ed.), *Proceedings of the Second International Conference on Mathematical and Numerical Aspects of Wave Propagation*. SIAM, Philadephia, PA, pp. 152–160.

Cohen, G., Joly, P., Roberts, J.E., Tordjman, N. (2001). Higher-order triangular finite elements with mass lumping for the wave equation. *SIAM J. Numer. Anal.* **38** (6), 2047–2078.

Collino, F., Monk, P. (1998). Optimizing the Perfectly Matched Layer. *Comput. Methods Appl. Mech. Engrg.* **164**, 157–171.

Collino, F., Tsogka, C. (2001). Application of the PML absorbing layer model to the linear elastodynamic problem in anisotropic heterogeneous media. *Geophysics* **66** (1), 294–307.

Cornou, C., Bard, P.-Y., Dietrich, M. (2003). Contribution of dense array analysis to the identification and quantification of basin-edge-induced waves, Part II: Application to the Grenoble basin (French Alps). *Bull. Seismol. Soc. Am.* **93** (6), 2624–2648.

Courant, R., Friedrichs, K.O., Lewy, H. (1928). Über die partiellen Differenzengleichungen der Mathematischen Physik. *Math. Ann.* **100**, 32–74 (in German).

Cowling, T.G. (1941). The non-radial oscillations of polytropic stars. *Mon. Not. R. Astron. Soc.* **101**, 369–373.

Dahlen, F.A., Tromp, J. (1998). Theoretical Global Seismology. Princeton Univ. Press, Princeton.

Day, S.M. (1998). Efficient simulation of constant Q using coarse-grained memory variables. *Bull. Seismol. Soc. Am.* **88**, 1051–1062.

Day, S.M., Bradley, C. (2001). Memory-efficient simulation of anelastic wave propagation. *Bull. Seismol. Soc. Am.* **91**, 520–531.

Deville, M.O., Fischer, P.F., Mund, E.H. (2002). High-Order Methods for Incompressible Fluid Flow. Cambridge Univ. Press, Cambridge, United Kingdom.

Dhatt, G., Touzot, G. (1984). The Finite Element Method Displayed. John Wiley and Sons, New York.

Dziewonski, A.M., Anderson, D.L. (1981). Preliminary reference Earth model. *Phys. Earth Planet. Int.* **25**, 297–356.

Emmerich, H., Korn, M. (1987). Incorporation of attenuation into time-domain computations of seismic wave fields. *Geophysics* **52**, 1252–1264.

Faccioli, E., Maggio, F., Paolucci, R., Quarteroni, A. (1997). 2-D and 3-D elastic wave propagation by a pseudo-spectral domain decomposition method. *J. Seismol.* **1**, 237–251.

Favier, N., Chevrot, S., Komatitsch, D. (2004). Near-field influences on shear wave splitting and traveltime sensitivity kernels. *Geophys. J. Int.* **156**, 467–482.

Festa, G., Nielsen, S. (2003). PML absorbing boundaries. *Bull. Seismol. Soc. Am.* **93**, 891–903.

Festa, G., Vilotte, J.P. (2005). The Newmark scheme as velocity–stress time-staggering: An efficient PML implementation for spectral element simulations of elastodynamics. *Geophys. J. Int.* **161**, 789–812.

Frankel, A. (1993). Three-dimensional simulations of ground motions in the San Bernardino valley, California, for hypothetical earthquakes on the San Andreas fault. *Bull. Seismol. Soc. Am.* **83**, 1020–1041.

Frankel, A., Vidale, J. (1992). A three-dimensional simulation of seismic waves in the Santa Clara valley, California, from the Loma Prieta aftershock. *Bull. Seismol. Soc. Am.* **82**, 2045–2074.

Funaro, D. (1992). Polynomial Approximation of Differential Equations. *Lecture Notes in Physics*, vol. 8. Springer-Verlag, Heidelberg, Germany.

Furumura, M., Kennett, B.L.N., Furumura, T. (1999). Seismic wavefield calculation for laterally heterogeneous Earth models-II. The influence of upper mantle heterogeneity. *Geophys. J. Int.* **139** (3), 623–644.

Furumura, T., Kennett, B.L.N., Furumura, M. (1998). Seismic wavefield calculation for laterally heterogeneous whole Earth models using the pseudospectral method. *Geophys. J. Int.* **135** (3), 845–860.

Giraldoa, F., Warburton, T. (2005). A nodal triangle-based spectral element method for the shallow water equations on the sphere. *J. Comput. Phys.* **207** (1), 129–150.

Givoli, D. (1992). Numerical Methods for Problems in Infinite Domains. Elsevier, Amsterdam, The Netherlands.

Graves, R.W., Day, S.M. (2003). Stability and accuracy of coarse-grain viscoelastic simulations. *Bull. Seismol. Soc. Am.* **93**, 283–300.

Gropp, W., Lusk, E., Skjellum, A. (1994). Using MPI, Portable Parallel Programming with the Message-Passing Interface. MIT Press, Cambridge, USA.

Hamdi, M.A., Ousset, Y., Verchery, G. (1978). A displacement method for the analysis of vibrations of coupled fluid-structure systems. *Int. J. Numer. Methods Engrg.* **13**, 139–150.

Hughes, T.J.R. (1987). The Finite Element Method, Linear Static and Dynamic Finite Element Analysis. Prentice Hall International, Englewood Cliffs, New Jersey, USA.

Igel, H., Weber, M. (1995). SH-wave propagation in the whole mantle using high-order finite differences. *Geophys. Res. Lett.* **22**, 731–734.

Igel, H., Weber, M. (1996). P-SV wave propagation in the whole mantle using high-order finite differences: Application to lowermost mantle structure. *Geophys. Res. Lett.* **23**, 415–418.

Jobert, N., Gaulon, R., Dieulin, A., Roult, G. (1977). Sur les ondes de très longue période, caractéristiques du manteau supérieur. *C. R. Acad. Sci. Paris, Série B* **285**, 49–51 (in French).

Kane, C., Marsden, J., Ortiz, M., West, M. (2003). Variational integrators and the Newmark algorithm for conservative and dissipative mechanical systems. *Int. J. Numer. Methods Engrg.* **49** (10), 1295–1325.

Karypis, G., Kumar, V. (1998a). A fast and high-quality multilevel scheme for partitioning irregular graphs. *SIAM J. Sci. Comput.* **20** (1), 359–392.

Karypis, G., Kumar, V. (1998b). Multilevel k-way partitioning scheme for irregular graphs. *J. Parallel Distributed Comput.* **48** (1), 96–129.

Kelly, K.R., Ward, R.W., Treitel, S., Alford, R.M. (1976). Synthetic seismograms: A finite difference approach. *Geophysics* **41**, 2–27.

Knupp, P.M. (1999). Applications of mesh smoothing: Copy, morph, and sweep on unstructured quadrilateral meshes. *Int. J. Numer. Methods Engrg.* **45** (1), 37–45.

Komatitsch, D. (1997). Méthodes spectrales et éléments spectraux pour l'équation de l'élastodynamique 2-D et 3-D en milieu hétérogène (Spectral and spectral-element methods for the 2-D and 3-D elastodynamics equations in heterogeneous media), Ph.D. Thesis, Institut de Physique du Globe, Paris, France, 187 pages (in French).

Komatitsch, D., Tromp, J. (1999). Introduction to the spectral-element method for 3-D seismic wave propagation. *Geophys. J. Int.* **139**, 806–822.

Komatitsch, D., Tromp, J. (2002a). Spectral-element simulations of global seismic wave propagation-I. Validation. *Geophys. J. Int.* **149**, 390–412.

Komatitsch, D., Tromp, J. (2002b). Spectral-element simulations of global seismic wave propagation-II. 3-D models, oceans, rotation, and self-gravitation. *Geophys. J. Int.* **150**, 303–318.

Komatitsch, D., Tromp, J. (2003). A perfectly matched layer absorbing boundary condition for the second-order seismic wave equation. *Geophys. J. Int.* **154**, 146–153.

Komatitsch, D., Vilotte, J.P. (1998). The spectral-element method: An efficient tool to simulate the seismic response of 2-D and 3-D geological structures. *Bull. Seismol. Soc. Am.* **88** (2), 368–392.

Komatitsch, D., Barnes, C., Tromp, J. (2000). Wave propagation near a fluid–solid interface: A spectral element approach. *Geophysics* **65** (2), 623–631.

Komatitsch, D., Martin, R., Tromp, J., Taylor, M.A., Wingate, B.A. (2001). Wave propagation in 2-D elastic media using a spectral element method with triangles and quadrangles. *J. Comput. Acoust.* **9** (2), 703–718.

Komatitsch, D., Ritsema, J., Tromp, J. (2002). The spectral-element method, Beowulf computing, and global seismology. *Science* **298**, 1737–1742.

Komatitsch, D., Tsuboi, S., Ji, C., Tromp, J. (2003). A 14.6 billion degrees of freedom, 5 teraflops, 2.5 terabyte earthquake simulation on the Earth Simulator. In: *Proceedings of the ACM/IEEE Supercomputing SC'2003 Conference*. Published on CD-ROM and at http://www.sc-conference.org/sc2003.

Komatitsch, D., Liu, Q., Tromp, J., Süss, P., Stidham, C., Shaw, J.H. (2004). Simulations of ground motion in the Los Angeles basin based upon the spectral-element method. *Bull. Seismol. Soc. Am.* **94**, 187–206.

Kristek, J., Moczo, P. (2003). Seismic-wave propagation in viscoelastic media with material discontinuities: A 3-D fourth-order staggered-grid finite-difference modeling. *Bull. Seismol. Soc. Am.* **93** (5), 2273–2280.

Kumazawa, M., Anderson, O.L. (1969). Elastic moduli, pressure derivates and temperature derivates of single-crystal olivine and single-crystal forstite. *J. Geophys. Res.* **74**, 5961–5972.

Li, X.D., Romanowicz, B. (1995). Comparison of global waveform inversions with and without considering cross–branch modal coupling. *Geophys. J. Int.* **121**, 695–709.

Li, X.D., Tanimoto, T. (1993). Waveforms of long-period body waves in a slightly aspherical Earth model. *Geophys. J. Int.* **112**, 92–102.

Liu, H.P., Anderson, D.L., Kanamori, H. (1976). Velocity dispersion due to anelasticity: Implications for seismology and mantle composition. *Geophys. J. R. Astron. Soc.* **47**, 41–58.

Liu, Q., Polet, J., Komatitsch, D., Tromp, J. (2004). Spectral-element moment-tensor inversions for earthquakes in Southern California. *Bull. Seismol. Soc. Am.* **94**, 1748–1761.

Lysmer, J., Drake, L.A. (1972). A finite element method for seismology. In: *Methods in Computational Physics*, vol. 11. Academic Press, New York, USA.

Madariaga, R. (1976). Dynamics of an expanding circular fault. *Bull. Seismol. Soc. Am.* **65**, 163–182.

Maday, Y., Patera, A.T. (1989). Spectral element methods for the incompressible Navier–Stokes equations. In: Noor, A.K., Oden, J.T. (Eds.), *State of the Art Survey in Computational Mechanics*, pp. 71–143.

Maday, Y., Rønquist, E.M. (1990). Optimal error analysis of spectral methods with emphasis on nonconstant coefficients and deformed geometries. *Comput. Methods Appl. Mech. Engrg.* **80**, 91–115.

Marfurt, K.J. (1984). Accuracy of finite-difference and finite-element modeling of the scalar wave equation. *Geophysics* **49**, 533–549.

Mégnin, C., Romanowicz, B. (2000). The 3-D shear velocity structure of the mantle from the inversion of body, surface and higher modes wave forms. *Geophys. J. Int.* **143**, 709–728.

Mercerat, E.D., Vilotte, J.P., Sánchez-Sesma, F.J. (2005). Triangular spectral-element simulation of 2-D elastic wave propagation using unstructured triangular grids. *Geophys. J. Int.*, submitted for publication.

Moczo, P., Kristek, J. (2005). On the geological models used for time-domain methods of seismic wave propagation. *Geophys. Res. Lett.* **32**, L01306, doi:10.1029/2004GL021598.

Newmark, N.M. (1959). A method of computation for structural dynamics. In: *Engineering Mechanics Division: Proceedings of the American Society of Civil Engineers*, pp. 67–93.

Nicoud, G., Royer, G., Corbin, J.-C., Lemeille, F., Paillet, A. (2002). Creusement et remplissage de la vallée de l'Isère au Quaternaire récent: Apports nouveaux du forage GMB1 (1999) dans la région de Grenoble (France). *Géologie de la France* **4**, 39–49 (in French).

Olsen, K.B. (2000). Site amplification in the Los Angeles basin from three-dimensional modeling of ground motion. *Bull. Seismol. Soc. Am.* **90**, S77–S94.

Olsen, K.B., Archuleta, R.J. (1996). 3-D simulation of earthquakes on the Los Angeles fault system. *Bull. Seismol. Soc. Am.* **86** (3), 575–596.

Pasquetti, R., Rapetti, F. (2004). Spectral element methods on triangles and quadrilaterals: Comparisons and applications. *J. Comput. Phys.* **198** (1), 349–362, doi:10.1016/j.jcp.2004.01.010.

Patera, A.T. (1984). A spectral element method for fluid dynamics: Laminar flow in a channel expansion. *J. Comput. Phys.* **54**, 468–488.

Phinney, R.A., Burridge, R. (1973). Representation of elastic-gravitational excitation of a spherical Earth model by generalized spherical harmonics. *Geophys. J. R. Astron. Soc.* **34**, 451–487.

Pitarka, A., Irikura, K., Iwata, T., Sekiguchi, H. (1998). Three-dimensional simulation of the near-fault ground motion for the 1995 Hyogo-ken Nanbu (Kobe), Japan, earthquake. *Bull. Seismol. Soc. Am.* **88**, 428–440.

Priolo, E., Carcione, J.M., Seriani, G. (1994). Numerical simulation of interface waves by high-order spectral modeling techniques. *J. Acoust. Soc. Am.* **95** (2), 681–693.

Ritsema, J., Van Heijst, H.J., Woodhouse, J.H. (1999). Complex shear velocity structure imaged beneath Africa and Iceland. *Science* **286**, 1925–1928.

Ronchi, C., Ianoco, R., Paolucci, P.S. (1996). The "Cubed Sphere": A new method for the solution of partial differential equations in spherical geometry. *J. Comput. Phys.* **124**, 93–114.

Rosat, S., Hinderer, J., Rivera, L. (2003). First observation of $_2S_1$ and study of the splitting of the football mode $_0S_2$ after the June 2001 Peru earthquake of magnitude 8.4. *Geophys. Res. Lett.* **30** (21), 2111, doi:10.1029/2003GL018304.

Sadourny, R. (1972). Conservative finite-difference approximations of the primitive equations on quasi-uniform spherical grids. *Monthly Weather Rev.* **100**, 136–144.

Seriani, G. (1998). 3-D large-scale wave propagation modeling by a spectral element method on a Cray T3E multiprocessor. *Comput. Methods Appl. Mech. Engrg.* **164**, 235–247.

Seriani, G., Priolo, E. (1994). A spectral element method for acoustic wave simulation in heterogeneous media. *Finite Elements Anal. Design* **16**, 337–348.

Tautges, T.J. (2004). MOAB-SD: Integrated structured and unstructured mesh representation. *Engrg. Comput.* **20** (3), 286–293, doi:10.1007/s00366-004-0296-0.

Taylor, M., Tribbia, J., Iskandarani, M. (1997). The spectral element method for the shallow water equation on the sphere. *J. Comput. Phys.* **130**, 92–108.

Taylor, M.A., Wingate, B.A., Vincent, R.E. (2000). An algorithm for computing Fekete points in the triangle. *SIAM J. Numer. Anal.* **38** (5), 1707–1720.

Teixeira, F.L., Chew, W.C. (1999). On causality and dynamic stability of Perfectly Matched Layers for FDTD simulations. *IEEE Trans. Microwave Theor. Tech.* **47** (6), 775–785.

Tessmer, E., Kosloff, D. (1994). 3-D elastic modeling with surface topography by a Chebyshev spectral method. *Geophysics* **59** (3), 464–473.

Thouvenot, F., Fréchet, J., Jenatton, L., Gamond, J.-F. (2003). The Belledonne border fault: Identification of an active seismic strike-slip fault in the western Alps. *Geophys. J. Int.* **155** (1), 174–192.

Tordjman, N. (1995). Éléments finis d'ordre élevé avec condensation de masse pour l'équation des ondes (High-order finite elements with mass lumping for the wave equation), Ph.D. Thesis, Université Paris IX Dauphine, Paris, France (in French).

Toshinawa, T., Ohmachi, T. (1992). Love-wave propagation in a three-dimensional sedimentary basin. *Bull. Seismol. Soc. Am.* **82** (4), 1661–1677.

Tromp, J., Tape, C., Liu, Q. (2005). Seismic tomography, adjoint methods, time reversal and banana-doughnut kernels. *Geophys. J. Int.* **160** (1), 195–216, doi:10.1111/j.1365-246X.2004.02453.x.

Tsuboi, S., Komatitsch, D., Ji, C., Tromp, J. (2003). Spectral-element simulations of the November 3, 2002, Denali, Alaska earthquake on the Earth Simulator. *Phys. Earth Planet. Int.* **139**, 305–313.

Valette, B. (1986). About the influence of prestress upon the adiabatic perturbations of the Earth. *Geophys. J. R. Astron. Soc.* **85**, 179–208.

Valette, B. (1989). Spectre des vibrations propres d'un corps élastique, auto-gravitant, en rotation uniforme et contenant une partie fluide (Free oscillations spectrum of an elastic, self-gravitating, uniformly rotating body with a fluid inclusion). *C. R. Acad. Sci. Paris* **309**, 419–422 (in French).

Vallon, M. (1999). Estimation de l'épaisseur d'alluvions et sédiments quaternaires dans la région grenobloise par inversion des anomalies gravimétriques (Estimation of the thickness of alluvial and quaternary deposits in the Grenoble area by inverting gravimetric anomalies). Tech. Rep., LGGE, IPSN/CNRS, Université Joseph Fourier (in French).

Vilotte, J.-P., Festa, G. (2004). Spectral element simulation of rupture dynamics on curvilinear faults. *EOS* **85** (47). Abstract S32B–08.

Virieux, J. (1986). P-SV wave propagation in heterogeneous media: Velocity–stress finite-difference method. *Geophysics* **51**, 889–901.

Winton, S.C., Rappaport, C.M. (2000). Specifying PML conductivities by considering numerical reflection dependencies. *IEEE Trans. Antennas and Propagation* **48** (7), 1055–1063.

Woodhouse, J.H., Dziewonski, A.M. (1984). Mapping the upper mantle: Three-dimensional modeling of Earth structure by inversion of seismic waveforms. *J. Geophys. Res.* **89**, 5953–5986.

Zeng, X. (1996). Finite difference modeling of viscoelastic wave propagation in a generally heterogeneous medium in the time domain and a dissection method in the frequency domain, Ph.D. Thesis, University of Toronto, Canada.

Zingg, D.W. (2000). Comparison of high-accuracy finite-difference methods for linear wave propagation. *SIAM J. Sci. Comput.* **22** (2), 476–502, doi:10.1137/S1064827599350320.

Zingg, D.W., Lomax, H., Jurgens, H. (1996). High-accuracy finite-difference schemes for linear wave propagation. *SIAM J. Sci. Comput.* **17** (2), 328–346, doi:10.1137/S1064827594267173.

THE FINITE-DIFFERENCE TIME-DOMAIN METHOD FOR MODELING OF SEISMIC WAVE PROPAGATION

PETER MOCZO[1], JOHAN O.A. ROBERTSSON[2] AND LEO EISNER[3]

[1] *Faculty of Mathematics, Physics and Informatics, Comenius University, Bratislava, Slovak Republic*
[2] *Western Geco Oslo Technology Center, Schlumberger House, Asker, Norway*
[3] *Schlumberger Cambridge Research Ltd., Cambridge, United Kingdom*

ABSTRACT

We present a review of the recent development in finite-difference time-domain modeling of seismic wave propagation and earthquake motion. The finite-difference method is a robust numerical method applicable to structurally complex media. Due to its relative accuracy and computational efficiency it is the dominant method in modeling earthquake motion and it also is becoming increasingly more important in the seismic industry and for structural modeling. We first introduce basic formulations and properties of the finite-difference schemes including promising recent advances. Then we address important topics as material discontinuities, realistic attenuation, anisotropy, the planar free surface boundary condition, free-surface topography, wavefield excitation (including earthquake source dynamics), non-reflecting boundaries, and memory optimization and parallelization.

Keywords: Anisotropy, Attenuation, Earthquake motion, Earthquake source dynamics, Finite-difference method, Free surface, Non-reflecting boundaries, Numerical modeling, Optimally accurate operators, Seismic waves

1. INTRODUCTION

Faithfully synthesizing observed seismic data requires simulation of seismic wave propagation in realistic computational models which can include anisotropic media, non-planar interfaces between layers and blocks, velocity/density/quality-factor gradients inside layers, and often with free-surface topography. In particular, the rheology of the medium should allow for realistic attenuation. Anisotropy increasingly becomes more important particularly in structural studies and is essential in many applications in seismic exploration. The modeling of large earthquakes requires at least kinematic modeling of the rupture propagation but the dynamic modeling is likely to provide more realistic simulations in the near future. Since analytical methods do not provide solutions for realistic (structurally complex) models, approximate methods are necessary. Among them, the finite-difference (FD) method is still the dominant method in modeling earthquake motion and it also is becoming increasingly more important in the seismic industry and structural modeling. This is because the FD method can handle relatively complex models, provides the "complete" solution as waves interact with the model, can be relatively accurate, and, at the same time, relatively computationally efficient. In addition to this, the FD method can be relatively easily parallelized. It is important to stress the word relatively—this chapter should explain why.

In the FD method, a computational domain is covered by a space–time grid, that is, by a set of discrete grid positions in space and time. The functions describing a wavefield as well as those describing material properties of the medium are represented by their values at the grid positions. In principle, the space–time grid may be arbitrary and usually no assumption is made about the function values in-between the grid points. Spatial and time derivatives of a function at a given grid position are approximated by the so-called FD formulae, the derivative being expressed using the function values at a specified set of the grid positions in a neighborhood of the given position. The original differential equation is thus replaced by a system of algebraic (FD) equations. The system of FD equations and their numerical solution have three basic properties—consistency of the FD equations with the original differential equations, stability and convergence of the numerical solution. These properties have to be analyzed prior to the numerical calculation.

In principle, the FD method can be applied either in the time or frequency domain. While for the forward modeling the time-domain formulation requires less calculations (at least for simple simulations), the frequency-domain formulation may be more efficient in inverse problems when simulations for multiple source locations are required, at least in 2-D (Pratt, 1990; Pratt *et al.*, 1998). As most of the FD applications to date focus on the forward modeling, this chapter only addresses the FD time-domain (FDTD) formulation.

The FD method belongs to the domain methods together with, for example, the finite-element, spectral element or the pseudospectral method (Chaljub *et al.*, 2006). Therefore, in general, it is less accurate than boundary methods but much more efficient when applied to complex models (for comparison see, e.g., Takenaka *et al.*, 1998).

Because, formally, the FD method almost always can provide some numerical results, those who are not familiar with the properties of the FD method and particularly with its inherent limitations, sometimes overestimate the capability and accuracy of the FD method, and especially some simple, user-friendly looking FD schemes and codes. An improper application of the FD method can give very inaccurate results. On the other hand, when properly treated, the FD method shows accuracy similar to that of other full waveform techniques, and can be a strong modeling tool applicable to many important problems of recent seismology and seismic prospecting.

In this chapter we try to provide a limited review of the recent progress in FDTD modeling of seismic wave propagation and earthquake motion as well as partial tutorial describing in detail specific FDTD techniques that can be used to solve real practical problems. Because explicit, heterogeneous, staggered-grid FD schemes clearly have dominated in the recent FDTD modeling we focus on them.

In the following subsections we briefly introduce basic concepts and properties of the FD method, some of them directly in relation to solving the elastodynamic equation. A more detailed introduction to the FD method would require considerably large space. Basics of the FD method can be found in applied mathematical textbooks such as, for example, Forsythe and Wasow (1960), Isaacson and Keller (1966), Richtmyer and Morton (1967), Marchuk (1982), Anderson et al. (1984), Mitchell and Griffiths (1994), Morton and Mayers (1994), Durran (1999), Cohen (2002). Detailed introductory texts on the application of the FD method to seismic wave propagation and seismic motion modeling can be found in Boore (1972), Levander (1989), Moczo (1998), Carcione et al. (2002), and Moczo et al. (2004b). Though focused on the computational electrodynamics, the extensive book of Taflove and Hagness (2005) is a good reference to the application of the FDTD method to partial differential equations in physics.

1.1. The Grid

Consider a Cartesian coordinate system (x, y, z) and a computational domain in the four-dimensional space of variables (x, y, z, t) with t meaning time. Consider a set of discrete points (x_I, y_J, z_K, t_m) given by $x_I = x_0 + I\Delta x$, $y_J = y_0 + J\Delta y$, $z_K = z_0 + K\Delta z$, $t_m = t_0 + m\Delta t$; $I, J, K, m = 0, 1, 2, \ldots$. The spatial increments Δx, Δy and Δz are usually referred to grid spacings, while Δt is the time step. The set of points (positions) defines a space–time grid. In many applications, the regular (uniform) rectangular grid with the grid spacings $\Delta x = \Delta y = \Delta z = h$ is a natural and reasonable choice. The value of a function u at a grid position (x_I, y_J, z_K, t_m), that is $u(I, J, K, m)$ or $u_{I,J,K}^m$, is approximated by a grid function $U_{I,J,K}^m = U(x_I, y_J, z_K, t_m)$.

Depending on the particular problem, other than Cartesian coordinate systems can be used to define a grid. For instance, spherical coordinates can be used for the whole Earth's models, and cylindrical for modeling boreholes. The choice of the grid determines the structure and properties of the FD approximations to derivatives and consequently the properties of the FD equations. Here we focus on FD schemes constructed for grids corresponding to the Cartesian coordinate systems.

In some problems it may be advantageous to define a non-uniform grid. Examples are grids with irregularly varying size of the grid spacing or discontinuous (combined) grids with a sudden change in size of the grid spacing. Such grids can better accommodate geometry of the model or reduce the total number of grid points covering the computational space. These grids belong to the structured grids: at a grid point the neighbor grid points are always known (for example, they are defined using some mathematical rule). On the other hand, at a grid point of the unstructured grid some additional information is needed about the neighbor grid points. Most FD techniques use structured grids because the algorithms are faster than those on the unstructured grids.

Another and perhaps more important aspect is whether all functions are approximated at the same grid position or not. In a conventional grid, all functions are approximated at the same grid positions. In a partly-staggered grid, displacement or particle-velocity components are located at one grid position whereas the stress-tensor components are located at another grid position. In a staggered grid, each displacement and/or particle-velocity component and each shear stress-tensor component has its own grid position. The normal stress-tensor components share another grid position. The staggered distribution of quantities in space is related (through the equation of motion) to the staggered distribution of quantities in time. In all types of grids, an effective density is assigned to a grid position of each displacement or particle-velocity component while an effective elastic modulus is assigned to each grid position of the stress-tensor components. The so-called grid cells of the conventional, partly-staggered and staggered grids are illustrated in Fig. 1.

1.2. The FD Approximations to Derivatives

Let the function $\Phi(x)$ have a continuous first derivative. The forward-difference formula

$$\frac{d\Phi}{dx}(x_0) \doteq \frac{\Phi(x_0+h) - \Phi(x_0)}{h}, \tag{1}$$

the backward-difference formula

$$\frac{d\Phi}{dx}(x_0) \doteq \frac{\Phi(x_0) - \Phi(x_0-h)}{h}, \tag{2}$$

and the central-difference formula

$$\frac{d\Phi}{dx}(x_0) \doteq \frac{\Phi(x_0+h) - \Phi(x_0-h)}{2h} \tag{3}$$

are three different approximations to the 1st derivative of function $\Phi(x_0)$. Substituting Taylor expansions of functional values $\Phi(x_0+h)$ and $\Phi(x_0-h)$ about the point x_0 in Eqs. (1) and (2) shows that the difference between the first derivative and the value of the right-hand side expression, that is, the truncation error, has the leading term proportional to h^1. The FD formulae (1) and (2) are the 1st-order approximations to the first derivative. Similarly, it is easy to check that the FD formula (3) is the 2nd-order approximation to the first derivative because the truncation error is proportional to h^2. For a chosen derivative, set of the grid points and order of approximation, it is possible to find a FD formula by construct-

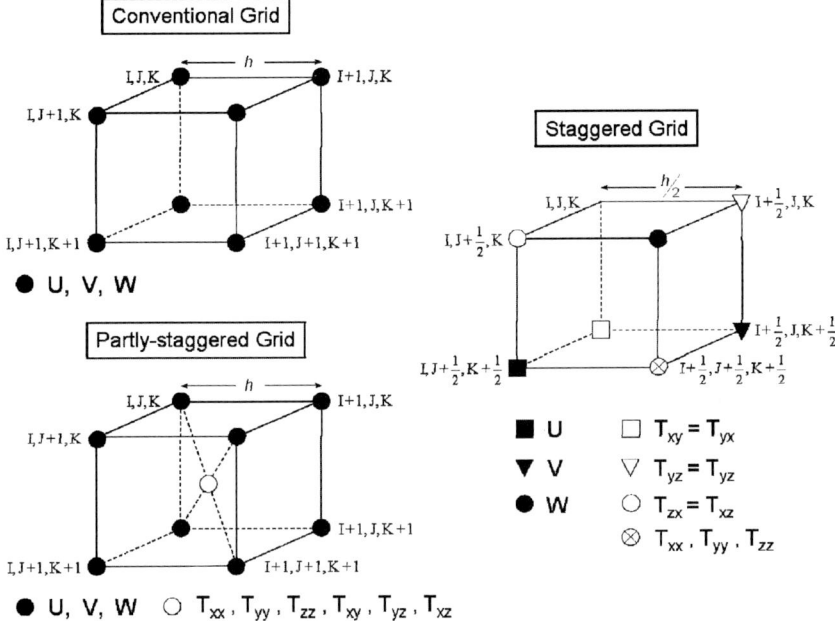

FIG. 1. Spatial grid cells in the conventional, partly-staggered and staggered grids. All displacement-vector components U, V and W are located at each grid position in the conventional grid. Either displacement or particle-velocity components U, V and W share the same grid positions whereas stress-tensor components T_{xx}, T_{yy}, T_{zz}, T_{xy}, T_{yz}, and T_{zx} share other grid positions in the partly-staggered grid. Displacement and/or particle-velocity components U, V and W are located at different grid positions as well as stress-tensor components T_{xx}, T_{yy}, T_{zz}, T_{xy}, T_{yz}, and T_{zx} are in the staggered grid. Because the normal stress-tensor components are determined by the same spatial derivatives of the displacement components, they share one grid position.

ing a system of algebraic equations based on Taylor expansions and equating the coefficients of identical powers of the grid spacing h (e.g., Durran, 1999; Moczo et al., 2004b). This is also true for approximating a derivative in a plane or in a volume.

A frequently used 2nd-order approximation to the second derivative is

$$\frac{d^2\Phi}{dx^2}(x_0) \doteq \frac{\Phi(x_0 - h) - 2\Phi(x_0) + \Phi(x_0 + h)}{h^2}. \quad (4)$$

The approximation is used in the conventional-grid displacement FD schemes.

An important 4th-order approximation to the 1st derivative is

$$\frac{d\Phi}{dx}(x_0) \doteq \frac{1}{h}\left\{a\left[\Phi\left(x_0 + \frac{3}{2}h\right) - \Phi\left(x_0 - \frac{3}{2}h\right)\right] \right.$$
$$\left. + b\left[\Phi\left(x_0 + \frac{1}{2}h\right) - \Phi\left(x_0 - \frac{1}{2}h\right)\right]\right\} \quad (5)$$

with $a = -1/24$, $b = 9/8$. The approximation is used in the staggered-grid FD schemes.

For other FD approximations, higher-order FD approximations and FD approximations on arbitrary spaced grids see also Anderson *et al.* (1984), Dablain (1986), Fornberg (1988), Geller and Takeuchi (1995, 1998), Klimeš (1996), and Cohen (2002).

1.3. Basic Properties of the FD Equations and their Solution

Denote a partial differential equation as *PDE* and a FD equation(s) as *FDE* (a FD scheme may be used instead of *FDE*). A *FDE* is consistent with the PDE if the difference between the *FDE* and the *PDE* (the truncation error) vanishes as the sizes of the time step and spatial grid spacing go to zero independently, that is, $|PDE - FDE| \to 0$ if $\Delta t \to 0$, $h \to 0$. If this is true only when a certain relationship is satisfied between Δt and h, the *FDE* is conditionally consistent.

A *FDE* is stable if it produces a bounded solution when the exact solution is bounded, and is unstable if it produces an unbounded solution when the exact solution is bounded. If the solution of the *FDE* is bounded for all values of Δt and h, the *FDE* is unconditionally stable. If the solution of the *FDE* is bounded only for certain values of Δt and h, the *FDE* is conditionally stable. If the solution of the *FDE* is unbounded for all values of Δt and h, the *FDE* is unconditionally unstable. The stability analysis can be performed only for linear *PDE*. A nonlinear *PDE* must be first linearized locally. The *FDE* of the linearized *PDE* can be analyzed for stability. The most commonly used method for the stability analysis is the von Neumann method. The basic idea of the von Neumann method is to represent a discrete solution at a time $m \Delta t$ and spatial point Ih, that is at one grid point, by a finite Fourier series, and examine stability of the individual Fourier components. Thus, the method investigates the local stability. The discrete solution is stable if and only if each Fourier component is stable. Von Neumann analysis is applicable to linear *FDE* with constant coefficients. Though a spatial periodicity is assumed for the finite Fourier series, the analysis can give a useful result even if this is not the case.

A *FDE* is convergent if the solution of the *FDE* approaches the exact solution of the *PDE* as the sizes of the time step and spatial grid spacing go to zero. Denoting the solutions obtained by the *PDE* and *FDE* as $u_{I,J,K}^m$ and $U_{I,J,K}^m$, respectively, the convergence means that $U_{I,J,K}^m \to u_{I,J,K}^m$ if $\Delta t \to 0$, $h \to 0$.

Whereas the consistency is the property of the *FDE* because it relates the *FDE* to the *PDE*, stability and convergence are properties of the numerical solution of the *FDE*. In general, while it is easy to analyze the consistency, proving convergence can be a very difficult mathematical problem. Therefore, it is very helpful that the convergence is related to the consistency and the stability: It follows from the Lax equivalence theorem that if the *FDE* is consistent and stable, it is also convergent.

Due to the discrete nature of the FD solution, the phase and group velocity in the grid differ from the true velocities in the medium. The grid velocities depend on the spatial sampling ratio $s = h/\lambda$, where λ is the wavelength that is to be propagated in the grid, and also on Courant number $c\Delta t/h$, where c is the velocity. The grid dispersion is a very important grid phenomenon. It has a cumulative effect on the wave propagation—the longer the travel distance, the larger the effect of the difference between the grid and true velocity. Therefore, the grid dispersion has to be analyzed prior to the numerical calculations. This is especially important if the medium is viscoelastic. The viscoelastic medium is intrinsically dispersive and thus a possible superposition of two dispersion effects has to be prevented by minimizing the grid dispersion; see Robertsson et al. (1994). The grid-dispersion relation can be obtained from the stability analysis. The grid dispersion has to be taken into account in planning the numerical calculation. For a detailed analysis of stability, grid dispersion and accuracy of the FD schemes solving the equation of motion on the conventional and staggered-grid schemes in 2-D and 3-D problems in homogeneous media see, for example, papers by (Marfurt (1984); Crase et al. (1992); Igel et al. (1995); Geller and Takeuchi (1995, 1998); Klimeš (1996); Takeuchi and Geller (2000); Mizutani et al. (2000); Moczo et al. (2000)).

1.4. Explicit and Implicit FD Schemes

In an explicit scheme, the motion at any (one) spatial grid point can be updated for the next time level using an explicit FD formula which uses only values of motion at previous time levels (and, obviously, using also material grid parameters). In the case of an implicit scheme, there is no explicit formula for updating motion only in one grid point. In an implicit scheme, the motion at a given time level is calculated simultaneously at all spatial grid points from the motion values at previous time levels using the inverse of a matrix. Obviously, the explicit schemes are computationally simpler. A vast majority of earthquake ground motion modeling and exploration seismology studies use explicit FD schemes. For the implicit schemes see, e.g., Emerman et al. (1982), Mufti (1985).

1.5. Homogeneous and Heterogeneous FD Schemes

Motion in a smoothly heterogeneous elastic or viscoelastic continuum is governed by the equation of motion. The equation can be solved using a proper FD scheme. If the medium contains a material discontinuity, i.e., an interface between two homogeneous or smoothly heterogeneous media, at which density and elastic moduli change discontinuously, the equation of motion still governs motion away from the discontinuity but boundary conditions apply at the discontinuity. There are two approaches to deal with this situation—the homogeneous and heterogeneous approaches. In the homogeneous approach, a FD scheme for the smoothly heterogeneous medium is applied at grid points away from the material discontinuity while a FD scheme obtained by a proper discretization of the boundary conditions is applied at grid points at or near the material discontinuity. In the alternative heterogeneous approach only one FD scheme is used for all interior grid points (points not lying on boundaries of a grid) no matter what their positions are with respect to the material discontinuity. The presence of the material discontinuity is accounted for only by assigning appropriate values of elastic moduli and density. Therefore, except for treating the free surface, the heterogeneous approach has been much more popular since the beginning of the seventies.

A homogeneous FD scheme is specific for a particular problem. Its application to complex models with curved material discontinuities is difficult and therefore impractical. Moreover, finding a stable and sufficiently accurate FD approximation to the boundary conditions is not a trivial problem, see, e.g., Kummer and Behle (1982), Slawinski and Krebes (2002).

While widely used, the heterogeneous approach has not been addressed properly until very recently. The point is that the heterogeneous approach can be justified only when a heterogeneous formulation of the equation of motion and Hooke's law, that is, the same form of the equations for both a point away from the material discontinuity and a point at the material discontinuity, can be found. This question will be analyzed in the next section.

1.6. The Equation of Motion, Hooke's Law, and FD Schemes

Given a model, the equation of motion and Hooke's law (the stress-strain relation, constitutive law) together with the initial and boundary conditions fully describe a problem of seismic wave propagation and motion. Consider a Cartesian coordinate system (x_1, x_2, x_3), for example, with the x_1-axis horizontal and positive to the right, and the x_3-axis positive downward. Let $\rho(x_i); i \in \{1, 2, 3\}$ be density, $c_{ijkl}(x_q)$ tensor of elastic coefficients, $\kappa(x_i)$ bulk modulus, $\mu(x_i)$ shear modulus, $\vec{u}(x_i, t)$ displacement vector, $\vec{v}(x_i, t)$ particle-velocity vector, $\vec{f}(x_i, t)$ body force per unit volume, $\sigma_{ij}(x_k, t)$ and $\varepsilon_{ij}(x_k, t); i, j, k \in \{1, 2, 3\}$ stress- and strain-tensors (from now on, x_1, x_2, x_3 and x, y, z may be used interchangeably; similarly, 1, 2, 3 and x, y, z in the subscripts of the displacement and stress-tensor components), δ_{ij} Kronecker delta. A partial time derivative will be denoted by a dot above the symbol or the operator ∂_t; for example, $\partial u_i/\partial t = \dot{u}_i = \partial_t u_i$. A derivative with respect to the coordinate x_j will be denoted by a comma followed by x_j or the operator ∂_j; for example, $\partial \sigma_{ij}/\partial x_j = \sigma_{ij,j} = \partial_j \sigma_{ij}$. The Einstein summation convention for repeated indices will be assumed unless stated otherwise.

With reference to the FD schemes developed during the last decades, the following alternative formulations of the equation of motion and Hooke's law for anisotropic or isotropic media can be used as a starting point for deriving the FD schemes:

displacement-stress

$$\rho \ddot{u}_i = \sigma_{ij,j} + f_i,$$

$$\sigma_{ij} = c_{ijkl}\varepsilon_{kl} \quad \text{or} \quad \sigma_{ij} = \kappa \varepsilon_{kk}\delta_{ij} + 2\mu\left(\varepsilon_{ij} - \frac{1}{3}\varepsilon_{kk}\delta_{ij}\right), \tag{6}$$

displacement-velocity-stress

$$\rho \dot{v}_i = \sigma_{ij,j} + f_i, \qquad v_i = \dot{u}_i,$$

$$\sigma_{ij} = c_{ijkl}\varepsilon_{kl} \quad \text{or} \quad \sigma_{ij} = \kappa \varepsilon_{kk}\delta_{ij} + 2\mu\left(\varepsilon_{ij} - \frac{1}{3}\varepsilon_{kk}\delta_{ij}\right), \tag{7}$$

velocity-stress

$$\rho \dot{v}_i = \sigma_{ij,j} + f_i,$$

$$\dot{\sigma}_{ij} = c_{ijkl}\dot{\varepsilon}_{kl} \quad \text{or} \quad \dot{\sigma}_{ij} = \kappa \dot{\varepsilon}_{kk}\delta_{ij} + 2\mu\left(\dot{\varepsilon}_{ij} - \frac{1}{3}\dot{\varepsilon}_{kk}\delta_{ij}\right), \tag{8}$$

displacement

$$\rho \ddot{u}_i = \left[\left(\kappa - \frac{2}{3}\mu\right)u_{k,k}\right]_{,i} + (\mu u_{i,j})_{,j} + (\mu u_{j,i})_{,j} + f_i \quad \text{or}$$

$$\rho \ddot{u}_i = (c_{ijkl}u_{k,l})_{,j} + f_i. \tag{9}$$

Because it is not differentiated with respect to the spatial coordinates, the strain tensor $\varepsilon_{ij} = \frac{1}{2}(u_{i,j} + u_{j,i})$ was used here in the first three formulations for brevity.

The above are so-called strong formulations of the equation of motion. However, a FD method can also be applied to a (integral) weak form of the equation of motion. For example, a weak form of the Galerkin type is typical for the standard finite-element method (e.g., Zienkiewicz and Taylor, 1989; Ottosen and Petersson, 1992).

When solving the strong form of the equation of motion, the boundary conditions at material discontinuities must be explicitly satisfied. The traction continuity at internal discontinuities and vanishing traction at the free surface are automatically satisfied by the weak-form solution (this is an advantage of the weak form). In contrast, continuity of displacement is an essential continuity condition that must be explicitly satisfied by the weak-form solutions.

Whereas most of the recent FD schemes solve one of the strong forms, Geller and Takeuchi (1995, 1998) developed their optimally accurate FD schemes in application to the weak form of Strang and Fix (1973) and Geller and Ohminato (1994).

In principle, any formulation can be used with any of the three types of the grids (conventional, partly-staggered, staggered). However, it is obvious that the displacement formulation is naturally connected with the conventional grid. This is because only displacement values are explicitly present both in the equations and grid. Similarly, the velocity-stress formulation is naturally connected with the staggered grid as all field quantities in the equation of motion and Hooke's law are explicitly present in the grid. The particular structure (i.e., relative positions of the field quantities in space and time) of the staggered grid is unambiguously determined by the structure of the equation of motion and Hooke's law, that is, by the temporal and spatial derivatives of the field quantities.

In the early days of the FD method applied to seismology and seismics, the displacement formulation and conventional grid were used; for example, Alterman and Karal (1968), Boore (1970, 1972), Kelly *et al.* (1976). Because the conventional-grid displacement FD schemes had problems with instabilities in models with high-velocity contrasts and with grid dispersion in media with high Poisson's ratio, Virieux (1984, 1986) introduced the staggered-grid velocity-stress FD schemes for modeling seismic wave propagation. Virieux followed Madariaga (1976) who introduced the staggered-grid formulation in his dynamic modeling of the earthquake rupture. Bayliss *et al.* (1986) and Levander (1988) introduced the 4th-order staggered-grid FD schemes which in 2-D and 3-D need at least four and eight times less memory, respectively, compared to the 2nd-order schemes. In terms of CPU the improvement is 5–8 times in 2-D and 10–16 in 3-D. This is related to the grid dispersion. The staggered-grid FD schemes have become the dominant type of schemes in the FD modeling of seismic wave propagation and earthquake motion. In order to further reduce the memory requirements, Luo and Schuster (1990) suggested a staggered-grid displacement-stress 2-D P-SV FD scheme which they called a parsimonious scheme. Because the scheme does not integrate stress in time, the stress-tensor components are only temporary quantities. Thus, the displacement-stress scheme in 3-D needs only 75% of the memory needed by the velocity-stress scheme. Rodrigues (1993), and Yomogida and Etgen (1993) used the 8th-order 3-D displacement-stress FD schemes, Ohminato and Chouet (1997) applied the 2nd-order while Moczo *et al.* (2000, 2002) the 4th-order approximations. Moczo *et al.* (2000) analyzed the grid dispersion of the displacement-stress schemes (4th and 2nd order) and pointed out that the stability and grid dispersion of the displacement-stress, displacement-velocity stress and velocity-stress schemes are the same. The advantage of the displacement-velocity-stress scheme is that both displacement and particle-velocity are calculated at no extra cost.

The partly-staggered grid was probably first used in seismology by Andrews (1973) who applied it in modeling the fault rupture using his traction-at-split-node method. Magnier *et al.* (1994) realized disadvantages of the staggered-grid schemes in treating the anisotropic media and used the partly-staggered grid. Zhang (1997) used the partly-staggered grid in his 2-D velocity-stress FD modeling. However, the developed schemes have not attracted much attention. Recently, the use of the partly-staggered grid was promoted by Saenger *et al.* (2000) and Saenger and Bohlen (2004). They called the grid rotated staggered grid since they obtained the spatial FD operator by the rotation of the standard staggered-grid operator. The term "rotated staggered grid" is somewhat unfortunate because assuming one spatial grid position for the stress tensor and another position for the displacement vector, it is possible to find a variety of FD schemes—depending on the order of approximation. Only in one particular case can the spatial FD operator be obtained by the rotation of the standard staggered-grid operator though it is also easy to obtain it without explicit consideration of the rotation. In fact, the particular scheme used by Saenger et al. (2000) is a simple consequence of requirement of the same truncation error with respect to all coordinate axes.

While the reason to use the partly-staggered grids for anisotropic media is obvious (all stress-tensor components located at the same grid position thus requiring no interpolation of particle velocities or strains), the application of the partly-staggered grid to account for the material heterogeneity has not been analyzed. Therefore, this application still needs to be theoretically more rigorously analyzed.

A special aspect of the development is the FD schemes formulated for non-uniform grids. These will be mentioned in the section on the memory optimization and parallelization.

Apart from the recent dominance of the staggered-grid FD schemes, Geller and his co-workers developed another important approach to the FD modeling of seismic wave propagation based on the weak form of the equation of motion. In their schemes displacement is the sole dependent variable, as opposed to the staggered-grid schemes. Geller and Takeuchi (1995) derived a general criterion for optimally accurate numerical operators, and used it to derive an optimally accurate frequency-domain scheme. Geller and Takeuchi (1998) used the criterion to derive an optimally accurate FDTD scheme for 1-D problems. Takeuchi and Geller (2000) then developed optimally accurate FDTD operators for 2-D and 3-D problems. These derivations yield implicit schemes which are solved using a predictor-corrector algorithm, so that the actual computational schemes are explicit. Whereas optimally accurate FDTD schemes require at least twice the CPU time per grid point and time step compared to 2nd-order staggered-grid FD schemes, they yield accuracy improvements on the order of 10 (for 1-D), 50 (for 2-D), or 100 (for 3-D). From this point of view they are cost-effective. The optimally accurate schemes have not yet been widely used in practical FD modeling. The likely reasons are (a) the theory might appear relatively complicated compared to that of the standard staggered-grid schemes, (b) the fact that quantification and minimization of computational error have not heretofore been widely viewed as high priorities, (c) inertia with respect to traditional approaches and the lack of user-friendly codes for optimally accurate schemes. However, it is likely that optimally accurate FD schemes will be more widely used in the future.

An interesting approach to develop alternative FD scheme was presented by Holberg (1987). Instead of minimizing the error measured in terms of higher-order derivatives he minimized the relative error in group velocity caused by the grid dispersion within a specific frequency band emitted by active sources.

Therefore he did not develop the FD operator with the predetermined order of the truncation error using the Taylor expansions. Holberg defined the FD operator as a differentiator realized by a convolutory (FIR) operator.

Whereas most of the recent FD schemes are 2nd-order accurate in time, it is possible to increase the order of accuracy by employing the Lax–Wendroff correction (Lax and Wendroff, 1964; Dablain, 1986). The leading term in the truncation error for the 2nd-order FD approximation to time derivative is replaced by a term with only spatial derivatives using the equation of motion. Although the form of the Lax–Wendroff schemes is quite different from the optimally accurate schemes of Geller and co-workers, Mizutani *et al.* (2000) have shown that these two types of scheme are in fact essentially identical. An efficient implementation of the approach for viscoelastic media was presented by Blanch and Robertsson (1997). An alternative approach for the time stepping is to use the Chebychev expansion method by Tal-Ezer *et al.* (1990) which combines computational efficiency with spectral accuracy in time.

1.7. Algorithms for Enhancing Computational Performance

A significant part of the literature has been devoted to enhancing the computational efficiency of FDTD by taking a more holistic view as to how FDTD are applied to solve a computational problem. Several methods for hybrid modeling were developed in the eighties and nineties where different computational techniques are used either for temporal/spatial dependences (e.g., Alekseev and Mikhailenko, 1980) or in different parts of the model appropriate for the local wave propagation regime (e.g., Shtivelman, 1984, 1985; Kummer *et al.*, 1987; Stead and Helmberger, 1988; Emmerich, 1989, 1992; Fäh, 1992; Robertsson *et al.*, 1996; Zahradník and Moczo, 1996; Moczo *et al.*, 1997; Lecomte *et al.*, 2004). For instance, for a reflection seismic problem, in a smoothly varying overburden a ray-based solution may be appropriate whereas FDTD are needed only in the vicinity of a complex target where the energy reflects and is reverberated back towards the surface. Such an approach can enable computational savings of several orders of magnitude depending on the specific application and model and will be particularly significant in 3-D. Problems with implementing and using such methods are related to model generation and interfacing the different methods which tend to be different from problem to problem making the process very difficult to implement in an automatic fashion.

A somewhat different approach was taken by, for instance, Robertsson and Chapman (2000) and Robertsson *et al.* (2000) who proposed to use a technique, referred to as FD-injection, fully based on finite differences for regenerating the FD response in a model following local model alterations. The computational savings can be very significant, particularly for applications such as full waveform inversion or time-lapse seismic analyses.

Recently, van Manen *et al.* (2005) have proposed a method based on concepts from time-reversal acoustics (Fink, 1997), that may become an important tool in synthesizing seismic data. The approach relies on a representation theorem of the wave equation to express the Green's function between points in the interior as an integral over the response in those points due to sources on a surface surrounding the medium. Following a predictable initial computational effort, Green's functions between arbitrary points in the medium can be computed as needed using a simple cross-correlation algorithm.

2. Optimally Accurate FD Operators

A linear mechanical system or a finite volume of elastic continuum preferably supports oscillatory motion at certain frequencies which are called eigenfrequencies or resonant frequencies, i.e. at normal modes. In other words, the oscillatory motion is naturally most amplified at these frequencies. The same is also true about a numerical error in a discrete numerical system which is a numerical approximation to the true physical system. The basic idea of Geller and Takeuchi (1995) therefore seems quite obvious: to minimize the error of the numerical solution first of all at eigenfrequencies.

Geller and Takeuchi (1995) used first-order Born theory and a normal mode expansion to obtain formal estimates of the relative error of the numerical solution and a general criterion for what they called optimally accurate operators. The criterion requires that the inner product of an eigenfunction and the net error of the discretized equation of motion should be approximately equal to zero when the operand is the eigenfunction and the frequency is equal to the corresponding eigenfrequency. An important aspect of the approach is that it is not necessary to know the actual values of the eigenfrequencies and eigenfunctions to use the criterion to derive optimally accurate operators. Geller and Takeuchi (1995) showed that in the case of a heterogeneous medium the criterion is the logical extension of the criterion to minimize grid dispersion of phase velocity for a homogeneous medium. Based on this criterion, Geller and Takeuchi (1998) derived optimally accurate 2nd-order weak-form FDTD scheme for the elastic 1-D case. Takeuchi and Geller (2000) then generalized their approach to the 2-D and 3-D cases.

Though the optimally accurate FD operators are not applicable to the staggered-grid schemes (they would lead to apparently intractable implicit schemes—according to Geller and Takeuchi, 1998), we consider them important and assume their wider applications in future FDTD modeling. Therefore, we briefly present the basics of Geller and Takeuchi's approach, closely following Geller and Takeuchi (1995, 1998).

2.1. General Criterion for Optimally Accurate FD Operators

In the Direct Solution Method (DSM; Geller and Ohminato, 1994) the equation of motion for an anelastic solid is transformed into its Galerkin weak form (Strang and Fix, 1973),

$$(\omega^2 \mathbf{T} - \mathbf{H})\vec{c} = -\vec{g}, \tag{10}$$

where ω is the angular frequency, $\mathbf{T}$ mass matrix, $\mathbf{H}$ stiffness matrix, $\vec{c}$ vector of expansion coefficients for the trial functions, $\vec{g}$ force vector,

$$T_{rs} = \int_V [\phi_i^{(r)}]^* \rho \phi_i^{(s)} \, dV, \qquad H_{rs} = \int_V [\phi_{i,j}^{(r)}]^* c_{ijkl} \phi_{k,l}^{(s)} \, dV,$$

$$g_r = \int_V [\phi_i^{(r)}]^* f_i \, dV, \tag{11}$$

$\phi_i^{(r)}$ is the ith component of the rth trial function, and $*$ means complex conjugate quantity. The displacement is represented as

$$u_i = \sum_r c_r \phi_i^{(r)}. \tag{12}$$

If an infinite trial function expansion were used, Eq. (10) would yield exact solutions. In any practical application the trial function expansion will be finite and there will be some numerical error. The exact equation of motion can be formally written as

$$(\omega^2 \mathbf{T}^e - \mathbf{H}^e)\vec{c}^e = -\vec{g}. \tag{13}$$

The relation between the numerical and exact quantities is assumed in a form

$$\mathbf{T} = \mathbf{T}^e + \delta\mathbf{T}, \qquad \mathbf{H} = \mathbf{H}^e + \delta\mathbf{H}, \qquad \vec{c} = \vec{c}^e + \vec{\delta c}, \tag{14}$$

where $\delta\mathbf{T}$, $\delta\mathbf{H}$ and $\vec{\delta c}$ are errors of the numerical operators and solution, respectively.

The normal modes satisfy equation

$$(\omega_p^2 \mathbf{T}^e - \mathbf{H}^e)\vec{c}_p = 0, \tag{15}$$

where ω_p is an eigenfrequency of the pth mode and $\vec{c}_p$ is the eigenvector. Orthonormalization is assumed in a form

$$\vec{c}_p^* \mathbf{H}^e \vec{c}_q = \omega_p^2 \vec{c}_p^* \mathbf{T}^e \vec{c}_q = \omega_p^2 \delta_{pq}. \tag{16}$$

Substituting Eqs. (14) into the l.h.s. of Eq. (10), replacing the r.h.s. of Eq. (10) by the l.h.s. of Eq. (13), and neglecting terms with products of errors (the first-order Born approximation) leads to

$$(\omega^2 \mathbf{T}^e - \mathbf{H}^e)\vec{\delta c} = -(\omega^2 \delta\mathbf{T} - \delta\mathbf{H})\vec{c}^e. \tag{17}$$

Equation (17) enables to determine the error of the numerical solution, $\vec{\delta c}$, if the exact solution and errors of the operators, i.e., $\vec{c}^e$, $\delta \mathbf{T}$ and $\delta \mathbf{H}$ are known.

The solution of Eq. (13) can be represented in terms of an eigenfunction expansion

$$\vec{c}^e = \sum_p d_p^e \vec{c}_p. \tag{18}$$

Substituting Eq. (18) into Eq. (13), and using Eq. (16) leads to

$$d_p^e = \vec{c}_p^* \vec{g} / (\omega^2 - \omega_p^2). \tag{19}$$

The expansion coefficient d_p^e will be large, when ω is close to ω_p. Otherwise, it will be negligible.

The solution of Eq. (17) can also be represented in terms of an eigenfunction expansion

$$\vec{\delta c} = \sum_p \delta d_p \vec{c}_p. \tag{20}$$

Substituting expansions (18) and (20) into Eq. (17), and using Eqs. (16) leads to

$$\delta d_p = -\sum_q (\omega^2 \vec{c}_p^* \delta \mathbf{T} \vec{c}_q - \vec{c}_p^* \delta \mathbf{H} \vec{c}_q) d_q^e / (\omega^2 - \omega_p^2). \tag{21}$$

The expansion coefficient δd_p will be large only when ω is close to ω_p. In such a case obviously only d_p^e will be large. Therefore, in the vicinity of $\omega = \omega_p$, the $q \neq p$ terms in Eq. (21) can be neglected, i.e., the relative error of the numerical solution in the vicinity of ω_p will approximately be

$$\frac{\delta d_p}{d_p^e} = -(\omega^2 \vec{c}_p^* \delta \mathbf{T} \vec{c}_p - \vec{c}_p^* \delta \mathbf{H} \vec{c}_p) / (\omega^2 - \omega_p^2). \tag{22}$$

It follows from Eq. (22) that the relative error will in general greatly increase with $\omega \to \omega_p$. However, if the numerator of Eq. (22) is also proportional to $\omega - \omega_p$, the relative error will remain approximately constant as $\omega \to \omega_p$. Such proportionality can be achieved if and only if

$$\omega_p^2 \vec{c}_p^* \delta \mathbf{T} \vec{c}_p - \vec{c}_p^* \delta \mathbf{H} \vec{c}_p \doteq 0 \tag{23}$$

for each mode. If Eq. (23) is approximately satisfied, then Eq. (22) can be simplified:

$$\left| \frac{\delta d_p}{d_p^e} \right| \approx \vec{c}_p^* \delta \mathbf{T} \vec{c}_p. \tag{24}$$

This means that the relative error for a given grid can be reliably estimated in advance of calculation.

Geller and Takeuchi (1995) defined optimally accurate operators, say $\mathbf{T}'$ and $\mathbf{H}'$, as operators that satisfy Eq. (23):

$$\vec{c}_p^*(\omega_p^2 \delta\mathbf{T}' - \delta\mathbf{H}')\vec{c}_p \doteq 0. \tag{25}$$

Substituting first two of Eqs. (14) for operators $\mathbf{T}'$ and $\mathbf{H}'$ into Eq. (25) and using Eq. (15) leads to equivalent equation

$$\vec{c}_p^*(\omega_p^2 \mathbf{T}' - \mathbf{H}')\vec{c}_p \doteq 0. \tag{26}$$

Equation (25) will be satisfied if the leading term of the truncation error of the discretized equation is zero when the operand is an eigenfunction and the frequency is equal to the corresponding eigenfrequency, in other words if

$$(\omega_p^2 \delta\mathbf{T}' - \delta\mathbf{H}')\vec{c}_p \doteq 0. \tag{27}$$

On the other hand, however, it is not necessary for Eq. (27) to be satisfied in order for Eq. (25) to be satisfied, because even if the quantity on the l.h.s. of Eq. (27) is non-zero, its inner product with $\vec{c}_p$ can still be approximately zero. Geller and Takeuchi (1995, 1998) and Takeuchi and Geller (2000) take advantage of this fact to derive optimally accurate operators using a two-step methodology. First, for interior points, they derived optimally accurate operators that satisfy Eq. (27). Second, to complete the derivation of the optimally accurate operators, they "fill in" the few remaining degrees of freedom (corresponding to boundary points or their neighbors) so that Eq. (25) is approximately satisfied, even though Eq. (27) is not necessarily satisfied.

In this review, for simplicity, we discuss only the simplest case of optimally accurate 1-D TDFD operators for interior points only. For a discussion of the treatment of the boundary operators, see, for example, Sections 3 and 4 of Geller and Takeuchi (1995).

Consider the equation

$$\text{exact } LHS(\omega, u) = \vec{f} \tag{28}$$

and such its discretization which gives

$$\text{discretized } LHS(\omega, u)$$
$$= \text{exact } LHS(\omega, u) + \frac{h^2}{a}\left[\text{exact } LHS(\omega, u)\right]'' + \cdots, \tag{29}$$

where the primes denote spatial differentiation. The normal modes satisfy the equation

$$\text{exact } LHS(\omega_p, u_p) = 0, \tag{30}$$

which implies

$$\left[\text{exact } LHS(\omega_p, u_p)\right]'' = 0. \tag{31}$$

Considering normal modes in Eq. (29), and substituting Eqs. (30) and (31) into Eq. (29) leads to

$$\text{discretized } LHS(\omega_p, u_p) \doteq 0, \tag{32}$$

which corresponds to condition (26). This shows that the leading term of the truncation error of each FD approximation used to discretize the l.h.s. of Eq. (28) has the same coefficient, h^2/a, and the displacement 2 times more differentiated than in the approximated term. Thus, we have an indication for constructing optimally accurate discretization.

2.2. Optimally Accurate FD Operators for 1-D problem in a Homogeneous Medium

In order to illustrate the above theory, we closely follow the 1-D problem for a homogeneous medium in Geller and Takeuchi (1998). Consider 1-D equation

$$\rho \ddot{u} = C \frac{\partial^2 u}{\partial x^2} + f. \tag{33}$$

A weak-form FDTD approximation to the equation can be written as

$$[A_I^m(M, i) - K_I^m(M, i)]U(M, i) = F_I^m, \tag{34}$$

where m is the time level, at which Eq. (34) is approximated, I index of the spatial position at which Eq. (34) is approximated, M time summation index, i spatial summation index. Matrix $\mathbf{A}_I^m$ has the form

$$\mathbf{A}_I^m(M, i) = \begin{bmatrix} A_I^m(m+1, I-1) & A_I^m(m+1, I) & A_I^m(m+1, I+1) \\ A_I^m(m, I-1) & A_I^m(m, I) & A_I^m(m, I+1) \\ A_I^m(m-1, I-1) & A_I^m(m-1, I) & A_I^m(m-1, I+1) \end{bmatrix}, \tag{35}$$

matrix $\mathbf{K}_I^m$ has an analogous structure. The conventional operators, corresponding to the standard difference formula (4), are

$$\mathbf{A}_I^m = \frac{\rho}{\Delta^2 t}\begin{bmatrix} 0 & 1 & 0 \\ 0 & -2 & 0 \\ 0 & 1 & 0 \end{bmatrix}, \quad \mathbf{K}_I^m = \frac{C}{h^2}\begin{bmatrix} 0 & 0 & 0 \\ 1 & -2 & 1 \\ 0 & 0 & 0 \end{bmatrix}. \tag{36}$$

The operator error at a single point in space and time approximately is (only the leading term is given)

$$[\delta A_I^m(M, i) - \delta K_I^m(M, i)]U(M, i) \doteq \left[\frac{\Delta^2 t}{12}\rho \frac{\partial^4 u}{\partial t^4} - \frac{h^2}{12}C\frac{\partial^4 u}{\partial x^4}\right]_{m,I}, \tag{37}$$

where $\delta \mathbf{A}_I^m$ and $\delta \mathbf{K}_I^m$ are differences between the conventional and exact operators.

In order to derive an optimally accurate FDTD scheme, Geller and Takeuchi (1998) take the Fourier transform of Eq. (34), to obtain the FD equation in the frequency domain for purposes of error analysis:

$$[\tilde{A}_I(i) - \tilde{K}_I(i)]\tilde{U}(i) = \tilde{F}_I. \tag{38}$$

Because Eq. (38) has essentially the same form as Eq. (10), the above theory of the error of the numerical solution can be applied. The difference is that the operator errors here depend on frequency and wavenumber.

The Fourier transform of Eq. (37) is

$$[\delta\tilde{A}_I(i) - \delta\tilde{K}_I(i)]\tilde{U}(i) \doteq \left[\frac{\omega^2 \Delta^2 t}{12}\rho\omega^2 u - \frac{h^2}{12}\frac{d^2}{dx^2}\left(C\frac{d^2 u}{dx^2}\right)\right]_I. \tag{39}$$

It is clear from Eq. (39) that the error is not equal to zero when the operand is an eigenfunction and frequency is equal to eigenfrequency. This would be the case if

$$[\delta\tilde{A}_I(i) - \delta\tilde{K}_I(i)]\tilde{U}(i) \doteq \left(\frac{\omega^2 \Delta^2 t}{12} - \frac{h^2}{12}\frac{d^2}{dx^2}\right)\left[\rho\omega^2 u + C\frac{d^2 u}{dx^2}\right]_I \tag{40}$$

because the expression inside the brackets is the l.h.s. of the exact homogeneous equation of motion in the frequency domain, which is zero when the operand is an eigenfunction and the frequency is the corresponding eigenfrequency. A time-domain equivalent to Eq. (40) can be obtained by applying the inverse Fourier transform to Eq. (40):

$$[\delta A_I^m(M,i) - \delta K_I^m(M,i)]U(M,i)$$
$$\doteq \left\{\frac{\Delta^2 t}{12}\frac{\partial^2}{\partial t^2}\left[\rho\frac{\partial^2 u}{\partial t^2} - C\frac{\partial^2 u}{\partial x^2}\right] + \frac{h^2}{12}\frac{\partial^2}{\partial x^2}\left[\rho\frac{\partial^2 u}{\partial t^2} - C\frac{\partial^2 u}{\partial x^2}\right]\right\}_{m,I}. \tag{41}$$

The operators that yield error (41) are

$$\mathbf{A}_I^m = \frac{\rho}{\Delta^2 t}\begin{bmatrix} 1/12 & 10/12 & 1/12 \\ -2/12 & -20/12 & -2/12 \\ 1/12 & 10/12 & 1/12 \end{bmatrix},$$

$$\mathbf{K}_I^m = \frac{C}{h^2}\begin{bmatrix} 1/12 & -2/12 & 1/12 \\ 10/12 & -20/12 & 10/12 \\ 1/12 & -2/12 & 1/12 \end{bmatrix}. \tag{42}$$

Though, obviously, the operators (42) come from the condition (41), which corresponds to conditions (25) and (26), it is now easy to see relation to the conventional operators. Consider, e.g., approximation to the second time derivative at time level m and spatial position I. While the conventional approximation, formula (4), is

$$\left.\frac{\partial^2 u}{\partial t^2}\right|_{m,I} \doteq \frac{1}{\Delta^2 t}(U_I^{m-1} - 2U_I^m + U_I^{m+1}), \tag{43}$$

the approximation corresponding to the optimally accurate operator $\mathbf{A}_I^m$ in Eq. (42) is

$$\left.\frac{\partial^2 u}{\partial t^2}\right|_{m,I} \doteq \frac{1}{12}\left(\left.\frac{\partial^2 u}{\partial t^2}\right|_{m,I-1} + 10\left.\frac{\partial^2 u}{\partial t^2}\right|_{m,I} + \left.\frac{\partial^2 u}{\partial t^2}\right|_{m,I+1}\right) \quad (44)$$

where the conventional approximation (43) is used to approximate derivatives at time level m at all three spatial positions $I-1$, I and $I+1$. Similarly, the optimally accurate spatial second derivative is approximated over three time levels $m-1$, m and $m+1$.

It is obvious that optimally accurate operators yield an implicit FD scheme. In order to avoid necessity to solve a system of simultaneous linear equations at each time step, Geller and Takeuchi (1998) applied a predictor-corrector scheme. Note that the predictor-corrector optimally accurate scheme and the Lax–Wendroff scheme are essentially the same (Mizutani et al., 2000).

As already mentioned, Takeuchi and Geller (2000) generalized the approach based on the optimally accurate operators to solve 2-D and 3-D problems.

3. Complex Viscoelastic Media with Material Discontinuities

As already mentioned, accounting for realistic attenuation and heterogeneity of the medium, particularly in the presence of material discontinuities in the Earth's interior, is of crucial importance in modeling seismic wave propagation and earthquake motion. While seismologists tried to model material discontinuities from the very beginning of the application of the FDM in seismology, the incorporation of the realistic attenuation was made possible considerably later. We first briefly review the problem of material heterogeneity and material discontinuities in the strong formulation of the equation of motion.

3.1. A Material Discontinuity in the Elastic Medium

In one of the pioneering efforts on the application of the FDM to seismic wave propagation, Alterman and Karal (1968) introduced the concept of fictitious grid points in order to approximate boundary conditions on material discontinuities in their displacement FD scheme. Difficulties in application of the homogeneous approach to curved discontinuities led Boore (1972) to explicitly include a stress-continuity condition on discontinuities differently from the homogeneous and heterogeneous approaches. Due to poor numerical properties of his explicit continuous stress method, Boore (1972) applied the heterogeneous approach in his SH modeling. In order to follow detailed variation of the torsion modulus and, at the same time, to avoid derivatives of the modulus, he used the mathematical trick of Tikhonov and Samarskii (see, e.g., Mitchell, 1969, p. 23) and calculated effective grid moduli as integral harmonic averages along a grid line between two neighboring grid points. Ilan *et al.* (1975) and Ilan and Loewenthal (1976) applied the homogeneous approach to the 2-D P-SV problem on horizontal and vertical planar discontinuities. They used Taylor expansions of displacement to couple the equation of motion with the boundary conditions. Kelly *et al.* (1976) presented their heterogeneous P-SV schemes with simple intuitive averaging of material parameters. They numerically compared the heterogeneous approach with the homogeneous one and showed unacceptable difference between the two approaches in the case of the corner-edge model. Kummer and Behle (1982) followed the approach of Ilan *et al.* (1975) and derived the 2nd-order SH schemes for grid points lying on different types of segments of the step-like polygonal discontinuity. Virieux (1984, 1986) used the velocity-stress formulation and staggered grid introduced to seismology by Madariaga (1976). His P-SV FD scheme did not suffer from stability problems caused by large values of Poisson's ratio, which was the case of all displacement schemes on conventional grids. Since the work by Virieux, the staggered-grid FD schemes became the dominant schemes applied to seismic wave propagation in heterogeneous media. Virieux also discussed the discrepancy between the homogeneous and heterogeneous approaches found by Kelly *et al.* (1976). He found it difficult to explain features of the solution obtained by the homogeneous approach.

An attempt to incorporate internal boundary conditions into a displacement FD scheme was made by Sochacki *et al.* (1991) who integrated the equation of motion over a grid cell that could possibly contain a material discontinuity. The integrated equation of motion was then discretized. Schoenberg and Muir (1989) developed a calculus to replace a stack of thin flat elastic anisotropic homogeneous layers by an equivalent (in the long-wavelength limit) homogeneous anisotropic medium. As a result, they found that stress-strain relation for an averaged medium satisfies the boundary conditions at interfaces. Applying the Schoenberg–Muir calculus simplifies modeling of wave propagation and, at the same time, accounts for transversal anisotropy. Muir *et al.* (1992) applied the Schoenberg–Muir calculus to a grid cell containing material discontinuity, i.e., in general, they treated contents of the cell as a stack of thin flat layers that can be averaged by the Schoenberg–Muir calculus. The two papers did not have the impact on the heterogeneous FD schemes that they deserved—likely because the authors did not make explicit reference to the question of the heterogeneous FD schemes. Zahradník and Priolo (1995) explicitly pointed out the problem of heterogeneous FD schemes. Assuming discontinuous material parameters in the equation of motion they obtained an expression whose dominant term is equivalent to the traction continuity condition. As other developers of the heterogeneous FD schemes they were not aware of the work by Schoenberg and Muir. Graves (1996) intuitively suggested a formula for determination of effective material grid parameters in the 3-D 4th-order velocity-stress staggered-grid schemes and numerically demonstrated the good level of accuracy. Zhang and Symes (1998) developed a 2-D 4th-order full-stencil immersed interface technique to account for a curved material discontinuity. In the first step, all grid points are solved using a standard FD scheme. In the second step, each grid point whose stencil includes grid points from both sides of the discontinuity is recalculated using previous time step's values with a special 25-point scheme determined using local boundary conditions. Moczo et al. (2002) analyzed the 1-D problem in a medium consisting of two halfspaces. They demonstrated their method on a simple physical model of the contact of two media, found a heterogeneous formulation of the equation of motion and Hooke's law, and derived a heterogeneous FD scheme. Finally they analyzed the 3-D problem, suggested a 4th-order heterogeneous staggered-grid FD scheme, and demonstrated its accuracy compared to the standard staggered-grid FD schemes. Their analysis will be followed in the next sections.

3.1.1. *1-D problem*

Consider two elastic halfspaces with a welded interface in the plane $x = 0$. The wave propagation in the halfspaces is described by equations

$$\rho^{\pm}\ddot{u}^{\pm} = \sigma^{\pm}_{,x} + f^{\pm}, \qquad \sigma^{\pm} = C^{\pm}u^{\pm}_{,x} \tag{45}$$

where the superscript $^+$ refers to one halfspace and superscript $^-$ to the other. Either $u(x, t)$ is the x-component of the displacement $\vec{u}(u_x, 0, 0)$, $\sigma(x, t)$ the xx-component of the stress tensor, $f(x, t)$ the x-component of the body force per unit volume $\vec{f}(f_x, 0, 0)$ and $C(x) = \lambda(x) + 2\mu(x)$ in the case of P wave, or $u(x, t)$ is the y-component of the displacement $\vec{u}(0, u_y, 0)$, $\sigma(x, t)$ the xy-component of the stress tensor, $f(x, t)$ the y-component of the body force per unit volume $\vec{f}(0, f_y, 0)$ and $C(x) = \mu(x)$ in the case of the SH wave (the coordinate system can always be rotated so that the S wave could be the SH wave). At the welded interface the continuity of displacement and traction applies: $u^-(0) = u^+(0)$, $\sigma^-(0) = \sigma^+(0)$. Because a heterogeneous FD scheme should be nothing else than a discrete approximation to a differential problem, a heterogeneous formulation of the differential problem is to be found. This means the same form of the equation of motion and Hooke's law for a point at a material discontinuity as a point away from the material discontinuity.

Let $\varphi^{\pm}(x)$, $c^{\pm}(x)$ and $g^{\pm}(x)$ be real functions of a real argument x such that

$$\varphi^{\pm}(x) = c^{\pm}(x)g^{\pm}(x) \tag{46}$$

and

$$\varphi^-(0) = \varphi^+(0). \tag{47}$$

Functions $c^{\pm}(x)$ and $g^{\pm}(x)$ may have discontinuities of the first order at $x = 0$. Define

$$\bar{g}(0) = 0.5 \cdot \left[g^-(0) + g^+(0)\right]. \tag{48}$$

Then it follows that

$$\varphi^-(0) = \varphi^+(0) = \bar{c}(0)\bar{g}(0), \quad \bar{c}(0) = 2/[1/c^-(0) + 1/c^+(0)]. \qquad (49)$$

If $c^\pm(x) = 1/r^\pm(x)$, then

$$\varphi^-(0) = \varphi^+(0) = \frac{1}{\bar{r}(0)}\bar{g}(0), \quad \bar{r}(0) = 0.5 \cdot [r^-(0) + r^+(0)]. \qquad (50)$$

It follows from Eqs. (46) to (50) that the equation of motion and Hooke's law for a point at the material discontinuity have the form

$$\bar{\rho}(0)\ddot{u}(0) = \overline{\sigma_{,x}}(0) + \bar{f}(0) \qquad (51)$$

and

$$\sigma(0) = \bar{C}(0)\overline{u_{,x}}(0), \qquad (52)$$

respectively, with a density equal to the arithmetic average of the densities in the two halfspaces, and elastic modulus equal to the harmonic average of the moduli in the two halfspaces:

$$\bar{\rho}(0) = 0.5 \cdot [\rho^-(0) + \rho^+(0)],$$
$$\bar{C}(0) = 2/[1/C^-(0) + 1/C^+(0)]. \qquad (53)$$

The average spatial derivatives of the stress and displacement are

$$\overline{\sigma_{,x}}(0) + \bar{f}(0) = 0.5 \cdot [\sigma_{,x}^-(0) + \sigma_{,x}^+(0) + f^-(0) + f^+(0)],$$
$$\overline{u_{,x}}(0) = 0.5 \cdot [u_{,x}^-(0) + u_{,x}^+(0)]. \qquad (54)$$

It is obvious that Eqs. (51) and (52) for a point at the material discontinuity have the same form as the equation of motion and Hooke's law (45), at a point away from the material discontinuity. This provides a basis for the heterogeneous 1-D FD scheme.

Moczo et al. (2002) also showed that two Hooke elements (elastic springs) connected in series make an appropriate rheological model for considering traction continuity at the welded interface of two elastic materials.

Consider, for example, the velocity-stress formulation. Let V_I^m, T_I^m and F_I^m be the discrete approximations to particle velocity $v_I^m = v(Ih, m\Delta t)$, stress $\sigma_I^m = \sigma(Ih, m\Delta t)$ and body force $f_I^m = f(Ih, m\Delta t)$. One possible heterogeneous 4th-order staggered-grid FD scheme for the 1-D problem is

$$T_{I+1/2}^m = T_{I+1/2}^{m-1} + C_{I+1/2}^H \frac{\Delta t}{h}[a(V_{I+2}^{m-1/2} - V_{I-1}^{m-1/2})$$
$$+ b(V_{I+1}^{m-1/2} - V_I^{m-1/2})],$$
$$V_I^{m+1/2} = V_I^{m-1/2} + \frac{1}{\rho_I^A}\frac{\Delta t}{h}[a(T_{I+3/2}^m - T_{I-3/2}^m) + b(T_{I+1/2}^m - T_{I-1/2}^m)]$$
$$+ \frac{\Delta t}{\rho_I^A}F_I^m \qquad (55)$$

with

$$\rho_I^A = \frac{1}{h} \int_{x_{I-1/2}}^{x_{I+1/2}} \rho(x)\,dx, \qquad C_{I+1/2}^H = \left[\frac{1}{h}\int_{x_I}^{x_{I+1}} \frac{1}{C(x)}\,dx\right]^{-1}. \tag{56}$$

The coefficients in Eqs. (55) are $a = -1/24$ and $b = 9/8$. In the scheme, the averaging of the spatial derivatives of the functions at the material discontinuity is neglected while the harmonic averaging of the elastic moduli and arithmetic averaging of densities at the material discontinuity is taken into account (see Moczo et al., 2002 for details). In general, the integrals are evaluated numerically. It is easy to check that the scheme yields very good accuracy in smoothly and/or discontinuously heterogeneous media. The scheme is capable to sense a true position of the material discontinuity no matter what the position of the discontinuity is with respect to the grid points.

While Tikhonov and Samarski (e.g., Mitchell, 1969, p. 23) obtained the harmonic averaging as a result of the mathematical 'trick' to avoid spatial derivatives of the coefficients in the 2nd-order displacement formulation, the harmonic average in the heterogeneous formulation (52) is due to traction-continuity condition at the material discontinuity.

3.1.2. 3-D problem

Define stress and strain vectors

$$\vec{\sigma} = [\sigma_{xx}, \sigma_{yy}, \sigma_{zz}, \sigma_{xy}, \sigma_{yz}, \sigma_{zx}]^T,$$
$$\vec{\varepsilon} = [\varepsilon_{xx}, \varepsilon_{yy}, \varepsilon_{zz}, \varepsilon_{xy}, \varepsilon_{yz}, \varepsilon_{zx}]^T. \tag{57}$$

Hooke's law for an isotropic medium can be written as

$$\vec{\sigma} = \mathbf{E}\vec{\varepsilon} \tag{58}$$

where

$$\mathbf{E} = \begin{bmatrix} \kappa + \tfrac{4}{3}\mu & \kappa - \tfrac{2}{3}\mu & \kappa - \tfrac{2}{3}\mu & 0 & 0 & 0 \\ \kappa - \tfrac{2}{3}\mu & \kappa + \tfrac{4}{3}\mu & \kappa - \tfrac{2}{3}\mu & 0 & 0 & 0 \\ \kappa - \tfrac{2}{3}\mu & \kappa - \tfrac{2}{3}\mu & \kappa + \tfrac{4}{3}\mu & 0 & 0 & 0 \\ 0 & 0 & 0 & 2\mu & 0 & 0 \\ 0 & 0 & 0 & 0 & 2\mu & 0 \\ 0 & 0 & 0 & 0 & 0 & 2\mu \end{bmatrix} \tag{59}$$

is the elasticity matrix. Let moduli κ and μ have a discontinuity of the first-order across a surface S with normal vector $\vec{n}$. The surface S defines the geometry of the material discontinuity (interface). The welded-interface boundary conditions are continuity of displacement $\vec{u}(\vec{\eta})$ and traction $\vec{T}(\vec{\eta}, \vec{n})$ across the surface:

$$\vec{u}^+(\vec{\eta}) = \vec{u}^-(\vec{\eta}), \qquad \vec{T}^+(\vec{\eta}, \vec{n}) = \vec{T}^-(\vec{\eta}, \vec{n}). \tag{60}$$

For simplicity, consider first the planar surface S parallel to the xy-coordinate plane with a normal vector $\vec{n} = (0, 0, 1)$. The conditions (60) imply

$$\sigma_{zx}^+ = \sigma_{zx}^-, \qquad \sigma_{zy}^+ = \sigma_{zy}^-, \qquad \sigma_{zz}^+ = \sigma_{zz}^-,$$
$$\varepsilon_{xx}^+ = \varepsilon_{xx}^-, \qquad \varepsilon_{yy}^+ = \varepsilon_{yy}^-, \qquad \varepsilon_{xy}^+ = \varepsilon_{xy}^-. \tag{61}$$

The components $\sigma_{xx}, \sigma_{yy}, \sigma_{xy}, \varepsilon_{zx}, \varepsilon_{zy}$ and ε_{zz} may be discontinuous across the material discontinuity. Define averaged stress and strain vectors at the material discontinuity:

$$\vec{\sigma}^A = \frac{1}{2}(\vec{\sigma}^+ + \vec{\sigma}^-), \qquad \vec{\varepsilon}^A = \frac{1}{2}(\vec{\varepsilon}^+ + \vec{\varepsilon}^-). \tag{62}$$

Due to the boundary conditions,

$$\vec{\sigma}^A = \left[\sigma_{xx}^A, \sigma_{yy}^A, \sigma_{zz}, \sigma_{xy}^A, \sigma_{yz}, \sigma_{zx}\right]^T,$$
$$\vec{\varepsilon}^A = \left[\varepsilon_{xx}, \varepsilon_{yy}, \varepsilon_{zz}^A, \varepsilon_{xy}, \varepsilon_{yz}^A, \varepsilon_{zx}^A\right]^T. \tag{63}$$

Then Hooke's law for a point on the material discontinuity is

$$\vec{\sigma}^A = \tilde{\mathbf{E}} \vec{\varepsilon}^A \tag{64}$$

with the elasticity matrix

$$\tilde{\mathbf{E}} = \begin{bmatrix} \Lambda + 2\mu^A & \Lambda & \Psi & 0 & 0 & 0 \\ \Lambda & \Lambda + 2\mu^A & \Psi & 0 & 0 & 0 \\ \Psi & \Psi & \left(\kappa + \frac{4}{3}\mu\right)^H & 0 & 0 & 0 \\ 0 & 0 & 0 & 2\mu^A & 0 & 0 \\ 0 & 0 & 0 & 0 & 2\mu^H & 0 \\ 0 & 0 & 0 & 0 & 0 & 2\mu^H \end{bmatrix} \tag{65}$$

and

$$\Lambda = \left[\left(\frac{\kappa - \frac{2}{3}\mu}{\kappa + \frac{4}{3}\mu}\right)^A\right]^2 \cdot \left(\kappa + \frac{4}{3}\mu\right)^H + 2 \cdot \left(\frac{(\kappa - \frac{2}{3}\mu)\mu}{\kappa + \frac{4}{3}\mu}\right)^A,$$
$$\Psi = \left(\frac{\kappa - \frac{2}{3}\mu}{\kappa + \frac{4}{3}\mu}\right)^A \cdot \left(\kappa + \frac{4}{3}\mu\right)^H. \tag{66}$$

Superscripts A and H denote arithmetic and harmonic averages, respectively.

Relation (64) means that for a point on the material discontinuity it is possible to find the same form of Hooke's law as for a point inside a homogeneous or smoothly heterogeneous medium, given by Eq. (58). Considering the point on the discontinuity as a point of the averaged medium characterized by matrix $\tilde{\mathbf{E}}$ assures traction continuity at the point. There is, however, an important difference between laws (64) and (58). The matrix $\tilde{\mathbf{E}}$ for the averaged medium given by Eq. (65) has 5 independent nonzero elements and the averaged medium is transversely isotropic. Matrix **E** for any of the two isotropic media in contact has only 2 independent nonzero elements. This means that the exact heterogeneous formulation for a planar material discontinuity parallel with a coordinate plane increases the number of elastic coefficients necessary to describe the medium.

Next let us consider a planar material discontinuity in a general position in a Cartesian coordinate system. Let the normal vector be $\vec{n} = (n_x, n_y, n_z)$ with all non-zero elements. Find a Cartesian coordinate system $x'y'z'$ in which $\vec{n}$ is parallel to the z'-axis. Then it is possible to find a matrix $\tilde{\mathbf{E}}'$ with 5 independent non-zero elements. Transforming the matrix $\tilde{\mathbf{E}}'$ into a matrix $\tilde{\mathbf{E}}$ in the original coordinate system xyz yields a symmetric elasticity matrix $\tilde{\mathbf{E}}$ which has, in general, all elements non-zero (5 being independent). This means that all strain-tensor components are necessary to calculate each stress-tensor component at a point of the interface (not the case with the standard staggered grid), and 21 non-zero elastic coefficients are necessary at the point.

If the geometry of a material discontinuity is defined by a non-planar smooth surface S, the surface may be locally approximated by a planar surface tangential to surface S at a given point.

It is obvious that finding a heterogeneous formulation of the differential problem as a basis for a heterogeneous FD scheme for a medium with a material discontinuity in a general 3-D problem is far more complicated than that in the 1-D problem. A non-simplified treatment and a corresponding FD scheme would lead to a substantial increase in memory requirement.

Therefore, Moczo et al. (2002) suggested a simplified approach: (a) They wanted to keep the structure, number of operations and memory requirements of the standard 4th-order staggered-grid scheme. (b) At the same time, they chose to determine an effective grid elastic modulus (κ or μ) at each grid position of the stress-tensor components as volume harmonic average of the modulus within a volume of the grid cell centered at the grid position. The latter choice was based on the fact that harmonic averaging is exact in the 1-D case, see Eq. (49), and is a part of exact averaging in the 3-D case, see Eqs. (64)–(66).

At each position of the displacement or particle-velocity component an effective grid density is determined as a volume arithmetic average of density within a volume of the grid cell centered at the grid position. The averaging applies to both smoothly and discontinuously heterogeneous media. The averages are evaluated by numerical integration.

Let $T^{xx,m}_{I+1/2,J+1/2,K+1/2}$ be the discrete approximation to the stress-tensor component $\sigma_{xx}[(I+1/2)h, (J+1/2)h, (K+1/2)h, m\Delta t]$. Similarly, let $V^{x,m}_{I,J+1/2,K+1/2}$, $V^{y,m}_{I+1/2,J,K+1/2}$, $V^{z,m}_{I+1/2,J+1/2,K}$ and $F^{x,m}_{I,J+1/2,K+1/2}$ be discrete approximations to the particle-velocity and body-force components. Examples of the FD schemes for the stress-tensor and particle-velocity components σ_{xx} and v_x are

$$\begin{aligned}
T^{xx,m}_{I+1/2,J+1/2,K+1/2} &= T^{xx,m-1}_{I+1/2,J+1/2,K+1/2} \\
&+ \frac{\Delta t}{h}\bigg\{\left(\kappa^H_{I+1/2,J+1/2,K+1/2} + \frac{4}{3}\mu^H_{I+1/2,J+1/2,K+1/2}\right) \\
&\quad\times [a(V^{x,m-1/2}_{I+2,J+1/2,K+1/2} - V^{x,m-1/2}_{I-1,J+1/2,K+1/2}) \\
&\quad+ b(V^{x,m-1/2}_{I+1,J+1/2,K+1/2} - V^{x,m-1/2}_{I,J+1/2,K+1/2})] \\
&+ \left(\kappa^H_{I+1/2,J+1/2,K+1/2} - \frac{2}{3}\mu^H_{I+1/2,J+1/2,K+1/2}\right) \\
&\quad\times [a(V^{y,m-1/2}_{I+1/2,J+2,K+1/2} - V^{y,m-1/2}_{I+1/2,J-1,K+1/2}) \\
&\quad+ b(V^{y,m-1/2}_{I+1/2,J+1,K+1/2} - V^{y,m-1/2}_{I+1/2,J,K+1/2}) \\
&\quad+ a(V^{z,m-1/2}_{I+1/2,J+1/2,K+2} - V^{z,m-1/2}_{I+1/2,J+1/2,K-1}) \\
&\quad+ b(V^{z,m-1/2}_{I+1/2,J+1/2,K+1} - V^{z,m-1/2}_{I+1/2,J+1/2,K})]\bigg\}
\end{aligned} \quad (67)$$

with

$$\kappa^H_{I+1/2,J+1/2,K+1/2} = \left[\frac{1}{h^3}\int_{x_I}^{x_{I+1}}\int_{y_J}^{y_{J+1}}\int_{z_K}^{z_{K+1}}\frac{1}{\kappa}\,dx\,dy\,dz\right]^{-1}, \quad (68)$$

$$\mu^H_{I+1/2,J+1/2,K+1/2} = \left[\frac{1}{h^3}\int_{x_I}^{x_{I+1}}\int_{y_J}^{y_{J+1}}\int_{z_K}^{z_{K+1}}\frac{1}{\mu}\,dx\,dy\,dz\right]^{-1} \quad (69)$$

and

$$\begin{aligned}
V^{x,m+1/2}_{I,J+1/2,K+1/2} &= V^{x,m-1/2}_{I,J+1/2,K+1/2} + [\Delta t/(h\rho^A_{I,J+1/2,K+1/2})] \\
&\times [a(T^{xx,m}_{I+3/2,J+1/2,K+1/2} - T^{xx,m}_{I-3/2,J+1/2,K+1/2}) \\
&+ b(T^{xx,m}_{I+1/2,J+1/2,K+1/2} - T^{xx,m}_{I-1/2,J+1/2,K+1/2}) \\
&+ a(T^{xy,m}_{I,J+2,K+1/2} - T^{xy,m}_{I,J-1,K+1/2}) + b(T^{xy,m}_{I,J+1,K+1/2} - T^{xy,m}_{I,J,K+1/2}) \\
&+ a(T^{zx,m}_{I,J+1/2,K+2} - T^{zx,m}_{I,J+1/2,K-1}) + b(T^{zx,m}_{I,J+1/2,K+1} - T^{zx,m}_{I,J+1/2,K})] \\
&+ (\Delta t/\rho^A_{I,J+1/2,K+1/2})F^{x,m}_{I,J+1/2,K+1/2}
\end{aligned} \quad (70)$$

with

$$\rho^A_{I,J+1/2,K+1/2} = \frac{1}{h^3}\int_{x_{I-1/2}}^{x_{I+1/2}}\int_{y_J}^{y_{J+1}}\int_{z_K}^{z_{K+1}}\rho\,dx\,dy\,dz. \quad (71)$$

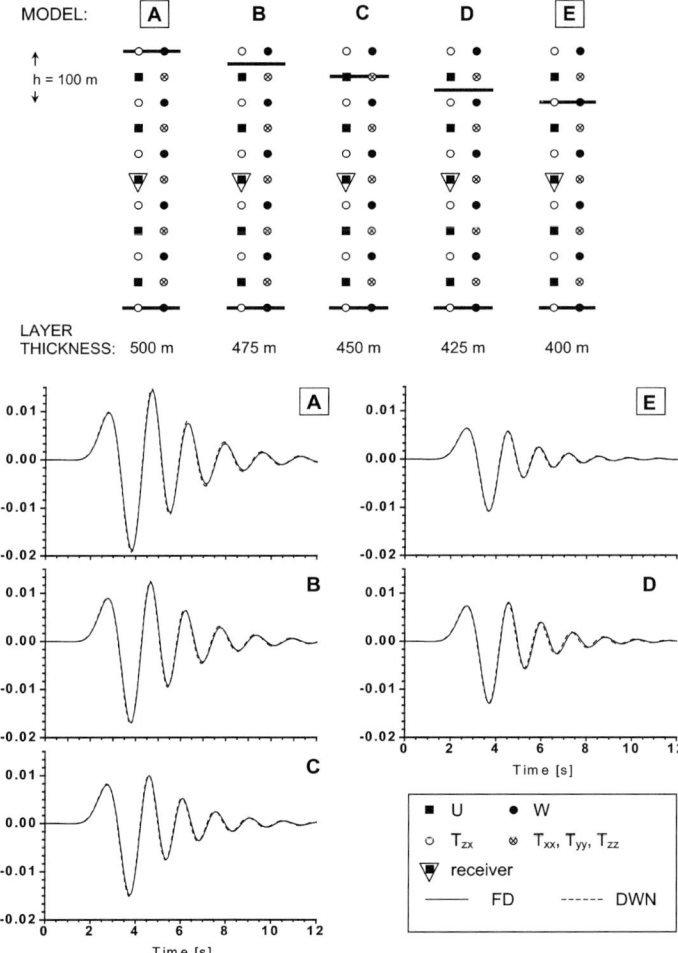

FIG. 2. The upper part: Positions of the upper and lower layer-halfspace interfaces in five models of a layer (P-wave and S-wave velocities 1125 m/s and 625 m/s) between two identical halfspaces (P-wave and S-wave velocities 5468 m/s and 3126 m/s)—shown schematically in one vertical grid plane. The five models differ from each other by position of the upper layer-halfspace interface in the spatial grid (the same grid for all models) and thus by the layer thickness. The lower part: Comparison of our FD and DWN synthetics for the five models. Note very good accuracy of the FD synthetics for any position of the layer-halfspace interface with respect to the spatial grid. Also note considerable differences between synthetics due to variations in the layer thickness that is smaller than one grid spacing. Reproduced from Moczo et al. (2002).

Moczo et al. (2002) used a set of models to numerically test the scheme. The tests demonstrated its very good numerical accuracy. Here we illustrate an important property of the scheme—the capability to sense the position of a material discontinuity regardless of its position with respect to the spatial grid. Five different models of a single horizontal homogeneous layer located in between two homogeneous halfspaces differ from each other by the thickness of the layer. The spatial grid is one and the same in all five models (see the upper part of Fig. 2). A double-couple point source was located in the lower halfspace. The FD synthetics (u_x component) are compared with those calculated by the discrete wavenumber (DWN) method (Bouchon, 1981; computer code Axitra by Coutant, 1989), see the lower part of Fig. 2. The FD and DWN synthetics agree very well regardless of the position of the upper layer-halfspace interface with respect to the spatial grid. It is also clear from Fig. 2 that differences in thickness of the layer—smaller than one grid spacing—cause considerable changes in seismic motion. This often is underestimated by many modelers who consider the size of one grid cell as "atom of resolution" within which a FD scheme cannot see differences. This and other examples given by Moczo et al. (2002) clearly show that this is not the case if the scheme is sufficiently accurate.

3.2. Incorporation of the Realistic Attenuation

3.2.1. Stress-Strain Relation in Viscoelastic Medium—the 1-D Case

The behavior of real Earth's material can be described as a combination of elastic solids and viscous fluids. The stress-strain relation therefore also should depend on time. The rheology of a viscoelastic medium seems appropriate for quantitative description of seismic wave propagation. Observations show that the internal friction in the Earth is nearly constant over the seismic frequency range, e.g. McDonal et al. (1958), Liu et al. (1976), Spencer (1981), Murphy (1982).

The stress-strain relation in a linear isotropic viscoelastic material is given by the Boltzmann superposition principle. For the 1-D problem this is

$$\sigma(t) = \int_{-\infty}^{t} \psi(t - \tau)\dot{\varepsilon}(\tau)\,d\tau, \tag{72}$$

where $\sigma(t)$ is the stress, $\dot{\varepsilon}(t)$ the time derivative of the strain, and $\psi(t)$ the stress relaxation function defined as a stress response to a Heaviside unit step function in strain. According to Eq. (72), the stress at a given time t is determined by the entire history of the strain until time t. The integral in Eq. (72) represents a time convolution of the relaxation function and the strain rate. Using the symbol $*$ for the convolution, Eq. (72) can be written as

$$\sigma(t) = \psi(t) * \dot{\varepsilon}(t). \tag{73}$$

It follows from the definition of the relaxation function that its time derivative

$$M(t) = \dot{\psi}(t) \tag{74}$$

is the stress response to the Dirac δ-function in strain and that

$$\sigma(t) = M(t) * \varepsilon(t). \tag{75}$$

Let $\mathcal{F}$ and $\mathcal{F}^{-1}$ denote the direct and inverse Fourier transforms

$$\mathcal{F}\{x(t)\} = \int_{-\infty}^{\infty} x(t) \exp(-i\omega t)\, dt,$$

$$\mathcal{F}^{-1}\{X(\omega)\} = \frac{1}{2\pi} \int_{-\infty}^{\infty} X(\omega) \exp(i\omega t)\, d\omega,$$

where ω is the angular frequency. Relation (75) can be Fourier-transformed into the frequency domain:

$$\sigma(\omega) = M(\omega) \cdot \varepsilon(\omega). \tag{76}$$

In general, $M(\omega)$ is a complex, frequency-dependent viscoelastic modulus. Due to properties of the Fourier transform

$$\psi(t) = \mathcal{F}^{-1}\left\{\frac{M(\omega)}{i\omega}\right\}. \tag{77}$$

Equation (76) indicates the correspondence principle in the linear theory of viscoelasticity—in the frequency-domain, relations for the viscoelastic medium are obtained by replacing real frequency-independent moduli by complex, frequency-dependent quantities. Thus, the incorporation of the attenuation into the frequency-domain computations is much easier than that in the time domain.

The time derivative of the stress is

$$\dot{\sigma}(t) = M(t) * \dot{\varepsilon}(t). \tag{78}$$

An instantaneous elastic response of the viscoelastic material is given by the so-called unrelaxed modulus M_U, a long-term equilibrium response is given by the relaxed modulus M_R:

$$M_U = \lim_{t \to 0} \psi(t) = \lim_{\omega \to \infty} M(\omega), \quad M_R = \lim_{t \to \infty} \psi(t) = \lim_{\omega \to 0} M(\omega). \tag{79}$$

The modulus defect or relaxation of modulus is

$$\delta M = M_U - M_R. \tag{80}$$

Given the viscoelastic modulus, the quality factor $Q(\omega)$ is defined as

$$Q(\omega) = \operatorname{Re} M(\omega) / \operatorname{Im} M(\omega). \tag{81}$$

Due to large computer time and memory requirements, the stress-strain relation (72) in some cases only allowed simplified $Q(\omega)$ laws, e.g. linear $Q(\omega)$.

3.2.2. Conversion of the Convolutory Stress-Strain Relation into a Differential Form

If $M(\omega)$ is a rational function, the inverse Fourier transform of Eq. (76) yields the nth-order differential equation for $\sigma(t)$, which can be numerically solved much more easily than the convolution integral. Day and Minster (1984) assumed that, in general, the viscoelastic modulus is not a rational function. Therefore they suggested approximating a viscoelastic modulus by an nth-order rational function and determining its coefficients by the Padé approximant method. They obtained n ordinary differential equations for n additional internal variables, which replace the convolution integral. The sum of the internal variables multiplied by the unrelaxed modulus gives an additional viscoelastic term to the elastic stress. The work of Day and Minster not only developed one particular approach but, in fact, indirectly suggested the future evolution—a direct use of the rheological models whose $M(\omega)$ is a rational function of $i\omega$. In response to work by Day and Minster (1984); Emmerich and Korn (1987) realized that an acceptable relaxation function corresponds to a rheology of what they defined as the generalized Maxwell body—n Maxwell bodies and one Hooke element (elastic spring) connected in parallel; Fig. 3. Because in the rheological literature the generalized Maxwell body is defined without the additional single Hooke element, the abbreviation GMB-EK will hereafter be used for the model defined by Emmerich and Korn. Because the viscoelastic modulus of the GMB-EK has a form of a rational function, Emmerich and Korn (1987) obtained similar differential equations as Day

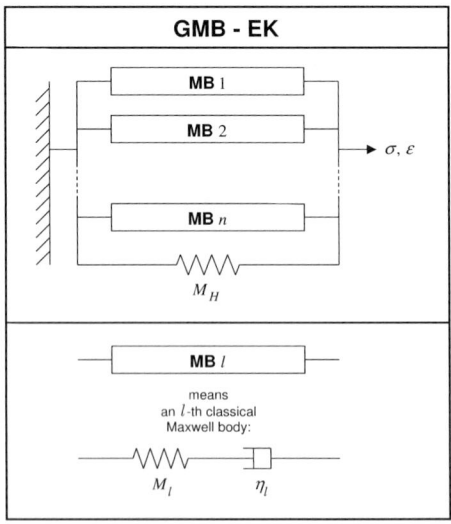

FIG. 3. Rheological model of the Generalized Maxwell Body (GMB-EK) defined by Emmerich and Korn (1987). M_H and M_l denote elastic moduli, η_l viscosity.

and Minster (1984). In order to fit an arbitrary $Q(\omega)$ law they chose the relaxation frequencies logarithmically equidistant over a desired frequency range and used the least-square method to determine weight factors of the relaxation mechanisms (classical Maxwell bodies). Emmerich and Korn (1987) demonstrated that their approach is better than the approach based on the Padé approximant method both in accuracy and computational efficiency. Independently, Carcione *et al.* (1988a, 1988b), in accordance with the approach by Liu *et al.* (1976), assumed the generalized Zener body (GZB)—n Zener bodies (ZB, standard linear bodies), connected in parallel; Fig. 4. Carcione *et al.* (1988a, 1988b) developed a theory for the GZB and introduced the term memory variables for the obtained additional variables.

After publications by Emmerich and Korn (1987) and Carcione *et al.* (1988a, 1988b) different authors chose to use either the GMB-EK (for example, Emmerich, 1992; Fäh, 1992; Moczo and Bard, 1993; Moczo *et al.*, 1997; Kay and Krebes, 1999) or GZB (for example, Robertsson *et al.*, 1994; Blanch *et al.*, 1995;

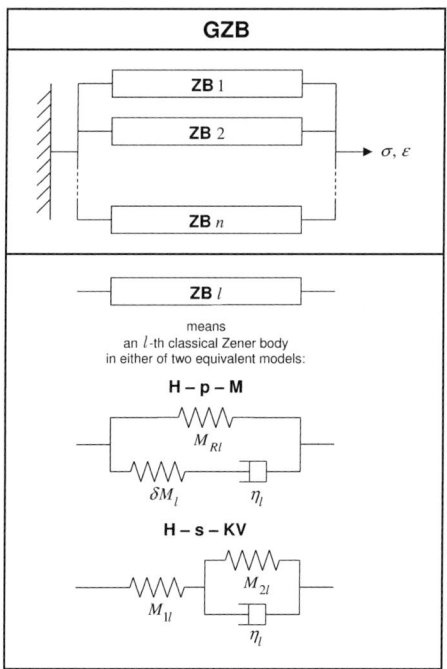

FIG. 4. Rheological model of the Generalized Zener Body (GZB). For a classical Zener body (standard linear body) there are two equivalent models: H-p-M, that is, Hooke element connected in parallel with Maxwell body, and H-s-KV, that is, Hooke element connected in series with Kelvin–Voigt body. In the H-p-M model it is easier to recognize the relaxed modulus M_{Rl} and modulus defect δM_l. M_{1l} and M_{2l} in the H-s-KV model denote elastic moduli. In both models η_l stands for viscosity.

Xu and McMechan, 1995; Robertsson, 1996; Hestholm, 1999). In both cases the authors followed the corresponding mathematical formalisms. Moczo *et al.* (1997) applied the GMB-EK approach also in the finite-element method and hybrid FD-finite-element method. Emmerich and Korn (1987), Emmerich (1992), Fäh (1992), and Moczo and Bard (1993) defined one memory variable for one displacement component. Robertsson *et al.* (1994) introduced the memory variables based on the GZB rheology into the staggered-grid velocity-stress FD scheme. Blanch *et al.* (1995) suggested an approximate single-parameter method, τ-method, to approximate the constant $Q(\omega)$ law. Xu and McMechan (1998) used simulated annealing for determining a best combination of relaxation mechanisms to approximate a desired $Q(\omega)$ law.

There appears to have been no or little comments by the authors using the GZB on the rheology of the GMB-EK and the corresponding algorithms, and vice versa. Thus, two parallel sets of publications and algorithms had been developed during years. Therefore, Moczo and Kristek (2005) addressed this development and showed relation between the two rheologies.

3.2.3. Rheologies of the GMB-EK and GZB

There are simple rules in the time and frequency domains for the mathematical representation of the linear rheological models consisting of Hooke and Stokes elements (springs and dashpots) connected in parallel or series. In the frequency domain, the stress-strain relations for the Hooke and Stokes elements are $\sigma(\omega) = M \cdot \varepsilon(\omega)$ and $\sigma(\omega) = i\omega\eta \cdot \varepsilon(\omega)$, respectively, where M is the elastic modulus, and η viscosity. If two elements are connected in series, stresses are equal while strains additive. If two elements are connected in parallel, stresses are additive while strains are equal.

The application of the frequency-domain rules to the GMB-EK yields

$$M(\omega) = M_H + \sum_{l=1}^{n} \frac{i M_l \omega}{\omega_l + i\omega}, \quad \omega_l = \frac{M_l}{\eta_l}, \ l = 1, \ldots, n, \tag{82}$$

where ω_l is a relaxation frequency. The relaxed and unrelaxed moduli are

$$M_R \equiv \lim_{\omega \to 0} M(\omega) = M_H, \quad M_U \equiv \lim_{\omega \to \infty} M(\omega) = M_R + \sum_{l=1}^{n} M_l. \tag{83}$$

Since $M_U = M_R + \delta M$, we get $M_l = \delta M_l$, and it is possible to assume

$$\delta M_l = a_l \delta M, \quad \sum_{l=1}^{n} a_l = 1 \tag{84}$$

without any simplification. Then

$$M(\omega) = M_R + \delta M \sum_{l=1}^{n} \frac{i a_l \omega}{\omega_l + i\omega}. \tag{85}$$

Using relation (77) it is straightforward to obtain the relaxation function

$$\psi(t) = \left[M_R + \delta M \sum_{l=1}^{n} a_l e^{-\omega_l t} \right] \cdot H(t), \tag{86}$$

where $H(t)$ is the Heaviside unit step function. The above formulae were presented by Emmerich and Korn (1987).

From the two equivalent models of the ZB, shown in Fig. 4, we choose the H-p-M type to obtain $M(\omega)$. This is because it is easy to recognize the relaxed modulus and modulus defect in the ZB. The application of the frequency-domain rules to the GZB results in

$$M(\omega) = \sum_{l=1}^{n} M_{Rl} \frac{1 + i\tau_{\varepsilon l}\omega}{1 + i\tau_{\sigma l}\omega} \tag{87}$$

with relaxation times

$$\tau_{\varepsilon l} = \frac{\eta_l}{\delta M_l} \frac{M_{Ul}}{M_{Rl}}, \qquad \tau_{\sigma l} = \frac{\eta_l}{\delta M_l}, \qquad \frac{\tau_{\varepsilon l}}{\tau_{\sigma l}} = \frac{M_{Ul}}{M_{Rl}} \tag{88}$$

and

$$M_{Ul} = M_{Rl} + \delta M_l. \tag{89}$$

The unrelaxed and relaxed moduli are

$$M_R = \sum_{l=1}^{n} M_{Rl}, \qquad M_U = \sum_{l=1}^{n} M_{Rl} \frac{\tau_{\varepsilon l}}{\tau_{\sigma l}} = M_R + \sum_{l=1}^{n} \delta M_l. \tag{90}$$

Relations (77) and (87) yield the relaxation function

$$\psi(t) = \left\{ \sum_{l=1}^{n} M_{Rl} \left[1 - \left(1 - \frac{\tau_{\varepsilon l}}{\tau_{\sigma l}} \right) \exp\left(-\frac{t}{\tau_{\sigma l}} \right) \right] \right\} \cdot H(t). \tag{91}$$

Assumption of simplification (Carcione, 2001)

$$M_{Rl} = \frac{1}{n} M_R \tag{92}$$

leads to

$$M(\omega) = \frac{M_R}{n} \sum_{l=1}^{n} \frac{1 + i\tau_{\varepsilon l}\omega}{1 + i\tau_{\sigma l}\omega},$$

$$\psi(t) = M_R \left[1 - \frac{1}{n} \sum_{l=1}^{n} \left(1 - \frac{\tau_{\varepsilon l}}{\tau_{\sigma l}} \right) \exp\left(-\frac{t}{\tau_{\sigma l}} \right) \right] \cdot H(t). \qquad (93)$$

Formulae (92) and (93) were presented by Carcione (2001). Unfortunately, all papers dealing with the incorporation of the attenuation based on the GZB, starting from Liu *et al.* (1976) until now, despite the book by Carcione (2001), have the same error—the missing factor $1/n$ in the viscoelastic modulus and relaxation function.

3.2.4. The Relation Between the GZB and GMB-EK

Following Moczo and Kristek (2005), consider again the ZB (H-p-M) model. The application of the frequency-domain rules to the *l*th ZB yields

$$\sigma_l(\omega) \cdot \left(\frac{1}{\delta M_l} + \frac{1}{i\eta_l \omega} \right) = \left(1 + \frac{M_{Rl}}{\delta M_l} + \frac{M_{Rl}}{i\eta_l \omega} \right) \cdot \varepsilon(\omega). \qquad (94)$$

Defining

$$\omega_l = \frac{\delta M_l}{\eta_l} \qquad (95)$$

and rearranging Eq. (94) gives

$$\sigma_l(\omega) = M_l(\omega) \cdot \varepsilon(\omega), \quad M_l(\omega) = M_{Rl} + \frac{i\delta M_l \omega}{\omega_l + i\omega}. \qquad (96)$$

For n ZB connected in parallel, that is, for the GZB (Fig. 4), the stress is

$$\sigma(\omega) = \sum_{l=1}^{n} \sigma_l(\omega) = \left[\sum_{l=1}^{n} M_l(\omega) \right] \cdot \varepsilon(\omega) \qquad (97)$$

and thus

$$M(\omega) = \sum_{l=1}^{n} M_{Rl} + \sum_{l=1}^{n} \frac{i\delta M_l \omega}{\omega_l + i\omega}. \qquad (98)$$

Since

$$M_R = \sum_{l=1}^{n} M_{Rl}, \quad M_U = M_R + \sum_{l=1}^{n} \delta M_l, \quad M_U = M_R + \delta M, \qquad (99)$$

it is possible to define

$$\delta M_l = a_l \delta M, \quad \sum_{l=1}^{n} a_l = 1 \qquad (100)$$

without loss of generality and obtain

$$M(\omega) = M_R + \delta M \sum_{l=1}^{n} \frac{i a_l \omega}{\omega_l + i\omega}. \tag{101}$$

The viscoelastic modulus (101) obtained for the GZB (H-p-M), Fig. 4, is exactly the same as it has been obtained by Emmerich and Korn (1987) for their GMB-EK, Fig. 3. Obviously, $M(\omega)$ for the GZB (H-s-KV) would be the same. It is also easy to rewrite the non-simplified $\psi(t)$ for the GZB, Eq. (91), into the form of $\psi(t)$ for the GMB-EK, Eq. (86), without any simplification. In other words, the rheology of the GMB-EK and GZB is one and the same. As a consequence, the GMB-EK will be used in the following.

3.2.5. Introduction of the Anelastic Functions (Memory Variables)—the 1-D Case

In order to focus on the essential aspects of the implementation of the realistic attenuation in the time-domain computations, we continue by considering the 1-D case with one stress and one strain component. Using the unrelaxed modulus, the viscoelastic modulus (101) and relaxation function (86) are rewritten as

$$M(\omega) = M_U - \delta M \sum_{l=1}^{n} \frac{a_l \omega_l}{\omega_l + i\omega},$$

$$\psi(t) = \left[M_U - \delta M \sum_{l=1}^{n} a_l \left(1 - e^{-\omega_l t} \right) \right] \cdot H(t). \tag{102}$$

The time derivative of the relaxation function (time-dependent modulus) is

$$M(t) = \dot{\psi}(t)$$
$$= -\delta M \sum_{l=1}^{n} a_l \omega_l e^{-\omega_l t} \cdot H(t)$$
$$+ \left[M_U - \delta M \sum_{l=1}^{n} a_l \left(1 - e^{-\omega_l t} \right) \right] \cdot \delta(t). \tag{103}$$

Inserting Eq. (103) into Eq. (75) yields

$$\sigma(t) = M_U \cdot \varepsilon(t) - \delta M \sum_{l=1}^{n} a_l \omega_l \int_{-\infty}^{t} \varepsilon(\tau) \cdot e^{-\omega_l (t-\tau)} \, d\tau. \tag{104}$$

The convolution integral can be replaced by additional functions (internal variables, new variables, memory variables, anelastic functions). While Day and Minster (1984), Emmerich and Korn (1987), and Carcione *et al.* (1988a, 1988b) defined the additional functions as material-dependent, Kristek and Moczo (2003) defined their anelastic functions as material-independent (the reason will be explained later):

$$\zeta_l(t) = \omega_l \int_{-\infty}^{t} \varepsilon(\tau) \cdot e^{-\omega_l(t-\tau)} \, d\tau, \quad l = 1, \ldots, n. \tag{105}$$

The stress-strain relation then becomes

$$\sigma(t) = M_U \cdot \varepsilon(t) - \sum_{l=1}^{n} \delta M_{al} \zeta_l(t). \tag{106}$$

Equations necessary to solve for the anelastic functions are easily obtained by taking the time derivative of Eq. (105):

$$\dot\zeta_l(t) = \omega_l \frac{d}{dt} \int_{-\infty}^{t} \varepsilon(\tau) \cdot e^{-\omega_l(t-\tau)} \, d\tau = \omega_l \bigl[-\zeta_l(t) + \varepsilon(t)\bigr] \tag{107}$$

and

$$\dot\zeta_l(t) + \omega_l \zeta_l(t) = \omega_l \varepsilon(t), \quad l = 1, \ldots, n. \tag{108}$$

Equations (106) and (108) define the time-domain stress-strain relation for the viscoelastic medium whose rheology corresponds to the rheology of the GMB-EK (and to its equivalent—the GZB). If the staggered-grid velocity-stress FD scheme is to be used, then the time derivative of the stress is needed. It is easy to obtain

$$\dot\sigma(t) = M_U \cdot \dot\varepsilon(t) - \sum_{l=1}^{n} \delta M_{al} \xi_l(t) \tag{109}$$

and

$$\dot\xi_l(t) + \omega_l \xi_l(t) = \omega_l \dot\varepsilon(t), \quad l = 1, \ldots, n. \tag{110}$$

As mentioned before, formalism developed specifically for the GZB was used in many papers. Therefore we give here equations equivalent to those presented by Robertsson *et al.* (1994). Using Eqs. (91), (74), (78) and (90) it is easy to obtain

$$\dot\sigma(t) = M_U \cdot \dot\varepsilon(t) - \sum_{l=1}^{n} r_l(t), \tag{111}$$

$$r_l(t) = \frac{M_{Rl}}{\tau_{\sigma l}} \left(1 - \frac{\tau_{\varepsilon l}}{\tau_{\sigma l}}\right) \int_{-\infty}^{t} \dot\varepsilon(\tau) \cdot \exp[-(t-\tau)/\tau_{\sigma l}] \, d\tau, \quad l = 1, \ldots, n, \tag{112}$$

$$\dot r_l(t) + \frac{1}{\tau_{\sigma l}} r_l(t) = \frac{M_{Rl}}{\tau_{\sigma l}} \left(1 - \frac{\tau_{\varepsilon l}}{\tau_{\sigma l}}\right) \dot\varepsilon(t), \quad l = 1, \ldots, n. \tag{113}$$

Note that anelastic functions (memory variables) $r_l(t)$ are material-dependent.

Defining anelastic coefficients (different from those used by Emmerich and Korn, 1987)

$$Y_l = a_l \delta M / M_U, \quad l = 1, \ldots, n, \tag{114}$$

the stress-strain relations (106) and (109) become

$$\sigma(t) = M_U \cdot \varepsilon(t) - \sum_{l=1}^{n} M_U Y_l \zeta_l(t),$$

$$\dot\sigma(t) = M_U \cdot \dot\varepsilon(t) - \sum_{l=1}^{n} M_U Y_l \xi_l(t). \tag{115}$$

The related Eqs. (108) and (110) remain unchanged.

It is clear that the stress or its time derivative can be calculated if the anelastic coefficients and unrelaxed modulus are known.

The anelastic coefficients Y_l, $l = 1, \ldots, n$ have to be determined from $Q(\omega)$-law. Using the anelastic coefficients, the viscoelastic modulus and quality factor (81) are

$$M(\omega) = M_U \left[1 - \sum_{l=1}^{n} Y_l \frac{\omega_l}{\omega_l + i\omega} \right] \tag{116}$$

and

$$\frac{1}{Q(\omega)} = \left[\sum_{l=1}^{n} Y_l \frac{\omega_l \omega}{\omega_l^2 + \omega^2} \right] \Big/ \left[1 - \sum_{l=1}^{n} Y_l \frac{\omega_l^2}{\omega_l^2 + \omega^2} \right]. \tag{117}$$

Equation (117) yields

$$Q^{-1}(\omega) = \sum_{l=1}^{n} \frac{\omega_l \omega + \omega_l^2 Q^{-1}(\omega)}{\omega_l^2 + \omega^2} Y_l. \tag{118}$$

Equation (118) can be used to numerically fit any $Q(\omega)$-law. A sufficiently accurate approximation to nearly constant $Q(\omega)$ is obtained if the relaxation frequencies ω_l are distributed logarithmically equidistant over the frequency range of interest. If, for example, $Q(\omega)$ values are known at frequencies $\tilde\omega_k$; $k = 1, \ldots, 2n - 1$, with $\tilde\omega_1 = \omega_1$, $\tilde\omega_{2n-1} = \omega_n$, Eq. (118) can be solved for $Y_l, l = 1, \ldots, n$ using the least square method. A more detailed discussion of the frequency range and its sampling at frequencies $\tilde\omega_k$ can be found in the papers by Blanch et al. (1995) and Graves and Day (2003; Eqs. (13) and (14)).

If an elastic P-wave velocity α or S-wave velocity β is known, then, in the considered 1-D problem, $M_U = \rho\alpha^2$ for the P-wave or $M_U = 2\rho\beta^2$ for the S-wave. In practice, a phase velocity at a certain reference frequency ω_r can be measured or estimated. The P- or S-wave phase velocity $c(\omega)$ is given by

$$\frac{1}{c(\omega)} = \text{Re}\left\{\left[\frac{M(\omega)}{\rho}\right]^{-1/2}\right\}. \tag{119}$$

It follows (Moczo et al., 1997) from Eqs. (116) and (119) that

$$M_U = \rho c^2(\omega_r)\frac{R + \Theta_1}{2R^2}, \qquad R = \left(\Theta_1^2 + \Theta_2^2\right)^{1/2}, \tag{120}$$

$$\Theta_1 = 1 - \sum_{l=1}^{n} Y_l \frac{1}{1 + (\omega_r/\omega_l)^2}, \qquad \Theta_2 = \sum_{l=1}^{n} Y_l \frac{\omega_r/\omega_l}{1 + (\omega_r/\omega_l)^2}. \tag{121}$$

As already pointed out, a constant or almost constant Q is of great importance. Therefore, Blanch et al. (1995) addressed the question of an efficient and sufficiently accurate curve-fitting procedure in the case of constant Q. Their τ-method is based on the fact that the level of attenuation caused by a ZB can be determined by a dimensionless variable $\tau = (\tau_\varepsilon - \tau_\sigma)/\tau_\sigma$. Blanch et al. (1995) derived explicit closed formula to determine parameters of the GZB for a desired constant Q, for P- and S-waves, respectively. The GZB obtained by tuning through a single parameter τ yields a very good constant-Q approximation.

3.2.6. A FD Scheme for the Anelastic Functions in the 1-D Case

The 2nd-order approximations to the anelastic functions ζ_l and $\dot{\zeta}_l, l = 1, \ldots, n$ give

$$\begin{aligned}\zeta_l(t_m) &\doteq \frac{1}{2}\big(\zeta_l(t_{m+1/2}) + \zeta_l(t_{m-1/2})\big), \\ \dot{\zeta}_l(t_m) &\doteq \frac{1}{\Delta t}\big(\zeta_l(t_{m+1/2}) - \zeta_l(t_{m-1/2})\big),\end{aligned} \tag{122}$$

where t_m denotes the mth time level. Then each of the equations for the anelastic functions can be solved by

$$\zeta_l(t_{m+1/2}) = \frac{2\omega_l \Delta t}{2 + \omega_l \Delta t}\varepsilon(t_m) + \frac{2 - \omega_l \Delta t}{2 + \omega_l \Delta t}\zeta_l(t_{m-1/2}). \tag{123}$$

The value of $\zeta_l(t_m)$ needed in the stress-strain relation

$$\sigma(t_m) = M_U \cdot \varepsilon(t_m) - \sum_{l=1}^{n} M_U Y_l^M \zeta_l(t_m), \tag{124}$$

is obtained from $\zeta_l(t_{m-1/2})$ and $\zeta_l(t_{m+1/2})$ using Eq. (122). This means that two values have to be kept in memory for one spatial position at one time. It is, however, possible (Kristek and Moczo, 2003) to avoid the necessity to keep in memory both values. It follows from Eqs. (123) and (122) that

$$\zeta_l(t_m) = -\frac{\omega_l \Delta t}{2 - \omega_l \Delta t}\varepsilon(t_m) + \frac{2}{2 - \omega_l \Delta t}\zeta_l(t_{m+1/2}). \quad (125)$$

Then the stress-strain relation (124) can be obtained in the form

$$\sigma(t_m) = \tilde{M}\varepsilon(t_m) - \sum_{l=1}^{n}\tilde{Y}_l^M \zeta_l(t_{m+1/2}) \quad (126)$$

where

$$\tilde{M} = M_U\left(1 + \sum_{l=1}^{n} G_{1l} Y_l^M\right), \quad \tilde{Y}_l^M = G_{2l} M_U Y_l^M,$$

$$G_{1l} = \frac{\omega_l \Delta t}{2 - \omega_l \Delta t}, \quad G_{2l} = \frac{2}{2 - \omega_l \Delta t}. \quad (127)$$

Using scheme (123) and a proper scheme for Eq. (126) it is sufficient to have only one variable for one anelastic function at one grid position at one time. In the case of the staggered-grid velocity-stress FD scheme, the form of equations is the same; only ζ_l and ε have to be replaced by ξ_l and $\dot{\varepsilon}$, respectively.

3.2.7. A Material Discontinuity in the Viscoelastic Medium—the 1-D Case

It is not a trivial task to find a heterogeneous formulation to the differential problem if the stress is given in the form of Eq. (115). Kristek and Moczo (2003) suggested an approximate approach which has been shown sufficiently accurate using numerical tests against the discrete wavenumber method (Bouchon, 1981; Coutant, 1989).

Consider a contact of two viscoelastic media with the GMB-EK rheology. Each of the two media is described by a real density and complex frequency-dependent modulus given by Eq. (116). The question is how to determine density, elastic (unrelaxed) modulus $\bar{M}_U$, and anelastic coefficients $Y_l^{\bar{M}}, l = 1, \ldots, n$ for an averaged medium that should represent the contact of two media (that is the boundary conditions at the interface between the two media) if a material discontinuity goes through a grid cell. There is no reason to consider other than volume arithmetic averaging for the density using formula (56). An averaged viscoelastic modulus $\bar{M}$ can be obtained by numerical averaging in the frequency domain over the grid cell. From the averaged viscoelastic modulus, the quality factor corresponding to this modulus can be determined, Eq. (81), for example, at frequencies $\tilde{\omega}_k$, $k = 1, \ldots, 2n - 1$; $Q_{\bar{M}}(\tilde{\omega}_k) = \text{Re } \bar{M}(\tilde{\omega}_k)/ \text{Im } \bar{M}(\tilde{\omega}_k)$. Assuming that the rheology of the averaged medium can be approximated by the GMB-EK rheology, the anelastic coefficients $Y_l^{\bar{M}}, l = 1, \ldots, n$ for the averaged medium can be obtained using Eq. (118).

It follows from Eq. (79) that $\bar{M}_U = \lim_{\omega \to \infty} \bar{M}(\omega)$. An implication is that, in the limit, the averaging of the viscoelastic modulus gives the averaging of the unrelaxed modulus. This means that the unrelaxed (elastic) modulus $\bar{M}_U$ for the averaged viscoelastic medium can be obtained in the same way as in the perfectly elastic medium.

3.2.8. A Summary of Equations in the 3-D Case

In the 3-D case it is assumed that the rheology of the medium is described by one GMB-EK (or, equivalently, GZB) for the complex frequency-dependent bulk modulus and one GMB-EK for the complex frequency-dependent shear modulus. The stress-strain relation is (Kristek and Moczo, 2003)

$$\sigma_{ij} = \kappa \varepsilon_{kk} \delta_{ij} + 2\mu \left(\varepsilon_{ij} - \frac{1}{3} \varepsilon_{kk} \delta_{ij} \right)$$
$$- \sum_{l}^{n} \left[\kappa Y_l^\kappa \zeta_l^{kk} \delta_{ij} + 2\mu Y_l^\mu \left(\zeta_l^{ij} - \frac{1}{3} \zeta_l^{kk} \delta_{ij} \right) \right] \quad (128)$$

where $i, j, k \in \{1, 2, 3\}$, the equal-index summation convention does not apply to l, $\kappa(x_i)$ and $\mu(x_i)$ are unrelaxed (elastic) bulk and shear moduli, and Y_l^κ and Y_l^μ are the corresponding anelastic coefficients. Assuming a measured or estimated $Q_\alpha(\omega)$ for the P- and $Q_\beta(\omega)$ for the S-waves, the corresponding anelastic coefficients Y_l^α and Y_l^β are obtained using Eq. (118). Then the anelastic coefficients Y_l^κ and Y_l^μ are

$$Y_l^\kappa = \left(\alpha^2 Y_l^\alpha - \frac{4}{3} \beta^2 Y_l^\beta \right) \Big/ \left(\alpha^2 - \frac{4}{3} \beta^2 \right),$$
$$Y_l^\mu = Y_l^\beta, \quad l = 1, \ldots, n. \quad (129)$$

There are n material-independent anelastic functions ζ_l^{ij} for each of 6 strain-tensor components satisfying equations

$$\dot{\zeta}_l^{ij} + \omega_l \zeta_l^{ij} = \omega_l \varepsilon_{ij}, \quad l = 1, \ldots, n, \quad (130)$$

where the equal-index summation convention does not apply to l.

While Eqs. (128) and (130) are applicable to the conventional displacement, or staggered-grid displacement-stress and displacement-velocity-stress FD schemes, equations

$$\dot{\sigma}_{ij} = \kappa \dot{\varepsilon}_{kk} \delta_{ij} + 2\mu \left(\dot{\varepsilon}_{ij} - \frac{1}{3} \dot{\varepsilon}_{kk} \delta_{ij} \right)$$
$$- \sum_{l}^{n} \left[\kappa Y_l^\kappa \xi_l^{kk} \delta_{ij} + 2\mu Y_l^\mu \left(\xi_l^{ij} - \frac{1}{3} \xi_l^{kk} \delta_{ij} \right) \right] \quad (131)$$

and

$$\dot{\xi}_l^{ij} + \omega_l \xi_l^{ij} = \omega_l \dot{\varepsilon}_{ij}, \quad l = 1, \ldots, n \tag{132}$$

are needed for the staggered-grid velocity-stress FD schemes.

In analogy to the 1-D case, it is possible to obtain

$$\zeta_l^{ij}(t_{m+1/2}) = \frac{2\omega_l \Delta t}{2 + \omega_l \Delta t} \varepsilon_{ij}(t_m) + \frac{2 - \omega_l \Delta t}{2 + \omega_l \Delta t} \zeta_l^{ij}(t_{m-1/2}),$$
$$l = 1, \ldots, n \tag{133}$$

and

$$\sigma_{ij}(t_m) = \tilde{\kappa} \varepsilon_{kk}(t_m) \delta_{ij} + 2\tilde{\mu} \left(\varepsilon_{ij}(t_m) - \frac{1}{3} \varepsilon_{kk}(t_m) \delta_{ij} \right)$$
$$- \sum_{l=1}^{n} \left[\tilde{Y}_l^{\kappa} \zeta_l^{kk}(t_{m+1/2}) \delta_{ij} \right.$$
$$\left. + 2\tilde{Y}_l^{\mu} \left(\zeta_l^{ij}(t_{m+1/2}) - \frac{1}{3} \zeta_l^{kk}(t_{m+1/2}) \delta_{ij} \right) \right], \tag{134}$$

where

$$\tilde{\kappa} = \kappa \left(1 + \sum_{l=1}^{n} G_{1l} Y_l^{\kappa} \right), \quad \tilde{\mu} = \mu \left(1 + \sum_{l=1}^{n} G_{1l} Y_l^{\mu} \right),$$
$$\tilde{Y}_l^{\kappa} = G_{2l} \kappa Y_l^{\kappa}, \quad \tilde{Y}_l^{\mu} = G_{2l} \mu Y_l^{\mu},$$
$$G_{1l} = \frac{\omega_l \Delta t}{2 - \omega_l \Delta t}, \quad G_{2l} = \frac{2}{2 - \omega_l \Delta t}. \tag{135}$$

Equations (133) to (135) are ready for programming. In the case of the staggered-grid velocity-stress FD scheme, the form of equations is the same; only ζ_l^{ij} and ε_{ij} are replaced by ξ_l^{ij} and $\dot{\varepsilon}_{ij}$.

3.2.9. Coarse Spatial Sampling

The incorporation of realistic attenuation considerably increases the number of operations and variables/parameters that have to be kept in computer (core) memory. In order to reduce the increased memory requirements and also computational time, Zeng (1996), independently Day (1998) and Day and Bradley (2001) introduced coarse spatial sampling of the anelastic functions and coefficients. In Day's (1998) approach, one anelastic function ξ_l^{ij} for one relaxation frequency ω_l is distributed with a spatial period of $2h$, h being a grid spacing. Consequently, $n = 8$. Considering, for example, location of the stress-tensor component T^{zx} at 8 corners of a grid cube $h \times h \times h$, only one of the 8 ξ_l^{zx} anelastic functions is assigned to one of the 8 corners (say, ξ_1^{zx} is assigned to one position, ξ_2^{zx} to other position, and so on). Consequently, the total number of ξ_l^{zx}, $l = 1, 2, \ldots, 8$ in the whole grid is $\frac{MX}{2} \cdot \frac{MY}{2} \cdot \frac{MZ}{2} \cdot 8 = MX \cdot MY \cdot MZ$, MX, MY and MZ being the numbers of the grid cells in the three Cartesian directions. Because there are 6 independent stress-tensor components, the total number of all the anelastic functions in the whole grid is $MX \cdot MY \cdot MZ \cdot 6$. Since the anelastic coefficients Y_l^κ and Y_l^μ at the grid positions of the normal stress-tensor components, and $Y_l^{\mu xy}$, $Y_l^{\mu yz}$ and $Y_l^{\mu zx}$ at the grid positions of the shear stress-tensor components are distributed in the same coarse manner, the total number of the anelastic coefficients in the grid is $MX \cdot MY \cdot MZ \cdot 5$. Thus, the additional memory due to attenuation in Day's (1998) approach in the staggered-grid scheme in the case of 8 relaxation frequencies is equivalent to the case of just one relaxation frequency without coarse sampling, which is significant.

Graves and Day (2003) analyzed stability and accuracy of the scheme with the coarse spatial sampling and defined the effective modulus and the quality factor necessary to achieve sufficient accuracy.

As discussed earlier, Moczo et al. (2002) demonstrated that a position of a material discontinuity within one grid cell can be sensed by a sufficiently accurate FD scheme. In a structurally complex model there are material discontinuities going through grid cells in different orientations with respect to the coordinate system. In such a case and with the originally suggested spatial sampling (Day, 1998; Day and Bradley, 2001) it can happen that the medium from one side of the material discontinuity is characterized over one half of the whole considered frequency range while the medium from the other side of the discontinuity is characterized over the other half of the considered frequency range. Since the behavior of the two media in contact is characterized in two disjunctive frequency sub-intervals, the two media cannot physically interact. Consequently, the two media cannot be averaged.

In principle, the geometry of the coarse spatial sampling shown in the papers by Day (1998) and Day and Bradley (2001) is not the only one possible. Keeping the same spatial periodicity of the anelastic quantities, it is possible to avoid division of a grid cell into two parts characterized in two disjunctive frequency sub-intervals. Still the best possible alternative situation would be characterization of one medium in contact using, for example, relaxation frequencies ω_1, ω_3, ω_5, ω_7 and characterization of the other medium in contact using ω_2, ω_4, ω_6, ω_8, which again is not satisfactory.

In evaluating the sum that makes an anelastic term in the stress-strain relation, Eqs. (128), (131) or (134), at a given spatial grid position, it would be possible to account for the anelastic coefficients and functions which are not located at that grid point by their properly weighted values. Such averaging, however, poses a problem: Because the anelastic functions (that is, internal variables or memory variables) introduced by Day and Minster (1984), Emmerich and Korn (1987), Carcione et al. (1988a, 1988b) and Robertsson et al. (1994) are material-dependent, any such spatial averaging (accounting for the functions missing at the

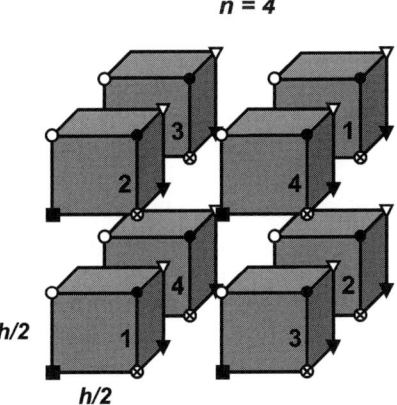

FIG. 5. Coarse spatial distribution of grid cells and anelastic functions. The number on a cell face indicates the relaxation frequency of the anelastic functions localized in the cell. For example, grid cell 1 contains $\xi_1^{xx}, \xi_1^{yy}, \xi_1^{zz}, \xi_1^{xy}, \xi_1^{yz}, \xi_1^{zx}$. Reproduced from Kristek and Moczo (2003).

considered grid point) would introduce and additional artificial averaging of the material parameters. There is no reason for such an additional averaging.

There would be no problem with the coarse spatial sampling and at the same time with weighted spatial averaging of the anelastic functions at a grid point with only one of the all anelastic functions if the anelastic functions were material-independent. Therefore Kristek and Moczo (2003) introduced material-independent anelastic functions (as given above). Moreover, they also suggested an alternative coarse spatial distribution of the anelastic functions which only requires $n = 4$ relaxation frequencies, keeping the same memory requirements as in Day (1998), and Day and Bradley (2001). The distribution is shown in Fig. 5. Kristek and Moczo (2003) demonstrated accuracy of their FD scheme with the material-independent anelastic functions and new coarse distribution of the anelastic functions.

3.2.10. Incorporation of Attenuation in Anisotropic Media

FD modeling in anisotropic media is addressed in the following section. Here we only point out the main difference between incorporating attenuation in the isotropic and anisotropic media. In the isotropic medium it is possible to consider that the rheology of the medium is described by two separate viscoelastic bodies—one for the complex frequency-dependent bulk modulus (corresponding to the dilatational part of the strain) and one for the complex frequency-dependent shear modulus (corresponding to the deviatoric part of the strain). In terms of the quality factors for P- and S-waves, the quality factors can be strictly separated. This makes the incorporation of the attenuation in the isotropic medium much easier compared to the anisotropic medium, where the two quality factors cannot be simply separated if they are not equal.

Robertsson and Coates (1997) presented a 2-D velocity-stress staggered-grid FD scheme for modeling qP- and qS-wave propagation in anisotropic media based on the rheological model described by Carcione and Cavallini (1994). Through eigenvalue decomposition of the stress and stiffness tensors, relaxation functions and memory variables are associated with the so-called eigenstiffness and eigenstresses. The decomposition of stresses and stiffness in this fashion was first observed by Lord Kelvin (Thomson, 1856).

4. Anisotropic Media

FD modeling of wave propagation in anisotropic media has attracted limited attention as the main focus of both applied and academic studies has been on wave propagation in isotropic heterogeneous models. However, a need for modeling of anisotropic wave propagation arises not only from modeling of wave propagation in truly anisotropic materials such as for instance aligned anisotropic crystals in the upper mantle or shales, but also from wave propagation in an equivalent anisotropic medium, see Eq. (65). An equivalent anisotropic medium is a low-frequency approximation for wave propagation in heterogeneous isotropic media (Backus, 1962; Helbig, 1984). An equivalent anisotropic medium can be used to represent for instance fine-scale layering or heterogeneity on a coarse scale appropriate for the scale of discretization of the FD modeling grid. Methods to compute equivalent anisotropic media are well known in 1-D (Schoenberg and Muir, 1989) and can applied to FD modeling of smooth interfaces in 2-D and 3-D (Muir et al., 1992; Moczo et al., 2002). An equivalent anisotropic medium can also be used for modeling of wave propagation through stress induced cracked materials, where equivalent homogeneous anisotropic material represents average orientation and size of cracks (e.g., Crampin and Chastin, 2003). A general equivalent medium theory for 3-D heterogeneous media is a formidable challenge and still a topic of research (and likely to remain one for quite some time).

Another reason for the limited use of FD modeling in anisotropic media is the lack of suitable formulation of anisotropic finite differencing. Whereas the conventional-grid schemes allow modeling of arbitrary anisotropic propagation, they may become unstable at fluid/solid boundaries (among other problems). The staggered-grid schemes are stable at fluid/solid boundaries but wave propagation in general anisotropic media is significantly more difficult. For example, Hooke's law for the general anisotropic medium has the form

$$\sigma_{ij} = c_{ijkl} u_{k,l}.$$

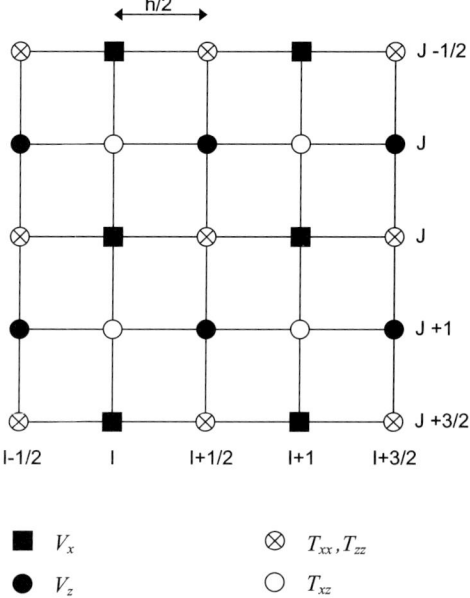

FIG. 6. Staggered grid in 2-D. V_x and V_z are the particle displacement/velocity components, T_{xx}, T_{zz} and T_{zx} are the stress-tensor components.

Tensor of elastic constants c_{ijkl} is symmetric in i and j, k and l, ij and kl indexes. The stress tensor σ_{ij} requires evaluation of terms such as $c_{1112}u_{1,2}$. Figure 6 illustrates that the derivative of u_1 along the 2nd axis is not centered on the stress component σ_{11}. In fact, in 3-D anisotropic media all derivatives with coefficients

$$c_{1112},\ c_{1113},\ c_{1123},\ c_{2212},\ c_{2213},\ c_{2223},\ c_{3312},\ c_{3313},\ c_{3323},\ c_{2313},$$
$$c_{2312},\ c_{1312}$$

are not centered on the corresponding stress components with the standard staggered-grid formulation. Therefore only wave propagation in an orthorhombic anisotropic medium with axis of symmetry aligned along the Cartesian coordinate system does not require any interpolation and is easily implemented. Neglecting the shift of the derivatives for a generally anisotropic medium introduces significant errors. Igel et al. (1995) therefore proposed interpolation operators that shift the derivatives to the corresponding stress positions on the staggered grid. Analogously to the differential operators, the higher-order interpolation operators are more accurate, but require more floating point operations. In fact the interpolation significantly increases the number of operations at every time-step. In general anisotropic media we must carry out 15 interpolations (12 for every coefficient shown above and three of these coefficients require interpolations in two directions), which leads to a substantial increase in terms of computational cost. For example, the 3-D second-order staggered pressure-velocity scheme, with no attenuation and no source representation, requires 16 floating-point operations per time-step for acoustic media. Analogously, the partly-staggered grid (discussed below) formulation (stress-velocity) requires 51 operations for an isotropic elastic medium and 81 operations per time-step for an anisotropic medium (without the cost interpolation required for staggered grids). The cost of interpolation operators in anisotropic media leads to an increase in the number of operations by 500 or more operations per time step (the number of float operations depends on particular implementation as some of the intermediate variables can be stored and reused for further calculations; if no intermediate variables are reused, each interpolation requires 55 operations for the fourth-order interpolation operator). The total number of operations per grid points is therefore on the order of 600 for the 3-D anisotropic case. In other words, anisotropic finite differences on staggered grids are approximately 50 times slower (per time step) than acoustic modeling, and more than 15–20 times slower than isotropic modeling. Note that in addition as elastic simulations typically require shorter time steps and larger simulation times to propagate S-waves thus making acoustic modeling comparably even more computationally inexpensive.

A FD formulation on a partly-staggered grid, originally developed for isotropic wave propagation (Andrews, 1973 and Zhang, 1997), was successfully applied to anisotropic wave propagation by Saenger and Bohlen (2004) (they used term rotated staggered). The formulation combines features of the conventional and staggered-grid schemes. In a grid cell, all stress-tensor components are located at one grid position, all displacement or particle-velocity components are located at other grid position. Numerical analysis of the stability condition shows that the formulation is stable for fluid/solid boundaries. For anisotropic finite-differences on the partly-staggered grid the calculation is roughly only 60% slower than isotropic modeling and the relative cost further decreases with the higher-order space derivative operators that are used as both types of media require the same number (18) of spatial derivatives (e.g., for the popular 4th-order in space and 2nd-order in time formulation, the anisotropic calculation requires only 35% additional computations). This comparison illustrates that there is no significant saving in computational cost by modeling only restricted degrees of anisotropy on the partly-staggered grid (e.g., modeling only orthorhombic versus fully anisotropic media). Although the partly-staggered grid appears to have some quite attractive properties it is not widely used today and the properties of the partly-staggered-grid FD schemes, such as stability for various qP-to-qS wave speeds ratio and interfaces, have to our knowledge not been fully analyzed and tested. Further research is therefore required before adapting the partly-staggered grid formulation instead of the standard staggered-grid formulation introduced by Madariaga (1976) and Virieux (1984) that we focus on in our chapter.

Next we shall derive numerical stability conditions for a FD approximation that is 2nd-order accurate in time and of arbitrary order accuracy in space (von Neumann condition; O'Brien et al., 1951) for wave propagation in a fully anisotropic homogeneous medium. We shall follow the derivation of numerical stability by Crase (1990), Igel et al. (1995), and Saenger and Bohlen (2004). The stability conditions for anisotropic heterogeneous media are evaluated by stability of the equivalent homogeneous medium.

The velocity-stress formulation of the equation of motion for a general anisotropic medium is (without the body-force term), Eqs. (8),

$$\rho \dot{v}_i = \sigma_{ij,j}, \qquad \dot{\sigma}_{ij} = c_{ijkl} v_{k,l}.$$

An approximation to the time derivatives yields

$$v_i\left(I, J, K, m+\frac{1}{2}\right) - v_i\left(I, J, K, m-\frac{1}{2}\right) \doteq \Delta t \frac{1}{\rho} \partial_j [\sigma_{ij}(I, J, K, m)],$$

$$\sigma_{ij}(I, J, K, m) - \sigma_{ij}(I, J, K, m-1) \doteq \Delta t\, c_{ijkl} \partial_l \left[v_k\left(I, J, K, m-\frac{1}{2}\right) \right]. \tag{136}$$

Here, $\partial_i = \partial/\partial x_i$. Taking the plane-wave ansatz, i.e., a harmonic wave $u_i(t, k) = A_i \exp[i(\omega t - k_j x_j)]$, $k = |\vec{k}|$ being a wavenumber, which satisfies the elastodynamic equation, we can rewrite Eq. (136) to contain only particle velocity at one time:

$$-4\sin^2\left(\frac{\omega \Delta t}{2}\right) v_i(I, J, K, m)$$
$$= -\Delta t^2 \frac{1}{\rho} \partial_j \{ c_{ijkl} \partial_l [v_k(I, J, K, m)] \}. \tag{137}$$

The spatial operators on r.h.s. of Eq. (137) are also discretized. The ansatz solution for the harmonic wave in a homogeneous anisotropic medium allows analytical evaluation of the FD approximation to the real wavenumber (note that we ignore the issue of interpolation here):

$$\partial_i \doteq \tilde{k}_i = \frac{2}{\Delta x_i} \sum_{n=1}^{N_S/2} p_n \sin\left(k_i \frac{2n-1}{2} \Delta x_i \right), \tag{138}$$

where $\tilde{k}$ is the numerical wavenumber, Δx_i grid spacing in the x_i-direction, N_S the order of the spatial operator, and p_n the coefficients of the spatial operator. Now we can evaluate complete discretized Eq. (137) for the ansatz solution and obtain the following expression for the angular frequency:

$$\omega = \frac{2}{\Delta t} \arcsin\left[\frac{1}{2} \frac{\Delta t^2}{2} \lambda_q(\tilde{k}_r, c_{ijkl}) \right]^{1/2}. \tag{139}$$

Here λ_p are diagonal elements of the matrix

$$D_{ik} = \frac{1}{\rho} \tilde{k}_j c_{ijkl} \tilde{k}_l. \tag{140}$$

If the expression under the square root in Eq. (139) is greater than one or less than zero, the numerical angular frequency is not real and the FD scheme becomes unstable (because the operator on particle velocity of Eq. (137) is unstable). Thus a general condition for stability of the FD approximation is

$$0 \leq \frac{\Delta t^2}{4} \lambda_q(\tilde{k}_r, c_{ijkl}) \leq 1. \tag{141}$$

To evaluate stability condition (141) we need the eigenvalues of the matrix (140). To evaluate these we bound the FD wavenumbers of Eq. (138) by

$$\tilde{k}_i \leq \frac{2}{\Delta x_i} \sum_{n=1}^{N_S/2} |p_n|, \tag{142}$$

and evaluate eigenvalues of the FD operator (140). In isotropic media the conditions (139)–(140) can be evaluated analytically and they correspond to the CFL condition (e.g., Mitchell and Griffiths, 1994, p. 167; Taflove and Hagness, 2005, p. 42). The analytic solution exists for isotropic homogeneous media where stability condition (141) is

$$0 \leq -2 \sum_{n=1}^{N_t/2} (-1)^n \frac{\Delta t^{2n} \alpha^{2n}}{\Delta x_i^{2n}(2n)!} \left(\sum_{j=1}^{N_S/2} |p_j| \right)^{2n} \leq 1, \tag{143}$$

where α is the P-wave velocity. The condition is valid for both P- and S-wave velocities, but P-wave velocity imposes a stricter constraint as it is larger than the S-wave velocity. However, for a general anisotropic medium the stability condition (141) has to be evaluated numerically.

5. Free Surface

5.1. Traction-Free Boundary Condition

In exploration and earthquake seismology recordings are often made at or close to the Earth's surface. In most seismological applications, the air/fluid (ocean) or air/solid (land) interface can be thought of as a free surface where a fluid or solid abruptly terminates and is replaced by vacuum, that is, the Earth's surface may be approximated by a surface with vanishing traction. Let the surface S with normal vector $\vec{n}$ define the geometry of the Earth's surface. Let $\vec{T}(\vec{u}, \vec{n})$ be the traction vector corresponding to the displacement vector $\vec{u}$ and normal vector $\vec{n}$. The traction-free condition is

$$\vec{T}(\vec{u}, \vec{n}) = 0. \tag{144}$$

A real location for the seismic recordings can be affected by at least two significant factors causing a large impact on wave propagation that must be modeled or understood in many applications. Apart from the fact that the near-surface medium structure may be highly complex with large velocity variations (e.g., Eisner and Clayton, 2002), the free surface itself may display significant topographic variations. At the same time, the FD method inherently has difficulty to implement traction boundary condition. Therefore the accuracy and efficiency of the free-surface approximation are key issues in FD modeling of seismic wave propagation and earthquake motion.

Though, in principle, the same free-surface boundary condition should be simulated for planar and non-planar (topographic) free surfaces, due to the definition of the FD method and unlike, for example, the finite-element method, the implementation of the traction-free condition in the case of topography is a considerably more difficult problem. This is also reflected in the recent development. Therefore, the two geometries will be addressed separately.

5.2. Planar Free Surface

Let the surface S be a planar free surface at $z = 0$ with the unit normal vector $\vec{n} = (0, 0, -1)$. Then

$$\sigma_{z\eta} = 0, \quad \eta \in \{x, y, z\} \tag{145}$$

is the desired boundary condition at the horizontal (flat) free surface.

Recall the 4th-order approximation to the 1st derivative used in the staggered-grid FD schemes for interior grid points. The 1st derivative of a function $\Phi(\xi)$ at $\xi = \xi_0$ is approximated by Eq. (5)

$$\begin{aligned}\Phi'(\xi_0) &= \Phi_{,\xi}(\xi_0) \\ &= \frac{1}{h}\left\{a\left[\Phi\left(\xi_0 + \frac{3}{2}h\right) - \Phi\left(\xi_0 - \frac{3}{2}h\right)\right] \right. \\ &\quad \left. + b\left[\Phi\left(\xi_0 + \frac{1}{2}h\right) - \Phi\left(\xi_0 - \frac{1}{2}h\right)\right]\right\}\end{aligned}$$

with $a = -1/24$, $b = 9/8$ and h being a grid spacing. Evaluation of the z-derivative of a function at $z = 0$ then requires the function values at positions $-h/2$ and $-3h/2$ above the free surface. Similarly, evaluation of the z-derivatives of a function at $z = h/2$ and $z = h$ requires values at $z = -h$ and $z = -h/2$, respectively. This implies two principal possibilities:

- the application of the FD scheme for the interior grid points with the field or material parameters somehow defined above the free surface,
- the application of a different, say, adjusted, FD scheme which does not require any values above the free surface.

Obviously, for the same reason which makes the heterogeneous schemes easier to use than the homogeneous ones, the first approach is preferable.

The first approach either leads to the so-called vacuum formalism, medium taper or imaging method. The vacuum formalism applies zero moduli above the free surface. While this approach gives good accuracy in the displacement formulation, e.g., Zahradník and Priolo (1995) in 2-D, Moczo et al. (1999) in 3-D, Graves (1996) and other authors did not find the approach satisfactory in the staggered-grid modeling. A density taper was used by Frankel and Leith (1992) in their conventional-grid displacement FD scheme.

5.2.1. Stress Imaging

The stress imaging was introduced by Levander (1988) in his 2-D P-SV 4th-order staggered-grid velocity-stress FD scheme. The stress-imaging technique applies explicit boundary conditions to the stress-tensor component(s) located at the grid plane coinciding with the free surface, and uses imaged values of the stress-tensor components above the free surface assuming their antisymmetry about the free surface. The antisymmetry

$$\sigma_{z\eta}(-z) = -\sigma_{z\eta}(z), \quad \eta \in \{x, y, z\} \tag{146}$$

ensures that the boundary condition given by Eqs. (145) are satisfied.

As summarized by Robertsson (1996), there are three possibilities for treating the displacement or particle-velocity values formally required by the FD scheme:

1. The values are calculated using the 2nd-order approximations to the boundary condition and imaged stress-tensor components. The approach was used by Levander (1988), Graves (1996), Kristek et al. (2002) and others.
2. The values are mirrored as even values with respect to the free surface. The approach was used by Crase (1990) and Rodrigues and Mora (1993). As pointed out by Robertsson (1996), the even values of the particle velocity values violate the boundary conditions.
3. The values are set to zero. This was proposed by Robertsson (1996).

Given the staggered grid, there are two natural options for locating the free surface. In one, say H formulation, the horizontal displacement or particle-velocity components, and stress-tensor components T_{xx}, T_{yy}, T_{zz} and T_{xy} are located at the free surface. In the other, say W formulation, the vertical displacement or particle-velocity component and T_{zx} and T_{zy} are at the free surface. Rodrigues (1993) developed a 3-D 8th-order staggered-grid displacement-stress scheme and used the stress-imaging technique in the H formulation. He found that it is necessary to use more than twice the number of grid points compared to inside the medium in order to avoid a significant numerical dispersion. Therefore, he combined the stress-imaging technique with a vertically refined grid near the free surface and achieved good accuracy. Kristek et al. (2002) numerically tested both the H and W formulations of the stress-imaging technique against the discrete-wavenumber method. They demonstrated that in the 3-D case the stress-imaging technique in the 4th-order FD modeling requires at least twice as many grid points per wavelength compared to what is sufficient inside the medium if the Rayleigh waves are to be propagated without significant grid dispersion even in the case of the simple homogeneous halfspace. They also tested the 4th-order version of the Rodrigues (1993) approach. While sufficiently accurate, the approach needs three times smaller time step (the factor of 3 is due to the most natural refinement of the staggered-grid).

It is obvious that either at least twice denser spatial sampling or three times smaller time step degrade the efficiency of the 4th-order staggered-grid modeling inside the medium.

5.2.2. Adjusted FD Approximations (AFDA)

The principle of the AFDA technique used by Kristek *et al.* (2002) is simple. The technique

1. Directly prescribes zero values of σ_{zz} at the free surface in the H formulation or σ_{zx} and σ_{zy} in the W formulation.
2. Applies adjusted FD approximations to calculate the z-derivatives at the grid points at the free surface and depths $h/2$ and h; the adjusted approximation uses only function values in the medium.

As a consequence, no imaged (virtual) values above the free surface are needed. Kristek *et al.* (2002) showed that while H-AFDA results in slightly better phases, W-AFDA results in better amplitudes. They concluded with the recommendation to use W-AFDA for earthquake ground motion modeling. The calculation of the stress-tensor and displacement components in W-AFDA can be summarized as follows (if the velocity-stress formulation is considered, displacement components are simply replaced by the particle-velocity components):

1. Direct application of the boundary condition:

$$T_{zx}(0) = 0, \qquad T_{zy}(0) = 0.$$

2. The following 4th-order FD approximations are used to calculate the stress-tensor and displacement vector components:

$$\Phi'(z_0) = \frac{1}{h}\left[-\frac{352}{105}\Phi(z_0) + \frac{35}{8}\Phi\left(z_0 + \frac{h}{2}\right) - \frac{35}{24}\Phi\left(z_0 + \frac{3}{2}h\right)\right.$$
$$\left. + \frac{21}{40}\Phi\left(z_0 + \frac{5}{2}h\right) - \frac{5}{56}\Phi\left(z_0 + \frac{7}{2}h\right)\right] + O(h^4), \quad (147)$$

$$\Phi'(z_0) = \frac{1}{h}\left[-\frac{11}{12}\Phi\left(z_0 - \frac{h}{2}\right) + \frac{17}{24}\Phi\left(z_0 + \frac{h}{2}\right) + \frac{3}{8}\Phi\left(z_0 + \frac{3}{2}h\right)\right.$$
$$\left. - \frac{5}{24}\Phi\left(z_0 + \frac{5}{2}h\right) + \frac{1}{24}\Phi\left(z_0 + \frac{7}{2}h\right)\right] + O(h^4), \quad (148)$$

$$\Phi'(z_0) = \frac{1}{h}\left[-\frac{h}{22}\Phi'(z_0 - h) - \frac{577}{528}\Phi\left(z_0 - \frac{h}{2}\right) + \frac{201}{176}\Phi\left(z_0 + \frac{h}{2}\right)\right.$$
$$\left. - \frac{9}{176}\Phi\left(z_0 + \frac{3}{2}h\right) + \frac{1}{528}\Phi\left(z_0 + \frac{5}{2}h\right)\right] + O(h^4),$$
$$(149)$$

$$\Phi'(z_0) = \frac{1}{h}\left[\frac{16}{105}\Phi(z_0 - h) - \frac{31}{24}\Phi\left(z_0 - \frac{h}{2}\right) + \frac{29}{24}\Phi\left(z_0 + \frac{h}{2}\right)\right.$$
$$\left. - \frac{3}{40}\Phi\left(z_0 + \frac{3}{2}h\right) + \frac{1}{168}\Phi\left(z_0 + \frac{5}{2}h\right)\right] + O(h^4).$$
$$(150)$$

2.1. The calculation of the stress-tensor components:

$T_{xx}(h/2)$ is obtained from the 4th-order FD approximation to Hooke's law for σ_{xx}; derivative $u_{z,z}$ is approximated by formula (148);

$T_{yy}(h/2)$ and $T_{zz}(h/2)$—similar to $T_{xx}(h/2)$;

$T_{zx}(h)$ is obtained from the 4th-order FD approximation to Hooke's law for σ_{zx}; derivative $u_{x,z}$ is approximated by formula (149) in which $u_{x,z}(0)$ is replaced by $u_{z,x}$ due to condition $\sigma_{zx}(0) = 0$;

$T_{zy}(h)$ is obtained from the 4th-order FD approximation to Hooke's law for σ_{zy}; derivative $u_{y,z}$ is approximated by formula (149) in which $u_{y,z}(0)$ is replaced by $u_{z,y}$ due to condition $\sigma_{zy}(0) = 0$.

2.2. The calculation of the displacement-vector components:

$W(0)$ is obtained from the 4th-order FD approximation to the equation of motion for u_z; derivative $\sigma_{zz,z}$ is approximated by formula (147) in which condition $\sigma_{zz}(0) = 0$ is used;

$U(h/2)$ is obtained from the 4th-order FD approximation to the equation of motion for u_x; derivative $\sigma_{zx,z}$ is approximated by formula (148);

$V(h/2)$ is obtained from the 4th-order FD approximation to the equation of motion for u_y; derivative $\sigma_{zy,z}$ is approximated by formula (148);

$W(h)$ is obtained from the 4th-order FD approximation to the equation of motion for u_z; derivative $\sigma_{zz,z}$ is approximated by formula (150) in which condition $\sigma_{zz}(0) = 0$ is used.

In the W formulation, displacement component W, and stress-tensor components σ_{zx} and σ_{yz} are located at the free surface. The corresponding grid material parameters are evaluated as integral averages in the half grid-cell volumes, that is, the upper half of the volume located above the free surface is not taken into account. For example, density and unrelaxed moduli are evaluated as

$$\rho_W^A = \rho_{I+1/2,J+1/2,0}^A = \frac{2}{h^3} \int_{x_I}^{x_{I+1}} \int_{y_J}^{y_{J+1}} \int_{z_0}^{z_{1/2}} \rho \, dx \, dy \, dz, \tag{151}$$

$$\mu_{zx}^H = \mu_{I,J+1/2,0}^H = \left[\frac{2}{h^3} \int_{x_{I-1/2}}^{x_{I+1/2}} \int_{y_J}^{y_{J+1}} \int_{z_0}^{z_{1/2}} \frac{1}{\mu} \, dx \, dy \, dz \right]^{-1}, \tag{152}$$

$$\mu_{yz}^H = \mu_{I+1/2,J,0}^H = \left[\frac{2}{h^3} \int_{x_I}^{x_{I+1}} \int_{y_{J-1/2}}^{y_{J+1/2}} \int_{z_0}^{z_{1/2}} \frac{1}{\mu} \, dx \, dy \, dz \right]^{-1}. \tag{153}$$

Using numerical comparisons against the DWN method, Kristek et al. (2002) demonstrated that with the W-AFDA technique it is possible to apply the same spatial sampling as inside the medium. Because in many realistic models of the Earth's interior lateral material discontinuities reach the free surface, Moczo et al. (2004a) tested the accuracy of the W-AFDA technique against the finite-element method for which it is easy and natural to satisfy boundary condition at the free surface. Detailed numerical tests demonstrated the sufficient accuracy of the W-AFDA technique in models with near-surface material discontinuities and the capability of the FD scheme to 'see' the true position of the material discontinuities in the spatial grid.

5.3. Free-Surface Topography

5.3.1. Approaches to Model Free-Surface Topography

The classification of approaches is similar to that given in the previous section. The approaches based on modifying material properties at or in the vicinity of the free surface to implicitly satisfy the boundary conditions (e.g., Mittet, 2002; Zahradník and Hron, 1992; Zahradník et al., 1993) lend themselves to be generalized to incorporate topographic variations without much difficulty. However, they tend to require significant spatial oversampling. Frankel and Leith (1992) modeled the topographic effects on seismic waves generated at a Russian test site by using a technique where a smoothly varying density taper was used to model the transition from vacuum to the elastic sub-surface. Ohminato and Chouet (1997) describe one of the first techniques implemented in 3-D where the exact location of the free surface is chosen such that it follows a staircase approximating the topographic surface. In their technique normal stresses are never located on the free surface. Instead, the location of the surface is chosen such that only the shear stresses which should be zero are at the free surface. The free-surface condition is simulated by setting the shear modulus to zero at the free surface and all elastic moduli to zero in the vacuum above the free surface. In the 2nd-order FD scheme at least 25 grid spacings per minimum wavelength are needed to achieve an accurate and stable solution. Pitarka and Irikura (1996) also developed a similar method which was applied in 3-D to study wave propagation site effects at the Kobe-JMA station.

Robertsson (1996) presented a method of simulating the free surface in staggered-grid FD modeling. The method is based on stress imaging and also results in a staircase-shaped surface approximation to the free-surface topography. By splitting the update at each time step in two iterations (in 2-D) that contain spatial derivatives in one direction only, the imaging conditions can be satisfied also in the case of surface topography. The method is relatively simple to implement and yields sufficiently accurate results if at least 15–20 grid points are used per minimum wavelength. The method has been successfully applied both to land seismic applications (e.g., Robertsson and Holliger, 1997; Holliger and Robertsson, 1998) as well as for modeling scattering from a rough sea surface (Laws and Kragh, 2002; Robertsson *et al.*, 2006).

As mentioned above the problem with spatial oversampling the wavefield in the vicinity of the free surface by roughly a factor of three can be circumvented by introducing a simple grid-refinement scheme in the vicinity of the free surface (Rodrigues, 1993; Robertsson and Holliger, 1997). A similar approach was also taken by Hayashi *et al.* (2001). The technique by Robertsson (1996) and its extension by Robertsson and Holliger (1997) will be described in the next subsection.

Several authors developed approaches to avoid a staircase-shaped free surface. Ilan (1977) considered an arbitrary polygonal free surface. Ilan's technique did not address the transition points between the segments of various slopes and required a non-uniform grid that decreased accuracy. An improved representation of the arbitrary polygonal free surface was developed by Jih *et al.* (1988). Following a predefined classification scheme, the different segments were treated using a one-sided approximation of the free-surface condition. Unfortunately, the one-sided difference approximations reduce the accuracy of the method such that a significant spatial oversampling is needed. Another approach to overcome the "staircase problem", initially developed by Fornberg (1988) and then further developed by a number of authors (Tessmer *et al.*, 1992; Carcione and Wang, 1993; Carcione, 1994; Nielsen *et al.*, 1994; Hestholm and Ruud, 1994; Tessmer and Kosloff, 1994; Hestholm, 1999; Hestholm and Ruud, 2002) is to solve the wave equation on a curved grid whose line/surface coincides with the topographic surface (or another internal surface). This is achieved by solving the equation of motion written in Cartesian coordinates and involves first computing the spatial derivatives in the new (conformally mapped) coordinate system (curved grid) and then applying the chain rule to calculate the required Cartesian spatial derivatives. The main drawbacks of the method are stability problems associated with modeling very rough surface topography as well as the computational overhead caused by the chain rule (25% in 2-D and 50% in 3-D according to Komatitsch *et al.*, 1996). Komatitsch *et al.* (1996) therefore proposed solving the equation of motion directly on the curved grid and thus avoiding the chain-rule computations. Although the approach addresses the computational efficiency, it requires additional memory.

As the FD method inherently has problems when incorporating the traction-free condition, particularly in the case of an arbitrary surface topography, Moczo *et al.* (1997) instead used a hybrid approach where a finite-element solution was used in the vicinity of the free-surface topography and a FD solution in the rest of the model. The method was shown to work very well and to be accurate and stable. Obviously, in order not to loose the computational efficiency it is desirable to minimize the region where the finite-element method is applied.

5.3.2. A Method for Modeling Surface Topography in 2-D

To summarize the discussion above, a completely accurate and efficient technique applicable in general to model free-surface topography does not exist. Here we describe a relatively robust technique proposed by Robertsson (1996). The method can be viewed as a generalization of the stress-imaging method of Levander (1988) (H formulation) with one important modification. Instead of updating the particle velocities in the vicinity of the free surface such that the free surface condition is explicitly satisfied by using second-order accurate difference approximations, the particle velocities are simply set to zero above the free surface. Stresses on the other hand are imaged such that tractions perpendicular to the free surface always are zero at the free surface. The generalization to the viscoelastic case is straightforward since no spatial derivatives of the memory variables occur.

A description of the staggered grid in the vicinity of the free surface is shown in Fig. 7, for a crest and a trough. The free surface is discretized such that all grey grid-cells belong to it. It is critical where the boundary is located within the staggered grid-cells (see Robertsson, 1996, for a discussion). Numerically, the free surface itself is located along the thick black line. The grid-points along the discretized boundary belong to one and only one of the seven following categories:

(1) horizontal boundary (H);
(2) vertical boundary with vacuum to the left (VL);
(3) inner corner with vacuum above to the left (IL);
(4) outer corner with vacuum below to the left (OL);
(5) vertical boundary with vacuum to the right (VR);
(6) inner corner with vacuum above to the right (IR); and
(7) outer corner with vacuum below to the right (OR).

It is worthwhile to recall the physical meaning of the imaging technique. The imaging is carried out to ensure that the normal and shear stresses perpendicular to the boundary under consideration (σ_{zz} and σ_{zx} for the 2-D case of a flat horizontal

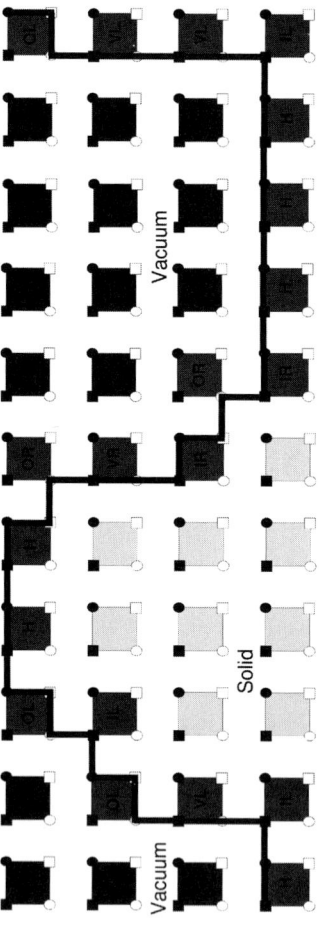

FIG. 7. Staggered FD grid in the vicinity of the free-surface boundary where it forms a crest or (left) and a trough (right). The light large squares represent the locations of the grid-cells in the sub-surface. The grey grid-cells are located along the free-surface boundary, which runs exactly along the thick black line. All boundary grid-points are classified H-, VL-, IL-, OL-, VR-, IR-, or OR-points as described in the text. Within the grid-cells, the solid squares represent the σ_{xx} and σ_{zz} components, the light squares the σ_{zx} component, the solid circles the v_x component, and the light circles represent the v_z component. Reproduced from Robertsson (1996).

surface) are zero. Notice that within all seven categories of boundary grid-cells, the free surface is always parallel to the grid. Imaging, therefore, only takes place in the directions of the x- and z-coordinate axes.

What happens in the vicinity of inner corners, where grid-points are subject to imaging from different directions? Briefly, by calculating separately the vertical and horizontal derivatives of the stress components in the update of the particle velocities, such problems are readily avoided (see below).

Horizontal boundary grid-point (H-point): An H-point has one neighbor on either horizontal side that is either an H-, OL-, OR-, IL- or IR-point. The H-points are treated identically to the flat free-surface approximation. Particle velocities are set to zero in the vacuum. Imaging of stresses only takes place in the vertical direction. Just as for the flat free surface, the σ_{xx} component is updated using the fourth-order accurate central FD approximation along the surface.

Vertical boundary grid-point with vacuum to the left (VL-point): A VL-point has one neighbor on either vertical side that is either a VL-, IL- or OL-point. The VL-points are treated similar to the H-points with the exception that here it is the σ_{xx} component that should be zero at the boundary, and the imaging therefore only takes place in the horizontal direction. The σ_{zz} component is updated in the same way that the σ_{xx} component is updated for the H-points.

Inner corner grid-point with vacuum above to the left (IL-point): An IL-point has one neighbor vertically above that is either an OL- or VL-point, and one horizontally to the left that is either an H- or OL-point. No imaging takes place around the IL-points. The σ_{xx} and σ_{zz} components are located along the boundary, perpendicular to parts of it, and are therefore set to zero.

Outer corner grid-point with vacuum to the left (OL-point): An OL-point has either an IL- or H-point immediately to the right and an IL- or VL-point immediately below. The free surface is located through the v_x, v_z, and σ_{zx} components, while the normal stresses are located in the vacuum above the free surface (Fig. 7). The σ_{zx} component is therefore set to zero. The σ_{xx} and σ_{zx} components are imaged horizontally with respect to the rightmost vertical part of the free surface, and the σ_{zz} and again the σ_{zx} components are imaged vertically with respect to the lowermost horizontal part of the free-surface (Fig. 7). If the OL-point is adjacent to an IL-point, the particle velocity component in between must be set to zero to obtain a stable and accurate solution.

Vertical boundary grid-point with vacuum to the right (VR-point): A VR-point has one neighbor on either vertical side that is either a VR-, IR- or OR-point. The VR-points are treated analogously to the VL-points, with the exception that the imaging takes place in the opposite horizontal direction.

Inner corner grid-point with vacuum above to the right (IR-point): An IR-point has one neighbor vertically above that is either an OR- or VR-point, and one horizontally to the right that is either an H- or OR-point. Only vertical imaging of the σ_{zx} component takes place for the IR-points. The σ_{xx} and σ_{zz} components are located along the boundary, perpendicular to parts of it, and are therefore set to zero.

Outer corner grid-point with vacuum to the right (OR-point): An OR-point has either an IR- or H-point immediately to the left and an IR- or VR-point immediately below. The free-surface makes a step immediately to the left of the OR-point and only intersects with its v_z component, whereas the other velocity- and stress-fields are located in the vacuum above the free surface (Fig. 7). The σ_{zx} component immediately to the left of the boundary point is set to zero. The σ_{xx} and σ_{zx} components are imaged horizontally with respect to the vertical segment of the free-surface boundary to the left of the grid-point (Fig. 7). The σ_{zz} component is imaged vertically with respect to the step at the v_z component in the OR-point. If the OR-point is adjacent to an IR-point, the particle velocity component in between must be set to zero to obtain a stable and accurate solution.

The free-surface boundary points have to be classified prior the FD. The maximum computational efficiency is obtained by setting the reciprocal values of the densities to zero everywhere in the FD grid above the free surface. The particle velocities will then automatically be zero above the free surface after every update. The imaging algorithm thus does not add a substantial amount of the computational cost. However, computations are wasted in the region occupied by vacuum above the topography in the FD grid.

Each of the equations for the particle velocities, consist of two derivatives (one vertical and one horizontal) of the stress-tensor components. Here we do not show explicitly equations for the considered 2-D case as they can be easily obtained from Eqs. (8) for the 3-D case. By first performing all vertical imaging of the stress-tensor components (H-, OR-, OL-, and IR-points) and then calculating and adding only the vertical derivatives in equations, the free-surface condition is satisfied completely in all grid-cells along the boundary. Next, the horizontal imaging of the stress components (VL-, VR-, OR-, and OL-points) is carried out followed by an update with the remaining horizontal derivatives in equations, again satisfying the free-surface boundary condition. Following the updates of equations it is necessary to set the v_x component to zero in the VR- and OR-points, since the reciprocal velocity is not zero in these points. Subsequently, the stress-fields are updated. This action does not involve any imaging of the variable fields. Following the update, the correction of the normal stress parallel to the surface must be made at the H- (σ_{xx}), VL- and VR-points (σ_{zz}), as described above for the horizontal free-surface.

There is a modeling limit as to how narrow the "troughs" or "crests" in the topography can be. A crest cannot be narrower than the number of grid-points imaged around its horizontal sides. Fig. 7 shows the narrowest crest and trough that are allowed when using fourth-order accurate spatial central-difference approximations.

Since the technique requires approximately three times dense spatial sampling than that inside the medium, Robertsson and Holliger (1997) used a grid-

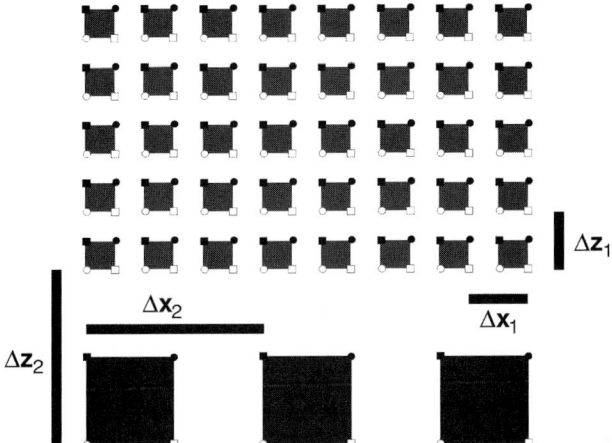

FIG. 8. Grid-refinement technique (Robertsson and Holliger, 1997) used together with the method for modeling surface topography (Robertsson, 1996). The upper part of the grid (light grid cells) is three times more densely sampled than the lower part (dark grid cells). Reproduced from Robertsson and Holliger (1997).

refinement as illustrated in Fig. 8. For the update at grid points of the finer grid close to the coarser grid a simple linear interpolation in the vertical direction is sufficient between the wavefield components in the coarse and fine parts of the grid. However, in the horizontal direction, some care should be taken since waves that propagate close to parallel to the transition region will undergo reflection due to very slight numerical errors in the interpolation between the two regions. An excellent performance can be obtained by using a more accurate sinc interpolation (Martin Musil, personal communication).

To illustrate the accuracy of the method for modeling surface topography we show an acoustic example that has been published elsewhere (Robertsson *et al.*, 2006). The example addresses the problem of modeling scattering from a rough sea surface modeled as a self-similar surface using the method by Pierson and Moskowitz (1964). The so-called Significant Wave Height (SWH) was set to 2 meters (a moderately rough sea as marine reflection seismic data typically are acquired in seas up to 4 m SWH).

Figure 9 shows the reflected response due to a plane wave incident vertically from below as well as at 30° incidence angle with respect to the vertical, recorded 6 m below the average level of the sea surface. We show solutions computed using three different methods. First, the FDTD method by Robertsson (1996) described here. Second, a solution computed using the spectral element method (Chaljub *et al.*, 2006) and finally a solution computed using the Kirchhoff approximation (Laws and Kragh, 2002).

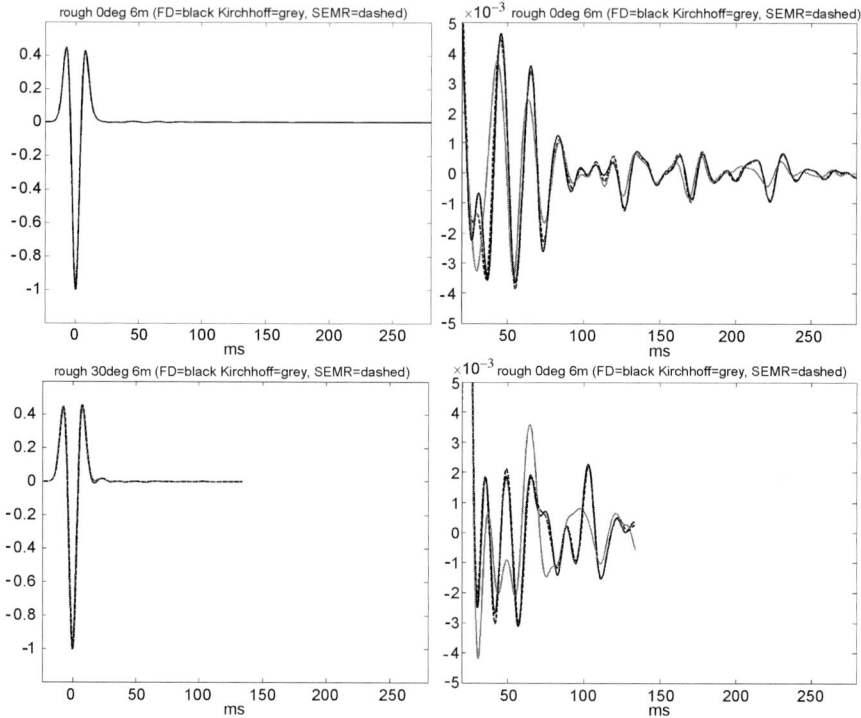

FIG. 9. Rough sea surface reflected response for a plane wave incident vertically from below (top row) and at an angle of 30° to the vertical (bottom row). The reflected/scattered field is recorded at a point placed at 6 m depth. The plots on the right are enlargements of the scattered coda in the plots on the left. Black: FDTD response. Grey: Response computed using a Kirchhoff method (Laws and Kragh, 2002). Dashed: Response computed using a spectral element method (Chaljub et al., 2006). Figure reproduced from Robertsson et al. (2006).

Careful convergence tests were carried out to ensure that details of the reflected and scattered response were not contaminated by numerical artifacts in the respective method. The FDTD method determined the spatial discretization of the sea surface used for all three methods, since it required the densest sampled sea surface (15–20 grid-points per minimum wavelength).

As can be seen from Fig. 9, there is very good match between the FDTD and the spectral element solutions both in terms of amplitude and phase far into the coda. Although the much simpler (and more efficient Kirchhoff method) provides a reasonably accurate solution, the synthesized coda differs significantly from that computed using the two other methods.

6. WAVEFIELD EXCITATION

6.1. Direct Modeling of the Point Sources

The average properties of complex seismic sources are usually represented as point sources in a continuous elastic medium (see, for example, Chapter 3 of Aki and Richards, 1980). The two simplest point sources are body forces and moment-tensor sources. A point source representation of a body-force type of source can be implemented directly as the increment of the corresponding components as given by the equation of motion, e.g. Eqs. (7):

$$\rho \dot{v}_i = \sigma_{ij,j} + f_i.$$

Note, that the body force source-time function is not differentiated with respect to time. The usual implementation of the body-force point source at the time t_m is then

$$v_i(t_m) = v_i(t_m) + \frac{1}{\rho}\Delta t f_i(t_m).$$

Implementation of the moment-tensor source can be included either by stress (Virieux, 1986; Coutant et al., 1995) or by particle velocity (Frankel, 1993; Yomogida and Etgen, 1993; Graves, 1996). The implementations are equivalent due to the body-force equivalent theorem (e.g., pages 40–44 of Aki and Richards, 1980). In the particle-velocity implementation, each component of the moment tensor is implemented by corresponding couple of the body forces with discrete arm length between the forces. For example, in the staggered-grid velocity-stress formulation the M_{xx} component of the moment tensor is equivalent to couple of the forces ($M_{xx}/\Delta x$) acting along the x-axis in the opposite directions. Because these forces are applied at one grid point, the appropriate volume is one grid cell, $\Delta x \cdot \Delta y \cdot \Delta z$. Therefore, the particle-velocity update for the M_{xx} component of the moment tensor at grid node (I, J, K) is

$$v_x\left(I+\frac{1}{2}, J, K, m\right) = v_x\left(I+\frac{1}{2}, J, K, m\right) + \frac{1}{\rho}\Delta t \frac{M_{xx}(I, J, K, m)}{\Delta x^2 \Delta y \Delta z},$$

$$v_x\left(I-\frac{1}{2}, J, K, m\right) = v_x\left(I-\frac{1}{2}, J, K, m\right) - \frac{1}{\rho}\Delta t \frac{M_{xx}(I, J, K, m)}{\Delta x^2 \Delta y \Delta z}.$$

However, the equivalent body forces for the representation of the M_{xy} component of the moment tensor are not located along the grid line I and they must be averaged from four equivalent body forces ($M_{xy}/2\Delta y$) acting along the x-axis in the opposite directions with a force arm of length $2\Delta y$ as illustrated in Fig. 10. Therefore the particle velocity update for the M_{xy} component of the moment tensor at

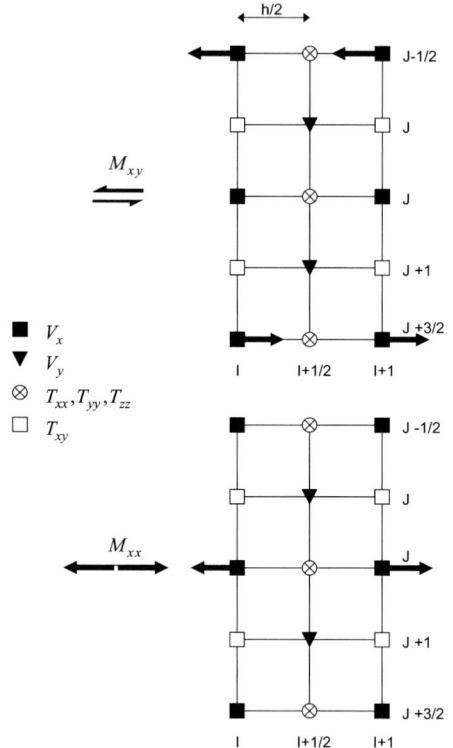

FIG. 10. An example of equivalent body force implementation of moment tensor source component M_{xy} (top) and M_{xx} (bottom). The FD grid shows only the XY plane of the 3-D grid as all body forces are implemented in the same plane (plane K).

grid node (I, J, K) is

$$v_x\left(I \pm \frac{1}{2}, J+1, K, m\right)$$
$$= v_x\left(I \pm \frac{1}{2}, J+1, K, m\right) + \frac{1}{\rho}\Delta t \frac{M_{xy}(I, J, K, m)}{4\Delta x \Delta y^2 \Delta z},$$

$$v_x\left(I \pm \frac{1}{2}, J-1, K, m\right)$$
$$= v_x\left(I \pm \frac{1}{2}, J-1, K, m\right) - \frac{1}{\rho}\Delta t \frac{M_{xy}(I, J, K, m)}{4\Delta x \Delta y^2 \Delta z}.$$

Analogously the remaining components of the moment tensor can be implemented as equivalent body forces centered at grid node (I, J, K). Note that this moment tensor implementation allows modeling of the explosive sources (equal diagonal elements and vanishing non-diagonal elements of the moment tensor), pure shear slip (if the moment tensor is determined from the dip, strike and rake and seismic moment, pages 117–118 of Aki and Richards, 1980), or a compensated linear vector dipole (CLVD) source.

6.2. Introducing the Source Wavefield Along a Boundary Internal to the Grid

Alterman and Karal (1968), Kelly *et al.* (1976) and Levander (1989) describe how to introduce a source field into a FD grid by "injecting" an analytical source solution on an internal artificial surface surrounding the source to for instance avoid a (point) source singularity in the FD computation. The source wavefield is introduced so that it radiates from the outside of the surface surrounding the source. In principle, all we need is to satisfy the principle of superposition and the continuity of the superimposed wavefields across the surface. In practice because the FD calculation is discontinuous at the injection surface S, some care must be taken in the FD calculations where the spatial FD stencil intersects the surface.

Figure 11 shows a 2-D staggered FD grid in the vicinity of the injection surface S. The region outside the surface is referred to as the external region V_e

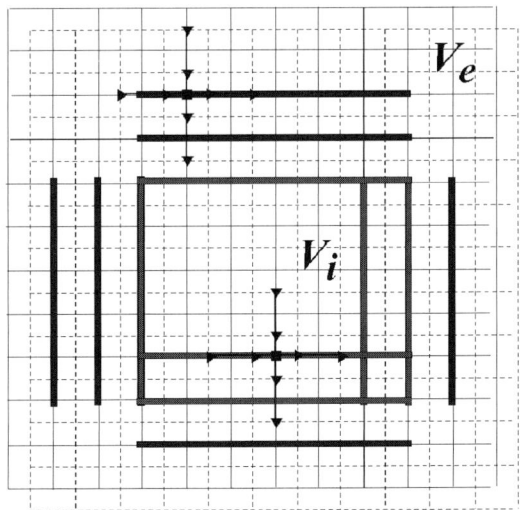

FIG. 11. Introduction of a source wavefield along an internal artificial surface in the staggered grid illustrating the update of normal stresses. Normal stresses are located on the solid grid whereas shear stresses are located on the dashed grid. The region with grey background corresponds to V_i whereas the remaining part of the grid (white background) corresponds to V_e. The so-called injection surface S separates the two regions. Two FD stencils are shown in the picture (one updating a point in V_i and one a point in V_e). Bold lines in the grey region correspond to grid points where the source wavefield should be subtracted at the appropriate points during the update whereas bold lines in the white region show grid points where it should be added. Reproduced from Robertsson and Chapman (2000).

(white in Fig. 11) whereas the region inside the surface is referred to as the internal region V_i (grey in Fig. 11). The process of introducing the source wavefield is straightforward and involves manipulating the update of the wavefield only in the vicinity of S for the update of points where the spatial extent of the FD stencil intersects S (where the wavefield in the grid is discontinuous). In Fig. 11 these points have been marked with by the thick bold lines.

The source field to be injected is known (e.g., analytically) in the vicinity of the surface S for all times (this is possible at least if the surface is located entirely within a homogeneous part of the model). The source wavefield will interact with the model throughout the FD grid within V_i and V_e and generate a scattered wavefield. In V_e both the source wavefield and the scattered wavefield is present whereas in V_i only the scattered wavefield is present. Since the source wavefield is known (e.g., analytically) this can be added and subtracted as appropriate for the update of points along the bold lines for the parts of the spatial stencils that intersect S (subtract the source wavefield at the appropriate points when updating points along the bold lines in the grey region and adding the source wavefield at the appropriate points when updating points along the bold lines in the white region).

In each FD time step the stresses (and corresponding memory variables for the viscoelastic case) are first updated in the entire grid. When the update is complete, we go back and correct the update at the points where the spatial FD stencil intersected the surface S. Here we describe how this is done for the normal stresses (illustrated in Fig. 11). Outside S in V_e, the wavefield is updated as if the injected wavefield were propagating through the entire grid. Therefore we must add the injected source wavefield to the particle-velocity components corresponding to the parts of the stencil that are inside S in V_i. In Fig. 11 this occurs when normal stresses along the bold lines in the white region along each side of V_e are updated. For a 4th-order accurate scheme we therefore need to know the source wavefield along the two closest grid points inside S for the upper and left edges of the rectangular injection region shown in Fig. 11, and along the closest grid point for the lower and right edges. Inside S in V_i, the wavefield is updated as if no wavefield were injected. Therefore we must subtract the injected source wavefield from the particle-velocity components corresponding to the parts of the stencil that are outside S in V_e (for update of normal stresses located along the bold lines in the grey region along each side in V_i).

Next, we advance the calculation by half a FD time step and update the particle velocities in the entire grid using equation. The wavefield injection is performed using the same procedure by adding and subtracting the stress components of the injection wavefield at the three grid points around S. In total, we therefore need to know the values at all times of stresses and particle velocities at three grid points around S, staggered appropriately both in time and space. By iterating these two steps of the update, the entire FD simulation is stepped through and the wavefield is injected along the surface S.

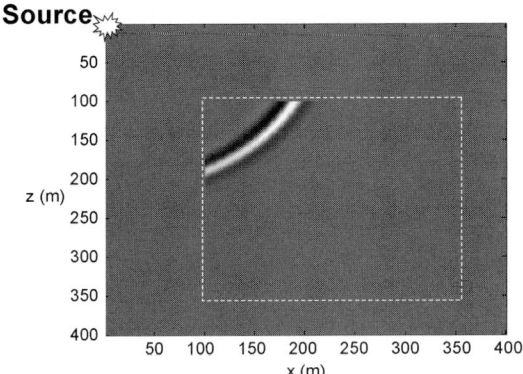
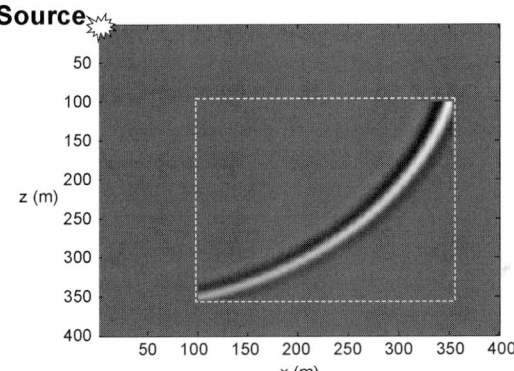
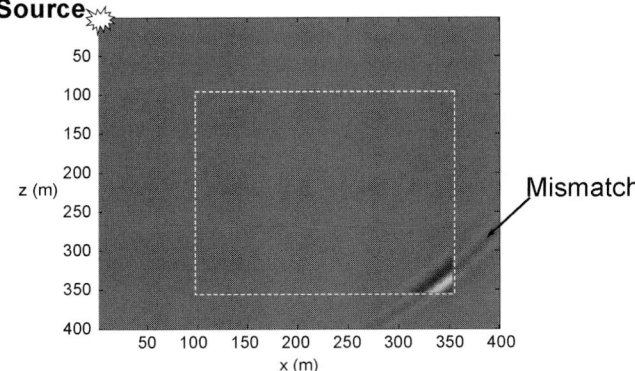

FIG. 12. Example of a hybrid simulation in a homogeneous medium, where the FD computation is driven by an analytical wavefield for a point source in a homogeneous medium. The source is located at a fictitious position in the upper corner of the FD grid. The injection surface is outlined as a dashed box. Snapshots from 0.14 s, 0.24 s and 0.30 s are shown.

As noted by for instance, Fäh (1992), Zahradník (1995), Robertsson et al. (1996), Zahradník and Moczo (1996), Moczo et al. (1997), Robertsson and Chapman (2000), and Takeuchi and Geller (2003), the same technique for introducing the source wavefield along an internal boundary can be used for hybrid modeling purposes where the wavefield from another computational technique (e.g., discrete-wavenumber or ray method) is introduced inside a FD grid. In that case we reverse the regions V_i and V_e so that the source wavefield is injected inside the surface S.

Figure 12 shows snapshots from tests of such a hybrid technique. Here the source wavefield is given by an analytical solution for a point source in a homogeneous medium. Since the FD grid also corresponds to a homogeneous medium we would expect the scattered wavefield in V_e to be zero. In the snapshots from 0.14 s and 0.24 s, we see no evidence of any wavefield being present in V_e. However, in the snapshot from 0.30 s, part of the wavefield leaks through into V_e. This is because the wavefront that has propagated in the FD grid has suffered from numerical dispersion and no longer exactly match the wavefield from the analytical solution which ideally should destructively interfere with the FD wavefront as it reaches the injection surfaced S.

In the so-called FD-injection technique (Robertsson and Chapman, 2000), the source wavefield is generated by a FD solution in an unperturbed model to drive the update on small FD sub-grids surrounding regions of change to compute the wavefield in a perturbed model. This can provide a powerful method in for instance waveform inversion applications.

6.3. Dynamic Modeling of Earthquake Rupture

6.3.1. Fault Boundary Condition

In many seismological problems an earthquake fault may be represented by a surface embedded in heterogeneous elastic or viscoelastic pre-stressed medium. A non-zero initial equilibrium stress is due to tectonic loading and residual stress after previous earthquakes on the fault. An earthquake itself may be modeled as spontaneous rupture propagation along the fault. The rupture generates seismic waves which then propagate from the fault into the embedding medium. In general, several ruptures can propagate along the fault at one time. Inside the rupture displacement and particle-velocity vectors are discontinuous across the fault. At the same time traction is continuous. Let $\vec{n}(x_i)$ be a unit normal vector to the fault surface pointing from the '−' to '+' side of the surface (Fig. 13), $D\vec{u}(x_i, t) = \vec{u}^+(x_i^+, t) - \vec{u}^-(x_i^-, t)$ slip (discontinuity in displacement vector across the fault), $D\vec{v}(x_i, t) = \vec{v}^+(x_i^+, t) - \vec{v}^-(x_i^-, t)$ slip rate (discontinuity in the particle-velocity vector across the fault), $\vec{T}(\vec{n}; x_i, t) = \vec{T}^0(\vec{n}; x_i) + \Delta\vec{T}(\vec{n}; x_i, t)$ total traction on the fault, $\vec{T}^0(\vec{n}; x_i)$ initial traction, and $\Delta\vec{T}(\vec{n}; x_i, t)$ traction variation. The latter

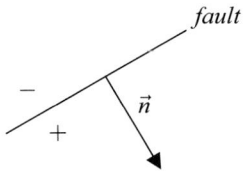

FIG. 13. Fault surface and the normal vector $\vec{n}$.

is due to the rupture propagation. Inside the rupture the total traction is related to slip at the same point through the friction law $\vec{T} = \vec{T}^f(D\vec{u}, D\vec{v}, \theta)$, where $\vec{T}^f$ is frictional traction and θ represents a set of state variables. Given the initial traction and (visco)elastic material parameters of the fault, it is the friction law which controls initialization, propagation and healing of the rupture.

Consider further only shear faulting: there is no opening of the fault and no interpenetrating of the fault materials. Assume a locked fault. If the magnitude of the shear traction is smaller than the frictional strength (product of the normal traction and the friction coefficient) at a point, the fault remains locked and slip rate zero at the point. Should the shear traction exceed the frictional strength, slip occurs. The shear traction decreases over a finite distance from the static down to the dynamic frictional level, following the friction law. The slipping is opposed by the friction.

Let subscripts sh and n denote the shear and normal components. Let S denote frictional strength. The boundary conditions on the fault can be formulated as follows (Day, 1982, 2005; Day et al., 2005).

Shear faulting:

$$D\vec{u}_n = 0, \qquad D\vec{v}_n = 0, \qquad D\vec{u}_{sh} \neq 0, \qquad D\vec{v}_{sh} \neq 0. \tag{154}$$

Shear traction bounded by the frictional strength:

$$|\vec{T}_{sh}| \leq S. \tag{155}$$

Colinearity of the shear traction and slip rate:

$$SD\vec{v}_{sh} - \vec{T}_{sh}(\vec{n})|D\vec{v}_{sh}| = 0. \tag{156}$$

The fact that the frictional traction opposes the slipping is consistent with the colinearity requirement because we consider vector $\vec{n}$ oriented in the direction from the '−' to '+' side of the fault and slip as the relative motion of the '+' side with respect to the '−' side of the fault: both $\vec{T}(\vec{n})$ and $D\vec{v}$ are viewed from the same side of the fault. If slip was defined as the relative motion of the '−' side with respect to the '+' side of the fault, requirement of the antiparallelism with the '+' sign in Eq. (156) would be consistent with the frictional traction opposing the relative motion of the fault faces.

While semi-analytical boundary integral equation (BIE) method is perhaps the most accurate method to account for the fault boundary conditions, especially

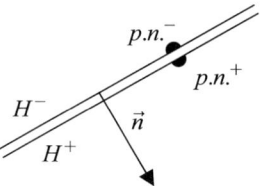

FIG. 14. Halfspaces H^- and H^+, partial nodes $p.n.^+$ and $p.n.^-$, and the normal vector $\vec{n}$.

on non-planar faults (e.g., Aochi and Fukuyama, 2002), its application is limited because it cannot include heterogeneity of the medium. Because the FDM can account for material heterogeneity and is computationally more efficient, it has been extensively applied to study source dynamics particularly on planar faults parallel to grid planes.

The FDM has been applied to the dynamic rupture propagation independently by Andrews (1973, 1976), and Madariaga (1976), and then by many others. Two main approaches have been developed and applied. In the split-node approach the fault is represented by a grid surface of split (partial) nodes. At a grid point, each of the two partial nodes belongs to only one side of the fault and the two nodes may experience a relative motion (slip) along the fault. In the zone approach the fault is represented by a "thick" zone and slip is evaluated either using inelastic strain or as a difference between displacements at grid points separated by the fault zone. Before we mention other approaches we explain the split-node and zone methods in some detail.

6.3.2. Traction-at-Split-Nodes (TSN) Method

The TSN method has been developed independently by (Andrews (1973, 1976, 1999); Day (1977, 1982, 2005)) see also Day *et al.* (2005). Consider a halfspace H^- covered by a FD grid and a grid node $p.n.^-$ on the free surface of the halfspace. Similarly, consider halfspace H^+ and a grid node $p.n.^+$ on its free surface (Fig. 14). Define an outer normal vector $\vec{n}$ to the surface of the halfspace H^- pointing to the halfspace H^+ (i.e., $\vec{n}$ is in the '$p.n.^- \to p.n.^+$' direction).

Let M^- and M^+ be masses of the two partial nodes. The partial node $p.n.^-$ is accelerated by a force $\vec{F}^-$ which is due to deformation in the halfspace H^- and, possibly, by body forces acting in the halfspace. Similarly, the partial node $p.n.^+$ is accelerated by a force $\vec{F}^+$. Thus, the accelerations are $\vec{a}^\pm = \vec{F}^\pm / M^\pm$. Couple the halfspaces H^- and H^+ along their surfaces in order to simulate a fault. The coupling can be accomplished by a constraint surface traction acting at the contact. Consider a traction $\vec{T}^c(\vec{n})$ quantifying a contact force with which material in H^+ acts upon material in H^-. Let A be an area of the fault surface associated with each partial node. Then the acceleration $\vec{a}^-$ of the partial node $p.n.^-$ is contributed by the force $\vec{F}^-$ (which is due to deformation in the halfspace H^-)

and by the constraint force $\vec{F}^{c,-} = \vec{F}^c = A \cdot \vec{T}^c(\vec{n})$ (which is due to the action of the halfspace H^+): $\vec{a}^- = (\vec{F}^- + \vec{F}^{c,-})/M^- = (\vec{F}^- + A \cdot \vec{T}^c)/M^-$. Similarly, the acceleration $\vec{a}^+$ of the partial node $p.n.^+$ is contributed by the force $\vec{F}^+$ (due to deformation in the halfspace H^+) and by the constraint force $\vec{F}^{c,+} = -\vec{F}^c = -A \cdot \vec{T}^c(\vec{n})$ (due to the action of the halfspace H^-): $\vec{a}^+ = (\vec{F}^+ + \vec{F}^{c,+})/M^+ = (\vec{F}^+ - A \cdot \vec{T}^c)/M^+$. Consider some initial equilibrium state with traction $\vec{T}^0(\vec{n})$. The traction does not contribute to the acceleration of the partial node $p.n.^-$. If $\vec{T}^c$ is the total traction, only the difference $\vec{T}^c - \vec{T}^0$ contributes to the acceleration. Considering the accelerations at time t,

$$\vec{a}^\pm(t) = \{\vec{F}^\pm(t) \mp A \cdot [\vec{T}^c(t) - \vec{T}^0]\}/M^\pm. \tag{157}$$

Though the initial traction is nonzero, the initial strain is considered zero. Then forces $\vec{F}^\pm$ correspond to deformations caused only by the dynamic changes due to rupture. The particle velocities and displacements of the partial nodes in the 2nd-order approximation are then

$$\vec{v}^\pm\left(t + \frac{dt}{2}\right) = \vec{v}^\pm\left(t - \frac{dt}{2}\right) + \frac{dt\{\vec{F}^\pm(t) \mp A \cdot [\vec{T}^c(t) - \vec{T}^0]\}}{M^\pm} \tag{158}$$

and

$$\vec{u}^\pm(t + dt) = \vec{u}^\pm(t) + dt \cdot \vec{v}^\pm\left(t + \frac{dt}{2}\right). \tag{159}$$

For the slip rate we obtain from Eq. (158)

$$D\vec{v}\left(t + \frac{dt}{2}\right) \\ \doteq D\vec{v}\left(t - \frac{dt}{2}\right) + dt\, B\left\{\frac{M^-\vec{F}^+(t) - M^+\vec{F}^-(t)}{A \cdot (M^- + M^+)} - [\vec{T}^c(t) + \vec{T}^0]\right\}, \tag{160}$$

where $B = A(M^- + M^+)/(M^- M^+)$. Find a constraint traction $\vec{T}^c(t) = \vec{T}^{ct}(t)$ that assures zero slip rate before two partial nodes start slipping as well as vanishing slip rate when the slipping ceases. The question is how to time condition $D\vec{v} = 0$. If $D\vec{v}(t) = 0$ is required, the trial traction acts for the interval from $t - dt/2$ to $t + dt/2$ and can reverse the slipping (i.e., produce back-slip) by the time it is integrated all the way up to $t + dt/2$. This results in the traction driving slip rather than opposing it and thus in violating conservation of energy. Therefore, $D\vec{v}(t + dt/2) = 0$ has to be required. Assume $D\vec{v}(t + dt/2) = 0$ in Eq. (160) and obtain the trial traction

$$\vec{T}^{ct}(t) \doteq \vec{T}^0 + \frac{dt^{-1} M^- M^+ D\vec{v}(t - dt/2) + M^- \vec{F}^+(t) - M^+ \vec{F}^-(t)}{A \cdot (M^- + M^+)}. \tag{161}$$

Find a constraint traction during the slip, that is, frictional traction $\vec{T}_{sh}^c(t) = \vec{T}_{sh}^f(t)$ such that $D\vec{v}(t+dt/2) \neq 0$. Assuming first $D\vec{v}(t+dt/2) \neq 0$ for $\vec{T}_{sh}^c(t) = \vec{T}_{sh}^f(t)$ in Eq. (160), and then $D\vec{v}(t+dt/2) = 0$ for $\vec{T}^c(t) = \vec{T}^{ct}(t)$ leads to

$$D\vec{v}_{sh}\left(t + \frac{dt}{2}\right) \doteq dt\, B\left[\vec{T}_{sh}^{ct}(t) - \vec{T}_{sh}^f(t)\right]. \tag{162}$$

Recalling the colinearity condition (156)

$$S(t)D\vec{v}_{sh}(t) - \vec{T}_{sh}^f(t)|D\vec{v}_{sh}(t)| = 0. \tag{163}$$

Using approximation $D\vec{v}_{sh}(t) = \frac{1}{2}[D\vec{v}_{sh}(t - dt/2) + D\vec{v}_{sh}(t + dt/2)]$ and Eq. (162) we obtain from the colinearity (163)

$$\left[|D\vec{v}_{sh}(t)| + S(t)\frac{dt}{2}B\right]\vec{T}_{sh}^f(t) \doteq \frac{S(t)}{2}\left[D\vec{v}_{sh}\left(t - \frac{dt}{2}\right) + dt\, B\vec{T}_{sh}^{ct}(t)\right]. \tag{164}$$

Define an auxiliary vector $\vec{\gamma}$

$$\vec{\gamma} = D\vec{v}_{sh}\left(t - \frac{dt}{2}\right) + dt\, B\vec{T}_{sh}^{ct}(t). \tag{165}$$

Equations (164) and (165) imply that $\vec{T}_{sh}^f(t)$ has the direction of the vector $\vec{\Upsilon} = \vec{\gamma}/|\vec{\gamma}|$. Therefore, enforcing the boundary conditions on the fault can be formulated as follows:

If $|\vec{T}_{sh}^{ct}(t)| \leq S(t)$ then $\vec{T}^c(t) = \vec{T}^{ct}(t)$. (166)

If $|\vec{T}_{sh}^{ct}(t)| > S(t)$ then $\vec{T}_{sh}^c(t) = S(t)\vec{\Upsilon}$, $\vec{T}_n^c(t) = \vec{T}_n^{ct}(t)$. (167)

The above approach, based on finding a trial traction $\vec{T}^{ct}(t)$ ensuring $D\vec{v}(t + dt/2) = 0$, and colinearity requirement at time t, can cause in some rare cases large oscillations of rake direction just around the time of rupture arrest (Day, 2005). It is possible to avoid this problem by the modification of the colinearity condition (Day, 2005):

$$S(t)D\vec{v}_{sh}\left(t + \frac{dt}{2}\right) - \vec{T}_{sh}^f(t)\left|D\vec{v}_{sh}\left(t + \frac{dt}{2}\right)\right| \doteq 0. \tag{168}$$

Inserting Eq. (162) into Eq. (168) yields

$$\left[S(t) + |\vec{T}_{sh}^{ct}(t) - \vec{T}_{sh}^f(t)|\right]\vec{T}_{sh}^f(t) \doteq S(t)\vec{T}_{sh}^{ct}(t). \tag{169}$$

Equation (169) means that $\vec{T}_{sh}^f(t)$ has the same direction as $\vec{T}^{ct}(t)$. Then condition (167) is replaced by

If $|\vec{T}_{sh}^{ct}(t)| > S(t)$ then $\vec{T}_{sh}^c(t) = S(t)\vec{T}_{sh}^{ct}(t)/|\vec{T}_{sh}^{ct}(t)|$, $\vec{T}_n^c(t) = \vec{T}_n^{ct}(t)$.
(170)

The modified approach behaves always well.

An assumption of the small displacements is necessary for the TSN method. The assumption means that the accumulating slip does not change the configuration of the partial nodes adjacent to each other. This means that $h \gg |D\vec{u}_{sh}|$, where h is a spatial grid spacing. Also, a necessary condition is that the time-stepping algorithm is explicit and a force at a node accelerates only that node.

Accuracy of the TSN implementation heavily depends on the accuracy of calculation of the body forces $\vec{F}^{\pm}$ due to deformations in the halfspaces. Formally, at each time the surfaces of the halfspaces are the free surfaces. In fact, both Andrews (1973, 1976, 1999) and Day (1982) used an FD formulation on the partly-staggered grids in which the spatial differencing is equivalent to the finite-element method.

6.3.3. Stress Glut (SG) Method

Andrews (1999) describes the SG method that he used in his 1976 paper (Andrews, 1976). Instead of a surface with an explicit displacement discontinuity in the TSN method, an inelastic zone of a finite thickness is used in the SG method to represent the fault surface. The method can be implemented in the partly-staggered or staggered grid. Consider for simplicity a horizontal zone (perpendicular to the z-axis oriented positive downward) centered at a grid plane with grid positions of the stress tensor and bounded in the vertical direction by grid planes with grid positions of the displacement and particle-velocity vectors. The thickness of the fault zone is thus equal to one grid spacing h. Assume the velocity-stress formulation. At each time level stresses are updated as if there were no fault zone. Denote shear stress-tensor components located at the central horizontal grid plane as trial values (superscript t) and evaluate a magnitude of the shear trial traction vector $|\vec{T}_{sh}^t| = [(\tau_{zx}^t)^2 + (\tau_{zy}^t)^2]^{1/2}$. If, at a grid point, $|\vec{T}_{sh}^t| \leq S$, S being the frictional strength of the fault, then $\tau_{zi} = \tau_{zi}^t$, $i \in \{x, y\}$ at that grid point. If $|\vec{T}_{sh}^t| > S$, then, due to the colinearity, $\tau_{zi} = (S/|\vec{T}_{sh}^t|)\tau_{zi}^t$, $i \in \{x, y\}$. The latter means an inelastic stress adjustment. Assuming that the corresponding inelastic strain is distributed over the fault thickness, the offset in stress, i.e., stress glut, is related to the increment of the seismic moment tensor in the volume of the grid cell:

$$\Delta M_{zi} = \left(\tau_{zi}^t - \tau_{zi}\right) \cdot h^3, \quad i \in \{x, y\}. \tag{171}$$

At the same time, the increment in the seismic moment tensor can be interpreted under the assumption of a shear slip along the central grid plane of the fault zone. Assuming a normal vector $\vec{n}(0, 0, 1)$ and slip rate at a given grid position on the central grid plane $Dv_i = v_i^+(z^+) - v_i^-(z^-)$,

$$\Delta M_{zi} = \mu Dv_i h^2 \Delta t, \quad i \in \{x, y\}. \tag{172}$$

Equating Eqs. (171) and (172) we obtain for the slip rate

$$Dv_i = \frac{h}{\Delta t} \frac{\tau_{zi}^t - \tau_{zi}}{\mu}, \quad i \in \{x, y\}. \tag{173}$$

Andrews (1999) discusses computational details as well as the more complicated implementation of the method on the staggered grid.

6.3.4. Thick-Fault-Zone (TFZ) Method

Madariaga *et al.* (1998) described a velocity-stress staggered-grid method to study dynamic faulting. As in the SG method, no surface with explicit displacement and particle-velocity discontinuity is considered. Instead, an alternative fault zone of the finite thickness is centered at the grid plane with grid positions of the normal stress-tensor components. Considering again, e.g., a horizontal fault zone, frictional boundary conditions are applied on the two nearest grid planes with grid positions of the shear stress-tensor components, i.e., on the grid planes half grid spacing below and above the central grid plane. Because the horizontal displacement components are located on the central grid plane, the slip rate is evaluated as the difference between particle-velocity values at grid planes one grid spacing above and one grid spacing below the central grid plane, that is, over a thickness of two grid spacings. Such a configuration preserves symmetry of stresses and particle velocities about the fault plane. Madariaga *et al.* (1998) apply standard 4th-order FD formulae to calculate spatial derivatives at all grid positions.

6.3.5. Comparison of the TSN, SG and TFZ Methods

Dalguer and Day (2004, 2006) performed an extensive numerical comparison of the three methods. Calculations were performed with the 2nd-order TSN and 4th-order staggered-grid SG and TFZ formulations using the same grid spacing h and uniform grid. The rupture propagation velocity in the fault-zone model is lower than that in the split model—likely due to the blunting of the stress concentration on the rupture front in the fault-zone models. The TSN solution converges substantially better than the SG solution, and far better than the TFZ solution. The SG method reaches convergence with a grid spacing $h/2$, while grid spacing smaller than $h/4$ is necessary for the TFZ method. Dalguer and Day therefore tested modified grid configurations for the SG and TFZ methods. With the fault thickness reduced to $0.75h$ in the SG and $0.5h$ in the TFZ modeling, the rupture velocities approach that in the TSN modeling. At the same time, the behaviors of both fault zones depend on the ratio between h and the fault thickness which has to be adjusted ad hoc. In any case, the theoretical efficiency of the 4th-order fault-zone formulations is lost if sufficiently accurate results are to be obtained.

6.3.6. Alternative Approaches

An interesting approach was presented by Ionescu and Campillo (1999) who studied the 2-D problem of slip instability under slip-dependent friction. They used a combination of two space–time FD grids. An interior (or body) grid is used to solve the equation of motion away from the fault. The other grid is used to implement frictional boundary conditions on the fault. The first-order hyperbolic system (velocity-stress formulation) is reduced on the fault to a first-order system in one space variable (orthogonal to the fault) for the shear velocity and shear stress-tensor components. Using the integration on the characteristic lines and frictional boundary conditions a nonlinear unstable ordinary differential equation for slip is obtained. The equation depends on the computed values of the stresses and particle velocities away from the fault. A small local time step is used to solve the equation during the larger time step applied in the body grid.

Nielsen modified a standard application of the 4th-order staggered-grid FD scheme to investigate rupture propagation (e.g., Nielsen et al., 2000; Nielsen and Carlson, 2000). In order to satisfy the frictional boundary conditions with a slip rate and tractions evaluated at the same time, he approximates a field variable at a given time using a 2nd-order interpolation between their values at consecutive time levels. Thus he reformulates the velocity-stress formulation of the equation of motion and Hooke's law with the same time-derivative operator applied to the particle-velocity and stress-tensor components. In addition to this, Nielsen avoids application of the spatial-derivative operator across the fault if the operator should apply to a discontinuous field. He achieves this by reducing the spatial-derivative operator down to the 2nd-order in evaluating shear stress-tensor components at a grid point on the grid planes one grid spacing above and one grid spacing below the fault, if the grid point one grid spacing away is slipping.

6.3.7. Modeling of Non-Planar Faults Using Partly-Staggered Grids

Cruz-Atienza and Virieux (2004) developed a new FD approach to model the dynamic rupture propagation on non-planar faults. They used a 2-D velocity-stress formulation on a partly staggered grid. Their formulation is based on a new application of the fault boundary conditions. No split nodes are considered, that is, there is no explicit discontinuity at any grid point. A finite source, rupturing fault, is represented by a set of neighboring numerical cells placed alongside a fault without sharing any stress grid point. Because this is the first FD approach enabling both medium heterogeneity and non-planar fault geometry, we describe its essential features in some detail.

Consider a Cartesian coordinate system and a corresponding spatial rectangular grid. A square $h \times h$ FD grid cell has particle-velocity vector positions at four corners and stress-tensor position at the center. Then one n-point square numerical cell is a set of $\sqrt{n} \times \sqrt{n}$ neighboring square FD cells. This means that one

n-point square numerical cell has n stress-tensor grid positions (points). Considering a fixed spatial support S (square area), the scaling relation for a numerical cell is $S = n \cdot h_n^2$, where n is the number of FD square $h_n \times h_n$ grid cells making one n-point numerical cell. The greater n the smoother the fault-geometry discretization is.

Boundary conditions on the fault are applied locally in each numerical cell. The application means the following. A planar fault tangential to the discretized fault geometry is considered inside the cell. Let ϑ be an angle between the x-axis and the fault plane. The Cartesian stress-tensor components at a stress grid point are rotated into shear and normal stresses with respect to the fault plane. The same boundary condition, following a considered friction law, is applied at each stress grid point within the numerical cell. After imposing the boundary condition, the stresses are rotated back to the original coordinate system connected with the grid.

The fault plane always passes through the center of the numerical cell. Velocity grid points in a numerical cell belong to either positive or negative fault block. At a velocity grid point I, displacement u_I parallel to the fault plane is obtained from the projected particle velocity by simple time integration. Because no split nodes are considered, displacement u_I depends on a distance between the point and the fault. For any time t, u_I will be at its maximum, if the point lies on a line passing through the center of the cell and perpendicular to the fault. It will be zero, if the grid point lies on the fault. For a given ϑ, it may happen that no velocity grid point lies on a line passing through the center of the cell and perpendicular to the fault. A weight function $H_I(\vartheta)$ for a grid point I is defined as a ratio between the maximum displacement and displacement, that is, $H_I(\vartheta) = \max_\vartheta \{u_I(t, \vartheta)\}/u_I(t, \vartheta)$. $H_I(\vartheta)$ does not depend on time and has to be determined for all grid points in a given type of a numerical cell. Then the weighted displacements $\tilde{u}_I(t) = u_I(t, \vartheta) \cdot H_I(\vartheta)$ do not depend on the fault orientation in the spatial grid. The relative displacement of the, for example, positive fault block, D^+ is then determined as $D^+ = (\sum_{I=1}^{p} \tilde{u}_I^2(t))^{1/2}/p$, where p is the number of the velocity grid points. Points within an angular vicinity γ around the fault (two opposed sectors of a circle centered at the cell's center) are not considered because their displacements are too small. The angle is determined as $\gamma = \arctan(\sqrt{n} - 1)$.

Cruz-Atienza and Virieux (2004) found that the zigzag discrete shape of the fault associated with a low number n of grid points leads to unwanted destructive dynamic interaction of cells. At the same time, as the grid spacing is smaller, the dynamic interaction is stronger. As a consequence, in order to increase numerical resolution, it is necessary both to reduce grid spacing and enlarge n. To guarantee the low-frequency equivalence between different discretizations of a fault length L, the scaling relationship $L \geq 30 h_n \sqrt{n}$ has to be followed.

6.4. Non-Reflecting Boundaries

6.4.1. Absorbing Boundary Conditions

For computational reasons it is almost always necessary to limit the computational domain to the part of the Earth in the vicinity of the seismic source and receivers as well as the Earth structure contributing to form the seismic response. This is achieved by applying so-called non-reflecting boundaries or absorbing boundary conditions (ABC) around the tractable truncated computational domain such that no energy is transmitted or reflected from the boundary of the computational domain back into its interior. Thus from the perspective of an observer located inside the computational domain the signals appear to have been perfectly absorbed by the boundary.

A significant proportion of the literature on time-domain FD modeling is devoted to the design of efficient ABC but most published approaches can be divided into two groups. The first group, including the popular Clayton and Engquist (1977); Higdon (1991); Lindman (1975) and Liao et al. (1984), attempt to extrapolate the wavefield beyond the edge of the computational domain and then use this extrapolated value in the spatial discretization operator used to update locations inside the computational domain. An interesting approach was suggested by Peng and Toksöz (1994, 1995).

Collino (1993) and later other authors attempted to apply so-called high-order ABCs which, in general, is an infinite sequence of ABCs with increasing accuracy. An advantage of the high-order ABCs is that they are implementable for an arbitrary high order (Givoli, 2004).

The second group, including the approach of Kosloff and Kosloff (1986) and Perfectly Match Layers (PML; Bérenger, 1994), gradually attenuate the amplitude of the wavefield within a "sponge" layer within the boundary.

For completeness, we should also mention work on complementary boundary conditions and operators (Smith, 1974; Schneider and Ramahi, 1998), which does not fit within either of the two groups. Dirichlet and Neumann boundary conditions are complementary as they have opposite reflection coefficients. By adding the response from two simulations with different complementary boundary conditions, the first-order boundary reflections will destructively interfere. Unfortunately, higher-order interactions will not necessarily cancel and the methods have therefore not found widespread application.

Each ABC approach has it own characteristics, advantages and disadvantages. In the first group mentioned above, for example, the Clayton–Engquist and Higdon methods have the advantage of requiring relatively little memory but only work well within a limited range of angles of incidence. Lindman's approach works well over a wide range of angles but only works when the material along the boundary is homogeneous. Liao's ABC works well over most angles and for some types of heterogeneity adjacent to the boundary but requires double precision implementation to maintain stability. It also requires more computer memory than the other methods listed above.

In the second group of ABCs, Kosloff's method (Kosloff and Kosloff, 1986) amounts to applying an exponential damping function of the wavefield within a "sponge" region surrounding the computational domain. Kosloff's method is effective at normal incidence and is very easy to implement but requires a relatively large amount of memory and can generate significant reflection artifacts away from normal incidence. A related method known as perfectly matched layers (PML) was first introduced for electromagnetic wave propagation by Bérenger (1994), and later generalized for elastic wave propagation by (Chew and Liu (1996); Collino and Tsogka (1998, 2001); Komatitsch and Tromp (2003); Marcinkovich and Olsen (2003); Festa and Nielsen (2003)), and for viscoelastic anisotropic media by Chen et al. (2000). PMLs have now become established as the most efficient absorbing boundary condition available, offering simplicity of implementation, stability, good absorption over a wide range of incident angles and frequencies and less memory requirements than Kosloff's ABC. In this section we will therefore focus on PMLs, review the theory and discuss issues related to their implementation.

6.4.2. The Perfectly Matched Layer (PML)

Following Chen et al. (2000), we define the split particle velocities $v_i^{(\gamma)}$ such that the equation of motion in a general heterogeneous anisotropic elastic medium is decomposed into three equations

$$\rho \dot{v}_i^{(\gamma)} = \sigma_{i\gamma,\gamma} \tag{174}$$

with

$$v_i = \sum_{\gamma=1}^{3} v_i^{(\gamma)} \tag{175}$$

where the Einstein summation convention is not assumed for the Greek letter γ. Similarly, the constitutive equation for the split stiffness tensors may be written as

$$\dot{\sigma}_{ij}^{(\gamma)} = c_{ij\gamma q} v_{q,\gamma} \tag{176}$$

with

$$\sigma_{ij} = \sum_{\gamma=1}^{3} \sigma_{ij}^{(\gamma)}. \tag{177}$$

Following Chew and Liu (1996), we introduce a modified spatial differencing operator

$$\tilde{\partial}_\gamma = \frac{1}{e_\gamma} \partial_\gamma, \tag{178}$$

where the denominator e_γ has been interpreted as a coordinate stretching variable. Outside the PML layers, vector $\vec{e} = (1, 1, 1)$. Inside the PML layers, a complex e_α is needed in order to attenuate the split $(v^{(\gamma)}, \sigma^{(\gamma)})$ in the γ direction. Assuming a time dependence of $e^{-i\omega t}$ we choose

$$e_\gamma = a_\gamma + \frac{i\Omega_\gamma}{\omega} \tag{179}$$

where a_γ and Ω_γ are the so-called PML medium profiles. In the time domain, this leads to the following equations for the partial variables:

$$\rho(a_\gamma \partial_t + \Omega_\gamma) v_i^{(\gamma)} = \partial_\gamma \sigma_{i\gamma} \tag{180}$$

and

$$(a_\gamma \partial_t + \Omega_\gamma) \sigma_{ij}^{(\gamma)} = c_{ij\gamma q} \partial_\gamma v_q. \tag{181}$$

The selection of the PML material profile a_γ and Ω_γ is key to the effective implementation of the PML ABC. Theoretically, these parameters can be arbitrary in the continuum space and the PML layers would perfectly absorb any incident wave. This is, however, not true in the discretized model. The discretization error is proportional to the product of the spatial discretization size and the contrast of these parameters (Chew and Weedon, 1994; Chew and Liu, 1996). A PML sponge of width $N = 10$ with the following profile at each layer $i = 1, 2, \ldots, N$ has been found to result in excellent absorption for many seismic applications for a wide range of frequencies, Courant numbers and wave types (Chen et al., 2000):

$$a_\gamma(i) = 1 + \frac{1}{2} a_{\max} \left(1 + \sin\left(\frac{N/2 - i + 1}{N} \pi \right) \right)^{2.8} \tag{182}$$

and

$$\Omega_\gamma(i) = \frac{1}{2} \Omega_{\max} \left(1 + \sin\left(\frac{N/2 - i + 1}{N} \pi \right) \right)^{2.8} \tag{183}$$

with $a_{\max} = 0.075$ and $\Omega_{\max} = 0.91/\Delta t$.

We note that the PML ABC differs fundamentally from the traditional material sponge ABCs such as the method by Kosloff and Kosloff (1986) as a result of the split wavefield formulation. For each set of split wavefields, only the propagation vector component normal to the boundary is complex. The propagation vector components parallel to the boundary remain the same across the boundary layers. As a result, a perfect match can be obtained while the wavefields are attenuated by the perfectly matched layers, at least in theory in the non-discretized model.

Finally, the extension of PMLs to viscoelastic media is straightforward (Chen et al., 2000). The memory variables governing the viscoelastic behavior of the medium due to propagating waves are updated from the change in stress and record the history of the stress. The memory variables in the PML region are therefore split in exactly the same way as the stresses. The equation for updating the split stress fields becomes, see Eq. (111),

$$(a_\gamma \partial_t + \Omega_\gamma)\sigma_{ij}^{(\gamma)} = c_{ij\gamma q}\tilde{\partial}_\gamma v_q + \sum_{l=1}^{n} r_l^{ij(\gamma)} \qquad (184)$$

and the equation for updating the split memory variable fields is, see Eq. (113),

$$\dot{r}_l^{ij(\gamma)}(t) + \frac{1}{\tau_{\sigma l}} r_l^{ij(\gamma)}(t) = \frac{1}{\tau_{\sigma l}}\left(1 - \frac{\tau_{\varepsilon l}}{\tau_{\sigma l}}\right)[c_{ij\gamma q}]_R \tilde{\partial}_\gamma v_q, \quad l=1,\ldots,n, \qquad (185)$$

where $[c_{ij\gamma q}]_R$ means relaxed moduli. The equations for updating the particle velocities in split form remain the same, Eq. (180).

6.4.3. PML Reflection Coefficients

In Fig. 15 we show reflection and conversion coefficients for incident P- and S-waves. The curves were obtained from 2-D simulations in a homogeneous isotropic elastic medium with a P-velocity of 2500 m/s, an S-velocity of 1500 m/s and a density of 1100 kg/m³ where a point source was positioned close to the PML boundary (Chen et al., 2000). For incident P-waves the source was explosive, for S-waves the source was a force oriented normal to the PML boundary. In each case the source was a 50 Hz Ricker wavelet. The reflections were corrected for propagation distance and radiation pattern to obtain reflection and conversion coefficients.

Simulations were carried out for PML boundary thicknesses of 10, 20, 30, 40 and 50 cells, and compared to a 40 cell thick Kosloff ABC (shown in the dashed lines). First, the results show that the choice of a 10 grid-point wide PML boundary is roughly as efficient as the 40 grid-point wide boundary (justifying the values we used for cost comparisons in the previous section). However, the poor performance at low grazing angles may not be acceptable for many applications. Increasing the thickness of the PML boundaries drastically improves the effectiveness so that we are close to single-precision machine accuracy for the 50 grid-point wide boundaries. The PS-plot (top right of Fig. 15) reveals very little P-to-S conversion. Machine precision is achieved with a 20 cell thick PML.

The SS-reflection coefficient (bottom left in Fig. 15) shows a behavior similar to the PP-results. The SP-plot (bottom right in Fig. 15) does not show the same remarkable performance as the PS-converted wave reflection although the performance is still satisfactory. It is possible that the SS-and SP-results could be degraded by P-wave energy generated by the source used.

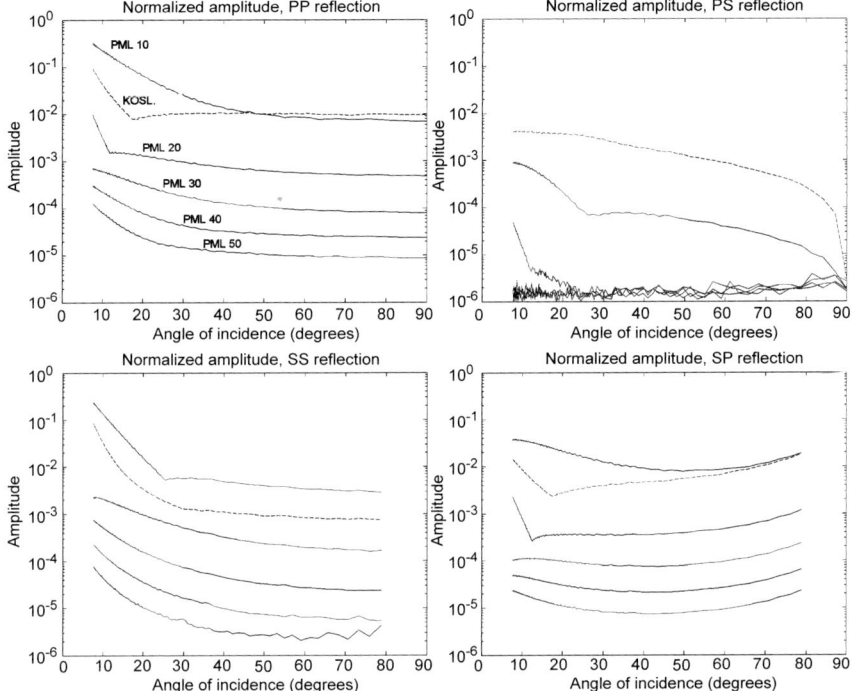

FIG. 15. Amplitude reflection coefficient as a function of incidence angle for PML boundaries of width 10, 20, 30, 40 and 50 grid-points (solid) as well as for a 40 grid-points wide Kosloff sponge (dashed). Isotropic model. Top left: Normalized PP reflection. Top right: Normalized PS reflection. Bottom left: Normalized SS reflection. Bottom right: Normalized SP reflection. Reproduced from Chen et al. (2000).

Figure 16 shows the corresponding results for an anisotropic model with the same density and vertical velocities as the isotropic model but with a vertical to horizontal P-velocity ratio of 1.12 and an anellipticity of 0.16 as defined by Carrio et al. (1992). The results do not differ significantly from the ones shown in Fig. 15 for the isotropic case.

6.4.4. PML Implementation Issues and Computational Cost

In general Eq. (180) is a system of nine equations: three partial, or split, velocities for each velocity component. The split elastic constitutive relation (181) is in general a system of eighteen equations: three partial stresses for each of six independent elements in the stress tensor. For the viscoelastic case, Eq. (181) translates into Eqs. (184) and (185) which, in general, are systems of eighteen equations. If all of these variables were required on the complete 3-D FD grid, the resulting

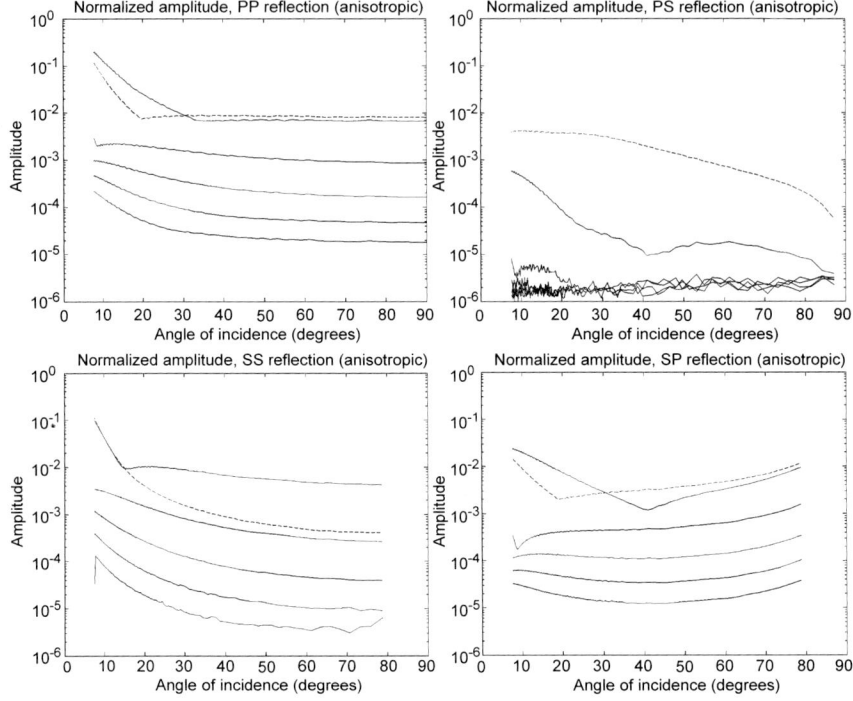

FIG. 16. Same as Fig. 15 but for an anisotropic model. Reproduced from Chen *et al.* (2000).

memory requirements would make this approach unattractive. However, because updating split field variables require only derivative of the full velocities and stresses, the domain in which the split field variables are defined (and stored) is restricted to a relatively narrow region around the boundaries.

The PML formulation presented here requires almost the same number of calculations per grid point as the interior of the computational domain since most of the calculation required for updating each split field is nothing but a constituent of the conventional time-domain FD calculation which must be evaluated in any case (e.g., the spatial derivatives). However, because of the splitting of variables, there is a significant price to pay in storage per grid point (approximately a factor three depending on dimension, model properties, and other factors). Nevertheless, with a 10 grid-points wide PML sponge the PML conditions are still computationally significantly more efficient both in terms of CPU and memory compared to other ABC (Chen *et al.*, 2000).

Recently Wang and Tang (2003) proposed an alternative PML implementation which avoids splitting the wavefields. The storage requirements per grid point are therefore equivalent to those in the interior of the grid. However, the CPU requirements increase to solve for a new set of differential equations with similar characteristics to the viscoelastic equations presented above.

We note that it is straightforward to interface PML boundary conditions with a free-surface condition.

As we have seen, despite what the name might suggest, PMLs are still far from perfect. Because of inaccuracy related to discrete numerical implementations of PMLs, a finite sponge width is required to avoid artificial reflections. The cost in terms of memory is therefore not insignificant, particularly for hybrid-modeling scenarios or grids that are narrow in one direction (e.g., cross-line), when the size of the absorbing boundary regions actually may be comparable to the size of the computational domain of interest. Moreover, if energy is incident under very low grazing angles, the PMLs will cause significant reflections. The search for a truly perfect absorbing boundary condition is therefore likely to continue as demands for absorbing boundary conditions that are effective in small (deformed) time-domain FD grids and extreme low angles of incidence is increasing with the interest in 3-D applications and hybrid modeling scenarios on the rise. One could then imagine solving the wave equation in 3-D on grids that were only one or a few grid-points wide in the cross-line direction, at a cost that would be comparable to that of a 2-D simulation. This would enable synthesizing seismic data with true 3-D amplitudes such that multiples, surface waves, guided waves and body waves all had the correct relative phases and amplitudes.

7. MEMORY OPTIMIZATION AND PARALLELIZATION

7.1. Memory Optimization

7.1.1. Memory Requirements in the Staggered-Grid Schemes

Many examples of FD modeling of earthquake motion in large sedimentary basins or seismic wave propagation in complex models in seismic exploration require very large computer memory and CPU. In order to solve the most challenging problems of seismology, it is necessary to reduce both memory and CPU requirements. Thus, memory optimization, sophisticated programming and parallelization are practically necessary. Here we briefly review several possibilities to reduce memory requirements and the number of operations. We also briefly mention parallelization.

Consider staggered-grid FD schemes. Assume that the material parameters of the medium can vary between any two grid positions. Such heterogeneity can be referred to as the point-to-point heterogeneity. Though it is sometimes an improper simplification, a homogeneous medium inside a grid cell is assumed by some modelers. Such material parameterization can be referred to as homogeneous material cells.

Denote the displacement and particle-velocity components at time level m as U^m, V^m, W^m, and $\dot{U}^m$, $\dot{V}^m$, $\dot{W}^m$, respectively. In the case of the point-to-point heterogeneity, each grid position of the displacement or particle-velocity components is assigned its value of density, that is, there are ρ^U, ρ^V and ρ^W assigned to a grid cell. Similarly, the elastic (unrelaxed) moduli μ^{xy}, μ^{yz} and μ^{zx} are assigned to grid positions of the shear stress-tensor components, while κ and μ are assigned to the joint grid position of the normal stress-tensor components. Corresponding to the elastic moduli, the anelastic coefficients Y_l^κ, Y_l^μ, $Y_l^{\mu xy}$, $Y_l^{\mu yz}$ and $Y_l^{\mu zx}$, $l = 1, \ldots, n$, are assigned to the grid positions of the stress-tensor components.

An assumption of homogeneous material cells reduces the material parameters in a grid cell to ρ, κ, μ, Y_l^κ and Y_l^μ, $l = 1, \ldots, n$. Consider displacement-stress (DS), displacement-velocity-stress (DVS) and velocity-stress (VS) staggered-grid FD schemes (for the equations and simplified schemes see, e.g., Moczo et al., 2001) which are 2nd-order accurate in time. The three schemes require the following variables to be stored in primary memory for a grid cell:

DS: $\quad U^m, V^m, W^m, U^{m-1}, V^{m-1}, W^{m-1},$ \hfill (186)

DVS: $\quad U^m, V^m, W^m, \dot{U}^{m-1/2}, \dot{V}^{m-1/2}, \dot{W}^{m-1/2},$ \hfill (187)

VS: $\quad \dot{U}^{m-1/2}, \dot{V}^{m-1/2}, \dot{W}^{m-1/2}, T_{xx}^{m-1}, T_{yy}^{m-1}, T_{zz}^{m-1}, T_{xy}^{m-1},$

$\quad T_{yz}^{m-1}, T_{zx}^{m-1}.$ \hfill (188)

As discussed in the section on incorporation of the realistic attenuation, the coarse spatial sampling of the anelastic functions (memory variables) can be considered. In the spatial distribution used by Kristek and Moczo (2003), $n = 4$ and each grid cell accommodates all six anelastic functions for just one of the relaxation frequencies, that is ξ_l^{xx}, ξ_l^{yy}, ξ_l^{zz}, ξ_l^{xy}, ξ_l^{yz} and ξ_l^{zx}, $l \in \{1, 2, 3, 4\}$.

Consider a computational region with dimensions XL, YL and ZL. Let $\beta_{\min}$ and $\alpha_{\max}$ be the minimum S-wave and maximum P-wave velocity, respectively. Let f_{ac} be the frequency up to which the FD computation should be sufficiently accurate. The maximum spatial grid spacing is the $h = (\beta_{\min}/f_{ac}) \cdot s$. Here, s is the spatial sampling ratio $s = h/\lambda_{\min}$ which has to be chosen based on the grid dispersion in the considered FD scheme and the wave propagation distance. The time step is $\Delta t = q \cdot p \cdot (h/\alpha)_{\min}$, $0 < p \leq 1$, where p is the stability ratio and q depends on the FD scheme. For example, for the 2nd-order in time, 4th-order in space staggered-grid, $q = 6/(7\sqrt{3})$; see, e.g. Moczo et al. (2000).

Assuming a uniform grid with the grid spacing h in the three Cartesian directions, the numbers of the grid cells in the three directions are $MX = (XL/h) + 1$, $MY = (YL/h) + 1$, $MZ = (ZL/h) + 1$.

In summary, the number of material parameters and field variables in the staggered-grid FD schemes are:

$$DS, DVS: \quad MX \cdot MY \cdot MZ \cdot [(3 + 5 + 5 \cdot 4) + (6 + 6)]$$
$$= MX \cdot MY \cdot MZ \cdot 40 \qquad (189)$$

and

$$VS: \quad MX \cdot MY \cdot MZ \cdot [(3 + 5 + 5 \cdot 4) + (9 + 6)]$$
$$= MX \cdot MY \cdot MZ \cdot 43. \qquad (190)$$

Here, $(3 + 5 + 5 \cdot 4)$ stands for 3 densities, 5 unrelaxed moduli and $5 \cdot 4$ anelastic coefficients, while $(6 + 6)$ for 6 displacement/particle-velocity components and 6 anelastic functions in one grid cell. In the VS scheme, $(9 + 6)$ stand 3 for particle-velocity components, 6 stress-tensor components and 6 anelastic functions. If the anelastic coefficients are spatially distributed in the same manner as the anelastic functions, $5 \cdot 4$ is replaced by 5 in formulae (189) and (190). In the case of the homogeneous grid cells, $3 + 5 + 5 \cdot 4$ is replaced by $1 + 2 + 2 \cdot 4$.

Clearly, depending on the total number of the grid cells, $MX \cdot MY \cdot MZ$, the memory requirements can be very large.

The total number of the grid cells and thus the memory requirements can be reduced by using a higher-order approximation in space (for example, Dablain, 1986), grid with varying size of the grid spacing or discontinuous grid. The 4th-order staggered-grid schemes were introduced by Bayliss et al. (1986) and Levander (1988). In recent FD modeling, the 4th-order accuracy in space is almost necessary. Yomogida and Etgen (1993) and Rodrigues (1993) used the 8th-order displacement-stress schemes. The rectangular grid with a varying size of the grid spacings was first used by Boore (1970) in the 1-D problem. Mikumo and Miyatake (1987) applied the varying size of the grid spacing in the 3-D case in a homogeneous medium. Moczo (1989) applied the grid with the varying size of the grid spacing to the 2-D SH problem in the laterally heterogeneous medium, Pitarka (1999) presented the 3-D velocity-stress scheme. Jastram and Behle (1992), Jastram and Tessmer (1994), Falk et al. (1996), Moczo et al. (1996), Kristek et al. (1999), Aoi and Fujiwara (1999), Hayashi et al. (2001), and Wang et al. (2001) introduced discontinuous grids.

Given some spatial grid, the core memory can be reduced using core memory optimization. While simple variants were used before, Graves (1996) described an optimized procedure. Moczo et al. (1999) presented a combined memory optimization which naturally combines core and disk memory optimizations. One other possibility to reduce the core memory is the use of the material cell types (e.g., Moczo et al., 2001). Here we briefly characterize the corresponding reductions.

7.1.2. Material Cell Types

In most models it is possible to efficiently describe heterogeneity of the medium by a spatial distribution of integer numbers (a look-up table). In this fashion each grid cell is assigned an integer number which represents a type of a material grid cell, that is, a set of material parameters characterizing the medium inside the grid cell. Such a description is efficient if there are sub-volumes of the computational model that can be covered by many grid cells of the same material type. Let K be the total number of different material types necessary to characterize heterogeneity of the whole model. Then the number of the material parameters and variables needed by the schemes is reduced to

$$\text{DS, DVS:} \quad MX \cdot MY \cdot MZ \cdot (6 + 6 + 1) + (3 + 5 + 5 \cdot 4) \cdot K \quad (191)$$

and

$$\text{VS:} \quad MX \cdot MY \cdot MZ \cdot (9 + 6 + 1) + (3 + 5 + 5 \cdot 4) \cdot K \quad (192)$$

assuming that $K < MX \cdot MY \cdot MZ$ which is the case even in relatively complex models. The number one added to the numbers of the field variables represents the distribution of the integers corresponding to the material cell types. In the case of the coarsely distributed anelastic coefficients or homogeneous material grid cells, the same reductions as in formulae (189) and (190) apply.

7.1.3. Core Memory Optimization

In core memory optimization method (Graves, 1996) such that only limited number of grid planes are kept in core memory all possible time updates for these planes are carried out. The subset of planes, say, NP planes, repeatedly moves throughout the whole model space, and the displacement or particle-velocity components, stress-tensor components and anelastic functions are successively (plane by plane) and periodically overwritten in disk. The key aspect in that all possible time updates are performed for the planes in the core memory means that the procedure is split into three parts—roll-in, cascade, and roll-out. Consider, for example, that the subset is made of horizontal grid planes. Then in the formulae (191) and (192) MZ is replaced by NP. The smaller the NP, the larger the reduction is.

7.1.4. Combined Memory Optimization

It is obvious that whereas the requirement for core memory can be significantly reduced, the needed amount of disk space can become very large. Although available disk space becomes larger and access to disk memory becomes faster, still, in principle, it is possible to reduce also the disk memory needed in the core-memory-optimization procedure. In the disk memory optimization (Moczo *et al.*,

1999), the wavelet transform is applied first to 2-D array of each displacement or particle-velocity component and each anelastic function. The transform decreases the information entropy. Then the sets of the wavelet coefficients are compressed by a standard compression procedure. Thus, for example, instead of the $MX \cdot MY$ displacement-component values in one grid plane only a relatively short stream of zeros and ones has to be written in disk. The total number of the material parameters and field variables stored in disk is then

$$\text{DS, DVS:} \quad MX \cdot MY \cdot MZ \cdot [(6+6)/CR + 1] + (3 + 5 + 5 \cdot 4) \cdot K \tag{193}$$

and

$$\text{VS:} \quad MX \cdot MY \cdot MZ \cdot [(9+6)/CR + 1] + (3 + 5 + 5 \cdot 4) \cdot K \tag{194}$$

where CR is the compression ratio. A reasonable value is about 10. Moczo et al. (1999) found that the increase of the CPU time due to one passage of the subset of planes with compression was always smaller than 0.75% of the time for one passage without compression. This is because the increase due to compression itself is partly compensated by the smaller number of the I/O operations.

7.1.5. Spatial Discontinuous Grid

In many models the S- and P-wave velocities are lower near the Earth's surface. In such a case it is advantageous to cover the lower part with a coarser spatial grid. Due to the structure of the staggered grid, the most natural combined (discontinuous) grid is the one whose upper part is the $h \times h \times h$ grid and the lower part the $3h \times 3h \times 3h$ grid (Aoi and Fujiwara, 1999; Kristek et al., 1999; Robertsson and Holliger, 1997). Let MZH be the number of the grid cells in the z-direction in the upper grid. Let $MZ3H$ be the number of the grid cells in the z-direction in the lower grid. Assuming only the material cell types, the number of material parameters and variables needed by the schemes is

$$\text{DS, DVS:} \quad \left[MX \cdot MY \cdot MZH + \left(\frac{MX - 1}{3} + 1 \right) \cdot \left(\frac{MY - 1}{3} + 1 \right) \right.$$
$$\left. \times MZ3H \right] \cdot \left[6 + 6 + 1 + (3 + 5 + 5 \cdot 4) \cdot K \right] \tag{195}$$

and

$$\text{VS:} \quad \left[MX \cdot MY \cdot MZH + \left(\frac{MX - 1}{3} + 1 \right) \cdot \left(\frac{MY - 1}{3} + 1 \right) \right.$$
$$\left. \times MZ3H \right] \cdot \left[9 + 6 + 1 + (3 + 5 + 5 \cdot 4) \cdot K \right]. \tag{196}$$

Formulae (195) and (196) are easy to modify if the core or combined memory optimization are applied.

A combination of the material cell types, combined memory optimization and discontinuous grid typically can reduce the memory requirements by more than one order of magnitude. An example is given by Moczo *et al.* (2001) for the modeling of the 1995 Kobe, Japan, earthquake.

7.1.6. Spatially Varying Time Step

Falk *et al.* (1998) and Tessmer (2000) introduced a combined grid in which a smaller time step is applied to the upper part of the grid while a larger time step is applied to the lower part. Their techniques considerably reduce CPU cost.

7.1.7. Discontinuous Space–Time Grids

Kang and Baag (2004a, 2004b) developed efficient techniques for 2-D and 3-D 4th-order staggered-grid modeling. They combined discontinuous grid in space with a discontinuous time step. While time integration in a finer grid with grid spacing h is performed with time step Δt, time integration in a coarser grid with grid spacing $3h$ is performed with time step $3\Delta t$. The finer grid covers 2-D or 3-D rectangular subregion which may have a planar free surface. This enables efficient modeling of localized surface sedimentary structures. Proportionality of the time step to the grid spacing is due to the fact that the two spatial grids have to overlap in the medium with a higher speed. The technique considerably reduces both the number of grid points and the number of operations.

7.2. Parallelization

Because the FD operators are local, the FD algorithms are suitable for parallelization. Over the past decade, several distinct approaches have been applied to parallelize FD codes.

7.2.1. Message Passing Libraries

Message-passing libraries, such as the Message Passing Interface (MPI, Gropp *et al.*, 1994), represent an approach suitable for shared or distributed memory architectures. The MPI typically requires the involvement of the sender-receiver communication: the source process makes a call to send data and the destination process makes a call to receive it. While the scaling of the MPI parallelized codes can be very good, the preparation of the code can require considerable time and effort. The resulting MPI parallelized code often differs substantially from the original source code.

7.2.2. Parallelizing Compilers

Some compilers are capable of producing a parallel code that is portable to shared and distributed memory machines. The compiler analyzes the inside-code dependences and can produce a parallelized code with or without additional user's directives and/or language extensions. The High-performance Fortran language (HPF, Koelbel *et al.*, 1994) is one of the best known examples. In some cases the resulting parallel code is quite efficient (e.g., Caserta *et al.*, 2002) but deficiencies of the approach are also known. In an automatic parallelization regime, compilers often make conservative assumptions on data dependence, which usually yields lower efficiency.

7.2.3. Interactive Parallelization Tools

There are parallelization tools which provide a user with the possibility of combining the automatic parallelization analysis with user's knowledge of the code. Whereas such interactive tools usually lead to quick parallelization, the efficiency can be limited due to non-optimized nesting, multiple decompositions and Fortran 90 constructs. Often additional manual modifications of the source code are necessary to obtain good performance.

7.2.4. High-Level Library-Based Tools

Library-based tools exist which are designed to help the user with the application of the lower-level libraries, such as MPI. Particularly, a user does not need to handle all details of the MPI parallelization. Some tools enable the use of additional optimizations specific for the machine architecture. The preparation of the parallel code may still be very time consuming and invasive.

7.2.5. Directive-Based Parallelization

Some computer manufacturers, for example Cray and SGI, introduced the possibility to manually supplement parallelization directives in the source code. In the beginning, the directives were mainly used to support loop-level shared memory parallelization. Recently SGIs OpenMP has become the best known and standard tool for relatively efficient and quick parallelization. One problem is that the OpenMP is most efficient only with shared-memory architectures.

ACKNOWLEDGEMENTS

The authors would like to thank Y.H. Chen, L.A. Dalguer, S.M. Day, R.T. Coates, M. Gális, R.J. Geller, I. Ionescu, J. Kristek, V. Maupin, S. Nielsen, and J. Virieux for useful discussions and comments on selected topics. Thanks also go to D. Lafleur and W. Kimman for misprint corrections. The work was supported in part also by the Geophysical Institute, Slovak Academy of Sciences.

REFERENCES

Aki, K., Richards, P.G. (1980). Quantitative Seismology. Theory and Methods, vols. I and II. Freeman, San Francisco.

Alekseev, A.S., Mikhailenko, B.G. (1980). The solution of dynamic problems of elastic wave propagation in inhomogeneous media by a combination of partial separation of variables and finite-difference methods. *J. Geophys.* **48**, 161–172.

Alterman, Z., Karal, F.C. (1968). Propagation of elastic waves in layered media by finite-difference methods. *Bull. Seismol. Soc. Am.* **58**, 367–398.

Anderson, D.A., Tannehill, J.C., Pletcher, R.H. (1984). Computational Fluid Mechanics and Heat Transfer. Hemisphere Publishing Corporation.

Andrews, D.J. (1973). A numerical study of tectonic stress release by underground explosions. *Bull. Seismol. Soc. Am.* **63**, 1375–1391.

Andrews, D.J. (1976). Rupture propagation with finite stress in antiplane strain. *J. Geophys. Res.* **81**, 3575–3582.

Andrews, D.J. (1999). Test of two methods for faulting in finite-difference calculations. *Bull. Seismol. Soc. Am.* **89**, 931–937.

Aochi, H., Fukuyama, E. (2002). Three-dimensional non-planar simulation of the 1992 Landers earthquake. *J. Geophys. Res.* **107**, doi:10.1029/2000JB000061.

Aoi, S., Fujiwara, H. (1999). 3-D finite-difference method using discontinuous grids. *Bull. Seismol. Soc. Am.* **89**, 918–930.

Backus, G.E. (1962). Long-wave elastic anisotropy produced by horizontal layering. *J. Geophys. Res.* **67**, 4427–4440.

Bayliss, A., Jordan, K.E., LeMesurier, B.J., Turkel, E. (1986). A fourth-order accurate finite-difference scheme for the computation of elastic waves. *Bull. Seismol. Soc. Am.* **76**, 1115–1132.

Bérenger, J.-P. (1994). A perfectly matched layer for the absorption of electromagnetic waves. *J. Comput. Phys.* **114**, 185–200.

Blanch, J.O., Robertsson, J.O.A. (1997). A modified Lax–Wendroff correction for wave propagation in media described by Zener elements. *Geophys. J. Int.* **131**, 381–386.

Blanch, J.O., Robertsson, J.O.A., Symes, W.W. (1995). Modeling of a constant Q: Methodology and algorithm for an efficient and optimally inexpensive viscoelastic technique. *Geophysics* **60**, 176–184.

Boore, D.M. (1970). Love waves in nonuniform waveguides: Finite difference calculations. *J. Geophys. Res.* **75**, 1512–1527.

Boore, D. (1972). Finite-difference methods for seismic wave propagation in heterogeneous materials. In: Bolt, B.A. (Ed.), In: *Methods in Computational Physics*, vol. 11. Academic Press, New York.

Bouchon, M. (1981). A simple method to calculate Green's functions for elastic layered media. *Bull. Seismol. Soc. Am.* **71**, 959–971.

Carcione, J.M. (1994). The wave equation in generalized coordinates. *Geophysics* **59**, 1911–1919.

Carcione, J.M. (2001). Wave Fields in Real Media: Wave Propagation in Anisotropic, Anelastic and Porous Media. Pergamon.

Carcione, J.M., Cavallini, F. (1994). A rheological model for anelastic anisotropic media with applications to seismic wave propagation. *Geophys. J. Int.* **119**, 338–348.

Carcione, J.M., Herman, G.C., ten Kroode, A.P.E. (2002). Seismic modeling. *Geophysics* **67**, 1304–1325.

Carcione, J.M., Kosloff, D., Kosloff, R. (1988a). Wave propagation simulation in a linear viscoacoustic medium. *Geophys. J.* **93**, 393–407.

Carcione, J.M., Kosloff, D., Kosloff, R. (1988b). Wave propagation simulation in a linear viscoelastic medium. *Geophys. J.* **95**, 597–611.

Carcione, J.M., Wang, J.P. (1993). A Chebyshev collocation method for the elastodynamic equation in generalized coordinates. *Comput. Fluid Dyn. J.* **2**, 269–290.

Carrio, P., Costa, J., Ferrer-Pinheiro, J.E., Schoenberg, M. (1992). Cross-borehole tomography in anisotropic media. *Geophysics* **57**, 1194–1198.

Caserta, A., Ruggiero, V., Lanucara, P. (2002). Numerical modelling of dynamical interaction between seismic radiation and near-surface geological structures: a parallel approach. *Comput. Geosci.* **28**, 1069–1077.

Chaljub, E., Komatitsch, D., Vilotte, J.P., Capdeville, Y., Valette, B., Festa, G. (2006). Spectral Element Analysis in Seismology. In: Wu, R.-S., Maupin, V. (Eds.), *Advances in Wave Propagation in Heterogeneous Earth*. In: *Advances in Geophysics*, vol. 48. Dmowska, R. (Ed.). Elsevier, p. 365 (this book).

Chen, Y.H., Coates, R.T., Robertsson, J.O.A. (2000). Extension of PML ABC to Elastic Wave Problems in General Anisotropic and Viscoelastic Media: *Schlumberger OFSR research note*.

Chew, W.C., Liu, Q.H. (1996). Perfectly matched layers for elastodynamics: a new absorbing boundary condition. *J. Comput. Acoustics* **4**, 341–359.

Chew, W.C., Weedon, W.H. (1994). A 3-D perfectly matched medium from modified Maxwell's equations with stretched coordinates. *Microwave Opt. Tech. Lett.* **7**, 599–604.

Clayton, R., Engquist, B. (1977). Absorbing boundary conditions for acoustic and elastic wave equations. *Bull. Seismol. Soc. Am.* **67**, 1529–1540.

Cohen, G.C. (2002). *Higher-Order Numerical Methods for Transient Wave Equations*. Springer.

Collino, F. (1993). High order absorbing boundary conditions for wave propagation models. In: Kleinman, R., et al. (Eds.), *Proceedings of the Second International Conference on Mathematical and Numerical Aspects of Wave Propagation*. SIAM, Delaware, pp. 161–171.

Collino, F., Tsogka, C. (1998). Application of the PML absorbing layer model to the linear elastodynamic problem in anisotropic heterogeneous media. *INRIA, Rapport de Recherche, No. 3471*.

Collino, F., Tsogka, C. (2001). Applications of the PML absorbing layer model to the linear elastodynamic problem in anisotropic heterogeneous media. *Geophysics* **66**, 294–305.

Coutant, O. (1989). Program of Numerical Simulation AXITRA. *Research Reports LGIT*. Université Joseph Fourier, Grenoble (in French).

Coutant, O., Virieux, J., Zollo, A. (1995). Numerical source implementation in a 2-D finite difference scheme for wave propagation. *Bull. Seismol. Soc. Am.* **85**, 1507–1512.

Crampin, S., Chastin, S. (2003). A review of shear wave splitting in the crack-critical crust. *Geophys. J. Int.* **155**, 221–240.

Crase, E. (1990). High-order (space and time) finite-difference modeling of the elastic wave equation. In: *60th Annual International Meeting, Society of Exploration Geophysicits*, pp. 987–991. Expanded Abstracts.

Crase, E., Wideman, Ch., Noble, M., Tarantola, A. (1992). Nonlinear elastic waveform inversion of land seismic reflection data. *J. Geophys. Res.* **97**, 4685–4703.

Cruz-Atienza, V.M., Virieux, J. (2004). Dynamic rupture simulation of non-planar faults with a finite-difference approach. *Geophys. J. Int.* **158**, 939–954.

Dablain, M.A. (1986). The application of high-order differencing to the scalar wave equation. *Geophysics* **51**, 54–66.

Dalguer, L.A., Day, S.M. (2004). Split nodes and fault zone models for dynamic rupture simulation. *EoS Trans. Am. Geophys. Union* **85** (47). Fall Meet. Suppl., Abstract S41A-0944.

Dalguer, L.A., Day, S.M. (2006). Comparison of fault representation methods in finite difference simulations of dynamic rupture. *Bull. Seismol. Soc. Am.*, in press.

Day, S.M. (1977). Finite element analysis of seismic scattering problems. Ph.D. Dissertation, *University of California, San Diego*.

Day, S.M. (1982). Three-dimensional simulation of spontaneous rupture: the effect of nonuniform prestress. *Bull. Seismol. Soc. Am.* **72**, 1881–1902.
Day, S.M. (1998). Efficient simulation of constant Q using coarse-grained memory variables. *Bull. Seismol. Soc. Am.* **88**, 1051–1062.
Day, S.M. (2005). Personal communication.
Day, S.M., Bradley, C.R. (2001). Memory-efficient simulation of anelastic wave propagation. *Bull. Seismol. Soc. Am.* **91**, 520–531.
Day, S.M., Dalguer, L.A., Lapusta, N., Liu, Y. (2005). Comparison of finite difference and boundary integral solutions to three-dimensional spontaneous rupture. *J. Geophys. Res. B* **110**, 12307, doi:10.1029/2005JB003813.
Day, S.M., Minster, J.B. (1984). Numerical simulation of wavefields using a Padé approximant method. *Geophys. J. R. Astron. Soc.* **78**, 105–118.
Durran, D.R. (1999). Numerical Methods for Wave Equations in Geophysical Fluid Dynamics. Springer.
Eisner, L., Clayton, R. (2002). Equivalent medium parameters for numerical modeling in media with near-surface low velocities. *Bull. Seismol. Soc. Am.* **92**, 711–722.
Emerman, S.H., Schmidt, W., Stephen, R.A. (1982). An implicit finite-difference formulation of the elastic wave equation. *Geophysics* **47**, 1521–1526.
Emmerich, H. (1989). 2-D wave propagation by a hybrid method. *Geophys. J. Int.* **99**, 307–319.
Emmerich, H. (1992). PSV-wave propagation in a medium with local heterogeneities: A hybrid formulation and its application. *Geophys. J. Int.* **109**, 54–64.
Emmerich, H., Korn, M. (1987). Incorporation of attenuation into time-domain computations of seismic wave fields. *Geophysics* **52**, 1252–1264.
Falk, J., Tessmer, E., Gajewski, D. (1996). Tube wave modelling by the finite-differences method with varying grid spacing. *Pure Appl. Geophys.* **148**, 77–93.
Falk, J., Tessmer, E., Gajewski, D. (1998). Efficient finite-difference modelling of seismic waves using locally adjustable time steps. *Geophys. Prosp.* **46**, 603–616.
Fäh, D. (1992). A hybrid technique for the estimation of strong ground motion in sedimentary basins. Diss. ETH Nr. 9767, *Swiss Federal Institute of Technology*, Zurich.
Festa, G., Nielsen, S. (2003). PML absorbing boundaries. *Bull. Seismol. Soc. Am.* **93**, 891–903.
Fink, M. (1997). Time reversed acoustics. *Phys. Today* **50**, 34–40.
Fornberg, B. (1988). Generation of finite difference formulas on arbitrary spaced grids. *Math. Comput.* **51**, 699–706.
Forsythe, G.E., Wasow, W.R. (1960). Finite Difference Methods for Partial Differential Equations. Wiley and Sons, New York.
Frankel, A. (1993). Three-dimensional simulations of ground motions in the San Bernardino Valley, California, for hypothetical earthquakes on the San Andreas fault. *Bull. Seismol. Soc. Am.* **83**, 1020–1041.
Frankel, A., Leith, W. (1992). Evaluation of topographic effects on P- and S-waves of explosions at the Northern Novaya Zemlya test site using 3-D numerical simulations. *Geophys. Res. Lett.* **19**, 1887–1890.
Geller, R.J., Ohminato, T. (1994). Computation of synthetic seismograms and their partial derivatives for heterogeneous media with arbitrary natural boundary conditions using the Direct Solution Method. *Geophys. J. Int.* **116**, 421–446.
Geller, R.J., Takeuchi, N. (1995). A new method for computing highly accurate DSM synthetic seismograms. *Geophys. J. Int.* **123**, 449–470.
Geller, R.J., Takeuchi, N. (1998). Optimally accurate second-order time-domain finite difference scheme for the elastic equation of motion: One-dimensional case. *Geophys. J. Int.* **135**, 48–62.
Givoli, D. (2004). High-order local non-reflecting boundary conditions: A review. *Wave Motion* **39**, 319–326.
Graves, R.W. (1996). Simulating seismic wave propagation in 3-D elastic media using staggered-grid finite differences. *Bull. Seismol. Soc. Am.* **86**, 1091–1106.

Graves, R.W., Day, S.M. (2003). Stability and accuracy analysis of coarse-grain viscoelastic simulations. *Bull. Seismol. Soc. Am.* **93**, 283–300.

Gropp, W., Lusk, E., Skjellum, A. (1994). Using MPI, Portable Parallel Programming with the Message Passing Interface. MIT Press.

Hayashi, K., Burns, D.R., Toksöz, M.N. (2001). Discontinuous-grid finite-difference seismic modeling including surface topography. *Bull. Seismol. Soc. Am.* **91**, 1750–1764.

Helbig, K. (1984). Anisotropy and dispersion in periodically layered media. *Geophysics* **49**, 364–373.

Hestholm, S. (1999). Three-dimensional finite difference viscoelastic wave modelling including surface topography. *Geophys. J. Int.* **139**, 852–878.

Hestholm, S., Ruud, B.O. (1994). 2-D finite difference elastic wave modeling including surface topography. *Geophys. Prosp.* **42**, 371–390.

Hestholm, S., Ruud, B.O. (2002). 3-D free-boundary conditions for coordinate-transform finite-difference seismic modeling. *Geophys. Prosp.* **50**, 463–474.

Higdon, R.L. (1991). Absorbing boundary conditions for elastic waves. *Geophysics* **56**, 231–241.

Holberg, O. (1987). Computational aspects of the choice of operator and sampling interval for numerical differentiation in large-scale simulation of wave phenomena. *Geophys. Prosp.* **35**, 629–655.

Holliger, K., Robertsson, J.O.A. (1998). Effects of the shallow subsurface on the upper crustal seismic reflection images. *Tectonophysics* **286**, 161–169.

Igel, H., Mora, P., Riollet, B. (1995). Anisotropic wave propagation through finite-difference grids. *Geophysics* **60**, 1203–1216.

Ilan, A. (1977). Finite-difference modeling for P-pulse propagation in elastic media with arbitrary polygonal surface. *J. Geophys.* **43**, 41–58.

Ilan, A., Loewenthal, D. (1976). Instability of finite-difference schemes due to boundary conditions in elastic media. *Geophys. Prosp.* **24**, 431–453.

Ilan, A., Ungar, A., Alterman, Z.S. (1975). An improved representation of boundary conditions in finite difference schemes for seismological problems. *Geophys. J. R. Astron. Soc.* **43**, 727–745.

Ionescu, I.R., Campillo, M. (1999). Influence of the shape of the friction law and fault finiteness on the duration of initiation. *J. Geophys. Res.* **104**, 3013–3024.

Isaacson, E., Keller, H.B. (1966). Analysis of Numerical Methods. Wiley and Sons, New York.

Jastram, C., Behle, A. (1992). Acoustic modeling on a grid of vertically varying spacing. *Geophys. Prosp.* **40**, 157–169.

Jastram, C., Tessmer, E. (1994). Elastic modelling on a grid with vertically varying spacing. *Geophys. Prosp.* **42**, 357–370.

Jih, R.-S., McLaughlin, K.L., Der, Z.A. (1988). Free-boundary conditions of arbitrary polygonal topography in a two-dimensional explicit elastic finite-difference scheme. *Geophysics* **53**, 1045–1055.

Kang, T.-S., Baag, Ch.-E. (2004a). Finite-difference seismic simulation combining discontinuous grids with locally variable timesteps. *Bull. Seismol. Soc. Am.* **94**, 207–219.

Kang, T.-S., Baag, Ch.-E. (2004b). An efficient finite-difference method for simulating 3-D seismic response of localized basin structures. *Bull. Seismol. Soc. Am.* **94**, 1690–1705.

Kay, I., Krebes, E.S. (1999). Applying finite element analysis to the memory variable formulation of wave propagation in anelastic media. *Geophysics* **64**, 300–307.

Kelly, K.R., Ward, R.W., Treitel, S., Alford, R.M. (1976). Synthetic seismograms: A finite-difference approach. *Geophysics* **41**, 2–27.

Klimeš, L. (1996). Accuracy of elastic finite differences in smooth media. *Pure Appl. Geophys.* **148**, 39–76.

Koelbel, C., Loverman, D., Shreiber, R., Steele Jr., G., Zosel, M. (1994). The High-Performance Fortran Handbook. MIT Press.

Komatitsch, D., Coutel, F., Mora, P. (1996). Tensorial formulation of the wave equation for modeling curved interfaces. *Geophys. J. Int.* **127**, 156–168.

Komatitsch, D., Tromp, J. (2003). A Perfectly Matched Layer (PML) absorbing condition for the second-order elastic wave equation. *Geophys. J. Int.* **154**, 146–153.

Kosloff, R., Kosloff, D. (1986). Absorbing boundaries for wave propagation problems. *J. Comput. Phys.* **63**, 363–376.

Kristek, J., Moczo, P. (2003). Seismic wave propagation in viscoelastic media with material discontinuities—a 3-D 4th-order staggered-grid finite-difference modeling. *Bull. Seismol. Soc. Am.* **93**, 2273–2280.

Kristek, J., Moczo, P., Archuleta, R.J. (2002). Efficient methods to simulate planar free surface in the 3-D 4th-order staggered-grid finite-difference schemes. *Studia Geophys. Geod.* **46**, 355–381.

Kristek, J., Moczo, P., Irikura, I., Iwata, T., Sekiguchi, H. (1999). The 1995 Kobe mainshock simulated by the 3-D finite differences. In: Irikura, K., Kudo, K., Okada, H., Sasatani, T. (Eds.), *The Effects of Surface Geology on Seismic Motion*, vol. 3. Balkema, Rotterdam, pp. 1361–1368.

Kummer, B., Behle, A. (1982). Second-order finite-difference modeling of SH-wave propagation in laterally inhomogeneous media. *Bull. Seismol. Soc. Am.* **72**, 793–808.

Kummer, B., Behle, A., Dorau, F. (1987). Hybrid modelling of elastic-wave propagation in two-dimensional laterally inhomogeneous media. *Geophysics* **52**, 765–771.

Laws, R., Kragh, E. (2002). Rough seas and time-lapse seismic. *Geophys. Prosp.* **50**, 195–208.

Lax, P.D., Wendroff, B. (1964). Difference schemes for hyperbolic equations with high order accuracy. *Commun. Pure Appl. Math.* **27**.

Lecomte, I., Gjøystdal, H., Maaø, F., Bakke, R., Drottning, Å., Johansen, T.-A. (2004). Efficient and flexible seismic modelling of reservoirs: The HybriSeis concept. *The Leading Edge* **23**, 432–437.

Levander, A.R. (1988). Fourth-order finite-difference P-SV seismograms. *Geophysics* **53**, 1425–1436.

Levander, A.R. (1989). Finite-difference forward modeling in seismology. In: James, D.E. (Ed.), *The Encyclopedia of Solid Earth Geophysics*. Van Nostrand Reinhold, pp. 410–431.

Liao, Z., Wong, H.L., Baipo, Y., Yifan, Y. (1984). A transmitting boundary for transient wave analysis. *Sci. Sinica A* **27**, 1063–1076.

Lindman, E.L. (1975). Free space boundary conditions for the time dependent wave equation. *J. Comput. Phys.* **18**, 66–78.

Liu, H.-P., Anderson, D.L., Kanamori, H. (1976). Velocity dispersion due to anelasticity; implications for seismology and mantle composition. *Geophys. J. R. Astron. Soc.* **47**, 41–58.

Luo, Y., Schuster, G. (1990). Parsimonious staggered grid finite-differencing of the wave equation. *Geophys. Res. Lett.* **17**, 155–158.

Madariaga, R. (1976). Dynamics of an expanding circular fault. *Bull. Seismol. Soc. Am.* **67**, 163–182.

Madariaga, R., Olsen, K., Archuleta, R. (1998). Modeling dynamics rupture in a 3-D earthquake fault model. *Bull. Seismol. Soc. Am.* **88**, 1182–1197.

Magnier, S.-A., Mora, P., Tarantola, A. (1994). Finite differences on minimal grids. *Geophysics* **59**, 1435–1443.

Marchuk, G.I. (1982). Methods of Numerical Mathematics. Springer Verlag.

Marcinkovich, C., Olsen, K. (2003). On the implementation of perfectly matched layers in a three-dimensional fourth-order velocity-stress finite difference scheme. *J. Geophys. Res. B* **108** (5), 2276, doi:10.1029/2002JB002235.

Marfurt, K.J. (1984). Accuracy of finite-difference and finite-element modeling of the scalar and elastic wave equations. *Geophysics* **49**, 533–549.

McDonal, F.J., Angona, F.A., Mills, L.R., Sengbush, R.L., van Nostrand, R.G., White, J.E. (1958). Attenuation of shear and compressional waves in Pierre shale. *Geophysics* **23**, 421–439.

Mikumo, T., Miyatake, T. (1987). Numerical modeling of realistic fault rupture processes. In: Bolt, B.A. (Ed.), *Seismic Strong Motion Synthetics*. Academic Press, pp. 91–151.

Mitchell, A.R. (1969). Computational Methods in Partial Differential Equations. Wiley and Sons, London.

Mitchell, A.R., Griffiths, D.F. (1994). The Finite Difference Method in Partial Differential Equations. Wiley and Sons, New York.

Mittet, R. (2002). Free-surface boundary conditions for elastic staggered-grid modeling schemes. *Geophysics* **67**, 1616–1623.

Mizutani, H., Geller, R.J., Takeuchi, N. (2000). Comparison of accuracy and efficiency of time-domain schemes for calculating synthetic seismograms. *Phys. Earth Planet. Int.* **119**, 75–97.
Moczo, P. (1989). Finite-difference technique for SH-waves in 2-D media using irregular grids—application to the seismic response problem. *Geophys. J. Int.* **99**, 321–329.
Moczo, P. (1998). Introduction to Modeling Seismic Wave Propagation by the Finite-Difference Method. Lecture Notes. Kyoto University. Available in pdf format at ftp://ftp.nuquake.sk/pub/Papers.
Moczo, P., Bard, P.-Y. (1993). Wave diffraction, amplification and differential motion near strong lateral discontinuities. *Bull. Seismol. Soc. Am.* **83**, 85–106.
Moczo, P., Bystrický, E., Kristek, J., Carcione, J.M., Bouchon, M. (1997). Hybrid modeling of P-SV seismic motion at inhomogeneous viscoelastic topographic structures. *Bull. Seismol. Soc. Am.* **87**, 1305–1323.
Moczo, P., Kristek, J. (2005). On the rheological models used for time-domain methods of seismic wave propagation. *Geophys. Res. Lett.* **32**, L01306, doi:10.1029/2004GL021598.
Moczo, P., Kristek, J., Bystrický, E. (2001). Efficiency and optimization of the 3-D finite-difference modeling of seismic ground motion. *J. Comput. Acoust.* **9**, 593–609.
Moczo, P., Kristek, J., Gális, M. (2004a). Simulation of planar free surface with near-surface lateral discontinuities in the finite-difference modeling of seismic motion. *Bull. Seismol. Soc. Am.* **94**, 760–768.
Moczo, P., Kristek, J., Halada, L. (2000). 3-D 4th-order staggered-grid finite-difference schemes: Stability and grid dispersion. *Bull. Seismol. Soc. Am.* **90**, 587–603.
Moczo, P., Kristek, J., Halada, L. (2004b). The Finite-Difference Method for Seismologists. An Introduction. Comenius University, Bratislava. Available in pdf format at ftp://ftp.nuquake.sk/pub/Papers.
Moczo, P., Kristek, J., Vavryčuk, V., Archuleta, R.J., Halada, L. (2002). 3-D heterogeneous staggered-grid finite-difference modeling of seismic motion with volume harmonic and arithmetic averaging of elastic moduli and densities. *Bull. Seismol. Soc. Am.* **92**, 3042–3066.
Moczo, P., Labák, P., Kristek, J., Hron, F. (1996). Amplification and differential motion due to an antiplane 2-D resonance in the sediment valleys embedded in a layer over the halfspace. *Bull. Seismol. Soc. Am.* **86**, 1434–1446.
Moczo, P., Lucká, M., Kristek, J., Kristeková, M. (1999). 3-D displacement finite differences and a combined memory optimization. *Bull. Seismol. Soc. Am.* **89**, 69–79.
Morton, K.W., Mayers, D.F. (1994). Numerical Solution of Partial Differential Equations. Cambridge Univ. Press.
Mufti, I.R. (1985). Seismic modeling in the implicit mode. *Geophys. Prosp.* **33**, 619–656.
Muir, F., Dellinger, J., Etgen, J., Nichols, D. (1992). Modeling elastic fields across irregular boundaries. *Geophysics* **57**, 1189–1193.
Murphy III, W.F. (1982). Effects of partial saturation on attenuation in Massilon sandstone and Vycor porous glass. *J. Acoust. Soc. Am.* **71**, 1458–1468.
Nielsen, S., Carlson, J.M. (2000). Rupture pulse characterization: Self-healing, self-similar, expanding solutions in a continuum model of fault dynamics. *Bull. Seismol. Soc. Am.* **90**, 1480–1497.
Nielsen, S., Carlson, J.M., Olsen, K. (2000). Influence of friction and fault geometry on earthquake rupture. *J. Geophys. Res. B* **105**, 6069–6088.
Nielsen, P., If, F., Berg, P., Skovgaard, O. (1994). Using the pseudospectral technique on a curved grid for 3-D acoustic forward modeling. *Geophys. Prosp.* **42**, 321–341.
O'Brien, G., Hyman, M., Kaplan, S. (1951). A study of the numerical solution of partial differential equations. *J. Math. Phys.* **29**, 233–251.
Ohminato, T., Chouet, B.A. (1997). A free-surface boundary condition for including 3-D topography in the finite-difference method. *Bull. Seismol. Soc. Am.* **87**, 494–515.
Ottosen, N.S., Petersson, H. (1992). Introduction to the Finite Element Method. Prentice Hall.
Peng, C., Toksöz, M.N. (1994). An optimal absorbing boundary condition for finite difference modeling of acoustic and elastic wave propagation. *J. Acoust. Soc. Am.* **95**, 733–745.

Peng, C., Toksöz, M.N. (1995). An optimal absorbing boundary condition for elastic wave modeling. *Geophysics* **60**, 296–301.

Pierson, W.J., Moskowitz, L. (1964). A proposed spectral form for fully developed wind seas based on the similarity theory of S.A. Kitaigorodskii. *J. Geophys. Res.* **69**, 5181–5190.

Pitarka, A. (1999). 3-D elastic finite-difference modeling of seismic motion using staggered grids with nonuniform spacing. *Bull. Seismol. Soc. Am.* **89**, 54–68.

Pitarka, A., Irikura, K. (1996). Modeling 3-D surface topography by finite-difference method: Kobe-JMA station site, Japan, case study. *Geophys. Res. Lett.* **23**, 2729–2732.

Pratt, R.G. (1990). Inverse theory applied to multi-source cross-hole tomography. Part 2: Elastic wave-equation method. *Geophys. Prosp.* **38**, 311–329.

Pratt, R.G., Shin, C., Hicks, G.J. (1998). Gauss–Newton and full Newton methods in frequency-space seismic waveform inversion. *Geophys. J. Int.* **133**, 341–362.

Richtmyer, R.D., Morton, K.W. (1967). Difference Methods for Initial Value Problems. Wiley and Sons, New York. (reprinted by Kreiger, New York, 1994).

Robertsson, J.O.A. (1996). A numerical free-surface condition for elastic/viscoelastic finite-difference modeling in the presence of topography. *Geophysics* **61**, 1921–1934.

Robertsson, J.O.A., Blanch, J.O., Symes, W.W. (1994). Viscoelastic finite-difference modeling. *Geophysics* **59**, 1444–1456.

Robertsson, J.O.A., Chapman, C.H. (2000). An efficient method for calculating finite-difference seismograms after model alterations. *Geophysics* **65**, 907–918.

Robertsson, J.O.A., Coates, R.T. (1997). Finite-difference modeling of Q for qP- and qS-waves in anisotropic media. In: *67th Annual International Meeting, Society of Exploration Geophysicists*, pp. 1846–1849. Expanded Abstracts.

Robertsson, J.O.A., Holliger, K. (1997). Modeling of seismic wave propagation near the Earth's surface. *Phys. Earth Planet. Int.* **104**, 193–211.

Robertsson, J.O.A., Laws, R., Chapman, C.H., Vilotte, J.-P., Delavaud, E. (2006). Modelling of scattering of seismic waves from a corrugated rough sea surface: A comparison of three methods. *Geophys. J. Int.*, in press.

Robertsson, J.O.A., Levander, A., Holliger, K. (1996). A hybrid wave propagation simulation technique or ocean acoustic problems. *J. Geophys. Res.* **101**, 11225–11241.

Robertsson, J.O.A., Ryan-Grigor, S., Sayers, C., Chapman, C.H. (2000). A finite-difference injection approach to modeling of seismic fluid flow monitoring. *Geophysics* **65**, 896–906.

Rodrigues, D. (1993). Large scale modelling of seismic wave propagation. Ph.D. Thesis, Ecole Centrale Paris.

Rodrigues, D., Mora, P. (1993). An efficient implementation of the free-surface boundary condition in 2-D and 3-D elastic cases. In: *63th Annual International Meeting, Society of Exploration Geophysicists*, pp. 215–217. Expanded Abstracts.

Saenger, E.H., Gold, N., Shapiro, S.A. (2000). Modeling the propagation of elastic waves using a modified finite-difference grid. *Wave Motion* **31**, 77–92.

Saenger, E.H., Bohlen, T. (2004). Finite-difference modeling of viscoelastic and anisotropic wave propagation using the rotated staggered grid. *Geophysics* **69**, 583–591.

Schneider, J.B., Ramahi, O.M. (1998). The complementary operators method applied to acoustic finite-difference time-domain simulations. *J. Acoust. Soc. Am.* **104**, 686–693.

Schoenberg, M., Muir, F. (1989). A calculus for finely layered anisotropic media. *Geophysics* **54**, 581–589.

Shtivelman, V. (1984). A hybrid method for wave field computation. *Geophys. Prosp.* **32**, 236–257.

Shtivelman, V. (1985). Two-dimensional acoustic modelling by a hybrid method. *Geophysics* **50**, 1273–1284.

Slawinski, R.A., Krebes, E.S. (2002). The homogeneous finite-difference formulation of the P-SV-wave equation of motion. *Studia Geophys. Geod.* **46**, 731–751.

Sochacki, J.S., George, J.H., Ewing, R.E., Smithson, S.B. (1991). Interface conditions for acoustic and elastic wave propagation. *Geophysics* **56**, 168–181.

Smith, W.D. (1974). A non-reflecting plane boundary for wave propagation problems. *J. Comput. Phys.* **15**, 492–503.
Spencer Jr., J.W. (1981). Stress relaxation at low frequencies in fluid-saturated rocks. *J. Geophys. Res.* **86**, 1803–1812.
Stead, R.J., Helmberger, D.V. (1988). Numerical-analytical interfacing in two dimensions with applications to modeling NTS seismograms. In: Aki, A., Wu, R.S. (Eds.), *Scattering and Attenuation of Seismic Waves*. Birkhauser, Basel, Switzerland, pp. 157–193.
Strang, G., Fix, G.J. (1973). An Analysis of the Finite Element Method. Prentice Hall, Englewood Cliffs, NJ.
Taflove, A., Hagness, S.C. (2005). Computational Electrodynamics: The Finite-difference Time-Domain Method. Artech House.
Takenaka, H., Furumura, T., Fujiwara, H. (1998). Recent developments in numerical methods for ground motion simulation. In: Irikura, K., Kudo, K., Okada, H., Sasatani, T. (Eds.), *The Effects of Surface Geology on Seismic Motion*, vol. 2. Balkema, Rotterdam, pp. 91–101.
Takeuchi, N., Geller, R.J. (2000). Optimally accurate second order time-domain finite difference scheme for computing synthetic seismograms in 2-D and 3-D media. *Phys. Earth Planet. Int.* **119**, 99–131.
Takeuchi, N., Geller, R.J. (2003). Accurate numerical methods for solving the elastic equation of motion for arbitrary source locations. *Geophys. J. Int.* **154**, 852–866.
Tal-Ezer, H., Carcione, J.M., Kosloff, D. (1990). An accurate and efficient scheme for wave propagation in linear viscoelastic media. *Geophysics* **55**, 1366–1379.
Tessmer, E. (2000). Seismic finite-difference modeling with spatially varying time steps. *Geophysics* **65**, 1290–1293.
Tessmer, E., Kosloff, D. (1994). 3-D elastic modeling with surface topography by a Chebyshev spectral method. *Geophysics* **59**, 464–473.
Tessmer, E., Kosloff, D., Behle, A. (1992). Elastic wave propagation simulation in the presence of surface topography. *Geophys. J. Int.* **108**, 621–632.
Thomson (Lord Kelvin), W. (1856). Elements of a mathematical theory of elasticity. *Philos. Trans. R. Soc.* **166**, 481–498.
van Manen, D.J., Robertsson, J.O.A., Curtis, A. (2005). Modeling of wave propagation in inhomogeneous media. *Phys. Rev. Lett.* **94**, 164301–164304.
Virieux, J. (1984). SH-wave propagation in heterogeneous media: Velocity-stress finite-difference method. *Geophysics* **49**, 1933–1957.
Virieux, J. (1986). P-SV wave propagation in heterogeneous media: Velocity-stress finite-difference method. *Geophysics* **51**, 889–901.
Wang, Y., Xu, J., Schuster, G.T. (2001). Viscoelastic wave simulation in basins by a variable-grid finite-difference method. *Bull. Seismol. Soc. Am.* **91**, 1741–1749.
Wang, T., Tang, X. (2003). Finite-difference modeling of elastic wave propagation: A nonsplitting perfectly matched layer approach. *Geophysics* **68**, 1749–1755.
Xu, T., McMechan, G.A. (1995). Composite memory variables for viscoelastic synthetic seismograms. *Geophys. J. Int.* **121**, 634–639.
Xu, T., McMechan, G.A. (1998). Efficient 3-D viscoelastic modeling with application to near-surface land seismic data. *Geophysics* **63**, 601–612.
Yomogida, K., Etgen, J.T. (1993). 3-D wave propagation in the Los Angeles basin for the Whittier–Narrows earthquake. *Bull. Seismol. Soc. Am.* **83**, 1325–1344.
Zahradník, J. (1995). Comment on 'A hybrid method for estimation of ground motion in sedimentary basins: Quantitative modeling for Mexico City' by D. Fäh, P. Suhadolc, St. Mueller and G.F. Panza. *Bull. Seismol. Soc. Am.* **85**, 1268–1270.
Zahradník, J., Hron, F. (1992). Robust finite-difference scheme for elastic waves on coarse grids. *Studia Geophys. Geod.* **36**, 1–19.
Zahradník, J., Moczo, P., Hron, F. (1993). Testing four elastic finite-difference schemes for behaviour at discontinuities. *Bull. Seismol. Soc. Am.* **83**, 107–129.

Zahradník, J., Moczo, P. (1996). Hybrid seismic modeling based on discrete-wavenumber and finite-difference methods. *Pure Appl. Geophys.* **148**, 21–38.

Zahradník, J., Priolo, E. (1995). Heterogeneous formulations of elastodynamic equations and finite-difference schemes. *Geophys. J. Int.* **120**, 663–676.

Zeng, X. (1996). Finite difference modeling of viscoelastic wave propagation in a generally heterogeneous medium in the time domain, and a dissection method in the frequency domain. Ph.D. Thesis, University of Toronto.

Zhang, J. (1997). Quadrangle-grid velocity-stress finite-difference method for elastic-wave-propagation simulation. *Geophys. J. Int.* **131**, 127–134.

Zhang, Ch., Symes, W.W. (1998). Fourth order, full-stencil immersed interface method for elastic waves with discontinuous coefficients. *1998 SEG Expanded Abstracts*.

Zienkiewicz, O.C., Taylor, R.L. (1989). The Finite Element Method, vol. 1, 4th edition. McGraw–Hill.

A LATTICE BOLTZMANN APPROACH TO ACOUSTIC-WAVE PROPAGATION

LIANJIE HUANG

Geophysics Group, Mail Stop D443, Los Alamos National Laboratory, Los Alamos, NM 87545, USA

ABSTRACT

It is difficult for simplified or approximated numerical modeling methods to accurately simulate complex wave phenomena. I present a lattice Boltzmann-based method for accurately modeling acoustic-wave propagation in strongly heterogeneous media. Rather than solving a partial-differential wave equation in a continuous medium, the lattice Boltzmann-based method directly simulates the physical processes of wave propagation: transportation, reflection, transmission, and collision of quasi-particles carrying pressure on a discrete lattice. It can accurately handle sharp interfaces at any locations along lattice links, and total reflections from free surfaces or empty bubbles, which are particularly difficult for classical finite-difference wave-equation methods to handle. The lattice Boltzmann-based method provides an accurate numerical modeling tool for simulating complex wave phenomena in strongly heterogeneous media containing complex geometrical shapes of sharp interfaces.

Keywords: Acoustic-wave propagation, Heterogeneous media, Lattice Boltzmann, Modeling, Phononic lattice solid, Sharp interface

1. INTRODUCTION

Only some physical parameters of a system can be measured directly using a given apparatus, but many of the unmeasured parameters may be important or even crucial for understanding the physical properties of the system. Observations of waves that have propagated through complex media are often used to estimate the physical properties of the media by interpreting or fitting the observed wave fields. Diffraction tomography and inversion are used to solve for the physical properties of this kind of complex and unreachable system (such as the solid Earth) from observations on their surfaces. The results of non-linear inversion rely on the ability to accurately simulate wave propagation in heterogeneous and complex media (e.g. Tarantola, 1984, 1986, 1987, 1988; Mora, 1987, 1988, 1989; Crase *et al.*, 1990; Sevink and Herman, 1996; Papazachos and Nolet, 1997; Bleistein *et al.*, 2001; Khan and Mosegaard, 2002).

To study wave phenomena observed on the surface of a system, the macroscopic approaches such as one-way and one-return approximations (see Chapters 5 and 6 of this book by Wu *et al.*) or classical finite differences (e.g. Dablain, 1986; Virieux, 1986; Holberg, 1987, 1988; Igel and Weber, 1996; Hestholm and Ruud, 1998; Hustedt *et al.*, 2004; Kang and Baag, 2004, see also Chapter 8 of this book by Moczo *et al.*) are generally used to solve acoustic- or elastic-wave equations, the partial-differential equations describing the behavior/physics of wave propagation in a continuous medium. Rather than solving a partial-differential equation in continuous time and space, a finite-difference equation is solved numerically for a continuous-valued wavefield at discrete time steps in discrete space in these approaches.

Real media are generally heterogeneous and complex, and involve a large number of sharp interfaces whose shapes can also be complex. The properties of heterogeneities in a complex medium may change drastically in space. However, to meet the condition of a continuous medium in wave equations, interfaces of media with different physical properties must be smoothed in finite-difference approaches and consequently, numerical errors arise in numerical experiments. The most rapid variation in properties can occur over one grid spacing in finite-difference methods. However, the medium is intrinsically smooth because it is assumed to be continuous along with its space derivatives. The more discontinuities present in the media, the greater are the numerical error due to error accumulation. The error could be significant for a highly heterogeneous and complex media such as fractured and fluid-filled porous media. In addition, classical finite-difference approaches suffer from instability problems in some cases depending on the medium properties and discretization parameters.

The important aspects of a numerical modeling method are its efficiency and its ability to simulate the desired (wave) phenomena. We are now in the era of supercomputers which could run at speeds approaching 100 Teraflops in the near future. Therefore, the most important aspect for a modeling method is its ability to simulate with essentially no approximations all wave phenomena resulting from the complexity of the media.

Lattice gas and lattice Boltzmann methods in which space, time, and dynamical variables are all discrete have successfully simulated complex hydrodynamical phenomena without essential approximations (Frisch *et al.*, 1986; Wolfram, 1986; d'Humières and Lallemand, 1987; Higuera, 1988; Rothman, 1988a, 1988b; Appert and Zaleski, 1990; Gunstensen and Rothman, 1991, 1992; Holme and Rothman, 1992; Succi and Benzi, 1993; Rothman and Zaleski, 1997; Chen and Doolen, 1998; He *et al.*, 1999; Kang *et al.*, 2002; Ubertini *et al.*, 2003; Xu and He, 2003; Inamuro *et al.*, 2004). The time evolution of the dynamical variables depends only on the values of the dynamical variables at nearest neighboring locations of a lattice site, and the stability of the methods is unconditional. Therefore, these microscopic/mesoscopic approaches are well suited for massively parallel computers to simulate complex phenomena.

An attempt at developing the corresponding fully discrete (or microscopic) methods for modeling wave propagation has been made since the birth of lattice gas methods. Rothman (1987) presented a cellular automaton method for modeling acoustic waves in a homogeneous medium. Huang et al. (1988) tried to model acoustic waves in inhomogeneous media by adjusting lattice spacings. However, their method requires simple velocity and density contrasts, and it could not handle irregular interfaces easily. Both methods evolve Boolean variables on a regular square lattice as is the case in the Hardy–Pomeau–Pazzis (HPP) lattice gas model (Hardy et al., 1973), and therefore, they lead to a non-physical anisotropic damping of the waves being modeled (see Huang et al., 1988).

A Boolean model updating the variables with a frequency proportional to the desired particle speed was proposed by Muir (1987b) and later by Mora and Maillot (1990) for the simulation of acoustic waves in inhomogeneous media. This approach has some numerical problems such as the growth of fine vertical perturbation bands from the interfaces. Four models based on lattice gases for modeling acoustic waves in heterogeneous media proposed by Muir (1987a, 1987b, 1991) differ in their data types, which are Boolean, probabilistic, real, and conventional conservation field variables, respectively. Muir introduced heterogeneity both in velocity and impedance, and observed the macroscopic anisotropy of waves in a finely layered isotropic medium. However, these models cannot handle heterogeneities with arbitrary velocity or impedance contrasts.

Mora (1992) introduced a method based on the Bose-like lattice Boltzmann method for the modeling of acoustic waves in heterogeneous media. In the method, a Boltzmann-like transport equation, which is a partial-differential equation describing the behavior of quasi-particles related with waves in continuous media, is solved using a finite-difference scheme. The disadvantages of the method include: (a) It requires a very small lattice spacing to handle (non-sharp) interfaces of small impedance contrasts; (b) It becomes inaccurate when the impedance contrast is large; and (c) It has a slow convergence rate in homogeneous regions due to finite-difference errors.

Huang and Mora (1994a) developed a lattice Boltzmann-based method for modeling acoustic-wave propagation in highly heterogeneous and complex media such as fractured and fluid-filled porous media. The method, termed *the phononic lattice solid by interpolation* (PLSI), is capable of simulating wave phenomena in complex media without essential approximations. The method is discrete in the dynamical variables themselves (i.e., wavefields), as well as space and time. It directly models the behavior of quasi-particle number densities (carrying pressure) on a discrete lattice rather than solving the corresponding macroscopic wave equation using finite differences. It employs the collision rules of the Bose-like Boltzmann method because the quasi-particles in the PLSI carry pressure and can superpose like bosons, rather than being exclusive like Fermi–Dirac-type particles. The PLSI is capable of handling sharp interfaces at any location along lattice links, and is comparable with lattice gas and lattice Boltzmann approaches from the point of view of discretization and numerical precision. It has been shown theoretically and numerically that the PLSI simulates acoustic waves in heterogeneous media at the macroscopic scale. The PLSI has been used for the simulation of acoustic-wave propagation in finely layered media, fractured media, and media with empty bubbles to demonstrate its ability to handle strong heterogeneities (Huang and Mora, 1996).

Theoretical studies of wave propagation in fluid-filled porous media are difficult due to the presence of fluids whose behavior is non-linear as well as viscous. Lattice site movements (i.e., finite deformations) induced by the passage of a macroscopic wave are particularly important for modeling wave propagation in a fluid-filled porous medium considering that non-linear solid–fluid interactions are thought to play a role in attenuation mechanisms. Huang and Mora (1994b) developed a lattice Boltzmann approach for studying effects of non-linear interactions of two acoustic media on acoustic-wave propagation. A lattice Bhatnager–Gross–Krook (BGK) method for modeling acoustic-wave propagation in strongly heterogeneous viscous media was developed by Huang et al. (1999). Ultimately, a fully discrete approach to elastic-wave propagation is required to properly simulate non-linear solid–fluid effects on elastic-wave propagation in fluid-filled porous media. Very few attempts have been made to model full elastic waves using a fully discrete/microscopic method. Maillot (1994) developed a semi-microscopic approach to elastic-wave propagation.

I first present here the history of the use of discretization and an overview of the lattice gas and lattice Boltzmann approaches. Then I present a lattice Boltzmann-based method for modeling acoustic-wave propagation in complex media (Huang and Mora, 1994a), absorbing boundary and free-surface conditions for the method (Huang et al., 2000), followed by some numerical modeling examples.

2. The History of the Use of Discretization

In 1866, Maxwell considered a gas as a set of particles, and founded the classical Kinetic Theory, in which the matter is discretized but the positions and velocities of particles vary continuously (Maxwell, 1890). This is the first use of discretization. The basis of the kinetic theory—the Boltzmann equation—was derived in 1872. It is particularly complex and therefore many simplified models were subsequently constructed.

Broadwell established the Discrete Kinetic Theory in 1964 and made the first step in the second stage of discretization in which both the matter and velocity are discretized (Broadwell, 1964). The general equations of the discrete kinetic theory were derived by Gatignol in 1970 (Gatignol, 1970). Several discrete models of the Boltzmann equation were developed later (e.g. Toscani, 1985).

The development of lattice gas methods for fluid dynamics is the third step of discretization. In the initial lattice gas model introduced by Hardy et al. (1973), particles at a lattice site on a 2-D regular lattice move to adjacent sites at each time step. They collide with each other if two particles reach the same lattice site at the same time, and move away such that the mass and momentum of particles are conserved. This first fully deterministic lattice gas is a model with discrete time, positions, and velocities, and it is named the HPP (Hardy–Pomeau–Pazzis) model. However, the HPP model has a non-isotropic character of the tensors related to the gases due to the use of a square lattice, and hence the modeled gases are different from reality.

The difficulties of the HPP model in coping with real fluid dynamics were overcome by replacing the square lattice with the triangular lattice. This was first achieved by Frisch *et al.* (1986) for the 2-D Navier–Stokes equations whose development was later extended to the 3-D case (d'Humières *et al.*, 1986; Frisch *et al.*, 1987). Since then, lattice gas methods have developed rapidly and have been used to simulate hydrodynamical phenomena such as fluid flow (d'Humières and Lallemand, 1987; Rivet *et al.*, 1988; Chen and Doolen, 1998), flow in porous media (Burges and Zaleski, 1987; Rothman, 1988a; Rothman and Keller, 1988; Appert and Zaleski, 1990; Gunstensen and Rothman, 1991; Holme and Rothman, 1992; Rothman and Zaleski, 1997; Kang *et al.*, 2002), magnetohydrodynamics (Chen and Matthaeus, 1987; Chen *et al.*, 1991; Succi *et al.*, 1991b; Martinez *et al.*, 1994; Schaffenberger and Hanslmeier, 2002), and other complex phenomena. Boghosian and Taylor (1995) formulated an exact kinetic theory for lattice gasses to ensure correct transport coefficients. The lattice gas approach is recognized as a new and powerful computational tool for the study of complex systems governed by related partial-differential equations (Doolen, 1990).

The lattice Boltzmann approach is an alternative to lattice gas methods for the study of hydrodynamic and other problems. It simulates the evolution of particle number-densities instead of Boolean variables that describe the presence or absence of a particle (McNamara and Zanetti, 1988; Higuera, 1988; Gunstensen *et al.*, 1991; Succi *et al.*, 1991a; Gunstensen and Rothman, 1992; Holme and Rothman, 1992; Succi and Benzi, 1993; Rothman and Zaleski, 1997; Chen and Doolen, 1998; He *et al.*, 1999; Kang *et al.*, 2002; Ubertini *et al.*, 2003; Xu and He, 2003; Inamuro *et al.*, 2004). A set of kinetic equations is solved in discrete space and discrete time. The Navier–Stokes equation is obtained in the limit of long wavelength and low frequency. Compared with original lattice gas models, one crucial feature of the continuum description of the lattice Boltzmann equation method is that it eliminates most of the noise of the system (McNamara and Zanetti, 1988; Higuera *et al.*, 1989). In addition, the lattice Boltzmann method has considerable flexibility in the choice of the local equilibrium particle distribution. In contrast, the Fermi–Dirac equilibrium is the only distribution usually considered for the lattice gas automata. This additional freedom for the lattice Boltzmann method allows us to choose an equilibrium distribution to achieve desired physical properties such as the Galilean-invariant convection and a velocity-independent pressure (Chen *et al.*, 1992;

Qian et al., 1992). The model proposed by Chen et al. (1992) is called *the pressure-corrected lattice Boltzmann equation (PCLBE)* and the model of Qian et al. (1992) is termed *the lattice Bhatnager–Gross–Krook (BGK) model*. Both models lead to recovery of the Navier–Stokes fluid equations in the macroscopic limit.

3. LATTICE GAS APPROACH

3.1. Particle Movement

A lattice gas is an analogic microscopic world to a real gas for the simulations of fluid dynamics (Frisch et al., 1986; Wolfram, 1986; Frisch et al., 1987). The lattice gas particles with unit mass move on a discrete lattice at a constant (unit) speed and collide at lattice sites. They are subject to an exclusion rule which requires that no more than one particle at a given lattice site can have a given momentum, namely, no more than one particle can move along the same lattice link in the same direction at the same time. Therefore, the maximum number of particles at each lattice site is 6 for the case of a 2-D triangular lattice. The particle moving along a specified direction at a lattice site will move to its adjacent site during the next time step, and consequently the particle movement step is exact.

3.2. Particle Collision Rules

Lattice gas particles arriving at lattice sites from different directions at the same time interact (i.e., collide) with each other such that both mass and particle momentum are conserved. For the six-particle FHP-I (Frisch–Hasslacher–Pomeau) model (Frisch et al., 1986), there are two kinds of collision rules for binary head-on collisions and triple collisions, respectively (Fig. 1). The labels of the unit vectors for a lattice site on a 2-D triangular lattice are given in the upper panel of Fig. 1. As shown in the upper panel of Fig. 1, there are two possible output configurations with probabilities of a (for particles departing along directions 2 and 5) and $(1 - a)$ (for particles departing along directions 3 and 6) for the given binary head-on collision with particles arriving at a lattice site along directions 1 and 4 at the same time. Equal probabilities for both output configurations are most commonly used. In this case, each output configuration has a probability of $1/2$.

Besides the laws for conservation of mass and momentum, there is an additional non-physical spurious conservation law for the head-on collisions on a 2-D triangular lattice (Frisch et al., 1986). The macroscopic hydrodynamical behavior of such a model with only head-on collisions would differ drastically from ordinary hydrodynamics, and therefore this non-physical spurious law must be removed so that the only conserved quantities are mass and momentum. For this purpose, the triple collisions rule was introduced (see the lower

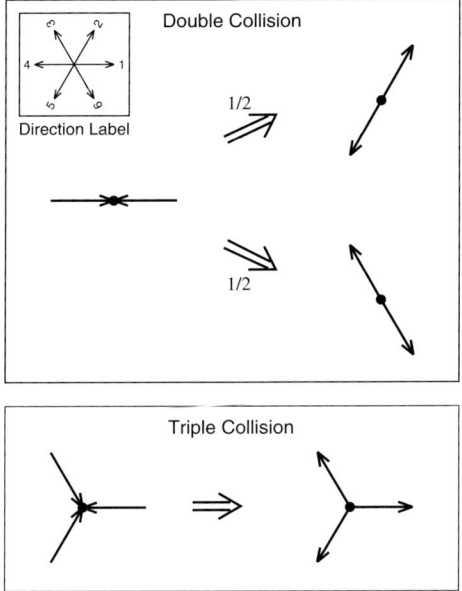

FIG. 1. FHP-I collision rules: head-on double collision (upper panel) and symmetric triple collision (lower panel).

panel of Fig. 1), in which particles arrive at the same lattice site from three symmetric directions (e.g., directions 2, 4 and 6) at the same time, and collide with each other, departing along the other three symmetric directions (i.e., directions 1, 3 and 5). The FHP-I model is the simplest set of collision rules without the non-physical spurious conservation law. Frisch *et al.* (1986) also introduced two other models termed the FHP-II and FHP-III models, both of which do not contribute to momentum transfer. The FHP-II model is a seven-bit variant of FHP-I model with an additional zero-velocity "rest particle" as well as some additional collision rules (Frisch *et al.*, 1986, 1987). The FHP-III model is a collision-saturated version of FHP-II (d'Humières and Lallemand, 1987; Frisch *et al.*, 1987).

A 3-D cubic lattice does not have enough symmetry to ensure macroscopic isotropy (Frisch *et al.*, 1986; d'Humières *et al.*, 1986; Wolfram, 1986) and consequently, the face-centered-hypercubic (FCHC)4-D model was introduced by d'Humières *et al.* (1986). To model 3-D fluids and maintain the required isotropy, they define the pseudo 4-D model as the 3-D projection of a FCHC model with unit periodicity in a particular direction.

After the collision process, lattice gas particles continue to move to adjacent sites during the next time step.

3.3. Microdynamical Equation

In the following, we consider the case of a 2-D triangular lattice. The number of lattice gas particles moving in the α-direction at a lattice site is the sum of the contribution of the movement step and the contribution of the collision step. The microdynamical equation governing the evolution of the lattice gas is therefore,

$$n_\alpha(\mathbf{x}, t + \Delta t) = n_\alpha(\mathbf{x} - \Delta \mathbf{x}_\alpha, t) + \Lambda_\alpha\big(n(\mathbf{x}, t)\big), \quad \alpha = 1, 2, \ldots, 6, \quad (1)$$

where $n = (n_1, n_2, \ldots, n_6)$ and $n_\alpha(\mathbf{x}, t + \Delta t)$ is the Boolean variable of lattice gas particles moving in the α-direction at lattice site $\mathbf{x}$ at time $t + \Delta t$ where Δt is time step selected to be unity. The value of the Boolean variable is either 0 or 1 because of the exclusion rule. $\Delta \mathbf{x}_\alpha = \Delta x \mathbf{e}_\alpha$ represents the vector in the α-direction between lattice sites with a lattice spacing Δx, where $\mathbf{e}_\alpha$ is the unit vector pointing in the α-direction. The lattice spacing Δx is selected for simplicity to be unity and hence $\Delta \mathbf{x}_\alpha$ is equal to the unit vector $\mathbf{e}_\alpha$ given by

$$\mathbf{e}_\alpha = \left(\cos \frac{(\alpha - 1)\pi}{3}, \sin \frac{(\alpha - 1)\pi}{3} \right), \quad \alpha = 1, 2, \ldots, 6. \quad (2)$$

For the FHP-I collision rules, the collision term Λ_α is given by

$$\begin{aligned}
\Lambda_\alpha = &\, a n_{\alpha+1} n_{\alpha+4}(1 - n_\alpha)(1 - n_{\alpha+2})(1 - n_{\alpha+3})(1 - n_{\alpha+5}) \\
&+ (1 - a) n_{\alpha+2} n_{\alpha+5}(1 - n_\alpha)(1 - n_{\alpha+1})(1 - n_{\alpha+3})(1 - n_{\alpha+4}) \\
&- n_\alpha n_{\alpha+3}(1 - n_{\alpha+1})(1 - n_{\alpha+2})(1 - n_{\alpha+4})(1 - n_{\alpha+5}) \\
&+ n_{\alpha+1} n_{\alpha+3} n_{\alpha+5}(1 - n_\alpha)(1 - n_{\alpha+2})(1 - n_{\alpha+4}) \\
&- n_\alpha n_{\alpha+2} n_{\alpha+4}(1 - n_{\alpha+1})(1 - n_{\alpha+3})(1 - n_{\alpha+5}),
\end{aligned} \quad (3)$$

where a is commonly selected to be $1/2$. The first three terms of the right-hand side of Eq. (3) refer to the head-on collisions, and the last two terms refer to the triple collisions. The terms with plus (or minus) sign represent populating (or depopulating) contributions to $n_\alpha(\mathbf{x}, t + \Delta t)$. In Eq. (3), the circular assumption for the subscripts is used. For instance, when $\alpha = 3$, then $\alpha + 4 = 7 \to 1$, $\alpha + 5 = 8 \to 2$, etc. The general form of the collision term (Frisch et al., 1986) is given by

$$\Lambda_\alpha(n) = \sum_{S, S'} (S'_\alpha - S_\alpha) A(S \to S') \prod_\beta n_\beta^{S_\beta} (1 - n_\beta)^{(1 - S_\beta)}, \quad (4)$$

where $S = (S_1, S_2, \ldots, S_6)$ and $S' = (S'_1, S'_2, \ldots, S'_6)$ represent the local states before and after collision processes, respectively, and $A(S \to S')$ is the transition probability from state S to S'. The Boolean variables S_α and S'_α are respectively for before and after collision processes.

A lattice gas obeying the microdynamical equation (1) has an equilibrium state and the equilibrium distribution is of the Fermi–Dirac type due to the exclusion rule. It evolves according to the Navier–Stokes equations in the incompressible hydrodynamical limit.

3.4. Conservation Relations

Mass and momentum of lattice gas particles conserve during the movement step and the collision step, respectively. Conservation of mass and momentum at each lattice site during the collision processes implies that

$$\sum_\alpha \Lambda_\alpha = 0 \tag{5}$$

and

$$\sum_\alpha \mathbf{e}_\alpha \Lambda_\alpha = 0. \tag{6}$$

Equations (5) and (6) lead to the following conservation relations for the Boolean field

$$\sum_\alpha n_\alpha(\mathbf{x}, t + \Delta t) = \sum_\alpha n_\alpha(\mathbf{x} - \Delta \mathbf{x}_\alpha, t), \tag{7}$$

$$\sum_\alpha \mathbf{e}_\alpha n_\alpha(\mathbf{x}, t + \Delta t) = \sum_\alpha \mathbf{e}_\alpha n_\alpha(\mathbf{x} - \Delta \mathbf{x}_\alpha, t). \tag{8}$$

4. Lattice Boltzmann Approaches

The Fermi–Dirac-type and Bose-like lattice Boltzmann approaches are respectively presented in the following.

4.1. I: A Fermi–Dirac-Type Approach

4.1.1. Lattice Boltzmann Equation

Rather than using the Boolean variable n_α, the mean population N_α defined by an ensemble average

$$N_\alpha(\mathbf{x}, t) = \langle n_\alpha(\mathbf{x}, t) \rangle \tag{9}$$

is used in lattice Boltzmann approaches. Making the ensemble average of the microdynamical equation (1) leads to

$$N_\alpha(\mathbf{x}, t + \Delta t) = N_\alpha(\mathbf{x} - \Delta \mathbf{x}_\alpha, t) + \Omega_\alpha(N(\mathbf{x}, t)),$$
$$\alpha = 1, 2, \ldots, 6, \tag{10}$$

where $N = (N_1, N_2, \ldots, N_6)$ and

$$\Omega_\alpha = \langle \Lambda_\alpha \rangle, \tag{11}$$

which, for the FHP-I collision rules, can be written as

$$\begin{aligned}\Omega_\alpha = &\, a\langle n_{\alpha+1}n_{\alpha+4}(1-n_\alpha)(1-n_{\alpha+2})(1-n_{\alpha+3})(1-n_{\alpha+5})\rangle \\ &+ (1-a)\langle n_{\alpha+2}n_{\alpha+5}(1-n_\alpha)(1-n_{\alpha+1})(1-n_{\alpha+3})(1-n_{\alpha+4})\rangle \\ &- \langle n_\alpha n_{\alpha+3}(1-n_{\alpha+1})(1-n_{\alpha+2})(1-n_{\alpha+4})(1-n_{\alpha+5})\rangle \\ &+ \langle n_{\alpha+1}n_{\alpha+3}n_{\alpha+5}(1-n_\alpha)(1-n_{\alpha+2})(1-n_{\alpha+4})\rangle \\ &- \langle n_\alpha n_{\alpha+2}n_{\alpha+4}(1-n_{\alpha+1})(1-n_{\alpha+3})(1-n_{\alpha+5})\rangle.\end{aligned} \quad (12)$$

In Eq. (12), $\langle \cdots \rangle$ denotes an ensemble average. Applying the Boltzmann approximation (the factorization hypothesis) to Eq. (12) gives

$$\begin{aligned}\Omega_\alpha = &\, a N_{\alpha+1} N_{\alpha+4}(1-N_\alpha)(1-N_{\alpha+2})(1-N_{\alpha+3})(1-N_{\alpha+5}) \\ &+ (1-a) N_{\alpha+2} N_{\alpha+5}(1-N_\alpha)(1-N_{\alpha+1})(1-N_{\alpha+3})(1-N_{\alpha+4}) \\ &- N_\alpha N_{\alpha+3}(1-N_{\alpha+1})(1-N_{\alpha+2})(1-N_{\alpha+4})(1-N_{\alpha+5}) \\ &+ N_{\alpha+1} N_{\alpha+3} N_{\alpha+5}(1-N_\alpha)(1-N_{\alpha+2})(1-N_{\alpha+4}) \\ &- N_\alpha N_{\alpha+2} N_{\alpha+4}(1-N_{\alpha+1})(1-N_{\alpha+3})(1-N_{\alpha+5}).\end{aligned} \quad (13)$$

Making the ensemble average over Eq. (4) and using the Boltzmann approximation yields the general form of the collision term

$$\Omega_\alpha(N) = \sum_{S,S'} (S'_\alpha - S_\alpha) A(S \to S') \prod_\beta N_\beta^{S_\beta}(1-N_\beta)^{(1-S_\beta)}. \quad (14)$$

Equation (10) with the collision term (13) or (14) is the lattice Boltzmann equation.

4.1.2. Fermi–Dirac Distribution

If $A(S \to S')$ in Eq. (14) satisfies the semi-detailed balance condition

$$\sum_S A(S \to S') = 1, \quad (15)$$

it can be shown that the collisions specified by Eq. (4) will force the system to a local equilibrium described by the Fermi–Dirac distribution (Frisch *et al.*, 1987)

$$N_\alpha = \frac{1}{1+\exp(\gamma_1 + \gamma_2 \mathbf{e}_\alpha \cdot \mathbf{v})}, \quad (16)$$

where γ_1 and γ_2 are the Lagrange multipliers determined by mass and momentum conservation laws and $\mathbf{v}$ is the mean velocity defined through the momentum

$$\rho \mathbf{v} = \sum_\alpha N_\alpha \mathbf{e}_\alpha, \quad (17)$$

where the particle (mass) density ρ is given by

$$\rho = \sum_\alpha N_\alpha. \tag{18}$$

4.1.3. Conservation Relations

It follows from Eqs. (5) and (6) by making ensemble averages that

$$\sum_\alpha \Omega_\alpha = 0, \tag{19}$$

and

$$\sum_\alpha \mathbf{e}_\alpha \Omega_\alpha = 0. \tag{20}$$

Averaging Eqs. (7) and (8) in the same way leads to the following conservation relations for the mean population and the momentum

$$\sum_\alpha N_\alpha(\mathbf{x}, t + \Delta t) = \sum_\alpha N_\alpha(\mathbf{x} - \Delta\mathbf{x}_\alpha, t), \tag{21}$$

and

$$\sum_\alpha \mathbf{e}_\alpha N_\alpha(\mathbf{x}, t + \Delta t) = \sum_\alpha \mathbf{e}_\alpha N_\alpha(\mathbf{x} - \Delta\mathbf{x}_\alpha, t). \tag{22}$$

4.1.4. Differential Boltzmann Transport Equation

In the macroscopic limit, i.e., when the macroscopic characteristic time and length are respectively much larger than the time step and lattice spacing, both of which are selected to be unity, Eq. (10) leads to the differential Boltzmann transport equation

$$\partial_t(N_\alpha) + \mathbf{e}'_\alpha \cdot \nabla N_\alpha = \Omega'_\alpha, \tag{23}$$

where ∂_t represents $\partial/\partial t$, $\mathbf{e}'_\alpha \equiv \mathbf{e}_\alpha/\Delta t = \mathbf{e}_\alpha$ when time step $\Delta t = 1$, and $\Omega'_\alpha \equiv \Omega_\alpha/\Delta t$ is the rate of change of the mean population N_α due to collision processes. Equation (23) is the continuum version of the kinetic equation.

4.1.5. Navier–Stokes Equations

Making use of a Chapman–Enskog expansion (Wolfram, 1986), one can obtain the fluid flow equations (Frisch et al., 1987)

$$\partial_t(\rho) + \nabla \cdot (\rho \mathbf{v}) = 0, \tag{24}$$

and

$$\partial_t(\rho\mathbf{v}) + \nabla \cdot [\rho g(\rho)\mathbf{v}\mathbf{v}] = -\nabla P + \nu\nabla^2(\rho\mathbf{v}), \tag{25}$$

where

$$P = \frac{1}{2}[\rho - g(\rho)\mathbf{v}^2], \tag{26}$$

ν is the viscosity and $g(\rho)$ is a function of density which should be unity but it is not unity for lattice gas models, resulting in non-Galilean effects. The incompressible Navier–Stokes equations could be recovered only in the low-Mach-number limit when time, pressure and viscosity are rescaled by the factor $g(\rho)$. Because $g(\rho)$ depends on density, this rescaling is consistent only for problems that have nearly constant density.

4.2. II: A Bose-Like Approach

Removing the exclusion principle from a Fermi–Dirac-type lattice Boltzmann method leads to a Bose-like lattice Boltzmann method in which the mean populations N_α's are unrestricted, non-negative, real-valued numbers. The lattice Boltzmann equation of the Bose-like approach still has the form of Eq. (10), but the change of the mean population due to collision processes is given by

$$\Omega_\alpha = \sum_{S,S'} (S'_\alpha - S_\alpha) A(S \to S') \prod_\beta N_\beta^{S_\beta}(1 + N_\beta)^{S'_\beta}. \tag{27}$$

It can be shown that the equilibrium distribution characterized by the conditions $\Omega_\alpha = 0$ for all α is the Bose–Einstein distribution function (Higuera, 1988)

$$N_\alpha = \frac{1}{\exp(\gamma_1 + \gamma_2 \mathbf{e}_\alpha \cdot \mathbf{v}) - 1}, \tag{28}$$

where γ_1 and γ_2 are the Lagrange multipliers determined by mass and momentum conservation laws.

5. A Lattice Boltzmann Method for Modeling Acoustic Waves: Phononic Lattice Solid by Interpolation

The phononic solid by interpolation (PLSI) is a lattice Boltzmann-based approach to wave propagation in heterogeneous acoustic media (Huang and Mora, 1994a). It is similar to the lattice gas method that is used to model idealized gas particles, but differs fundamentally in that quasi-particles in the PLSI method carry pressure wavefields rather than mass, and they propagate through a heterogeneous medium rather than a homogeneous medium. The PLSI simulates the physical processes of wave propagation. The number density of quasi-particles at each time step is obtained by calculating the contributions of four steps:

(1) transportation step (movement along the links between lattice nodes),
(2) transmission step,
(3) reflection step, and
(4) collision step.

5.1. Transportation Step

When a lattice link is homogeneous, the transmission and reflection steps are not required. For the 2-D PLSI method, we use a triangular lattice with each node on the lattice connected to other nodes located at equally spaced 60° azimuths. For the transportation step of quasi-particles carrying pressure along a homogeneous lattice link in the α-direction ($\alpha = 1, 2, \ldots, 6$), there is a group of subnodes between each pair of nodes as shown in Fig. 2a. The spacing of the subnodes is chosen such that quasi-particles move from one sub-node to the next in each time step Δt. Thus, the sub-node spacing varies spatially as velocity varies. In Fig. 2a, the lattice nodes are located at $\mathbf{x} - \Delta \mathbf{x}_\alpha$ and $\mathbf{x}$ and the subnodes are located at $\mathbf{x}_1$, $\mathbf{x}_2, \ldots$, etc. Quasi-particles carrying wavefields at $\mathbf{x} - \Delta \mathbf{x}_\alpha$ move to $\mathbf{x}_1$ in one time step. During subsequent time steps, quasi-particles move to $\mathbf{x}_2$, $\mathbf{x}_3$, etc. Therefore, at time t, the number density at $\mathbf{x}_j$ ($j = 1, 2, \ldots$) is given by

$$\tilde{N}_\alpha(\mathbf{x}_j, t) = N_\alpha(\mathbf{x} - \Delta \mathbf{x}_\alpha, t - j\Delta t). \tag{29}$$

The quasi-particles at $\mathbf{x}_*$ will move to lattice node at $\mathbf{x}$ during the next time step, and their number density $\tilde{N}_\alpha(\mathbf{x}_*, t)$ is calculated by interpolating the number densities of quasi-particles at the surrounding subnodes using an algorithm such as the Lagrange interpolation method (Zwillinger, 1996, p. 676). The contribution of the transportation step to the number density of quasi-particles moving in the α-direction at $\mathbf{x}$ at time $t + \Delta t$ is therefore given by

$$\tilde{N}_\alpha(\mathbf{x}, t + \Delta t) = \tilde{N}_\alpha(\mathbf{x}_*, t). \tag{30}$$

For the movement of quasi-particles in the α-direction along a homogeneous lattice link between lattice nodes at $\mathbf{x}$ and $\mathbf{x} + \Delta \mathbf{x}_\alpha$, a new set of subnodes is set up along the lattice link, and the number density at $\mathbf{x} + \Delta \mathbf{x}_\alpha$ is calculated in the same manner described above for calculating the number density at $\mathbf{x}$ using the number density at $\mathbf{x} - \Delta \mathbf{x}_\alpha$. For each node on the triangular lattice, such calculations are conducted for each of homogeneous lattice links along six directions.

5.2. Transmission and Reflection Steps

For the case when a lattice link is inhomogeneous (namely there is an interface at any location along the lattice link), transmission and reflection steps must be

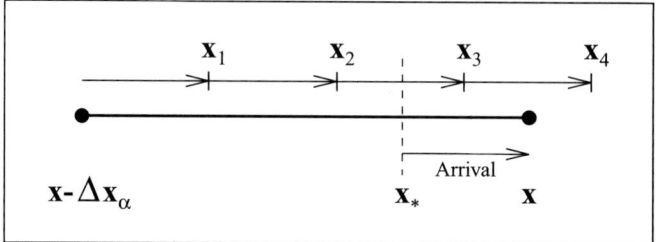

(a) Quasi-particle movement along a homogeneous link

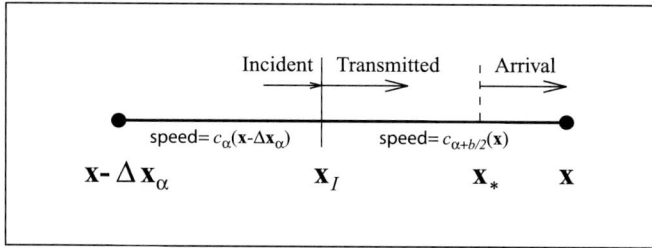

(b) Quasi-particle transmission along a heterogeneous link

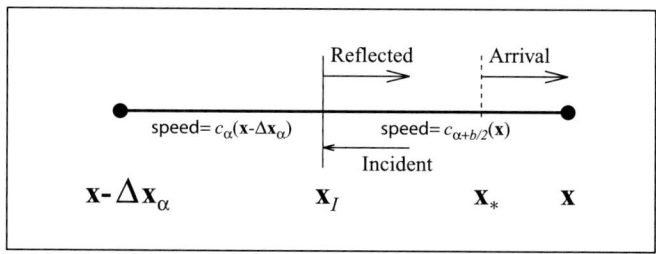

(c) Quasi-particle reflection along a heterogeneous link

FIG. 2. Illustration of quasi-particle movement along a lattice link. In (b) and (c), the interface is located at x_I that can be at any location between lattice nodes $\mathbf{x} - \Delta\mathbf{x}_\alpha$ and $\mathbf{x}$.

taken into account during quasi-particle movements along the lattice link. These steps account for the changes caused by spatial variations in velocity and/or density (Figs. 2b and c). As illustrated in Fig. 2b, quasi-particles carrying pressure move from $\mathbf{x} - \Delta\mathbf{x}_\alpha$ in the α-direction to the interface at x_I with a speed of $c_\alpha(\mathbf{x} - \Delta\mathbf{x}_\alpha)$, transmit through the interface where the number density of quasi-particles is multiplied by the transmission coefficient, then move to lattice node at $\mathbf{x}$ with a speed of $c_{\alpha+b/2}(\mathbf{x})$, which is the speed in the direction opposite to the α-direction, namely the speed on the other side of the interface. The transmission coefficient of quasi-particles is obtained by

$$T_\alpha(\mathbf{x} - \Delta\mathbf{x}_\alpha; \mathbf{x}) = \frac{2Z_{\alpha+b/2}(\mathbf{x})}{Z_{\alpha+b/2}(\mathbf{x}) + Z_\alpha(\mathbf{x} - \Delta\mathbf{x}_\alpha)}, \qquad (31)$$

where $Z_\alpha(\mathbf{x}) = \rho(\mathbf{x})c_\alpha(\mathbf{x})/\sqrt{D}$ is the impedance, $\rho(\mathbf{x})$ is the medium density, $c_\alpha(\mathbf{x})/\sqrt{D}$ the macroscopic wave speed of the medium, and D is the number of space dimensions. Figure 2c shows that quasi-particles move from $\mathbf{x}$ in the $\alpha + b/2$-direction to the interface at $\mathbf{x}_I$ with a speed of $c_{\alpha+b/2}(\mathbf{x})$, are reflected by the interface where the number density of quasi-particles is multiplied by the reflection coefficient, then move with the same speed to lattice node at $\mathbf{x}$. The reflection coefficient is given by

$$\begin{aligned}R_{\alpha+b/2}(\mathbf{x}; \mathbf{x} - \Delta\mathbf{x}_\alpha) &= -R_\alpha(\mathbf{x} - \Delta\mathbf{x}_\alpha; \mathbf{x}) \\ &= -\frac{Z_{\alpha+b/2}(\mathbf{x}) - Z_\alpha(\mathbf{x} - \Delta\mathbf{x}_\alpha)}{Z_{\alpha+b/2}(\mathbf{x}) + Z_\alpha(\mathbf{x} - \Delta\mathbf{x}_\alpha)}.\end{aligned} \qquad (32)$$

The movements of quasi-particles along links between a lattice node and the interface are the same as that along a homogeneous link (Fig. 2a). Therefore, the same interpolation algorithm used for a homogeneous link can be used to obtain the number densities of the incident quasi-particles and those of the quasi-particles at $\mathbf{x}_*$ (i.e. $\tilde{N}_\alpha^T(\mathbf{x}_*, t)$ for the transmission and $\tilde{N}_\alpha^R(\mathbf{x}_*, t)$ for the reflection), which will arrive at $\mathbf{x}$ during the next time step. The transmission and reflection steps require two interpolations, respectively. Huang and Mora (1994a) proposed two alternative approaches for obtaining the number densities of quasi-particles at $\mathbf{x}_*$ using only one interpolation for the transmission and reflection steps, respectively. In those approaches, a few fictitious values of quasi-particle number density along a homogeneous link are used. They can handle the cases when $|\mathbf{x} - \mathbf{x}_*| > |\mathbf{x} - \mathbf{x}_I|$ for the transmission and reflection steps.

5.3. Collision Step

Quasi-particles arriving at a lattice node from different directions at the same time collide with each other. After collision processes, the number density $N_\alpha(\mathbf{x}, t + \Delta t)$ of quasi-particles moving in the α-direction at $\mathbf{x}$ at time $t + \Delta t$ is given by

$$N_\alpha(\mathbf{x}, t + \Delta t) = \tilde{N}_\alpha(\mathbf{x}_*, t) + \Omega_\alpha(N(\mathbf{x}, t)), \qquad (33)$$

where $\alpha = 1, 2, \ldots, 6$, and $N = (N_1, N_2, \ldots, N_6)$. For an inhomogeneous lattice link, $\tilde{N}_\alpha(\mathbf{x}_*, t)$ in Eq. (33) is given by

$$\tilde{N}_\alpha(\mathbf{x}_*, t) = \tilde{N}_\alpha^T(\mathbf{x}_*, t) + \tilde{N}_\alpha^R(\mathbf{x}_*, t), \qquad (34)$$

where the terms $\tilde{N}_\alpha^T$ and $\tilde{N}_\alpha^R$ are the transmitted and reflected number densities, respectively. In Eq. (33), the term $\Omega_\alpha(N(\mathbf{x}, t))$ is the rate of change of the quasi-particle number density due to the collision process (Higuera, 1988)

$$\Omega_\alpha = \sum_{S,S'} (S'_\alpha - S_\alpha) A(S \to S') \prod_\beta n_\beta^{S_\beta} (1 + n_\beta)^{S'_\beta}, \tag{35}$$

with

$$n_\beta = d + N_\beta(\mathbf{x}, t), \quad \beta = 1, 2, \ldots, b, \tag{36}$$

where the term d is the background number density, the number density N_β of quasi-particles carrying wavefields is a small perturbation relative to the background number density d, and the Boolean variables S_α and S'_α define input and output states with a transition probability of $A(S \to S')$. The input configuration is characterized by a set of numbers $S = \{S_\alpha\}$ with $S_\alpha = 0$ or 1 for $\alpha = 1, 2, \ldots, 6$ to determine whether direction α contributes any particles to the collision ($S_\alpha = 1$) or not ($S_\alpha = 0$). Analogously, $S' = \{S'_\alpha\}$ with $S'_\alpha = 0$ or 1 for $\alpha = 1, 2, \ldots, 6$ determines the distribution of the outgoing particles among the directions of the lattice.

We use the first set of the FHP-I collision rules for the 2-D case (Fig. 1) (Frisch et al., 1986), which account for the head-on double collisions and the symmetric triple collisions. For instance, the collision term for direction $\alpha = 1$ is given by

$$\Omega_1 = \frac{1}{2}[-n_1 n_4 (h_2 h_5 + h_3 h_6) + h_1 h_4 (n_2 n_5 + n_3 n_6)]$$
$$+ \frac{1}{3}(n_2 n_4 n_6 h_1 h_3 h_5 - h_2 h_4 h_6 n_1 n_3 n_5), \tag{37}$$

with

$$h_\alpha = 1 + n_\alpha, \quad \alpha = 1, 2, \ldots, 6. \tag{38}$$

The terms in Eq. (37) account for the collision process described in the following. The first term is the result of the head-on collision of quasi-particles arriving in 1- and 4-directions and leaving both in 2- and 5-directions and 3- and 6-directions with a transition probability of 1/2 after the collision (see Fig. 1). Since this collision process removes particles from 1-direction, the term is negative. The second term is the increase in number density in 1-direction due to the head-on collisions of particles arriving from 2- and 5-directions and 3- and 6-directions. The third and fourth terms account for symmetric triple collisions. The third term is the collision of particles arriving in 2-, 4-, and 6-directions that scatter into 1-, 3-, and 5-directions (see Fig. 1). This term is positive since it adds number density in 1-direction. The last term is for particles arriving from 1-, 3-, and 5-directions that scatter into 2-, 4-, and 6-directions. This removes particles from 1-direction. The transition probability for the FHP-I collision rules are 1/2 for head-on collisions and 1/3 for symmetric triple collisions. The rate of change of the number density in other directions due to the FHP-I collisions can be obtained by rotating the subscripts of physical quantities in Eq. (37).

Higuera (1988) showed that the viscosity ν of the medium is related to the value of d by

$$\nu = \frac{1}{12d(1+d)} - \frac{1}{8}. \tag{39}$$

Equation (39) indicates that the viscosity vanishes for $d = \sqrt{11/12} - 1/2 \approx 0.457427$. The value of d is usually selected appropriate for zero viscosity. A nonzero viscosity is used for the absorbing boundary condition.

The advantage of the PLSI method is that it can easily handle models with sharp interfaces, complex surface topography, and strong velocity-contrast inclusions, including cavities. These features are very difficult for classical finite-difference solutions to the wave equation to reliably simulate

5.4. The First-Order Chapman–Enskog Expansion of the Quasi-Particle Number Density N_α

For the case without intrinsic anisotropy (i.e., quasi-particle speed $c_\alpha = c$), the macroscopic pressure P and velocity v_i are, respectively, defined as

$$P \equiv \frac{1}{b} \sum_\alpha N_\alpha, \tag{40}$$

and

$$v_i \equiv \frac{D}{b\rho c} \sum_\alpha N_\alpha e_{\alpha i}, \tag{41}$$

where b is the total number of directions in the lattice (i.e., $b = 6$ for a 2-D triangular lattice), and D is the number of space dimensions.

The expansion of N_α in term of the macroscopic velocity to the first order (i.e., the first order Chapman–Enskog expansion) has the form

$$N_\alpha = A_0 + A_1 v_i e_{\alpha i}, \tag{42}$$

where the undetermined coefficients A_0 and A_1 are independent of direction α. Summing Eq. (42) over α and making use of the relation

$$\sum_\alpha e_{\alpha i} = 0 \tag{43}$$

and the definition (40) yields

$$A_0 = P. \tag{44}$$

Multiplying Eq. (42) by $e_{\alpha j}$, summing it over direction α and making use of the relation

$$\sum_\alpha e_{\alpha i} e_{\alpha j} = \frac{b}{D} \delta_{ij} \qquad (45)$$

and the definition (41) yields

$$A_1 = \rho c. \qquad (46)$$

Substituting Eqs. (44) and (46) into Eq. (42) yields

$$N_\alpha = P + \rho c v_i e_{\alpha i}. \qquad (47)$$

This is the first order Chapman–Enskog expansion of N_α.

5.5. The Acoustic-Wave Equation

For the case without intrinsic anisotropy, the differential transport equation for the PLSI method can be written as (Huang and Mora, 1994a)

$$\frac{\partial N_\alpha}{\partial t} + c e_{\alpha j} \frac{\partial N_\alpha}{\partial x_j} = \frac{\partial \Omega_\alpha}{\partial t} + \frac{\partial N_\alpha^S}{\partial t}, \qquad (48)$$

where $\partial N_\alpha^S / \partial t$, the rate of change of particle number density due to scattering processes, is given by

$$\frac{\partial N_\alpha^S}{\partial t} = \frac{1}{2Z(\mathbf{x})} \frac{\partial Z}{\partial x_j} c(\mathbf{x}) e_{\alpha j} \left[N_\alpha(\mathbf{x}, t) - N_{\alpha+b/2}(\mathbf{x}, t) \right] \qquad (49)$$

with the impedance

$$Z(\mathbf{x}) = \rho(\mathbf{x}) \frac{c(\mathbf{x})}{\sqrt{D}}. \qquad (50)$$

Summing Eq. (49) over direction α and making use of the relation

$$e_{(\alpha+b/2)j} = -e_{\alpha j} \qquad (51)$$

gives

$$\sum_\alpha \frac{\partial N_\alpha^S}{\partial t} = \frac{1}{\rho} \frac{\partial (\rho c)}{\partial x_j} \sum_\alpha N_\alpha e_{\alpha j}. \qquad (52)$$

Multiplying Eq. (49) by $e_{\alpha i}$ and then summing it over direction α yields

$$\sum_\alpha \frac{\partial N_\alpha^S}{\partial t} e_{\alpha i} = 0, \qquad (53)$$

where the identity $e_{(\alpha+b/2)i} e_{(\alpha+b/2)j} = e_{\alpha i} e_{\alpha j}$ was used.

On the other hand, the energy and quasi-momentum conservation of particles during their collision processes implies that

$$\sum_\alpha \frac{\partial \Omega_\alpha}{\partial t} = 0, \tag{54}$$

and

$$\sum_\alpha \frac{\partial \Omega_\alpha}{\partial t} e_{\alpha i} = 0. \tag{55}$$

Summing the transport equation (48) over direction α and making use of the pressure definition (40) as well as the relations (54) and (52) yields

$$\frac{\partial P}{\partial t} + \frac{\rho c^2}{D}\frac{\partial v_j}{\partial x_j} = 0, \tag{56}$$

which can be integrated to give the classical relationship between pressure P and displacement u_j

$$P = P_0 - K \frac{\partial u_j}{\partial x_j}, \tag{57}$$

where

$$K = \rho \left(\frac{c}{\sqrt{D}}\right)^2 \tag{58}$$

is the bulk modulus of the medium which is defined by

$$K \equiv \rho C^2, \tag{59}$$

where C is the wave speed. It follows from Eqs. (58) and (59) that

$$C = \frac{c}{\sqrt{D}}. \tag{60}$$

This is the relationship between the wave speed and quasi-particle speed.

Equation (48) multiplied by $\rho[D/(b\rho c)]e_{\alpha i}$ and summed over α provides another equation. By applying the velocity definition (41) and the relations (53) and (55), it can be written as

$$\rho \frac{\partial v_i}{\partial t} + \frac{D}{b}\frac{\partial}{\partial x_j}\left(\sum_\alpha N_\alpha e_{\alpha i} e_{\alpha j}\right) = 0. \tag{61}$$

Using the Chapman–Enskog expansion (47) and the relations (45) and

$$\sum_\alpha e_{\alpha i} e_{\alpha j} e_{\alpha k} = 0, \tag{62}$$

Eq. (61) becomes

$$\rho \frac{\partial v_i}{\partial t} + \frac{\partial P}{\partial x_i} = 0. \tag{63}$$

Combining Eqs. (56) and (63) yields

$$\rho \frac{\partial^2 v_i}{\partial t^2} - \frac{\partial}{\partial x_i}\left(K \frac{\partial v_j}{\partial x_j}\right) = 0, \tag{64}$$

or

$$\frac{1}{K} \frac{\partial^2 P}{\partial t^2} - \frac{\partial}{\partial x_j}\left(\frac{1}{\rho} \frac{\partial P}{\partial x_i}\right) = 0. \tag{65}$$

Equations (64) and (65) are different forms of the acoustic-wave equation in heterogeneous media.

6. Absorbing Boundary Condition

6.1. Zero-Valued Reflection Coefficients at Absorbing Boundaries

For a lattice node at a model boundary, no quasi-particles exist outside of the model, and therefore there are no transmission and transportation steps at boundaries. Boundary artifacts arise only from the reflection and collision steps. To eliminate the contribution from the reflection step to these artifacts, the microscopic reflection-coefficients at boundaries are set to be zero.

A homogeneous model defined on a 160×160 triangular lattice was used to investigate the effectiveness of zero-valued microscopic reflection-coefficients. The quasi-particle speed is 0.4 and the density of the medium is 1.0 during the PLSI simulation. A pressure source with a first derivative of Gaussian time history was introduced at lattice node (80, 80). It had an amplitude of 0.010 and a time delay of 60 time steps. The fundamental frequency of the source was such that it generated waves with a wavelength of approximately 16 lattice spacings. Receivers were located along a horizontal line at a depth of 40 vertical lattice spacings from the upper boundary of the model. The background number density of quasi-particles is 0.457427 for a medium without viscosity (see Eq. (39)). Zero-valued reflection coefficients were set at all boundaries of the model. Computations were made using the PLSI method for 900 time steps. Figure 3a depicts the seismograms (i.e., pressure fields) recorded at the receivers, amplified by a factor of $t^{0.7}$ where t is the time step. In addition to the direct wave, one can see from Fig. 3a that some weak boundary reflections are still visible after the direct wave. This indicates that setting zero-valued reflection coefficients at boundaries is not adequate for eliminating the artificial boundary reflections.

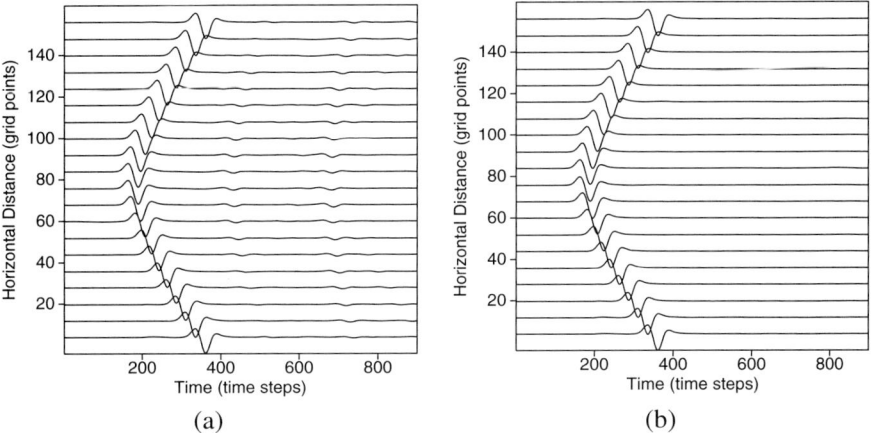

FIG. 3. The amplified seismograms recorded during the PLSI calculations for a homogeneous model with (a) zero-valued microscopic reflection-coefficients at the boundaries and (b) zero-valued microscopic reflection-coefficients plus viscous boundary layers.

6.2. Viscous Absorbing Boundary Layers

The only source of the artificial boundary reflections in Fig. 3a is the contribution of the quasi-particle collision process at the boundaries. A viscous layer is padded to each absorbing boundary in an attempt to further reduce the amplitude of the artificial boundary reflections. The viscosity of a lattice solid model may be adjusted by selection of the background number density of quasi-particles. For a specified set of collision rules, a particular value of the background number density can be chosen such that the viscosity approaches zero (see Eq. (39)). When the background number density is less than this particular value, the viscosity becomes positive and its value increases as the background number density decreases. Beyond this value, the viscosity becomes negative (non-physical) and its absolute value increases with the background number density. Therefore, a viscous layer (i.e., a layer with positive viscosity) with an appropriate choice of background number density can be padded to the boundaries of a model to absorb the artificial boundary reflections. This layer is called *a viscous absorbing boundary layer*. To avoid the reflection from the "interface" between the edge of a model and the edge of a viscous absorbing boundary layer (this interface is henceforth called a boundary interface), the viscosity of the absorbing boundary layer increases smoothly from zero (or the value of the medium) at the boundary interface up to a given value at the outer edge of the absorbing boundary layer. Numerically, the background number density of the boundary layer is varied using some smooth function such as a Hanning (also called Hann) taper (Lyons, 2004, pp. 7–81). In the calculations presented here, a Hanning taper was used to vary the number density smoothly from 0.457427 in the interior region to zero at the boundary of the computational domain. The microscopic reflection-coefficients at the outer edge of the boundary layer are set to zero.

A viscous boundary layer with a thickness of 40 vertical or horizontal lattice spacings was added to each boundary of the model used in the above numerical test, and consequently the PLSI calculations were carried out on a 240×240 triangular lattice. The same pressure source used above was introduced at lattice node (120, 120). Receivers were located along a horizontal line at a depth of 80 vertical lattice spacings from the upper boundary of the model. No receivers were placed in absorbing boundary layers. All other parameters used in the PLSI simulation remained the same as those used in the first numerical test. Computations were made using the PLSI method for 900 time steps. The corresponding seismograms (i.e., pressure) recorded at the receivers are displayed in Fig. 3b at the same scale as Fig. 3a. Artificial boundary reflections are no longer visible.

In summary, the absorbing boundary condition for the PLSI method is designed by

(1) setting the microscopic reflection-coefficients at boundaries of a model to zero, and
(2) adding viscous layers to the model boundaries.

Figure 4 shows a comparison between the above seismograms obtained with the absorbing boundary condition and those obtained using an analytic solution of the 2-D acoustic wave equation for a homogeneous medium. No time amplification was used in the figure. It demonstrates that the PLSI solution agrees very well with the analytic solution.

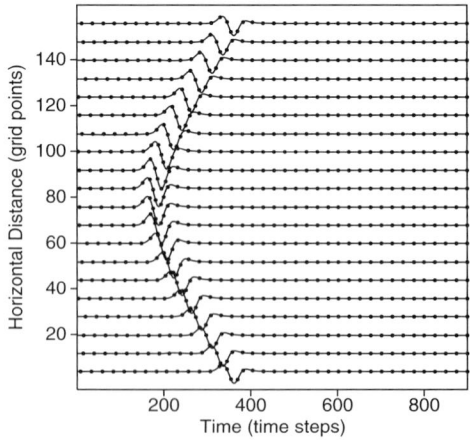

FIG. 4. Comparison between the seismograms as shown in Fig. 3b but without time amplification and those obtained using an analytic solution. Solid lines are for the PLSI solutions and dots are for the analytic solution.

6.3. Incidence Angle Dependence of the Absorbing Boundary Condition

In the following, we study the incidence angle dependence of reflections from the absorbing boundary condition introduced above. The incidence angle β is defined as the angle between the boundary normal direction and the ray path, as shown in Fig. 5. A homogeneous model defined on an 80×320 triangular lattice without viscous absorbing boundary layers was used for the PLSI simulation of wave propagation. A pressure source with the same characteristics as the source used in the previous simulation was introduced at lattice node (40, 280). All other parameters were the same as those of the previous numerical examples. Receivers were located along horizontal lines at lattice depths of 280, 250, 200, 150 and 10 from the upper boundary of the model. The corresponding incidence angles β for boundary reflections recorded at receivers located at the right (or left) boundary of the model are 0.0°, 33.0°, 60.0°, 70.4° and 80.3°, respectively. Computations were made using the PLSI method for 1200 time steps. The seismograms are displayed in Figs. 6(a1)–(e1) using a time-step-dependent gain of $t^{0.7}$.

Next, viscous boundary layers with the same thickness and properties as used in the PLSI simulation corresponding to Fig. 3b were added to all model boundaries. The PLSI computations with the absorbing boundary condition were made again for 1200 time steps. The corresponding amplified seismograms are displayed in Figs. 6(a2)–(e2) at the same scale as used in Figs. 6(a1)–(e1). Comparing each pair of figures (such as (a1) and (a2), (b1) and (b2), etc.), one can see that there are no visible boundary reflections in Figs. 6(a2)–(c2) which correspond to the

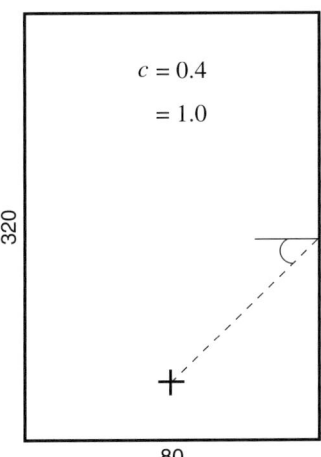

FIG. 5. Schematic illustration of a homogeneous model without viscous absorbing boundary layers used for the study of the incidence angle dependence of the absorbing boundary condition. The plus sign represents the position of the pressure source and the angle β is the incidence angle at the right boundary of the model.

540 HUANG

incidence angles of 0.0°–60°. For the case of the maximum incidence angle of 70.4° (Fig. 6(d2)), weak boundary reflections are visible but much smaller in amplitude than those in Fig. 6(d1). When the incidence angle increases up to 80.3° (Fig. 6(e2)), the boundary reflections from the left and right boundaries of the model become visible.

7. FREE-SURFACE REFLECTIONS

Classical finite-difference methods to solve partial differential wave equations describing wave phenomena in continuous media often have difficulty in handling free-surface boundary conditions, particularly when the free surface is irregular (Komatitsch et al., 1996; Ohminato and Chouet, 1997). Huang et al. (2000) proposed four methods for modeling free-surface reflections in PLSI simulations (Fig. 7). In each panel of Fig. 7, only a small part of a free surface in a 2-D triangular lattice is depicted. In the first method, quasi-particles arriving at a free surface in a certain direction will be specularly bounced back to another direction. This is referred to as the *specular bouncing* method. For example, as shown in Fig. 7a, quasi-particles moving in direction 2 will be bounced to direction 6 (cf. lattice node A) and those moving in direction 3 will be bounced to direction 5

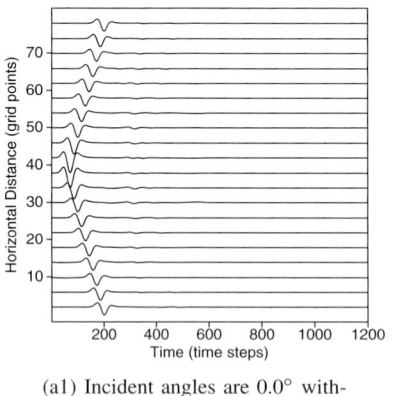

(a1) Incident angles are 0.0° without viscous layers

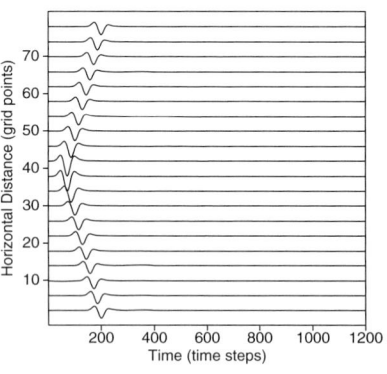

(a2) Incident angles are 0.0° with viscous layers

FIG. 6. Comparisons of the amplified seismograms (i.e., pressure) recorded during the PLSI calculations to simulate wave propagation from a pressure source in a homogeneous model. (a1)–(e1) are results obtained using zero-valued microscopic reflection-coefficients for a model defined on an 80 × 320 triangular lattice with a pressure source at lattice node (40, 280). (a2)–(e2) are the corresponding results obtained using viscous boundary layers in addition to zero-valued microscopic reflection-coefficients. The maximum incidence angles are for boundary reflections recorded at receivers located at the left or right boundaries of the model.

A LATTICE BOLTZMANN APPROACH

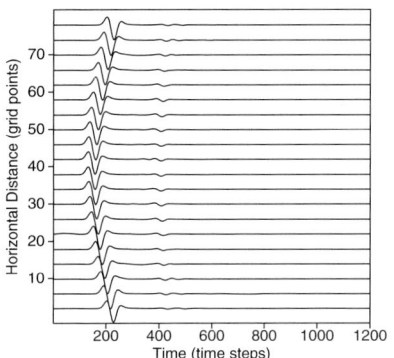

(b1) Maximum incidence angle = 33.0°
without viscous layers

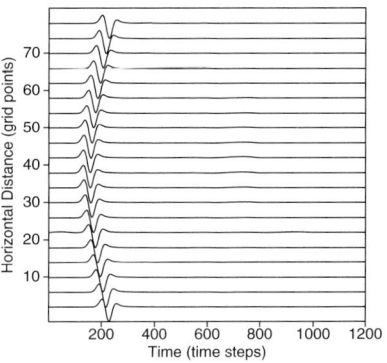

(b2) Maximum incidence angle = 33.0°
with viscous layers.

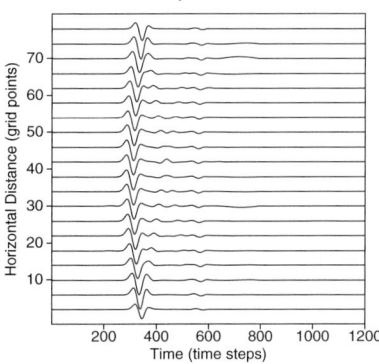

(c1) Maximum incidence angle = 60.0°
without viscous layers

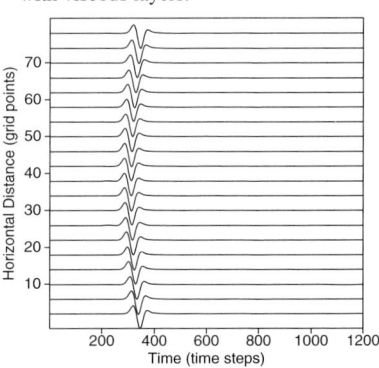

(c2) Maximum incidence angle = 60.0°
with viscous layers.

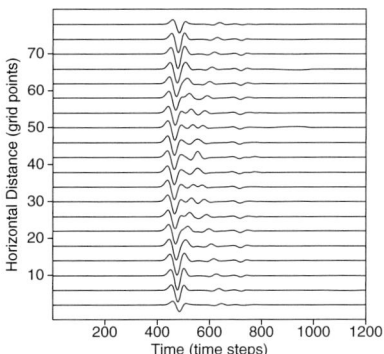

(d1) Maximum incidence angle = 70.4°
without viscous layers

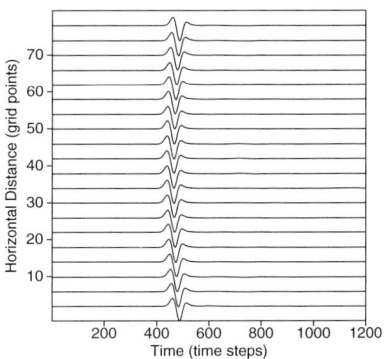

(d2) Maximum incidence angle = 70.4°
with viscous layers

FIG. 6. (Continued).

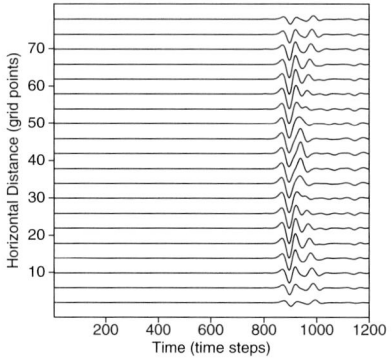
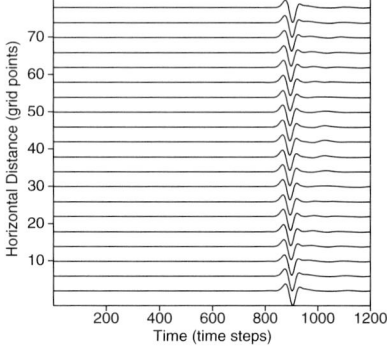

(e1) Maximum incidence angle = 80.3° without viscous layers

(e2) Maximum incidence angle = 80.3° with viscous layers

FIG. 6. (Continued).

(cf. lattice node B). These reflections are performed with a change of sign in the number density perturbation.

Another alternative method is termed the *backward bouncing-I* (cf. Fig. 7b). In this case, quasi-particles will be bounced back along the opposite direction to the incidence direction, coupled with a change in sign of the number density perturbation. For instance, quasi-particles moving in direction 2 will be bounced to direction 5 and those moving in direction 3 will be bounced to direction 6, as shown at lattice node A and B in Fig. 7b, respectively.

For the simulation of free-surface reflections, it seems that the specular bouncing method should be more suitable than the backward bouncing-I method.

In lattice gas simulations, the corresponding so-called "bounce-back" reflection method is commonly used for the study of interactions of fluid flows with solid boundaries (Lavallée *et al.*, 1991). However, Lavallée *et al.* (1991) pointed out that there are good reasons to examine other types of reflections from solid boundaries. One reason stems from consideration of experiments at the molecular level. Knudsen (1934) conducted an experiment where molecules were directed towards a wall at a fixed incidence angle, and observed that molecules were randomly scattered into all directions. In a lattice gas, this would correspond to a combination of the specular and bounce-back reflection methods. The second reason given by (Lavallée *et al.*, 1991) is that, with purely deterministic interactions with the wall, the Boltzmann assumption that no correlation exists between particles prior to collision is not valid. Note, however, that the hypothesis is correct in a statistical sense for a combination of 50% probability of the bounce-back method and 50% probability of the specular reflection method.

When modeling free-surface reflections using the PLSI method, one way to account for the molecular observations of Knudsen (1934) is to combine the

A LATTICE BOLTZMANN APPROACH 543

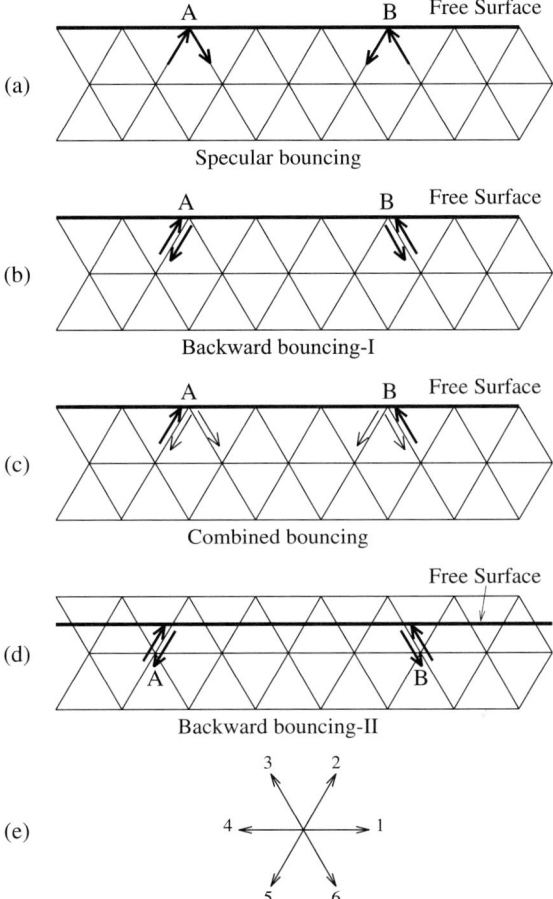

FIG. 7. Illustration of four methods for modeling free-surface reflections in PLSI simulations. The specular bouncing (a), backward bouncing (b) and combined bouncing (c) methods are used when a free surface is located exactly along lattice nodes. The backward bouncing-II (d) method is used when the free surface is located at any position on a lattice. All the bouncing processes are performed with a change in sign of the number density perturbation. (e) depicts the labelling of the lattice directions for a node on a 2-D.

specular bouncing method and the backward bouncing-I method to yield a third method (Fig. 7c), which is termed the *combined bouncing* method. In this case, quasi-particles arriving at the free surface in direction 2 or in direction 3 will be equally bounced back to directions 5 and 6 with a sign change of the number-density perturbation, as shown, respectively, at lattice nodes A and B in Fig. 7c. The above three methods can be considered as special collision rules for quasi-particles at a free surface.

When a free surface is not exactly located along lattice nodes, it is handled in the same way as interfaces within a model (see Fig. 7d). For example, as shown in Fig. 7d, quasi-particles moving in direction 2 at lattice node A move to the free surface and are reflected back along direction 5 with a reflection coefficient of -1. Similarly, quasi-particles moving in direction 3 to the free surface are reflected backwards with a reflection coefficient of -1, as shown at lattice node B in Fig. 7d. For this case, the process describing the quasi-particle behavior at the free surface is similar to the backward bouncing-I method. Therefore, this method for modeling free-surface reflections is termed the *backward bouncing-II*.

In each of the above methods for modeling free-surface reflections, the process of quasi-particle interaction with a free surface can occur at any location along lattice links. Therefore, these methods do not have any particular restrictions on the shape of a free surface. In other words, these methods can be applied to plane or irregular free surfaces.

7.1. Numerical Simulations of Free-Surface Reflections

A homogeneous medium, defined on a 240×200 triangular lattice with a free surface at the upper boundary of the model and the absorbing boundary condition at the left, right and lower boundaries, was used in all the following numerical examples to simulate free-surface reflections. Each viscous absorbing boundary layer has a thickness of 40 vertical or horizontal lattice spacings and the same properties as those used to calculate Fig. 6(a2)–(e2). A pressure source with a first derivative Gaussian time history, an amplitude of 0.010, and a time delay of 60 time steps was introduced at lattice node (120, 80). It generated waves with a wavelength of approximately 16 lattice spacings. The quasi-particle speed was 0.4 and the density of the medium was 1.0. The background number density of quasi-particles was 0.457427 so the viscosity approaches zero. Seismograms were recorded at receivers located along a horizontal line at a depth of 40 vertical lattice spacings from the upper boundary of the model. No receivers were located in absorbing boundary layers. PLSI computations were run for 800 time steps.

First, the specular bouncing method was used during the PLSI simulation. The direct and reflected waves from the free surface can be identified in a snapshot of pressure at 350 time steps shown in Fig. 8, where the plus sign represents the position of the pressure source. The figure shows that the wave reflected from the free surface has changed its polarity relative to the incidence wave. The solid lines in Fig. 9a depict the seismograms recorded at the receivers during the above PLSI simulation with the absorbing boundary condition. The lines show the direct waves and the reverse polarity waves reflected from the free surface as expected.

In the next two PLSI simulations, the backward bouncing-I and the combined bouncing methods were used respectively to simulate free-surface reflections.

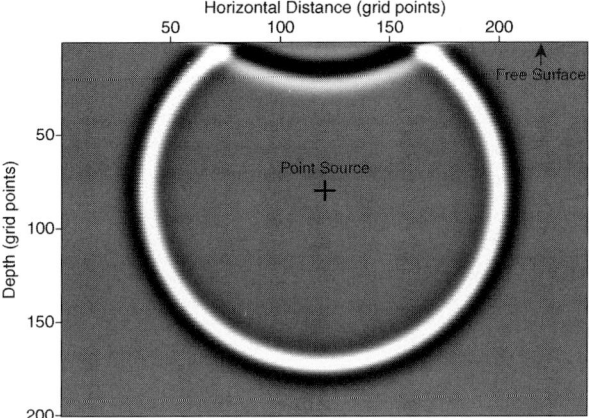

FIG. 8. A snapshot at 350 time steps during the PLSI simulation to model free-surface reflections. The plus sign '+' represents the position of the pressure source. The homogeneous model was defined on a 240×200 triangular lattice. The specular bouncing method was used for modeling the free-surface reflections.

The corresponding seismograms showing pressure at the receivers are displayed in Figs. 9b and c with solid lines.

For the case where the free surface was located along a horizontal line below the first row of the lattice with a distance of 1/2 vertical lattice spacing in the above model, the backward bouncing-II method was used to simulate free-surface reflections. The resulting seismograms are displayed in Fig. 9d with solid lines.

Analytic solutions for all the above simulations are shown in Figs. 9a–d with dots. Comparisons between the PLSI solutions and the analytic solutions indicate that the above four methods for modeling total reflections can accurately model free-surface reflections during PLSI simulations.

7.2. Numerical Simulations of Reflections From Free-Surface Topography

Figure 10 shows two snapshots obtained during PLSI simulation for models having free-surface topography. When the lateral variation of free-surface topography is much larger than the wavelength, reflections from different regions of free-surface topography in Fig. 10a can be clearly identified. When the lateral variation of surface topography is in the order of a wavelength, pressure wavefields reflected from free-surface topography become complicated, as shown in Fig. 10b.

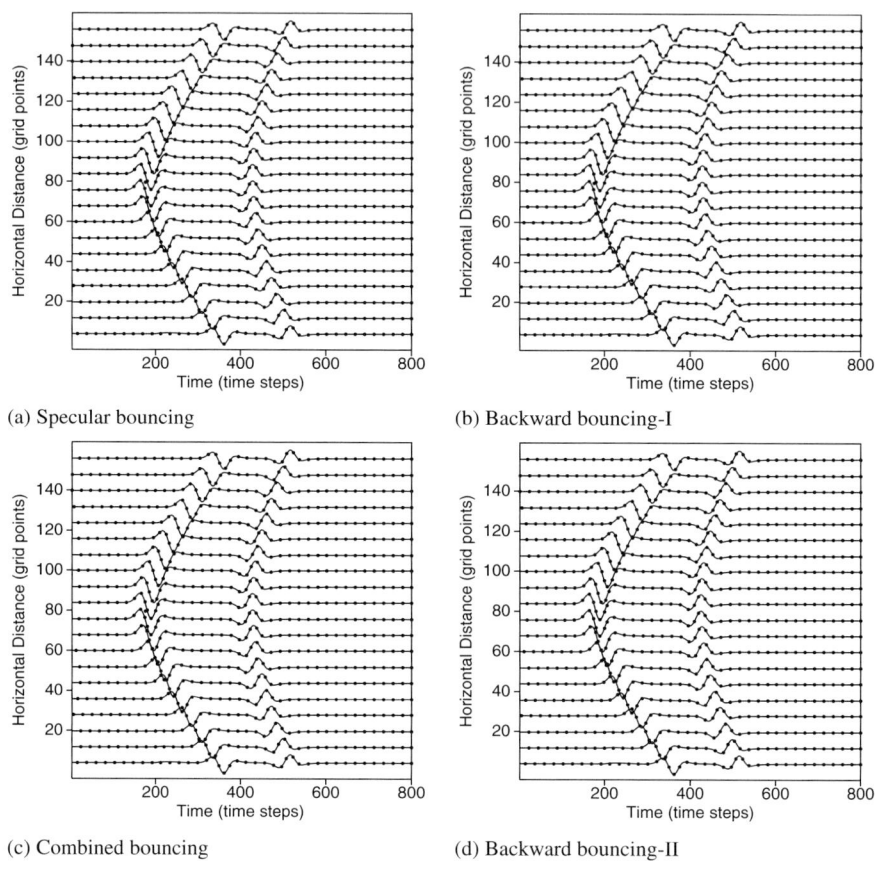

FIG. 9. Seismograms (solid lines) recorded at receivers located at a depth of 40 vertical lattice spacings from the free surface during the PLSI simulations. The dots represent the corresponding analytic solutions. For the free-surface reflections during the PLSI simulations, the specular bouncing (a), backward bouncing-I (b), combined bouncing (c), and backward bouncing-II (d) methods were used.

8. Acoustic Waves in Strongly Heterogeneous Media

The anisotropic properties of finely layered media have been studied theoretically, and numerical studies of wave propagation in such media are of interest (Backus, 1962; Schoenberg and Muir, 1989).

Wave propagation in fractured media is also an interesting subject for the study of anisotropic properties of such media (Crampin, 1985). It is difficult to analyze the anisotropy of such media if there is a high level of geometric complexity of the fractures. Rather than directly simulating wave propagation in such media, some

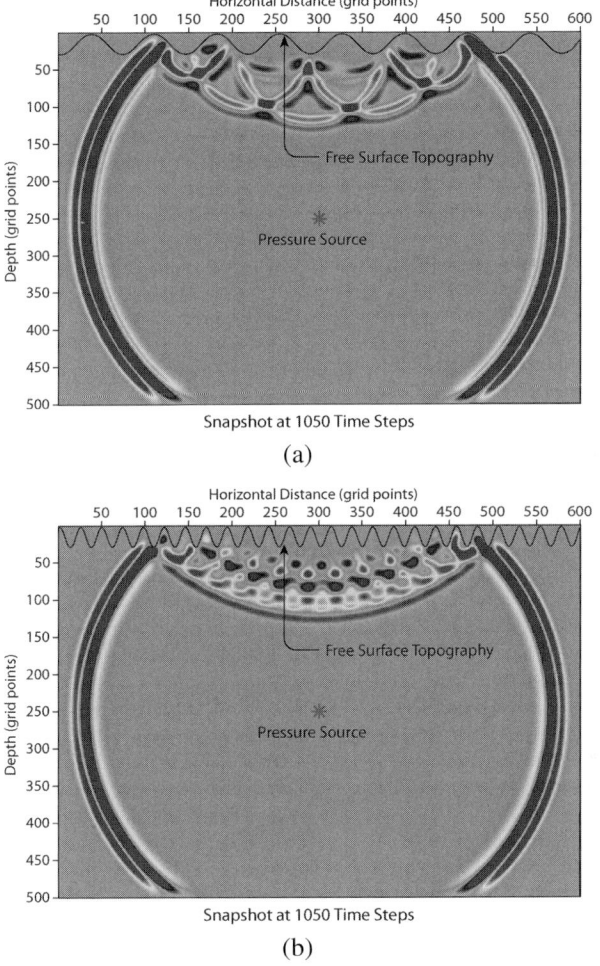

FIG. 10. Snapshots at 1050 time steps obtained during PLSI simulations to model reflections from free-surface topography. Because the free-surface topography in (b) changes more rapidly in the lateral direction than that in (a), wavefields reflected from free-surface topography in (b) are more complicated than those in (a).

effective parameters characterizing the properties of the media may be used if the density of the aligned pore space in the media is not too high (Hudson, 1981; Crampin, 1984).

A typical example of sharp interfaces is boundaries of empty pores or cavities in a medium (or boundaries of empty bubbles in acoustic media). The density and wave speed in the regions of empty bubbles are both zero, and hence, waves are completely reflected back to the medium from boundaries of empty bubbles.

The geometrical shapes of boundaries of empty bubbles may also be complex. Therefore, it is difficult for finite-difference methods to model wave propagation in such media.

Strongly heterogeneous media with sharp interfaces such as those with fine layers and/or aligned heterogeneities and/or empty pores/bubbles may pose problems for classical finite-difference methods to simulate wave propagation. This is because partial-differential wave-equations describe wave propagation in a *continuous* medium rather than discontinuous media and consequently, finite-difference errors resulting from smoothing of interfaces can accumulate and become significant if the medium contains a large number of heterogeneities. It is also difficult for finite-difference methods to handle total reflections from boundaries of empty pores/bubbles.

The PLSI is capable of handling sharp interfaces, and therefore provides a powerful tool for modeling acoustic-wave propagation in the above-mentioned heterogeneous and complex media. Numerical PLSI modeling examples to simulate acoustic-wave propagation in a finely layered medium, a medium with aligned heterogeneities, a medium with a circular empty bubble, and a medium with random empty bubbles (a bubbly material), are presented in the following.

8.1. Wave Propagation in a Finely Layered Medium

A finely layered isotropic medium defined on a 512×512 triangular lattice was used for the PLSI simulation to study fine-layer-induced anisotropy. The model consists of 256 horizontally aligned fine layers, each of which has the same thickness of two vertical lattice spacings (see the background of Fig. 11). All the odd fine layers had a quasi-particle speed of 0.5 and a medium density of 0.4, while all the even fine layers had a quasi-particle speed of 0.1 and a medium density of 2.0. Consequently, the impedance contrasts between fine layers are zero.

The choice of zero impedance-contrasts here as well as in the next example was made for efficiency. A smaller layer spacing or thickness of aligned heterogeneities relative to wavelength when impedance contrasts were not zero, would yield the same effect of eliminating visible scattering but take more CPU time. This more CPU-intensive method is nonessential for the purposes of testing the PLSI approach and studying the anisotropic properties induced by fine layers or aligned heterogeneities.

The absorbing-boundary condition was used during the PLSI calculation. The thicknesses of viscous absorbing boundary layers at the boundaries of the model are either 40 vertical lattice spacings or 40 horizontal lattice spacings. A pressure source with a first derivative Gaussian time history and an amplitude of 0.010, was introduced at lattice site (256, 256). The source with a time delay of 150 time steps generated waves with wavelengths of 80 and 16 lattice spacings in the fine layers with the higher and lower quasi-particle speeds, respectively. The mean number density of quasi-particles was 0.457427 and hence the viscosity approaches zero.

The same value of the mean number density was used in all of the following PLSI simulations.

Figure 11 is a snapshot at 1200 time steps obtained during the PLSI simulation. Its elliptic shape demonstrates the fine-layer-induced anisotropy. The direct wave propagates faster in the direction parallel to the orientation of fine layers than in the direction perpendicular to the orientation, as expected theoretically (Backus, 1962; Schoenberg and Muir, 1989). The high-frequency noise following the wavefront is due to scattering from the fine layers because one wavelength in the medium with the slower quasi-particle speed contains only eight fine layers.

8.2. Wave Propagation in a Medium with Aligned Heterogeneities

The next PLSI modeling example is to simulate acoustic-wave propagation in medium with aligned heterogeneities to demonstrate their induced anisotropic properties. A model with aligned heterogeneities was defined on a 512×512 triangular lattice (see the background of Fig. 12). The heterogeneities are aligned horizontally and located randomly. The thicknesses of the heterogeneities are two vertical lattice spacings. Their lengths range randomly from 7 to 13 horizontal lattice spacings. The number of lattice sites occupied by the heterogeneities is approximately 36.9 percent of the total number of lattice sites. The quasi-particle speeds of the background medium and the horizontal aligned heterogeneities are 0.5 and 0.1, respectively. Their corresponding medium densities are 0.4 and

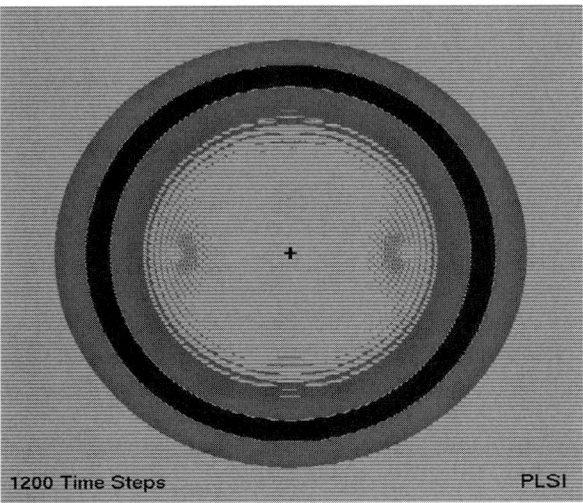

FIG. 11. An elliptic snapshot at 1200 time steps obtained during the PLSI simulation for a finely layered isotropic medium defined on a 512×512 triangular lattice. The pressure source was introduced at the position of the plus sign.

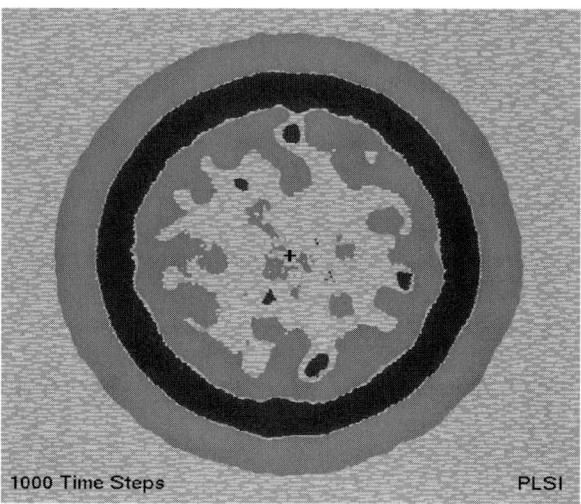

FIG. 12. A snapshot at 1000 time steps during the PLSI simulation for a model with aligned heterogeneities. The model was defined on a 512×512 triangular lattice. The pressure source was introduced at the position of the plus sign.

2.0, respectively. Consequently, the impedance contrasts between the background medium and the heterogeneities are zero. The same pressure source as above with a time delay of 190 time steps was introduced at lattice site (256, 256). It generated waves with wavelengths of 100 and 20 lattice spacings in the background medium and the aligned heterogeneities, respectively. A snapshot at 1000 time steps (see Fig. 12) obtained during the PLSI simulation for such a model demonstrates the anisotropic properties of the medium with aligned heterogeneities. Again, the fast propagation direction is parallel to the orientation of the aligned heterogeneities. The noisy appearance within the wavefront is the results of scattering/diffraction of the small heterogeneities that are not small enough relative to the wavelength.

8.3. Wave Propagation in a Medium with a Circular Empty Bubble

The third PLSI modeling example presented here is to demonstrate the PLSIs capability to handle total reflections from an empty bubble. A circular empty bubble with a diameter of 17 lattice spacings was embedded at the center of a homogeneous background medium defined on a 280 × 280 triangular lattice. Both the quasi-particle speed and medium density within the empty bubble are zero. The background medium has a quasi-particle speed of 0.4 and a medium density of 1.0. A pressure source with a first derivative Gaussian time history, an amplitude of 0.010 and a time delay of 60 time steps was introduced at lattice site (140, 100). It generated waves with a wavelength of 16 lattice spacings in the

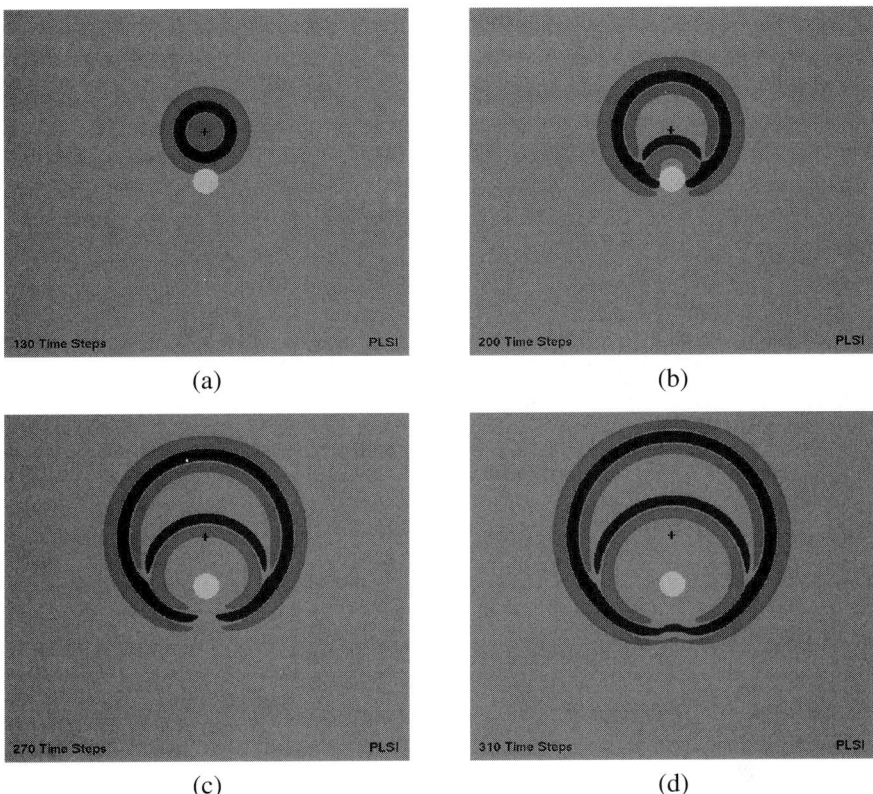

FIG. 13. Snapshots at 130, 200, 270 and 310 time steps during the PLSI simulation for a circular-empty-bubble model defined on a 280 × 280 triangular lattice. In each panel, the white region is the circular empty bubble with a diameter of 17 lattice spacings and the plus sign represents the position of the pressure source.

homogeneous region. A PLSI calculation with the absorbing boundary condition was made for 800 time steps. In addition to the direct wave, the reflected and diffracted waves can be identified from the four snapshots at 130, 200, 270 and 310 time steps during the PLSI simulation as shown in Fig. 13.

For comparison, the circular empty bubble was filled with a medium that has a quasi-particle speed of 0.3 and a medium density of 1.0. Figure 14 shows the corresponding four snapshots of the PLSI simulation at 130, 200, 270 and 310 time steps for the model. They are very different from those in Fig. 13 except for parts of the direct wave. Waves were reflected by the upper part of the boundary of the circular, non-empty heterogeneous region and transmitted through the boundary, propagated into the heterogeneous region and then transmitted through. Waves were also reflected by the lower part of the boundary of the heterogeneous

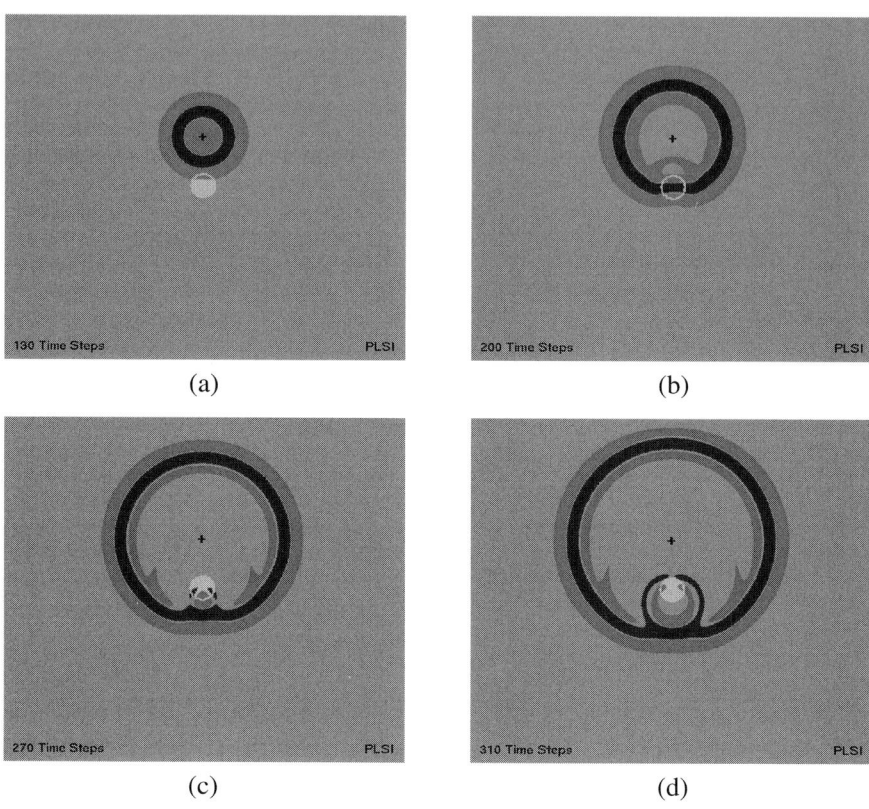

FIG. 14. Snapshots at 130, 200, 270 and 310 time steps during the PLSI simulation for a model containing a circular, non-empty heterogeneous region. The model was defined on a 280 × 280 triangular lattice. In each panel, the white region is the circular, non-empty heterogeneous region with a diameter of 17 lattice spacings and the plus sign represents the position of the pressure source.

region, as shown in the snapshots in Fig. 14. Figures 13 and 14 show that, due to the total reflections of the upper part boundary of the circular empty bubble, the backward scattering of the empty bubble is much stronger than the circular, non-empty heterogeneous region.

For both above PLSI calculations, receivers were located along two horizontal lines at the depths of 60 and 220 vertical lattice spacings, from the upper boundary of the model. The recorded seismograms for both calculations are displayed in Fig. 15. Figure 15(a1) shows the strong reflection from the upper part of the boundary of the circular empty bubble while Fig. 15(b1) shows the two reflections from the upper and lower parts of the boundary of the non-empty heterogeneous region. The forward propagating fields for both models are also very different from one another (see Figs. 15(a2) and (b2)).

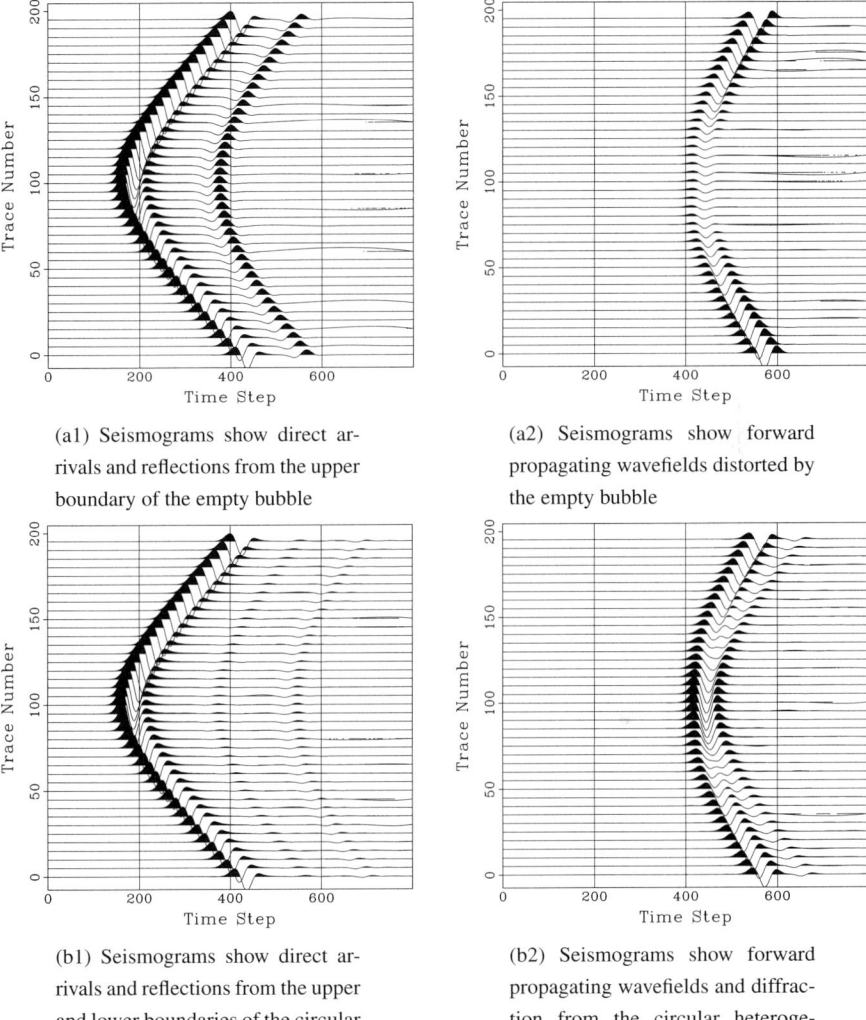

(a1) Seismograms show direct arrivals and reflections from the upper boundary of the empty bubble

(a2) Seismograms show forward propagating wavefields distorted by the empty bubble

(b1) Seismograms show direct arrivals and reflections from the upper and lower boundaries of the circular heterogeneous region

(b2) Seismograms show forward propagating wavefields and diffraction from the circular heterogeneous region

FIG. 15. Seismograms obtained by the PLSI for models with a circular empty bubble (the upper two panels) and a circular, non-empty heterogeneous region (the lower two panels) at the centers of the models. The models were defined on a 280 × 280 triangular lattice. The diameters of the circular bodies are both 17 lattice spacings. All seismograms were displayed with the same scale.

8.4. Wave Propagation in a Medium with Random Empty Bubbles

The last PLSI modeling example presented here is to demonstrate the capability of the PLSI method to simulate strong scattering effects of empty bubbles with different sizes. Empty bubbles were located randomly in a homogeneous background medium with a quasi-particle speed of 0.4 and a medium density of 1.0 (see the background of Fig. 16a). The model was defined on a 280×280 triangular lattice. The dimensions of the empty bubbles ranged randomly from 2 to 5 lattice spacings. The same type of pressure source with a wavelength of 32 lattice spacings in the homogeneous background medium and a time delay of 120 time steps was introduced at lattice site (140, 140). A PLSI simulation with the absorbing boundary condition was made for 1200 time steps. In order to compare the strong effects of the random empty bubbles on wave propagation with the effects of non-empty heterogeneities located in the same regions (see the background of Fig. 16b), a second PLSI calculation was run. In this simulation, the quasi-particle speed and the medium density of the non-empty heterogeneous regions were set to 0.3 and 1.0, respectively. Two snapshots at 400 time steps for both models are displayed in Fig. 16. The results demonstrate that the effects of the empty bubbles on the wave fields are much stronger than those of the non-empty heterogeneities. The noisy appearance within the wavefront in Fig. 16b is due to scattering of the heterogeneities that are not small enough relative to the minimum wavelength.

Seismograms recorded along a horizontal line at the depth of 60 vertical lattice spacings are displayed in Fig. 17. In Fig. 17a, the zero value trace is because the corresponding receiver was located in the region of an empty bubble and hence no waves propagated into that region. It is obvious that the direct wave field in Fig. 17a is much different from that in Fig. 17b due to the total reflections from the boundaries of the random empty bubbles. In addition, the later arrivals in Fig. 17a are much stronger than those in Fig. 17b even though there was not a high density of empty bubbles in the medium. Figure 17 demonstrates that the empty bubbles generate much stronger scattering effects than non-empty heterogeneities.

9. Conclusions

Rather than solving the partial-differential wave-equation using a numerical algorithm such as a finite-difference scheme, the lattice Boltzmann-based method termed "the phononic lattice solid by interpolation" directly simulates the physical processes of acoustic-wave propagation in strongly heterogeneous media with essentially no approximations. The method has a high numerical precision for complex media because the transportation step of the method makes use of an accurate interpolation algorithm, while the transmission and reflection steps are exact to the numerical precision of the computer for any velocity or impedance contrasts. It can accurately handle sharp interfaces at any positions along lattice links, and total reflections from free surfaces or empty bubbles. Numerical simulations demonstrate the ability of the method to simulate acoustic-wave propagation

A LATTICE BOLTZMANN APPROACH   555

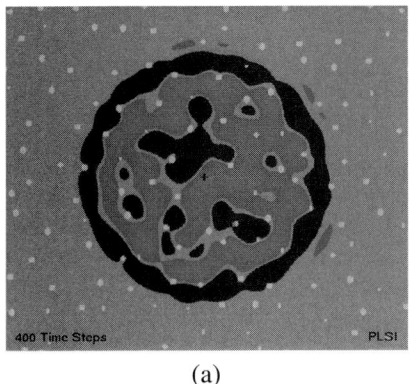 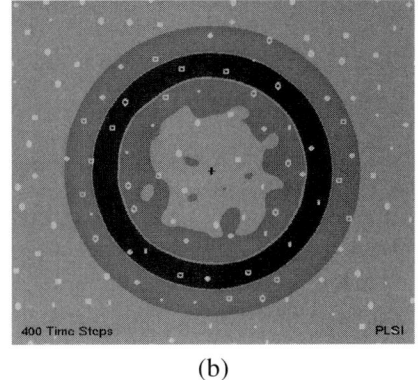

(a) (b)

FIG. 16. Snapshots at 400 time steps during the PLSI simulations for a random empty bubble model (a) and the same model but with non-empty heterogeneities (b). The white curves in (b) are the boundaries of the heterogeneities. The plus signs represent the positions of the pressure sources.

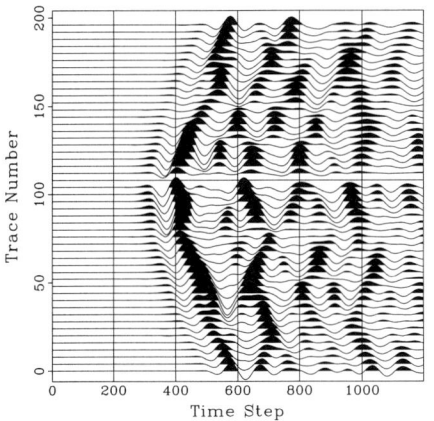 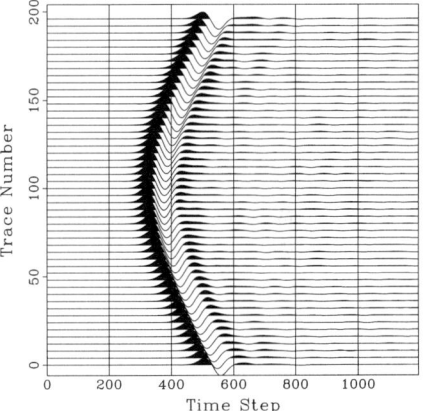

(a) Seismograms show severely-distorted direct arrivals and later arrivals of strong scattering from empty bubbles

(b) Seismograms show nearly-perfect direct arrivals and weak scattering from non-empty heterogeneities

FIG. 17. Seismograms recorded during the PLSI simulations for a random empty bubble model (a) and the same model but with non-empty heterogeneities (b). The same scale was used in (a) and (b).

in strongly heterogeneous media, which are not easily handled by classical finite-difference methods due to the geometrical complexity and sharpness of interfaces. Ultimately, a fully discrete approach to elastic wave propagation in heterogeneous and complex media is required to properly handle not only compressional waves, but also shear waves.

ACKNOWLEDGEMENTS

This work was partially supported by the US Department of Energy through contract W-7405-ENG-36 to Los Alamos National Laboratory. The author thanks Bertrand Maillot and an anonymous reviewer for their helpful comments.

REFERENCES

Appert, C., Zaleski, S. (1990). Lattice gas with a liquid–gas transition. *Phys. Rev. Lett.* **64**, 1–4.
Backus, G. (1962). Long wave elastic anisotropy produced by horizontal layering. *J. Geophys. Res.* **67**, 4427–4440.
Bleistein, N., Cohen, J.K., Stockwell Jr., J.W. (2001). Mathematics of Multidimensional Seismic Imaging, Migration and Inversion. Springer-Verlag, New York.
Boghosian, B., Taylor, W. (1995). Correlations and renormalization in lattice gases. *Phys. Rev. E* **52**, 510–514.
Broadwell, J. (1964). Study of rarefied shear flow by discrete velocity method. *J. Fluid Mech.* **19**, 401–414.
Burges, C., Zaleski, S. (1987). Buoyant mixtures of cellular-automaton fluids. *Complex System* **1**, 31–50.
Chen, H., Chen, S., Matthaeus, W. (1992). Recovery of the Navier–Stokes equation using a lattice gas Boltzmann method. *Phys. Rev. A* **45**, R5339–R5342.
Chen, H., Matthaeus, W. (1987). New cellular automaton model for magnetohydrodynamics. *Phys. Rev. Lett.* **58**, 1845–1848.
Chen, S., Chen, H., Martinez, D., Matthaeus, W. (1991). Lattice Boltzmann model for simulation of magnetohydrodynamics. *Phys. Rev. Lett.* **67**, 3776–3779.
Chen, S., Doolen, G.D. (1998). Lattice Boltzmann method for fluid flows. *Annu. Rev. Fluid Mech.* **30**, 329–364.
Crampin, S. (1984). Effective anisotropic elastic constants for wave propagation through cracked solids. *Geophys. J. R. Astron. Soc.* **76**, 135–145.
Crampin, S. (1985). Evidence for aligned cracks in the earth's crust. *First Break* **3**, 12–15.
Crase, E., Pica, A., Noble, M., McDonald, J., Tarantola, A. (1990). Robust elastic nonlinear inversion: Application to real data. *Geophysics* **55**, 527–538.
Dablain, M. (1986). The application of high-order differencing to the scalar wave equation. *Geophysics* **51**, 54–66.
d'Humières, D., Lallemand, P. (1987). Numerical simulations of hydrodynamics with lattice gas automata in two dimensions. *Complex System* **1**, 599–632.
d'Humières, D., Lallemand, P., Frisch, F. (1986). Lattice gas models for 3-D hydrodynamics. *Europhys. Lett.* **2**, 291–297.
Doolen, G.E. (1990). Lattice Gas Methods for Partial Differential Equations. Addison–Wesley, Redwood City, USA.
Frisch, U., d'Humières, D., Hasslacher, B., Pomeau, Y., Rivet, J. (1987). Lattice gas hydrodynamics in two and three dimensions. *Complex System* **1**, 649–707.
Frisch, U., Hasslacher, B., Pomeau, Y. (1986). Lattice-gas automata for the Navier–Stokes equation. *Phys. Rev. Lett.* **56**, 1505–1508.
Gatignol, R. (1970). Theéorie cinfique d'un gaz à répartition discrète de vitesses. *Z. Flügwissenschaften* **18**, 93–97.
Gunstensen, A., Rothman, D. (1991). A lattice-gas model for three immiscible fluids. *Physica D* **47**, 47–52.

Gunstensen, A., Rothman, D. (1992). Microscopic modeling of immiscible fluids in three dimensions by a lattice Boltzmann method. *Europhys. Lett.* **18**, 157–161.
Gunstensen, A., Rothman, D., Zaleski, S., Zanetti, G. (1991). Lattice Boltzmann model of immiscible fluids. *Phys. Rev. A* **43**, 4320–4327.
Hardy, J., Pomeau, Y., de Pazzis, O. (1973). Time evolution of two-dimensional model system, i: Invariant states and time correlation functions. *J. Math. Phys.* **14**, 1746–1759.
He, X., Chen, S., Zhang, R. (1999). A lattice Boltzmann scheme for incompressible multiphase flow and its application in simulation of Rayleigh–Taylor instability. *J. Comput. Phys.* **152**, 642–663.
Hestholm, S., Ruud, B. (1998). 3-D finite-difference elastic wave modeling including surface topography. *Geophysics* **63**, 613–622.
Higuera, F. (1988). Lattice gas simulation based on the Boltzmann equation. In: Monaco, R. (Ed.), *Discrete Kinetic Theory, Lattice Gas Dynamics and Foundations of Hydrodynamics*. World Scientific, Singapore, pp. 162–177.
Higuera, F., Succi, S., Benzi, R. (1989). Lattice gas dynamics with enhanced collisions. *Europhys. Lett.* **9**, 663–668.
Holberg, O. (1987). Computational aspects of the choice of operator and sampling interval for numerical differentiation in large-scale simulation of wave phenomena. *Geophys. Prosp.* **35**, 629–655.
Holberg, O. (1988). Towards optimum one-way wave propagation. *Geophys. Prosp.* **36**, 99–114.
Holme, R., Rothman, D. (1992). Lattice-gas and lattice-Boltzmann methods of miscible fluids. *J. Stat. Phys.* **68**, 409–429.
Huang, J.-I., Chu, Y.-H., Yin, C.-H. (1988). Lattice-gas automata for modeling acoustic wave propagation in inhomogeneous media. *Geophys. Res. Lett.* **15**, 1239–1241.
Huang, L., Fehler, M.C., He, X. (1999). Lattice BGK method for modeling acoustic wave propagation in strongly heterogeneous viscous media. *Seismol. Res. Lett.* **70**, 263.
Huang, L., Mora, P. (1994a). The phononic lattice solid by interpolation for modelling p waves in heterogeneous media. *Geophys. J. Int.* **119**, 766–778.
Huang, L., Mora, P. (1994b). The phononic lattice solid with fluids for modelling non-linear solid–fluid interactions. *Geophys. J. Int.* **117**, 529–538.
Huang, L., Mora, P. (1996). Numerical simulation of wave propagation in strongly heterogeneous media using a lattice solid approach. In: Hassanzadeh, S. (Ed.), *Mathematical Methods in Geophysical Imaging IV*, vol. 2282. The International Society for Optical Engineering, Bellingham, Washington, pp. 170–179.
Huang, L., Mora, P., Felher, M. (2000). Absorbing boundary and free-surface conditions in the phononic lattice solid by interpolation. *Geophys. J. Int.* **140**, 147–157.
Hudson, J.A. (1981). Wave speeds and attenuation of elastic waves in material containing cracks. *Geophys. J. R. Astron. Soc.* **64**, 133–150.
Hustedt, B., Operto, S., Virieux, J. (2004). Mixed-grid and staggered-grid finite-difference methods for frequency-domain acoustic wave modelling. *Geophys. J. Int.* **157**, 1269–1296.
Igel, H., Weber, M. (1996). P-SV wave propagation in the earth's mantle using finite differences: Application to heterogeneous lowermost mantle structure. *Geophys. Res. Lett.* **23**, 415–418.
Inamuro, T., Ogata, T., Tajima, S., Konishi, N. (2004). A lattice Boltzmann method for incompressible two-phase flows with large density differences. *J. Comput. Phys.* **198**, 628–644.
Kang, Q., Zhang, D., Chen, S. (2002). Unified lattice Boltzmann method for flow in multiscale porous media. *Phys. Rev. E* **66**, 056307.
Kang, T., Baag, C. (2004). An efficient finite-difference method for simulating 3-D seismic response of localized basin structures. *Bull. Seismol. Soc. Am.* **94**, 1690–1705.
Khan, A., Mosegaard, K. (2002). An inquiry into the lunar interior: A nonlinear inversion of the Apollo lunar seismic data. *J. Geophys. Res. Planets* **107**, 5036.
Knudsen, M. (1934). *The Kinetic Theory of Gases*. Methuen Monographs, London.
Komatitsch, D., Coutel, F., Mora, P. (1996). Tensorial formulation of the wave-equation for modeling curved interfaces. *Geophys. J. Int.* **127**, 156–168.

Lavallée, P., Boon, J., Noullez, A. (1991). Boundaries in lattice gas flows. *Physica D* **47**, 233–240.
Lyons, R.G. (2004). Understanding Digital Signal Processing. Pearson Education, New Jersey.
Maillot, B. (1994). Semi-microscopic models of elastic waves. Ph.D. Thesis, University of Paris VII.
Martinez, D., Chen, S., Matthaeus, W. (1994). Lattice Boltzmann magnetohydrodynamics. *Phys. Plasmas* **1**, 1850–1867.
Maxwell, J. (1890). On the dynamical theory of gases. *Sci. Papers* **2**, 26–78.
McNamara, G., Zanetti, G. (1988). Use of the Boltzmann equation to simulate lattice gas automata. *Phys. Rev. Lett.* **61**, 2332–2335.
Mora, P. (1987). Nonlinear 2-D elastic inversion of multi-offset seismic data. *Geophysics* **52**, 1211–1228.
Mora, P. (1988). Elastic wavefield inversion of reflection and transmission data. *Geophysics* **53**, 750–759.
Mora, P. (1989). Inversion = migration + tomography. *Geophysics* **54**, 1575–1586.
Mora, P. (1992). The lattice Boltzmann phononic lattice solid. *J. Stat. Phys.* **68**, 591–609.
Mora, P., Maillot, B. (1990). Cellular automata acoustic waves in inhomogeneous media. In: *52nd EAEG Meeting*. EAEG, Copenhagen.
Muir, F. (1987a). M3 to M4: From Huygens' to conservation variables. In: *Stanford Exploration Proj. Tech. Rep.*, vol. 56. Stanford University, Stanford, CA, pp. 19–22.
Muir, F. (1987b). Three experimental modeling system. In: *Stanford Exploration Proj. Tech. Rep.*, vol. 51. Stanford University, Stanford, CA, pp. 119–128.
Muir, F. (1991). Experience with cm Fortran. In: *Stanford Exploration Proj. Tech. Rep.*, vol. 70. Stanford University, Stanford, California, pp. 53–58.
Ohminato, T., Chouet, B. (1997). A free-surface boundary-condition for including 3-D topography in the finite-difference method. *Bull. Seismol. Soc. Am.* **87**, 494–515.
Papazachos, C., Nolet, G. (1997). P and S deep velocity structure of the hellenic area obtained by robust nonlinear inversion of travel times. *J. Geophys. Res. B* **102** (4), 8349–8367.
Qian, Y., d'Humières, D., Lallemand, P. (1992). Lattice BGK models for Navier–Stokes equation. *Europhys. Lett.* **17**, 479–484.
Rivet, J., Hénon, M., Frisch, U., d'Humières, D. (1988). Simulating fully three-dimensional external flow by lattice gas methods. In: Monaco, R. (Ed.), *Discrete Kinetic Theory, Lattice Gas Dynamics and Foundations of Hydrodynamics*. World Scientific, Singapore, pp. 276–285.
Rothman, D. (1987). Modeling seismic P-waves with cellular automata. *Geophys. Res. Lett.* **14**, 17–20.
Rothman, D. (1988a). Cellular-automaton fluids: A model for flow in porous media. *Geophysics* **53**, 509–518.
Rothman, D. (1988b). Lattice-gas automata for immiscible two-phase flow. In: Monaco, R.B. (Ed.), *Discrete Kinetic Theory, Lattice Gas Dynamics and Foundations of Hydrodynamics*. World Scientific, Singapore, pp. 286–299.
Rothman, D., Keller, J. (1988). Immiscible cellular-automaton fluids. *J. Stat. Phys.* **52**, 1119–1127.
Rothman, D., Zaleski, S. (1997). Lattice-Gas Cellular Automata. Cambridge Univ. Press, New York.
Schaffenberger, W., Hanslmeier, A. (2002). Two-dimensional lattice Boltzmann model for magnetohydrodynamics. *Phys. Rev. E* **66**, 46702-1-7.
Schoenberg, M., Muir, F. (1989). A calculus for finely layered anisotropic media. *Geophysics* **54**, 581–589.
Sevink, A., Herman, G. (1996). Three-dimensional, nonlinear, asymptotic seismic inversion. *Inverse Problems* **12**, 757–777.
Succi, S., Benzi, R. (1993). Lattice Boltzmann equation for quantum mechanics. *Physica D* **69**, 327–332.
Succi, S., Benzi, R., Higuera, F. (1991a). The lattice Boltzmann equation: A new tool for computational fluid-dynamics. *Physica D* **47**, 219–230.
Succi, S., Vergassola, M., Benzi, R. (1991b). Lattice Boltzmann scheme for two-dimensional magnetohydrodynamics. *Phys. Rev. A* **43**, 4521–4524.

Tarantola, A. (1984). The seismic reflection inverse problem. In: Santosa, F., Pao, Y., Symes, W., Holland, C. (Eds.), *Inverse Problems of Acoustic and Elastic Waves*. SIAM, Philadelphia.
Tarantola, A. (1986). A strategy for nonlinear elastic inversion of seismic reflection data. *Geophysics* **51**, 1893–1903.
Tarantola, A. (1987). Inverse Problem Theory: Methods for Data Fitting and Model Parameter Estimation. Elsevier, Amsterdam, The Netherlands.
Tarantola, A. (1988). Theoretical background for the inversion of seismic waveforms including elasticity and attenuation. *Pure Appl. Geophys.* **128**, 365–399.
Toscani, G. (1985). Global existence and asymptotic behavior for the discrete models of the Boltzmann equation. *J. Math. Phys.* **26**, 2918–2921.
Ubertini, S., Bella, G., Succi, S. (2003). Lattice Boltzmann method on unstructured grids: Further developments. *Phys. Rev. E* **68**, 16701.
Virieux, J. (1986). P-SV wave propagation in heterogeneous media: Velocity-stress finite-difference method. *Geophysics* **51**, 889–891.
Wolfram, S. (1986). Cellular automaton fluids 1: Basic theory. *J. Stat. Phys.* **45**, 471–526.
Xu, K., He, X. (2003). Lattice Boltzmann method and gas-kinetic BGK scheme in the low-Mach number viscous flow simulations. *J. Comput. Phys.* **190**, 100–117.
Zwillinger, D. (Ed.) (1996). *CRC Standard Mathematical Tables and Formulae*. CRC Press, Boca Raton, Florida.

SYNTHESIS OF SEISMOGRAM ENVELOPES IN HETEROGENEOUS MEDIA

HARUO SATO[1] AND MICHAEL C. FEHLER[2]

[1]*Tohoku University, Sendai, Japan*
[2]*Los Alamos National Laboratory, Los Alamos, New Mexico, USA*

ABSTRACT

Wave trains in high-frequency seismograms of local earthquakes are mostly composed of incoherent waves that are scattered by distributed heterogeneities. Their waveforms are very complex and significantly different from those computed for conventional layered structures; however, their envelopes are repeatable, frequency dependent, and vary regionally. Well-logs obtained from deep boreholes show random fluctuations of medium properties superposed over a layered background structure. Recognizing the complexity of seismograms and Earth inhomogeneity, seismologists often focus on understanding envelopes of band-pass filtered traces rather than on unfiltered waveforms. Stochastic approaches are superior to deterministic wave-theoretical approaches for modeling wave envelopes in random media. There are several methods to predict how envelopes vary with travel distance and frequency depending on the power spectra of random media. As the most tractable case, this chapter precisely examines scalar wave propagation in 2-D random media for an impulsive wavelet isotropically radiated from a point source. We mainly discuss three methods: the Markov approximation method, the isotropic scattering model based on the radiative transfer theory, and a hybrid method. By using reference envelopes simulated by stochastic averaging of waveforms calculated using the finite difference method, the three methods for direct envelope simulation are tested. Random medium having Gaussian auto-correlation function with correlation distance longer than the seismic wavelength is a simple case to analyze because it is dominated by forward scattering. For this case, we mathematically introduce the Markov approximation method, which directly and reliably predicts wave envelopes based on the parabolic wave equation and an extension of the phase screen method. Then, we examine the validity of the Markov approximation for the case of von Kármán-type random media having realistic power-law spectrum as an asymptote. The envelopes predicted by the Markov approximation satisfactorily explain the reference envelopes for random media having weak short wavelength spectra, which are also dominated by forward scattering. For the case of media with strong short wavelength spectra, however, the coincidence is good around the peak amplitude but becomes poor for the coda portion because of wide-angle scattering. In that case, the isotropic scattering model based on radiative transfer theory well explains the coda portion of the reference envelopes, where the momentum transfer scattering coefficient is used as the effective isotropic scattering coefficient. Introducing a hybrid method using the momentum transfer scattering coefficient and the envelope predicted by the Markov approximation as a propagator in the radiative transfer integral equation for isotropic scattering process, we successfully simulate wave envelopes well explaining the reference envelopes from onset to coda for the case of rich short wavelength spectra. It will be necessary to develop envelope simulation methods for vector elastic waves in 3-D random media for a wide range of frequencies for the study of Earth inhomogeneity from the analysis of seismograms. The hybrid method proposed here could be one of mathematical bases for these developments.

Keywords: Seismology, Coda waves, Heterogeneity, Scattering, Random media

1. INTRODUCTION

When we examine seismograms of local earthquakes for frequencies higher than 1 Hz, we observe complex wave trains in addition to direct waves. The complexity of such seismograms cannot be modeled by using a conventional layered structure. The apparent duration of micro-earthquake seismograms observed at distant stations are often much larger than the source duration time. Figure 1a shows velocity seismograms of a micro-earthquake of local magnitude 4.4 in Tohoku, Japan. Traces in Fig. 1b are root mean square (RMS) envelopes of band pass filtered traces. There is a lag between the peak arrival and the S-wave

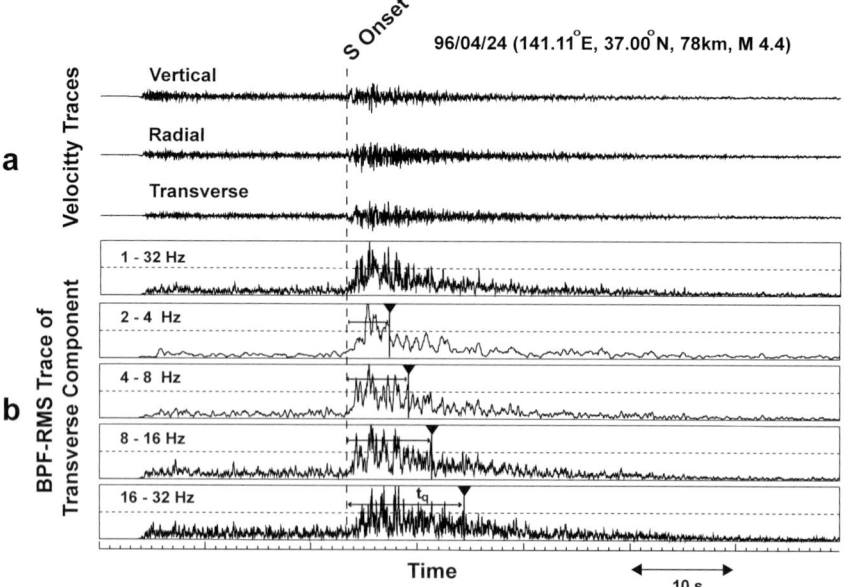

FIG. 1. Velocity seismograms of a microearthquake of M4.4 observed in Tohoku, Japan: (a) Raw traces. (b) Band-pass filtered RMS traces (Saito *et al.*, 2002).

onset in each frequency band. The apparent duration indicated by a horizontal arrow of length t_q in Fig. 1b increases with increasing central frequency. In general, the duration of envelopes broaden with increasing propagation distance. Interpreting the envelope broadening as a result of multiple forward scattering and diffractions due to medium inhomogeneity, Sato (1989) proposed to use the Markov approximation for the synthesis of plane-wave envelopes in media having random velocity inhomogeneity characterized by Gaussian spectra. The Markov approximation is based on the parabolic wave equation and an extension of the phase screen method. This approximation is an attractive stochastic method that was developed for light propagation through upper atmosphere and acoustic sound propagation through internal waves in ocean (e.g. Ishimaru, 1978; Flatté et al., 1979). Analyzing S-wave seismograms recorded in Kanto, Japan based on this model, Sato (1989) and Scherbaum and Sato (1991) estimated the ratio of mean square (MS) fractional fluctuation ε^2 of velocity to the correlation distance a to be of the order of 10^{-3} km^{-1}. Later, Obara and Sato (1995) found regional differences in the frequency dependence of envelope broadening in Kanto-Tokai, Japan. They interpreted the difference in frequency dependence to be due to the difference in spectral structure of random inhomogeneity.

There is other, clear evidence of Earth inhomogeneity. Wave trains following the direct S wave are called coda waves. Figure 2 shows band pass fil-

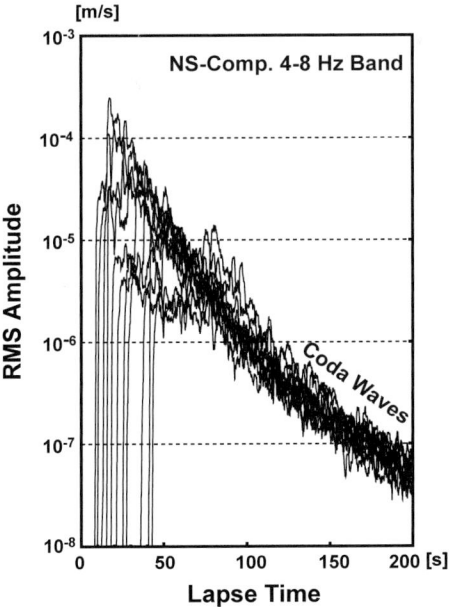

FIG. 2. Coda envelopes of a microearthquake of M 4.8 at different epicentral distances in Tohoku, Japan.

tered seismogram envelopes of a microearthquake recorded at various epicentral distances in Tohoku, Japan. Coda envelopes decay smoothly with increasing lapse time irrespective of epicentral distance. Interpreting coda waves of local earthquakes as scattered waves due to distributed random inhomogeneity in the lithosphere, Aki and Chouet (1975) proposed a simple scattering model to explain the smooth decay of coda amplitude. The existence of a common decay curve for coda envelopes has been used as the basis for the magnitude determination in addition to the measurement of attenuation and site amplification factors (Aki, 1980). Coda envelopes have been investigated theoretically by using radiative transfer theory (Shang and Gao, 1988; Zeng et al., 1991; Sato, 1993). Radiative transfer theory was thought to be rather heuristic and phenomenological; however, it can be statistically derived from the ensemble average equation for the wave field moment in random media (e.g. Foldy, 1945; Frisch, 1968; Ryzhik et al., 1996; Margerin, 2005). Numerical simulations based on the Monte Carlo method have been developed to solve radiative transfer equation (e.g. Hoshiba, 1991; Gusev and Abubakirov, 1996; Yoshimoto, 2000; Margerin et al., 2000).

The spectral structure of the Earth inhomogeneity has been measured from seismological analyses of coda excitation with lapse time and amplitude attenuation with distance (e.g. Wu and Aki, 1985). Acoustic and density logs in deep wells suggest that the fractional fluctuation of Earth inhomogeneity is random and its power spectrum obeys a power-law in wavenumber (Shiomi et al., 1997; Goff and Holliger, 2000). Figure 3 shows the power spectral density function (PSDF) of the fractional fluctuation of acoustic well log obtained in Kyushu, Japan as an example. The power-law spectrum is seen for a wide range of wavelengths from a few tens of centimeters to a hundred meters.

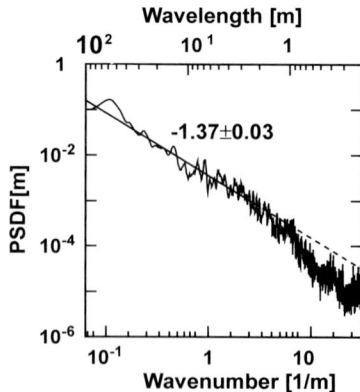

FIG. 3. PSDF of the fractional fluctuation of acoustic well log obtained in Kyushu, Japan (Shiomi et al., 1997).

Scattering phenomena appear over a very wide length scale in materials constituting the Earth. By using a laser Doppler vibrometer, Nishizawa et al. (1997) examined how ultrasonic waves of MHz range are scattered in rock samples. Comparing envelopes of ultrasonic shear waves through a steel block and rock samples of gabbro and granite, Fukushima et al. (2003) precisely measured how envelopes of ultrasonic shear waves decrease in amplitude and have longer duration with increasing propagation distance in rock samples. Figure 4 shows the variation of envelopes of ultrasonic shear waves in different samples and different frequencies, where steel (ST) is uniform and the grain size of Oshima granite (OS3) is larger than that of gabbro (GB). Fukushima et al. (2003) found that the envelope broadening becomes larger and coda excitation becomes stronger with increasing frequencies in rock samples. They reported that the autocorrelation function (ACF) of velocity inhomogeneity of these rock samples is exponential, that is, the PSDF obeys a power law.

If the velocity structure of a medium is precisely known, wave traces can be numerically simulated to a desired resolution within the constraints of computer memory and power. There was a pioneering work of Frankel and Clayton (1986), which simulated wave traces in random media as a model of crustal inhomogeneity by using the finite difference method. They showed how waves attenuate with increasing travel distance and coda waves are excited with increasing lapse time in random media. Frankel and Wennerberg (1987) proposed an empirical model for coda energy distribution based on their numerical simulation results. There have been many attempts based on the finite difference method or the boundary integral method to simulate wave propagation through media containing various kinds of inclusions. But, there have been few studies about how to describe the characteristics of wave envelopes in such media by using a small number of parameters that statistically describe a random cavity/crack distribution. Amplitude attenuation

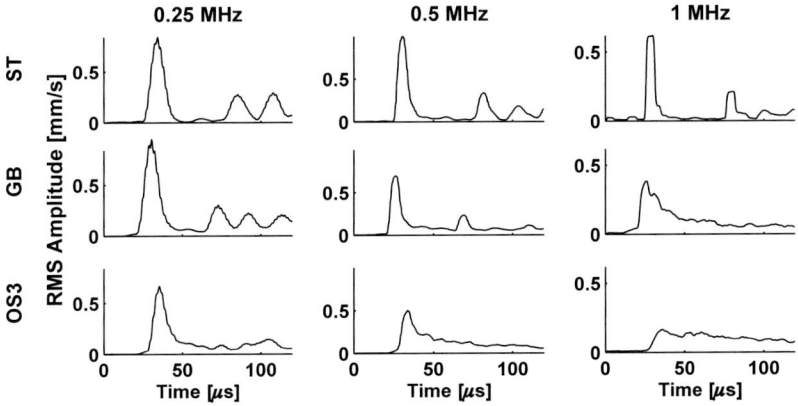

FIG. 4. RMS envelopes of ultrasonic shear waves propagated through steel (ST), gabbro (GB) and granite (OS3) for three different frequencies (Fukushima et al., 2003).

of waves is the only connection with the statistical characterization of medium inhomogeneity: the frequency dependence of amplitude attenuation is well predicted from the scattering loss due to distributed cavities/cracks (Yamashita, 1990; Yomogida and Benites, 1995; Kelner et al., 1999).

Knowing that the real velocity structure appears to be random and rich in short-wavelength spectral components, the full waveform inversion for precise velocity structure including small-scale structure will not be possible even if computer power increases because of insufficient wave field sampling due to the limited number of seismometers and the restriction of observation geometry on the Earth's surface. Therefore, we may change the target from the inversion for the spatial variation of deterministic structure to the inversion for the statistical characteristics of medium inhomogeneity. We know that waveforms can be calculated by a convolution of the medium response with the source time function. If we disregard phase information, we cannot deterministically invert for the inhomogeneous velocity structure from observed waveforms, but we may statistically invert for the power spectrum of velocity inhomogeneity from the frequency dependence of observed wave envelopes. For that purpose, we need to establish a forward method to simulate envelopes in randomly inhomogeneous media characterized by a small set of statistical parameters. The resultant envelopes should be characterized by these statistical parameters.

There have been many attempts to directly simulate wave envelopes in random media for an impulsive radiation from a point source using methods such as the Born approximation with radiative transfer theory and/or the Markov approximation; however, most of them are limited to forward modeling. Fehler *et al.* (2000) first examined the validity of the Markov approximation by making a comparison between the ensemble averaged envelopes of numerically simulated waves for a suite of Gaussian-type random media in 2-D. Later, Saito *et al.* (2003) and Sato *et al.* (2004) examined the case of von Kármán-type random media in 2-D, and developed hybrid methods for the synthesis of whole envelopes from onset to coda. There have also been several attempts to derive envelopes in random media in 3-D, too (e.g. Saito *et al.*, 2002; Lambert and Rickett, 1999); however, they involve forward modeling and a comparison with finite difference simulation has not been done.

As the most tractable case, this chapter gives an exposition of the approaches for direct envelope simulation in 2-D random media for a Ricker wavelet isotropically radiated from a point source by using the methods developed in our papers (Sato, 1993; Fehler *et al.*, 2000; Saito *et al.*, 2003; Sato *et al.*, 2004). By using waveforms simulated by the finite difference method as reference envelopes, the reliability of direct envelope simulation methods can be examined. Taking random media having Gaussian spectra as the simplest case, we first mathematically introduce the Markov approximation method that directly predicts wave envelopes based on the parabolic wave equation and an extension of the phase screen method. Next, we examine the validity of the Markov approximation for the case of von Kármán-type random media, which have realistic power-law spectra as asymptotes. For media with strong short wavelength spectra, we will find a disagreement between the envelopes synthesized by the Markov approximation and the reference envelopes. In such cases, the isotropic scattering model based on radiative transfer theory well explains only the coda portion of the reference envelopes, where the momentum transfer scattering coefficient is used as the effective isotropic scattering coefficient. Then, we introduce a hybrid method using the envelopes predicted by the Markov approximation as a propagator in the radiative transfer integral equation. At last, we will review recent developments in envelope studies.

2. Scalar Wave Equation in 2-D Random Media

We study scalar wave propagation through randomly inhomogeneous two-dimensional media without intrinsic absorption. Scalar waves u are governed by the wave equation,

$$\left(\frac{\partial^2}{\partial x^2} + \frac{\partial^2}{\partial y^2} - \frac{1}{V(\mathbf{x})^2}\frac{\partial^2}{\partial t^2}\right)u(\mathbf{x}, t) = 0. \tag{1}$$

The medium inhomogeneity is characterized by wave velocity

$$V(\mathbf{x}) \equiv V_0 + \delta V(\mathbf{x}) = V_0\bigl(1 + \xi(\mathbf{x})\bigr), \tag{2}$$

where $\xi(\mathbf{x})$ is a fractional fluctuation. We imagine an ensemble $\{\xi(\mathbf{x})\}$, where $\xi(\mathbf{x})$ is assumed to be a statistically homogeneous and isotropic random function of space coordinate $\mathbf{x}$. All the quantities averaged over the ensemble denoted by angular brackets should be compared with observed values. The mean velocity is given by an ensemble average $V_0 = \langle V(\mathbf{x}) \rangle$, where $\langle \xi(\mathbf{x}) \rangle = 0$. The statistical measure of fractional fluctuation is given by the autocorrelation function (ACF)

$$R(\mathbf{x}) \equiv R(r) = \bigl\langle \xi(\mathbf{x}')\xi(\mathbf{x}' + \mathbf{x}) \bigr\rangle, \tag{3}$$

which depends on lag distance $r \equiv |\mathbf{x}|$ only. The function R is also called autocovariance function. The magnitude of the fractional fluctuation is given by the MS fractional fluctuation, $\varepsilon^2 \equiv R(0) = \langle \xi(\mathbf{x})^2 \rangle$. The spectral structure of random media is given by the power spectral density function (PSDF)

$$P(\mathbf{m}) \equiv P(m) = \int_{-\infty}^{\infty}\int_{-\infty}^{\infty} R(\mathbf{x})e^{-i\mathbf{m}\mathbf{x}}\,d\mathbf{x}, \tag{4}$$

where wavenumber $m = |\mathbf{m}|$.

3. Finite Difference Simulation in Gaussian-Type Random Media

Simulating wave traces for a suite of random media by using the finite difference method, Fehler *et al.* (2000) quantitatively evaluated the average envelope of waves over an ensemble of random media. Following their work, we introduce a synthesis of wave envelopes in random media which are statistically characterized by a Gaussian ACF.

3.1. Realization of Random Media

The Gaussian ACF is given by

$$R(\mathbf{x}) = R(r) = \varepsilon^2 e^{-\frac{x^2+y^2}{a^2}}. \tag{5}$$

The length parameter a is called the correlation distance and is the scale over which the random media are smooth. The corresponding PSDF is also Gaussian,

$$P(\mathbf{m}) \equiv P(m) = \varepsilon^2 \pi a^2 e^{-\frac{m^2 a^2}{4}}. \tag{6}$$

To generate a random medium, we construct a function that has the PSDF given by (6), take the square root, and randomize the phase. Figure 5 schematically illustrates an example of a random medium with correlation distance $a = 5$ km and RMS fractional fluctuation $\varepsilon = 0.05$. The value of ε is selected to keep minimum and maximum velocities within a range required for reliable numerical

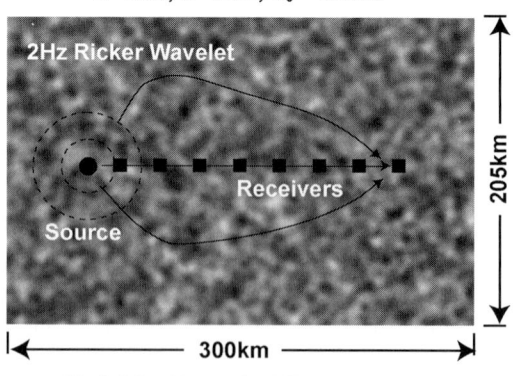

FIG. 5. Model for the FD simulation of scalar waves in 2-D random media and the layout of a source and receivers (Fehler *et al.*, 2000).

wave field calculations. We choose a model with dimensions 300 km long and 205 km wide. A source and receivers are put in the model space as illustrated and the maximum source-receiver distance is 200 km. Choosing $V_0 = 4$ km/s, the travel time for a direct arrival from the source to the receiver at 200 km distance in the background medium is 50 s and the minimum time for a pulse to travel from the source to a lateral boundary and then to the receiver is 71.6 s. Therefore, we conclude that at least the initial 20 s of the first arriving wave packet at 200 km distance has no influence from reflections off the model boundaries.

For the study of high frequency wave propagation, we use an isotropic radiation of 2 Hz Ricker wavelet from the point source, where the frequency and the medium dimension are simply limited by the computer resources. In this case, the dominant wavelength is 2 km and forward scattering dominates since $ak_0 \approx 16$, where wavenumber k_0 corresponds to 2 Hz.

3.2. Finite Difference Simulation

The finite difference (FD) code used has fourth order accuracy in space and second order accuracy in time. The code uses Holberg coefficients (Holberg, 1987), which are optimal for minimizing grid dispersion for a given number of grid points per wavelength. We use Higdon absorbing boundaries (Higdon, 1991). In FD modeling, we wish to have small grid dispersion over the propagation distance being modeled. For modeling to distances of 100 wavelengths from the source, the number of grid points required per wavelength is 40. We thus used grid spacing 50 m. To stabilize the finite difference calculation, we eliminate random inhomogeneity in the model that has spatial wavelength smaller than 200 m. Elimination of such small-scale inhomogeneity from Gaussian-type random medium having $a = 5$ km has a negligible effect on the wave propagation. Time step is chosen to be 4 ms by using criteria for numerical stability. The FD method provides a reliable estimation of the entire wave field because it includes all wave phenomena including reverberations. As an example, Fig. 6 shows wave traces with increasing travel distance in one realization of a random medium, where the abscissa is reduced time with a moveout velocity $V_0 = 4$ km/s. Waveforms are similar to the Ricker wavelet at receivers near the source; however, this section shows clearly the increasing complexity of the waveform as propagation distance increases. Waveforms were simulated for a total of 100 realizations of random media. FD waveforms at a propagation distance of 200 km calculated for 15 realizations are shown in Fig. 7a as an example. There is considerable variation in waveform character among the various realizations. Note also that there is considerable fluctuation in first arrival times among the realizations.

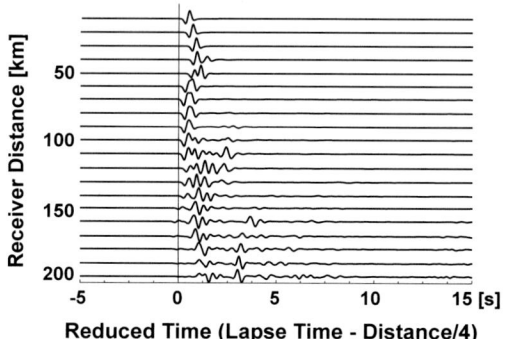

FIG. 6. Variation of wave traces with increasing travel distance simulated by the FD method in one realization of a random medium characterized by the Gaussian ACF, where the source is a 2 Hz Ricker wavelet and each trace is normalized by the maximum amplitude (Fehler et al., 2000).

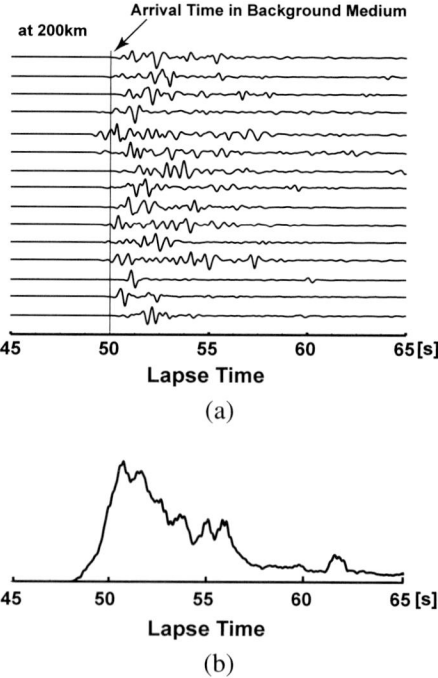

FIG. 7. (a) Wave traces at 200 km distance from the source simulated by the FD method in 15 realizations of a random medium characterized by the Gaussian ACF. (b) RMS envelope averaged over an ensemble of 100 realizations (Fehler et al., 2000).

3.3. FD Envelopes as Reference Envelopes

Ensemble average envelopes were calculated from a total of 100 realizations of random media. Ensemble average envelopes were calculated by squaring waveforms from each realization, averaging results from all realizations, taking the square root, and smoothing over a 0.32 s window. Figure 7b shows an ensemble-average RMS envelope at 200 km distance. In Fig. 8, gray curves show RMS ensemble average envelopes calculated by the FD method for a 2 Hz Ricker wavelet source for travel distance ranging from 50 km to 200 km. As travel distance increases, the peak delay from the onset and the broadening of time width increase. We find a small excitation of coda waves, which are late-arriving waves that have been scattered at wide angles by short wavelength spectra of random inhomogeneities since the FD method includes the contribution of wide-angle scattering. We refer to these envelopes simulated by the FD method as FD envelopes. We use FD envelopes as reference envelopes for the evaluation of direct envelope simulation methods in the following.

4. MARKOV APPROXIMATION FOR GAUSSIAN-TYPE RANDOM MEDIA

We study the long travel distance propagation of quasi-monochromatic waves radiated isotropically from a point source. Instead of modeling individual waveforms, we directly model the envelope. When the wavelength is much smaller than the correlation distance $ak_0 \gg 1$, diffraction and scattering around the forward direction dominate and we may use the parabolic wave equation, which describes wave propagation in one direction. We will derive the master equation for the two-frequency mutual coherence function (TFMCF), which is the correlation of the wave field between two locations and two frequencies. As shown in the following, the TFMCF is a Fourier transform of the MS envelope for a given central angular frequency. In this case, master equation can be used to extrapolate the TFMCF recursively in distance away from the source in a manner that the

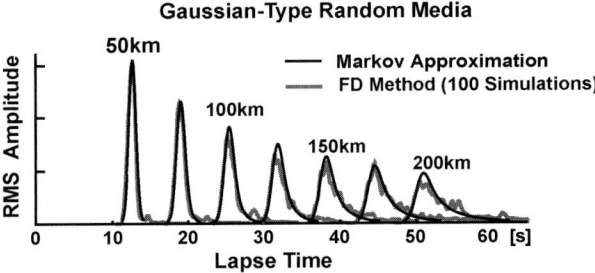

FIG. 8. RMS FD envelopes (gray curves) and RMS Markov envelopes (solid curves) in random media characterized by the Gaussian ACF (Fehler et al., 2000). FD envelopes are reference envelopes.

TFMCF at some distance can be determined once the TFMCF at some slightly smaller distance is known. The term Markov approximation has its roots in the concept of Markov process, which is one where the probability of future events is dependent only on most recent events (Tatarskii, 1971). The following derivation is taken from Fehler et al. (2000).

In the case of small fractional fluctuation $|\xi| \ll 1$, substituting (2) into (1), we have

$$\left(\frac{\partial^2}{\partial x^2} + \frac{\partial^2}{\partial y^2} - \frac{1}{V_0^2}\frac{\partial^2}{\partial t^2}\right)u(\mathbf{x},t) + \frac{2}{V_0^2}\xi(\mathbf{x})\frac{\partial^2}{\partial t^2}u(\mathbf{x},t) = 0. \tag{7}$$

At a source-receiver distance r, which is longer than the wavelength $(r \gg 1/k_0)$ and correlation distance $(r \gg a)$, we may write outgoing wave radiated from a source at the origin using the Fourier transform with respect to angular frequency as

$$u(r,\theta,t) = \frac{1}{2\pi}\int_{-\infty}^{\infty} d\omega \frac{U(r,\theta,\omega)}{\sqrt{r}} e^{i(k_0 r - \omega t)}, \tag{8}$$

where $r \equiv |\mathbf{x}|$, θ is the polar angle, and U is the amplitude of an outgoing harmonic wave. Substituting (8) into (7), we obtain the parabolic wave equation

$$2ik_0\partial_r U + \frac{\partial_\theta^2 U}{r^2} - 2k_0^2 \xi U = 0, \tag{9}$$

where we neglected a second derivative term with respect to radius $\partial_r^2 U$ since $a \gg 1/k_0$ and a term proportional to the inverse square of distance U/r^2 since $r \gg 1/k_0$. Note that, contrary to the wave equation, the parabolic wave equation contains only a first derivative in the spatial coordinate corresponding to distance from the source. This means that we can find the wave field at r by integrating (9) over the interval $r - \Delta r$ to r. This is the approach that is implemented when doing wave equation migration imaging (e.g. Claerbourt, 1985; Fehler and Huang, 2002).

4.1. Coherent Wave Field

If we consider an ensemble of random media and take the ensemble average of field U, we get the coherent wave field, or mean field $\langle U \rangle$. We introduce the Markov approximation by deriving the equation for the coherent wave field in the following. Taking the ensemble average of (9), we have

$$2ik_0\partial_r \langle U \rangle + \frac{\partial_\theta^2 \langle U \rangle}{r^2} - 2k_0^2 \langle \xi U \rangle = 0. \tag{10}$$

Following Lee and Jokipii (1975), we evaluate the last term as follows. Integrating (9), we can write the wave field at r by using the wave field at $r - \Delta r$, where $\Delta r > 0$:

$$U(r, \theta, \omega) \approx U(r - \Delta r, \theta, \omega) + \frac{i}{2k_0} \frac{\Delta r}{r^2} \partial_\theta^2 U(r - \Delta r, \theta, \omega)$$

$$- ik_0 U(r - \Delta r, \theta, \omega) \int_{r-\Delta r}^{r} dr'' \, \xi(r'', r''\theta). \tag{11}$$

One argument of ξ is taken as $r\theta$, which is the distance on the transverse axis perpendicular to the radial direction in the vicinity of $\theta \approx 0$. We suppose the existence of an intermediate scale Δr, which is larger than the correlation distance a but smaller than the scale of variation of U. Multiplying (11) by $\xi(r, r\theta')$ and taking the ensemble average, we have

$$\langle \xi(r, r\theta') U(r, \theta'', \omega) \rangle$$

$$\approx \langle \xi(r, r\theta') U(r - \Delta r, \theta'', \omega) \rangle + \frac{i}{2k_0} \frac{\Delta r}{r^2} \langle \xi(r, r\theta'') \partial_\theta^2 U(r - \Delta r, \theta'', \omega) \rangle$$

$$- ik_0 \int_{r-\Delta r}^{r} dr'' \, \langle \xi(r, r\theta') U(r - \Delta r, \theta'', \omega) \xi(r'', r''\theta'') \rangle$$

$$= -ik_0 \langle U(r - \Delta r, \theta'', \omega) \rangle \int_{r-\Delta r}^{r} dr'' \, \langle \xi(r, r\theta') \xi(r'', r''\theta'') \rangle$$

$$\approx -ik_0 \langle U(r - \Delta r, \theta'', \omega) \rangle \int_{0}^{\infty} dy \, R(y, r\theta' - r\theta'')$$

$$\approx -i \frac{k_0}{2} A\big(r(\theta' - \theta'')\big) \langle U(r, \theta'', \omega) \rangle, \tag{12}$$

where we used that the variation of U is small, $\langle U(r, \theta'', \omega) \rangle \approx \langle U(r - \Delta r, \theta'', \omega) \rangle$, and there is no contribution of the inhomogeneity at r to the wave field at $r - \Delta r$, $\langle \xi(r, \theta') U(r - \Delta r, \theta'', \omega) \rangle \approx 0$, which means no backward scattering. Function A in (12) is a longitudinal integral of the ACF of ξ as defined by

$$A(r\theta_d) = \int_{-\infty}^{\infty} dy \, R(y, r\theta_d), \tag{13}$$

where $r\theta_d = |x' - x''| = r\theta' - r\theta''$. Substituting (12) into (10), we get

$$2ik_0 \partial_r \langle U \rangle + \frac{\partial_\theta^2 \langle U \rangle}{r^2} + ik_0^3 A(0) \langle U \rangle = 0. \tag{14}$$

This is the equation for coherent wave field. The coherent wave field at some distance can be determined once the coherent wave field at some slightly smaller distance is known. It is just like an extension of the phase screen method (e.g. Jensen et al., 1994). Alternative derivations based on a functional formulation are given in Tatarskii (1971) and/or Ishimaru (1978). The range of application conditions for this approximation is discussed in detail by Barabanenkov et al. (1971). The coherent wave field decays exponentially with increasing propagation distance as

$$\langle U(r,\theta,\omega)\rangle \propto e^{-\frac{A(0)k_0^2 r}{2}}. \tag{15}$$

We note that the term $A(0)k_0^2/2$ gives scattering loss per distance for the coherent wave field.

4.2. Two-Frequency Mutual Coherence Function

We define the two-frequency mutual coherence function (TFMCF) as the correlation of the field U between two locations θ' and θ'' on the transverse axis at distance r, at different angular frequencies at ω' and ω'' (Ishimaru, 1978),

$$\Gamma_2(r,\theta',\theta'',\omega',\omega'') \equiv \langle U(r,\theta',\omega')U(r,\theta'',\omega'')^*\rangle, \tag{16}$$

where an asterisk indicates complex conjugate. Multiplying U^* by (9) and U by the complex conjugate of (9), and taking the ensemble average of their sum, we obtain

$$2i\partial_r \Gamma_2(r,\theta',\theta'',\omega',\omega'') + \left(\frac{\partial_{\theta'}^2}{r^2 k_0'} - \frac{\partial_{\theta''}^2}{r^2 k_0''}\right)\Gamma_2(r,\theta',\theta'',\omega',\omega'')$$
$$- 2\langle(k_0'\xi(r,r\theta') - k_0''\xi(r,r\theta''))U(r,\theta',\omega')U(r,\theta'',\omega'')^*\rangle = 0. \tag{17}$$

Using the same procedure for the derivation of (12), we have

$$\langle(k_0'\xi(r,r\theta') - k_0''\xi(r,r\theta''))U(r,\theta',\omega')U(r,\theta'',\omega'')^*\rangle$$
$$= -\frac{i}{2}\left[(k_0'^2 + k_0''^2)A(0) - 2k_0'k_0''A(r\theta_d)\right]\Gamma_2(r,\theta',\theta'',\omega',\omega''). \tag{18}$$

Substituting (18) into (17), we have

$$2i\partial_r \Gamma_2 + \left(\frac{1}{k_0'} - \frac{1}{k_0''}\right)\frac{1}{r^2}\partial_{\theta_d}^2 \Gamma_2$$
$$+ i\left[(k_0'^2 + k_0''^2)A(0) - 2k_0'k_0''A(r\theta_d)\right]\Gamma_2 = 0, \tag{19}$$

where we used $\partial_{\theta'}^2\Gamma_2 = \partial_{\theta''}^2\Gamma_2 = \partial_{\theta_d}^2\Gamma_2$ since Γ_2 depends only on the difference between locations $r\theta'$ and $r\theta''$ because of the statistical homogeneity of random media.

4.3. Quasi-Monochromatic Waves

For waves having angular frequency centered around $\omega_c = (\omega' + \omega'')/2$, we use the center-of-mass coordinate ω_c and define the difference coordinate $\omega_d = \omega' - \omega''$ in the angular-frequency domain, and $k_{0c} = \omega_c/V_0$ and $k_{0d} = \omega_d/V_0$. For quasi-monochromatic waves ($|\omega_d| \ll |\omega_c|$), we can derive the equation governing the TFMCF as

$$\partial_r \Gamma_2 + i\frac{k_{0d}}{2k_{0c}^2 r^2}\partial_{\theta_d}^2 \Gamma_2 + k_{0c}^2[A(0) - A(r\theta_d)]\Gamma_2 + \frac{k_{0d}^2}{2}A(0)\Gamma_2 = 0, \quad (20)$$

where we used $k_{0c} = (k_0' + k_0'')/2$, $k_{0d} = k_0' - k_0''$, $k_0'^2 + k_0''^2 \approx 2k_{0c}^2 + k_{0d}^2/2$, $k_0' k_0'' \approx k_{0c}^2 - k_{0d}^2/4$, and $1/k_0' - 1/k_0'' \approx -k_{0d}/k_{0c}^2$. The variation of TFMCF comes only from short offsets on the transverse axis as travel distance increases. If we choose Γ_2 to be of the form

$$\Gamma_2 = {}_0\Gamma_2 e^{-\frac{1}{2}k_d^2 A(0)z}, \quad (21)$$

where the exponential term represents the wandering effect, the equation for ${}_0\Gamma_2$ is given by

$$\partial_r \,{}_0\Gamma_2 + i\frac{k_{0d}}{2k_{0c}^2 r^2}\partial_{\theta_d}^2 \,{}_0\Gamma_2 + k_{0c}^2[A(0) - A(r\theta_d)]\,{}_0\Gamma_2 = 0. \quad (22)$$

The ensemble average of the wave intensity at a distance r and at lapse time t is written as

$$\langle |u(\mathbf{x}, t)|^2\rangle = \frac{1}{(2\pi)^2}\frac{1}{r}\int_{-\infty}^{\infty}d\omega'\int_{-\infty}^{\infty}d\omega''\langle U(r,\theta,\omega')U(r,\theta,\omega'')^*\rangle$$
$$\times e^{-i(\omega'-\omega'')(t-r/V_0)}$$
$$= \frac{1}{(2\pi)^2}\frac{1}{r}\int_{-\infty}^{\infty}d\omega_c\int_{-\infty}^{\infty}d\omega_d\,\Gamma_2(r,\theta_d=0,\omega_d,\omega_c)$$
$$\times e^{-i\omega_d(t-r/V_0)}$$
$$= \frac{1}{(2\pi)^2}\frac{1}{r}\int_{-\infty}^{\infty}d\omega_c\int_{-\infty}^{\infty}d\omega_d\,e^{-\frac{A(0)r}{2V_0^2}\omega_d^2}\,{}_0\Gamma_2(r,\theta_d=0,\omega_d,\omega_c)$$
$$\times e^{-i\omega_d(t-r/V_0)}$$
$$= \frac{V_0}{2\pi}\int_{-\infty}^{\infty}d\omega_c\int_{-\infty}^{\infty}dt'\,w(\mathbf{x},t-t')G_M(\mathbf{x},t';\omega_c), \quad (23)$$

where the Fourier transform of $\exp[-A(0)r\omega_d^2/2V_0^2]$ gives the wandering effect in the time domain,

$$w(\mathbf{x},t) = \frac{1}{2\pi}\int_{-\infty}^{\infty}d\omega_d\,e^{-\frac{A(0)r}{2V_0^2}\omega_d^2}e^{-i\omega_d t} = \frac{V_0}{\sqrt{2\pi A(0)r}}e^{-\frac{V_0^2 t^2}{2A(0)r}}. \quad (24)$$

The wandering term $w(\mathbf{x}, t)$ does not influence the broadening of individual wave packets but shows wandering effect from the statistical averaging of the phase fluctuations of different rays on the transverse line at distance r (Lee and Jokipii, 1975). The wave intensity at a distance r and at a central angular frequency ω_c is given by a convolution integral of the wandering effect w and function G_M, $G_M * w$. Function G_M is given by the inverse Fourier transform of ${}_0\Gamma_2$ with respect to difference angular frequency ω_d as

$$G_M(\mathbf{x}, t; \omega_c) = \frac{1}{2\pi r V_0} \frac{1}{2\pi} \int_{-\infty}^{\infty} d\omega_d \left[2\pi \, {}_0\Gamma_2(r, \theta_d = 0, \omega_d, \omega_c)\right] e^{-i\omega_d(t - r/V_0)}. \quad (25)$$

Function G_M is the MS envelope at central angular frequency ω_c. Function G_M must be real, which means that ${}_0\Gamma_2(r, \theta_d = 0, \omega_d, \omega_c) = {}_0\Gamma_2(r, \theta_d = 0, -\omega_d, \omega_c)^*$. We also note that causality gives $G_M = 0$ for $t < r/V_0$. As the initial condition for the coherent isotropic radiation from the point source, we take ${}_0\Gamma_2$ to be non-dimensional for a unit source radiation,

$$2\pi \, {}_0\Gamma_2(r = 0, \theta_d, \omega_d, \omega_c) = 1. \quad (26)$$

Then, G_M has a dimension of spatial density and satisfies

$$G_M(\mathbf{x}, t; \omega_c) \to \frac{1}{2\pi V_0 r} \delta\left(t - \frac{r}{V_0}\right) \quad \text{as } r \to 0. \quad (27)$$

Function G_M satisfies $\int_{-\infty}^{\infty} 2\pi r V_0 G_M(\mathbf{x}, t; \omega_c) \, dt = 1$, which means that all the energy radiated from the origin passes through a circle of radius r since we disregarded backward scattering in the derivation.

4.4. Analytic Solution for the Case of Gaussian ACF

For the Gaussian ACF (5), we have

$$A(r\theta_d) = \sqrt{\pi}\varepsilon^2 a e^{-\frac{r^2\theta_d^2}{a^2}} \approx \sqrt{\pi}\varepsilon^2 a \left[1 - \frac{r^2\theta_d^2}{a^2}\right] \quad \text{for } r\theta_d \ll a. \quad (28)$$

Taking $A(0) = \sqrt{\pi}\varepsilon^2 a$ from (28) and placing it in (15), we find that the coherent wave field decays as $\langle U \rangle \propto \exp(-\sqrt{\pi}\varepsilon^2 a k_0^2 r/2)$. For a 2 Hz wave in our model structure, the coherent wave field scattering loss per travel distance is 0.11 km^{-1}. This means that the coherent wave field dies away very rapidly. As illustrated in Fig. 6, individual wave packets look to keep their original form but the arrival times vary a lot, therefore, the average wave field over an ensemble becomes very small even at small distances. In this case, the wandering effect (24) is written as

$$w(\mathbf{x}, t) = \frac{V_0}{\sqrt{2\pi \sqrt{\pi}\varepsilon^2 a r}} e^{-\frac{V_0^2 t^2}{2\sqrt{\pi}\varepsilon^2 a r}}. \quad (29)$$

Substituting (28) into the third term of (22), we explicitly write the equation for ${}_0\Gamma_2$ as

$$\partial_r \, {}_0\Gamma_2 + i \frac{k_{0d}}{2k_{0c}^2 r^2} \partial_{\theta_d}^2 \, {}_0\Gamma_2 + \frac{\sqrt{\pi}\varepsilon^2 k_{0c}^2 r^2 \theta_d^2}{a} \, {}_0\Gamma_2 = 0. \quad (30)$$

By using the travel distance r_0 and the characteristic time

$$t_M \equiv \frac{\sqrt{\pi}}{2V_0}\frac{\varepsilon^2}{a}r_0^2, \qquad (31)$$

we define the non-dimensional longitudinal distance τ and the non-dimensional transverse distance χ as

$$\tau = \frac{r}{r_0} \quad \text{and} \quad \chi = \sqrt{2r_0 V_0 k_{0c}^2 t_M \theta_d}, \qquad (32)$$

where $\tau = 0$ at the source and $\tau = 1$ at the receiver. We can write Eq. (30) in non-dimensional form as

$$\partial_\tau {}_0\Gamma_2 + i\frac{t_M \omega_d}{\tau^2}\partial_\chi^2 {}_0\Gamma_2 + \tau^2 \chi^2 {}_0\Gamma_2 = 0. \qquad (33)$$

First, we assume that the solution has the following form:

$${}_0\Gamma_2(\tau, \chi) = \frac{e^{v(\tau)\tau^2\chi^2}}{w(\tau)}. \qquad (34)$$

Then, (33) reduces to

$$\left[\frac{dv}{d\tau} + \frac{2}{\tau}v + 4it_M\omega_d v^2 + 1\right]\tau^2\chi^2 + \left[2it_M\omega_d v - \frac{1}{w}\frac{dw}{d\tau}\right] = 0. \qquad (35)$$

Each term in brackets in (35) must be zero to satisfy the equation regardless of χ. The differential equation for $v(\tau)$ is a Riccati equation. Using the initial condition ${}_0\Gamma_2(\tau = 0, \chi) = 1/2\pi$, that is, $v(0) = 0$ and $w(0) = 2\pi$, we get

$$v(\tau) = \frac{1}{s_0}\cot s_0\tau - \frac{1}{s_0^2\tau} \quad \text{and} \quad w(\tau) = 2\pi\sqrt{\frac{\sin s_0\tau}{s_0\tau}}, \qquad (36)$$

where $s_0 = 2e^{\pi i/4}\sqrt{t_M\omega_d}$. Finally we obtain

$${}_0\Gamma_2(\tau, \chi) = \frac{1}{2\pi}\sqrt{\frac{s_0\tau}{\sin s_0\tau}}e^{(\frac{\tau^2}{s_0}\cot s_0\tau - \frac{\tau}{s_0^2})\chi^2}. \qquad (37)$$

At distance $r = r_0$ and difference angle $\theta_d = 0$ corresponding to $\tau = 1$ and $\chi = 0$, respectively, we get

$${}_0\Gamma_2(r_0, \theta_d = 0) = {}_0\Gamma_2(\tau = 1, \chi = 0) = \frac{1}{2\pi}\sqrt{\frac{2e^{\pi i/4}\sqrt{t_M\omega_d}}{\sin(2e^{\pi i/4}\sqrt{t_M\omega_d})}}. \qquad (38)$$

Substituting (38) into (25), we get

$$G_M(r_0, t; \omega_c)$$
$$= \frac{1}{2\pi V_0 r_0}\frac{1}{2\pi}\int_{-\infty}^{\infty} d\omega_d \sqrt{\frac{2e^{\pi i/4}\sqrt{t_M\omega_d}}{\sin(2e^{\pi i/4}\sqrt{t_M\omega_d})}}e^{-i\omega_d(t-r_0/V_0)}, \qquad (39)$$

where the RHS is practically independent of central angular frequency ω_c.

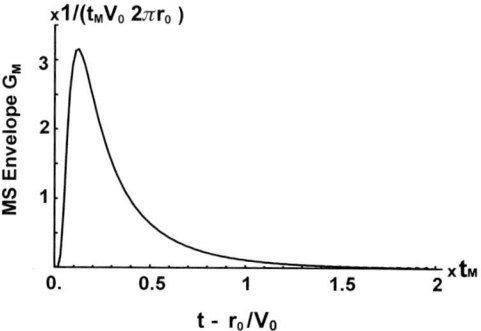

FIG. 9. MS envelope G_M derived from the Markov approximation for 2-D random media with Gaussian ACF, where t_M is defined by (31) (Fehler et al., 2000).

We can numerically evaluate (39) by using an FFT. When we scale the time and angular frequency by using the characteristic time as $\bar{\omega}_d = t_M \omega_d$, we find that $2\pi r_0 V_0 t_M G_M$ is a function of reduced time scaled by the characteristic time as illustrated in Fig. 9. Note that the envelope characteristics are functions of the scaled time t_M and thus depend on the ratio ε^2/a. The peak amplitude of 3.2 occurs at time $0.12 t_M$ after the direct arrival, and the quarter maximum, which corresponds to the half maximum of the RMS envelope, is at $0.45 t_M$. Since t_M in (31) is proportional to the square of the propagation distance, the envelope width for an impulsive source is proportional to the square of travel distance. We note that the characteristic time is independent of frequency in this case.

4.5. Markov Envelopes in Gaussian-Type Random Media

To calculate the MS envelope for the isotropic radiation of 2 Hz Ricker wavelet from the point source, we use the convolution of $G_M(r, t; \omega_c)$, the wandering effect $w(\mathbf{x}, t)$ and the temporal change in the power of 2 Hz Ricker wavelet source $W_R(t)$, $G_M * w * W_R$. We call these synthesized envelopes the Markov envelopes. Taking the square root of MS envelope, we have an RMS envelope. Solid traces in Fig. 8 show RMS Markov envelopes for the case of Gaussian ACF ($\varepsilon = 0.05$ and $a = 5$ km). The RMS envelopes calculated using the Markov approximation agree well with the reference envelopes. Our results indicate that the predictions made by the Markov approximation are reliable for the case of small velocity fluctuations ($\varepsilon \leq 0.05$). We note that it is valid to compare $G_M * w * W_R$ with the ensemble-averaged envelopes; however, it is better to exclude the wandering effect by using $G_M * W_R$ for a comparison with an individual wave envelope.

5. MARKOV APPROXIMATION FOR VON KÁRMÁN-TYPE RANDOM MEDIA

Figure 3 shows that the spectrum of random inhomogeneity may be appropriately described as a power law. However, a power law function goes to infinity at a

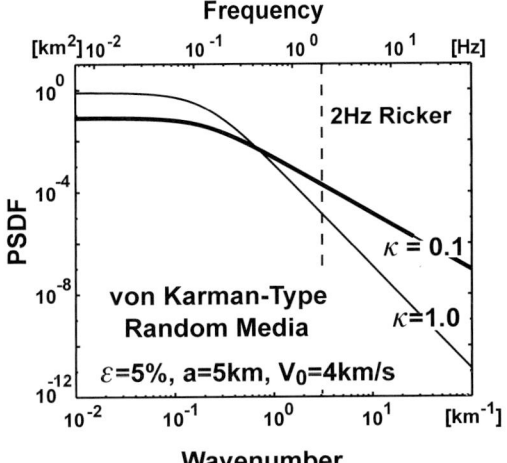

FIG. 10. PSDFs of von Kármán-type random media in 2-D for $\kappa = 0.1$ and $\kappa = 1.0$.

wavenumber of 0, so we cannot use a power law over all wavenumbers to describe a medium. Instead, the inhomogeneity of the Earth may be more appropriately described by using a von Kármán-type ACF that has a power-law spectrum as a high wavenumber asymptote. The von Kármán-type ACF of order κ in 2-D is written by

$$R(\mathbf{x}) = R(r) = \frac{\varepsilon^2 2^{1-\kappa}}{\Gamma(\kappa)} \left(\frac{r}{a}\right)^\kappa K_\kappa\left(\frac{r}{a}\right), \tag{40}$$

where Γ is a gamma function, and K_κ is the modified Bessel function of the second kind. The corresponding PSDF is

$$P(\mathbf{m}) = P(m) = \frac{4\pi \varepsilon^2 a^2 \kappa}{(1 + a^2 m^2)^{\kappa+1}}$$
$$\propto m^{-2\kappa - 2} \quad \text{for } am \gg 1, \tag{41}$$

where this PSDF obeys a power law for large wavenumbers with exponent $-2\kappa - 2$. Figure 10 shows PSDFs for $\kappa = 0.1$ and 1.0. We take $\kappa = 1.0$ to represent random media that are poor in short-wavelength spectral components and $\kappa = 0.1$ for a typical example of random media that are rich in short-wavelength spectral components.

5.1. FD Envelopes as Reference Envelopes

Using the same FD method as used for the case of Gaussian-type random media, we numerically simulate wave fields for the isotropic radiation of a 2 Hz Ricker wavelet from a point source through 50 realizations of a von Kármán-type

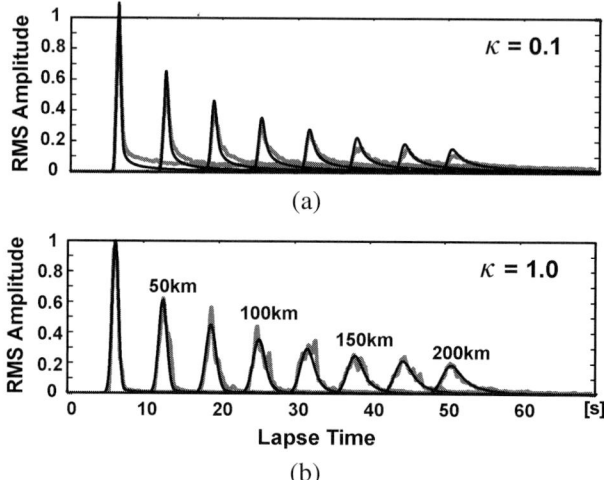

FIG. 11. RMS FD envelopes (gray curves) and RMS Markov envelopes (solid curves) in random media characterized by the von Kármán-type ACF ($\varepsilon = 0.05$, $a = 5$ km, $V_0 = 4$ km/s), where the source is a 2 Hz Ricker wavelet: (a) $\kappa = 0.1$; (b) $\kappa = 1.0$ (Saito et al., 2003). FD envelopes are reference envelopes.

random medium of dimension 205 km by 300 km having $V_0 = 4$ km/s, $\varepsilon = 5\%$, $a = 5$ km, for each case of $\kappa = 0.1$ and $\kappa = 1.0$. Gray curves in Figs. 11a and b show ensemble-average FD envelopes at eight source-receiver distances. The maximum peak amplitude of the envelope decreases, the lag between the maximum envelope amplitude and the onset increases, and the time width broadens as the travel distance increases. The excitation of coda waves in media with $\kappa = 0.1$ is larger than that in media with $\kappa = 1.0$ at each distance because of the difference in short wavelength spectral components, which are responsible for wide-angle scattering. Even though RMS fractional fluctuation ε is small as 0.05, the coda excitation is strong in the case of $\kappa = 0.1$. We use those FD envelopes as reference envelopes for the evaluation of direct envelope simulation methods for the case of von Kármán-type random media.

5.2. Markov Envelopes

For von Kármán-type random media, the longitudinal integral A (13) is directly calculated from the PSDF (41) as

$$A(r\theta_d) = \int_{-\infty}^{\infty} dy \frac{1}{(2\pi)^2} \iint_{-\infty}^{\infty} P(\mathbf{m}) e^{im_x r\theta_d + im_y y} \, dm_x \, dm_y$$

$$= \frac{2^{-\kappa+\frac{3}{2}} \sqrt{\pi} \varepsilon^2 a}{\Gamma(\kappa)} \left(\frac{r\theta_d}{a}\right)^{\kappa+\frac{1}{2}} K_{\kappa+\frac{1}{2}}\left(\frac{r\theta_d}{a}\right). \quad (42)$$

We may write the value of $A(0) - A(r\theta_d)$ in (22) at small transverse distance $r\theta_d \ll a$ by using integral formulas (Abramowitz and Stegun, 1970) as

$$A(0) - A(r\theta_d) \approx \begin{cases} \dfrac{-\pi^{\frac{3}{2}}\varepsilon^2 a}{2\Gamma(\kappa)\Gamma(\frac{3}{2}-\kappa)\cos\kappa\pi}\left(\dfrac{r\theta_d}{a}\right)^2 & \text{for } \kappa \gg \frac{1}{2}, \\[2mm] \dfrac{\pi^{\frac{3}{2}}\varepsilon^2 a}{2^{2\kappa}\Gamma(\kappa)\Gamma(\frac{3}{2}+\kappa)\cos\kappa\pi}\left(\dfrac{r\theta_d}{a}\right)^{2\kappa+1} & \text{for } \kappa \ll \frac{1}{2}. \end{cases} \quad (43)$$

These asymptotic solutions are not available when κ is close to 1/2. Using a numerical evaluation, Saito et al. (2003) got an approximation as a power of transverse distance as

$$A(0) - A(r\theta_d) \approx \varepsilon^2 a C(\kappa) \left(\frac{r\theta_d}{a}\right)^{p(\kappa)} \quad \text{for } r\theta_d/a \ll 1. \quad (44)$$

They numerically estimate $A(0)$, $p(\kappa)$ and $C(\kappa)$ for the range of $r\theta_d/a$ from 10^{-4} to 10^{-1} for different κ values within 15% error: $A(0) = 0.55$, $C = 0.56$ and $p = 1.19$ for $\kappa = 0.1$, and $A(0) = 3.14$, $C = 1.50$ and $p = 1.99$ for $\kappa = 1.0$.

The master equation of the Markov approximation for the quasi-monochromatic waves (22) is valid for von Kármán-type random media. Substituting (44) into (22), we get

$$\partial_r {}_0\Gamma_2 + i\frac{k_{0d}}{2k_{0c}^2 r^2}\partial_{\theta_d}^2 {}_0\Gamma_2 + k_{0c}^2 \varepsilon^2 a C(\kappa)\left(\frac{r\theta_d}{a}\right)^{p(\kappa)} {}_0\Gamma_2 = 0. \quad (45)$$

Using the characteristic time at a distance r_0 newly defined as

$$t_M = \frac{C(\kappa)^{\frac{2}{p(\kappa)}} \varepsilon^{\frac{4}{p(\kappa)}} a}{2V_0}\left(\frac{a\omega_c}{V_0}\right)^{\frac{-2p(\kappa)+4}{p(\kappa)}}\left(\frac{r_0}{a}\right)^{\frac{p(\kappa)+2}{p(\kappa)}} \quad (46)$$

in scaling the longitudinal distance and transverse distance in (32), we write the master equation in non-dimensional form as

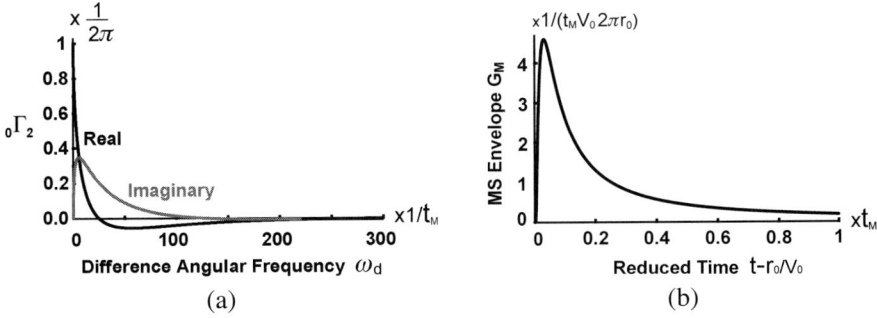

FIG. 12. (a) TFMCF calculated by the numerical integration for the von Kármán-type random media of $\kappa = 0.1$, where t_M is defined by (46). (b) MS envelope G_M derived from the Markov approximation (Sato et al., 2004).

$$\partial_\tau {}_0\Gamma_2 + it_M\omega_d \frac{1}{\tau^2}\partial_\chi^2 {}_0\Gamma_2 + \tau^p \chi^p {}_0\Gamma_2 = 0. \tag{47}$$

Comparing (47) with (33), we find a difference in the power of the third term. Numerically integrating (47), we obtain ${}_0\Gamma_2$ as a function of difference angular frequency. Figure 12a shows real and imaginary parts of ${}_0\Gamma_2$ at a distance r_0 for $\kappa = 0.1$. By using an FFT in (25), we can easily evaluate G_M from ${}_0\Gamma_2$. Figure 12b shows the temporal change in G_M against reduced time $t - r/V_0$ at $r = r_0$ for $\kappa = 0.1$. When we compare G_M in Figs. 9 and 12b, we note the difference in the definitions of characteristic time.

5.3. Comparison of Markov Envelopes with Reference Envelopes

In Fig. 11, solid curves show RMS Markov envelopes in von Kármán-type random media for a 2 Hz Ricker wavelet source. MS Markov envelopes are simulated by using a convolution integral $G_M * w * W_R$, where the wandering effect w is calculated from (24) with (42). We may say that Markov envelopes (solid curves) well explain reference envelopes (gray curves) around the peaks for both cases, $\kappa = 0.1$ and 1.0. The Markov approximation for the parabolic wave equation is able to predict at least the early wave-envelopes quantitatively even for random media with rich short-wavelength spectra, because the early part is mainly composed of waves scattered around the forward direction. For the case of $\kappa = 1.0$, Markov envelopes and reference envelopes coincide well with each other even in the coda; however, we find a departure of the Markov envelope from the reference envelope as lapse time increases at each receiver for the case of $\kappa = 0.1$. We may say that the coda part is mostly composed of waves that are scattered at wide angles from short wavelength spectra and they are correctly synthesized by the FD simulation; however, the wide-angle scattering is completely neglected in the derivation of the Markov envelope.

6. Radiative Transfer Theory for von Kármán-Type Random Media of $\kappa = 0.1$

Radiative transfer theory has been developed for describing the transport of energy in scattering media (Chandrasekhar, 1960). Wu (1985) proposed to use the stationary-state solution of radiative transfer theory in the analysis of S-seismogram envelopes. Shang and Gao (1988) proposed a correct formulation of the radiative transfer equation for the non-stationary state in the case of 2-D space. Zeng et al. (1991) formulated the non-stationary case in 3-D space for the case of impulsive source radiation.

We begin by introducing the concept of the momentum transfer scattering coefficient, which can be viewed as equivalent to an isotropic scattering coefficient for describing multiple non-isotropic scattering process. We then use the momentum transfer scattering coefficient in radiative transfer theory to synthesize seismogram envelopes for the case of $\kappa = 0.1$.

6.1. Momentum Transfer Scattering Coefficient

The Born (single-scattering) approximation well describes the scattering process when the medium fluctuation is small. For the incidence of plane wave $u^{\text{Inc.}}$ on a localized inhomogeneity with dimension L^2, cylindrically outgoing scattered waves $u^{\text{Scat.}}$ are generated. The ensemble average power of scattered waves of angular frequency ω_c at distance r and at angle ψ is given by

$$\langle |u^{\text{Scat.}}(r,\psi,\omega_c)|^2 \rangle \approx \frac{L^2}{2\pi r} g(\psi,\omega_c) \langle |u^{\text{Inc.}}|^2 \rangle, \tag{48}$$

where $g(\psi, \omega_c)$ is the scattering coefficient, which has dimensions of inverse length, characterizing the scattering power per unit area of 2-D random media. Applying the Born approximation, we may write the scattering coefficient as a function of the PSDF of random media as (Frankel and Clayton, 1986; Sato and Fehler, 1998),

$$g(\psi;\omega_c) = \left(\frac{\omega_c}{V_0}\right)^3 P\left(2\frac{\omega_c}{V_0}\sin\frac{\psi}{2}\right). \tag{49}$$

In Fig. 13a, a fine solid curve shows the angular dependence of scattering coefficient at 2 Hz in von Kármán-type random media with $V_0 = 4$ km/s, $\varepsilon = 5\%$, $a = 5$ km and $\kappa = 0.1$. The scattering pattern is far from isotropic, and a large lobe in the forward direction represents strong forward scattering due to long

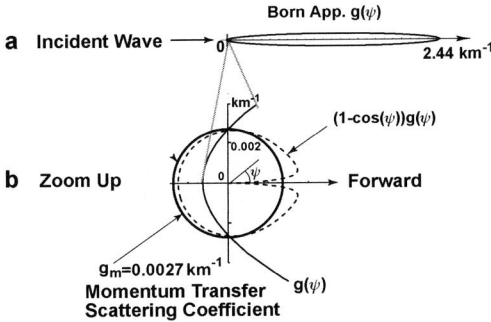

FIG. 13. (a) Scattering coefficient $g(\psi)$ based on the Born approximation at 2 Hz in random media of von Kármán-type with $\kappa = 0.1$. (b) Comparison of the momentum transfer scattering coefficient g_m with the scattering coefficient $g(\psi)$ (Sato et al., 2004).

wavelength spectral components of the random media. The total scattering coefficient is defined as the angular average of the scattering coefficient,

$$g_0(\omega_c) = \frac{1}{\pi}\int_0^\pi g(\psi;\omega_c)\,d\psi = \frac{1}{\pi}\int_0^\pi \left(\frac{\omega_c}{V_0}\right)^3 P\left(\frac{2\omega_c}{V_0}\sin\frac{\psi}{2}\right) d\psi. \quad (50)$$

Since forward scattered waves arrive at the receiver at nearly the same time as the direct wave, the scattering process can be effectively well described by isotropic scattering when multiple scattering is dominant. From the derivation of the diffusion equation from the transport equation for non-isotropic scattering, the momentum transfer scattering coefficient,

$$g_m(\omega_c) = \frac{1}{\pi}\int_0^\pi (1-\cos\psi)g(\psi;\omega_c)\,d\psi$$

$$= \frac{1}{\pi}\int_0^\pi 2\left(\sin^2\frac{\psi}{2}\right)\left(\frac{\omega_c}{V_0}\right)^3 P\left(\frac{2\omega_c}{V_0}\sin\frac{\psi}{2}\right) d\psi \quad (51)$$

is found to represent the effective isotropic scattering coefficient (e.g. Morse and Feshbach, 1953), where factor $(1-\cos\psi)$ eliminates scattering around the forward direction. The reciprocal of g_m gives the transport mean free path. For this case, $g_m = 0.00273$ km^{-1} at 2 Hz. In Fig. 13b, a bold solid curve shows a circle of radius g_m representing the effective isotropic scattering coefficient and a fine solid curve shows the scattering coefficient predicted by the Born approximation. Scattering at wide angles is three orders of magnitude smaller than forward scattering. A broken curve shows the plot of the integrand of (51), $(1-\cos\psi)g(\psi;\omega_c)$, which is very close to the circle of radius g_m. The factor $(1-\cos\psi)$ works as a filter that eliminates small wavenumber spectra (long wavelength spectra) of the random media. That is, the momentum transfer scattering coefficient represents the effective isotropic scattering, which is the average contribution of wide-angle scattering due to short wavelength spectra of random media. In the 3-D case, we note that g_m for S waves estimated from the coda excitation is from 2×10^{-3} km^{-1} to 5×10^{-2} km^{-1} in the lithosphere (e.g. Fig. 3.10 of Sato and Fehler, 1998).

6.2. Radiative Transfer Integral Equation for Isotropic Scattering Process

We introduce the mathematical concept of radiative transfer theory for envelope synthesis. Energy is assumed to propagate through random media with the background velocity V_0 and the scattering power per unit area of random media is characterized by the effective isotropic scattering coefficient g_m. Isotropic impulsive radiation of unit energy from a source located at the origin is described by the delta function in time,

$$G_\delta(\mathbf{x},t) = \frac{e^{-V_0 g_m t}}{2\pi V_0 r} H(t)\delta\left(t - \frac{r}{V_0}\right), \quad (52)$$

which includes the causal propagation with a geometrical spreading factor and the effect of exponential scattering loss. By using $G_\delta(\mathbf{x}, t)$ as a propagator, we may describe the multiple isotropic scattering process for energy density G_I by the following radiative transfer integral equation:

$$G_I(\mathbf{x}, t) = G_\delta(\mathbf{x}, t)$$
$$+ V_0 g_m \int_{-\infty}^{\infty} \int_{-\infty}^{\infty} \int_{-\infty}^{\infty} G_\delta(\mathbf{x} - \mathbf{x}', t - t') G_I(\mathbf{x}', t') \, dt' \, d\mathbf{x}'. \quad (53)$$

Shang and Gao (1988) solved the scattering process in 2-D to get the analytical solution of the spatiotemporal distribution of energy density. The 1-D case was solved by Hemmer (1961). The following derivation is taken from Sato (1993).

The Laplace transform of (52) with respect to time is

$$\hat{G}_\delta(\mathbf{x}, s) \equiv \int_0^\infty G_\delta(\mathbf{x}, t) e^{-st} \, dt = \frac{e^{-(s+g_m V_0) r / V_0}}{2\pi V_0 r}. \quad (54)$$

The Fourier transform of (54) with respect to space coordinates is

$$\hat{\tilde{G}}_\delta(\mathbf{k}, s) \equiv \int_{-\infty}^{\infty} \int_{-\infty}^{\infty} \hat{G}_\delta(\mathbf{x}, s) e^{-i\mathbf{k}\mathbf{x}} \, d\mathbf{x}$$
$$= \frac{1}{V_0} \int_0^\infty e^{-(s+g_m V_0) r / V_0} J_0(kr) \, dr$$
$$= \frac{1}{\sqrt{(s + V_0 g_m)^2 + V_0^2 k^2}}. \quad (55)$$

The Fourier–Laplace transform of the radiative transfer integral equation (53) is written by

$$\hat{\tilde{G}}_I(\mathbf{k}, s) = \frac{\hat{\tilde{G}}_\delta(\mathbf{k}, s)}{1 - V_0 g_m \hat{\tilde{G}}_\delta(\mathbf{k}, s)}$$
$$= \hat{\tilde{G}}_\delta(\mathbf{k}, s) + \sum_{n=1}^{\infty} (V_0 g_m)^n \hat{\tilde{G}}_\delta(\mathbf{k}, s)^{n+1}, \quad (56)$$

where the last line is a formal expansion with respect to the power of g_m. Taking the inverse Fourier transform of (56) and using (55), we get

$$\hat{G}_I(\mathbf{x}, s) = \hat{G}_\delta(\mathbf{x}, s) + \frac{1}{(2\pi)^2} \sum_{n=1}^{\infty} (V_0 g_m)^n \int_{-\infty}^{\infty} \int_{-\infty}^{\infty} \hat{\tilde{G}}_\delta(\mathbf{k}, s)^{n+1} e^{-i\mathbf{k}\mathbf{x}} \, d\mathbf{k}$$
$$= \hat{G}_\delta(\mathbf{x}, s) + \frac{1}{(2\pi)^2} \sum_{n=1}^{\infty} (V_0 g_m)^n$$
$$\times \int_0^\infty [(s + V_0 g_m)^2 + V_0^2 k^2]^{-\frac{n}{2} - \frac{1}{2}} J_0(kr) e^{-ikr} k \, dk$$

$$= \hat{G}_\delta(\mathbf{x}, s) + \frac{1}{2\pi V_0^2} \sum_{n=1}^{\infty} \frac{(V_0 g_m)^n}{\Gamma(\frac{n}{2} + \frac{1}{2})} \left[\frac{r}{2V_0(s + V_0 g_m)}\right]^{\frac{n}{2}-\frac{1}{2}}$$

$$\times K_{\frac{n}{2}-\frac{1}{2}}\left[\frac{(s + V_0 g_m)r}{V_0}\right]$$

$$= \hat{G}_\delta(\mathbf{x}, s) + \frac{1}{2\pi V_0^2}$$

$$\times \sum_{n=1}^{\infty} \frac{(V_0 g_m)^n}{\Gamma(n)} \int_{r/V_0}^{\infty} \left(t^2 - \frac{r^2}{V_0^2}\right)^{\frac{n}{2}-1} e^{-(s+V_0 g_m)t} \, dt, \qquad (57)$$

where we used an integral formula and a Laplace transformation formula (see Gradshteyn and Ryzhik, 1980, p. 322 and p. 686) and $K_\nu = K_{-\nu}$. Taking the inverse Laplace transform of (57), we obtain

$$G_I(\mathbf{x}, t) = G_\delta(\mathbf{x}, t) + \frac{e^{-V_0 g_m t}}{2\pi V_0^2} H\left(t - \frac{r}{V_0}\right) \sum_{n=1}^{\infty} \frac{(V_0 g_m)^n}{\Gamma(n)} \left(t^2 - \frac{r^2}{V_0^2}\right)^{\frac{n}{2}-1}$$

$$= G_\delta(\mathbf{x}, t) + \frac{V_0 g_m e^{-V_0 g_m t}}{2\pi V_0^2 \sqrt{t^2 - \frac{r^2}{V_0^2}}}$$

$$\times H\left(t - \frac{r}{V_0}\right) \sum_{n=1}^{\infty} \frac{(V_0 g_m)^n}{n!} \sqrt{t^2 - \frac{r^2}{V_0^2}}^n$$

$$= G_\delta(\mathbf{x}, t) + \frac{g_m}{2\pi \sqrt{V_0^2 t^2 - r^2}} H\left(t - \frac{r}{V_0}\right) e^{g_m \sqrt{V_0^2 t^2 - r^2} - V_0 g_m t}. \qquad (58)$$

The first term represents the direct propagation and the second term represents scattering contribution. Solving the radiative transfer equation in differential form, Paaschens (1997) derived the solution (58). Figure 14 shows the spatiotemporal change in G_I. In Fig. 14a, the energy density decreases rapidly after the direct arrival at a small lapse time; however, the decay rate becomes smaller and the coda portion becomes larger due to multiple scattering at large lapse times. In Fig. 14b, we find a smooth spatial distribution of scattered energy density at a short lapse time. We note that there is no violation of causality since energy density does not exist beyond the direct arrival. As lapse time increases, the second term dominates over the first term and converges to the solution of a diffusion equation as

$$G_I(\mathbf{x}, t) \approx \frac{e^{-\frac{r^2}{2(V_0 t/g_m)}}}{2\pi (V_0 t/g_m)} \quad \text{for } V_0 t \gg r. \qquad (59)$$

At the source location $r = 0$, the second term in (58) decreases according to the reciprocal of travel distance as $g_m/(2\pi V_0 t)$.

In Fig. 15a we show a comparison of RMS envelopes synthesized by using G_I (solid curves) with reference envelopes (gray curves) at four receivers. Each solid trace is the square root of the convolution $G_I * W_R$. We find a good coincidence between solid and gray curves for the latter coda portion at each receiver; however, there is some discrepancy for the early coda. Coda excitation is well explained as a cumulative contribution of wide-angle scattering even though scattering coefficient at wide angles is much smaller than that in the forward direction. We note that Jannaud *et al.* (1991) numerically confirmed that the excitation of coda waves

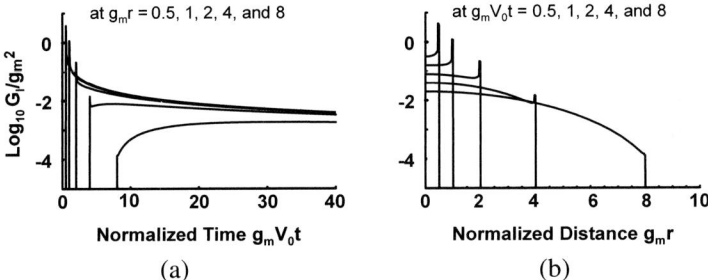

FIG. 14. (a) Temporal variations of normalized energy density G_I/g_m^2 at selected distances calculated from the analytical solution (58) of the radiative transfer equation for isotropic scattering process in 2-D. (b) Spatial variations of normalized energy density at selected lapse times.

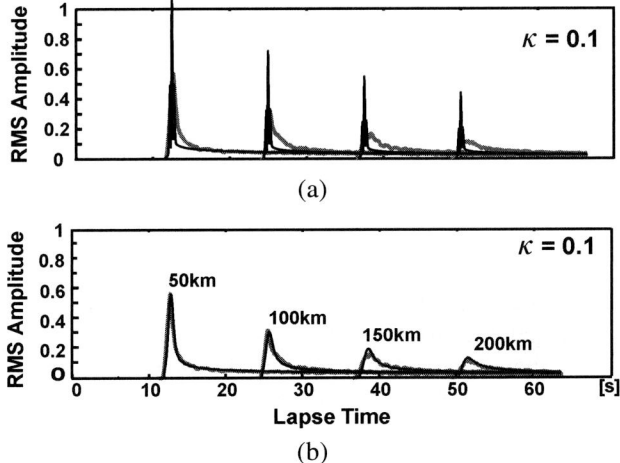

FIG. 15. (a) RMS envelopes (solid curves) synthesized based on the analytical solution (58) of the radiative transfer equation for isotropic scattering process (Saito *et al.*, 2003). (b) RMS envelopes (solid curves) synthesis based on the hybrid model (60) (Sato *et al.*, 2004). Random media are von Kármán-type of $\kappa = 0.1$ and the source is a 2 Hz Ricker wavelet. Gray curves are reference envelopes an ensemble average of FD envelopes in the same random media as illustrated in Fig. 11a.

is proportional to the source factor and the PSDF of velocity inhomogeneities in a weak scattering regime.

7. Hybrid Envelope Synthesis for von Kármán-Type Random Media Having $\kappa = 0.1$

The conventional radiative transfer theory using the momentum transfer scattering coefficient has been shown to yield envelopes for the codas that are similar to those obtained from the FD method. However, for the case of $\kappa = 0.1$, radiative transfer theory derived envelopes are too impulsive as illustrated in Fig. 14a. The impulsive character of the initial pulse is due to the use of the delta function propagator G_δ in the radiative transfer equation.

In order to introduce diffraction effects into the formulation, Saito et al. (2003) proposed to replace the first term of the solution of (58) with the solution of the Markov approximation G_M including a scattering loss term. Their simple model well explains both envelope broadening around the peak and coda excitation in each trace.

Recently, Sato et al. (2004) proposed a hybrid model to use the solution of the Markov approximation G_M with exponential scattering loss $e^{-\mu g_m V_0 t}$ as a propagator in the radiative transfer integral equation (53). The radiative transfer equation for a unit isotropic source radiation is written as

$$G_H(\mathbf{x}, t) = G_M(\mathbf{x}, t) e^{-\mu g_m V_0 t}$$
$$+ V_0 g_m \int_{-\infty}^{\infty} \int_{-\infty}^{\infty} \int_{-\infty}^{\infty} G_M(\mathbf{x} - \mathbf{x}', t - t') e^{-\mu g_m V_0 (t-t')}$$
$$\times G_H(\mathbf{x}', t') \, dt' \, d\mathbf{x}'. \tag{60}$$

This integral equation contains diffraction effects during propagation between wide-angle scattering locations. From the requirement of total energy conservation $\int_{-\infty}^{\infty} \int_{-\infty}^{\infty} G_H(\mathbf{x}, t) \, d\mathbf{x} = 1$, we numerically evaluate the value of μ for a given lapse time range. At 2 Hz, in the case of von Kármán-type random media of $\kappa = 0.1$, we have to take $\mu = 0.50$ with accuracy of 99% for a lapse time up to 60 s. In Fig. 15b, we show a comparison of RMS envelopes synthesized by the hybrid method (solid curves) and reference envelopes (gray curves). In the synthesis of MS envelopes, we used the convolution $G_H * w * W_R$. At all travel distances, the hybrid envelopes are in good agreement with reference envelopes from the onset through coda. As schematically illustrated in Fig. 16a, the impulsive propagator G_δ describing causality with a constant velocity is used in conventional radiative transfer theory for isotropic scattering (Shang and Gao, 1988; Zeng et al., 1991; Yoshimoto, 2000) and/or non-isotropic scattering (Sato, 1994; Hoshiba, 1995; Gusev and Abubakirov, 1996). The concept of the hybrid method is schematically

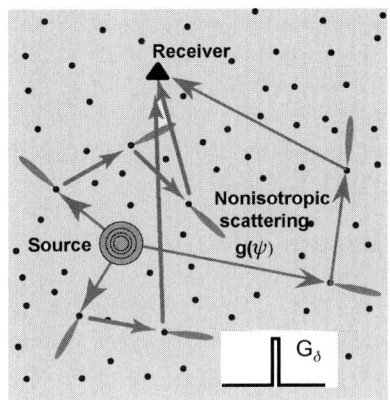 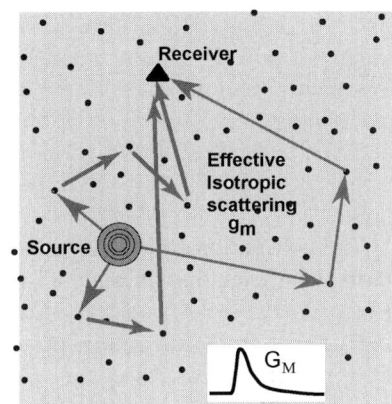

FIG. 16. Concept of (a) the conventional radiative transfer theory and (b) the hybrid method (Sato et al., 2004).

shown in Fig. 16b, where we assume the effective isotropic scattering due to short wavelength components but include envelope broadening due to long wavelength components during propagation between isotropic scatterers.

8. Overview of Envelope Studies

8.1. Envelope Broadening Studies

Saito et al. (2002) solved the Markov approximation for the spherically outgoing scalar waves in 3-D von Kármán-type random media as an extension of Shishov (1974) for the Gaussian-type random media. They found that envelopes are independent of frequency in random media with poor short wavelength spectra ($\kappa = 1.0$); however, higher frequency envelopes decay more rapidly than lower frequency envelopes in random media with rich short wavelength spectra ($\kappa = 0.1$). Applying this model to S-seismogram envelopes of microearthquakes observed in northern Honshu, Japan, they estimated that random inhomogeneity of the lithosphere can be described by a von Kármán-type ACF with $\kappa = 0.6$ and the ratio $\varepsilon^{2.2} a^{-1} \approx 10^{-3.6}$ km^{-1}. Short wavelength inhomogeneity decreases as wavenumber to the power -4.2 in this PSDF. Gusev and Abubakirov (1996) used the radiative transfer theory to model scalar wave propagation including forward scattering in random media characterized by power-law spectrum. They used a Monte Carlo method. Their goal was to explain the qualitative character of S-waves observed in Kamchatka including both envelope broadening around the direct arrivals and the excitation of coda. Later, Petukhin and Gusev (2003) averaged seismogram envelopes of small earthquakes recorded in Kamchatka and compared the shapes with those calculated for various types of random media. They concluded that random media whose short-wavelength inhomogeneity power spectrum decreases as wavenumber to the power of -3.5 to -4 are appropriate.

Extension of the Markov approximation to a vector wave case is necessary for the practical interpretation of seismograms. Using a concept of angular spectra, Korn and Sato (2005) proposed a method to synthesize vector wave envelopes in 2-D random elastic media. For the incidence of an impulsive plane wavelet from a homogeneous medium onto a random medium having a Gaussian ACF, they confirmed the validity of the formulation by comparing theoretical vector envelopes with wave envelopes of finite difference simulations.

8.2. Diffusion Studies

Seismic waves within volcanoes are strongly distorted by scattering from heterogeneities. Analyzing seismograms having spindle-like envelopes, Wegler and Lühr (2001) and Wegler (2003) estimated the scattering power at Merapi volcano in Java and Vesuvius volcano in Italy. They modeled strong scattering using a diffusion equation and estimated that mean free path in the near-surface region is 100 m and 200 m at Merapi and Vesuvius, respectively. Margerin *et al.* (1998) studied the influence of the layered structure of scattering properties on coda Q.

8.3. Three-Component Envelope Studies

Sato (1984) proposed a method to simulate three-component seismogram envelopes for a point shear-dislocation source using the radiative transfer equation in the single scattering regime. His method was based on the use of scattering amplitudes derived from the Born approximation for random elastic media characterized by an exponential ACF. He found the dominance of S-to-S scattering in S coda and the appearance of pseudo P and S arrivals in seismograms at a receiver even in the null direction. Yoshimoto *et al.* (1997a) extended the study by simulating three-component seismogram envelopes recorded on a free surface. Comparing the simulated envelopes to observed records of micro-earthquakes located in the shallow crust of Nikko, northern Kanto, Japan, Yoshimoto *et al.* (1997b) estimated the ratio $\varepsilon^2/a = 8 \times 10^{-3}$ km^{-1} for the exponential ACF. The effects of multiple scattering on S-wave polarization have been recently studied by Margerin *et al.* (2000). They used Stokes vectors and scattering amplitudes due to distributed cavities and simulated vector-wave envelopes using Monte Carlo simulation. Bal and Moscoso (2000) also modeled wave polarization using the radiative transfer equation for the case of multiple isotropic scattering.

8.4. Envelopes of Teleseismic Seismograms

Envelope modeling of teleseismic P waves was first done by using the energy-flux model by Korn (1993) for the study of lithospheric inhomogeneity. Hock *et al.* (2004) used the model results as a basis for measuring lithospheric inhomogeneity in central Europe. Stimulated by studies of scattering in the lithosphere, a number of investigators have studied the envelopes of seismic waves that have propagated through the mantle. Nishimura *et al.* (2002) studied the spatial distribution of lateral inhomogeneity in the upper mantle by analyzing transverse component envelopes of teleseismic P-waves in the frequency range 0.04–2.56 Hz. Scattering in the whole mantle was studied by Lee *et al.* (2003), who estimated the scattering coefficient in the lower and upper mantle from the analysis of seismogram envelopes for times before and after the ScS arrival. Envelopes simulated by using the radiative transfer theory and the Born approximation were used in the analyses of precursors of PKP by Margerin and Nolet (2003). They suggested that whole mantle scattering may be significant, contrary to previous suggestions that mantle scattering occurs primarily in the vicinity of the D'' layer. Analyzing stacked envelopes of teleseismic P waves and subsequent modeling using a Monte Carlo simulation, Shearer and Earle (2004) pointed out the importance of scattering due to lower mantle inhomogeneity even though it is smaller than that in the lithosphere and the upper mantle.

9. Conclusion and Discussion

Scattering phenomena play an important role in the propagation of high frequency seismic waves through the Earth. Focusing on the systematic variation in envelopes and disregarding their phase information, we have developed stochastic methods for the direct synthesis of wave envelopes. We precisely examined the scalar wave propagation in 2-D random media for an isotropic radiation of a Ricker wavelet from a point source as the most tractable case. By using envelopes simulated by stochastic averaging of waveforms calculated using the finite difference method as reference envelopes, various methods for direct envelope simulation were tested. For random media having Gaussian spectra with correlation length longer that the wavelength, we mathematically introduced the Markov approximation, which means an approach with both the parabolic approximation and a stochastic extension of the phase screen method. Envelopes predicted by this method well agree with the reference envelopes. Then, we examined the validity of the Markov approximation for the case of von Kármán-type random media which have power-law spectra at large wavenumbers. The envelopes predicted by the Markov approximation well explain the reference envelopes for random media having weak short wavelength spectra. For the case of random media with strong short wavelength spectra; however, the coincidence is good around the peak amplitude but becomes poor for the coda portion because of wide-angle scattering. In that case, the isotropic scattering model based on radiative transfer theory explains satisfactorily the coda portion of the reference envelopes, where the momentum transfer scattering coefficient is used as the effective isotropic scattering coefficient. Introducing a hybrid method using the momentum transfer scattering coefficient and the envelope predicted by the Markov approximation as a propagator in the radiative transfer integral equation for isotropic scattering process, we successfully simulated wave envelopes that explain the reference envelopes from onset to coda for the case of rich short wavelength spectra.

Comparisons with the FD simulation results were limited only for scalar waves at 2 Hz in a few set of statistical parameters characterizing random media because of computer capacity; however, the hybrid method proposed here could be one of several mathematical bases for probing the Earth inhomogeneity. It will be necessary to develop various simulation methods for vector wave envelopes in 3-D random elastic media for a wide range of frequencies to establish solid mathematical bases for the study of Earth inhomogeneity.

Wave theoretical approaches are, of course, basic and important for the quantification of the effects of velocity and/or density inhomogeneity on observed seismic traces; however, we would like to affirm also the importance of the description of scattering media in the framework of radiative transfer theory. Radiative transfer theory provides a framework to estimate the regional difference of the scattering coefficient and its frequency dependence without making assumptions about the physical origin of scattering. Our understanding of the inhomogeneity of the Earth will be enhanced if the spatial distribution of scattering coefficient is known along with the velocity structure.

Seismic observations using data from dense seismic networks such as the Hi-net and the K-net that were recently deployed in Japan will accelerate our understanding of the Earth inhomogeneity. Continued improvements in computer power will enable us to solve complex problems on the wave propagation through highly inhomogeneous elastic media. We also note recent developments in ultrasonic array measurements of three-component velocity wave propagation through rock samples by using a laser Doppler vibrometer as a physical simulation of wave propagation. Joint theoretical and observational collaborations using these laboratory results could lead to a breakthrough in our understanding of seismic wave propagation in the inhomogeneous Earth.

Acknowledgements

We thank L.J. Huang and T. Saito for their collaboration for the basics of this work. Comments by M. Korn and an anonymous reviewer and editor, Rushan Wu are very helpful for improving the manuscript.

References

Abramowitz, M., Stegun, I.A. (1970). Handbook of Mathematical Functions with Formulas, Graphs and Mathematical Tables. Dover, New York.
Aki, K. (1980). Attenuation of shear-waves in the lithosphere for frequencies from 0.05 to 25 Hz. *Phys. Earth Planet. Inter.* **21**, 50–60.
Aki, K., Chouet, B. (1975). Origin of coda waves: Source, attenuation and scattering effects. *J. Geophys. Res.* **80**, 3322–3342.
Bal, G., Moscoso, M. (2000). Polarization effects of seismic waves on the basis of radiative transfer theory. *Geophys. J. Int.* **142**, 571–585.
Barabanenkov, Yu.N., Kravtsov, Yu.A., Rytov, S.M., Tamarskii, V.I. (1971). Status of the theory of propagation of waves in randomly inhomogeneous medium. *Sov. Phys. Usp.* **13**, 551–680 (English Translation).
Chandrasekhar, S. (1960). Radiative Transfer. Dover, New York.
Claerbourt, J.F. (1985). Imaging the Earth's Interior. Blackwell Scientific Publication, Oxford.
Fehler, M., Huang, L.-J. (2002). Modern Imaging using seismic reflection data (invited book chapter). *Annu. Rev. Earth Planet. Sci.* **30**, 259–284.
Fehler, M., Sato, H., Huang, L.-J. (2000). Envelope broadening of outgoing waves in 2-D random media: A comparison between the Markov approximation and numerical simulations. *Bull. Seismol. Soc. Am.* **90**, 914–928.
Flatté, S.M., Dashen, R., Munk, W.H., Watson, K.M., Zachariasen, F. (1979). Sound Transmission through a Fluctuating Ocean. Cambridge Univ. Press, New York.
Foldy, L.L. (1945). The multiple scattering of waves-I. General theory of isotropic scattering by randomly distributed scatterers. *Phys. Rev.* **67**, 107–119.
Frankel, A., Clayton, R.W. (1986). Finite difference simulations of seismic scattering: Implications for the propagation of short-period seismic waves in the crust and models of crustal heterogeneity. *J. Geophys. Res.* **91**, 6465–6489.
Frankel, A., Wennerberg, L. (1987). Energy-flux model of seismic coda: Separation of scattering and intrinsic attenuation. *Bull. Seismol. Soc. Am.* **77**, 1223–1251.
Frisch, U. (1968). In: Bharucha-Reid, A.T. (Ed.), *Wave Propagation in Random Media*. In: *Probabilistic Method in Applied Mathematics*, vol. I. Academic Press, New York, pp. 76–198.
Fukushima, Y., Nishizawa, O., Sato, H., Ohtake, M. (2003). Laboratory study on scattering characteristics of shear waves in rock samples. *Bull. Seismol. Soc. Am.* **93**, 253–263.
Goff, J.A., Holliger, K. (2000). Nature and origin of upper crustal seismic velocity fluctuations and associated scaling properties: Combined stochastic analyses of KTB velocity and lithology logs. *J. Geophys. Res.* **104**, 13169–13182.
Gradshteyn, I.S., Ryzhik, I.M. (1980). Table of Integrals, Series, and Products. Academic Press, London. (Corrected and enlarged edition).
Gusev, A.A., Abubakirov, I.R. (1996). Simulated envelopes of non-isotropically scattered body waves as compared to observed ones: Another manifestation of fractal heterogeneity. *Geophys. J. Int.* **127**, 49–60.
Hemmer, P.C. (1961). On a generalization of Smoluchowski's diffusion equation. *Physica A* **27**, 79–82.

Higdon, R.L. (1991). Absorbing boundary conditions for elastic waves. *Geophysics* **56**, 231–241.

Hock, S., Korn, M., Ritter, J., Rothert, E. (2004). Mapping random lithospheric heterogeneities in Northern and Central Europe. *Geophys. J. Int.* **157**, 251–264.

Holberg, O. (1987). Computational aspects of the choice of operator and sampling interval for numerical differentiation in large-scale simulation of wave phenomena. *Geophys. Prosp.* **35**, 629–655.

Hoshiba, M. (1991). Simulation of multiple-scattered coda wave excitation based on the energy conservation law. *Phys. Earth Planet. Int.* **67**, 123–136.

Hoshiba, M. (1995). Estimation of nonisotropic scattering in western Japan using coda wave envelopes: Application of a multiple nonisotropic scattering model. *J. Geophys. Res.* **100**, 645–657.

Ishimaru, A. (1978). Wave Propagation and Scattering in Random Media, vols. 1 and 2. Academic, San Diego.

Jannaud, L.R., Adler, P.M., Jacquin, C.G. (1991). Spectral analysis and inversion of codas. *J. Geophys. Res.* **96**, 18215–18231.

Jensen, F., Kuperman, W., Porter, M., Schmidt, H. (1994). Computational Ocean Acoustics. American Institute of Physics, New York.

Kelner, S., Bouchon, M., Coutant, O. (1999). Numerical simulation of the propagation of P waves in fractured media. *Geophys. J. Int.* **137**, 197–206.

Korn, M. (1993). Determination of site-dependent scattering Q from P-wave coda analysis with an energy-flux model. *Geophys. J. Int.* **113**, 54–72.

Korn, M., Sato, H. (2005). Synthesis of plane vector-wave envelopes in 2-D random elastic media based on the Markov approximation and comparison with finite difference simulations. *Geophys. J. Int.* **161**, 839–848.

Lambert, H.C., Rickett, B.J. (1999). On the theory of pulse propagation and two-frequency field statistics in irregular interstellar plasmas. *Astrophys. J.* **517**, 299–317.

Lee, L.C., Jokipii, J.R. (1975). Strong scintillations in astrophysics. II. A theory of temporal broadening of pulses. *Astrophys. J.* **201**, 532–543.

Lee, W.S., Sato, H., Lee, K.-W. (2003). Estimation of S-wave scattering coefficient in the mantle from envelope characteristics before and after the ScS arrival. *Geophys. Res. Lett.* **30**, 2248, doi:10.1029/2003GL018413.

Margerin, L. (2005). Introduction to radiative transfer of seismic waves. In: Levander, A., Nolet, G. (Eds.), *Seismic Earth: Array Analysis of Broad-band Seismograms*. In: *Geophysical Monograph Series*, vol. 157. AGU, Washington, pp. 229–252. Chapter 14.

Margerin, L., Nolet, G. (2003). Multiple scattering of high-frequency seismic waves in the deep Earth: PKP precursor analysis and inversion for mantle granularity. *J. Geophys. Res. B* **108** (11), 2514, doi:10.1029/2003JB002455.

Margerin, L., Campillo, M., van Tiggelen, B. (1998). Radiative transfer and diffusion of waves in a layered medium: Insight into coda Q. *Geophys. J. Int.* **134**, 596–612.

Margerin, L., Campillo, M., van Tiggelen, B. (2000). Monte Carlo simulation of multiple scattering of elastic waves. *J. Geophys. Res.* **105**, 7873–7893.

Morse, P.M., Feshbach, H. (1953). Methods of Theoretical Physics, vols. I and II. McGraw–Hill, New York.

Nishimura, T., Yoshimoto, K., Ohtaki, T., Kanjo, K., Purwana, I. (2002). Spatial distribution of lateral heterogeneity in the upper mantle around the western Pacific region as inferred from analysis of transverse components of teleseismic P-coda. *Geophys. Res. Lett.* **29** (23), 2137, doi:10.1029/2002GL015606.

Nishizawa, O., Satoh, T., Lei, X., Kuwahara, Y. (1997). Laboratory studies of seismic wave propagation in inhomogeneous media using a laser Doppler vibrometer. *Bull. Seismol. Soc. Am.* **87**, 809–823.

Obara, K., Sato, H. (1995). Regional differences of random inhomogeneities around the volcanic front in the Kanto-Tokai area, Japan, revealed from the broadening of S wave seismogram envelopes. *J. Geophys. Res.* **100**, 2103–2121.

Paaschens, J.C.J. (1997). Solution of the time-dependent Boltzmann equation. *Phys. Rev. E* **56**, 1135–1141.

Petukhin, A.G., Gusev, A.A. (2003). The duration-distance Relationship and average envelope shapes of small Kamchatka earthquakes. *Pure Appl. Geophys.* **160**, 1717–1743.

Ryzhik, L.V., Papanicolaou, G.C., Keller, J.B. (1996). Transport equations for elastic and other waves in random media. *Wave Motion* **24**, 327–370.

Saito, T., Sato, H., Ohtake, M. (2002). Envelope broadening of spherically outgoing waves in three-dimensional random media having power-law spectra. *J. Geophys. Res.* **107**, doi:10.1029/2001JB000264.

Saito, T., Sato, H., Fehler, M., Ohtake, M. (2003). Simulating the envelope of scalar waves in 2-D random media having power-law spectra of velocity fluctuation. *Bull. Seismol. Soc. Am.* **93**, 240–252.

Sato, H. (1984). Attenuation and envelope formation of three-component seismograms of small local earthquakes in randomly inhomogeneous lithosphere. *J. Geophys. Res.* **89**, 1221–1241.

Sato, H. (1989). Broadening of seismogram envelopes in the randomly inhomogeneous lithosphere based on the parabolic approximation: Southeastern Honshu Japan. *J. Geophys. Res.* **94**, 17735–17747.

Sato, H. (1993). Energy transportation in one- and two-dimensional scattering media: Analytic solutions of the multiple isotropic scattering model. *Geophys. J. Int.* **112**, 141–146.

Sato, H. (1994). Formulation of the multiple non-isotropic scattering process in 2-D space on the basis of energy transport theory. *Geophys. J. Int.* **117**, 727–732.

Sato, H., Fehler, M. (1998). Seismic Wave Propagation and Scattering in the Heterogeneous Earth. AIP Press, Springer Verlag, New York, pp. 1–308.

Sato, H., Fehler, M., Saito, T. (2004). Hybrid synthesis of scalar wave envelopes in 2-D random Media having rich short wavelength spectra. *J. Geophys. Res. B* **109**, 06303, doi:10.1029/2003JB002673.

Scherbaum, F., Sato, H. (1991). Inversion of full seismogram envelopes based on the parabolic approximation: Estimation of randomness and attenuation in southeast Honshu Japan. *J. Geophys. Res.* **96**, 2223–2232.

Shang, T., Gao, L. (1988). Transportation theory of multiple scattering and its application to seismic coda waves of impulsive source. *Sci. Sinica B* **31**, 1503–1514 (China).

Shearer, P.M., Earle, P.S. (2004). The global short-period wavefield modeled with a Monte Carlo seismic phonon method. *Geophys. J. Int.* **158**, 1103–1117.

Shiomi, K., Sato, H., Ohtake, M. (1997). Broad-band power-law spectra of well-log data in Japan. *Geophys. J. Int.* **130**, 57–64.

Shishov, V.L. (1974). Effect of refraction on scintillation characteristics and average pulsars. *Sov. Astron.* **17**, 598–602.

Tatarskii, V.I. (1971). The Effects of the Turbulent Atmosphere on Wave Propagation. Israel Program for Science translations, Jerusalem.

Wegler, U. (2003). Analysis of multiple scattering at Vesuvius volcano, Italy, using data of the TomoVes active seismic experiment. *J. Volcanol. Geotherm. Res.* **128**, 45–63.

Wegler, U., Lühr, B.-G. (2001). Scattering behaviour at Merapi volcano (Java) revealed from an active seismic experiment. *Geophys. J. Int.* **145**, 579–592.

Wu, R.S., Aki, K. (1985). The fractal nature of the inhomogeneities in the lithosphere evidenced from seismic wave scattering. *Pure Appl. Geophys.* **123**, 805–818.

Wu, R.S. (1985). Multiple scattering and energy transfer of seismic waves—separation of scattering effect from intrinsic attenuation-I. Theoretical modeling. *Geophys. J. R. Astron. Soc.* **82**, 57–80.

Yamashita, T. (1990). Attenuation and dispersion of SH waves due to scattering by randomly distributed cracks. *Pure Appl. Geophys.* **132**, 545–568.

Yomogida, K., Benites, R. (1995). Relation between direct wave Q and coda Q: A numerical approach. *Geophys. J. Int.* **123**, 471–483.

Yoshimoto, K., Sato, H., Ohtake, M. (1997a). Three-component seismogram envelope synthesis in randomly inhomogeneous semi-infinite media based on the single scattering approximation. *Phys. Earth Planet. Int.* **104**, 37–61.

Yoshimoto, K., Sato, H., Ohtake, M. (1997b). Short-wavelength crustal inhomogeneities in the Nikko area, central Japan, revealed from the three-component seismogram envelope analysis. *Phys. Earth Planet. Int.* **104**, 63–73.

Yoshimoto, K. (2000). Monte-Carlo simulation of seismogram envelope in scattering media. *J. Geophys. Res.* **105**, 6153–6161.

Zeng, Y., Su, F., Aki, K. (1991). Scattering wave energy propagation in a random isotropic scattering medium I. Theory. *J. Geophys. Res.* **96**, 607–619.

SUBJECT INDEX

1-D problem
– finite difference method   431, 433, 436, 437, 441, 448, 455, 458, 459
2-D problem
– finite difference method   427, 431, 433, 475
2nd-order approximation
– finite difference method   424, 425, 432, 433, 458, 489
3-D problem
– finite difference method   427, 431, 433, 443, 460
4th-order approximation
– finite difference method   425, 469, 471

A

Absorbing boundary conditions (ABC)
– finite difference method   495
Acoustic wave
– lattice Boltzmann approach   519, 520, 528, 546
Acoustic-wave equation
– lattice Boltzmann approach   534, 536
Acoustic-wave propagation
– lattice Boltzmann approach   517, 519, 520, 548, 549, 554
– one-way and one-return method   280
Adiabatic mode   140
Aki–Larner method   176, 215–222
Alterman–Karal decomposition   483
Amplitude   see Ray-theory amplitude
Amplitude   see AVA, AVO
Anelastic coefficients   457, 462
Anelastic functions (memory variables)   455, 463
Anelasticity in one-return method   290
Angular spectra
– in synthesis of seismogram envelope   590
Anisotropic
– surface waves   142
Anisotropic common ray approximation   60
Anisotropy
– surface waves   142
– finite difference method   428, 463, 468
Asymptotic ray series   5
– amplitude coefficients of   5
– higher-order approximation   110–112
– zero-order approximation   5, 110

Attenuation   393, 448, 465
– intrinsic   339
– leakage   343, 358
– Lg wave   339, 343
– Q 339–343
– scattering   339–343
Autocorrelation function   565, 567
AVA (amplitude variation with angle)   266, 318
AVO (amplitude variation with offset)   267, 306
– dipping sandstone reservoir   309
– laterally varying   316
– sedimentary interface   307
– thin-layers   315
– with heterogeneous overburden   311

B

Backscattering (back scattering)   269, 272–274
– single backscattering   268, 276
– single scattering   281, 284, 289, 297, 300, 301
Backward-difference formula   424
Basin
– SH waves   191, 193, 221–226, 232, 233, 251
Body-force equivalent   481
Boltzmann superposition principle   448
Born approximation   52, 96, 267, 328, 332, 566, 583, 584, 590, 591
Born approximation (weak scattering approximation)   267–271
– distorted Born approximation   276
– in global tomography   402
– local Born approximation   273, 277, 278, 280, 283, 299
Born scattering   52, 96, 118
Born series   272–274
Boundary condition   132, 134
Boundary element method (BEM)   158, 346, 366
– direct BEM   167
– indirect BEM (IBEM)   162, 167
– indirect BIE   158
– comparison with spectral element method   162
Boundary integral equation (BIE)   157

– direct BIE   157, 159
– hybrid method   179
– indirect BIE   157, 161
– time domain implementation   174

C

Causality   576, 586, 588
Causality (effective) of the ray method   50
Central-difference formula   424
Chapman–Enskog expansion   527, 533–535
Characteristic frequency   220, 238
Characteristic time   577, 578, 581, 582
Christoffel matrix   6, 7, 57
– degenerate case   7, 9
– eigenvalue of   7–9
– eigenvector of   7, 8, 11, 57
Coarse spatial sampling   461, 463
Coda waves   561, 563, 565, 571, 587
– in surface waves   136
Coherent wave field   572, 573, 576
Coherent-state transform   89–92
Collision rule
– in lattice Boltzmann approach   519, 522–524, 526, 532, 533, 537, 543
Common ray approximation   60
– anisotropic   60
– isotropic   61
Complex velocity   291
Constant Q 457, 458
Constraint traction   488, 490
Continental margin   135
Controlled initial-value ray tracing   14, 16–23, 86
– applications of   21–23
– for the asymptotic summation of Gaussian beams   21
– for the asymptotic summation of Gaussian packets   21
– for the interpolation within ray cells   21
Convolution
– in envelope synthesis   566, 575, 578, 582, 587, 588
Correlation distance   561, 568, 571–573
Coupling equation for S waves   57, 58
Coupling ray theory for S waves   3, 4, 56–68
– common ray approximation of   60
– coupling equation   57, 58
– propagator matrix of   58, 59

– quasi-isotropic approximation of   59–62, 66
Courant number   427
Courant–Friedrichs–Lewy (CFL) condition   468
Cowling approximation in the gravito-elastodynamic equations   370
Crustal wave guide   328, 338
Cubed sphere transformation   383

D

De Wolf approximation   265, 272–275, 277, 283
– comparison of Born, Rytov and De Wolf approximation   268, 271, 272
De Wolf series (DWS)   275, 276
Diffraction   571, 588
Diffusion   584, 586, 590
Direct solution method of equation of motion   434
Dirichlet-to-Neumann
– DtN coupling   378, 388, 412
– DtN operator   376, 379, 410
Discretization
– in lattice Boltzmann approach   518, 520
Dispersion equation in surface waves   235, 237
Distorted mode in surface waves   241–243, 246, 249–251
D'' layer
– synthesis of seismogram envelope   591
Dual domain in one-way propagator   332, 334
Dual-domain propagator   267, 268, 277, 280, 297, 302, 318
– for acoustic wave   281
– for elastic wave   284
Duration in envelope synthesis   562, 563, 565
Dynamic ray tracing (DRT)   2, 29–34
– in Cartesian coordinates   30, 31
– in ray-centred coordinates   31–34
Dynamic ray tracing propagator matrix   2, 29, 34–36
– applications of   38, 39
– Chain rule of   35
– interface propagator matrix   35
– symplectic property of   35

E

Effective isotropic scattering coefficient   561, 584

SUBJECT INDEX

Eigen-value in surface waves   235–238, 249
Eigen-vector in surface waves   235–239, 244, 249, 259–261
Eigenfrequency in finite difference approach   433, 438
Eigenfunction in finite difference approach   433, 438
Eikonal equation   1, 7–9, 69, 75
Elementary ray seismograms   49
Elementary wave in ray theory   1, 5
– code of   13
Embedding technique for surface waves   145
Energy-flux conserved Green's function   276
Energy-flux model in envelope modelling   591
Ensemble in waves propagation in random media   564, 566–568, 570–576, 578, 580, 583, 587
Envelope broadening   563, 565, 571, 575, 588, 590
Equation of motion   428, 432
– displacement formulation   429
– displacement-stress formulation   429
– displacement-velocity-stress formulation   429
– Galerkin type   434
– heterogeneous formulation   428, 442
– strong form   429
– velocity-stress formulation   429, 467
– weak form   431, 434
Excitation formula of Love wave   241, 246, 249, 250
Explosive sources in finite difference method   482
Exponential ACF (auto correlation function)   565, 575, 585, 588, 590
Extensions of the ray method   3, 118
– asymptotic diffraction theory   118
– coherent-state transform   89–92
– coupling ray theory for S waves   3, 4, 56–68
– geometric theory of diffraction   118
– hybrid ray methods   116–118
– Kirchhoff–Helmholtz integrals   118
– linear canonical transform   87, 92
– Maslov methods   91–94
– quasi-isotropic approximation   59, 60, 66

– ray method with a complex eikonal   114, 115
– space–time ray method   118
– summation of Gaussian beam   2, 82–84
– summation of Gaussian packet   2, 84–86

F

Fault
– boundary condition   486, 488
– non-planar   488, 493
– planar   488
FD envelopes
– in synthesis of seismogram envelopes   571, 579, 580, 587
FD (finite difference) method
– in comparison with spectral element method   366, 407
FD-injection technique   483, 486
Finite difference
– in SH wave problems   192, 215, 222
– in synthesis of seismogram envelopes   561, 565, 566, 568, 569, 590, 591
Finite element in SH wave problems   192, 215, 222
Finite-difference equations (FDE)   426
Finite-difference (FD) approximation   424, 426, 438
– adjusted   471, 473
Finite-difference (FD) scheme
– consistency   426
– convergence   426
– explicit   427
– heterogeneous   427
– homogeneous   427
– implicit   427
– material grid parameters   427, 443, 445, 459, 463
– parsimonious   430
– stability   426, 466
Finite-difference method in crustal waveguides   325
Finite-element methods (FEM)
– relation with spectral element method   366
First-arrival travel times   112, 113
– direct computation of   112, 113
– finite difference travel time tracing   113
– network shortest-path ray tracing   113
First-order Born theory in finite difference approximation   433
Focusing/defocusing in surface waves   138

Forescattering (forward scattering) in one-way and one-return approximation   265, 269, 275
Forward multiple-scattering
– surface waves   148
Forward-difference formula in finite difference method   424
Fractional fluctuation   563, 564, 567, 568, 572, 580
Free surface
– in generalized screen propagators   330, 331, 350, 353
– planar, in finite difference method   469, 473
– topography, in generalized screen propagators   473, 480
Free-surface reflections in lattice Boltzmann approach   540, 542–546
Free-surface topography in lattice Boltzmann approach   545, 547
French model   293, 294
Fresnel zone   150
Friction law, in finite difference method   487

G

Gaussian ACF (auto correlation function)   561, 563, 566, 568–571, 576, 578, 579, 589–591
Gaussian beam   2, 68–73
– diffracted   68
– higher-order   68
– optimization of the shape of   81, 82
– paraxial   68, 70, 71
– surface-wave   118
Gaussian packet   2, 74–81
– optimization of the shape of   81, 82
– paraxial   74, 76, 77
Generalized amplitude in irregular multi-layered media   201–203, 205, 206, 235
Generalized Bremmer series (GBS)   265, 266, 276
Generalized Galerkin approximation   379
Generalized Maxwell body   450, 461
Generalized reflection/transmission coefficients   201, 246, 247, 249
Generalized reflection/transmission matrices   191, 201, 249
Generalized screen propagator (GSP)   266, 267, 280–331

Generalized Zener body (GZB)
– in rheological model   451, 455
Geometrical spreading   2, 29, 41
– reciprocity of relative   36
– relative   36
Global seismology
– wave equation in   367
Global tomography
– Born approximation   402
– great-circle approximation   402
– non-linear asymptotic coupling theory (NACT)   402
– path average approximation (PAVA)   402
Green function   see Ray-theory Green function
Green's function   146, 160, 166, 194, 252, 253
– 2-D   167
– 2.5-D   168, 171
– 3-D   171
– discrete wavenumber representation of   167
– for homogeneous media in BIE   161
– for non-homogeneous media in BIE   171
Green's function in wavenumber domain   279, 283
– elastic wave   283
– scalar wave   279
Grid   422, 423, 425
– conventional   425
– discontinuous space–time   506
– partly-staggered   425
– spatial discontinuous   505
– staggered   425
Grid dispersion   427, 430, 433
Grid spacings   423

H

Half space screen propagator   329, 331, 356
Hamiltonian   8–10, 52
– averaged, for S waves   9
– for anisotropic media   9
– for isotropic media   8
Heterogeneity
– crustal   326
– random   323, 326, 330, 336
– volume   325, 343, 350
High-frequency asymptotic approximation   265, 266, 276
Higher-order ray approximation   110–112
– additional component of   111
– principal component of   111

SUBJECT INDEX

History function of the ray 15, 23
– boundaries 15, 16
– demarcation belts 16, 20
Hooke element (elastic spring)
– in rheological model 442, 450–452
Hooke's law 428, 429, 442–444, 464
Huygens' principle 161, 164
Hybrid envelopes in synthesis of seismogram envelope 588
Hybrid method in synthesis of seismogram envelope 561, 562, 567, 588, 589, 592
Hybrid modeling
– in connection with FD method 432, 486
Hybrid ray methods 116–118
– ray-finite-difference 116
– ray-matrix 117
– ray-mode 116

I

Inclusion in SH wave problems 191, 193, 225–229
Initial condition in envelope synthesis 576, 577
Initial traction 486
Initial-value ray tracing 8–13
– controlled 14, 16–23, 86
Intensity 575
Interface 14, 42–44, 110
Internal variable 450, 455, 462
Interpolation within ray cells 24–29
Intrinsic attenuation
– in spectral element method 407
Inversion
– in ray method 55, 56, 107, 108, 148
Irregular interface 191–193, 196, 197, 201, 208, 218, 222, 226, 233, 237, 249, 250
Isotropic common ray approximation 61
Isotropic medium 428
Isotropic scattering
– in synthesis of seismogram envelope 561, 562, 567, 582–585, 587–589, 591, 592
Isotropic scattering coefficient
– in synthesis of seismogram envelope 582

K

Kelvin–Voigt body 451
KMAH index 15, 20, 23, 46, 47
– reciprocity of 47

L

Lattice Boltzmann 517–521, 525, 526, 528, 554
– lattice Boltzmann equation 525
Lattice Boltzmann equation 522, 526
Lattice gas 518–525, 528, 542
Lax equivalence theorem 426
Lax–Wendroff scheme 432, 439
Lg wave 325, 326, 328
Linear canonical transform 87–89, 92
– symplectic property of 88
Lippmann–Schwinger integral equation 269, 274
Local modes 130, 131, 152
Longitudinal distance
– in synthesis of seismogram envelope 577, 581
Longitudinal integral of the ACF 573, 580
Love wave 233, 234, 236–238, 241, 244–246, 248–251
Lower mantle inhomogeneity 591
Lyapunov exponent 97–110
– approximation in 2-D 100, 101
– average for the model 103
– directional 103
– for a system of finite rays 102, 103
– for ray tracing 98, 99

M

Markov approximation 561–563, 566, 567, 571, 572, 578, 581, 582, 588–591
Markov envelopes 582
Maslov methods 91–94
Master equation 571, 581
Material discontinuity
– boundary conditions 439, 445, 459
Maxwell body 450, 451
Mean field 572
Medium inhomogeneity 563, 566, 567
Memory optimization in FD method 501, 506
– combined 504
– core 504
– disk 504
Memory variables 451, 452, 455, 456, 462, 498
Mesher (Mesh generation package) 384
MFSB (multiple-forescattering–single-backscattering) 265, 273, 277, 286, 302, 318
Migration 96, 97

– Gaussian beam 97
– Gaussian packet 96
Modal solution
– in surface waves 234–236, 239, 241, 251
Mode coupling
– in surface waves 128
Mode-matching
– in surface waves 136
Mode-matching methods
– in surface waves 135
Modulus
– bulk 428
– relaxed 451
– shear 428
– torsion 439
– unrelaxed 449
– viscoelastic 449, 455
Modulus defect 449, 451, 453
Moho discontinuity 325, 334
Momentum transfer scattering coefficient 561, 567, 582–584
Monte Carlo method
– in envelope synthesis 564, 590
MS envelope 571, 576, 578, 581
Multi-layered media 191, 192, 201, 202, 208, 210, 218, 226, 233, 234, 237, 241, 246, 249–251
Multiple-scattering
– in surface waves 131, 146

N

Navier–Stokes equation 521, 522, 524, 527, 528
Non-isotropic scattering
– in envelope synthesis 588
Non-reflecting boundaries 495, 496
Non-vertical boundaries
– in surface waves 136
Normal mode 234, 237, 241–243, 246, 248, 250, 251

O

One return approximation (method) (De Wolf approximation) 266, 268, 273, 277, 278, 286, 290, 318
One-way approximation (method) 266
One-way propagator 266, 267, 280, 292, 326, 327, 331, 357
– asymptotic phase-matching 280, 284
Optimally accurate operators 431, 433, 439

Optimally accurate schemes 431, 437, 439
Orthogonality relation 132

P

P-SV wave 350, 355
Padé approximant method 450
Parabolic wave equation 561, 563, 566, 571, 572, 582
Parallelization
– in FD method 506, 507
Paraxial ray methods 2, 15, 16, 29–34, 38, 39
– paraxial rays 2, 29
– paraxial slowness vector 29
– paraxial travel times 2, 29
Paraxial ray tracing see Dynamic ray tracing
Perfectly matched layer (PML) 370, 393, 409, 496, 501
Perturbation derivative
– in ray theory 53
– of travel time 53
– of travel time, applications 53, 55, 56
– of travel time, first-order 51, 54, 115
– of travel time, second-order 51, 55
– of travel time gradient 54, 55
Perturbation expansion
– in ray theory 15, 16, 51, 53, 54
Perturbation function (scattering potential)
– in wave equation 278
Perturbation parameter
– in ray theory 52, 53
Phase screen method 563, 566, 573, 591
Phase screen propagator 334
Phase shift due to caustics 41, 46, 47
– example of 47
– KMAH index 15, 20, 23, 46, 47
Phase space 52, 53, 57, 87, 91, 97–100
Phase velocity 6, 7, 12
Phononic lattice solid 528, 554
PLSI (phononic lattice solid by interpolation) 519, 520, 528, 529, 533, 534, 536–540, 542–555
Poisson's ratio 430, 440
Polarization 138, 142, 590, 591
Polarization of elementary waves 41, 42, 56, 57, 60
Potential formulation of elastic wave equation
– in fluid regions 371
Power spectral density function 564, 567

SUBJECT INDEX 603

Power-law spectrum, random media 561, 564, 567, 579, 589, 592
Predictor-corrector in FD approach 431, 439
PREM, earth model 128
Primary reflected wave 277
Primary reflection 265, 276, 309, 318
Primary transmitted wave 277
Propagator
– dual-domain propagator 267, 268, 277, 280, 297, 302, 318
– forward scattering propagator 265, 269, 275
– one-way propagator 266, 267, 280, 292
– transpared propagator 276
Pseudo-spectral method 366

Q

Q-factor 290, 291, 339–343
Quality factor 449
Quasi-isotropic approximation 59, 60, 66–68
– common ray approximation 60
– of the Christoffel matrix 61, 62
– of the Green function 61
– of travel time 62
Quasi-isotropic perturbation of travel time 62
Quasi-isotropic projection of the Green function 61
Quasi-monochromatic waves 571, 574, 581
Quasi-particle 517, 519, 528–532, 536, 537, 540, 542–544, 548
Quasi-particle movement 530

R

R/T (reflection/transmission) 191, 192, 201, 203–205, 208, 209, 226, 233, 234, 241, 250, 251, 258
Radiative transfer integral equation 562, 584, 585, 588, 592
Radiative transfer theory 561, 564, 566, 567, 582–584, 588, 589, 591, 592
Random media 561, 564, 566, 568, 571, 578, 579
Ray 138
– anisotropic common S-wave reference 9, 53, 60
– anisotropic-ray-theory reference 60
– boundary 17

– boundary-value 20
– complex 115
– isotropic common S-wave reference 61
– reference 51, 52, 54, 57, 60
– successful 16
– two-point 20
– two-point diffracted 21, 22
Ray cell 2, 13, 15, 25
– degenerate 25
– interpolation within 24–29
– regular 25
Ray chaos 4, 13, 97–110
– Lyapunov exponent 97–110
– rotation number 97–103
Ray coordinates 13, 30
Ray history 2, 14–16, 23, 24
– boundaries 15, 16
– demarcation belts 16, 20
Ray Jacobian 40
Ray parameters 13, 16
– normalized 16
Ray theory 150, 152
– for anisotropic media 4–8
– for isotropic media 6, 8
Ray theory for S waves
– anisotropic 9, 56, 57
– coupling 9, 56–68
– isotropic 9, 56, 57
Ray tracing 2, 131, 139
– boundary-value 14
– controlled initial-value 14, 16–23, 86
– first-order (FORT) 52
– in the vicinity of S-wave singularities 11, 12
– initial-value 8–13
– models suitable for 14, 15, 106–110
– paraxial see Dynamic ray tracing
– surface-wave 118
– two-point 2, 13, 20, 21
Ray-centred coordinates 31–34, 36, 72, 73, 78, 79, 99, 100
– basis vectors 32, 33
– dynamic ray tracing in 31–34
Ray-parameter domain 2, 13, 16, 22
– homogeneous subdomains of 20
– metric tensor in 17–20
– normalized 16, 22
– secondary metric tensor in 20, 23, 86
– triangulation of 16–20
Ray-theory amplitude 39–42
– complex-valued scalar 2, 7, 40, 41

– complex-valued vectorial 1–3, 5–8
– polarization of vectorial 41, 42
Ray-theory Green function 2, 45–49
– reciprocity of 46
Ray-theory perturbations 51–56
Ray-theory seismograms 49–51
Ray-tracing system 2, 10
– for anisotropic media 11
– for isotropic media 10
– in Hamiltonian form 10
Ray-velocity vector 8, 10, 12
Rayleigh ansatz 215, 216, 218
Rayleigh wave 329, 350, 357
Realization of random media 568–571
Reciprocities in the ray methods
– of KMAH index 47
– of normalized R/T coefficients 46
– of ray-theory Green function 46
– of relative geometrical spreading 36
Reference envelopes
– in envelope synthesis 561, 566, 571, 578–580, 582, 587, 591, 592
Reference Green's function 266
Reference modes 130, 143
Reference ray 51, 52, 54, 57, 60
– anisotropic common S-wave 9, 53, 60
– anisotropic-ray-theory 60
– isotropic common S-wave 61
Reference surface
– in ray method 15, 20–23
– reference coordinates 20
Reflected, surface waves 132
Reflection, surface waves 132, 145
Reflection/transmission (R/T) coefficients
– in ray method 3, 39–44
– normalized 39, 45
– reciprocity of 46
Regional wave 324
Relaxation frequency
– in rheological models 452, 463
Relaxation function
– in rheological models 448, 453, 455
Relaxation times
– in rheological models 453
Representation theorem 164
Riccati equation 36, 37, 577
– for Gaussian beam 71, 72, 79
– for Gaussian packet 78, 79
Ricker wavelet 211, 216, 220, 222, 233, 566, 569, 570, 578–580, 582, 587, 591
Rotation number

– in ray theory 97–103
– approximation in 2-D 102
– average for the model 103
– for a system of finite rays 102, 103
– for ray tracing 99, 100
Rupture 486
Rytov approximation 265, 267, 268, 270–272, 276
– comparison with the Born approximation 271
Rytov transform 334

S

S-wave singularities
– in ray theory 9, 11
S-wave splitting 66
Scattering
– back 332, 339–343
– backscattering (back scattering) 269, 272–274
– forescattering (forward scattering) 265, 269, 275
– forward 332–343
– multiple-forescattering–single-backscattering 326
– surface waves 152
– topography 348, 350
– volume heterogeneity 343, 350
Scattering coefficient 583, 584, 587, 592
Scattering pattern 583
Screen approximation (method) 297
– complex screen approximation (method) 265, 267, 289, 302, 304, 306
– phase-screen approximation (method) 280, 284, 302
– pseudo-screen 280
Secular function 235–237
Sedimentary basins 396
Seismic risk analysis 136
Seismic wave excitation 191, 192
Selfgravitation 367, 368, 379, 391
Sensitivity kernels 150, 151
SH wave 191, 192, 212, 234, 241, 244–247, 256, 331, 334, 343
Sharp interfaces
– in lattice Boltzmann approach 517–519, 547, 548, 554
Single scattering
– surface waves 131, 148
– in envelope synthesis 583, 590
Singularity, boundary integral method 161
– hyper 161

SUBJECT INDEX 605

– strong 161
– weak 161
Slip, rapture modelling 486, 487
Slip rate, rapture modelling 486
Slowness vector 2, 6
Small-angle approximation 265, 267, 272, 297, 300, 302, 303, 315, 318, 335, 346
Snell's law 2, 43, 132
Spatial sampling ratio
– in FD method 427
Spectral-element method (SEM) 365, 367, 379
– as quadranglar 380
– collocation points 385, 388
– for global seismology 402
– for regional seismology 371
– for surface waves 366
– Gauss–Lobatto–Legendre (GLL) points 394, 395, 397
– generalized Galerkin approximation 388
– hexahedra 380, 383
– modeling for sedimentary basins 383
– modeling in the D'' region 403
– newmark time stepping method 389
– parallel implementation 395
– tetrahedra 380
– triangular 381
Statistical characterization 566
Statistical homogeneity 574
Stokes element
– in rheological model 452
Stokes vectors 590
Stress glut method
– in finite difference method 491, 492
Stress imaging
– for free-surface boundary condition 470, 471
Stress-strain relation 428, 448
Strongly heterogeneous media 517, 520, 546, 548, 554, 555
Summation of Gaussian beams 2–4, 21, 82, 91, 92
– discretization error of 86, 87
Summation of Gaussian packets 2–4, 21, 84, 91, 92
Surface wave 191, 192, 211, 228, 232–234, 244, 247, 366, 393, 397
Synthetic seismogram 192, 216–218, 220, 222–224, 227, 228, 230, 231, 233, 247, 250, 251, 309–317, 335–338, 346–349, 354–356

T

T-matrix 192
Taylor expansion 424, 425
Tensorization 384, 385, 387
Thick-fault-zone method 492
Thin-slab approximation (method) 265, 267, 277, 282, 289–291, 297, 303
– applications 306, 307, 309, 311, 315, 316
– formulation 278, 287, 289
Time step, finite difference method 506
Topography 191, 193, 208, 210, 211, 215, 218, 220, 255, 257, 343, 350
Total scattering coefficient 584
Traction-at-split nodes (TSN) method 488, 492
Traction-free boundary condition 468, 469
Transmission, surface waves 132, 145
Transmission and reflection, surface waves 134
Transparent propagator
– in one-way and one-return methods 276
Transport equation
– in ray theory 2, 8, 40
Transverse distance
– in synthesis of seismogram envelope 577, 581
Travel distance
– in synthesis of seismogram envelope 561, 565, 569–571, 575–578, 580, 586
Travel time 1, 3, 6
– first-arrival 112, 113
– perturbation derivatives of 53
– ray-theory 2, 112
– spatial derivatives of 36–38
Travel-time perturbations 53–55
Travel-time spatial derivatives 36–38
– applications of 37, 38
Trial traction
– in rapture modelling 489
Truncation error
– in FD method 424, 426, 436
Twersky approximation 273
Two-point ray tracing 2, 13, 20, 21
– reference surface for 15, 20–23

V

Vacuum formalism
– for free surface boundary condition 470
Variational method 365, 367, 377
Vector wave envelopes 590, 592
Visco-elastic medium 290, 448

von Kármán (ACF)   566, 567, 578–582, 588, 589, 591
Von Neumann method   426
Voronoi polygon
– in Gaussian beam summation   86

W

Wandering effect
– in envelope synthesis   575, 576, 582
Wave envelopes   561, 563, 565, 566, 568, 590–592
Wave equation
– absorbing boundary   368, 377
– acoustic wave equation (acoustic media)   280
– coupling boundary   375, 377
– elastic wave equation (elastic media)   282
– gravito-elastodynamic equations   369
– in regional and global seismology   367
– initial conditions   374
– internal boundary condition   375
– potential formulation   373
– scalar wave equation (scalar media)   278
– weak form   377, 378
Wave-envelopes   582
Wavefield decomposition
– asymptotic into Gaussian beams   82–84
– asymptotic into Gaussian packets   84–86
– of a general wavefield into Gaussian beams   94
– of a general wavefield into Gaussian packets   94
Wavefront   5
Wavefront tracing   13, 15, 23, 24
Weak form of wave equation   367, 368, 377
Weak scattering approximation (Born approximation)   268, 269, 289
Wide-angle approximation   265, 280, 282, 318, 333, 335
Wide-angle scattering   561, 580, 582, 587, 588
WKBJ approximation   140, 152, 276, 318

Z

Zener body (standard linear body)
– in rheological model   451